SCIENCE
AND THE
AMERICAN
CENTURY

SCIENCE AND THE AMERICAN CENTURY

READINGS FROM
ISIS

EDITED AND WITH AN
INTRODUCTION BY
Sally Gregory Kohlstedt and David Kaiser

THE UNIVERSITY OF CHICAGO PRESS
Chicago and London

Sally Gregory Kohlstedt is professor in and chair of the History of Science, Technology, and Medicine Program at the University of Minnesota.

David Kaiser is the Germeshausen Professor in and department head of the Program in Science, Technology, and Society at the Massachusetts Institute of Technology.

The University of Chicago Press, Chicago 60637
The University of Chicago Press, Ltd., London
© 1996, 2000, 2001, 2002, 2003, 2006, 2007,
2008, 2009, 2010 by the History of Science Society
© 2013 by The University of Chicago
All rights reserved. Published 2013.
Printed in the United States of America

22 21 20 19 18 17 16 15 14 13   1 2 3 4 5

ISBN-13: 978-0-226-92514-1 (paper)
ISBN-13: 978-0-226-92515-8 (e-book)
ISBN-10: 0-226-92514-5 (paper)
ISBN-10: 0-226-92515-3 (e-book)

LIBRARY OF CONGRESS CATALOGING-IN-PUBLICATION DATA

Science and the American century : readings from Isis / edited and with
an introduction by Sally Gregory Kohlstedt and David Kaiser.
      pages cm
   Collection of articles previously published in the journal Isis.
   Includes bibliographical references and index.
      ISBN 978-0-226-92514-1 (paperback : alkaline paper) —
ISBN 978-0-226-92515-8 (e-book) — ISBN 0-226-92514-5 (paperback
: alkaline paper) — ISBN 0-226-92515-3 (e-book) 1. Science—United
States—History—20th century. I. Kohlstedt, Sally Gregory, 1943– editor.
II. Kaiser, David, editor. III. Isis (Chicago, Ill.)
   Q127.U5S35 2013
   509.73'0904—dc23

                                                          2012033424

♾ This paper meets the requirements of ANSI/NISO Z39.48-1992
(Permanence of Paper).

CONTENTS

III
SOCIAL POLICIES,
SCIENTIFIC PRACTICE, AND THE LAW

INTRODUCTION

Sally Gregory Kohlstedt and David Kaiser

A quarter century ago, historical studies of American science, technology, and medicine began to coalesce into a recognizable field. Classic early writings were complemented and challenged by a fast-growing literature. Back in 1980, among all the dissertations in the history of science completed in North America, only one in seven analyzed developments in the United States. In 1990 the proportion was more than half. With good reason the editors of one of the first compilations on the subject, published in 1985, concluded that "there seems to be room for optimism, perhaps for the first time."[1]

Historians of American science, technology, and medicine quickly fanned out to investigate all manner of topics, disciplines, and themes. They expanded from studies of the physical and biological sciences to the social sciences, science policy, science education, science in popular culture, and more. Historians in these thematic areas, like the profession in general, have moved progressively forward in time, coming very close to the present. The 1985 *Osiris* volume on "Historical Writing on American Science" largely reviewed books and articles on nineteenth-century topics, then prominent. Indeed, the editors noted at that time that certain topics—such as the relationship between science and the federal government during and after World War II—still remained largely unexamined. A decade later, a compilation of articles from *Isis* appeared under the title *The Scientific Enterprise in America*. Though that volume covered material dating back to 1820, the center of gravity had moved squarely to the early decades of the twentieth century. That collection also devoted substantial attention—fully one quarter of the chapters—to the great transformations of the midcentury, when science and technology became thoroughly enmeshed with wartime exigencies, and the federal government emerged as the most significant patron for research. Nonetheless, *The Scientific Enterprise in America* had little to say about developments after 1945.[2]

The present collection is based on articles published in *Isis* since the publication of *The Scientific Enterprise in America*. With essays selected from among the diverse and substantive contributions to the journal, the editors sought to identify and reflect more recent trends in the historiography of science, technology, and medicine in the United States. Two-thirds of the chapters treat episodes since the end of World War II. In fact, the enormous changes wrought by the war in the institutions, funding, and policies for science serve as a backdrop for much historical writing today. Alongside the now widespread focus on the sprawling projects that marked the Cold War years—an era of technoscientific gigantism, when everything from nuclear reactors and particle accelerators to computer mainframes and electron microscopes seemed to double in capacity and complexity every few years—most of the selections in the volume chart a more subtle trend, as many less dramatic efforts in science, technology, and medicine seeped ever more concertedly into everyday life.[3] Scientists' ambi-

[1] Margaret W. Rossiter and Sally Gregory Kohlstedt, introduction to *Historical Writing on American Science*, ed. Kohlstedt and Rossiter, published as *Osiris* (2nd series) 1 (1985): 9–16, quotation on 15. On the evolution of the field, see also Charles Rosenberg, "Science in American Society: A Generation of Historical Debate," *Isis* 74 (1983): 356–67. On dissertation trends, see David Kaiser, "Training and the Generalist's Vision in the History of Science," *Isis* 96 (2005): 244–51, esp. 249.

[2] Ronald L. Numbers and Charles E. Rosenberg, eds., *The Scientific Enterprise in America: Readings from Isis* (Chicago: University of Chicago Press, 1996).

[3] On Cold War gigantism in science and technology, see especially the recent review essay by Cyrus C. M.

tions grew by leaps and bounds over the twentieth century, as did their profile in society. They increasingly found themselves called upon to address matters sometimes considered private or beyond their expertise, from child rearing and civil rights to environmental management and medical ethics. Some scientists sought not just to unpack deep truths of the cosmos but were ready to refashion the built environment, engage in public policy decisions, and even design entirely new life forms for laboratory study and commercial profit.

This recent shift in the historical literature—not just to more recent topics but to ever more intimate incursions of scientific practice into Americans' daily routines—inspired our choice of the title, *Science and the American Century*. Journalist and *Life* magazine publisher Henry R. Luce coined the phrase "the American Century" in a spirited editorial back in February 1941. His goal was to goad the American public out of an isolationist stance and to embrace the war effort, several months before the surprise attack on Pearl Harbor instigated the country's formal declaration of war. Luce's rhetoric was by turns pleading, boastful, even jingoistic. He argued, for example, that "the fundamental trouble with America has been, and is, that whereas their nation became in the 20th Century the most powerful and the most vital nation in the world, nevertheless Americans were unable to accommodate themselves spiritually and practically to that fact. Hence they have failed to play their part as a world power." The time had come, Luce urged, for Americans to "accept wholeheartedly our duty and our opportunity as the most powerful and vital nation in the world and in consequence to exert upon the world the full impact of our influence, for such purposes as we see fit and by such means as we see fit." Luce anticipated a golden age of "health and vigor" for the world at large, to be brought forth and sustained by the benevolent leadership of the United States.[4]

The tumultuous decades that followed Luce's editorial—cataclysmic world war, long-simmering cold war, and sporadic proxy wars around the globe—did little to bolster Luce's vision of a coming era of peace and bounty. Critics have rightly complained that the term, "the American Century," blinded politicians and policymakers to the full ramifications of Luce's evocative phrase, and might well have contributed to missteps and overreaching at various moments in American foreign policy.[5] For better or worse, however, the term retains analytic power. The long arc of "the American Century"—from the Spanish-American War of 1898, with its implications for territorial expansion, through the end of the Cold War in the early 1990s—was indeed a period in which American influence on worldly matters grew to unprecedented scale. The hubris behind Luce's pronouncement might well have been misplaced, but as a descriptive term his ear-catching phrase remains useful. We adopt the phrase here not to celebrate Luce's nationalistic ambitions but to demarcate a particular historical period; indeed, one that may well be passing.

The phrase seems doubly appropriate for the present volume, given Luce's original formulation. In laying out his proposal for American leadership, Luce pointed to the "baffling, difficult, paradoxical, revolutionary" character of the twentieth century. First among those revolutionary characteristics, in Luce's accounting, were fast-paced changes "in science and in industry." The time would soon come, Luce hoped, when American involvement in world affairs would seem "as natural to us in our time as the airplane or the radio."[6] Luce drew inspiration from the fervent optimism of the increasingly technical age. The nation's

---

Mody, "How I Learned to Stop Worrying and Love the Bomb, the Nuclear Reactor, the Computer, Ham Radio, and Recombinant DNA," *Historical Studies in the Natural Sciences* 38 (2008): 451–61.

[4] Henry R. Luce, "The American Century," *Life*, February 17, 1941, 61–64, quotations on 63–64.

[5] See, e.g., Andrew Bacevich, "Farewell to the American Century," *Salon.com*, April 30, 2009, available at http://www.salon.com/news/opinion/feature/2009/04/30/bacevich (accessed March 7, 2011).

[6] Luce, "The American Century," 64.

growing influence on the world stage was by turns inspired and reinforced by a massive infrastructure for research and innovation in science, technology, and medicine.

The essays that follow are grouped into three sections, each presented in roughly chronological order. The first section, "Nature, Science, and Environmental Perspectives," charts several ways in which knowledge of nature was cultivated in constant dialogue with various publics over the course of the century, even as practitioners absorbed and refracted global information. In "Local Knowledge, Environmental Politics, and the Founding of Ecology in the United States," Daniel W. Schneider reveals how the origins of a new ecological science grew out of collaboration between naturalists with increasingly sophisticated methods and local fishermen who understood the ebb and flow of the Illinois River based on intimate experience with its wildlife and seasons. Other essays also reflect on the geographies of science and the productive tension between specific regional knowledge and an increasingly familiar global network linking investigators and ideas. The imperialist incentives that took Darwin on his significant voyage increased trade and exchanges of plants and insects, some deliberate and others not. The invaders moved the federal government into action. As Philip J. Pauly demonstrates in "The Beauty and Menace of the Japanese Cherry Trees," political and public opinion in the early twentieth century inflected expert reaction to the gift of inadvertently infected cherry trees from Japan. A conservative perspective in the U.S. Department of Entomology had been initially counterbalanced by the acclimatizing outlook of the Bureau of Plant Industry, but in the context of anti-immigration sentiment, nativism held sway and the trees were unceremoniously burned. Despite the incident, significant constituencies remained fascinated by nature, and scientists encouraged the rapid development of public botanical gardens, natural history museums, and zoos. Moving natural objects into somewhat static spaces while retaining something of their profile in nature proved challenging, as Michael Rossi documents in "Fabricating Authenticity." His account of the modeling of a "sulfur-bottom whale" at the American Museum of Natural History demonstrates the technical and philosophical issues that were negotiated over half a century. In the process, he also reveals shifting attitudes toward visitors as well as increasingly sophisticated techniques used to attract their attention to exotic (in this case enormous) creatures from the deep that few of them would ever encounter in nature.

The essays by Paul S. Sutter and Joshua Blu Buhs bookend the twentieth century with their studies of insects and ideas about them, as both migrated between North and South America. Their accounts reveal the entomologists' optimism, as they became convinced of the capacity of their expert knowledge and technology to control pests. Yet with the practical experience and growing ecological sensibility came ambivalence. Sutter's essay, "Nature's Agents or Agents of Empire?," documents how specialized staff who worked to manage disease-carrying insects during the building of the Panama Canal had mixed loyalties, as they sought to balance control with a sensibility about interventions in the natural habitat. As Joshua Blu Buhs points out in "The Fire Ant Wars," by the end of the century the imperative to control fire ants from South America in the American South engaged multiple experts, each group alert to layers of influence and a growing public reluctance to use strong and perhaps dangerous insecticides. Humans now had to be taken into account as part of the balance of nature.

Section II, "Patrons, Politics, and the Physical Sciences," investigates the changing places and practices of the physical sciences since World War II. The sprawling, top secret Manhattan Project to design and build the first nuclear weapons fundamentally altered relationships between the physical sciences, engineering, and the federal government. Funding from military patrons suddenly seemed limitless, while esoteric knowledge assumed a worldly importance that few had attributed to it before. In "Patenting the Bomb," Alex Wellerstein uncovers a long-forgotten chapter in that transformation: the wartime patent system for the Manhattan Project, seemingly the very antithesis of secrecy and classification, in which every component of the new weapons was described in exquisite detail, all in black and white. Wellerstein's essay reminds readers what the nuclear-weapons project looked like to policymakers *before* the dramatic bombings of Hiroshima and Nagasaki in August 1945, when atomic bombs were considered just one more military project to be slotted into existing procedures and protocols.

Immediately after the war, fears of losing "the atomic secret" fueled anticommunist hysteria and drove bitter legislative disputes over who should manage nuclear projects and those who worked on them. What had confidently been presumed by Luce to be "the American century" now seemed at risk. David Kaiser's essay, "Nuclear Democracy," follows how some nuclear physicists' political ideals—forged in response to the red scare and the new wave of "loyalty oaths"—inspired new ways to interpret the subatomic domain and to encourage participation in its ongoing study. Around the same time, the international rivalry of the Cold War shaped a different form of scientific participation, as W. Patrick McCray documents in "Amateur Scientists, the International Geophysical Year, and the Ambitions of Fred Whipple." Professional astronomers recruited amateur skywatchers to help scan the night sky, on the lookout for artificial satellites like Sputnik. Much like the tensions between military paymasters and academic scientists, however, professional and amateur astronomers often held conflicting views about how best to organize and control their work.

Even as academic astronomers tussled with amateur volunteers, a different group looked to the academy as a model for how best to foster creativity and innovation. As Scott G. Knowles and Stuart W. Leslie discuss in " 'Industrial Versailles,' " architects' designs during the 1950s and 1960s for new corporate research laboratories built upon specific notions of how basic scientific research leads to industrial applications. By the early 1970s, however, when the postwar boom years for science and technology had turned to bust, some stunning laboratories seemed like empty cathedrals from a bygone era, their heady promise unfulfilled. The shift is evident in Catherine Westfall's account of nuclear physicists' changed ambitions since the Sputnik days, "Rethinking Big Science." Westfall documents how more modest projects and an emphasis upon flexibility and coalition building came to replace the mammoth projects of postwar "big science."

The essays in section III, "Social Policies, Scientific Practice, and the Law," chart scientists' progressive involvement in once private domains, from the social reform-era 1930s to the politically charged exposés in recent years of the dangers of tobacco use. Since its formation as a discipline, sociology had attracted some scholars who sought to affect public policy in the light of their research. In "Families Made by Science," Ellen Herman demonstrates how Arnold Gesell's use of physical and intelligence tests during the early decades of the century were incorporated into adoption practices for young children. In the process of using scientific measurements, sociologists encouraged agencies to use Gesell's scales in order to socially engineer families, through their models of identity and normality. Also deliberately tackling a social problem, Kenneth and Mamie Clark undertook investigations of black children's racial self-definition during the 1940s. John P. Jackson, Jr., shows in "Blind Law and Powerless Science" how their experimental research and resulting quantitative data, in which African American children in Harlem chose white dolls as most like themselves, became part of the social-science evidence marshaled for the successful Supreme Court challenge to segregation in 1954, in the famous *Brown v. Board of Education* case.

By midcentury, however, questions about experiments and results, their ownership, and their importance to the public became more contested. In "Visions of a Cure," David S. Jones traces the unsettling debate over randomized control trials and outcome assessments in dealing with cardiac problems. Working to evaluate coronary bypass surgery and stents, cardiologists and surgeons literally viewed outcomes in different terms, complicating the decisions of doctors and patients. Such laboratory-intensive research in late twentieth-century biomedicine and biotechnology crisscrossed between university laboratories and clinics, private corporations, and the federal government. Who should benefit financially? Sally Smith Hughes's essay, "Making Dollars Out of DNA," explores the successful effort of Stanford biologist Stanley Cohen and coworkers to patent a process for joining and replicating DNA. In the process, they established a precedent for commercialization of molecular biology. In the DNA case, benefits accrued to the researchers and universities, but private corporations quickly adapted to the new research terrain as well, aggressively pursuing patent protections for biomedical processes and even challenging outside research that threatened their financial interests. Brianna Rego's study of the tobacco industry, "The Polonium Brief," based in significant ways on the Legacy Tobacco Documents Library at the University of California, San Francisco, reveals how corporate research on polonium over four decades—most of it unpublished—was stockpiled to serve as a counterweight to external investigations intent on showing the dangers posed by that radioactive substance. As this cluster of essays makes plain, relations between scientific investigation, medical therapeutics, and market forces have grown increasingly common and complex since the 1970s.

Various themes recur across the sections, blurring once tidy categories such as "science," "technology," and "medicine." The studies by Daniel W. Schneider, W. Patrick McCray, John P. Jackson, Jr., Ellen Herman, and Brianna Rego tease out the intricate interrelationships that have linked (and continue to link) specialized research with ordinary lives. Essays by Philip J. Pauly, Paul S. Sutter, Alex Wellerstein, Catherine Westfall, Joshua Blu Buhs, and Sally Smith Hughes illuminate different moments in the fraught evolution of federal science policy, from World War I through the closing decades of the twentieth century. A distinct subtheme that has recently garnered the attention of historians of science—and that animates Wellerstein's and Hughes's chapters in particular—concerns patents and intellectual property rights. Who owns the fruits of scientific research, and to what ends can that knowledge be put? Another thread connecting various essays concerns scientists' uses of political metaphors. Philip J. Pauly, Paul S. Sutter, David Kaiser, and Joshua Blu Buhs each study ways in which metaphors intentionally drawn from the social sphere were generative in scientific pursuits, suggesting new ideas or burnishing rhetorical positions. Lastly, several authors in this collection, like many of their colleagues in the field, have grown fascinated by the roles played by images, visualization, and shifting styles of argument in the sciences. Michael Rossi, David Kaiser, Scott G. Knowles and Stuart W. Leslie, and David S. Jones each ask how the practice of representing complicated scientific relationships reflects and refracts broader assumptions about science, politics, and culture.

Other themes that had once featured centrally in historical studies of American science, technology, and medicine are less prominent in the present collection. Few essays treat questions of women, gender, and science, for example.[7] Only the essay by John P. Jackson, Jr.,

---

[7] Mapping the subject of women in science are three volumes by Margaret W. Rossiter, whose third volume, *Forging a New World since 1972: Women Scientists in America* (Baltimore, MD: Johns Hopkins University Press, 2012), completes the account begun in her two prize-wining books, *Women Scientists in America: Struggles and Strategies to 1940* (Baltimore, MD: Johns Hopkins University Press, 1984), and *Women Scientists in America: Before Affirmative Action, 1940–1972* (Baltimore, MD: Johns Hopkins University Press, 1995). Also relevant are collections of essays by Sally Gregory Kohlstedt and Helen Longino, eds., *Women, Gender, and Science: New Direc-*

focuses squarely on the role of minorities and the category of race.[8] The years since the 1985 *Osiris* volume, meanwhile, saw an explosion of detailed studies of science, technology, and the federal government, especially its military branches.[9] If the question of science and the military receives less explicit attention in the present volume than one might expect, that is because the theme now serves as an unavoidable backdrop for understanding other relationships, such as science, industry, and commercialization. Another perennial topic—science and religion—receives scarcely any attention in this volume at all, though we hardly believe the topic has ceased to be important.[10] The fierce battles over teaching biological evolution versus "Intelligent Design" in public schools (to name but one major example) have still not played their course, and important historical work remains to be done.

Much as Luce had anticipated, the essays in this collection reveal that science in "the American century" proved to be influential well beyond the nation's borders. The notion of an isolated country has been challenged by the exigencies of war, the exchange of ideas and objects across national boundaries, and the sometimes self-conscious efforts toward imperialist expansion on the one hand and international collaboration on the other. Pace Luce's spirited rhetoric, American ambitions in science, technology, and medicine over the twentieth century were as often met with failure and frustration as with resounding success; periods of ambitious expansion were followed—perhaps inevitably—by eras of steep decline.

We selected the essays for this volume from among many excellent articles in *Isis* with several goals in mind. We want to showcase some of the best recent work in the flourishing literature on the history of science, technology, and medicine in the modern United States— with a special emphasis on effective material for the classroom. We also aim to strengthen bonds between the history of science and American history. Especially when considering recent decades, one simply cannot ignore the emergence of a science-infused world. From nukes to lasers, nylon to bioengineered mice, science, technology, and medicine have become inescapable elements of daily life. Historians' only hope for capturing the texture of mundane affairs or highbrow concerns in the United States over the long twentieth century is by merging insights from the history of science, technology, and medicine with the evolving narrative of the American experience.

---

tions, published as *Osiris* 12 (1997); and Angela N. H. Creager, Elizabeth A. Lunbeck, and Londa Schiebinger, eds., *Feminism in Twentieth-Century Science, Technology, and Medicine* (Chicago: University of Chicago Press, 2001).

[8] The quantity of research on African American inventors and scientists has lagged behind that of women in science, but there have been significant contributions. Kenneth Manning's biography of a prominent early twentieth-century biologist in *Black Apollo: The Life of Ernest Everett Just* (New York: Oxford University Press, 1983) has become a classic. See also Rayvon Fouché, *Black Inventors in the Age of Segregation* (Baltimore, MD: Johns Hopkins University Press, 2003), and the useful bibliographical resource in Amy Bix, "A Bibliography on the History of Technology and the African American Experience," in *Technology and the African American Experience*, ed. Bruce Sinclair (Cambridge, MA: MIT Press, 2004). On changing scientific perspectives on race, see especially John P. Jackson, Jr., and Nadine M. Weidman, eds., *Race, Racism, and Science* (New Brunswick, NJ: Rutgers University Press, 2006); Evelynn M. Hammonds and Rebecca M. Herzig, eds., *The Nature of Difference: Sciences of Race in the United States from Jefferson to Genomics* (Cambridge, MA: MIT Press, 2008); and Ian Whitmarsh and David S. Jones, eds., *What's the Use of Race? Modern Governance and the Biology of Difference* (Cambridge, MA: MIT Press, 2010).

[9] Classic studies include Paul Forman, "Behind Quantum Electronics: National Security as Basis for Physical Research in the United States, 1940–1960," *Historical Studies in the Physical Sciences* 18 (1987): 149–229; Everett Mendelsohn, M. Roe Smith, and Peter Weingart, eds., *Science, Technology, and the Military* (Boston, MA: Kluwer, 1988); Peter Galison and Bruce Hevly, eds., *Big Science: The Growth of Large-Scale Research* (Stanford, CA: Stanford University Press, 1992); and Stuart W. Leslie, *The Cold War and American Science: The Military-Industrial-Academic Complex at MIT and Stanford* (New York: Columbia University Press, 1993).

[10] Important introductions to major themes in the literature on science and religion in an American context include Ronald L. Numbers, *The Creationists: From Scientific Creationism to Intelligent Design* (Cambridge, MA: Harvard University Press, 2006 [1992]), and Edward Larson's *Summer for the Gods: The Scopes Trial and America's Continuing Debate over Science and Religion* (New York: Basic Books, 1997).

I

NATURE, SCIENCE, AND ENVIRONMENTAL PERSPECTIVES

Local Knowledge, Environmental Politics, and the Founding of Ecology in the United States

Stephen Forbes and "The Lake as a Microcosm" (1887)

By Daniel W. Schneider

ABSTRACT

Stephen Forbes's "The Lake as a Microcosm" is one of the founding documents of the science of ecology in the United States. By tracing the connections between scientists and local fishermen underlying the research on floodplain lakes presented in "The Lake as a Microcosm," this essay shows how the birth of ecology was tied to local knowledge and the local politics of environmental transformation. Forbes and the other scientists of the Illinois Natural History Survey relied on fishermen for manual labor, expertise in catching fish, and knowledge of the natural history of the fishes. As Forbes and his colleagues worked in close contact with fishermen, they also adopted many of their political concerns over the privatization of the floodplain and became politically active in supporting their interests. The close connection between scientists and local knowledge forced the ecologists to reframe the boundaries of ecology as objective or political, pure or applied, local or scientific.

Isis 91, no. 4 (December 2000): 681–705.

I thank Roberta Farrell and Meg Cederoth for their research assistance. Glenn Sandiford provided material on the Illinois and U.S. Fish Commissions. Leslie J. Reagan's suggestions and editing were invaluable. John Hoffman of the Illinois Historical Survey Library, Bob Bailey of the Illinois State Archives, and William Maher of the University of Illinois Archives were extremely helpful in making collections available. Thomas Rice of the Illinois Natural History Survey helped locate and copy photographs. I thank Chip Burkhardt, Robert Kohler, Eileen McGurty, Richard Sparks, and Ruth Sparks for discussions of this work. Geoff Bowker and Daniel Walkowitz read earlier drafts of this manuscript. Comments by four anonymous referees and Margaret Rossiter improved the manuscript greatly. Parts of this study were presented at the American Society for Environmental History; the Department of Integrative Biology, University of California, Berkeley; and the Program in Science, Technology, Information, and Medicine at the University of Illinois at Urbana-Champaign. This study was supported by the University of Illinois Campus Research Board.

O N 25 FEBRUARY 1887 Stephen A. Forbes delivered "The Lake as a Microcosm," one of the founding papers of the new science of ecology, to a small scientific society in the Illinois River town of Peoria. Forbes spoke of the ecology of "fluviatile" lakes, "situated in the river bottoms and connected with the adjacent streams by periodical over-flows." Less than thirty miles downstream, a battle brewed that same year as members of the Peoria elite began placing "No Trespassing" signs on several thousand acres of the same types of fluviatile lakes analyzed by Forbes. Local residents who had been hunting and fishing in these lakes for decades shot down signs in protest and claimed the land and resources as theirs by poaching fish, ducks, and muskrat. Over the next several decades the battles over the Illinois River floodplain escalated, culminating in gunfire, armed block-ades, and court fights.[1]

Forbes and the other scientists at the Illinois Natural History Survey had begun their systematic investigation of the Illinois River and its floodplain at the precise moment that the ecosystem became the focus of a struggle over rights to lands and waters. Indeed, the rise of the science of ecology in the United States, from the first use of the term "oekologie" in 1866 to the establishment of the Ecological Society of America in 1915, coincided with and was shaped by a dramatic transformation of society's relation to the natural world as subsistence, artisanal, and traditional patterns of resource use were supplanted by capital intensive resource extraction, on the one hand, and recreational fishing and hunting by an urban elite, on the other.[2] During the period that ecology was developing into a science, Americans fought intense battles over the new science's very object of study.

The development and practice of ecology was linked to the politics of environmental transformation from the beginnings of the science in the United States. Historians of sci-ence generally agree that the post–World War II politics of environmental degradation,

---

[1] Stephen A. Forbes, "The Lake as a Microcosm," *Bulletin of the Peoria Scientific Association,* 1887, pp. 77–87, on p. 77. I reference the original in this essay, but it is more easily available as reprinted in *Illinois Natural History Survey Bulletin,* 1925, *15:*537–550. On the 1887 conflict downstream of Peoria see testimony of Fer-dinand Luthy, transcript of *Duck Island Hunting and Fishing Club v. Chester L. Whitnah et al.,* 306 Ill. 291 (1923), Record Series 901, Supreme Court Trial Transcripts, Vault 39941, Illinois States Archives, Springfield, Abstract, pp. 109–112, Record, p. 736. For discussion of the conflict in general see Daniel W. Schneider, "Enclosing the Floodplain: Resource Conflict on the Illinois River, 1880–1920," *Environmental History,* 1996, *1:*70–96.

[2] The Natural History Society of Illinois, formed in 1858, became the State Laboratory of Natural History in 1877. In 1917 it was reorganized as the Illinois State Natural History Survey, its current name. I call the organization the Natural History Survey throughout this essay. For an institutional history see "A Century of Biological Research," *Illinois Nat. Hist. Surv. Bull.,* 1958, 27(2):85–234. On "oekologie" and the Ecological Society of America see Robert P. McIntosh, *The Background of Ecology: Concept and Theory* (Cambridge: Cambridge Univ. Press, 1985), pp. 2, 66. On the transformation of resource use see Richard White, *Land Use, Environment, and Social Change: The Shaping of Island County, Washington,* rpt. ed. (Seattle: Univ. Washington Press, 1992); William G. Robbins, *Colony and Empire: The Capitalist Transformation of the American West* (Lawrence: Univ. Press Kansas, 1994); and Richard W. Judd, *Common Lands, Common People: The Origins of Conservation in Northern New England* (Cambridge, Mass.: Harvard Univ. Press, 1997). For forests see James Willard Hurst, *Law and Economic Growth: The Legal History of the Lumber Industry in Wisconsin, 1836–1915,* rpt. ed. (Madison: Univ. Wisconsin Press, 1984); and Michael Williams, *Americans and Their Forests: A His-torical Geography* (New York: Cambridge Univ. Press, 1989). For fisheries see Arthur F. McEvoy, *The Fish-erman's Problem: Ecology and Law in the California Fisheries, 1850–1980* (Cambridge: Cambridge Univ. Press, 1986); Schneider, "Enclosing the Floodplain"; and Margaret Beattie Bogue, "To Save the Fish: Canada, the United States, the Great Lakes, and the Joint Commission of 1892," *Journal of American History,* 1993, *79:*1429–1454. For game see Louis Samuel Warren, *The Hunter's Game: Poachers, Conservationists, and Twentieth-Century America* (New Haven, Conn.: Yale Univ. Press, 1997). For grasslands see Donald Worster, *Dust Bowl; The Southern Plains in the 1930s* (Oxford: Oxford Univ. Press, 1979); Allan G. Bogue, *From Prairie to Corn Belt: Farming on the Illinois and Iowa Prairies in the Nineteenth Century,* rpt. ed. (Ames: Iowa State Univ. Press, 1994); and John Mack Faragher, *Sugar Creek: Life on the Illinois Prairie* (New Haven, Conn.: Yale Univ. Press, 1986).

and particularly the politics of nuclear weapons and energy and pesticides, strongly influenced the science of ecology. Scholars have also traced the influence of environmental politics on ecologists' response to the Depression and Dust Bowl in the United States in the late 1930s. Yet historians have debated the extent to which the early science of ecology reflected the environmental politics of forestry, agriculture, and fishing.[3]

Examination of the social and political context of Forbes's foundational paper "The Lake as a Microcosm" demonstrates the linkage between the science of ecology and environmental politics, formed through the intimate connections between ecologists and the local cultures of resource use at their study sites. This work builds on investigations into the "place of knowledge" in science studies. Scholars have increasingly investigated the role of the specific locations where knowledge is produced in the process of building scientific facts. Originally emphasizing the laboratory as a place of production, they have recently begun looking to the field in sciences such as geography, oceanography, anthropology, biology, and soil science, adapting approaches from laboratory studies to analyze field practices of sampling, collecting, sorting, and displaying material.[4]

This study shifts the focus to how scientists related to the social environment at their field sites and the importance of these social relations to ecology. Fieldwork is a distinguishing feature of ecological science. In the late nineteenth century ecology began to differentiate itself from the more established biological disciplines such as botany and zoology by stressing the study of the adaptations of organisms to environmental conditions in nature. This emphasis led ecologists out of museums and laboratories and into the field. Indeed, one leading early ecologist emphasized these aspects of the new science in defining

---

[3] On the role of environmental politics see Stephen Bocking, *Ecologists and Environmental Politics: A History of Contemporary Ecology* (New Haven, Conn.: Yale Univ. Press, 1997); McIntosh, *Background of Ecology,* pp. 289–323; Donald Worster, *Nature's Economy: A History of Ecological Ideas,* 2nd ed. (New York: Cambridge Univ. Press, 1994), pp. 218–253; Worster, *Dust Bowl;* Paolo Palladino, "On 'Environmentalism': The Origins of Debates over Policy for Pest-Control Research in America, 1960–1975," in *Science and Nature: Essays in the History of the Environmental Sciences,* ed. Michael Shortland (Oxford: British Society for the History of Science, 1993), pp. 181–212; and Joel B. Hagen, *An Entangled Bank: The Origins of Ecosystem Ecology* (New Brunswick, N.J.: Rutgers Univ. Press, 1992), pp. 100–121. On the debates over environmental politics in the early history of ecology see Eugene Cittadino, "Ecology and the Professionalization of Botany in America, 1890–1905," *Studies in the History of Biology,* 1980, *4*:171–198; and Ronald C. Tobey, *Saving the Prairies: The Life Cycle of the Founding School of American Plant Ecology, 1895–1955* (Berkeley: Univ. California Press, 1981), pp. 60–62. Richard Judd argues that local fishermen and foresters influenced scientific conservation in New England in *Common Lands, Common People,* esp. Ch. 9: "Tradition and Science in the Coastal Fisheries," pp. 229–262. Gregg Mitman, in *The State of Nature* (Chicago: Univ. Chicago Press, 1992), identifies the influence of the broader politics surrounding World War I on the Chicago school of ecology. Michael L. Smith connects the rapid transformation of the California environment to the history of the earth sciences in the United States in *Pacific Visions: California Scientists and the Environment, 1850–1915* (New Haven, Conn.: Yale Univ. Press, 1987), esp. pp. 143–185.

[4] On the "place of knowledge" see Adi Ophir and Steven Shapin, "The Place of Knowledge: A Methodological Survey," *Science in Context,* 1991, *4*:3–21; Charles W. J. Withers, "Reporting, Mapping, Trusting: Making Geographical Knowledge in the Late Seventeenth Century," *Isis,* 1999, *90*:497–521; David N. Livingstone, "The Spaces of Knowledge: Contributions towards a Historical Geography of Science," *Environment and Planning D: Society and Space,* 1995, *13*:5–34; and Jan Golinski, *Making Natural Knowledge: Constructivism and the History of Science* (Cambridge: Cambridge Univ. Press, 1998), pp. 79–102. On laboratory studies see Bruno Latour and Steve Woolgar, *Laboratory Life: The Construction of Scientific Facts* (1979; Princeton, N.J.: Princeton Univ. Press, 1986); and, for a recent review, Karin Knorr-Cetina, "Laboratory Studies: The Cultural Approach to the Study of Science," in *Handbook of Science and Technology Studies,* ed. Sheila Jasanoff, Gerald E. Markle, James C. Petersen, and Trevor Pinch (Thousand Oaks, Calif.: Sage, 1995), pp. 140–166. On the movement to the field see Henrika Kuklick and Robert E. Kohler, eds., *Science in the Field, Osiris,* 2nd Ser., 1996, *11*; Bruno Latour, *Pandora's Hope: Essays on the Reality of Science Studies* (Cambridge, Mass.: Harvard Univ. Press, 1999), pp. 24–79; Richard W. Burkhardt, Jr., "Ethology, Natural History, the Life Sciences, and the Problem of Place," *Journal of the History of Biology,* 1999, *32*:489–508; and Kohler, "Place and Practice in American Field Biology," unpublished MS.

ecology as "field physiology."[5] As ecologists moved into the field, they selected particular locations for detailed scientific observation and analysis. Scientists new to a particular area depended on local people to provide manual labor but also relied on their knowledge of its animals, plants, and habitats. Thus connected to a particular biological habitat, ecologists were also connected to the locale's myriad of other characteristics—its people, culture, and politics.

Understanding the science produced on the Illinois River floodplain requires an analysis of the local context of that knowledge, in all the senses of "local." The "local" has many meanings in the scholarship of science studies. The social constructivist approach views all scientific knowledge as locally produced, generated under specific circumstances in particular locations, such as a laboratory or field site. Locally specific knowledge can also mean the knowledge of particular locales: how one patch of forest or floodplain lake is different from another. Finally, local knowledge refers to the knowledge of local users of the resource, obtained through their everyday interactions with the ecosystem. All of these meanings of the "local" are important for understanding early ecology on the Illinois River. In addition, I extend the concept of local knowledge and argue that as ecologists absorbed the practices and knowledge of local resource users they came to adopt their political concerns as well. The local politics and social relations of a site can be as important a part of the specific circumstances of knowledge production as the local ecology or particular laboratory practices.[6]

Their engagement with the many meanings of "local" forced turn-of-the-century scientists to undertake what Thomas Gieryn has termed "boundary work."[7] Ecologists' reliance on local knowledge and involvement in local politics jeopardized their assumptions about the objectivity of science and its demarcation from other kinds of knowledge, particularly that held by fishermen and hunters. Forbes and the other survey scientists responded by mapping and navigating boundaries between local and scientific knowledge, between pure and applied ecology, and between science and politics. This boundary work, undertaken on the Illinois River floodplain, established the contours of the science of ecology as fundamentally connected to the politics of environmental transformation.

SCIENTIFIC UNDERSTANDING OF THE ILLINOIS RIVER

Stephen Forbes is a key figure in the history of ecology. (See Figure 1.) One of the first presidents of the Ecological Society of America, he was recognized by the National Academy of Sciences as a "founder of the science of ecology in the United States." In "The Lake as a Microcosm" Forbes described ecological communities, one of the central con-

---

[5] Frederic Edward Clements, *Research Methods in Ecology* (Lincoln, Nebr.: Univ. Printing Company, 1905), p. 7, cited in Hagen, *Entangled Bank* (cit. n. 3), p. 15. See also Cittadino, "Ecology and the Professionalization of Botany" (cit. n. 3), pp. 174–181.

[6] I thank an anonymous referee for helping me clarify the meanings of "local." For a review of the social constructivist approach see Golinski, *Making Natural Knowledge* (cit. n. 4), pp. 27–46. Local knowledge may also be understood as practical knowledge, indigenous knowledge, working knowledge, or folk wisdom. James Scott calls it "mētis," emphasizing both its practical aspects and its adaptability, in *Seeing Like a State: How Certain Schemes to Improve the Human Condition Have Failed* (New Haven, Conn.: Yale Univ. Press, 1998), pp. 309–341. See also Helen Watson-Verran and David Turnbull, "Science and Other Indigenous Knowledge Systems," in *Handbook of Science and Technology Studies*, ed. Jasanoff *et al.* (cit. n. 4), pp. 115–139.

[7] Thomas F. Gieryn, *Cultural Boundaries of Science: Credibility on the Line* (Chicago: Univ. Chicago Press, 1999); and Gieryn, "Boundaries of Science," in *Handbook of Science and Technology Studies*, ed. Jasanoff *et al.*, pp. 393–443.

Figure 1. *Stephen A. Forbes, from the 1880s, around the time he wrote "The Lake as a Microcosm." Photo courtesy of the Illinois Natural History Survey.*

cepts of the field. The concept of community extended the focus of ecology from the interaction of organisms with the physical environment to their interactions with each other. Forbes thought that lakes provided the ideal environment for investigating these ideas: he called a lake "a little world within itself—a microcosm within which all the elemental forces are at work and the play of life goes on in full." Within this microcosm existed an ecological community, a "complete and independent equilibrium of organic life and activity."[8]

Although the bulk of "The Lake as a Microcosm" concerned the glacial lakes of northern Illinois, the Illinois River floodplain was at the center of Forbes's ecological research and his concepts of the microcosm and the ecological community. Forbes considered the flood-plain lakes to be "much more numerous and important" than the glacial lakes, and, as the "most important breeding grounds and reservoirs of life," they supported a large and active fishery. Forbes first used the idea of the microcosm in an 1880 article on the food of fishes, part of a series of papers investigating the feeding relations of fishes, birds, and insects. Food had a prominent place in Forbes's work because predation was one of the clearest ways in which an animal could affect others in its habitat. Thus Forbes looked to feeding to observe the interactions of animals with each other and their effects on the community as a whole. "Whatever affects any species" in a lake, he wrote, "must speedily have its influence of some sort upon the whole assemblage." Forbes introduced this approach to examining communities by documenting the feeding relations of fishes. He sampled river and lake fishes and analyzed the contents of their stomachs to determine their ecological relations with other species. Most of the material analyzed in the 1880 paper in which he first developed the idea of a microcosm came from the Illinois River and floodplain.[9]

In "The Lake as a Microcosm," Forbes developed these ideas further to examine the role of natural selection in maintaining a harmonious balance among the organisms in-habiting an area. Such a balance might have been expected in the northern glacial lakes because they varied "but little in level with the change of the season, and scarcely at all from year to year," giving species time to equilibrate with each other. The floodplain, however, subject to a continuous cycle of flooding and drying, provided Forbes with "perhaps no better illustration of the methods by which the flexible system of organic life adapts itself, without injury, to widely and rapidly fluctuating conditions."[10]

Forbes significantly expanded on his ideas about the importance of cycles of distur-bances in aquatic ecology by establishing a year-round biological field station on the Illinois River at Havana in 1894. (See Figure 2.) The overall goal of the research program

[8] L. O. Howard, "Biographical Memoir of Stephen Alfred Forbes, 1844–1930," *Biographical Memoirs of the National Academy of Sciences,* 1932, *15*:2–54, on p. 16; and Forbes, "Lake as a Microcosm" (cit. n. 1), p. 77. For Forbes's role in the history of ecology see Stephen Bocking, "Stephen Forbes, Jacob Reighard, and the Emergence of Aquatic Ecology in the Great Lakes Region," *J. Hist. Biol.,* 1990, *23*:461–498; McIntosh, *Back-ground of Ecology* (cit. n. 2), pp. 58–60; Frank B. Golley, *A History of the Ecosystem Concept in Ecology: More Than the Sum of the Parts* (New Haven, Conn.: Yale Univ. Press, 1993), pp. 36–37; Hagen, *Entangled Bank* (cit. n. 3), pp. 7–10, 15; and Robert Allyn Lovely, "Mastering Nature's Harmony: Stephen Forbes and the Roots of American Ecology" (Ph.D. diss., Univ. Wisconsin–Madison, 1995). For the importance of "The Lake as a Microcosm" in particular see Sharon E. Kingsland, "Defining Ecology as a Science," in *Foundations of Ecology: Classic Papers with Commentaries,* ed. Leslie A. Real and James H. Brown (Chicago: Univ. Chicago Press, 1991), pp. 1–13. Forbes's landmark paper appears as the first selection in this collection. "The Lake as a Microcosm" is still required reading in graduate programs in ecology, and, one hundred years after its publication, it is still regularly cited in scientific publications; see Institute for Scientific Information, *Science Citation Index.*

[9] S. A. Forbes, "The Food of Fishes," *Bulletin of the Illinois State Laboratory of Natural History,* 1880, *1*(3):18–65, on p. 18. The exposition of the microcosm is on pp. 17–19; this material was repeated almost verbatim in "The Lake as a Microcosm." Other articles by Forbes appearing in this volume were "On Some Interactions of Organisms," pp. 3–17; "On the Food of Young Fishes," pp. 66–79; "The Food of Birds," pp. 80–148; "Notes upon the Food of Predaceous Beetles," pp. 149–152; and "Notes on Insectivorous Coleoptera," pp. 153–160. Of twelve species of fishes whose sampling location is given, nine were from the Illinois River or floodplain. Of these, three were specifically from the Illinois River bottoms and one from the Mississippi River bottoms as well.

[10] Forbes, "Lake as a Microcosm" (cit. n. 1), pp. 78–79. On Forbes and the balance of nature see Kingsland, "Defining Ecology as a Science" (cit. n. 8), pp. 1–4.

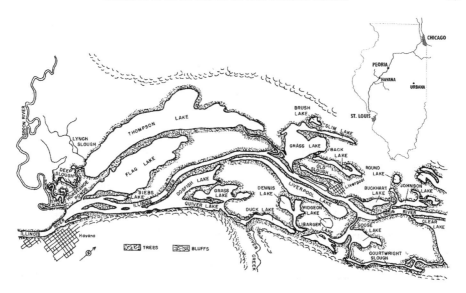

Figure 2. *"Map of the Illinois River and its adjoining bottomland lakes above Havana before 1912." The survey conducted detailed research on these lakes beginning in 1894. Most of the lakes in this map were drained for agriculture by the mid 1920s. Showing a continued reliance on local knowledge, this map, drafted for a 1965 survey bulletin, was reported as "checked by William Riley and Frank Rudolph, commercial fishermen residing in Havana; both had fished in the various lakes shown on the map." From William C. Starrett and Arnold W. Fritz, "A Biological Investigation of the Fishes of Lake Chautauqua, Illinois," Illinois Natural History Survey Bulletin, 1965, 29(1):14. The inset map, showing the location of the area in Illinois, is not in the original.*

there was to examine "the effect on the aquatic plant and animal life of a region produced by the periodical overflow and gradual recession of the waters of great rivers."[11] For the next three decades, under the direction of Forbes and the superintendents of the field station, Charles Kofoid and then Robert Richardson, scientists conducted studies of the plants, fish, and wildlife of the river and its floodplain. One of their main conclusions was that the productivity of the fishes was related to the area of the flooded land—the bottomland lakes, sloughs, and wetlands.

By 1910 Forbes could summarize the survey's scientific understanding of the Illinois River. The bottomland lakes contributed to fish productivity in two ways, he argued. First, they provided shallow, still water for breeding and feeding grounds. Second, they supplied food for the fishes in the river proper. Forbes's early work on fishes showed that "virtually all our young fishes, whatever their adult habits may be, live at first on the same kind of food. . . . This first food—the minute plant and animal life of the water, called its plankton—is produced almost wholly in the backwaters." In flowing water, without continuous replenishment, all of the plankton would eventually be washed downstream, leaving the river devoid of food. Periodic flooding, however, added plankton to the river from the backwater lakes. Fish, he argued, had adapted to use the resources provided by the flood. They moved on rising water into the backwater areas where they bred; there, too, the young fish found an abundance of food. With falling water, the fish moved back into the

[11] *Biennial Report of the Director, 1893–1894* (Chicago: Illinois State Laboratory of Natural History, 1894), p. 17.

main channel. "The longer the period and the larger the scale of the spring overflow," Forbes wrote, "the better is the prospect for a heavy annual contribution to the population of the stream." As a result, he concluded, "the fish-producing capacity of the stream is thus proportionate, other things being equal, to the extent and fertility of the backwaters."[12] The work of the survey provided the first quantitative estimates of the productivity of the backwaters and documented the importance of the floodplain to the maintenance of fish populations in the river.

Forbes's emphasis on the connections between the river and the floodplain differed from the work of contemporary fisheries scientists. At the time Forbes began his studies, biologists emphasized the longitudinal nature of rivers: how fishes moved up- or downstream rather than from the main channel to the backwater lakes. For instance, the eminent fish biologist David Starr Jordan classified fishes according to their usual location along a river, starting with those typical of "lowland" areas and moving upstream to "channel," "upland," and "mountain" fishes.[13] Even when scientists considered the effects of flooding, the focus was typically on how floods washed species out of upstream areas rather than on how fish used the habitats of the floodplain. This emphasis on the longitudinal nature of rivers was also a feature of the earliest European literature, which classified fishes by their location. The only other prominent work on floodplains, that of Grigore Antipa on the Danube, appeared over thirty years after Forbes began his work on the Illinois River.[14]

FISHERMEN AND THE DEVELOPMENT OF FORBES'S RIVER ECOLOGY

Forbes's divergence from fisheries scientists in his understanding of a river as the connected waters of the channel and its floodplain built on the local culture of fishing on the river. Although Forbes and his colleagues at the survey established the scientific evidence for the importance of the floodplain to the fishery, the relation of the backwaters to fish productivity was already well known to fishermen. Over twenty years before Forbes delivered "The Lake as a Microcosm," Isaiah Prickett used his knowledge of the backwaters

[12] Stephen A. Forbes, "The Investigation of a River System in the Interest of Its Fisheries," in *Biological Investigations of the Illinois River* (Urbana: Illinois State Laboratory of Natural History, 1910), pp. 11–12, 12.

[13] David Starr Jordan, "The Distribution of Fresh-Water Fishes," in *Transactions of the American Fish-Cultural Association: Seventeenth Annual Meeting* (1888), pp. 4–24. This classification scheme emphasizing the upstream–downstream axis in rivers was certainly known to Forbes, as Jordan's paper immediately preceded a paper given by Forbes at the meeting of the American Fish-Cultural Association. See also Fred Mather, "Poisoning and Obstructing the Waters," in *Proceedings of the American Fish Culturists' Association: Third Annual Meeting* (1888), pp. 14–19; Marshall Mc'Donald, "A New System of Fish-way Building," in *Transactions of the American Fish-Cultural Association: Twelfth Annual Meeting* (1883), pp. 57–62; and Mc'Donald, "Natural Causes Influencing the Movements of Fish in Rivers," in *Transactions of the American Fish-Cultural Association: Thirteenth Annual Meeting* (1884), pp. 164–170.

[14] For work on flooding see, e.g., Victor E. Shelford, "Ecological Succession, I: Stream Fishes and the Method of Physiographic Analysis," *Biological Bulletin of the Woods Hole Marine Biological Laboratory,* 1911, *21*:9–35; and Charles C. Adams, "Baseleveling and Its Faunal Significance, with Illustrations from Southeastern United States," *American Naturalist,* 1901, *35*:839–851. Classification of river fishes in the European literature is discussed in H. B. N. Hynes, *The Ecology of Running Waters* (Toronto: Univ. Toronto Press, 1970), pp. 383–397. For Antipa's work see Grigore Antipa, *Regiunea inundabilă a Dunării: Starea ei actuală si mijloacele de a o pune in valoare* (Bucharest, 1910), cited in Robin L. Welcomme, *Fisheries Ecology of Floodplain Rivers* (London: Longman, 1979), p. 276. Forbes was familiar with this work, as Antipa had sent him a 1912 German translation, and was intrigued by how it might apply to the Illinois River. He ultimately concluded that the specific conditions of the Danube River made Antipa's recommendations inappropriate for the Illinois. See S. A. Forbes to C. J. Dittmar, 9 Jan. 1914; Forbes to Lyman E. Cooley, 4 Feb. 1914; Forbes to Paulus Schiemenz, 18 Feb. 1914; and Forbes to Robert Richardson, 27 Feb. 1914, Natural History Survey, Chief's Office, 1912–1931, Record Series 43/1/5, Box 2, University of Illinois Archives, Urbana.

and the river's fishes to great effect. In the 1860s he built a fish trap in the narrow slough that connected the Illinois River with Thompson Lake, the largest floodplain lake along the river. When the river flooded, the fish would move into the slough from the river and then into the lake. Prickett's trap was simple in design, essentially a box with a door in the middle, placed across the entire width of the connecting slough. As the water levels dropped, the fish would try to move back into the river. Prickett would open the trap's door, "and it would fill up with fish. Then he would raise it up and shove the fish out and then lower it again." Fishermen called this and related techniques the "shutting of a slough"; because the fishermen understood fish behavior in relation to the flood, "every one of [the fish] can be taken."[15]

Richard White, in his history of the Columbia River, emphasized the importance of "knowing nature through labor."[16] Through their labor, the fishermen on the Illinois River knew the natural history of the river species, their habits, and how to catch them. This fishermen's knowledge, based on decades of observing fish in order to catch them, became part of the scientific understanding of the ecologists. Recognizing their value, Forbes worked with fishermen from the beginning of his studies on the rivers of Illinois. As he came to appreciate fishermen's ability to contribute to the scientific knowledge of the Illinois River, he also began to understand them as people. Because of this understanding, the scientists would eventually connect their ecological work to political efforts to protect the fishery.

Beginning in 1876, while he was investigating the food habits of the fishes, Forbes worked with local fishermen on the Illinois, Ohio, and Mississippi Rivers. He was following in a tradition established by scientists at the U.S. Fish Commission, who interviewed marine and freshwater fishermen in their efforts to understand the causes of declines in the fisheries.[17] Yet unlike the Fish Commission scientists, Forbes established ties and a mode of research that continued for the next fifty years of his professional life. At first, Forbes felt both repelled and intrigued by the fishermen, whom he saw as extremely strange, "others" to whom he had difficulty relating. Though initially taken aback by the fishermen, their customs, and their living conditions, Forbes continued to work with them. Through ongoing contact, he forged an understanding and empathy that influenced his future work and contributed to his support of their interests. His letters home during these research trips are filled with descriptions of the fishermen and their families, houses, food, music, and character.

A series of letters that Forbes sent to his wife while on a collecting trip in the Kentucky

[15] Testimony of Richard England, transcript of *State v. New*, 280 Ill. 393 (23 Oct. 1917), RS 901, Supreme Court Trial Transcripts, Vault 36505, Illinois State Archives (hereafter cited as **transcript of State v. New**), Abstract, pp. 276–277 (describing how the trap worked). Because traps of this sort blocked access by other fishermen trying to enter the lake, they were often destroyed. After describing the trap, England, a fisherman, continued, "We took an ax and cut it out." *Report of Board of Illinois State Fish Commissioners, to the Governor of Illinois: October 1, 1894 to September 30, 1896* (Springfield, Ill.: Phillips Bros., State Printers, 1897) ("every one . . . taken").

[16] Richard White, *The Organic Machine: The Remaking of the Columbia River* (New York: Hill & Wang, 1995), esp. pp. 3–29.

[17] S. A. Forbes to C. Forbes, 1 Nov. 1877, Folder I.B.1, Stephen Alfred Forbes Collection, Illinois Historical Survey Library, University of Illinois, Urbana (hereafter cited as **Forbes Collection**); and Dean C. Allard, *Spencer Fullerton Baird and the U.S. Fish Commission* (New York: Arno, 1978), pp. 92–93, 107. James W. Milner of the U.S. Fish Commission spent the summer of 1871 interviewing fishermen and fish dealers while investigating the Lake Michigan fisheries; see Milner, "Report on the Fisheries of the Great Lakes: The Result of Inquiries Prosecuted in 1871 and 1872," in *United States Commission of Fish and Fisheries: Report of the Commissioner for 1872 and 1873*, U.S. Senate, 42nd Cong., 3rd sess., 1874, Mis. Doc. 74, pp. 1–78. I thank Glenn Sandiford for directing me to these references. See also Bogue, "To Save the Fish" (cit. n. 2), p. 1439.

bottoms in 1879 documents his changing attitudes. At first horrified by the fishermen, Forbes wrote, "I am working alone on an indescribably dirty table . . . in the genuine Kentucky farm house—built up on stilts apparently to afford shelter for the pigs under it. . . . The boards of the floors are all loose and slip and rattle under our feet, and whatever is dropped falls through to the pigs." Three days later he continued, "My Kentucky friends with the best intentions, are fast becoming unendurable—but I shall endure them, never-theless. . . . When you are worried [about our children] thank heaven that they are not pigs and drunkards, and vagabonds, and dogs and burs and dirt and flies in the butter and that they don't smell of mingled smoke and bacon grease nor swear nor drink nor chew. Whatever happens to you, console yourself by remembering that you are not in the Ken-tucky bottoms."[18]

Yet as he continued to work with them, Forbes developed an understanding of and empathy with the bottomlanders. While he complained of "our pigs—Kentuckians," he also began to realize that they might think him strange, and his activities annoying. "Per-haps they are writing scalding reports of us 'Yanks' to their friends and sweethearts," he speculated, asking his wife, "What would you think of a crew of boarders from a foreign land who should convert your back porch into a slaughter house, dress . . . catfishes thereon day after day, and tip all their scraps over the railing to the pigs under your house? Perhaps they don't like it any better." Displaying a growing empathy, he decided to set up his tent as a workroom rather than foul his hosts' house. After another week, Forbes had begun to establish a personal relationship with the fishermen. "I succeed quite to my own admira-tion," he declared, "in affiliating with these bottom-landers. We work and talk together with a mutual confidence quite touching." Although he still considered the Kentucky bottomlanders as "others," he began to paint them more sympathetically.[19]

Local fishermen played a crucial role in the scientific studies conducted by Forbes and the survey on the Illinois River over the following decades. The establishment of the field station was of paramount importance. The biological station at Havana linked Forbes's river research to a particular place. Prior to this, he had collected throughout the state, primarily on the Illinois, Mississippi, Fox, and Ohio Rivers. But with the establishment of the station, he set up a series of fixed sampling sites that linked his work not only to a particular habitat but to a particular community, its culture, and its politics. First and foremost, Havana was selected because of its access to floodplain lakes. Yet other consid-erations were important as well. It was on a direct rail line from the university, was situated on a bluff overlooking the river—which minimized the danger of malaria—and had access to springs providing clean water.

But Forbes and Kofoid, in different ways, also emphasized the broader context of the particular location, its specific social relations and importance to the state's economy. From the initial selection of the field site, aspects of the social environment entered into research decisions. Charles Kofoid's thoughts on the importance of location in planning fieldwork are made clear in his report on European biological stations, in which he drew on his experience at Havana. Kofoid wrote that biological stations were particularly well suited to take an important role in the "conservation of the aesthetic and economic resources of lakes and streams." Giving "special attention . . . to the economic or applied scientific phases of their activities," he discussed the pertinent considerations in locating a station. Biological stations that emphasized these applied questions, he noted, were "obviously

[18] S. A. Forbes to C. Forbes, 13 Sept. 1879, 16 Sept. 1879, Folder I.B.1, Forbes Collection.
[19] S. A. Forbes to C. Forbes, 17 Sept. 1879, 24 Sept. 1879, Folder I.B.1, Forbes Collection.

best located in or near great fishing centers where contact with fisheries problems is most intimate." Havana was just such a center. It had the largest fishery on the river, accounting for over 20 percent of the river's catch in 1896, just after the station was established. In 1908 Havana's catch amounted to 10 percent of that produced by all the freshwater fisheries in the United States.[20]

In his lyrical vignette on the Havana station, "Midsummer at the Biological Station," Forbes emphasized not Havana's practicality but its exoticism. Although "Biological Station" appeared in the title, no science was done in this piece; rather, it was an evocation of a place and its inhabitants. Though just a hundred miles from the survey's headquarters in Urbana, Havana was presented in the most exotic of terms: "a river scene, glowing under the semi-tropical heat of a July day." The wind was "an exhalation of the torrid zone, and leads one's imagination back along its course to the Orinoco and the Amazon." By the time Forbes wrote this piece the local bottomlanders were no longer the "pigs" of his first encounters; now his lyrical description romanticized them. Describing a family of bottomlanders, he wrote, "A boat juts out below—as primitive a boat as any on the Nile in the time of the Pharaohs,—and in it a man and two boys—also as primitive as Moses' Hebrews in the wilderness—they are crossing the stream from the little town behind us to the opposite bottom lands where they have their home in a temporary hut among the trees."[21]

While the vignette may seem odd as a description of a scientific research program, its emphasis on the local inhabitants was appropriate, given their importance to the station's work. These bottomlanders—fishermen, hunters, and boatmen—were central players in the scientific enterprise of the station. They worked as field assistants, guiding survey scientists in unfamiliar locations. They helped the scientists as laborers, rowing boats and collecting fish in their seines, trammel nets, and fish traps. Beyond this manual labor, however, they provided knowledge—both on how to catch fish and on their natural history. Because the fishermen had developed their working techniques over many decades, their gear was highly suited to the conditions of the Illinois River and its floodplain. They used a variety of nets and fish traps and developed boats that could navigate with heavy loads of fish on the often shallow and weed-choked backwaters. As scientists discovered that the techniques they brought from other areas were not effective, they adopted those of the local fishermen. Much of the equipment the survey scientists used to investigate the ecology of the river was modeled on that of the fishermen. When Kofoid needed a research boat to begin his studies of the river's plankton, he turned to a local builder to provide a boat designed "after the pattern of the fish boats—with a model bow, square stern and flat bottom." The fisheries researchers initially used a small, 40-yard seine, but Kofoid realized that the net was inadequate when he found that "we miss many things that the fishermen get." (See Figure 3.) He modified the survey's nets to match the commercial gear, with larger seines and the addition of gill nets.[22]

Fishermen also taught the scientists about the natural history of fish. Forbes filled his

[20] Charles Atwood Kofoid, "The Biological Stations of Europe," *United States Bureau of Education Bulletin, 1910,* no. 4, pp. xiii, 4. For the statistics see John W. Alvord and Charles B. Burdick, *Report of the Rivers and Lakes Commission on the Illinois River and Its Bottomlands, with Reference to the Conservation of Agriculture and Fisheries and the Control of Floods* (Springfield, Ill., 1915), p. 66.

[21] S. A. Forbes, "Midsummer at the Biological Station," n.d., Folder II.A.4, Forbes Collection.

[22] C. A. Kofoid to Forbes, 6 July 1897, Natural History Survey, Chief's Office, Chief's Correspondence, 1871–1909, Record Series 43/1/1, Box 8, Univ. Illinois Archives; and Kofoid to Forbes, 3 May 1899, RS 43/1/1, Box 10, Univ. Illinois Archives.

Figure 3. *"Seining in the Illinois River, Beardstown, Ill." This view is from a postcard mailed in 1909. Most seining took place not in the river itself, as depicted here, but in the backwater lakes. Prior to working for the Natural History Survey, Miles Newberry, Henry Allen, and other locals had worked on seine crews like this one. Postcard in possession of author.*

reports with commentary on the habits of the fish that he learned about from talking with fishermen. For instance, he noted that he had been "repeatedly assured by fishermen that the catfish seizes the foot of the mollusk . . . and tears the animal loose by vigorously jerking and rubbing it about. One intelligent fisherman informed me that he was often first notified of the presence of catfishes in his seine, in making a haul, by seeing the fragments of clams floating on the surface, disgorged by the struggling captives." Forbes was skeptical; yet when he examined the stomach contents of the catfish the absence of any shell fragments supported the observations of the fishermen. "How these fishes manage to separate mollusks . . . from the shell, I am scarcely able to imagine," he admitted. Later, when survey ecologists were investigating the effects of levee construction and sewage on fish populations, they specifically sought out the knowledge of locals, canvassing commercial fishermen along the length of the river for information on changing fish populations.[23]

As research frequently confirmed local knowledge, Forbes developed a trust in his informants that led him to seek their advice throughout his studies. When the survey established the field station in 1894 one of the first employees was a local fisherman, Miles

[23] For just a few examples of commentary on the habits of fish see S. A. Forbes, "Studies of the Food of Fresh-Water Fishes," *Bull. Illinois State Lab. Nat. Hist.*, 1888, 2(7):433–473 (quotation from p. 458); and Forbes and Robert Earle Richardson, *The Fishes of Illinois*, 2nd ed. (Springfield: Illinois State Journal Company, 1920), p. 70. On canvass of fishermen see Richardson to Forbes, 15 Apr. 1913; and "Fishes, Illinois River, De Pue to Hennepin: Testimony of F. L. Powers, De Pue, Ill., April 16," RS 43/1/5, Box 1, Univ. Illinois Archives. These data were published in Forbes and Richardson, "Studies on the Biology of the Upper Illinois River," *Bull. Illinois State Lab. Nat. Hist.*, 1913, 9(10):481–574, on p. 537.

Newberry. (See Figure 4.) Newberry, born on a farm, had moved to the Illinois River at Havana in 1876, when he was eighteen. He was looking for farm work but soon got a job hauling a seine for a commercial fisherman on Thompson Lake. He acquired his knowledge of the Illinois River through his paid labor as a fisherman, deckhand, and engineer, as well as from subsistence activities: fishing in the summer, hunting in the spring and fall, and trapping in the winter. Newberry applied his skills for the survey, where he constructed and repaired nets, repaired the boats and engines, served as a river guide, helped with sampling, and acted as an unofficial liaison to other fishermen on the river. He also provided more specific advice on particular projects, recommending the sampling scheme for research on the distribution, migration, breeding habits, and food of the Illinois River fishes. Kofoid described Newberry as a "valuable" member of the survey staff; his experience as a fisherman and riverman played an important role in the development of their research program on the Illinois River.[24] Following the experience with Newberry, the survey continued to hire experienced fishermen to assist in its studies. When Newberry left after several years, his role was assumed by a young man, Hurley, who was the son of the foreman of a commercial fish crew. Hurley was initially hired as a laborer, to assist Newberry in the field sampling, but with Newberry's departure he became indispensable. When Hurley left in 1909 he helped train his replacement, Henry (Hank) Allen, the foreman of a commercial seine crew.

Not all local fishermen were regarded as experts. Kofoid, for instance, found some of them unreliable as to the location of good fishing grounds. The survey thus sought not just local knowledge, but expert knowledge. Some fishermen knew the habits of fishes better than others; some were more familiar with local waters. Discussions of the qualifications of the fishermen give an indication of the skills that survey scientists valued. Some fishermen, like Newberry, were hired for their local expertise—their knowledge of local waters and their skill at fishing and in piloting boats and repairing equipment. Allen, too, was "thoroughly experienced in handling fishing tackle, launches, and engines"; Richardson was a bit concerned that his "knowledge of local waters is perhaps not so minute . . . having moved here from Bath only 3 years ago." Others were hired simply for their labor, at least initially: Kofoid described Hurley as "a good steady stout lad of 19 who can do a man's work at seining and hauling boats."[25]

In addition to local knowledge and physical strength, the scientists also sought fishermen who would fit in with the more educated staff of the station. Some of the ambivalence that was revealed in Forbes's early letters from the field persisted in the dealings of the survey with the fishermen. Kofoid referred to some local fishermen as "river toughs," while Richardson wrote of "Havana's bed bugs, river rats and other human insects." Scientists described other fishermen with a backhanded compliment: "in intelligence far above ordinary fishermen." Other colleagues apparently shared Forbes's initial shock at the filthiness of the locals: one of the key criteria for hiring fishermen was personal cleanliness. "Personally he is neat and clean, above the ordinary for fishermen," wrote Richardson of Hank Allen. (See Figure 5.) Scientists were also concerned about maintaining proper authority. Allen "is a few years older than I am," Richardson worried. But he concluded that "considering

[24] Testimony of Miles Newberry, transcript of State v. New, Record, pp. 965–986; "University of Illinois Biological Station," Illini (Champaign), 28 Feb. 1896, 25(20):315–320; Kofoid to Forbes, 23 June 1898, RS 43/1/1, Box 8, Univ. Illinois Archives; Biennial Report of the Director for 1899–00 (Urbana: Illinois State Laboratory of Natural History, 1901), pp. 3–4; and Kofoid to Forbes, 31 July 1898, RS 43/1/1, Box 8, Univ. Illinois Archives.

[25] Kofoid to Forbes, 28 Apr. 1899 (?), 23 Apr. 1899, RS 43/1/1, Box 10, Univ. Illinois Archives.

Figure 4. *"Bottom-lands at High Water." Ecologist Charles Kofoid is in the bow, with fisherman Miles Newberry in the stern. Newberry was a local fisherman and hunter hired as a field assistant by the survey. This 1895 photograph appears staged to show the scientist as "local." Compare this scene with Stephen Forbes's vignette describing life on the Illinois River: "A boat juts out below—as primitive a boat as any on the Nile in the time of the Pharaohs,—and in it a man and two boys—also as primitive as Moses' Hebrews in the wilderness—they are crossing the stream from the little town behind us to the opposite bottom lands where they have their home in a temporary hut among the trees. One of them, standing in the bow, paddles the boat with a single oar, first on this side and then on that, another stands in the middle." (S. A. Forbes, "Midsummer at the Biological Research Station," n.d., Folder II.A.4, Forbes Collection.) Photo courtesy of the Illinois Natural History Survey.*

the facts as I have observed them . . . I think there is no good reason in sight for fearing we could not get along." Before hiring fishermen, survey scientists often asked around among the wealthier fish dealers for references. These fish dealers, among the "respectable" citizens of Havana, helped provide an entrée into the seamier world of the laborer and the independent fisherman.[26]

ECOLOGISTS AND THE STRUGGLE FOR THE FLOODPLAIN

Aside from local knowledge of fishes and their habits and habitats, scientific understanding of the river was also informed by the political struggles over resource use in which the fishermen were involved. Prior to 1880, the floodplain had been treated as a commons.

---

[26] Kofoid to Forbes, 21 July 1897, RS 43/1/1, Box 8; Richardson to Forbes, 16 Jan. 1913, n.d., RS 43/1/5, Box 1; and Richardson to Forbes, 3 Oct. 1909, RS 43/1/1, Box 10, Univ. Illinois Archives. The dealers reported, for instance, that Allen was regarded as "steady and well behaved, industrious and responsible."

Figure 5. *Robert Richardson (left) and Henry Allen (right) with a seine. Before working for the survey, Allen was a foreman of a fishing crew for one of the larger commercial fishing operations on the river. As a survey employee, he was to care for the boats and perform routine collecting. Richardson acknowledged Allen in a paper: "We had also as field helper, at the oars and in every service in which help was needed, an experienced, intelligent, and interested fisherman and mechanic, to whom is due no small part of the credit for whatever success attended the season's operations." (R. E. Richardson, "Observations on the Breeding Habits of Fishes at Havana, Illinois, 1910 and 1911,"* Bulletin of the Illinois State Laboratory of Natural History, *1913, 9[8]:405–416, on page 405.) Photo courtesy of the Illinois Natural History Survey.*

Up and down the river, people harvested the fish, hunted waterfowl, grazed their livestock on floodplain grasses, cut timber, and collected pecans. Beginning in the 1880s, wealthy sportsmen began buying up the floodplain and restricting access to these once-public areas. Independent commercial fishermen protested these changes by trespassing, poaching, and legal challenges. As the protests increased, hunting clubs and other landowners gave up on preserving flooded areas for the pleasure of private duck hunting and began looking for a new kind of profit. They converted their holdings to cropland by building levees and draining the enclosed lakes and wetlands. In response to the loss of these waters, fishermen increased their protests. Matters came to a head in 1908, when about fifty armed fishermen prevented a dredge from completing an agricultural levee that would drain an important fishing lake.[27]

Survey scientists quickly recognized the threat to the fishery that drainage represented. In the 1890s, early in the survey's studies, Kofoid wrote that "the development in recent

[27] Schneider, "Enclosing the Floodplain" (cit. n. 1).

years of extensive systems of levees in the bottoms of the Illinois River for the purpose of protecting farm lands from untimely floods increases the importance of, and necessity for, the reservoir backwaters." In 1910, as the pace of levee construction quickened and its harm to the fishery became apparent, survey scientists repeated their warnings against drainage. "Nothing can be more dangerous to the continued productiveness of these waters than a shutting of the river into its main channel and the drainage of the bottom-land lakes for agricultural purposes," declared Forbes to the American Fisheries Society. By 1910, two fifths of the floodplain had already been drained, and "in the face of the gigantic interests—agricultural, industrial, commercial, and political—which are now mustering along its course," Forbes worried that the remainder of the floodplain would be destroyed and the fishery permanently decimated.[28]

The impending loss of the floodplain suggested new research questions and experiments. Forbes's research began to focus on a new problem: how to protect the fishery of the river from drainage. This was both an ecological problem and a political one. "Since 1910," Forbes wrote, "we have given all our work a turn towards the fisheries interest." Forbes thought it would be exceedingly difficult to fight the "gigantic interests" threatening the floodplain; backwater lakes would be drained as long as drainage remained the most profitable use of the land. In an attempt to reverse the cost-benefit analysis, Forbes encouraged research to make the fishery more economically valuable so that it could compete with agriculture. One obstacle to protecting the fish industry was the low price commanded by the dominant commercial fish, the carp. While valued in Europe and among immigrant communities in the United States, the carp was not broadly popular in America. In collaboration with Forbes, Dr. Nellie Goldthwaite, a member of the household science department at the University of Illinois, developed recipes for carp that might increase demand and thus its price.[29] If carp sold at higher prices, floodplain areas might be more valuable as wetlands than drained and plowed into fields.

In addition, Forbes began investigating a method of growing both crops and fish on the floodplain. In this scheme, akin to crop rotation, the floodplain would be rotated between wetlands and crops as it was flooded one year and kept dry the next. Forbes outlined a set of experiments to determine whether aquatic plants and animals could remain dormant in the floodplain soils until reflooded. He instructed Richardson to collect earth "from places which were overflowed last spring and are now dry, and from other places which were overflowed two years ago but not since—possibly from some which have been now dry for three years"—and to place the earth in aquaria where it would be reflooded.[30] Forbes

[28] C. A. Kofoid, "Plankton Studies, IV: The Plankton of the Illinois River, 1894–1899, with Introductory Notes upon the Hydrography of the Illinois River and Its Basin, Pt. 1: Quantitative Investigations and General Results," *Bull. Illinois State Lab. Nat. Hist.,* 1903, *6*(2):95–629, on p. 568; and Forbes, "Investigation of a River System" (cit. n. 12), p. 14. Other ecologists also became concerned with the protection of their research sites. Henry Cowles was involved in efforts to protect the Indiana Dunes where he had done his pioneering work on succession; see J. Ronald Engel, *Sacred Sands: The Struggle for Community in the Indiana Dunes* (Middletown, Conn.: Wesleyan Univ. Press, 1983), pp. 79–84.

[29] S. A. Forbes, "Notes of Conference between the Illinois State Game and Fish Conservation Commission and the Director of the Natural History Survey, Urbana, Ill., November 11, 1913," p. 3, RS 43/1/5, Box 1; and Forbes to Rivers and Lakes Commission, 19 Mar. 1914, RS 43/1/5, Box 2, Univ. Illinois Archives. On the status of the carp—which fetched about a fifth its German price in U.S. markets—see Alvord and Burdick, *Report on the Illinois River and Its Bottomlands* (cit. n. 20), pp. 124–126. Regarding the efforts to increase demand see Forbes to Dittmar, State Game and Fish Conservation Commission, 12 Dec. 1913; and Forbes to N. E. Goldthwaithe [*sic*], Household Science Department, 20 Mar. 1914, RS 43/1/2, Box 2, Univ. Illinois Archives. For some recipes see N. E. Goldthwaite, "The Cooking of Carp," *University of Illinois Bulletin,* 1915, *13*(9):1–9.

[30] Forbes to Rivers and Lakes Commission, 19 Mar. 1914. This scheme was suggested by the work of Antipa.

was interested in which plants and animals would be present after various periods of desiccation. This information would be necessary to determine the potential availability of food for newly hatched fish larvae and thus the proper cycle of rotation between dry-land agriculture and fish culture.

Forbes's research priorities had political implications. Recreational and commercial fisheries interests were in conflict over this period. Research on rotational use of the floodplain and encouraging the use of carp was directed toward preserving the health of the "wild fishery" rather than developing methods of fish culture.[31] While fish culturists of the period emphasized the role of hatcheries in augmenting populations produced in the wild, Forbes worked to protect the floodplain lakes from drainage and so to preserve the ability of the river and floodplain to produce fish without augmentation. While hatcheries were primarily developed for increasing sport-fish production, the effort to preserve the wild fishery of the Illinois River supported the political interests of the commercial fishermen because it emphasized protection of the habitat—the floodplain lakes that produced the wild fishes they targeted.

In addition to concerns with overall productivity, conflict on the river centered on access: whether fishermen would have the right to fish on the privatized floodplain. Denial of access affected researchers as well and helped move the survey into a direct political role in the struggle. As privatization expanded, the survey could no longer get access to important study sites, and the scientists found themselves to be political allies of the fishermen in much more direct ways. Further, as the conflict over access propelled agriculturalists and speculators to levee the bottomlands, the very lakes the biologists were studying were destroyed. "Owing to the drainage of this lake our operations in this locality must cease with the present season," Kofoid wrote of Phelps Lake in 1898.[32]

The political interests of the fishermen and the survey coalesced in the struggle to prevent Thompson Lake from being drained. The survey's relationship to both fishermen and locale was epitomized by its long history of involvement with this lake. Thompson Lake was one of the largest of the backwater lakes on the Illinois River and had supported a commercial fishery since the mid-nineteenth century. In 1894 Forbes chose it as one of the regular sampling stations for his work on the backwaters of the Illinois River. Starting that year, every one to three weeks survey scientists sampled the fishes, plankton, and bottom organisms of Thompson Lake and five other nearby localities. In 1901, however, the owner sold the lake to the Thompson Lake Rod and Gun Club, which began restricting access.[33] Fishermen became militant in protecting their rights of access, while survey scientists first became explicitly involved in the politics of conservation. Individual fishermen and scientists who had begun working together in the field in the 1890s joined forces in the courts to involve the State of Illinois in the effort to preserve public access to the floodplain and protect the fishery interests.

As the club put up "No Trespassing" signs and hired wardens to patrol the lake, local fishermen and hunters fought back. In March 1907 wardens attempted to arrest William Cobb for poaching in Thompson Lake. A gun battle ensued in which Cobb was seriously

---

[31] Forbes to Richardson, 28 Feb. 1911, RS 43/1/5, Box 2, Univ. Illinois Archives. On the interest conflict see *Report of State Fish Commission to the Governor of Illinois: September 30, 1884* (Springfield, Ill.: Rokker, State Printer and Binder, 1884), p. 18; and *Report of State Fish Commissioners, from October 1, 1898, to September 30, 1900,* p. 2.

[32] *Biennial Report of the Director for 1897–98* (Urbana: Illinois State Laboratory of Natural History, 1898), p. 15.

[33] Transcript of *State v. New,* Abstract, p. 214.

wounded. That fall fishermen began to poach in earnest; this protest continued well into 1908 as the traditional users of the lake continued to assert their rights to harvest its resources. The club responded by seeking injunctions against trespass by the commercial fishermen. The federal court obliged and in 1908 granted an injunction preventing anyone from fishing, hunting, or boating on Thompson Lake without the permission of the club.[34]

These restrictions applied to the survey scientists as well as the belligerent fishermen. Without permission from the club, they could not conduct their research on Thompson Lake. At the time, Richardson was exploring new ideas about why fish populations were greater in some areas of the river than others. He was especially interested in how physical conditions, such as the characteristics of the bottom material and the size of inlets and outlets to the river, affected fish populations. To answer this question, he wanted to compare Thompson Lake, the most productive on the river, with Matanzas Lake, which supported far fewer fish. However, with the injunction in force, the club had forced Richardson to agree to a rigid set of conditions before he was allowed to conduct fieldwork on the lake. Rather than being allowed to design the sampling to answer the research questions, he had "been forced to it by conditions and events." The resultant sampling scheme left Richardson frustrated, "without any certain feeling . . . that it is the one we should continue to follow." In 1909 he complained bitterly to Forbes about these restrictions. Survey workers were prevented from taking any fish to sample; that right was reserved to John Schulte, a wealthy commercial fisherman who had signed a lease from the club. Instead, they had to examine Schulte's hauls to collect their data. They were allowed to sample plankton, but their access was severely proscribed and they could not use their motor launches, a restriction that turned what should have been short research trips into day-long endeavors. Even permission for this limited access was precarious. "For the purposes of keeping the ducks undisturbed," Richardson wrote, "even row boats are excluded, except in special cases, from the lake." At first Richardson merely asked Forbes to contact the club's manager to seek broader permission for sampling, noting sarcastically, "I suppose we must put up with such trivial inconveniences as that in the interests of a few full game bags." After a few days' consideration, however, he requested a more political intervention. He again wrote to Forbes, this time suggesting that the Illinois attorney general should challenge the injunctions against trespass on the survey's behalf.[35] These injunctions were originally filed to prevent poaching by poor fishermen. Suggesting that they be challenged put the survey firmly on the side of the local fishermen in their dispute with the wealthy sportsmen.

Richardson's concerns appeared to be primarily scientific: he wanted to address particular research questions but was prevented from doing so by the privatization of the floodplain. It was Forbes who placed these concerns in the broader context of conserving the river and its fisheries for both scientific and social reasons. While Forbes did not take the specific steps Richardson suggested, he became active in supporting public access to the floodplain. He recommended to Illinois's newly established Rivers and Lakes Commission that "the reservation of the most valuable feeding grounds and breeding grounds of fishes might well be undertaken by whatever legal process is necessary and possible. . . . Conservation of the fish and game of the state, and a permanent maintenance of the fertility of the reclaimed lands, must be taken into full account." When the commission asked for comments on a draft report concerning the future of the floodplain, Forbes suggested strengthening its conclusion regarding the importance of public waters: "I wish it might

[34] *Ibid.*, Abstract, p. 130.
[35] Richardson to Forbes, 19 Oct. 1909, 21 Oct. 1909, RS 43/1/1, Box 10, Univ. Illinois Archives.

be possible to add something further on the importance of retaining or obtaining for the public these waters."[36]

As pressure—from both poachers and the State of Illinois—to establish public rights to Thompson Lake increased, the Thompson Lake Rod and Gun Club gave up the plan to maintain a hunting and fishing reserve and reorganized itself into a levee district in order to drain the lake for agriculture. The affected fishermen, recognizing the permanent threat that drainage represented, went out on Thompson Lake in 1913, in deliberate defiance of the injunction and intending to be arrested, in order to test the ownership of the lake in the courts. Among them was Miles Newberry. This former survey employee, whom Forbes described as having "served the station very efficiently from the beginning," now tested the law in order to challenge the right of the gun club to drain Thompson Lake. The fishermen argued that "Thompson Lake is a public body of water owned by the State of Illinois." When the club sued them for trespassing, the State of Illinois joined the case on the fishermen's side.[37] Thompson Lake became a test case of the state's right to assert ownership and control on the Illinois River floodplain in the interests of conservation.

The state developed a strategy for establishing public rights to the floodplain that reflected the complicated nature of local and scientific knowledge developed on the river. Part of the strategy was based on testimony as to the half-century of use of the lake by the public: the state called upon local fishermen, hunters, and rivermen to document the navigability of the lake and its history as a public hunting and fishing area. Second, the state called upon survey scientists to testify as to the public importance of the lake in scientific terms. In 1914 the Rivers and Lakes Commission wrote to Forbes, inviting him to testify "regarding action against the encroachment on Thompson Lake." Forbes, in turn, instructed Richardson to "make a special point of attendance there, being especially prepared, of course, to testify concerning the value of the lake as a public fisheries ground." In the hearing on trespass, both Richardson and Newberry spoke in support of public access to Thompson Lake. Newberry testified to his knowledge of the river, drawing on his history of use of Thompson Lake from 1877 on, including his work for the Natural History Survey. Richardson testified concerning his scientific studies of the river.[38] The Thompson Lake case, which ultimately reached the Illinois Supreme Court, was decided in 1917 in favor of the Thompson Lake Rod and Gun Club, despite the combined testimony of fishermen and scientists. In 1922 the lake was drained and the land converted to agriculture.

THE BOUNDARIES OF ECOLOGY

Ecology's practical orientation and engagement with local resource politics stemmed from the nature of practitioners' work in the field. Working alongside local fishermen, hunters, and farmers, ecologists learned of an area's natural history; this knowledge was incorporated into the developing science. Similarly, the concerns of the local residents about

[36] Forbes to Rivers and Lakes Commission, 19 Mar. 1914, RS 43/1/5, Box 2; and Forbes to J. W. Alvord and C. B. Burdick, 22 June 1915, RS 43/1/5, Box 3, Univ. Illinois Archives.

[37] *Biennial Report of the State Laboratory and Special Report of the University Biological Station, 1895–1896* (Chicago: Illinois State Laboratory of Natural History, 1896), p. 19 (on Newberry); and Transcript of *State v. New,* Record, pp. 2059 (quotation), 2079.

[38] Forbes to Richardson, 5 Nov. 1914, RS 43/1/5, Box 4, Univ. Illinois Archives; and testimony of Miles Newberry, Record, pp. 965–986, and testimony of Robert Richardson, Record, pp. 1047–1060, 1068–1076, transcript of *State v. New.*

environmental change were also incorporated into ecology, moving scientists into the politics of conservation. In responding to the influence of local knowledge and politics, the scientists were forced to redefine ecology into what Thomas Gieryn terms a "hybrid," a science that could encompass expanding influences and aims.[39]

Two of Forbes's papers, which span his own career as well as the period of the birth of ecology in the United States, illustrate the development of this hybrid science and its connections to social relations. In the 1880 paper in which Forbes first expressed his concept of the microcosm, he also discussed the nature of applied ecology. He spoke of the need for a *"working knowledge"* of nature, an understanding of how to manipulate elements like "its edible fishes, its injurious and beneficial insects, and its parasitic plants" for the benefit of humans. Forbes argued that the only way for this working knowledge "to have an applicable value" was through a complete understanding of all an area's species and their interconnections, achieved via a "comprehensive survey of our entire natural history."[40] From his first thinking about the interdependence of life in an ecological community, Forbes was concerned with the applications of this work to issues of fisheries, forestry, and farming.

By the end of his career, Forbes had come to define his "working knowledge" of nature as "ecology." In his 1921 presidential address to the Ecological Society of America he discussed the basic nature of ecological science, which he defined not as "an academic science merely" but, rather, "that part of every other biological science which brings it into immediate relation to human kind."[41] Confronted with conflicts between the demands of a "pure" science and the need to establish and advance ecology and its institutions, Forbes had mapped the boundaries of the new science and developed a complex endeavor that straddled the borders of objective and political, pure and applied, local and scientific.

Local struggles over the floodplain brought the survey ecologists into the realm of environmental politics. Once there, the ecologists drew on locals for explicitly political ends that went beyond issues of the floodplain to questions concerning state fish policy and Chicago's sewage. This political involvement raised difficulties for the ecologists as they sought to maintain their reputations as objective scientists. Forbes, based at the Natural History Survey in Urbana, may have been physically removed from day-to-day interactions with the fishermen, but he was by no means politically distant. He relied on Richardson as his contact with the fishermen and other interests on the river. Just as ecologists had learned about the natural history of the river from the locals, they sought locals' perspective on the river's politics. In seeking Richardson's advice on how he should respond to proposed changes in the state fish and game laws, Forbes commented, "You have the great advantage that you have been in close contact with fishermen and fishing operations for several years."[42] Richardson lived on the river in Havana and was in close communication with many of its factions.

Forbes depended on Richardson's connections to help him fight the dumping of Chicago sewage into the river that began in 1900. George Soper, a member of a commission making recommendations on Chicago's sewage, wrote Forbes about his plan to recommend that the city invest in an expensive filtration plant to reduce its discharge to the Illinois River. He also warned Forbes that there was no chance that his plan would be implemented unless

---

[39] Gieryn, *Cultural Boundaries of Science* (cit. n. 7), esp. Ch. 5: "Hybridizing Credibilities: Albert and Gabrielle Howard Compost Organic Waste, Science, and the Rest of Society," pp. 233–335.

[40] Forbes, "Food of Fishes" (cit. n. 9), p. 19 (emphasis in original).

[41] Stephen A. Forbes, "The Humanizing of Ecology," *Ecology,* 1922, *3*:89–92, on p. 90.

[42] Forbes to Richardson, 29 Jan. 1913, RS 43/1/5, Box 1, Univ. Illinois Archives.

the politicians saw some evidence of downstream opposition to Chicago's waste. Forbes quickly wrote to Richardson, asking him to canvass his contacts on the river for information about political opposition to Chicago's sending sewage down the Illinois River. "Can you give or send me," he wrote, "any protests or other expressions, by people along its banks, of serious discontent with the condition of the river?"[43]

Forbes was convinced that preservation of the river depended on a marriage of scholarly science with practical politics. Having been warned by scientists in other states, however, about the dangers of political involvement, Forbes recognized that involvement in river politics could damage his and the survey's credibility.[44] He thus worked to create a public image of scientific objectivity even while striving to influence state policy. When a wealthy member of a hunting and fishing club asked Forbes to support his candidacy for fish commissioner, Forbes replied that even if he did have a preference, he was prevented by the new civil service law from supporting any candidate. This professed disinterest was belied, however, when Forbes quickly wrote to Richardson to ask who the commercial fishermen supported. He intended to send the governor a general list of qualifications for the position, thus influencing the makeup of the commission without explicitly supporting one candidate. As conflicts escalated and their political work became more crucial, Forbes sought to distance the survey from charges of playing politics. He defended attacks on Richardson for criticizing the Chicago Sanitary District in public lectures by describing him as "simply . . . a biological expert" whose research and conclusions were "public property and must be held at the disposal of any one concerned in their use and application."[45] This public face of scientific objectivity, however, was only a mask for the survey's deep political involvement.

The political goals of ecology were related to its practical applications, its influence on decisions concerning the harvesting of resources and the degradation of the environment. Working at the boundary of scientific and political ecology required simultaneously treading the border between ecology as pure and applied science. One of the reasons for Forbes's emphasis on the practical utility of ecology can be found in his struggles to fund the Natural History Survey. Scientific institutions in the nineteenth century often lacked secure funding, and administrators emphasized the practical benefits of their work to their patrons.[46] Forbes's annual reports and letters to the governor all stressed the benefits that accrued to the state from his work. As state entomologist, Forbes did a good deal of work with the farmers and the agricultural establishment of Illinois and frequently argued for the practical value of his basic entomological research. He saw agricultural experiment stations as an institutional model for supporting both basic and applied aquatic ecology. Forbes drew

[43] G. A. Soper to Forbes, 15 Mar. 1915, RS 43/1/5, Box 6; and Forbes to Richardson, 17 Mar. 1915, RS 43/1/5, Box 4, Univ. Illinois Archives.

[44] Tarleton H. Bean to Forbes, 21 Jan. 1914, RS 43/1/5, Box 2, Univ. Illinois Archives. The New York State Fish Culturist had written: "You know how nearly impossible it is to administer a scientific Bureau in politics. One of the prime requisites for success . . . is freedom from the handicaps which invariably exist wherever places depend upon political favor."

[45] Forbes to F. W. Shepardson, Director of Education, State Department of Registration and Education, 21 Feb. 1920, RS 43/1/5, Box 13, Univ. Illinois Archives (concerning Richardson). On the matter of the fish commissioner see B. G. Merrill to Forbes, 12 Nov. 1916; Forbes to Merrill, 14 Dec. 1916; Forbes to Richardson, 29 Dec. 1916; and Richardson to Forbes, 1 Jan. 1917, RS 43/1/5, Box 7, Univ. Illinois Archives.

[46] On efforts to fund the survey see Bocking, "Stephen Forbes" (cit. n. 8), pp. 471–472. On practical emphases more generally see Smith, *Pacific Visions* (cit. n. 3), pp. 107–122; Hugh Richard Slotten, "The Dilemmas of Science in the United States: Alexander Dallas Bache and the U.S. Coast Survey," *Isis*, 1993, *84*:26–49; and Sharon E. Kingsland, "An Elusive Science: Ecological Enterprise in the Southwestern United States," in *Science and Nature,* ed. Shortland (cit. n. 3), pp. 151–179, esp. pp. 175–179.

parallels between aquiculture and agriculture, referring to the Illinois River as a "flowing soil" and to plankton as its "crop," in an attempt to connect an agricultural perspective to river ecology. Kofoid also likened the benefits of aquatic field stations to those of agricultural experiment stations: "Biological stations may do in our country for aquiculture what experiment stations have done, and are doing, for agriculture."[47]

Forbes was concerned with applied issues not just because they could help fund the survey, however. He saw the utility of ecological work as a fundamental part of the science. One could no more study "pure" ecology than humans could remove themselves from nature. As the fledgling Ecological Society of America was beginning to deal with issues of conservation, pollution, the establishment of natural areas, and the like, Forbes took up the question of the "humanizing of ecology," its need to consider human welfare. If people were part of nature—which Forbes thought self-evident—then ecology, as the science that examines the interactions of animals and plants with each other, was uniquely able, of all the biological sciences, to address practical problems. "Ecology is . . . the humanistic science *par excellence,*" he argued. In fact, "the applied—the applicable—part of each of these sciences is simply and solely the ecological element which enters into its make-up."[48]

Yet Forbes resisted the notion that ecology was merely practical. When it was suggested that the survey be responsible for investigating fishing methods "with a view to the fullest utilization of remaining fisheries resources," he complained that such a task "would pull us completely off from our scientific program and put us into a field of purely practical experimentation." Forbes was driven to search for knowledge about the workings of nature as well as to provide practical advice for the exploitation of that nature. At the outset of the survey's studies, he emphasized that although the work they planned "should stand in the closest possible relation to the general public welfare," it should promote "pure science" and not be limited "to the economic field." In outlining to Richardson the series of experiments on floodplain rotation, Forbes emphasized the novelty of this research, stating that "nothing of the kind has been done in this country, at least." One of the attractive aspects of the work was that it would produce new knowledge, not simply that it would be useful. Forbes saw the two approaches to ecological study as complementary and always pursued both.[49]

Investigation of the Illinois River fisheries was one of the primary problems in applied ecology addressed by Forbes. The survey scientists developed a complicated relationship with the local knowledge they collected from inhabitants while doing this work: they appreciated its importance but sought to differentiate it from science. Perhaps in order to set his infant science of ecology apart from the local knowledge on which it was based, Forbes drew a distinction between the practices of ecology and fishing. "I have been at work on the Illinois River . . . as a biologist and not as a fisherman," he declared. Forbes

[47] Bocking, "Stephen Forbes," p. 472; Stephen A. Forbes and Robert Earle Richardson, "Some Recent Changes in Illinois River Biology," *Illinois Nat. Hist. Surv. Bull.,* 1919, *13*(6):147; and Kofoid, "Biological Stations of Europe" (cit. n. 20), p. xiii. See Charles E. Rosenberg, *No Other Gods: On Science and American Social Thought* (1976; Baltimore: Johns Hopkins Univ. Press, 1997), pp. 135–210, on the development of the agricultural experiment stations and tension between pure and applied research.

[48] Forbes, "Humanizing of Ecology" (cit. n. 41), p. 90. For a history of ecologists' thinking on the role of humanity in ecological systems see Eugene Cittadino, "The Failed Promise of Human Ecology," in *Science and Nature,* ed. Shortland (cit. n. 3), pp. 251–283.

[49] Forbes to Richardson, 8 Oct. 1915, RS 43/1/5, Box 6, Univ. Illinois Archives; *Biennial Report of the Director, 1893–1894* (cit. n. 11), p. 15; and Forbes to Richardson, 20 Jan. 1914, RS 43/1/5, Box 2, Univ. Illinois Archives. On the complementary roles of pure and applied in Forbes's science see Bocking, "Stephen Forbes" (cit. n. 8), pp. 471–473.

and his survey colleagues further distinguished their scientific knowledge from fishermen's knowledge by prefacing accounts of information provided by fishermen with qualifiers meant to assure scientific readers that the data were reliable. Richardson referred to one fisherman as "experienced and unusually intelligent," while Forbes and Richardson wrote of "reliable, experienced, and unusually well-informed fishermen of our acquaintance." At times Forbes commented on the fishermen's lack of ecological understanding. Claiming that they didn't understand the importance of noncommercial fishes like the gizzard shad as food for valuable species, he criticized them for leaving "long lines of this species to rot on the bank where the seines are hauled."[50]

These distinctions emphasized that scientific knowledge of the Illinois River, though based in the local culture of fishing, had different aims than the local knowledge of the fishermen. There were limits to the usefulness of fishermen's knowledge for the scientists. For instance, fishermen's information often pertained only to the time of commercial harvest, while the scientists wanted to sample fish from the river throughout the year. Further, the scientists often wanted live material for study, while fishermen generally didn't need to keep their catch alive. The set nets used by the commercial fishermen were not useful for the scientists because unless they were checked constantly the fish would be killed and unusable. Fishermen, on the other hand, sought efficient ways of catching the fish within the constraints of marketability. A good catch for the fishermen—one that made the effort expended worthwhile—was much larger than a good catch for the scientists, who simply wanted a representative sample of species composition and material for museum specimens and laboratory experiments. Given their different needs, fishermen and scientists regarded different areas of the river as interesting.[51]

Nonetheless, Forbes not only relied on the local knowledge of the fishermen in developing his science but began to make claims for the local nature of scientific expertise as well. In a summary of his research on the effects of pollution on the upper Illinois River, Forbes prefaced the scientific information with a discussion of his own and coauthor Richardson's qualifications. Forbes reported that he "began work, as a biologist, on Illinois River problems some thirty-six years ago." In contrast, he noted only one pertinent point about Richardson: "the junior author has virtually lived on the river for purposes of investigation during the last four years." Truly to understand the river and its ecology, Forbes implied, one needed not only the biologist's knowledge but also the insights gained by knowing the river in a different way: one had to live on the river, to understand it as a local. Richardson was uniquely able to write of the biology of the river because he had lived there, gaining knowledge from everyday experience as only a resident could. As Forbes emphasized the importance of local knowledge in understanding an ecological system, he was simultaneously claiming that Richardson could legitimately be considered a local—that his knowledge was of a similar nature to that of the fishermen. Richardson also saw himself as a local in Havana—albeit one placed in a particular class stratum. Working for over a month on algae in the Missouri Botanical Garden in St. Louis, he was eager to leave: "I am not seeking to stay here; on the contrary will be very glad when I

[50] Forbes,"Notes of Conference, November 11, 1913" (cit. n. 29), p. 2; R. E. Richardson, "Observations on the Breeding Habits of Fishes at Havana, Illinois, 1910 and 1911," *Bull. Illinois State Lab. Nat. Hist.*, 1913, 9(8):405–416, on p. 405; Forbes and Richardson, "Studies on the Biology of the Upper Illinois River" (cit. n. 23), p. 537; and S. A. Forbes, "The Food of Illinois Fishes," *Bull. Illinois State Lab. Nat. Hist.*, 1878, *1*(2):71–89, on p. 72.

[51] Kofoid to Forbes, 3 May 1899, 8 May 1899, and Richardson to Forbes, 19 Oct. 1909, RS 43/1/1, Box 10, Univ. Illinois Archives.

am ready to leave." Despite Havana's problems and the escalating conflict between fishermen and gun clubs, he felt at home there. "I am usually pretty well there, have a few good friends there, and when it come to a choice would much rather be there than here"—even, he added, "if the day comes when I have to go armed."[52]

But the scientists were not the only ones drawing boundaries between authoritative and nonauthoritative knowledge of the floodplain lakes. As scientists were taking on the mantle of "local," fishermen attempted to take on the mantle of "expert." In some cases fishermen claimed their own authority and undertook their own scientific investigations. Since one of the issues in the legal arena concerned water depth and the definition of navigability, a number of fishermen skated out on the frozen waters of a disputed lake to measure its depth. Drilling holes in the ice, they sounded the bottom. These measurements were used to support their claim to the navigability and, thus, state control of the waters. Rather than relying on other locals' vague recollections of depth during certain seasons ("I have seen it plumb dry except a little pond which is kind of muddy, but what you might call plumb dry"), these fishermen provided facts: "The deepest water we found was five feet, ten inches. The shallowest water was five feet, two inches."[53]

Yet even though they could rightly be regarded as experts on the natural history of the river's fishes, fishermen ultimately ceded authority to scientists. This shift in claims to authoritative knowledge is evident in the Thompson Lake court case. Both scientists and fishermen testified to the importance of Thompson Lake to the fishery. Asked to state his "opinion" on this matter, Richardson replied: "I should say we can give more than our opinion. We have facts to prove that the general condition is superior to almost all the other lakes for fish." In contrast, when asked a similar question by the lawyers, a commercial fisherman, Charles Rudolph, testified, "It is the best lake we have along the Illinois River." Asked for his "opinion for the reason of that," he continued, "It has got a better feeding ground." When asked to justify his "opinion," he did not allude to his thirty-seven years of experience as a commercial fisherman or to the habits of the "thousands" of fish he had observed on the Illinois River. Instead of presenting himself as an expert on the fishery, he pointed to the new experts on the Illinois River, the ecologists. He knew Thompson Lake was a better feeding ground, he replied, because "this Bug Man tests it." Asked if he meant the man "from the University of Illinois," Rudolph agreed. "That man that comes down here says it is the best."[54]

CONCLUSION: THE "LOCAL" IN "THE LAKE AS A MICROCOSM"

By analyzing the local context of Forbes's famous paper "The Lake as a Microcosm," this essay demonstrates how social relations on the river shaped the ecological work done there

---

[52] Forbes and Richardson, "Studies on the Biology of the Upper Illinois River" (cit. n. 23), p. 481; and Richardson to Forbes, 16 Jan. 1913, RS 43/1/5, Box 1, Univ. Illinois Archives (I thank Rob Kohler for calling my attention to this letter). Richardson had been on the river four years; but note his worries that Allen's knowledge—based on three years' experience at Havana—was not "minute" enough.

[53] Testimony of Philip Horchem, p. 61; testimony of Wilton Bull, pp. 163–164; and testimony of John Whitehead, pp. 159–163, on p. 162, Abstract, transcript of *Schulte v. Warren*, 218 Ill. 108 (24 Oct. 1905), RS 901, Supreme Court Trial Transcripts, Vault 29362, Illinois State Archives.

[54] Testimony of Robert E. Richardson, transcript of *State v. New*, Record, p. 1051; and testimony of Charles H. Rudolph, transcript of *State v. New*, Record, pp. 957–964, on p. 960. When challenged under cross-examination—"How do you go at it to make the study of fish and their habits?"—Rudolph replied, "I have been in the fish business for thirty-seven years, almost. And I know pretty near every move a fish makes I have seen them. I have seen thousands of them spawning and I have seen thousands of them that wasn't spawning" (p. 963).

and the development of fundamental theories of ecology. The many connections between scientists and "the local" were embodied in the research described in and flowing from this paper. Forbes's ecological studies were localized on the Illinois River floodplain in complex ways. At its simplest, "local knowledge" referred to the ecologists' scientific knowledge of particular floodplain lakes. Understanding why one lake supported a large fishery while another was home to only a few fish required a detailed, localized study.

Although the understanding of fish populations in key lakes was in itself important to the ecologists, they also sought to apply that knowledge to novel situations. How did the knowledge of these particular lakes become generalized in the theories of ecologists? Forbes used the narrative trope of the "microcosm" to universalize the scientific knowledge he derived from one lake. In a single isolated lake an ecologist could see "the play of life" at work on a scale amenable to observation and "mental grasp."[55] The interactions Forbes observed in these microcosms revealed the importance of interactions in ecology in general.

Yet other aspects of the local were also key to the understanding of Forbes's ecology. Fishermen, too, knew individual lakes: specific areas in a lake good for trapping fish, beaches useful for landing a seine, or spots where particular species of fish spawned. The knowledge of fishermen and hunters about specific habitats, the natural history of the fish, and how to catch them was essential to the developing research of the survey. Fishermen worked for the survey scientists, contributing their labor and expertise. Further, as the scientists learned about the river and its aquatic life from the fishermen, they also learned about the fishermen's lives and political struggles. The scientists forged friendships and came to identify with the people who lived on the river and depended on its resources. Ecology incorporated not only the local knowledge of the resident fishers and hunters but their perspectives on the changing environment. As a result, the survey scientists became strong advocates for the preservation of public access and became embroiled in a struggle with elites over control of the floodplain. Over the course of almost five decades of working in close contact with commercial fishermen, Forbes's scientific research reflected the culture of fishing on the river. As ecology established itself, its borders were mapped and remapped, and the knowledge and concerns of local people over environmental transformation became incorporated into the science. Through that incorporation, however, ecology had simultaneously devalued the local knowledge on which it was based.

---

[55] Forbes, "Lake as a Microcosm" (cit. n. 1), p. 77.

The Beauty and Menace of the Japanese Cherry Trees

Conflicting Visions of American Ecological Independence

By Philip J. Pauly

I N AUGUST 1909, Tokyo officials presented two thousand ornamental cherry trees to their "sister Capital City" of Washington, D.C. Incorporating exemplars of oriental horticulture into the landscape of official Washington (in place of a recently planted grove of American elms) would offer symbolic compensation for the recent American demand, in the so-called Gentlemen's Agreement, that Japanese immigration cease. The trees arrived in early January 1910, but within a few days Charles L. Marlatt, acting chief of the Bureau of Entomology of the United States Department of Agriculture (USDA), reported that they were infested with crown gall, root gall, two kinds of scale, a potentially new species of borer, and "six other dangerous insects." He urged that "the entire shipment should be destroyed by burning as soon as possible."[1] Marlatt's recommendation went all the way to President William Howard Taft. Taft acceded to his expert's views, and on 28 January federal employees took the dormant trees from storage sheds on the Washington Monument grounds, heaped them into piles, and reduced them to ashes (Figure 1).

At the time, as well as more recently, Marlatt's report was interpreted as an awkward but necessary technical interruption of the intricacies of cultural diplomacy. Yet this view is so incomplete as to be inaccurate. The recommendation to burn the trees was thoroughly political. Marlatt was willing to undercut presidential policy in order to further his campaign for plant quarantine legislation. This campaign was one element of a larger struggle within the USDA over the relative importance of introducing valuable "plant immigrants"

Isis 87, no. 1 (March 1996): 51–73.

Thanks to Bertram Zuckerman of Fairchild Tropical Gardens, to David Farr and Louise Russell of the U.S. Department of Agriculture, to Pamela Henson of the Smithsonian Institution, and to Kathryn Jacob of the National Historical Publications and Records Commission for assistance on sources. Florence Marlatt provided an illuminating picture of her father. Michele Bogart and Virginia Yans offered valuable suggestions. A preliminary version of this essay was presented at the biennial meeting of the International Society for the History, Philosophy, and Sociology of Biology, Brandeis University, Waltham, Massachusetts, 14 July 1993.

[1] Roland M. Jefferson and Alan E. Fusonie, *The Japanese Flowering Cherry Trees of Washington, D.C.: A Living Symbol of Friendship* (National Arboretum Contribution, 4) (Washington, D.C.: Agricultural Research Service, USDA, 1977), pp. 9, 49–54 (reprinting memos). See also Roger Daniels, *The Politics of Prejudice: The Anti-Japanese Movement in California and the Struggle for Japanese Exclusion* (Berkeley: Univ. California Press, 1962), pp. 31–45.

Figure 1. *Japanese cherry trees in flames near Agriculture Department greenhouses on the Washington Monument grounds, 28 January 1910. (Courtesy of the National Agricultural Library.)*

and excluding undesirable "alien crop enemies."[2] Viewed from a still broader perspective, the cherry tree incident lay at the center of a reorientation in the relations that the human population of temperate North America had with species and varieties originating in other parts of the world. The aim of this essay is to elucidate these implications of Marlatt's horticultural auto-da-fé.

The framework for my account is provided by Alfred Crosby's remarkable historical synthesis, *Ecological Imperialism.*[3] Crosby has interpreted the European conquest of the Americas, Australia, and New Zealand as the replacement, between 1492 and the early 1800s, of plants, animals, and humans resident in these areas by types that had evolved in the more competitive environments surrounding the Mediterranean Sea. The earliest European settlers were only dimly aware that their easy success had a biological dimension. Crosby's central metaphor is so striking just because it conveys the distance between the oblivious self-regard of European colonists and the profound ecological forces that made their hegemony in the temperate "neo-Europes" possible.

How did Europeans become aware of the bases and implications of ecological imperialism? In the late nineteenth century, small groups of specialists in the neo-European ecological "colonies" began to appreciate the dimensions of this phenomenon and to ex-

[2] "Destroy Tokio Gift Trees," *New York Times,* 29 Jan. 1910; editorial, *ibid.,* 30 Jan. 1910; and Jefferson and Fusonie, *Japanese Flowering Cherry Trees,* pp. 11–15. *Plant Immigrants* was the title of a bulletin published by the USDA in the 1910s; "enemies" were discussed in Beverly T. Galloway, "Protecting American Crop Plants against Alien Enemies," *Transactions of the Massachusetts Horticultural Society,* 1919, Pt. 1, p. 75.

[3] Alfred W. Crosby, *Ecological Imperialism: The Biological Expansion of Europe, 900–1900* (Cambridge: Cambridge Univ. Press, 1986); see also William H. McNeill, *Plagues and Peoples* (Garden City, N.Y.: Doubleday, 1976).

plore what might be done to gain more control over the flow of species and varieties into their territories. It is thus possible to extend Crosby's metaphor: in certain places, ecological imperialism prompted the descendants of human colonials to initiate movements for what can be called "ecological independence."

Taken as a whole, the United States led in these efforts. In contrast to the British dominions and the South American countries, it combined political independence, a developed scientific community, and a functioning national bureaucracy. Its resident organisms endured the continuing effects of ecological imperialism. Most important, however, at the end of the nineteenth century its leaders were rapidly establishing new kinds of relations with the rest of the world. Increased global trade, rapid steamship transport, and the creation of an American overseas empire combined to generate potentially far-reaching biological consequences. American agricultural explorers began to import large stocks of new germ plasm, thereby increasing the quality and diversity of crops both in North America and in the United States' new tropical dependencies. At the same time, there was fear that the new international contacts would lead to an increase in foreign pests and parasites entering the country. The consequence of empire could be the destruction of the American agricultural economy, and the native biota, from within.

The leaders of the American ecological independence movement were located in Washington at the USDA. At the turn of the century, the Bureaus of Plant Industry, Animal Industry, Forestry, Entomology, and Biological Survey comprised a massive biological science enterprise. We know very little about these bureaus by comparison with the much smaller and more diffuse agricultural experiment station programs.[4] Yet their employees were deeply engaged in shaping both the political and the natural economies of the nation. In particular, leaders of USDA biological science worked systematically to manage species and varieties whose presence or absence could affect the well-being of American citizens.

Investigation of Washington's agricultural scientists is particularly rewarding because it reveals a deep division among them over the biotic future of the country. Their differing perspectives were a function of both specialty and personality and were heightened by the interbureau rivalries that permeated the department at this time. On one side were botanists in the Bureau of Plant Industry, who were cooperative, sensitive, and sometimes diffident about their careers as bureaucrats. They sought to increase the number and variety of useful kinds of vegetation in the United States and were serene about the country's ability to prosper in a global biotic system. They can, appropriately, be described as ecological "cosmopolitans." On the other side were zoologists in the Bureaus of Entomology and Biological Survey, who tended more toward competitiveness, aggressive masculinity, and careerism. They worked to protect Americans from foreign organisms considered pests and were concerned to preserve the distinctive biotic elements they saw in the continental United States. Theirs was, in a word, a "nativist" perspective.[5]

The issues at stake were technical, though not abstruse. At the same time, they were bound, through language and imagery, to visceral ethnic sensibilities cultivated over cen-

---

[4] On USDA research see Thomas S. Harding, *Two Blades of Grass: A History of Scientific Development in the U.S. Department of Agriculture* (Norman: Univ. Oklahoma Press, 1947); on experiment stations see Charles E. Rosenberg, *No Other Gods: On Science and American Social Thought* (Baltimore: Johns Hopkins Univ. Press, 1976); and Alan I. Marcus, *Agricultural Science and the Quest for Legitimacy* (Ames: Iowa State Univ. Press, 1985).

[5] On American cosmopolitanism in the late nineteenth century see Howard Mumford Jones, *The Age of Energy: Varieties of American Experience, 1865–1915* (New York: Viking, 1971); on nativism see John Higham, *Strangers in the Land: Patterns of American Nativism, 1860–1925* (New Brunswick, N.J.: Rutgers Univ. Press, 1955). The term *nativist* was coined, without irony, in the 1840s to designate American-born opponents of (mostly Irish) immigration.

turies of political conflicts and ecological displacements. European-Americans identified with familiar and useful plants, whether the "native" spreading chestnut tree or the "introduced" amber waves of grain. Conversely, attitudes toward foreign pests merged with ethnic prejudices: the gypsy moth and the oriental chestnut blight both took on and contributed to characteristics ascribed to their presumed human compatriots. The cherry tree incident—with its contrast between strong, symmetrical, but dull American elms and effete, twisted, and spectacular oriental exotics, its subtexts of racial inequality and (horti)cultural sisterhood, and its implications of an insidious Yellow Peril hidden within beautiful packaging—was saturated with such meanings. USDA leaders' bureaucratic survival depended upon sensitivity to the political power of connotations.

The direction that the American ecological independence movement took was highly contingent. Predominance, in fact, shifted from nativism in the 1890s, to cosmopolitanism in the following decade, and then back to the nativists in the 1910s. Ultimately, the nativists prevailed. For more than eighty years, both government policy and informed opinion have granted the protection of indigenous species and varieties priority over the introduction of exotics. An introducer bears a significant burden of proof and, if problems arise, shoulders a large amount of blame. Moreover, past movements of organisms are retrospectively stigmatized as "ecological imperialism," a morally suspect and politically retrograde category.

The focus of this essay is on bugs, plants, and birds. As the cherry tree controversy implies, however, debates in this area were closely linked to attitudes about human groups. I will conclude by sketching how an appreciation of disputes over flora and fauna can lead to a significant reinterpretation of the efforts by "native" European-Americans in the 1910s and 1920s to limit human immigration. My aim is to show that a scientist-led ecological independence movement was a major element in both the environmental and the demographic history of modern America. Recovering the existence of an alternative vision of the country's biotic future, and recognizing the extent to which twentieth-century views on native and exotic organisms arose in a context permeated by mundanely illiberal prejudices, together raise historicist doubts about contemporary common wisdom.

PRELUDE: AMERICA'S ECOLOGICAL OPEN DOOR

The effects of ecological imperialism on eastern North America are well known. From the time of the first contacts, European explorers and colonists spread pathogenic bacteria and viruses among indigenous peoples. They introduced Old World plants and livestock and inadvertently imported rats and rabbits, weeds, and destructive insects and fungi. By the 1650s the Indians were disappearing from New England, and the makeup of the fauna and flora in settled areas had altered dramatically.[6]

A second wave of plant and animal introductions took place during the nineteenth century, as the rising standard of living, a new interest in horticultural novelties among gentleman farmers, and improved transportation combined to facilitate the movement of organisms. This phase of ecological imperialism was less dramatic than the first, and its initiators were groups long resident in North America. But, as had been the case in the 1600s, these introductions resulted from the haphazard activities of individuals with quite

---

[6] William Cronon, *Changes in the Land: Indians, Colonists, and the Ecology of New England* (New York: Hill & Wang, 1983); see also Carolyn Merchant, *Ecological Revolutions: Nature, Gender, and Science in New England* (Chapel Hill: Univ. North Carolina Press, 1989).

particular interests, and they had ecological consequences well beyond those conceived by their initiators.

Horticultural and agricultural improvers provided the main impetus behind the increasing flow of organisms into temperate North America during the nineteenth century. Nurserymen, tourists, and immigrants directed fruits, seeds, seedlings, and bulbs from Europe and Asia toward the United States. They substantially increased the diversity of crop varieties and, even more, of fruits and ornamentals. However, some introduced plants (such as honeysuckle and water hyacinth) soon became weedy in their new environments. Imported plant material also harbored weed seeds (notably the tumbleweed, or Russian thistle) and provided transportation for undesirable insects (including the codling moth, cabbageworm, and various scales and aphids) and a significant number of blights and rusts.[7]

More isolated, but, in retrospect, more remarkable, were the campaigns by enthusiasts to bring Eurasian animals, notably birds, to the United States. Between 1850 and 1870 local "acclimatization societies" imported thousands of English sparrows. By the 1880s bird lovers had attempted to introduce at least twenty exotic species. They soon succeeded in establishing both the ring-necked pheasant and the starling.[8]

The most notorious animal introduction of the nineteenth century was the gypsy moth. Etienne L. Trouvelot, an artist employed by Louis Agassiz at Harvard's Museum of Comparative Zoology, worked in his spare time breeding hybrid silkworms that could survive Massachusetts winters. In 1869 he imported gypsy moth eggs from France to his home in the Boston suburb of Medford. Some of the fuzzy egg masses blew out an open window, and the moths established themselves in brushland behind his house. After two decades of acclimatization, the Medford moth population suddenly exploded, producing the spectacular fouling and deforestation that northeasterners have experienced periodically ever since.[9]

The federal government participated in plant and animal introductions in essentially the same ways as private individuals. By turns, the State Department, the Patent Office, and the Department of Agriculture collected and distributed foreign seeds and plants. Spencer Baird's U.S. Fish Commission distributed the carp throughout the country at the same time that it redistributed such American fishes as the rainbow trout and striped bass from one coast to the other.[10] Although the scale of government work was at times substantial, it was never more than a series of discrete, essentially opportunistic, actions. Government officials did not attempt to systematize introductions. More important, with the notable but quite limited exception of carriers of epidemic disease, they made no effort to regulate the flow of organisms across the borders of the country.[11]

[7] Jack Ralph Kloppenburg, Jr., *First the Seed: The Political Economy of Plant Biotechnology, 1492–2000* (New York: Cambridge Univ. Press, 1988), pp. 50–57. On imported pests see Leland O. Howard, "Danger of Importing Insect Pests," in United States Department of Agriculture (USDA), *Yearbook,* 1897, pp. 529–539; and Lyster H. Dewey, "Migration of Weeds," in USDA, *Yearbook,* 1896, pp. 274–279.

[8] Theodore S. Palmer, "The Danger of Introducing Noxious Animals and Birds," in USDA, *Yearbook,* 1898, pp. 98–106; and Palmer, "A Review of Economic Ornithology in the United States," in USDA, *Yearbook,* 1899, p. 289.

[9] Edward H. Forbush and Charles H. Fernald, *The Gypsy Moth: Porthetria dispar (Linn.): A Report of the Work of Destroying the Insect in the Commonwealth of Massachusetts, Together with an Account of Its History and Habits Both in Massachusetts and Europe* (Boston: Wright & Potter, 1896), pp. 3–23.

[10] Norman Klose, *America's Crop Heritage: The History of Foreign Plant Introduction by the Federal Government* (Ames: Iowa State College Press, 1950); Kloppenburg, *First the Seed* (cit. n. 7), pp. 50–61; and Dean Conrad Allard, Jr., *Spencer Fullerton Baird and the U.S. Fish Commission* (New York: Arno, 1978), pp. 271–281.

[11] Quarantine measures dated back to colonial times, and the federal government began to get involved in the 1810s. Such procedures were extended to livestock around 1880. See Alan M. Kraut, *Silent Travellers: Germs,*

While Americans enthusiastically incorporated improved crop varieties and novel ornamentals into domesticated landscapes, they excoriated both introduced pests and those
seemingly responsible for them. The Hessian fly, Russian thistle, European corn borer,
English sparrow, and gypsy moth all generated hatred that easily encompassed those
deemed responsible for bringing them—whether the acclimatization societies, the scientific
community, or, most frequently, all members of the eponymous ethnic group.

Yet neither scientists nor mainstream political leaders argued for greater national control
over the movement of organisms. The invisible hands of nature and the market were
equally obscure and wonderful in their effects; the duty of Americans was to cope with
situations as they developed. The Harvard botanist Asa Gray reported dispassionately in
1879 on the continuing introduction of European weeds into the United States. The Cornell
horticulturist Liberty Hyde Bailey articulated the dominant perspective forcefully in 1894.
He argued that pests resulted inevitably from the disequilibrium introduced into nature by
cultivation. The only solutions were vigilance and time. Western farmers were plagued by
the Russian thistle because they had been too lazy to plant and plow properly. Government's only role was educational; the law, Bailey concluded, "cannot correct a vacancy
in nature."[12]

THE BEGINNINGS OF A FEDERAL RESPONSE TO PESTS

Two years after Bailey's speech, the USDA botanist Lyster Dewey prepared a bulletin
promoting "legislation against weeds."[13] Federal scientists were coming to believe that a
system driven by individual whims and by natural and market forces could be improved
and that they, as experts employed by the "general government," should hold the central
responsibility in this area. They began to consider how problems previously viewed in
isolation—not only the introduction of promising exotic organisms into the United States
and the exclusion of noxious alien species, but also the preservation of native species and
varieties, the cultivation of desirable forms in new regions of the country, the improvement
of wild types, and the control of pests—were in fact interrelated. But they held changing
and conflicting views on where to place the emphasis: on weeding out the bad, or on
cultivating the good.

Prior to the late 1880s, USDA scientists in Washington were small-scale operators with
little visibility. They gained a clear national role only around 1888, when the department
rose to cabinet status and the Hatch Act created a national network of agricultural experiment stations. With most local and routine tasks delegated to the provinces, Agriculture
Secretary Jeremiah Rusk was able to envision the scientific staff in Washington as the

Genes, and the "Immigrant Menace" (New York: Basic, 1994), pp. 24–25; and Ulysses G. Houck, *The Bureau of Animal Industry of the United States Department of Agriculture* (Washington, D.C.: Hayworth, 1924).

[12] Asa Gray, "The Pertinacity and Predominance of Weeds," *American Journal of Science,* 3rd Ser., 1879, *18*:161–167; and Liberty Hyde Bailey, "Coxey's Army and the Russian Thistle: An Essay on the Philosophy of Weediness," in *The Survival of the Unlike,* 4th ed. (New York: Macmillan, 1901), pp. 200–201. Bailey delivered this address to the American Association for the Advancement of Science in the nadir of the great depression following the Panic of 1893. Coxey's army was a group of unemployed men who had marched on Washington to demand public road-repair jobs (which would have included weed clearing). Bailey smugly contrasted the poor marchers themselves to weeds: while the former had been dispersed by federal police, the latter were "beyond the reach of the sheriff."

[13] Lyster H. Dewey, *Legislation against Weeds,* USDA Division of Botany, Bulletin No. 17, 1896; similarly, see Leland O. Howard, *Legislation against Injurious Insects,* USDA Division of Entomology, Bulletin No. 33 (O.S.), 1895.

department's "central station," the site for its most complex and abstract research and for work on problems of national scope.[14]

A cohort of young agricultural college graduates staffed the department's ramifying divisions, which by 1890 included Entomology, Vegetable Pathology, Economic Ornithology and Mammalogy, Botany, Forestry, and Pomology, in addition to the more autonomous Bureau of Animal Industry. Jammed together in a few buildings on the south side of the Mall, staffers formed a nascent technocratic culture organized around the well-funded production of official reports. Yet they knew that divisions were in competition for funds and that each group's success in the long run depended on producing work that would interest potentially influential outside "friends."[15]

Some of the young agricultural scientists also learned rapidly about the broader intellectual and national settings within which their work might be situated. In the 1880s Washington was the center for the most significant group of naturalists in the United States, dispersed among the Smithsonian Institution, the Fish Commission, and the Geological Survey, in addition to the Agriculture Department. Men such as John Wesley Powell, Lester Ward, and George Brown Goode expounded a sweeping evolutionary biosocial philosophy and projected natural science as part of a federal mission to comprehend the continent. These themes provided the common denominator of discussions at both the various Washington scientific societies and the aptly named Cosmos Club, to which many of the agriculturists gravitated.[16]

The initial structure of USDA central station work was modeled largely on that of the Smithsonian and the Geological Survey. Basic research, in the form of data collection leading toward national surveys, was mixed with opportunistic problem solving that would display the immediate value of a division's activities to influential interests. C. Hart Merriam's Division of Economic Ornithology and Mammalogy, for example, explored the distribution of North American mammals and birds, hoping ultimately to produce a continental map of "life zones"—both a contribution to pure science and a set of parameters that farmers could use when thinking about new crops. At the same time, division staffers assessed whether species such as bluejays and woodpeckers were truly "noxious" to agriculture, and they reviewed the efficacy of bounties for reducing the number of pests and of closed hunting seasons for protecting game species. Entomologists and vegetable pathologists pursued similar dual programs: taxonomy and mapping on the one hand, and tests of insecticides and fungicides on the other.[17]

---

[14] "Report of the Secretary," in USDA, *Annual Report,* 1888, p. 12; and "Report of the Secretary," in USDA, *Annual Report,* 1889, pp. 8–9.

[15] Other divisions included Microscopy, Gardens and Grounds, Chemistry, Soils, and Statistics, along with the essentially autonomous Weather Bureau. The entomologist Leland Howard recalled the world of the USDA in the 1880s in two autobiographies: Leland O. Howard, *History of Applied Entomology (Somewhat Anecdotal)* (Smithsonian Miscellaneous Collections, 84) (Washington, D.C.: Smithsonian Institution, 1930); and Howard, *Fighting the Insects: The Story of an Entomologist* (New York: Macmillan, 1933). See also two manuscripts by Beverly T. Galloway: "Fifty Years Ago in the Department of Agriculture," 14 pages; and "The Genesis of the Bureau of Plant Industry," 8 pages, 30 June 1926, Galloway File, National Fungus Collection, Systematic Botany and Mycology Laboratory, Beltsville Agricultural Research Center, Beltsville, Maryland. On the importance of scientific entrepreneurship and interbureau competition see A. Hunter Dupree, *Science in the Federal Government* (New York: Harper & Row, 1964) pp. 176–183; and William A. Niskanen, *Bureaucracy and Representative Government* (Chicago: Aldine, 1971).

[16] Dupree, *Science in the Federal Government;* Michael Lacey, "The Mysteries of Earthmaking Dissolve" (Ph.D. diss., George Washington Univ., 1979); and Philip J. Pauly, "The Potomac Formation: Building Big Biology in Gilded Age Washington," paper presented to History of Science Society, Washington, D.C., December 1992.

[17] C. Hart Merriam, "The Geographic Distribution of Animals and Plants in North America," in USDA, *Yearbook,* 1894, pp. 203–214; Keir B. Sterling, *The Last of the Naturalists: The Career of C. Hart Merriam*

Beginning in 1894, however, USDA scientists shifted their attention dramatically toward the problem of introduced pests. The possibility of active intervention in this area had first arisen in California, a novel environment where American settlers, European and Asian vegetation, and oriental insects had converged within a few years. In 1881 California fruit growers lobbied successfully for a state plant quarantine law; eight years later, the successful control of the cottony-cushion scale by a USDA-introduced predator made clear the value and credit that could come from government pest control successes.[18]

These issues rose to national significance, however, only in the early 1890s. Owing to happenstance and increased trade, tumbleweeds and wheat rust overwhelmed farmers in the northern plains region, gypsy moths devastated the Boston suburbs, the "Mexican" boll weevil crossed the Rio Grande, and the (Asian) San Jose scale spread from California to the East on infested nursery stock.[19] Panic about pests was greater than at any time since the Midwest's locust plagues of the 1870s. These problems in the natural economy were heightened by devastation to the nation's political economy in the wake of the Panic of 1893. The collapse of farm prices led rural leaders to demand immediate and comprehensive government action to save their livelihoods.[20]

Within this context, basic research on, for example, the natural history of peach yellows seemed trivial. Work to promote quarantine laws, by contrast, made department scientists visible while it moved responsibility for dealing with pests to the legislative arena. Such efforts affirmed the value of cooperation and identified specific causes of the malaise of the period. Finally, agitation on behalf of quarantines promised immediate results but was much cheaper than laboratory research—an important consideration, since the USDA budget had been cut one-quarter by the retrenching Cleveland administration.

These were the bases for Dewey's guidelines on weed control legislation. Entomology chief Leland Howard joined in with a bulletin on insect control laws, and he went a step further by issuing a stark warning on the enormous number of Mexican and Japanese insect species then poised to invade the United States. Entomologists proposed to stop the spread of the boll weevil by closing a fifty-mile-wide strip of Texas to cotton culture, and they argued for measures to keep the San Jose scale from spreading throughout the East. Howard and the plant pathologist Beverly Galloway capped these efforts by organizing the National Convention for the Suppression of Insect Pests and Plant Diseases by Legislation; this group of experiment station scientists and horticultural leaders met in Wash-

---

(New York: Arno, 1974), pp. 257–314; Palmer, "Review of Economic Ornithology" (cit. n. 8); Leland O. Howard, *Chinch Bug: General Summary of Its History, Habits, Enemies, and of Remedies and Preventives to Be Used against It,* USDA Division of Entomology, Bulletin No. 17 (O.S.), 1888; Howard and Charles L. Marlatt, *San Jose Scale, Its Occurrences in U.S., with Full Account of Its Life History and Remedies to Be Used against It,* USDA Division of Entomology, Bulletin No. 3 (N.S.), 1896, pp. 33–35; *Condensed Information Concerning Some of the More Important Insecticides,* USDA Division of Entomology, Circular No. 1 (2nd Ser.), 1891; and David G. Fairchild, *Bordeaux Mixture as a Fungicide,* USDA Division of Vegetable Pathology, Bulletin No. 6, 1894.

[18] Richard C. Sawyer, "To Make a Spotless Orange: Biological Control in California" (Ph.D. diss., Univ. Wisconsin–Madison, 1990), pp. 26–30, 76.

[19] Gustavus Weber, *The Bureau of Entomology* (Washington, D.C.: Brookings Institution, 1930), pp. 27–31, 46–47; Howard, *Fighting the Insects* (cit. n. 15), p. 53; Leland O. Howard, *The Mexican Cotton-Boll Weevil,* USDA Division of Entomology, Circular No. 6 (2nd Ser.), 1895; and Howard and Marlatt, *San Jose Scale* (cit. n. 17).

[20] Lawrence Goodwyn, *Democratic Promise: The Populist Movement in America* (New York: Oxford Univ. Press, 1976); and Robert C. McMath, Jr., *American Populism: A Social History, 1877–1898* (New York: Hill & Wang, 1993). The extent of disruption in the hinterlands was brought home to Washington staffers when Kansas State College of Agriculture, the academic home of a number of them, was taken over by Populists, and the father of Vegetable Pathology assistant David Fairchild was forced out as president; see James C. Carey, *Kansas State University: The Quest for Identity* (Lawrence: Regents Press of Kansas, 1977), pp. 67–86.

ington under USDA sponsorship the day after McKinley's inauguration, drafting legislation to establish a system for inspecting both importations and interstate shipments of
nursery stock for injurious insects and plant diseases.[21]

The quarantine movement entered a new phase in 1898, when the United States suddenly
acquired a global empire. For reasons to which I will return, this change undercut the
restrictionist activities of USDA entomologists and botanists. The department's zoologists,
by contrast, were energized by their new imperial outlook. Theodore Palmer began to put
North America's ecological position into a historical perspective, and he used that analysis
to establish, for the first time, the principle that the federal government should control the
nation's biotic borders.

As an adolescent naturalist in California, Palmer had been deeply interested in questions
of the geographic distribution of animals and, hence, in whether organisms were in their
proper places. After receiving his undergraduate degree at Berkeley in 1888, he joined C.
Hart Merriam in the "great adventure" of a biological survey of Death Valley. Under
Merriam's aegis, he rapidly advanced to become assistant chief of the Division of Biological Survey in 1896, with responsibility for its "economic" programs. A clubman and
collector, he established close and enduring ties with leading bird and sportsmen's organizations.[22]

In 1898 Palmer became concerned about the ecological backlash that might result from
American imperialism. On the one hand, he argued, the federal government was now
responsible for protecting "our island dependencies" of Hawaii and Puerto Rico from new
noxious species; on the other, he worried that the "increase in the means of communication" between these islands and North America, consequent on the establishment of imperial ties, could soon result in the "calamity" of the introduction of animals such as the
mongoose into the United States.[23]

In order to convey the depth and scope of these threats, Palmer explained to the half-
million recipients of the USDA *Yearbook* the animal aspects of what would later be called
ecological imperialism. He described how, since Columbus, Europeans had repeatedly
devastated biotas in America, Australia, and the Pacific Islands by introducing pigs, goats,
rabbits, rats, and cats. The recent importation of the mongoose into Jamaica had had similar
effects. Brought there in 1872 to control rats, the animals soon expanded their diets to
include chickens, fruits, and a number of indigenous animals. The decimation of these
species had led to a rapid increase in ticks and injurious insects. An erroneous report in
1892 that the USDA was planning to import mongooses to control gophers had led individuals "ignorant of the animal's past record and anxious to try some new method" to
obtain specimens privately; these importations had been prevented, but only "by the most
strenuous efforts."[24]

[21] Leland O. Howard, *Some Mexican and Japanese Injurious Insects Liable to Be Introduced into the United States,* USDA Division of Entomology, Technical Bulletin No. 4, 1896; Douglas Helms, "Technological Methods for Boll Weevil Control," *Agricultural History,* 1979, *53*:287; Leland O. Howard, *The Mexican Cotton-Boll Weevil,* USDA Division of Entomology, Circular No. 14 (2nd Ser.), 1896; Howard, *The Gipsy Moth in America,* USDA Division of Entomology, Bulletin No. 11 (N.S.), 1897; Beverly T. Galloway, ed., *Proceedings of the National Convention for the Suppression of Insect Pests and Plant Diseases by Legislation, Held at Washington, D.C., March 5 and 6, 1897* (Washington, D.C.: Government Printing Office, 1897); and Howard, "Danger of Importing Insect Pests" (cit. n. 7).
[22] On Palmer see W. L. McAttee, "In Memoriam: Theodore Sherman Palmer," *Auk,* 1956, *73*:367–377; and Sterling, *Last of the Naturalists* (cit. n. 17), pp. 129, 149. The Division of Economic Ornithology and Mammalogy was renamed the Division of Biological Survey in 1896.
[23] "Report of the Division of Biological Survey," in USDA, *Annual Report,* 1899, pp. 67–68; and Palmer, "Danger of Introducing Noxious Animals" (cit. n. 8).
[24] Palmer, "Danger of Introducing Noxious Animals," p. 96.

Against this backdrop, Palmer described the "ill-directed" efforts on the part of acclimatization societies to import European songbirds. He deplored the introduction of the English sparrow and the starling and noted that the recently introduced great titmouse "is said to attack small and weakly birds, splitting open their skulls with its beak to get at the brains, and doing more or less damage to fruit, particularly pears."[25]

Palmer drew two lessons. The particular vulnerability of island organisms to predatory exotics was such that "unusual care" needed to be taken when considering introductions to Puerto Rico and Hawaii. More generally, he argued, "some restriction" should be placed on the ability of private individuals and groups to introduce species that "may become injurious" in the continental United States. "Since," he pointed out, "it has been found necessary to restrict immigration and to have laws preventing the introduction of diseases dangerous to man or domesticated animals, is it not also important to prevent the introduction of any species which may cause incalculable harm?"[26] Power over exclusion, Palmer suggested, should be one of the duties of the Agriculture Department.

With the passage of the Lacey Act in early 1901, Palmer gained this power. The law was a striking example of the ability of scientific bureaucrats to reshape congressional initiatives. John F. Lacey, a duck-hunting Iowa congressman, had naively proposed that the Fish Commission expand its hatchery program to include game birds. Lacey's bill included an acclimatization provision: that the government "aid in the introduction of . . . foreign birds." Agriculture Secretary James Wilson, however, succeeded in completely reversing the bill's thrust. He suggested that it be "broadened" to include not only hatcheries but all kinds of preservation, and (pointing to Palmer's account of the problems caused by the sparrow and the mongoose) he persuaded Lacey that introductions of foreign species needed to be regulated, not aided.[27]

The leading purpose of the Lacey Act was stated clearly in its heading: "to enlarge the powers of the Department of Agriculture." The law curtailed the acclimatization societies, asserted political control over biotic borders, and gave the federal government—more specifically, the scientists working for the Biological Survey—the power to establish a comprehensive policy on faunal introductions. It ended four hundred years of biotic laissez-faire.

ECOLOGICAL COSMOPOLITANISM IN THE BUREAU OF PLANT INDUSTRY

The Lacey Act was nonetheless an isolated initiative. During the long tenure of James (Tama Jim) Wilson as secretary of agriculture (1897–1913), the USDA moved in a quite different direction. Wilson believed that American farmers had suffered so much from recent natural and economic disasters because they grew only a few varieties of a few crops. They needed new varieties of staples that were more resistant to disease and better

---

[25] *Ibid.*, p. 105.

[26] *Ibid.*, pp. 106, 107–108.

[27] James Wilson to John Fletcher Lacey, 15 Jan. 1900, John Fletcher Lacey Papers, State Historical Society of Iowa, Des Moines, Iowa. Environmental historians have emphasized the Lacey Act's provisions regulating interstate commerce in game. However, this entry of the federal government into wildlife conservation was less controversial, and less radical in principle, than its prohibitions on importations. For the legislative history see Jenks Cameron, *The Bureau of Biological Survey: Its History, Activities, and Organization* (Baltimore: Johns Hopkins Press, 1929), pp. 70–83; and Theodore Whaley Cart, "The Struggle for Wildlife Protection in the United States, 1870–1900: Attitudes and Events Leading to the Lacey Act" (Ph.D. diss., Univ. North Carolina at Chapel Hill, 1971), pp. 158–166, 188–189. More generally see John F. Reiger, *American Sportsmen and the Origins of Conservation*, rev. ed. (Norman: Univ. Oklahoma Press, 1986); and Thomas R. Dunlap, *Saving America's Wildlife* (Princeton, N.J.: Princeton Univ. Press, 1988).

Figure 2. Beverly T. Galloway, Chief of the Bureau of Plant Industry, 1912. (Courtesy of the National Fungus Collection, Systematic Botany and Mycology Laboratory, Beltsville Agricultural Research Center.)

adapted to the wide range of American climates. In addition, cultivation of new crops such as soybeans and sorghum would relieve price pressure on wheat, corn, and cotton. Lastly, Wilson hoped to expand the agricultural potential of the United States by enabling its climatically exceptional regions—Florida, the desert Southwest, and the tropical possessions—to produce citrus, dates, and other products previously imported from abroad.[28]

This program to restructure the country's domesticated flora reflected an ecologically cosmopolitan perspective. For Wilson, the world beyond the continental United States was a source of potential treasure, not danger. He thus initiated a vigorous search for unfamiliar foreign plants, a program whose realization depended upon the new activist foreign policy of the McKinley and Roosevelt administrations. It also benefited from the expanding global presence of American capitalists. Wilson sought, in a literal sense, to bring home the fruits of empire.

Wilson's chief implementer was the gifted scientific entrepreneur Beverly Galloway (Figure 2). Thin and myopic, prone to depression, and a fancier of violets, Galloway was prominent among the "hermaphrodites" that the manly bachelor chemist Harvey Wiley

[28] On Wilson's perception of the farm problem see "Report of the Secretary," in USDA, *Yearbook,* 1897, p. 10; and Barbara Kimmelman, "A Progressive Era Discipline: Genetics at American Agricultural Colleges and Experiment Stations, 1900–1920" (Ph.D. diss., Univ. Pennsylvania, 1987), pp. 25–41.

complained were in control of the USDA. He came to Washington in 1887 from the University of Missouri as an assistant in the Section of Mycology. He gradually gathered a young staff that combined initiative and camaraderie, and he was continually pushing to broaden his group's responsibilities, budget, and personnel roster.[29] In the mid 1890s Galloway coordinated his aims with those of other USDA divisions, most notably in his work with Leland Howard on the 1897 convention to control pests and plant diseases. Yet soon after Wilson became secretary, Galloway began to develop a broad program to promote new crops, breed new varieties, and extend the parameters of plant culture.[30] In late 1897 he asked Wilson to appoint David Fairchild, his former assistant, to organize a new office devoted to "systematic plant introduction."[31] It was a perfect, but fortuitous, convergence of organization and individual.

Fairchild, son of the president of Kansas State College of Agriculture, had resigned from the USDA four years earlier in order to pursue mycological graduate work in Italy and Germany. While crossing the Atlantic, however, he met Barbour Lathrop. Brother of the president of the Chicago Board of Trade (the most powerful force in the nation's agricultural economy), Lathrop, a forty-six-year-old bachelor, had diffuse passions for both intelligent young men and crop improvement. He gave Fairchild $1,000 for a trip to Java to study termite fungus gardens. He then followed the younger man there, pressured him to abandon his research, and invited him on a tour of the Far East, with the vague goal of finding new fruits to grow in California or Hawaii (Figure 3).[32]

Fairchild returned from this romantic odyssey in 1897 with polished manners, valuable social connections, a cosmopolitan outlook, and a love for the tropics. He took the position of head of the Section for Foreign Seed and Plant Introduction (SPI) enthusiastically. His office was immediately involved in coordinating the activities of official plant explorers: Neils Hansen and Mark Carleton traveled to Russia to collect wheat varieties and fruits adaptable to the northern plains, Walter Swingle went to the Middle East for figs and dates, and others fanned out to Japan, the tropics, and Europe.[33] Fairchild also refined the delicate

---

[29] Oscar E. Anderson, Jr., *The Health of a Nation: Harvey W. Wiley and the Fight for Pure Food* (Chicago: Univ. Chicago Press, 1958), p. 99. On Galloway see John A. Stevenson, "Beverly Thomas Galloway," *Dictionary of American Biography,* Vol. 10, p. 217; Galloway, "Fifty Years Ago" (cit. n. 15); Beverly T. Galloway, *Commercial Violet Culture* (New York: De la Mare, 1899); David G. Fairchild, *The World Was My Garden: Travels of a Plant Explorer* (New York: Scribner's, 1938), pp. 18–29; and Stevenson, "Plants, Problems, and Personalities: The Genesis of the Bureau of Plant Industry," *Agr. Hist.,* 1954, 28:155–162. Being a "hermaphrodite" was not equivalent to being homosexual—manners were more important than sexual partners. Galloway was in fact the married father of three. On late nineteenth-century gender categories see George Chauncey, *Gay New York: Gender, Urban Culture, and the Making of the Gay Male World, 1890–1940* (New York: Basic, 1994).

[30] "Report of the Secretary," in USDA, *Annual Report,* 1900, pp. xxxiii–xxxiv. Galloway was already skeptical about inspection for plant diseases and had broadened his division's domain to include environmental influences on disease susceptibility and the search for disease-resistant varieties. See USDA, *Annual Report,* 1893, p. 258; Beverly T. Galloway, "Plant Diseases and the Possibility of Lessening Their Spread by Legislation," in *Convention for Suppression of Insect Pests,* ed. Galloway (cit. n. 21), pp. 8–11; and Beverly T. Galloway to Walter T. Swingle, 10 Feb. 1897, Record Group 54, Entry 160, Vol. 17, National Archives, Washington, D.C. In 1895 Galloway broadened his division's name from Vegetable Pathology to Vegetable Physiology and Pathology.

[31] Klose, *America's Crop Heritage* (cit. n. 10); David G. Fairchild, "Report of the Special Agent in Charge of Seed and Plant Introduction," in USDA, *Annual Report,* 1898, pp. 3–4; USDA, *Annual Report,* 1899, pp. lxix–lxx; and Beverly T. Galloway, "Searching the World for New Crops," typescript, 15 pages, 1920, Beverly T. Galloway Papers, Special Collections, National Agricultural Library, Beltsville, Maryland.

[32] Fairchild, *World Was My Garden* (cit. n. 29), pp. 30–37, 79–84; and Marjory Stoneman Douglas, *Adventures in a Green World* (Miami: Field Research Projects, 1973).

[33] Fairchild, *World Was My Garden;* Klose, *America's Crop Heritage* (cit. n. 10), pp. 109–119; Isabel S. Cunningham, *Frank N. Meyer: Plant Hunter in Asia* (Ames: Iowa State Univ. Press, 1984); Rose Schuster Taylor, *To Plant the Prairies and the Plains: The Life and Work of Niels Ebbesen Hansen* (Mt. Vernon, Iowa: Bios, 1941); and Paul De Kruif, *Hunger Fighters* (New York: Harcourt, Brace, 1928). The Iowan Hansen had in fact been sent out independently by Wilson in early 1897.

Figure 3. Barbour Lathrop and David Fairchild on a steamer near Sumatra, Christmas 1895. From David G. Fairchild, The World Was My Garden: Travels of a Plant Explorer (New York: Scribner's, 1938), p. 84A.

task of appropriating foreign plants. He gave his "special agents" the less sinister title of "agricultural explorers," and he provided them with the elaborate official commissions, drawn by hand on heavy parchment with ribbons and gold seals, that were known vulgarly in the department as "Dago Dazzlers."[34]

[34] Fairchild, *World Was My Garden*, p. 119. After setting up the SPI Section, Fairchild left for a second multiyear tour with Lathrop. He reclaimed his government position in 1903, and he soon solidified his social standing by marrying Alexander Graham Bell's daughter. On hearing of the marriage, Lathrop wrote a deeply

From Fairchild's perspective, "systematic" plant introduction was quite different from previous casual attempts, private or governmental, to broadcast unfamiliar varieties. SPI staffers sought to discriminate among the myriad varieties they confronted and to select only those with particular promising characteristics. They established acclimatization gardens in different regions of the country to nurture unfamiliar types and built up an extensive network of academic and private cooperators willing to propagate and evaluate new specimens. In some cases they also worked—most notably with Carleton's durum wheat—to alter commercial processing and marketing techniques. Finally, they struggled to make Americans' tastes more cosmopolitan. Fairchild was continually pushing prominent people to consume artichokes, avocados, mangoes, and other tropical products regularly. Durum wheat succeeded so well because its primary end product, pasta, was soon popularized by the new Americans born in Italy.[35]

Plant introductions gained greater significance because they were enmeshed in Galloway's broader program for plant improvement. Division staffers Herbert J. Webber and Ernst Bessey emphasized that otherwise unimpressive foreign plants could be valuable raw material in the production of useful hybrids; small Russian apples, for example, could be bred with large American varieties to improve the latter's hardiness. As Barbara Kimmelman has shown, USDA scientists were at the center of American hybridization research in the late 1890s; with access to unprecedented stocks of germ plasm, and with great enthusiasm for Mendelism after 1900, they transformed the art of breeding into the science of genetics.[36] SPI provided the material on which the breeders were able to work.

Initially Galloway and his followers were hampered by the USDA's fragmentation. Wilson's interest in bureaucratic consistency had led him to locate plant exploration within the Division of Botany; but that organization's chief, Frederick Coville, was notably unenthusiastic about the work. Staffers continually worried that he was undermining their efforts.[37] Moreover, all the department's research divisions were hampered because the Washington-area hothouses and fields were controlled by the separate Division of Gardens and Grounds, run by the septuagenarian landscaper and Granger, William Saunders. When Saunders died in September 1900 Galloway immediately advised Wilson to merge the USDA's five plant science divisions into a new Bureau of Plant Industry. Consolidation, he argued, would lead to "more sympathetic union" among researchers and would enable them to tackle larger problems. Wilson agreed, and Galloway became head of a botanical establishment with more than two hundred employees.[38]

felt letter to "dear old Fairy" professing his acceptance of this action; after some initial tension, Marian Bell Fairchild established cordial relations with the gentleman she and her husband called "Uncle Barbour." See Barbour Lathrop to David Fairchild, 6 May 1905, David Fairchild Papers, Fairchild Tropical Garden, Miami, Florida; and Fairchild, *World Was My Garden,* pp. 457–458.

[35] David G. Fairchild, "The Plant Introduction Work of the Department of Agriculture, Memorandum for Conference in Secretary's Office, March 1915," typescript, File "Plant Introduction Reports," Fairchild Papers, p. 2; Fairchild, *Systematic Plant Introduction: Its Purposes and Methods,* USDA Division of Forestry, Bulletin No. 21, 1898; Fairchild, *World Was My Garden;* and Harvey Levenstein, *The Paradox of Plenty: A Social History of Eating in Modern America* (New York: Oxford Univ. Press, 1993), p. 29.

[36] Beverly T. Galloway, "Division of Vegetable Physiology and Pathology," in USDA, *Yearbook,* 1897, pp. 106–107; Herbert J. Webber and Ernst A. Bessey, "Progress of Plant Breeding in the United States," in USDA, *Yearbook,* 1899, p. 468; and Kimmelman, "Progressive Era Discipline" (cit. n. 28).

[37] See, e.g., Swingle to Fairchild, 8 Jan. 1899, Swingle File, Fairchild Papers; and David Fairchild, "Notes Prepared for an Informal Address to the Washington Staff of the Office of Foreign Seed and Plant Introduction . . . ," typescript, 9 pages, 9 Oct. 1922, File "Plant Introduction Reports," Fairchild Papers.

[38] "Report of the Secretary," in USDA, *Annual Report,* 1900, pp. lxxv–lxxvi; and "Report of the Secretary," in USDA, *Annual Report,* 1901, p. xxiii. On the politics of consolidation see Beverly T. Galloway, "The Genesis of the Bureau of Plant Industry," typescript, 8 pages, 30 June 1926, Galloway File, National Fungus Collection;

The scope and thrust of Galloway's vision in the early 1900s can be seen in presidential addresses he gave to the American Association for the Advancement of Science and the Botanical Society of America at the height of his scientific influence. Not surprisingly, he argued that the creation of the Bureau of Plant Industry marked a new era for botanists. It showed that governments, faced with the problems of population growth, urban concentration, and the international commercial system, were beginning to see in botany the potential key to the reorganization of the nation's food production system. Scientists could respond to this interest by turning their attention to such "applied" subjects as exploration, breeding, plant diseases, and the coadaptation between plants and environments. In rhetoric laden with the discipline's long-standing gender-role anxieties, he chided his academic colleagues for wanting to keep science "pure and undefiled" and for being "handicapped on every side by a sort of immaculateness." The "fully progressive" botanist, by contrast, was a "man of affairs" able to persuade "keen, analytic, practical men" that his work was worthwhile. These new kinds of botanists, Galloway hoped, would work toward "endless harmonious expansion," in both science and society.[39]

THE RETURN OF THE NATIVISTS

As Galloway's vision of botany became grander, his willingness to cooperate with the Division of Entomology declined. In 1895 Galloway and the entomologist Leland Howard each controlled budgets of about $30,000. Within ten years Galloway had acquired, through expansion and aggregation, a budget of $740,000, nearly ten times that of the entomologists. Galloway's protégé Fairchild reflected this change in interbureau relations. In 1898 he had expressed conspicuous concern about the problem of "objectionable introductions" and had pleaded that *"carefully conducted* Government importations" would not bring in pests. Within a few years that worry faded. By the mid 1900s Frank Meyer, his most intrepid explorer, was sending back a wide range of novelties from China and Siberia. Fairchild began to advertise acquisitions widely in the aptly-titled bulletin *Plant Immigrants,* and he distributed large amounts of exotic nursery stock to a network of private collaborators throughout the country. His office became the center of an international plant-exchange network.[40]

During these same years, however, federal entomologists became increasingly anxious about what they came to call "the insect menace." The economic and social devastation wrought by the boll weevil became evident around the turn of the century. Insects were being implicated in the spread of malaria, yellow fever, typhus, and other epidemic dis-

---

and Stevenson, "Plants, Problems, and Personalities" (cit. n. 29). On the structure and personnel of the new bureau see "Report of the Chief of the Bureau of Plant Industry," in USDA, *Annual Report,* 1901, pp. 43–45. The bureau combined the Divisions of Agrostology, Botany, Pomology, Vegetable Physiology and Pathology, and Gardens and Grounds. The Grangers were a secret society of farmers briefly prominent in the 1870s.

[39] Beverly T. Galloway, "Applied Botany, Retrospective and Prospective," *Science,* 1902, *16*:49–59, on pp. 55–57, 59; and Galloway, "The Twentieth Century Botany," *ibid.,* 1904, *19*:11–18, on pp. 12–13.

[40] Fairchild, *Systematic Plant Introduction* (cit. n. 35), pp. 20–21; Fairchild to Swingle, 28 Apr., 13 Oct. 1898, Walter T. Swingle Papers, Special Collections Department, Otto G. Richter Library, University of Miami, Coral Gables, Florida. On Meyer's adventures and Fairchild's efforts see Cunningham, *Frank N. Meyer* (cit. n. 33); Fairchild, *World Was My Garden* (cit. n. 29), pp. 289, 410–411; and Jefferson and Fusonie, *Japanese Flowering Cherry Trees* (cit. n. 1), pp. 6–7. On the budgets see Fred Wilbur Powell, *The Bureau of Plant Industry: Its History, Activities, and Organization* (Baltimore: Johns Hopkins Press, 1927), p. 101; and Weber, *Bureau of Entomology* (cit. n. 19), p. 157.

Figure 4. *Charles L. Marlatt, Chairman, Federal Horticultural Board, ca. 1923. (Courtesy of the National Archives.)*

eases.[41] The individual most concerned to defend the country's borders against pests was Charles Marlatt, the Bureau of Entomology's assistant chief (Figure 4). Recognizing that imported plant material was the major conduit for new insects, he directed his campaign against nurserymen and, ultimately, against the chief advocate for plant immigrants, David Fairchild.

It was a highly personal conflict. Marlatt and Fairchild had grown up together in Manhattan, Kansas, and had come to Washington in the same year; Marlatt was best man in Fairchild's wedding. Both had experienced the Orient. But while Fairchild traveled widely and easily with Lathrop as a tourist-collector, Marlatt used his honeymoon in 1901–1902

---

[41] Leland O. Howard, *The Insect Menace* (New York: Century, 1931); Howard, *Fighting the Insects* (cit. n. 15), pp. 118–137; and W. D. Hunter and W. E. Hinds, *The Mexican Cotton Boll Weevil,* USDA Division of Entomology, Bulletin No. 145, 1904, pp. 13–14. On the basis of their work against these enemies, the entomologists acquired bureau status in 1904.

to inspect Japan and eastern China for scale insect species that might endanger American fruit trees. He saw the devastation that had resulted from Chinese attacks on Westerners in the Boxer uprising, and on his return to America he witnessed the slow death of his young wife, Florence Brown Marlatt, from an infectious illness she contracted on their trip. He was thus deeply concerned about the dangers that foreign organisms posed to Americans. No one would stand in his way.[42]

Marlatt took charge of the USDA campaign to control plant imports in 1909. Leland Howard, a master of persuasion and a famous raconteur, had worked for over a decade to build a consensus in support of a plant quarantine law, but without success. Marlatt used different tactics. First he tried to shepherd a quarantine bill through Congress quietly. When that did not succeed (the nurserymen learned about it after House passage and sidetracked it in the Senate), he shifted to the opposite approach. He determined to raise the consciousness of both influential elements of the public and forces in the government about the danger posed by infested nursery stock. The Japanese cherry trees seemed heaven-sent for this purpose.[43] The fact that the originator of the cherry tree project was his old friend David Fairchild was unimportant.

Both Marian and David Fairchild were enthusiasts for *japonisme*. In 1907 they imported cherry trees (and a Japanese gardener) to adorn their Chevy Chase estate. The next year David Fairchild began to promote the idea of planting a "field of cherries" in the newly constructed West Potomac Park. When the Japanese government appropriated the idea and expanded it as part of their 1909 diplomatic offensive, he became the liaison, accepting the tree shipment from the Japanese in Seattle, arranging transport across the country in refrigerated railroad cars, and providing planting advice to federal landscapers. Marlatt undercut Fairchild's efforts completely. The Bureau of Entomology held authority to inspect the USDA's own plant introductions.[44] Marlatt used this power to make official Washington aware, once and for all, that exclusion of pests was the highest priority in horticultural commerce.

Japanese leaders recognized the long-term benefits of the tree-planting project and declined to take public offense when the cherry trees were burned. They immediately proposed a second shipment, and this time they countered experts with experts. Representatives of the Imperial Quarantine Service, the Imperial Horticultural Station, and the Imperial University supervised cultivation and assured the USDA that all specimens had been selected from an area free from scale insects, raised in ground free from nematodes, sprayed with fungicides and insecticides, and fumigated twice before packing. This time Agriculture Department officials approved every tree, and they were planted in 1912

---

[42] Charles L. Marlatt, *An Entomologist's Quest: The Story of the San Jose Scale* (Baltimore: Monumental Printing, 1953); and "Charles Lester Marlatt," *Proceedings of the Entomological Society of Washington,* 1955, *57*:37–43. In 1906 Marlatt was married a second time, to Helen McKey-Smith, daughter of the Episcopal bishop of Philadelphia. Money from his parents, the estate of his first wife, and his second wife's fortune combined to make Marlatt a wealthy man. He built a large house on fashionable 16th Street and was prominent in country club circles, joining foursomes with Franklin Roosevelt when Roosevelt was assistant secretary of the navy: Philip J. Pauly, telephone conversation with Florence Marlatt, daughter of Charles and Helen Marlatt, 11 July 1995.

[43] In 1908 the annual meeting of the American Association of Nurserymen voted to terminate negotiations about a plant quarantine law. See Richard P. White, *A Century of Service: A History of the Nursery Industry Associations of the United States* (Washington, D.C.: American Association of Nurserymen, 1975), pp. 168–171; and Marlatt, *Entomologist's Quest,* pp. 328–329.

[44] Fairchild, *World Was My Garden* (cit. n. 29), pp. 316–318, 326–327; Jefferson and Fusonie, *Japanese Flowering Cherry Trees* (cit. n. 1), pp. 5–6; and Gustavus A. Weber, *The Plant Quarantine and Control Administration: Its History, Activities, and Organization* (Washington, D.C.: Brookings Institution, 1930), p. 3.

around the Tidal Basin, along the Potomac, and on the White House grounds "as a living symbol of friendship between the Japanese and American peoples."[45]

In his heart, Marlatt was still not satisfied. A few years later he complained that this second shipment had introduced "the oriental fruit worm, which is now widespread in the eastern half of the United States, and is occasioning losses estimated well into the millions." At the time, however, he raised no objections. His focus was on the goal of gaining permanent control over commercial activity, and in 1912 he succeeded in engineering passage of the Plant Quarantine Act. This law gave the power to regulate plant imports to a new Federal Horticultural Board; for nearly two decades, that board was controlled by Marlatt.[46]

Between 1912 and 1920, the Federal Horticultural Board changed American policy regarding the continent's biotic future. The acquisitiveness and cosmopolitanism fostered by Galloway and Fairchild gave way to ideals of autonomy and isolation.

Initially, plant quarantine was a focused program, limited to specific infestations from particular countries. Marlatt, however, did not consider that approach sufficient. On the one hand, organisms innocuous in their natural settings could become rampant in North America; on the other, scientists were still unfamiliar with many dangerous parasites. Chestnut blight epitomized these dangers: American scientists realized only in the 1900s that a valuable native hardwood, a central symbol, after Longfellow, of the bountiful American countryside, was being exterminated by a strangling fungus. After identifying the pathogen for the first time, they traced it to China and concluded that enthusiasts for Chinese chestnut trees had unknowingly imported it a generation earlier. From Marlatt's perspective, "it is the unknown things that you cannot find that we have to protect this country from."[47] He believed that total exclusion was the only way to protect American plants from such a multitude of unknown, invisible enemies.

Marlatt's influence was limited while Galloway controlled USDA plant science. In 1914, however, after rising to the position of assistant secretary of agriculture, Galloway left the government for the less pressured setting of Cornell (where he had a nervous breakdown the next year).[48] His successor, William A. Taylor, was more interested in the systematic improvements that geneticists could induce with the material they had than in the hit-or-miss search for new germ plasm that he perceived exploration to offer. Marlatt was soon orchestrating a broad exclusion policy. At a convention of the sympathetic American Forestry Association in January 1917, he called for new security measures. He reviewed the now-standard list of foreign "plant enemies," adding the Japanese beetle, which had just appeared in New Jersey. He recalled that the "virgin lands of the New World" had originally been free from such pests. Protecting the "standard products of our soil," he

[45] Jefferson and Fusonie, *Japanese Flowering Cherry Trees*, pp. 16–21.

[46] Weber, *Plant Quarantine* (cit. n. 44), pp. 3 (quotation), 10–29. See Marlatt's testimony, 19 Feb. 1912, "A Bill to Regulate the Importation and Interstate Transportation of Nursery Stock (H.R. 18000)," *Hearings before the Committee on Agriculture on Miscellaneous Bills and Other Matters*, 62nd Cong., 2nd sess., pp. 51–106.

[47] Charles L. Marlatt, "Plant Quarantine No. 37," *American Florist*, 1919, 53:411; and Haven Metcalfe, "The Chestnut Bark Disease," *Journal of Heredity*, 1914, 5:8–17. On the early days of the plant quarantine program see Weber, *Plant Quarantine*, pp. 87, 61; for Marlatt's views see his testimony, 27 Apr. 1910, in "Inspection of Nursery Stock," *Hearings before the Committee on Agriculture*, 61st Cong., 2nd sess., 3 vols., Vol. 3, pp. 494–501.

[48] Galloway succeeded Liberty Hyde Bailey as dean of the College of Agriculture. His collapse was precipitated by attacks on his rationalizing policies by faculty and alumni, but it probably had deeper causes: he suffered intermittently from deep depression and ultimately took his own life. See Gould P. Colman, *Education and Agriculture: A History of the New York State College of Agriculture at Cornell University* (Ithaca, N.Y.: Cornell Univ. Press, 1963), pp. 250–260; Beverly T. Galloway to Fairchild, 16 May 1916, 31 Oct. 1916, and Agnes Galloway to Fairchild, 20 May 1921, Fairchild Papers; and Stevenson, "Beverly Thomas Galloway" (cit. n. 29).

asserted, was more important than accumulating the "novelties and curiosities of the plant world for our gardens, lawns, and parks." A few months later he proposed Quarantine #37, a rule that would end nearly all private importations of plants and bulbs.[49]

David Fairchild, as the chief government advocate for "novelties and curiosities," did what he could to prevent this change. Addressing the same elite forestry audience, he articulated the basic principles of ecological cosmopolitanism. He argued, first, that although any given foreign variety might seem to have minimal value, the total economic worth of plant imports was substantial—and would only be known in the future as a result of continued importation. His second appeal was to American principles of justice and charity: "it would be eminently unfair to assume that because we do not know that these little apple seedlings from the old world or from Japan are as clean and free from disease as any which we can produce in America, they represent undesirable immigrants and should be excluded from the country." He pointed to the inextricable commingling of evil and good: chestnut blight had led Americans to discover both new varieties of Chinese chestnuts and oriental pears resistant to blight. Finally, he argued—with rhetorical contrast between the archaic and the modern—that no one could hold back the future:

> We can say to ourselves, "let us be independent of foreign plant production. Let us protect our own by building a wall of quarantine regulations and keep out all the diseases which our agricultural crops are heir to and have the great advantage over the rest of the world." But the whole trend of the world is toward greater intercourse, more frequent exchange of commodities, less isolation, and a greater mixture of the plants and plant products over the face of the globe.[50]

Loss of some cherished species was unavoidable; the best that could be done was to foster research, ingenuity, and attention to individual diseases.

Fairchild's opposition was futile. In November 1918 the secretary of agriculture promulgated Quarantine #37. Ostensibly this prohibition on private imports had no impact on the Agriculture Department's own plant introduction program. In fact, however, the new regime subverted Fairchild's activities. It put the Office of Foreign Seed and Plant Introduction under suspicion as a potential source of disease, disrupted the delicate exchange networks that Fairchild had built up, and forced him to focus on the processing of commercially promising introductions. *Plant Immigrants* was terminated for lack of funds, but construction began on a new "Plant Detention Station" outside Washington, where "plant immigrants will be received and carefully grown, watched and propagated, to be sure that all alien enemies are excluded." In 1924, at age fifty-five, Fairchild gave up. He retired to Florida and joined a new private patron—the meat industry heir Allison Armour—on a yachting expedition in search of new plants. Galloway, who had quietly returned to Washington as Fairchild's assistant in 1916, remained behind to maintain a policy of "rational plant exclusion."[51]

Marlatt, on the other hand, ended his career in bureaucratic triumph. He succeeded

[49] Charles L. Marlatt, "Losses Caused by Imported Tree and Plant Pests," *American Forestry,* 1917, *23:*75–80, on pp. 75, 80; and Weber, *Plant Quarantine* (cit. n. 44), pp. 29–42. On Taylor's views see Klose, *America's Crop Heritage* (cit. n. 10), pp. 123–124.

[50] David G. Fairchild, "The Independence of American Nurseries," *Amer. Forestry,* 1917, *23:*213–216, on p. 216.

[51] Weber, *Plant Quarantine* (cit. n. 44), p. 29; Galloway, "Searching the World for New Crops" (cit. n. 31), p. 14 (quotation); and Beverly T. Galloway, "The Beginnings of the Bell, Maryland, Plant Introduction Garden," typescript, n.d., Galloway File, National Fungus Collection. On Fairchild's later career see Fairchild, *World Was My Garden* (cit. n. 29), pp. 425, 471–475; and David G. Fairchild, *Exploring for Plants* (New York: Macmillan, 1930). See also Galloway, "Rational Plant Exclusion," typescript, 8 pages, 1925, Galloway Papers, National Agricultural Library.

Leland Howard as chief of the Bureau of Entomology and in 1928 became head of the new Plant Quarantine and Control Administration as well. When the Mediterranean fruit fly appeared in Florida the next year, Marlatt obtained emergency funding of nearly $6 million, and near-totalitarian power, to fight it. He banned interstate shipment of all Florida produce, prohibited planting of vegetables in infested areas, required picking of all fruit prior to ripening, and organized a massive spraying campaign. Within a year the insect had been wiped out.[52]

To be sure, Quarantine #37 did not end ecological imperialism in North America. Biotic inequality and the potential for movement continued, resulting in introductions ranging from Dutch elm disease to the walking catfish and killer bees. Yet the convergence of scientific thinking and state power that occurred between 1895 and 1920 did mark the end of the United States' status as an ecological colony. Since that time ecological imperial forces have been hedged by at least some of the claims of political sovereignty.

ECOLOGICAL INDEPENDENCE AND IMMIGRATION RESTRICTION

Thus far this account has been limited to the nonhuman. Yet it should be clear that attitudes about foreign and native organisms were intimately linked, through both everyday experience and analogies of policy, to views on "alien" and "native" humans.

Commonplace symbolic connections between geographically identified organisms and humans were omnipresent and powerful. Americans perceived the English sparrow as an avian Cockney pushing aside larger but better-mannered American birds. The gypsy moth's devastation of respectable neighborhoods confirmed casual prejudices about Gypsies. The introduction of pulpy tropical fruits, carrying the aromas of the Far East, really did increase sensuality among Americans raised on tart apples and bland pears. These specific associations could operate at high cultural levels, and their implications could be molded consciously: the Japanese cherry trees were, by turns, tokens of equality, infectious interlopers, and sanitized ambassadors. Their successful introduction did, to some degree, neutralize the assertion of American racial superiority inherent in the Gentlemen's Agreement.

The density and pervasiveness of these thickets of meaning, however, make their overall import difficult to assess. As in the case of the nonhuman alone, we can gain a clearer conception of the efficacy of these modes of thought by focusing on the transformation of policy. The crucial arena in which ecological ideals intersected with demographic designs was, of course, immigration. The tendency in general histories has been to emphasize the opposition between those who welcomed newcomers and those who wanted to keep them out, and to view the Immigration Act of 1924 as the culmination of a decades-long narrow-minded campaign for the exclusion of those who were different. By looking at American immigration policy through the lens of the debates over ecological independence, we find a more complex, but more meaningful and relevant, narrative.

For a century after the founding of the Republic, Americans had the same mixture of interest and distrust, and the same laissez-faire policy, toward foreign humans as they did toward foreign plants and birds. At the end of the nineteenth century, however, federal

---

[52] Weber, *Plant Quarantine,* pp. 108–116. Marlatt complained wistfully to Fairchild that, given the American governmental system, he was not *really* a dictator—"all we can do is to come as near to [absolute power] as we can": Charles L. Marlatt to Fairchild, 10 Sept. 1929, Fairchild Papers. Their personal relations finally soured when, in his autobiography, Fairchild blamed Marlatt for the decline of SPI: Fairchild, *World Was My Garden,* p. 425. Marlatt was quite upset at this attack and attributed it to Fairchild's jealousy of his success. See Marlatt to Fairchild, 30 Oct. 1938, Fairchild Papers; and Pauly, conversation with Florence Marlatt (cit. n. 42).

passivity about immigration was challenged by the same two competing approaches discussed here. Like Palmer and Marlatt, nativists (in the original, political, meaning of the term) had little interest in the possible benefits that foreigners might bring to the United States. On the other hand, they were deeply concerned about the damage that aliens might wreak on American life. Like the ornithologists and entomologists, they sought to deal with these problems by excluding entire groups from the United States. The paradigm of the nativist approach was the Chinese Exclusion Act, passed at the insistence of "native" California workingmen in 1882, a year after the state's plant quarantine law. It kept out all but a trickle of Chinese laborers, and, by excluding women, it was designed to prevent the Chinese "race" from establishing itself permanently in North America.

Moderates on immigration were more cosmopolitan. They were comfortable with the entrance of some foreigners from most countries, and they recognized the value of immigrant merchants and laborers to the American economy. In place of blanket exclusions, they envisioned a policy of individual selection, similar in its fundamentals to that pursued by Galloway and Fairchild with plant immigrants. Botanists exercised discrimination constantly in their collecting choices, and they used plant introduction gardens to cull out less promising imported species and varieties. Bureaucrats dealing with humans had relatively little influence over the composition of the immigrant stream. As a consequence, they concentrated on sorting those who had arrived. They proposed to assess each entrant's inherent quality and then to welcome superior specimens, treat those with minor defects, and send inferior beings back to their homelands. A policy of selection was never instituted for Orientals. But it provided the rationale for the operation of Ellis Island, the United States' major immigration facility, during its most important years.

Ellis Island was built in the early 1890s to improve the government's ability to collect taxes from immigrants. Its mission was altered, however, by Theodore Roosevelt. As a nationalist and a naturalist, Roosevelt was deeply concerned about the biotic future of the country—including the preservation of large herbivores (on which he closely collaborated with Merriam and Palmer of the Biological Survey), the fecundity of European-Americans, and the quality of immigration.[53]

On becoming president in 1901, Roosevelt sought to transform Ellis Island into a site for immigrant selection. His image of this enlarged sandbar became clear in his choice of immigration commissioner for the Port of New York. William Williams was in most respects a typical patrician Wall Street lawyer. From Roosevelt's standpoint, however, he had important relevant experience: a decade earlier, as a State Department employee working in close consultation with the USDA Division of Economic Ornithology and Mammalogy, he had defended American management of the Pribiloff Islands fur seal herd before an international arbitration tribunal. He was thus familiar with problems of migration, understood the management of large semiconfined populations, appreciated the principles of selection, and knew how to use experts to implement and bolster policy.[54]

During his two terms as commissioner (1902–1905, 1909–1914), Williams developed procedures to cull the Ellis Island "herd." He believed that at least 25 percent of entrants were "of no benefit to the country" and would "lower our standards." The key to finding that bottom quartile lay in the system of medical inspection. Williams eliminated the

[53] Paul Russell Cutwright, *Theodore Roosevelt: The Making of a Conservationist* (Urbana: Univ. Illinois Press, 1985); and Thomas G. Dyer, *Theodore Roosevelt and the Idea of Race* (Baton Rouge: Louisiana State Univ. Press, 1980).

[54] "William Williams." *National Cyclopedia of American Biography*, Vol. 35, p. 196; and William Williams, "Reminiscences of the Bering Sea Arbitration," *American Journal of International Law*, 1943, 37:562–584.

informal "corrupt" arrangements whereby immigrants had previously bypassed the doctors. He doubled the medical staff, formalized the details of the inspection procedure, specified actionable defects, and introduced new tests—most notably the eversion of every entrant's eyelid with a buttonhook in the search for the common eye infection trachoma. In 1909, when Williams returned for his second tour, his staff introduced jigsaw puzzles as a means to pick out feebleminded entrants. The American Genetic Association, whose president was David Fairchild, strongly supported these measures to improve the quality of immigration.[55]

The disruptions of World War I ended American efforts to manage European immigration through individual selection. The interruption of transatlantic commerce made the absence of immigrants seem the norm for the first time in decades, and wartime "100% Americanism" heightened their alienness. When immigration suddenly recommenced in 1919, on a scale magnified by the war's devastation, a political consensus rapidly formed around the need for exclusionary measures similar to those already directed against Asians. In 1921 Congress established a cap on all European immigrant groups, limiting entry to 3 percent of the total already resident in the United States. This was refined and made permanent in the Immigration Act of 1924.[56]

Discussions about the biological aspects of immigration restriction have focused on the extent to which the details of the system established in the 1920s were racist.[57] They clearly were, but such racism was a historically secondary phenomenon. The more fundamental change in the biology of immigration was the introduction of a stringent limit on the total number of European entrants. In 1913, nearly 1.2 million foreigners came to the United States; twenty years later, only 23,000 arrived. Federal policy made the United States, for the first time, a nation made up almost entirely of "natives." This new demographic policy was part and parcel of the drive by scientific bureaucrats to make the United States ecologically independent.

In the 1890s American leaders, concerned to make the United States a successful global economic and imperial power, decided to manage the nation's ecological and demographic borders. The real controversies, which lasted more than two decades, concerned the relative importance of improvement via selection or protection through exclusion. These issues were eminently suitable for compromise. Yet in the late 1910s the nativists suddenly triumphed. Their victory was possible only because World War I interrupted international travel, generated unprecedented xenophobia, and narrowed participation in political decision making.

[55] William Williams, quoted in Prescott F. Hall, "Selection of Immigration," *Annals of the American Academy of Political and Social Science,* 1904, *24*:175. On the tests Williams introduced see Elizabeth Yew, "Medical Inspection of the Immigrant at Ellis Island, 1891–1943," *Bulletin of the New York Academy of Medicine,* 1980, *56*:488–510; and Kraut, *Silent Travellers* (cit. n. 11), pp. 50–77. See also "Second Report of the Committee on Immigration of the Eugenics Section of the American Genetic Association," *J. Hered.,* 1914, *5*:297–300. The impact of inspection on immigration is uncertain. Williams's procedures resulted in the exclusion of, at most, 2.5 percent of entrants, much less than he envisioned. Yet awareness of the inspection process certainly deterred many individuals with detectable illnesses or disabilities from making the trip; see Kraut, *Silent Travellers,* p. 66.

[56] For provisions of the laws see Marion T. Bennett, *American Immigration Policies: A History* (Washington, D.C.: Public Affairs Press, 1963); for the change in perspective see, e.g., Robert De C. Ward, "The Immigration Problem Today," *J. Hered.,* 1920, *11*:323–328.

[57] Stephen J. Gould, *The Mismeasure of Man* (New York: Norton, 1981), pp. 231–232; Daniel J. Kevles, *In the Name of Eugenics* (New York: Knopf, 1985), pp. 96–97; Elazar Barkan, "Reevaluating Progressive Eugenics: Herbert Spencer Jennings and the 1924 Immigration Legislation," *Journal of the History of Biology,* 1991, *24*:91–112; and Carl N. Degler, *In Search of Human Nature: The Decline and Revival of Darwinism in American Social Thought* (New York: Oxford Univ. Press, 1991).

Recovering the interpenetration of demographic and ecological policy, the range and vitality of cosmopolitan thinking, and the singular circumstances that made nativism the norm together provide significant intellectual resources for interpreting contemporary immigration issues. It leads to a distrust of absolutist solutions, a recognition that management does not necessarily mean exclusion, and, finally, an awareness that comparisons between the human and nonhuman are important, unavoidable, and often helpful symbolic resources. The liberal garden writer Michael Pollan emphasized this last point in a recent attack on the nativism implicit in the "native plants" movement of the last decade.[58] A more positive perspective on the same issue can be found through a final look at Washington's cherry trees.

Cherry trees have participated in American national symbolism from the early nineteenth century, when Parson Weems broadcast the story of George Washington's boyhood attack on his father's prize specimen (presumably a European variety). The Japanese, by sponsoring the planting of their own varieties within sight of the Washington Monument, were linking themselves to American stories in ways that no one could fully predict. The primary meaning imputed to the trees—that their growth and bloom, year after year, would symbolize the enduring friendship between the Japanese and American peoples—became, through Pearl Harbor and Hiroshima, ironic and even embarrassing. The trees hold a different significance, however, if they are viewed in their alternate guise as plant immigrants. Fewer than 10 percent of the original specimens are still extant. Yet the annual displays are more spectacular than ever, because the trees donated by Tokyo have been replaced and supplemented by specimens grown in American nurseries from stock imported in the early years of the century. The trees have, in a word, been naturalized. They can remind visitors that the United States, both ecologically and demographically, has been a place of mixture, and that this mixture is a major source of its beauty.[59]

---

[58] Michael Pollan, "Against Nativism," *New York Times,* 15 May 1994, sect. 6, pp. 52–55. He built his critique on histories of the Nazi enthusiasm for native plants; see Gert Groening and Joachim Wolschke-Bulmahn, "Some Notes on the Mania for Native Plants in Germany," *Landscape Journal,* 1992, *11*:116–126.

[59] On this theme see esp. William Cronon, *Nature's Metropolis: Chicago and the Great West* (New York: Norton, 1991); information on today's cherry trees comes from the author's telephone conversation with Robert DeFeo, Chief Horticulturist, National Park Service, 18 Oct. 1995.

Fabricating Authenticity

Modeling a Whale at the American Museum of Natural History, 1906–1974

By Michael Rossi

ABSTRACT

Historians of science have in recent years become increasingly attentive to the ways in which issues of process, matter, meaning, and value combine in the fabrication of scientific objects. This essay examines the techniques that went into the construction—and authentication—of one such scientific object: a model of a blue, or "sulfur-bottom," whale manufactured at the American Museum of Natural History in 1907. In producing their model, exhibitors at the American Museum employed a patchwork of overlapping discursive, procedural, and material techniques to argue that their fabrication was as authentic—as truthful, accurate, authoritative, and morally and aesthetically worthy of display—as an exhibit containing a real, preserved cetacean. Through an examination of the archival and published traces left by these exhibitors as they built their whale, I argue that the scientific meanings of authenticity at the American Museum were neither static nor timeless, but rather were subject to constant negotiation, examination, re-evaluation, and upkeep.

I N 1907, WILLIAM MUHLIG, a New York engineer, brought a lawsuit in Manhattan's municipal court over the matter of a fake whale. As the *New York Times* reported, two men, Christopher Rebhan and his partner August Brahn, had offered Muhlig a share in the "Only Preserved Greenland Whale Ever Seen in Captivity"—a specimen captured off the shores of Greenland and stuffed at great expense in Hamburg, Germany. For only $1,500 (roughly equivalent to $30,000 in 2009), Mr. Muhlig could purchase the chance to become the whale's manager as it toured the United States. He would collect a salary of twenty dollars a week, and a quarter of the profits from the show in return. Sensing a deal, Muhlig agreed, paying $500 up front and retaining Mr. Brahn's wife as a cashier. The whale went

Isis 101, no. 2 (June 2010): 338–61.

Thank you to David Jones, Stefen Helmreich, Jennifer Roberts, and *Isis's* anonymous referees for their invaluable advice and patient readings over the course of the essay's many iterations. Thanks also to the archivists Barbara Mathé and Pat Brunauer at the AMNH and Sigrund Nützsche at the Museum für Völkerkunde, Dresden. Deepest appreciation to William Broadhead for assistance navigating leviathan Latin locations.

on tour, and things went well until, in Harrisburg, Pennsylvania, Muhlig discovered to his surprise that the whale was not a preserved Greenland whale at all, but a phony—"a wooden dummy covered with canvas." The show was over and Muhlig put the model whale up for auction, netting a paltry ninety-nine dollars, which Mrs. Brahn seized as her salary. Back in New York, Muhlig took legal action to get his money back from Rebhan. Though he lost his original suit, the Appellate Term of New York's Supreme Court ruled for a new trial, finding that the lower court had subjected Muhlig to undue ridicule.[1]

During that same summer, in the same city, another fake whale went on display at the American Museum of Natural History near Central Park, and was met not with ridicule, but with widespread approbation. Roy Chapman Andrews, a young museum employee, and a team of twelve preparators had fabricated a life-sized model of a blue (or "sulfur-bottom") whale in the museum's Hall of the Biology of Mammals (see Figure 1 and Cover). Constructed from wood and angle iron and covered with *papier-mâché*, the model was lauded in the popular press. *Scientific American* dedicated an article in its supplement to the whale, emphasizing that "each detail" of the whale was a "close copy from life." The magazine continued: "The wonderful grooves along the lower jaw of the whale . . . are exact in number and position," and the construction of such a replica was "an achievement deserving of more than passing notice or credit."[2] A correspondent for *Outlook* magazine concurred, remarking that the seventy-six foot long model was an appropriate addition for an institution of the museum's stature, assuring his readers that the model was so accurate, it was as though the real whale "is on exhibition now, to all intents."[3]

How can it be that Andrews's wooden whale was scientifically authentic and admirable, while Muhlig's wooden whale was silly, and even actionable? Muhlig's model occasioned a brief snippet tucked among classified ads in the back of the *Times*; Andrews's model was a lasting sensation, garnering fond notice in popular media for the better part of the century. Muhlig was ridiculed in a court of law for believing his model whale to be real; visitors to the American Museum were actively encouraged to pretend that Andrews's whale was the real thing. Muhlig's whale was humbug; Andrews's whale was cetology—but how?

This paper examines the techniques—measuring, photographing, casting, and sculpting—marshaled by exhibitors at the American Museum to insure that their model was a *bona fide* representative of *Balaenoptera musculus*. Muhlig's reasons for believing in the veracity of his whale are lost, as are Rebhan and Brahn's methods for trafficking in fake cetaceans.[4] Andrews and his fellow museum exhibitors, on the other hand, left a faint but

---

[1] "Whale was a Dummy," *New York Times*, 7 June 1907, p. 18. Muhlig's appeal was heard by the Supreme Court of New York, Appellate Term: *Muhlig v. Rebhan*, 55 Misc. 305, 105 N.Y.S. 110 (1907).

[2] "The Whale in the American Museum of Natural History," *Scientific American Supplement*, 1907, *64*:161–162, on p. 162.

[3] Everett Wallace Smith, "Natural Science for the Every-Day Man," *Outlook*, 1908, *89*:183–191, on p. 185.

[4] Somewhat astonishingly, there was a precedent for peddling fake whales in traveling shows. As D. Graham Burnett reports, in the early nineteenth century, a "clique of flimflam artists" took to the road with an artificial cetacean, deceiving audiences around New York State until, in Waterloo, the fake was unmasked. Although Burnett cannot say for certain what the whale was made of, the material was flammable; the fake was set aflame, and its hucksters sent packing. See D. Graham Burnett, *Trying Leviathan: the Nineteenth-Century New York Court Case That Put the Whale on Trial and Challenged the Order of Nature* (Princeton: Princeton Univ. Press, 2008), p. 182. For more general thoughts on fakery see Mark Jones, ed., *Fake? the Art of Deception* (Berkeley: Univ. of California Press, 1990); Jones, ed., *Why Fakes Matter: Essays on Problems of Authenticity* (London: British Museum Press, 1992); and Jane MacLaren Walsh, "Crystal Skulls and Other Problems: or, 'Don't Look it in the Eye,'" in *Exhibiting Dilemmas: Issues of Representation at the Smithsonian,* ed. Amy Henderson and Adrienne L. Kaeppler (Washington, D.C.: Smithsonian Institution Press, 1997). For a general overview of

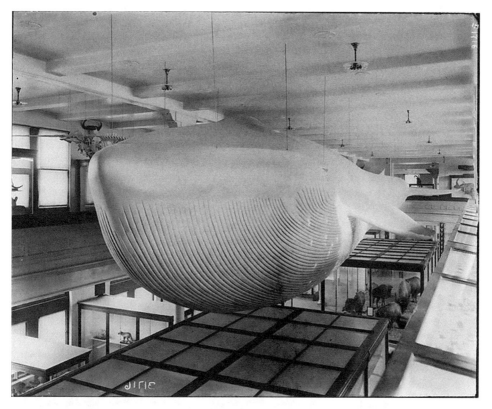

Figure 1. *The completed model sulfur-bottom whale hanging in the Hall of the Biology of Mammals, American Museum of Natural History, in 1907. American Museum of Natural History Photograph Collection, PPCA45.M88, neg. 031716. Reproduced with permission from the American Museum of Natural History Archives.*

well-preserved trail of preoccupations and apprehensions as they attempted to fashion replicas of the largest creatures in the world. Through these records, one finds that practices used by exhibitors to replicate whales were intimately tied up with questions of truthfulness, accuracy, authority, and moral and aesthetic worth—qualities that I will sum up under the rubric of *authenticity*.

Andrews and his peers attended carefully and explicitly to these questions as they assembled their model. For these exhibitors, authenticity was a quality to be laboriously produced and maintained, and each of the techniques they employed came with practical, moral, and aesthetic advantages and disadvantages. Measurements and photographs, for example, were easier to transport and study than whale corpses, but the ease with which they circulated could be problematic when the firsthand touch of the exhibitor counted. Casting insured that the model was point-by-point isometric with its referent whale, and vouched-for by a scientist, but the retouching work needed in order to achieve a cast that was not only epistemologically true but lifelike as well, was viewed by some exhibitors

various instances of fakery in science see Marcel C. LaFollette, *Stealing into Print: Fraud, Plagiarism, and Misconduct in Scientific Publishing* (Berkeley: Univ. California Press, 1992); and Horace F. Judson, *The Great Betrayal: Fraud in Science* (Orlando, Fla.: Harcourt, 2004). For a more sustained look at one particular incident see Daniel Kevles, *The Baltimore Case: A Trial of Politics, Science and Character* (New York: Norton, 1992).

as an unacceptable manipulation of natural form. The antidote to lackluster, lifeless casts was skillful sculpting, but sculpting needed to be underwritten by measurements, photographs, casts, and knowledgeable scientists in order for the work to count as authentic. These techniques provided a sort of epistemological coverage that ultimately could be utilized by exhibitors to argue that the model was just as viable, or just as authentic, as an exhibit made from a real whale.

In recent years, historians have devoted increasing attention to the ways in which these sorts of questions of fabrication, craft, aesthetics, and meaning-making apply, at different times and in different ways, to scientific practices as diverse as geometry, origami, molecular biology, and microphysics.[5] Natural history displays have proved a particularly fertile source of reflection on the historically situated characteristics of notions like nature and artifice, falsity and truthfulness, and ideology and materiality. Exhibitors had to produce displays that were scientifically rigorous (and therefore, it was assumed, rigorously truthful) and yet still attracted the broad audiences that were museums' *raisons d'être*.[6] In her examination of natural history displays in German museums in the 1900s,

[5] On the material practices of Hellenic geometry see Reviel Netz, *The Shaping of Deduction in Greek Mathematics: A Study in Cognitive History* (Cambridge: Cambridge Univ. Press, 1999). For work on craft and origami see Alma Steingart, "In Practice and On Paper: Embodiment and Materiality in the Twentieth-Century Convergence of Mathematics and Origami," paper presented at the Annual Conference for the Society for the Social Studies of Science, Washington D.C., 28–31 Oct. 2009. On molecular biology see Soraya de Chadarevian, *Designs for Life: Molecular Biology after World War II* (Cambridge: Cambridge Univ. Press, 2002), esp. pp. 98–164; de Chadarevian, "Models and the Making of Molecular Biology," in *Models: The Third Dimension of Science*, ed. Soroya de Chadarevian and Nick Hopwood (Stanford, Calif.: Stanford Univ. Press, 2004), pp. 339–368; and Natasha Myers, "Molecular Embodiments and the Body-Work of Modeling in Protein Crystallography," *Social Studies of Science*, 2008, *38*:163–199. Peter Galison's *Image and Logic: A Material Culture of Microphysics* (Chicago: Univ. Chicago Press, 1997) is a classic musing on the intersections between matter and theory in particle physics; for more recent work see David Kaiser, *Drawing Theories Apart: the Dispersion of Feynman Diagrams in Postwar Physics* (Chicago: Univ. Chicago Press, 2005). This is, of course, by no means an exhaustive list of the topics covered or approaches taken by historians of science interested in questions of epistemology and material culture.

[6] Historians interested in the materials, modes of production, and politics of natural history displays have offered a variety of interpretations of the underlying mission(s) of museums. On the ways in which museums were configured as organs of uplift and progress, differently construed see Douglas Sloan, "Science in New York City, 1867–1907," *Isis*, 1980, *71*:35–76; John Michael Kennedy, "Philanthropy and Science in New York City: The American Museum of Natural History, 1868–1968" 'Ph.D. diss., Yale Univ., 1969'; and James Livingston, *Pragmatism and the Political Economy of Cultural Revolution, 1850–1940* (Chapel Hill: Univ. of North Carolina Press, 1997), pp. 41–48. On the American Museum as an educational institution of debatable efficacy see Philip J. Pauly, "The Development of High School Biology: New York City, 1900–1925," *Isis*, 1991, *82*: 662–668. Sally Gregory Kohlstedt looks at natural history displays as vectors for gendered power norms in "Nature by Design: Masculinity and Animal Display in Nineteenth-Century America," in *Figuring it Out: Science, Gender and Visual Culture,* ed. Ann B. Shteir and Bernard Lightman (Lebanon, N.H.: Dartmouth College Press, 2006), pp. 110–139. Donna Haraway similarly thinks through questions of race, gender, class, and craft in the production of the American Museum's gorilla diorama; see Ch. 3 ("Teddy Bear Patriarchy") of *Primate Visions* (New York: Routledge, 1989), pp. 26–58. For his part, Tony Bennett challenges Haraway's interpretation, arguing that by focusing on just one display in one hall of the museum Haraway misses some of the institutional complexities of the American Museum's relationship with other power centers, particularly municipal and state education authorities in New York: Tony Bennett, *Pasts Beyond Memory: Evolution, Museums, Colonialism* (New York: Routledge, 2004), esp. pp. 114–135. For further reading on museum exhibits as conveyances for normative social values see Ronald Rainger, *An Agenda for Antiquity: Henry Fairfield Osborn & Vertebrate Paleontology at the American Museum of Natural History, 1890–1930* (Tuscaloosa: Univ. Alabama Press, 1991); Gregg Mitman, "Cinematic Nature: Hollywood Technology, Popular Culture, and the American Museum of Natural History," *Isis*, 1993, *84*:637–661; and Karen Wonders, *Habitat Dioramas: Illusions of Wilderness in Museums of Natural History* (Uppsala: Almqvist & Wiksell, 1993). Mieke Bal moves beyond the American Museum's diorama to examine the semiotics of the museum space itself in "Telling, Showing, Showing Off," *Critical Inquiry*, 1992, *18*:556–594. For more on museum space, see Sophie Forgan, "Building the Museum: Knowledge, Conflict, and the Power of Place," *Isis*, 2005, *96*:572–585. For deeper investigations into the identities and "thingness" of objects in museums see, for instance, Lorraine Daston, "The

Lynn Nyhart discusses the ways in which exhibitors struggled mightily to create authentic exhibits through the use of unadulterated, site-specific materials (including, in one instance, transporting a cliff face—rocks, plants, and all—from the seashore into the museum). These exhibits could be augmented when necessary by appealing touches like painted backgrounds and fake water. Victoria Cain makes a similar point about the delicate balance between notions of nature and artifice in American museum exhibits when she writes, "despite some early resistance by scientists, after the 1880s, [the] combination of authentic material and lifelike appearance allowed the public to categorize taxidermied specimens as real, rather than as Barnum-like fakery." This process of categorization—real or fake, flesh or *papier-mâché*, true or false—was persistently present for exhibitors, not least of all when they endeavored to construct honest replicas of tremendous creatures which spent most of their time obscured beneath the ocean. As Lorraine Daston notes, "matter constrains meanings, and vice versa."[7] Just so, the American Museum's whale modelers and their peers at other museums constantly juggled questions of materials and intentions, techniques and trust, and artifice and authenticity, when they endeavored to construct truthful and engaging reproductions of their subjects for public viewing.

DATA AND DERMA: ON MEASURING, PHOTOGRAPHING, REPLICATING, AND JUSTIFYING

In late summer, 1906, a "gray haired gentleman" named F. C. A. Richardson commenced work on the museum's model whale, assisted by the taxidermy department's new hire, Andrews, and a team of carpenters and blacksmiths.[8] In terms of gross mechanics, the American Museum's whale modelers employed many of the same methods used in constructing taxidermies of other very large animals. Working from paper templates derived from a scale model of a sulfur-bottom ("one inch to one foot"), the modelers first constructed a lateral cross-section of the whale of iron rods on the floor of the hall, creating an outline of a large whale in profile.[9] The outline was reinforced by parallel cross bars to which blacksmiths connected perpendicular sections of iron rods bent, like ribs, to constitute the model's girth. They then added iron outlines of the fins and flukes. To this iron skeleton, carpenters affixed wooden struts, also in the shape of the final replica, over which they placed strips of thin basswood to further refine the model's shape. (See Figure 2.) This technique had a notable precedent in the mounting of Jumbo, P. T. Barnum's famed elephant. Jumbo had run afoul of a locomotive in 1885 and was taxidermized at Ward's Natural Science Establishment in Rochester, New York, by William J. Critchley and Carl Akeley. Like the whale modelers, Critchley and Akeley constructed a manikin of wood and iron rods that they then covered with "inch square steamed basswood . . . bent and hewn so as to give the exact shape of the animal." This basswood substructure was so

---

Glass Flowers," in *Things that Talk: Object Lessons from Art and Science,* ed. Daston (New York: Zone Books, 2004), pp. 223–254; Sally Gregory Kohlstedt, "'Thoughts in Things': Modernity, History, and North American Museums," *Isis,* 2005, 96:586–601; and Sam Alberti, "Objects and the Museum," *Isis,* 2005, 96:559–571.

[7] Victoria Cain, "Nature Under Glass: Popular Science, Professional Illusion and the Transformation of American Natural History Museums, 1870–1940" 'Ph.D. diss., Columbia Univ., 2006', pp. 156–157; and Lorraine Daston, "Speechless," in *Things that Talk,* ed. Daston, pp. 9–24, on p. 17.

[8] Andrews remembered Richardson's hair in *Ends of the Earth* (New York: G. P. Putnam's Sons, 1929), p. 7.

[9] "The Whale" (cit. n. 2), p. 162.

Figure 2. *Two views of the model whale under construction. In the image on the left, Andrews poses before the model and a group of carpenters working on the model's armature. On the far side of the model, a line of basswood strips can be seen making its way up the side of the whale. In the image on the right, the model's cladding of basswood is complete, and awaits an epidermis of chicken wire and papier maché. American Museum of Natural History Collection, PPCA45.M88; on left, neg 31598; on right, neg 31601; American Museum of Natural History Archives. Reproduced by kind permission of the American Museum of Natural History.*

accurate, an article in a newsletter for Ward's insisted that the taxidermy was completed *"without any stuffing between the wood and the skin."*[10]

The denouement to Critchley and Akeley's model-making then, was to cover their subject in its own skin. For Richardson and Andrews, the use of actual whale skin was not an option—but that didn't mean that their model was any less authentic than Jumbo or his fellows in the Hall of the Biology of Mammals. The museum's 1911 *General Guide* summarized the salient points of the finished model for visitors: "The most striking object in the hall is the life-sized model of a sulfur-bottom whale, seventy-nine feet in length. The original of this specimen was captured in Newfoundland, and the model is accurately reproduced from careful measurements." What does the reader learn from this deceptively anodyne passage? For one thing, the model was striking. The sculpture stretched nearly from one end of the Hall of the Biology of Mammals to the other, dwarfing the herd of taxidermized bison grazing in their glass cases directly below, and looming over visitors on the lower floor of the hall. More importantly, the *General Guide* emphasized that the object in question was, to be clear, a replica; unlike Muhlig's whale, the American Museum's model wouldn't find itself discredited in a metaphysical scandal.[11] Yet this was by no means a generic model. It belonged to a specific animal, with a specific history, just like the bison below. It preserved the zero-sum relationship between referent and representation that underwrote the other taxidermies in the museum. The only difference between the model whale and the examples of mammals that surrounded it was that the medium of transfer between the natural world and the museum was "careful measurements" in the case of the former, and skin in the case of the latter.

James Greismer has argued that displays in natural history museums, which he terms

[10] "Jumbo," *Ward's Natural Science Bulletin*, 1 May 1886, pp. 10–11, on p.11, quoted in Oliver Davie, *Methods in the Art of Taxidermy* (Columbus, Ohio: Hann & Adair, 1894), p. 106.

[11] George Sherwood, *General Guide to the Exhibition Halls of the American Museum of Natural History*, no. 35 of the Guide Leaflet series (New York: American Museum of Natural History, 1911), p. 70. The museum encouraged visitors to watch the model's fabrication, and early on, several visitors evidently wondered why the institution was building a submarine: "Model Nearly Ready," *New York Tribune*, 27 Dec. 1906, p. 6.

"remnant models," should be thought of simply as "models of data constructed out of the physical entities being modeled." For Andrews's model whale, however, the exact opposite situation obtained. Instead of considering skin to be a special case of data, the guide leaflet in effect asked its readers to consider measurements as a special case of skin. *Outlook* magazine repeated the museum's approach when it wrote that the method the museum used to "*preserve*" its whale was to "send a corps of men who noted all its dimensions, contours and colorings" to the site of the original whale body. *Scientific American* similarly reassured its readers that the scale maquette that had been used to make the paper templates for the whale's skeleton was an "exact model" made from data collected in Newfoundland and "exactly transferred" to the museum's fabrication through square ruled paper. Moreover, in the *Los Angeles Times* in 1937, Andrews recalled: "I made a life-size model of a 78-foot long sulfur bottom whale . . . . When the specimen was killed it was weighed in sections with the following results." Andrews then dutifully recited the weight of the whale's flesh, blubber, blood, and bones.[12] When stripped of its claim to support real whale skin, Muhlig's purchase was just a hollow dummy. But underwritten by the authority of measurements, Andrews could lend metaphorical flesh, unctuous blubber, and a huge heart pumping warm blood to his wooden creation, even years after the original had decomposed into the ocean.

This delicate situation was as much a response to the vagaries of the creatures at hand as it was a statement of epistemological purpose. With respect to museum exhibition, whales, to borrow a phrase from Adele Clark and Joan Fujimura, are recalcitrant subjects at best.[13] "The study of whales," advised Andrews, "is beset with many and unusual difficulties." He added, "their great size alone is a serious obstacle"—and great size was only the beginning. Although it enshrouds creatures of mighty girth, whale skin is remarkably thin and fragile, tending to discolor and decay quickly upon its owner's death. It lends itself to neither treatment nor transportation, and certainly not to lifelike taxidermy or permanent display. Arnold Jacobi, director of the Königliches Zoologischen und Anthropologisch-Ethnographischen Museum in Dresden, complained in 1914 that even when the skin of (very small) whales could be removed and stuffed, the hide inevitably

[12] James R. Greisemer, "Modeling in the Museum: On the Role of Remnant Models in the Work of Joseph Grinnell," *Biology and Philosophy,* 1990, 5:3–36; Smith, "Natural Science for the Every-Day Man" (cit. n. 3), p. 185 (italics added); and Roy Chapman Andrews, "This Amazing Planet," *Los Angeles Times,* 17 Oct. 1937, p. 13J.

[13] Adele E. Clark, "What Tools? Which Jobs? Why Right?" in *The Right Tools for the Job,* ed. Adele E. Clark and Joan Fujimura (Princeton, N.J.: Princeton Univ. Press, 1992), pp. 3–46, esp. pp. 14–16. Recalcitrance, for Clark and Fujimura, indicates the ways in which particular things resist or complicate attempts to assimilate them into scientific discourse. Historians who address taxidermy often tend not to focus on the particular properties of individual animals that help or hinder attempts to preserve them as sculpture, though some exceptions can be found. For instance, in "Craft vs. Commodity, Mess vs. Transcendence," in *The Right Tools for the Job,* pp. 257–286, Susan Leigh Star urges her readers to "imagine trying to filet an elephant in the African summer heat!" and emphasizes the "back breaking labor (probably done by ill-paid assistants)" needed to transform the elephant's hide into a covering for a life-like replica" (pp. 263–264). In a somewhat more oblique fashion, Karen Wonders argues that, *contra* Haraway, masculine anxiety must be discounted as a key factor in motivating big game taxidermy during the early twentieth century because of the tremendous output of bird taxidermy during the same period—thus implying a macho differential between light, easily manipulated birds and heavy, recalcitrant mammals, like gorillas: Wonders, *Habitat Dioramas* (cit. n. 6), p. 224. This, of course, begs the question of masculinity, size, and effort—consider John Audubon's impromptu taxidermies and life-sized drawings of birds executed on site and *en plein aire*. Jennifer Roberts discusses the great efforts Audubon went through to achieve his pictures, noting also that he styled himself a rugged frontiersman for European audiences, his effete subjects notwithstanding: Jennifer Roberts, *Transporting Visions, The Transit of Images in Early America* (Berkeley: Univ. California Press, forthcoming).

"sweats blubber" for a long time afterwards. The history of exhibiting whale bodies for public observation tends towards brief and unappealing episodes.[14]

Whale bones, on the other hand, could be preserved and put on display, but as no less a cetological authority than Herman Melville noted, "the naked skeleton of the stranded whale ... gives very little idea of his shape." In 1937, Martin Hinton of the British Museum (Natural History) expressed an exhibitor's perspective on the matter: "the need for one life-sized model of a big Whale in the Gallery cannot be doubted; a great many members of the public go away disappointed at seeing only skeletons, small casts, and models." Without recourse to preserving the exterior of the animal itself, exhibitors had to deploy proxy methods. As Andrews proclaimed, "the ordinary methods used in the study of other animals" could not be applied to whales. "Instead of having actual specimens before one for comparison, a naturalist must depend almost entirely upon photographs, notes, measurements and descriptions."[15]

Photographs, notes, measurements, and descriptions—these were the technologies of cetology at the turn of the twentieth century. As difficult as whales themselves were to locate and study, however, these sorts of data were not especially easy to come by if one did not hunt whales for a living.[16] One indispensable source was Frederick William True's 1904 monograph, *Whalebone Whales of the Western North Atlantic*. Jacobi at the Dresden museum listed it as a foundational document for anyone who wished to model baleen whales. Frederick A. Lucas, a zoologist and exhibitor whose own work modeling whales would influence Andrews's project, wrote to the exhibition department at the American Museum to request a copy of True's monograph; at Lucas's own post at the Brooklyn Museum, "the taxidermists and modelers, after their unrighteous manner, have pretty much put my own copy out of use."[17]

To obtain his photographs, notes, measurements, and descriptions, True had contracted with the Cabot Steam Whaling company, an outfit whose station in Newfoundland was hauling ashore a plentiful catch in the early 1900s. In the preceding century, commercial whaling had largely been conducted using pelagic sailing vessels which put to sea for years at a time and refined their catch on the go, storing oil and whalebone on board and

---

[14] The sole exception to the apparent impossibility of preserving whale skin appears to be the "Malm whale," a juvenile blue whale successfully preserved by professor August Wilhelm Malm in the mid-nineteenth century. This exhibit—which was still on display at the Gothenburg Natural History Museum in 2009—is notable not only for being the only preserved blue whale in existence, but also for being the only known blue whale in existence with a pleasantly upholstered interior. See Cecilia Grönberg and Jonas J. Magnusson, "The Gothenburg Leviathan: *Into the Belly of the Malm Whale*," *Cabinet*, 2009, *33*:30–36. On Jacobi's difficulties with whale skin see A. Jacobi, "Modelle von Waltieren und ihre Herstellung," *Abhandlung und Bericht des Königlich Zoologischen und Anthropologisch-Ethnographischen Museums des Dresden*, 1914, *14*:1–8, on p 3. For more on the history of whale exhibitions see, for instance, "Vienna's Dead Whale," *New York Times*, 14 July 1889, p. 15. See also, Richard Ellis, *Men and Whales* (New York: Knopf, 1991); and Burnett, *Trying Leviathan* (cit. n. 4), p. 182.

[15] Herman Melville, *Moby Dick* (New York: Penguin, 1992), on p. 289; Martin Hinton, Letter to Trustees, 21 Apr. 1937, DF 1004/748, Whale Room, 1923–1955, The Natural History Museum Archives, London; and Roy Chapman Andrews, *Whale Hunting With Gun and Camera: A Naturalist's Account of the Modern Shore-Whaling Industry, of Whales and their Habits, and of Hunting Experiences in Various Parts of the World* (New York: D. Appleton, 1916), on p. 19.

[16] On whalers as cetologists see esp. Burnett, *Trying Leviathan* (cit. n. 4), pp. 95–144; and Lyndall Baker Landauer, "From Scoresby to Scammon: Nineteenth-Century Whalers in the Foundations of Cetology" (Ph.D. diss., International College, 1982).

[17] Jacobi, "Modelle von Waltieren," p. 4; Frederick William True, *The Whalebone Whales of the Western North Atlantic Compared with those Occurring in European Waters, with Some Observations on the Species of the North Pacific* (Washington, D.C.: Smithsonian Institution, 1904); and Lucas to Hermon C. Bumpus, 21 Nov. 1906, Box 21, Central Archives, American Museum of Natural History (hereafter cited as *AMNH*).

discarding the rest of their catch to the scavengers of the open ocean. By the late nineteenth century, however, this practice was increasingly eclipsed by shore whaling, in which smaller, steam-powered vessels conducted daily raids upon cetacean populations, using exploding harpoons fired from cannons to kill their prey. These kills were then dragged back to shore stations, where the whole whale corpse could be processed not only for oil and whalebone, but also for use as fertilizer or feed. In addition to achieving new economies of scale, this meant that for the first time, non-whalers could observe large numbers of large cetaceans without having to leave dry land. Furthermore, these novel techniques allowed the hunting of species of whales such as blue, fin, and minke whales, whose size, speed, and tendency to sink when dead had previously marked them as a species best left alone. To the list of technologies valued by modern cetologists, then, one had also to add exploding harpoons and steam-powered whaling ships, as well as a central feature of shore whaling stations, the "slip." True described the slip as "a large inclined platform. . .upon which whales were drawn up, one at a time, completely out of the water, thus affording excellent opportunities for close inspection."[18]

Andrews thought that *Whalebone Whales* was "superb," especially since, by 1907, "very little had been published on the external anatomy of whales."[19] True provided plentiful tables of measurements and numerous photographs of whale bodies, and explained his procedure for insuring consistency and accuracy: "as each individual was drawn up on the slip, I measured it, using a uniform schedule of measurements, and photographed it from one or more points of view, and made as copious notes as circumstances would permit on its color and other characters." In the book's section on *Balaenoptera musculus,* one could find notes on a sulfur-bottom's eye: "The iris is brown. The pupil is oblong with a straight superior margin." Also included were detailed charts setting out the color of the flukes, furrows, ridges, body, head, and pectoral fins of different specimens of whales.[20] True approvingly cited a comment by Georg Sars published in 1874 that the "whole ground color of the whale has a distinctly bluish cast." At the same time, True advised that the coloring of sulfur-bottom whales that he had personally encountered betrayed, "a large individual variation, no two specimens being precisely alike."[21] In addition to meticulous descriptions, the better to document as many variations as possible, *Whalebone Whales* provided nine plates of black and white photographs of sulfur-bottoms, honing in on details such as the patterns of light splotches across the whales' skin, and the grooves and furrows along their chins. A close-up photograph of whale #23's eye bears an unmistakable resemblance to the one finally modeled at the American Museum (see Figure 3).

Shortly after he completed work on the model sulfur-bottom in the Hall of the Biology

[18] For thorough coverage of the transition from sail-powered pelagic whaling to steam-powered shore whaling see, for instance, Lance E. Davis, Robert E. Gallman, and Karin Gleiter, *In Pursuit of Leviathan: Technology, Institutions, Productivity, and Profits in American Whaling, 1816–1906* (Chicago: Univ. of Chicago Press, 1997); John R. Bockstoce, with contributions by William A. Baker and Charles F. Batchelder, *Steam Whaling in the Western Arctic* (New Bedford, Mass.: Old Dartmouth Historical Society, 1977); J. N. Tønnessen and A. O. Johnsen, *History of Modern Whaling*, trans. R. I. Christophersen (Berkeley: Univ. of California Press, 1982); and James Travis Jenkins, *A History of the Whale Fisheries: From the Basque Fisheries of the Tenth Century to the Hunting of the Finner Whale at the Present Date* (London: H. F. & G. Witherby, 1921). For True on "slips" see *Whalebone Whales*, (cit. n. 17) p. 111.

[19] Roy Chapman Andrews, *Ends of the Earth* (New York: G. P. Putnam's Sons, 1929), pp. 21 and 7.

[20] True, *Whalebone Whales* (cit. n. 17), pp. 111, 175, 170.

[21] *Ibid.*, p. 162. True mentioned that Sven Foyn—often credited as the inventor of the harpoon cannon—recommended calling sulfur-bottoms "blue" whales on account of their color, an idea that True endorsed.

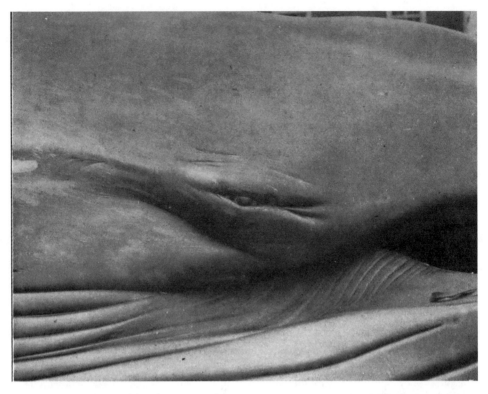

Figure 3. *The eye of whale #23, a sulfur-bottom killed in Newfoundland. From Frederick William True,* The Whalebone Whales of the Western North Atlantic Compared with those Occurring in European Waters, with Some Observations on the Species of the North Pacific *(Washington, D.C.: Smithsonian Institution, 1904), plate 16, figure 1.*

of Mammals, Andrews set out in True's footsteps. Where True had principally catalogued the whalebone whales of the western north Atlantic, Andrews persuaded director Hermon C. Bumpus to fund a solo expedition to study the whalebone whales of the eastern North Pacific. Over the course of eight years, Andrews journeyed to Vancouver, Alaska, Korea, and Japan, armed with a Graflex camera, notebooks, steel measuring tape, and "prepared lists, having all the important measurements to be obtained."[22] From the Pacific coast Andrews wrote to director Bumpus (echoing True's remarks) that his work "consisted of taking detailed measurements and descriptions of all the whales as they were brought in, besides photographing each individual from several points."[23] In his journals, Andrews was careful to describe particular metrical encounters between himself and his subjects; for example, on 10 April 1910, in Okinawa, whalers killed a young male sei whale: "this whale was drawn out upon the cutting slip so my measurements are all accurate & were taken with the greatest care." On occasion, the opposite was the case. Then a specimen might be inadequate for reliable measuring, as was a finback whale on 6 July 1908, which had been left in the water overnight and partially eaten by sharks. On 25 July 1910,

---

[22] Andrews, *Ends of the Earth* (cit. n. 19), pp. 29–30.
[23] Andrews to Bumpus, 12 Sept. 1908, Roy Chapman Andrews Folder II-3, H. C. Bumpus Correspondence, 1908–1920, Library of the Department of Mammalogy, AMNH.

Andrews prefaced his notes about a finback whale with a curt elegy to an old accomplice: "These measurements were taken with a cloth tape as all the succeeding ones will be. My steel tape is broken."[24]

His experience with literally hundreds of cetaceans allowed Andrews to retroactively declare that the model he had made was "a good whale." Andrews reflected that he had studied cetaceans "at sea with field glasses and camera," while on shore at the station he had investigated them "both inside and out as they were hauled from the water to be carved up." He concluded: "After all that, I can still be proud of our *papier maché* whale in the American Museum."[25]

As stated in the museum's guidebook, then, "careful measurements" made from within and without, could reinforce the provenance of the model. Indeed, as *Scientific American* pointed out, Andrews could take measurements from a giant whale in Newfoundland, construct a scale model with the data that could be transported with ease, and then create paper templates from which the life-sized replica could be constructed. Tables of measurements and collections of images—"inscription" technologies, to use Bruno Latour's notorious formulation—allowed exhibitors at the American Museum to "work on paper . . . but still manipulate three-dimensional objects" with confidence.[26] Nevertheless, even if whale data was more long-lived than whale skin in terms of *accuracy*, all the great power of these inscriptions could not insure the model's *authenticity*.

For one thing, whales formalized as tables of numbers bear little resemblance either to the creatures that cetologists were supposed to have encountered in nature, or to their representations. As Andrews pointed out, the model weighed "several tons," and he warned: "God help those below if he ever falls."[27] In his own accounts, Andrews made sure that his work abstracting cetaceans into numerical data remained closely associated with the visceral physicality of manipulating their huge bodies. Whereas True, concerned mainly with the task of shuffling cetaceans into formal categories, complained that skeletons and models in museums could only be studied "with much begriming of note-books, hands and clothes," Andrews went to much greater lengths to drive home the hardship of scientific inquiry, emphasizing the gory, strenuous nature of modern cetology. The specimens at whaling stations were available for analysis but only "if the investigator is not afraid of blood and grease," he told readers of his 1916 *Whale Hunting with Gun and Camera*. In the *American Museum Journal*, he wrote: "If one wishes to do such an ordinary thing as turn over a fin [of a whale] for observation of the color or markings of the other side, one must have the assistance of not only one man but several." After retrieving the bones of a beached right whale at Amagansett, Long Island (a task assigned to Andrews while he was in the midst of constructing the model sulfur-bottom), he wrote, "I never have suffered more in any experience of my life than I did then." At the end of all of the hard work, however, Andrews could look back upon his blue whale representation and declare it not simply an accurate whale, but a "*good*" whale—a morally sound representation, and one to be "proud" of.[28]

---

[24] Roy Chapman Andrews, Journal entries, 10 Apr. 1910, vol. 3; 6 July 1908, vol. 2; and 25 July 1910, vol. 3; Special Collections, AMNH.

[25] Andrews, *Ends of the Earth* (cit. n. 19), p. 9.

[26] Bruno Latour, "Drawing Things Together," in *Representation in Scientific Practice*, ed. Michael Lynch and Steve Woolgar (Cambridge, Mass.: MIT Press, 1990), pp. 19–68, on p. 46.

[27] Andrews, *Ends of the Earth* (cit. n. 19), p. 9.

[28] Indeed, True dismissed the hardships of the field with a rather mild statement: "As nearly every cetologist takes occasion to say, the investigation of animals so large as whales is surrounded with peculiar difficulties."

Andrews's virtuous pride in being able to vouch for his model, moreover, points to part of a broader relationship between measuring, modeling, and authenticity shared by some of Andrews's clique of workers in the taxidermy department of the American Museum. Certainly it was good scientific practice to base one's exhibits on data taken from real animals, cetacean or otherwise. At the same time, it mattered who took the measurements. Through his work at Ward's, and later at Chicago's Field Museum, Akeley became a powerful advocate for scientific, observational, and artistic rigor in natural history display. Even before he moved to the American Museum in 1909, Akeley's personality and ethos created a powerful impression on Andrews's cohort of young exhibitors, some of whom had apprenticed with him in Chicago. In the case of Jumbo, as in the case of the whale, "it was only by very painstaking and tedious work, together with a rigid adherence to the exact size of the animal, that the result was so successfully attained."[29] At the same time, Akeley cautioned that it wasn't enough for a conscientious museum man to rely on "careful measurements" to replicate nature; the exhibitor had to have had an encounter with his (or less often, her) subject in its natural environment in order to replicate the creature as it really was.

Akeley gave two related reasons for his insistence on firsthand experience with measurements. In the first place, he wrote, "other people's measurements are never very satisfactory, and actual study of the animals in their own environment is necessary in making natural groups." He emphasized, "one man's measurements are not often reliable guides for another man to work by." Measurements were not actually so easily transferred as one might expect. This in turn led to something of a more principled argument: without having experienced one's subject firsthand, the exhibitors—from heroic hunters to junior taxidermists—had vanishingly little authority to create his or her "true and faithful copy of nature." Akeley put it in no uncertain terms—without having experienced one's subject firsthand, "the exhibit is a lie and it would be nothing short of a crime to place it in one of the leading educational institutions of the country."[30]

In this sense, measurements alone couldn't underwrite Andrews's model. Although in 1907, *Scientific American* echoed the museum's guidebook, assuring its readers that the American Museum's whale project "was placed in the hands of Mr. Roy Chapman Andrews of the museum staff, who visited Newfoundland and was fortunate in securing his data from a large whale measuring 76 feet," this account was only half right. Data for the whale had, indeed, been secured in Newfoundland, but not by Andrews, who literally had difficulties telling a whale's head from a whale's tail until after he had begun work on the model.[31] In later years, Andrews would indeed encounter whales in the wild on his expeditions to the Pacific, and would in fact have the opportunity to shoot his own whale, placing him among game hunters like Akeley who had personally killed the animals they

See *Whalebone Whales* (cit. n. 17), p. 2. For Andrews's remarks on field investigations for whale reconstruction see Andrews, *Whale Hunting* (cit. n. 15), p. 20; "Pacific Coast Whalers," *American Museum Journal*, 1909, *9*:2; and *Ends of the Earth* (cit. n. 19), p. 13. For more on arduous research see, for instance, Naomi Oreskes, "Objectivity or Heroism? On the Invisibility of Women in Science," *Osiris*, 1996, *11*:87–113; and Rebecca Herzig, *Suffering for Science: Reason and Sacrifice in Modern America* (New Brunswick, N.J.: Rutgers Univ. Press, 2005).

[29] Quoted in Davie, *Methods in the Art of Taxidermy* (cit. n. 10), p. 106.

[30] Carl Akeley, *In Brightest Africa* (Garden City, N.Y.: Doubleday, 1924), pp. 15, 264, 254.

[31] "The Whale" (cit. n. 2), p. 162; and Andrews, *Ends of the Earth* (cit. n. 19), pp. 11–12. In fairness, the whale in question was surrounded by freezing cold surf and its flukes had been chopped off—doubtless a disorienting circumstance.

then preserved for eternity.[32] At the time Andrews was placed in charge of replicating a whale from "careful measurements," by his own admission he had never seen a whale before, and thus placed much of the responsibility for the authenticity of the whale in the hands of Frederick A. Lucas. An accomplished zoologist and exhibitor, Lucas's peripatetic career had taken him from the Smithsonian to the Brooklyn Museum, and finally to the American Museum of Natural History.

FIRST IMPRESSIONS: ON THE HEROICS OF CASTING AND THE PERILS OF ISOMORPHISM

Andrews didn't obscure Lucas's involvement with the model. In *Ends of the Earth,* he wrote of building the whale: "My lack of knowledge made little difference for we were actually only to enlarge a scale model of a sulphur-bottom whale, which was to be made by James L. Clark under the immediate supervision of Dr. F. A. Lucas, then director of the Brooklyn Museum." Indeed, it seems that Andrews attempted to apprentice himself to the more experienced cetologist during the period of the sulfur-bottom's construction but Lucas rebuffed him, albeit regretfully. In a letter to director Bumpus, Lucas wrote: "I advised Mr. Andrews, very much against my will, to stay at the American Museum. Thus do the rich and powerful, outside of the melodrama, ever triumph over the poor and lowly."[33]

The "poor and lowly" Lucas, of course, was exaggerating his lack of cache. The task facing Andrews and his crew at the American Museum was considerably more nuanced than merely enlarging a scale model, and Lucas had more direct knowledge of the whale at hand than simply the general credentials of having been to Newfoundland where he had "cut up whales and knew every inch of them from flukes to blow-hole," as Andrews put it.[34] While at the Smithsonian, Lucas had, in fact, overseen construction of a model blue whale of his own in 1903—a replica which could claim its own authenticity not from "careful measurements" as such, or simply from Lucas's experience cutting up cetaceans, but from the considerably more arduous isomorphism of casting. As he wrote in 1920 in *Museum Work* magazine,

> if it be not possible to restore a creature to the semblance of its appearance in life, it may be possible to reproduce it: the simplest method of doing this is naturally to make a picture of it, but this is unsatisfactory, and a museum of pictures would be monotonous and lifeless—most literally flat, stale and unprofitable; we need three dimensions to give a feeling of reality.[35]

Lucas's favored method for dealing with "animals that were never intended by Nature to be . . . mounted by ordinary methods" was to cast them—to cover them head to hind in plaster, then use the resulting mold to make a replica from more durable material than livid flesh.[36] Better than two-dimensional inscriptions, casting was like a three dimensional photo, recording precisely the whale's dimples and convexities, and properly enumerating the grooves which so enchanted the writers of *Scientific American.* Unlike a

---

[32] Gary Kroll, "Exploration in the *Mare Incognita:* Natural History and Conservation in Early Twentieth Century America" 'Ph.D. diss., Univ. of Oklahoma, 2000', pp. 100–102.

[33] Andrews, *Ends of the Earth* (cit. n. 19), p. 7; and Lucas to Bumpus, 21 Nov. 1908, Box 21, Central Archives, AMNH.

[34] Andrews, *Ends of the Earth* (cit. n. 19), p. 7.

[35] Frederick A. Lucas, "Wax and Other Casts," *Museum Work,* 1920, 2:114–118, on p. 114.

[36] *Ibid.*

photograph, however, casting captured not just a two-dimensional representation of the subject at a particular moment in time, but the subject's presence in proprioceptive space—its actuality, its authenticity, or, in Lucas's terms, its "feeling of reality."

In pursuit of this goal, in May of 1903, an expedition from the Smithsonian departed Washington D.C. for Balena, Newfoundland—the same spot to which True had journeyed in 1899—in order to add a cast of "the largest whale that swims" to the Smithsonian's display for the United States pavilion at the 1904 St. Louis Exposition. The team—consisting of Lucas and two colleagues, Joseph Scotlick and William Palmer, both of the museum's exhibitions department—were inspired by reports of "monsters" measuring up to eighty feet in length. They waited for weeks for the "wished for big one," as blue whale after blue whale arrived at the station, but with no luck. No single whale was either large enough or beautiful enough to represent the museum at the exposition. As Palmer took molds of various whale parts which caught his fancy—an eye here, a flipper there—Lucas fretted, at one point noting with alarm that whale size appeared to be decreasing by the day.[37]

Eventually, Lucas's luck changed. In the third week of June, a whaling boat brought in a specimen just two feet shy of the desired length. "While it was not the very largest of whales," Lucas commented, "it was a fine specimen," with a noticeably large head that contributed to its "character." The Smithsonian team quickly set to work, supplemented by a party of whalers dragooned as cheap labor. Positioning the whale on its side in shallow water, the group smeared its body with batch after batch of plaster, working in sections to yield more manageable molds; its head and extremities were decapitated and cast separately. After three days of grueling work, which left Lucas lamenting his "raw and bloody" fingers, their project was a success and the team returned to Washington bearing the whale mold in pieces on several railroad cars.[38] One year later, the Smithsonian's model—a gleaming, lacquered cetacean painted in shades of Prussian blue—posed high in the air in the United States Hall in St. Louis, accompanied by photographs of Lucas and his team laboring in the surf in Newfoundland.

The Smithsonian's whale was a sensation—"a chef d'oeuvre in the museum's portrait gallery of strange creatures!" trumpeted the *Washington Post*.[39] Newspapers lavished attention on impressive physical details of the model's construction: the materials used (thirty-five barrels of plaster, three bales of excelsior, four kegs of nails); the scale of the effort expended by the work crew (each of the 125 sections of the mold was only "to be with difficulty handled by two men"); the portents of lucre in the whale's construction (the final model was cast in linen, donated by the United States Treasury Department in the form of several million dollars worth of decommissioned, pulped bills); and the record-breaking nature of the artisanship (Palmer's was said to be the "largest plaster-of-paris mold and cast ever made in America").[40] Newspapers emphasized the achievement that the whale represented to the common man's observation of natural history. "For the first time an opportunity is given to the public to look upon the most gigantic of existing

---

[37] "Monsters of the Deeps," *Washington Post,* 26 July 1903, p. 10E; Lucas to True, 23 May 1903, and 8 June 1903, RU 70, Box 62, Folder 27, Frederick A. Lucas Papers, 1903–1904, Smithsonian Institution Archives.

[38] Lucas to True, 16 June 1903, RU 70, Box 62, Folder 27, Frederick A. Lucas Papers, 1903–1904, Smithsonian Institution Archives.

[39] "Living Snake Models," *Washington Post,* 6 Sept. 1904, p.12.

[40] Palmer seems to have broken his own record in this regard—in 1881, the *New York Times* reported he was responsible for "the biggest thing yet in plaster casts," with a whale mold that he took in Provincetown. See "Art Notes," *New York Times,* 16 Oct. 1881, p. 11.

creatures," proclaimed one article. The model was "the first and only authentic portrait of a whale in existence," affirmed another, "life sized and exact to the minutest detail."[41]

This "authentic portrait of a whale" could, in turn, lend its authenticity to the whale being constructed at the American Museum. Among Richardson's first acts as foreman of the American Museum's whale construction crew was to pay an unannounced, unapproved visit to the United States National Museum in Washington, where he snapped photographs of various views of the Smithsonian's whale (see Figure 4). This unlicensed adventure in cetacean espionage drew the displeasure of director Bumpus, who quickly wrote to Richard Rathbun, head of the National Museum, to apologize for Richardson's behavior.[42] Rathbun, for his part, was magnanimous, allowing the American Museum to keep the photos, and offering any additional assistance needed. However, the incident demonstrates that the model was imbued with more than simply morphological correctness—Richardson's photographs allowed both the anatomical precision assumed by the casting process as well as its narrative of personal expertise and backbreaking work to be transferred to the American Museum's whale. The "careful measurements" cited in the American Museum's guide as being taken from an animal in Newfoundland were, in fact, of the most careful variety—one hundred percent isomorphic with the animal under the plaster. Moreover, they were underwritten by the expertise of one who had been there, and who could be referenced immediately should the authenticity of the model be in doubt.

SMALL FRAUDS: ON SCULPTING AND LIFELIKE VERISIMILITUDE

From the standpoint of fashioning a model imbued with authority and rectitude, casting was a solid bet. It provided extraordinary assurance that viewers were standing in the presence of a replica which, if not made with real skin, was the product of the closest imaginable correspondence between the replica and the animal it represented. Moreover, casting necessitated an intimate contact between the modeler and the subject, supplying the moral grounding prescribed by Akeley. Nevertheless, casting alone was no magic bullet. From an exhibitor's perspective, casting could produce an incredibly *precise* mold of a dead whale's exterior, while nevertheless yielding a terribly *inaccurate* impression of the animal in life.

Among the most pressing problems (so to speak) that casting presented to exhibitors was the flattening that could occur in large animals when they were hauled onto dry land and covered with weighty plaster. As Lucas remarked, the subject might be so "'flabby' that an actual mould gives no correct idea of the form of the animal when in the water."[43] In Dresden, director Jacobi promoted an especially meticulous method to address just this problem, at least for small whales. The cetacean to be modeled was first floated in a large bath of water, adjusted to the desired position, and held tightly in place by a fabric sling. Exhibitors next made a rough cast of half of the whale. With this firm backing in place, the other half could be cast in plaster with less fear of distortion. The procedure was then undertaken for the first half of the whale that had been cast, to insure the utmost accuracy of the mold. Extremities such as fins and flukes would be cast separately. Rather than using *papier-mâché* to derive the positive, as was common practice, Jacobi was proud to

---

[41] Rene Bache, "Science Shows that Jonah Could Have Lived in Whale's Body," *Washington Post,* 6 May 1906, p. 1B; and "Living Snake Models" (cit. n. 39).

[42] Bumpus to Richard Rathbun, 24 Sept. 1906, Box 54, Central Archives, AMNH.

[43] Lucas to Sidney Harmer, 28 Sept. 1932, Box 1001, Central Archives, AMNH.

Figure 4. *Two photographs of the Smithsonian's National Museum of Natural History's model sulfur-bottom whale model, photographed by F. L. A. Richardson in 1906. American Museum of Natural History Photograph Collection, PPCA45.M88; on right, neg. 13258; on left, neg. 13263. Reproduced with permission from the American Museum of Natural History Archives.*

advertise his technique of applying alternating layers of plaster and linen to the well-greased inside of the mold in order to build up a sensitive yet sturdy skin which would take oil paint well, hoping for the most lifelike result possible. Of course, as Jacobi conceded, this technique could only be applied to the smallest of whales, leaving unanswered the question of how to cast larger cetaceans without such distortion. Jacobi himself readily admitted that the only really authentic part of a large finback whale which came through the Dresden museum as part of a traveling exhibition was the whalebone in its mouth—the rest was just a *papier-mâché* fabrication, based on the best available evidence. Nevertheless, he assured his readers that the model finback was so well-realized that the service done in the name of lay understanding of science far outweighed the "small fraud" of the model's construction.[44]

For Lucas, small frauds were both more present, and perhaps less troublesome, concerns than they were for his counterparts across the Atlantic. Rather than endorse Jacobi's arduous method of mummifying and disinterring porpoises to achieve undistorted casts that had not been manipulated, Lucas insisted to exhibitors at the British Museum (Natural History) that simply applying skillful sculptural techniques to the physical data one had at hand could more than make up for casting problems that marred a model's lifelike appearance.

Lucas called these models "composites" and used the term in two separate senses. On the one hand, as he noted during his stint in Newfoundland, it could be difficult to find an animal that was unblemished enough to make a good exhibit—a reasonable complaint when his subjects had all been dispatched by exploding harpoons. As Lucas wrote from Newfoundland to his superiors at the Smithsonian, "we shall probably be obliged to make a composite whale [from the multiple casts of whale parts that Palmer had made], not only on account of the difficulty of getting the whale wanted, but of . . . the great difficulty of securing . . . a specimen in good shape." This sort of mix-and-match wasn't a totally unheard of practice in natural history exhibition. As Cain points out, while exhibitors might shy away from altering a strictly scientific specimen, for public exhibition purposes a set of antlers could be moved from one deer to another, or a single passenger pigeon

[44] Jacobi, "Modelle von Waltieren" (cit. n. 14), pp. 3–4.

cobbled together from more than one specimen, without significantly harming the rectitude, authority, or perceived accuracy of the exhibit.[45] The presence of animal flesh in an exhibit was a strong argument for authenticity, and casting was the next best thing.

At the same time, a model might be "composite" in the sense of being "a cast whose body had been remodeled to do away with the flattening and sinking in of the body cavity that occurs, or is apt to occur, in making a mould over a heavy animal." Such distortion obviously called for correction, though such manipulation could be problematic. Exhibitors in London, for example, balked at the idea of allowing a sculptor's aesthetic judgment to supersede the three dimensional photograph implied by casting. William Thomas Calman, head of the Zoological Department of the British Museum (Natural History) expressed this view succinctly in a letter to the museum's trustees: "Dr. Calman does not think it desirable that any expenditure should be incurred on the preparation of full sized models if it is possible to obtain actual casts." Lucas, on the other hand, was more pragmatic in his methods, and wrote to skeptical counterparts in London that those at the American Museum "preferred a good model to a poor cast" because it gave "a better idea of the animal to the average visitor than an indifferent cast." Indeed, Lucas candidly admitted to his colleague: "Occasionally, in cases where it was not practicable to keep the specimen, we have relied entirely on photographs, measurements and descriptions." As he proudly insisted, the American Museum had "a number of skillful, well-trained modelers and preparators." Consequently, he thought that the results of freehand modeling instead of casting were "pretty good."[46]

Exhibitors in Germany and Britain however, weren't the only ones who were leery about the viability of simply sculpting an animal, and Lucas's opinions weren't unanimously shared even within the walls of the American Museum. Since his time in Newfoundland, Lucas had hoped to produce a life-sized model of a humpback whale, as he felt the creatures to be "so striking."[47] Having moved to the American Museum in 1923 as honorary director, Lucas argued strongly for the construction of a humpback whale

[45] Lucas to True, 8 June 1903, RU 70, Box 62, Folder 27, Frederick A. Lucas Papers, 1903–1904, Smithsonian Institution Archives. See also Cain, "Nature Under Glass" (cit. n. 7), p.159.

[46] Lucas to Harmer, 28 Sept. 1932, Box 1001, Central Archives, AMNH; W. T. Calman to Trustees, 25 July 1934, DF 1004/748, Whale Room, 1923–1955, The Natural History Museum Archives, London; and Lucas to William Pycraft, 1 Apr. 1925, Box 1209, Central Archives, AMNH.

[47] Lucas to True, 16 June 1903, RU 70, Box 62, Folder 27, Frederick A. Lucas Papers, 1903–1904, Smithsonian Institution Archives. Lucas likely also had cetacean conservation in mind when he advanced his scheme for the humpback whale model. See Mary Anne Andrei, *Nature's Mirror* (Chicago: Univ. Chicago Press, forthcoming). Whale skeletons and models were, of course, frequently displayed alongside whaling paraphernalia, which complicates the story, ensnaring whales, whalers, and whaling in a narrative of national memory. In his article, "The Passing of the Whale," for instance, Lucas remarked, "from a strictly American viewpoint the whale deserves serious consideration as it was half a century ago the basis of an industry which brought great wealth to the New England States": Lucas, "The Passing of the Whale," *Supplement to the New York Zoological Society Bulletin*, 1908, 16:445–448, on p. 445. Similarly, Bumpus wrote a letter to Madison Grant, haunted by a conversation the two men once had over "the imminent extinction of whales and the passing away of a characteristic American industry." Bumpus pointed out: "There are but few remains of the Cetacea in our museums, and practically nothing . . . which would illustrate the work of the sturdy Nantucket and New Bedford whalers." He urged that Grant help the museum to concentrate its efforts on obtaining "specimens of at least the more important Cetacea," as well as items of whaling equipment, lest it become "impossible for the institution to fill this chapter in America's industrial activity." See Bumpus to Grant, 20 Jan. 1908, Box 663B, Central Archives, AMNH. For an excellent discussion of the ways in which whales and their representations intersected with arguments about cetacean conservation, commercial whaling, and perceptions of national identity in early twentieth century Mexico see Serge Dedina *Saving the Gray Whale: People, Politics, and Conservation in Baja California* (Tucson: Univ. Arizona Press, 2000). On Andrews's and Lucas's attitudes towards cetacean conservation see Kroll, "Exploration in the *Mare Incognita*" (cit. n. 32), pp. 67–120.

model in the museum's new Hall of Ocean Life, then under construction. For a time, he seemed to be getting his way, prompting a ribald supporter to write expressing his excitement over a "humpback whale in the act of humping himself," as well as "all the other interesting things planned."[48] Lucas's model, however, was opposed by a powerful trustee, Childs Frick (the steel magnate's son), who felt that the $6,000 price tag was too much to pay "for an exhibit which would not be the original animal." Lucas explained in response: "The model would not be a hypothetical one, but would be constructed from exact data now on hand"—by which Lucas meant his own knowledge, as well as photographs, measurements, and Andrews's recently accumulated experience with ceta- ceans. Even so, the affirmation implied by casting was great, and some trustees argued that the humpback whale plans could be "considerably improved" by casting just the flippers and flukes of the animal itself.[49] Lucas readily agreed, and for a while, it looked as though the plan would come to fruition. Nevertheless, in spite of the museum director's own assurance that "nothing will give the visitor a better conception of what these great animals are like in life than this model," Frick resisted and the project foundered under the weight of apathy and an eventual budget pinch, much to the disappointment of Andrews and Lucas.[50] For trustees such as Frick, an exhibit that was not based on the most unimpeachable correspondence between an animal and its replica was simply "hypothet- ical"—an idealization, a work of fiction. For Lucas, the most critical task for the exhibitor was to secure a representation that mimicked the animal in life. Whether it was through stuffing prepared porpoise skins, or indexing whales through a cast, or simply through sculpting a sublimely convincing representation of a whale's exterior, the process used was ultimately in aid of achieving that "feeling of reality."

Through its elaborate provenance, the American Museum's 1907 sulfur-bottom model could not be considered "hypothetical," nor was it purely the product of an imaginative "closet naturalist," as Andrews (later) labeled those who knew their subjects only through books.[51] With the model's moral rectitude and metrical consistency vouched for by others, Andrews could at least claim a sort of knowing touch with which he—and not Richard- son—brought the model to fruition.

As the head of the whale project, Richardson was a decided failure. It was true that under his guidance the initial construction had proceeded without a hitch as the exhibitors scaled up Lucas's model and carpenters and blacksmiths worked on the model's frame. The next step in Richardson's plan, Andrews remembered later, was to cover the wooden "skeleton" with a paper skin, in lieu of real whale flesh. However, unlike Jumbo, whose skin fit its frame perfectly, Richardson's paper pelt evidently did not match its wooden

---

[48] Arthur Curtiss James to Lucas, 5 Sept. 1923, Box 1001, Central Archives, AMNH. Perhaps James's double entendre alluded to the curved pose of the proposed model.

[49] "Memorandum of Meeting of the Committee on the Decoration and Development of the Hall of Ocean Life and the Fish Hall," 2 Nov. 1927, Folder: Exhibits-Hall of Ocean Life, 1921–1955, Library of the Department of Mammalogy, AMNH. It should be noted that Frick's primary interest was paleontology—a field in which somewhat different standards of authenticity were historically relevant. In his guide to the Hall of Fossil Vertebrates, for instance, William Diller Matthew noted that replacement bones (or parts of bones) in the museum's reconstructed fossil skeletons would be clearly indicated: "Some of the skeletons are partly restored in plaster, indicated by a red cross (restored bones) or red lines (outlines of restored parts of bones)." See W. D. Matthew, "Hall of Fossil Vertebrates," Guide Leaflet, no. 3 (New York: American Museum of Natural History, 1902), on p. 4. On the trustees's argument for casting see "First Meeting of the Special Committee on Decoration and Development of the Fish and Oceanic Halls," 11 May 1925, Box 1178.1, Central Archives, AMNH. Also see Barton Warren Evermann to Lucas, 5 Sept. 1923, Box 233. Central Archives, AMNH.

[50] George H. Sherwood to Frick, 7 July 1927, Box 1178.1, Central Archives, AMNH.

[51] Andrews, Ends of the Earth (cit. n. 8), p. 19.

armature. "The paper wouldn't work," remembered Andrews. "It buckled and cracked and sank in between the ribs. Our whale looked awful. It seemed to be in the last stages of starvation."[52]

It's not entirely clear what went wrong with Richardson's approach to the model. Richardson was a seemingly accomplished manipulator of *papier-mâché*, and had prior experience in recreating tremendous, spectacular animals that he had never seen alive. Bumpus had hired him for the American Museum's whale project on the recommendation of Lucas, with whom Richardson had produced a striking *papier-mâché* stegosaurus for the Smithsonian's display at the St. Louis Exposition (a replica which, again, was vetted by "careful measurements taken by the bones of one of these reptiles in the National Museum"). Several years after Richardson was terminated following the whale debacle, Lucas received a concerned letter from a Harvard mycologist asking for information on a man who had unexpectedly dropped by his office claiming to be friends with Louis Agassiz and Asa Gray, and boasting of having once made a whale for the American Museum of Natural History. In his reply, Lucas confirmed for the scientist that Richardson had, indeed, started construction on the Museum's whale, adding that he did a "great part of the work extremely well." Lucas remembered Richardson to be a man of "very considerable" ability ("though by no means what he thinks it is") and commented that Richardson had previously "devised methods for making paper reproductions of various objects that were very successful." However, Lucas continued, "when it came to completing the model, he, so to speak, fell down entirely, because he considered that he knew all about it, when both Doctor Bumpus and myself had warned him that his proposed methods would be a failure." Richardson, moreover, was "a difficult man to work with," as he was thought "not easy to control." He was "somewhat unbalanced mentally, . . . a decidedly erratic individual." Though Lucas could reassure the scientist (who thought Richardson had threatened his family in an oblique fashion) that Richardson was harmless, he was nevertheless "continually losing money through overconfidence in himself."[53]

Perhaps Richardson had refused to cover the model in basswood before attempting one of his "novel paper solutions," to the annoyance of his employers. A photograph of the half-completed whale shows Andrews standing before a group of workers. (See Figure 2.) In the far distance of the photograph, a layer of wooden shims appears to be making its way up the side of the model, suggesting that Richardson was fired before the basswood strips were applied. Or perhaps Richardson had started the basswood strips, but his intractability convinced Bumpus to drop him in favor of the ambitious and enthusiastic, if untried, Andrews. In any case, in Andrews's retelling, he and another promising young recruit from the museum's taxidermy studio, James ("Jimmy") Clark, approached director Bumpus with a plan to finish the project using a layer of chicken wire which would be coated with *papier-mâché*—a procedure Andrews billed in his memoirs as the "Akeley method." By this, Andrews presumably meant to emphasize not only the portion of Akeley's method that called for the use of basswood in modeling large animals, but the idea that great displays called for great personal character and great sculptural adroitness. Clark was among the young museum hands who had studied with Akeley in Chicago and had proven himself to have a tremendous facility for rendering animal anatomy in clay and

[52] Roy Chapman Andrews, *Under a Lucky Star: A Lifetime of Adventure* (New York: Viking Press, 1943), p. 26.
[53] "Life-Size Stegosaurus," *Washington Post*, 3 May 1903, p. A7; and Lucas to William Gilson Farlow, 27 Jan. 1914, Box 213, Folder 905, Central Archives, AMNH.

other media. In any case, Bumpus accepted Andrews and Clark's proposal, relieving Richardson of his duty, and placing Andrews and Clark in charge of the project.

Akeley, that advocate of empirical honesty and lifelike taxidermy, had insisted that a one-on-one encounter between exhibitor and subject was necessary to cement the moral worth of an exhibit. However, it was equally clear that Akeley's technique called for a particular sort of *élan vital* from the naturalist/artist. After all, as Clark later wrote, by using the Akeley method it was possible to capture not just the form, but also the "spirit" of the living animal. "The artist had only to overcome himself," Clark mused, "since the Akeley method had freed him of all technical impediments."[54] In Andrews's remembrance, he and Clark were just the sorts who could overcome their own limitations to bring their model to life. As Andrews recalled, "It was amazing what a well-regulated diet of paper maché did for the beast. He lost that pitiful, starved, lost-on-dry-land appearance, his sides filled out and became smooth as a rubber boot; we could almost feel him roll and blow as we built him up with our new tonic."[55] Under Andrews's hands and those of his team—endorsed by Lucas's casts, True's measurements, and Akeley's methods—the "good whale" came to life; it began to "roll and blow." The museum's model did not simply reference an animal in fine detail; it assured viewers of the experience of the animal itself: "The model of the sulfur bottom whale," remarked the *American Museum Journal,* "represents an animal *in the act of swimming.*" In 1909, curators went one step further, repainting the Hall of the Biology of Mammals with a frieze resembling the ocean, to provide an "appropriate background for the marine mammals which are its chief exhibits."[56]

For their part, exhibitors at the British Museum (Natural History), it must be noted, eventually reversed their strong aversion to freehand modeling. In 1933, Calman of the Zoological Department, floated a scheme to secure a cast of a blue whale for the museum's new whale gallery. The plan called for the museum to either purchase a large whale, or else hunt one down by means of a chartered whaling vessel; the specimen would then be suspended in a graving dock in order to prepare molds. Even if this process only yielded casts of the "head, flippers, and tail," exhibitors anticipated that it would still result in a model constructed that would be "far more accurate than anything which has yet been attempted in any museum in the world." Indeed, part of the considerable cost of the casting was to be defrayed by selling copies of the cast whale to "some of the larger museums in America"—an offer towards which Lucas and Andrews would doubtless have taken a dim view. Francis Fraser, a young cetologist at the Zoological Department was to have overseen the suspending and casting, but the considerable cost of making the model delayed the plan's execution, giving Fraser time to change his mind. In 1937, he wrote to the museum's Keeper of Zoology, expressing confidence that if the construction of the plaster mold was eliminated, more attention could be paid to "the accuracy of the wood foundation" of the model, and the model's cost would be "cut down considerably." He was confident that the museum's modelers, having just completed a whale shark by a similar method, "could construct an accurate model in the way just mentioned."[57] The

[54] James L. Clark, *Good Hunting: Fifty Years of Collecting and Preparing Habitat Groups for the American Museum* (Norman: Univ. of Oklahoma Press, 1966), p. 16.

[55] Andrews, *Lucky Star* (cit. n. 52), p. 26.

[56] "The Museum Whales," *Am. Mus. J.,* 1907, 7:94–96, on p. 96; and "Museum News and Notes," *Am. Mus. J.,* 1909, 9:134–138, on p. 134.

[57] Francis Fraser to Trustees, 15 April 1937, DF 1004/748, Whale Room, 1923–1955, The Natural History Museum Archives, London. For more on the British Museum's whale see Susan Snell and Polly Tucker, *Life*

British Museum's blue whale—sculpted entirely *in situ* by Percy and Stuart Stammwitz under Fraser's supervision—was finished in 1938.

CONCLUSION: *BALAENA IPSA LOQUITOR*

It's impossible to know how Rebhan and Brahn argued their case before Manhattan's municipal court, or what reasons Muhlig gave for having thought that his whale was the real thing. We know that Rebhan and Brahn claimed their whale was caught in a particular location, and identified it as a particular—and particularly well-known—species. But what else made Muhlig so sure that he was getting a deal, or at least able to claim with a straight face that he was duped? Was something about the replica's appearance particularly confidence-inspiring? Did Rebhan and Brahn claim some sort of cetological expertise? Or could it be that there was nothing else to the story except a gullible engineer and a milieu in which stuffed whales were a potentially bankable item that made the claims of Rebhan and Brahn seem believable? It is impossible to say.

It is clear, on the other hand, that Andrews and his coworkers at the American Museum (as well as Lucas, Jacobi, and the others) were emphatically not attempting to defraud their publics. Far from it. Rather, they were attempting to imbue their model with the authority, accuracy, moral weight, and verisimilitude of a real whale that had once been swimming in the ocean.

In attempting to fabricate a viable model, exhibitors at the American Museum had to attend to these complicated issues because their program, as director Bumpus saw it, was one in which seeing and learning was directly facilitated by the actuality of the exhibits. "People are increasingly inclined to *observe* and then to draw their own conclusions," he wrote. "This is the real process of education; it is a function of intelligence; it is the activity of an independent being . . . . Moreover, it is the primitive and the almost universal method adopted by Nature—the animal is educated by experience, not by training." Whereas once, Bumpus continued, museums had simply served as repositories for undistinguished artifacts and stale ideas—"stuffed animals," "noisome alcoholics" (animals preserved in alcohol), and the "pedantry of scientific labels"—modern museums relied on "realistic [taxidermy] groups in an out-of-doors atmosphere" and "models of exquisite workmanship" to enable this true experience of the natural world.[58]

In order to produce an exhibit that could credibly ensure the ingenuousness of this pedagogical encounter, exhibitors at the American Museum relied on a complicated and unsystematic—but effective—bricolage of techniques. Judged on its own, Andrews's model sulfur-bottom was no more authentic or worthy of note than Muhlig's fake "captured Greenland whale." Mere illusionism—skillful sculpture without the backing of a sturdy explanatory apparatus—was, to paraphrase Jacobi, a "fraud." But well-vetted measurements enabled tremendous animals to be made fungible and abstract, accessible to exhibitors thousands of miles away. Casting certified the precision of the model's details, and bolstered the heroics of the model's creator. And with the tactile skill not simply of skillful sculptors, but of revitalizing hands, the model could approach the epistemological (if not ontological) actuality of a real whale, hunted in Newfoundland at a particular time by a particular group of people. As director Bumpus put it, the museum

---

Through a Lens: Photographs from the Natural History Museum, 1880 to 1950 (London: Natural History Museum, 2003).

[58] Bumpus to Walter Hines Page, 26 Mar. 1908, and 20 Jan. 1908, Box 123, Central Archives, AMNH.

was "an important agent of education, an active centre of scientific work, and a recognized standard of the culture of the community in which it is located," where the cultural "standard" touted by Bumpus specifically celebrated science, utility and rationality.[59] With a provenance derived from carefully rationalized techniques, Andrews's whale could be a powerful pedagogical tool, rather than a fly-by-night exhibit mainly worthy of amusement or scorn.

Nevertheless, just as the materials, rationales, and techniques that director Bumpus found convincing in the first years of the 1900s failed to sway Childs Frick into endorsing a similarly constructed whale in the 1920s, the capacity of Andrews's whale to honorably stand for its counterparts in the wild was subject to constant re-evaluation and reinterpretation. Though Andrews's carefully authenticated *papier-mâché* and wood whale was more durable than the stuffed skin of a once-living whale, it was nevertheless not so durable as to be able to stand up to shifting fashions in museum practice. As Victoria Cain and Karen Rader write, between the 1930s and 1980s, officials of natural history museums began—sometimes quite self-consciously—to reassess their exhibits in terms of laboratory science (especially biology) rather than field natural history. While this change implied different practices in different museums, broadly speaking, exhibitors turned from presenting static tableaux of taxidermized animals, to displays that more and more frequently integrated sound, moving pictures, and sometimes live animals into their spaces. Echoing Nyhart, Cain and Rader write: "Debates over authenticity, once conducted within the walls of museums and the pages of museum journals, exploded into broader discourse in this era. . . . Longstanding disputes about what constituted 'the real thing' continued to focus on specimens, but expanded to encompass the meaning of displays for science education and the kind of public interactions they evoked."[60] Although there was no decisive moment of epistemic shift—old techniques for insuring an exhibit's moral, scientific, and aesthetic worth continued to find use alongside more novel tactics—exhibitors increasingly tended to view scientific authenticity governed not by objects, but by the experiences of viewers as participants.

At the American Museum, this reappraisal of an exhibit's truth-value, accuracy, and worth increasingly focused on the senses of the viewer, rather than the (equally sensual) experience of the scientist/naturalist in the field, or an essential honesty that imbued in natural objects. A study submitted in April 1934 to planners of the museum's Hall of Ocean Life—a gallery which had been a perpetual work in progress for the better part of three decades—noted that museum visitors tended to pause for the longest periods of time in front of exhibits that featured "the mechanics of exploration." A particular example was a habitat group which showed a duo of divers fishing in a coral reef. Though the author of the report was careful to note that he was not in a position to recommend how the museum should "administrate" its displays, by the late 1960s the museum had made a decisive move towards exhibits that engaged a viewer's feelings and senses. Explaining a geological exhibit to the *New York Times*, Gordon Reekie of the museum's exhibition department noted: "We try to give the people an indoctrination . . . an emotional feeling before they get down to looking at rocks." The *Times* article went on to praise the way that the museum's exhibit on ecological conservation funneled patrons through an increasingly hot, crowded, and loud exhibit space—a vivid way of dramatizing industrial depredations

[59] Bumpus to Page, 20 Jan. 1908, Box 123, Central Archives, AMNH.

[60] Karen Rader and Victoria Cain, "From Natural History to Science: Display and the Transformation of American Museums of Science and Nature," *Museum and Society,* 2008, 6:152–171, on p. 163.

of the earth. Indeed, in some cases, the authenticity of the exhibits *and* the viewers' own authenticity could be seen to be co-productive. For instance, the *Times* praised the museum for providing African American viewers with a sense of their own true identity: the displays in the Hall of Man in Africa gave "Afro-Americans an individuality of their own," the paper reported.[61]

In the early 1960s, prompted by the museum's upcoming centennial, the museum announced that it would finally be completing the Hall of Ocean Life, replacing its early twentieth-century whale in the process. Instead of Andrews's whale, a spectacular, new model would represent the largest animal on the planet for museum audiences—a ninety-four foot long fiberglass and polyurethane creation, suspended in a dramatic, ready-to-dive pose from the ceiling of the Hall. The model was accurate; the man overseeing its construction—Richard Van Gelder, the museum's curator of mammalogy—paid close attention to detail, insisting, for instance, that twenty-eight tiny hairs be placed in the massive model's chin, in accordance with what was known about blue whale whiskers.[62]

But for exhibitors, the model's moral and aesthetic truth derived as much from viewers' judgments of their experiences as from the sorts of techniques of validation used by Andrews and his peers. The model was an engineering feat, and unabashedly so. To avoid suspending the model by wires from the gallery's ceiling—the mode of display used by exhibitors of Andrews's generation, now considered *déclassé*—its tremendous bulk was cantilevered in space from the point where its curved back gently grazed the ceiling. The fondest wish for the exhibit, as Alfred E. Parr wrote in 1963, was "to create the illusion of having joined the whale in its own domain." In his memoir of building the whale, Van Gelder painted the scene for his readers: "You are a skin diver without apparatus. You are one with the sea."[63]

As for the old whale, Van Gelder eulogized it as a part of the museum's past: "Three generations of New York school kids had grown up wondering of [Andrews's model], flinching from its amber-eyed stare, gazing at its striated belly from the floor below, and coming away with the awe and the assurance that, 'they have a real stuffed whale in there—I seen it with my own eyes.'" Such dated veracity was both the model's charm and its undoing. In Van Gelder's telling, people came to gaze *upon* the old whale. Its ability to educate was tied up in the authenticity of the object; its charisma derived from its correspondence with the natural world. But while in 1907 *Scientific American* praised Andrews's replica for its extraordinary verisimilitude, by 1963 Parr could compare the model to a "big boat hung in a barn."[64]

The question of the proper relationship between matter and meaning was never finally resolved either in Andrews's or Van Gelder's models—or, for that matter, in museum displays in general. Speaking of the museum's new whale, which went on display in May of 1969, Van Gelder complained to the *Times* that he had become accustomed to "complaints of visitors who expected [the new model] to be the real thing." Still, this did

[61] Abraham Krell, "A Report on the Hall of Ocean Life," April, 1934, Folder: Exhibits-Hall of Ocean Life, 1921–1955, Library of the Department of Mammalogy, AMNH; and Robert W. Stock, "At 100, the Museum of Natural History is No Fossil," *New York Times,* 29 June 1969, p. 8SM.

[62] Van Gelder to Gordon Reekie, 10 Mar. 1969, Folder: Gordon Reekie, 1957–1972, Library of the Department of Mammalogy, AMNH. For more on the construction of the 1969 model whale see Michael Rossi, "Modeling the Unknown: How to Make a Perfect Whale," *Endeavour,* 2008, *32*:58–63.

[63] A. E. Parr, "Concerning Whales and Museums," *Curator,* 1963, 6:65–76, on p. 75; and Richard Van Gelder, "Whale on My Back," *Curator,* 1970, *13*: 95-119, on p.104. In this instance, it should be specified, Van Gelder was talking about a proposed plan for the whale exhibit, but the final outcome had a similar effect.

[64] Van Gelder, "Whale on My Back," p. 95; and Parr, "Concerning Whales and Museums," p. 72.

not mean that the new model lacked authenticity, any more than it meant that Andrews's whale had *lost* its truth-value, aesthetic worth, or even accuracy. Indeed, in the case of the latter, these qualities had rather undergone a sea change—into something rich, if not slightly strange. The museum closed the Hall of the Biology of Mammals in 1969 in preparation for the opening of the Hall of Ocean Life. Andrews's model collected dust in the darkened hall of mammals for the next four years while museum officials tried to decide what to do with it. An announcement that the whale was available free of charge to anyone who could transport it from the hall yielded little in the way of serious offers. Finally, in 1973, a demolition crew removed the whale by means of sledgehammers and crowbars. Perhaps sensing a subtle modulation in the whale's value, museum officials saved one of the whale's eyes and one of its flippers, which the museum put up for auction in 1974 along with other ephemera from the museum's past. In spite of a press release written before the decommissioning of Andrews's replica whale in which Van Gelder specified that there were "no parts of a real whale in the model," it was nevertheless possible to understand the glass eye and chunks of plaster both as emotive *artifacts* and as parts of the "real stuffed whale" upon which Van Gelder waxed nostalgic. One woman bought the model whale's fin because it reminded her of her childhood. The model whale's glass eye, meanwhile, fetched $530, going to a woman who explained, "once you look in the eye of a whale, you never forget it."[65]

[65] Stock, "Museum no Fossil" (cit. n. 61), p. 8S; Richard Van Gelder, open letter, 3 Mar. 1969, Old Whale Folder, Library of the Department of Mammalogy, AMNH; and Lucida Franks, "Museum Sells Bits of Itself at Fund-Raiser," *New York Times,* 9 Mar. 1974, p. 31.

Nature's Agents
or Agents of Empire?

Entomological Workers and Environmental Change during the Construction of the Panama Canal

By Paul S. Sutter

ABSTRACT

This essay examines the role that entomological workers played in U.S. public health efforts during the construction of the Panama Canal (1904–1914). Entomological workers were critical to mosquito control efforts aimed at the reduction of tropical fevers such as malaria. But in the process of studying vector mosquitoes, they discovered that many of the conditions that produced mosquitoes were not intrinsic to tropical nature *per se* but resulted from the human-caused environmental disturbances that accompanied canal building. This realization did not mesh well with an American ideology of tropical triumphalism premised on the notion that the Americans had conquered unalloyed tropical nature in Panama. The result, however, was not a coherent counternarrative but a set of intra-administrative tensions over what controlling nature meant in Panama. Ultimately, entomological workers were loyal not just to the U.S. imperial mission in Panama but also to a modernist culture of science and to the workings of mosquito ecology as they understood them.

I
N HIS LAUDATORY "Introduction" to the 1916 book *Mosquito Control in Panama,* L. O. Howard, the chief of the U.S. Bureau of Entomology and longtime permanent secretary to the American Association for the Advancement of Science, celebrated the

Isis 98, no. 4 (December 2007): 725–54.

Much of the research for this article was completed while I was a postdoctoral fellow at the Smithsonian Institution, where Jeffrey Stine and Pamela Henson were my able mentors. I presented versions of this essay at the German Historical Institute, the José Martí International Colloquium in Cuba, the Smithsonian Institution, the Latin American Studies Association, the EcoHealth One conference, and the University of Georgia, and I want to thank all those who provided valuable feedback. For their comments and support I particularly want to thank Shane Hamilton, Neil Maher, Gregg Mitman, and Guillermo Castro. Finally, I want to express my gratitude to Bernie Lightman, the three anonymous referees, and the *Isis* staff.

mosquito control work done by the Isthmian Canal Commission (ICC) during the construction of the Panama Canal. Howard, who supported that effort from a distance, insisted that such work was "an object lesson for the sanitarians of the world and has demonstrated the vitally important fact that it is possible for the white race to live healthfully in the tropics."[1] With this assertion, Howard added his voice to the triumphalist chorus celebrating the U.S. conquest of tropical nature in Panama.

The technological and engineering feats that distinguished the canal—the giant locks and dams, the Bucyrus steam shovels, and the railroad systems that removed spoil from the famed Culebra Cut—were the most common subjects of such celebratory rhetoric. For many Americans of the time, they were the very stuff of progress, made all the more potent as an "object lesson" because they were achieved by an administrative arm of the U.S. government in a hostile natural environment beyond its borders. In an imperial context in which the technological mastery of nature helped define a nation's level of civilization, American observers crowed that their successful construction of a canal across the isthmus set them apart from the world's other great powers, to say nothing of the "primitive" peoples who increasingly occupied the American imagination.[2]

But next to those civil engineering wonders, on a similar plane of importance, many American observers placed the ICC's sanitary engineering—its local eradication of yellow fever and its control of malaria in particular. Scores of prominent Americans joined Howard in celebrating their apparent sanitary conquest of the tropics. Charles Francis Adams, Jr.— Civil War veteran, past president of the Union Pacific Railroad, amateur historian, grandson of John Quincy Adams, and brother of Henry Adams—called it "an epochal event in sanitation." Having taken control of what was "heretofore the most pestilential region on earth," Adams cheered, the U.S. administration had "proven that the adult male can, by following a prescribed mode of life and observing strict precautionary rules, live, and do a man's work, where he could not live safely or work effectively before." And the head of the U.S. sanitary effort in Panama, William Gorgas, engaged in similar rhetoric in his memoir, *Sanitation in Panama.* "The white man, of all the races of the human family, is the most eager in his pursuit of wealth," Gorgas theorized. "As it becomes generally known that he can live in the tropics and maintain his health, necessarily a large emigration will occur from the present civilized temperate regions to the tropics."[3] These were a few of the many voices—in scores of published books and articles on the subject—that toasted the apparent opening of the tropics to white male settlement and development as a result of American sanitary efforts.

But beyond joining this chorus, Howard made a specific case for the important work done in Panama by a diverse group of entomologists and sanitarians working with ento-

---

[1] L. O. Howard, "Introduction," in Joseph LePrince and A. J. Orenstein, *Mosquito Control in Panama: The Eradication of Malaria and Yellow Fever in Cuba and Panama* (New York: Putnam's, 1916) (hereafter cited as **LePrince and Orenstein, *Mosquito Control in Panama***), pp. iii–v, on p. iv. On Howard see E. O. Essig, *A History of Entomology* (New York: Macmillan, 1931), pp. 658–664; and L. J. Bruce-Chwatt, "Leland Ossian Howard (1857–1950) and Malaria Control: Then and Now," *Mosquito News,* June 1981, *41*(2):215–225.

[2] On technological dimensions of imperialism see Michael Adas, *Machines as the Measure of Men: Science, Technology, and Ideologies of Western Dominance* (Ithaca, N.Y.: Cornell Univ. Press, 1989); and Adas, *Dominance by Design: Technological Imperatives and America's Civilizing Mission* (Cambridge, Mass.: Harvard Univ. Press, 2006). On the American fascination with primitive peoples see Matthew Frye Jacobson, *Barbarian Virtues: The United States Encounters Foreign Peoples at Home and Abroad, 1876–1917* (New York: Hill & Wang, 2000).

[3] Charles Francis Adams, *The Panama Canal: An Epochal Event in Sanitation* (Boston, 1911), pp. 26–27; and William C. Gorgas, *Sanitation in Panama* (New York: Appleton, 1915), p. 289.

mological knowledge, a group I call "entomological workers." The discovery of the mosquito vectors for malaria and yellow fever, confirmed around the turn of the century in several far-flung parts of the globe, separated the successful U.S. effort to build a transisthmian canal (1904–1914) from the French failure two decades earlier. As those discoveries became accepted scientific truth—a messy and uneven process—Howard and his colleagues went to work. First in Cuba and then more thoroughly in Panama, U.S. entomological workers determined the breeding and feeding habits of vector species—for malaria, several species of the genus *Anopheles,* and for yellow fever, the species *Aedes aegypti*—and convinced high-ranking officials that precision mosquito control ought to be the centerpiece of the U.S. sanitary efforts against these diseases. In the process, entomologists became critical actors in the American administration of Panama and other tropical territories. Control of the Panamanian tropics and the vector-borne diseases closely associated with the region, Howard intimated, would not have been as efficient or successful without entomologists' intimate understanding of environmental conditions on the ground. What Howard did not suggest, and what his rhetoric hid, was that that understanding of the ecology of these diseases—in this essay I focus solely on malaria—came into tension with a dominant set of American, indeed Western, assumptions about the hostility of the tropics to white settlement. While most Americans, participants and observers, assumed that they needed to conquer a singular and unalloyed tropical nature to complete the canal, entomological workers discovered that what needed controlling was a debased hybrid landscape.[4]

This essay examines the place of entomological workers in the dynamic interplay between an environmental ideology of tropical triumphalism and the ecology of malaria during the U.S. construction of the Panama Canal. I divide the essay into several parts. First, I situate my effort at the intersection of several historiographies and grapple with theoretical questions arising from this nexus. Next, I briefly outline American thinking about the tropics, particularly emphasizing how this body of thought tended to naturalize diseases such as malaria. Then I examine how mosquito control became a central strategy of U.S. sanitation in Panama and how entomological workers emerged as lead actors in that process. The decision to focus on mosquito control rather than its alternatives, I argue, opened up an ecological line of inquiry that highlighted the tensions between malarial ecology and the logic of tropical triumphalism. Finally, I examine a mundane but telling debate—over how short to cut the grass around workers' quarters—in which these tensions played out. I argue that entomological workers gave voice to a set of environmental conditions that defied American expectations of nature in Panama and that their experiences offer an important model for rethinking the relationship between science, imperialism, and the environment.

HISTORIOGRAPHY AND THEORY

As an environmental historian, I am interested in how the interplay between ideology and ecology shaped the history of canal construction in Panama—and, for that matter, the larger history of U.S. activity in the Latin American, Asian, and African tropics—and in

---

[4] The best general history of both the French and U.S. efforts to build a canal is still David McCullough, *The Path between the Seas: The Creation of the Panama Canal, 1870–1914* (New York: Simon & Schuster, 1977). On the importance of hybrid landscapes to U.S. environmental historiography see Richard White, "From Wilderness to Hybrid Landscapes: The Cultural Turn in Environmental History," *Historian,* 2004, *66:*557–564.

how scientists influenced and navigated that interplay. Environmental historians of the United States have been slow to move beyond the nation's borders to chart the environmental implications of the United States as a hemispheric and global actor. The strong connections between nature and nation in U.S. historiography have kept us at home, focused on continental colonization and the wilderness discourse that dominated that process.[5] But just as that discourse reached a key transitional moment in the late nineteenth century, Americans entered into a transimperial conversation about the tropics as both a region and a problem to be solved. The tropics thus became the dominant environmental imaginary of extracontinental U.S. expansion, just as wilderness was for continental expansion. An analytical focus on the tropics, then, pulls U.S. environmental historiography in an international direction, putting it in dialogue with another vital historiography.

Students of U.S. empire, influenced by broader developments in colonial and postcolonial historiography, have moved away from monolithic analyses of imperial power, finding and mining veins of local accommodation and resistance, delineating the complex and intimate social and cultural territory of the imperial encounter, charting internal heterogeneity among the colonizers and the colonized, and demonstrating how agents and events on the periphery often reshaped core policies, practices, and cultures.[6] Particularly germane to this essay has been a move away from analyses of discursive and representational practices and toward the material dimensions of imperial control—to a focus on how power has been exercised on the ground.[7] But historians of U.S. empire have only just begun to recognize the centrality of environmental management to the nation's extracontinental expansion, and environment has only rarely been taken seriously as one of those material dimensions shaping imperial power. A central goal of this essay is to apply to American tropical triumphalism a rigorous analysis of what happened when that ideology met the environments that its practitioners sought to describe and control.[8]

---

[5] For a discussion of these issues see Paul S. Sutter, "What Can U.S. Environmental Historians Learn from Non-U.S. Environmental Historiography?" *Environmental History,* Jan. 2003, 8:109–129. Richard Tucker's *Insatiable Appetite: The United States and the Ecological Degradation of the Tropical World* (Berkeley/Los Angeles: Univ. California Press, 2000) and John Soluri's *Banana Cultures: Agriculture, Consumption, and Environmental Change in Honduras and the United States* (Austin: Univ. Texas Press, 2006) are important models.

[6] This is a growing literature, though it is well represented in three collections of essays: Gilbert M. Joseph, Catherine C. Legrand, and Ricardo D. Salvatore, eds., *Close Encounters of Empire: Writing the Cultural History of U.S.–Latin American Relations* (Durham, N.C.: Duke Univ. Press, 1998); Amy Kaplan and Donald Pease, eds., *Cultures of United States Imperialism* (Durham, N.C.: Duke Univ. Press, 1993); and Ann Laura Stoler, ed., *Haunted by Empire: Geographies of Intimacy in North American History* (Durham, N.C.: Duke Univ. Press, 2006). Among other important studies in this field are Mary Renda, *Taking Haiti: Military Occupation and the Culture of U.S. Imperialism, 1915–1940* (Chapel Hill: Univ. North Carolina Press, 2001); Kristen Hoganson, *Fighting for American Manhood: How Gender Politics Provoked the Spanish–American and Philippine–American Wars* (New Haven, Conn.: Yale Univ. Press, 1998); Laura Briggs, *Reproducing Empire: Race, Sex, Science, and U.S. Imperialism in Puerto Rico* (Berkeley: Univ. California Press, 2002); and Paul Kramer, *The Blood of Government: Race, Empire, the United States, and the Philippines* (Chapel Hill: Univ. North Carolina Press, 2006). An influential text in the broader historiographical reconsideration has been Frederick Cooper and Stoler, eds., *Tensions of Empire: Colonial Cultures in a Bourgeois World* (Berkeley: Univ. California Press, 1997).

[7] The classic text that set the stage for much of the postcolonial focus on the discursive and representational was Edward Said, *Orientalism* (New York: Vintage, 1979).

[8] Several scholars have recognized the power of tropical thinking to serve an agenda of imperial domination in the same way that racial discourses did, but those analyses have dwelled mostly on its social and cultural dimensions. See David Arnold, "Inventing the Tropics," in *The Problem of Nature: Environment, Culture, and European Expansion* (Cambridge, Mass.: Blackwell, 1996), pp. 141–168; Nancy Leys Stepan, *Picturing Tropical Nature* (Ithaca, N.Y.: Cornell Univ. Press, 2001); Stephen Frenkel, "Jungle Stories: North American Representations of Tropical Panama," *Geographical Review,* 2007, 86:317–333; and Frenkel, "Geographical Representations of the 'Other': The Landscape of the Panama Canal Zone," *Journal of Historical Geography,* 2002, 28:85–99. A growing number of scholars have shown how the human subjects of this discourse on tropicality

Such an approach raises a theoretical conundrum that has long plagued environmental historians. How does one, aware of the constructedness of past ideas of nature, give the material environment a voice (or voices) in speaking to those constructions without simply offering a construction of one's own? Environmental historians' most common response to this dilemma has been a pragmatic one: do the best we can with current science, explicating how nature as we now understand it works and using that understanding to expose the fallacious thinking of the past. Such an approach has its merits, but it treats past and present knowledge inconsistently, using current science uncritically while assuming that past science (and the broader intellectual enterprise of thinking about nature) was the constructed political, social, and cultural practice that historians of science have rightly seen it as. Attention to the historiography and sociology of science, then, ought to compel environmental historians to think more critically about how we use current scientific knowledge to portray the environment as an active historical force.

A more satisfying approach to this theoretical problem than using current science pragmatically is to observe how scientific workers in the past integrated their acquired knowledge of environmental conditions with the ideologies that put them into the field—or, in the case of Panama, how entomological workers squared their emerging understanding of mosquito ecology, gained through observation and experimentation, with reigning ideas of tropical nature.[9] Several approaches in the historiography of science are helpful here. One is the route laid out by the historian of science Robert Kohler, who has provided an intriguing model for what we might call the environmental history of science. In several books, Kohler has looked not only at how scientific practice has reworked nature but also at how the material environment has intruded upon and reshaped scientific practice. His work suggests that the environment is not only a mute subject of scientific study but also an intrusive force that contributes to the social construction of scientific activity.[10] Another promising approach is that offered by actor-network theory (ANT). The proponents of ANT offer a model of causation lodged in collective networks of human and nonhuman actors and structures. ANT rejects the categorical dualism of nature and humanity, allowing for the agency of the nonhuman environment while also calling into question the existence

were able to use its logic to bolster, or even construct, local authority, turning tropical identity into a point of pride. See Arnold, *Colonizing the Body: State Medicine and Epidemic Disease in Nineteenth-Century India* (Berkeley: Univ. California Press, 1993); Julyan Peard, *Race, Place, and Medicine: The Idea of the Tropics in Nineteenth-Century Brazilian Medicine* (Durham, N.C.: Duke Univ. Press, 1999); and Stepan, "Tropical Modernism: Designing the Tropical Landscape," *Singapore Journal of Tropical Geography,* 2000, *21:*76–91.

[9] There has been a hearty, if narrow, debate about how science and scientists functioned as agents of imperialism. Some scholars argue that scientists were merely agents of empire, applying an objectifying discourse of natural history exploration and collection in ways that had decidedly instrumentalist results. See, e.g., Lucille Brockway, *Science and Colonial Expansion: The Role of the British Royal Botanic Gardens* (1979; New Haven, Conn.: Yale Univ. Press, 2002); and Mary Louise Pratt, *Imperial Eyes: Travel Writing and Transculturation* (New York: Routledge, 1992). Other scholars, Richard Grove most notably, suggest that naturalists working on the periphery of empire were in a position to witness profound degradation and that their experience gave birth to modern environmental concern; see Grove, *Green Imperialism: Colonial Expansion, Tropical Island Edens, and the Origins of Environmentalism, 1600–1860* (New York: Cambridge Univ. Press, 1995). Stuart McCook's notion of botany in the Spanish Caribbean as a "creole science" is an excellent example of a growing scholarship that rejects this either/or debate; see McCook, *States of Nature: Science, Agriculture, and Environment in the Spanish Caribbean, 1760–1940* (Austin: Univ. Texas Press, 2002).

[10] Robert E. Kohler, *Lords of the Fly:* Drosophila *Genetics and the Experimental Life* (Chicago: Univ. Chicago Press, 1994); and Kohler, *Landscapes and Labscapes: Exploring the Lab–Field Border in Biology* (Chicago: Univ. Chicago Press, 2002). For a similar approach see Joshua Blu Buhs, "The Fire Ant Wars: Nature and Science in the Pesticide Controversies of the Late Twentieth Century," *Isis,* 2002, *93:*377–400; and Buhs, *Fire Ant Wars: Nature, Science, and Public Policy in Twentieth-Century America* (Chicago: Univ. Chicago Press, 2004).

of a singular "Nature" capable of independent action. While ANT sometimes lumps together in networks various agencies and structures—bodies, institutions, technologies, environmental entities and forces—that scholars might want to tease apart, it nonetheless provides an intriguing model for navigating between materialist and social constructivist perspectives. ANT suggests that, rather than seeing "Nature" as an actor apart from humanity, we need to conceptualize and examine hybrid environments as fields of agency and power in which the human and nonhuman intermingle and together shape change over time.[11]

The material environment that I seek to animate in this essay, then, is not a willful agent, a singular actor with intent or clear interests. Rather, it is an agglomeration of forces and structures that can be difficult to read in the service of agency. Environmental historians know full well that "Nature" does not speak for itself in any simple sense, and we have become acutely aware that speaking for nature is, no matter how solid our scientific backing, a subjective and fraught enterprise. Indeed, for some, environmental history has become the practice of documenting competing claims to speak for nature, all of which reveal social power but none of which adequately gets at a material environment beyond our reach. In this essay, and the larger project of which it is a part, I hope to reveal rather than repeat the power that the singular construct "Nature" has had to naturalize human actions and choices. But I am not fully ready to give up on finding and narrating an environment that acts in history beyond human subjectivity. Recent work in the historiography and sociology of science seems particularly helpful, then, in the project of challenging monolithic and naturalizing notions of "Nature" while also maintaining an insistence on material environmental agency.

I want to take these historiographical and theoretical insights and bring them to bear not just on scientific practice, but on the larger field of U.S. colonial administration, to show how the environment intruded upon that experiment as well. In Panama, entomologists worked at a critical juncture between dominant ideas of nature and environmental phenomena; they had intimate experiences with the material environment that had the potential to reinforce or destabilize ruling ideas of nature and to reveal tensions between their positions and those of other imperial actors. They were not mere captives of tropical ideology. My argument is not that scientists give us unmediated access to material environmental agency—that they are, in a sense, nature's agents. Nor do I intend to imply that they are the only group in the imperial field who work across this gap between the material environment and idealized nature. Rather, my aim is to suggest that material environmental influence can be seen quite clearly at the points of tension between ideological predisposition and empirical observation and that scientists are particularly fruitful subjects for examining such tensions. To paraphrase Richard White: the environment not only spoke back in Panama; the goals and methods of entomological workers invited it to do so.[12]

---

[11] In charting this approach I have been influenced by Gregg Mitman, "Where Ecology, Nature, and Politics Meet: Reclaiming *The Death of Nature*," *Isis,* 2006, *97:*494–504; Timothy Mitchell, "Can the Mosquito Speak?" in *The Rule of Experts: Egypt, Techno-Politics, Modernity* (Berkeley: Univ. California Press, 2002), pp. 19–53; and Scott Kirsch and Don Mitchell, "The Nature of Things: Dead Labor, Non-Human Actors, and the Persistence of Marxism," *Antipode,* 2004, *36:*687–705. On ANT see Bruno Latour, *Reassembling the Social: An Introduction to Actor-Network Theory* (New York: Oxford Univ. Press, 2005); Latour, *Politics of Nature: How to Bring the Sciences into Democracy* (Cambridge, Mass.: Harvard Univ. Press, 2004); and John Law, "Technology and Heterogeneous Engineering: The Case of Portuguese Expansion," in *The Social Construction of Technological Systems: New Directions in the Sociology and History of Technology,* ed. Wiebe E. Bijker, Thomas P. Hughes, and Trevor J. Pinch (Cambridge, Mass.: MIT Press, 1987), pp. 111–134.

[12] Richard White, "Discovering Nature in North America," *Journal of American History,* 1992, *79:*874–891, paraphrase drawn from quote on p. 889.

What entomological workers in Panama observed subtly challenged a basic premise of American tropical thinking: that Panama was, by nature, hostile to white people from temperate regions and thus had to be tightly controlled for whites to live healthy lives there. Entomologists in Panama never ceased to be agents of empire; as Howard's opening comment suggests, they were eager triumphalists when highlighting their role in controlling tropical diseases. But to see them merely as agents of empire, and to fail to disaggregate that category, is to miss how the environment they observed, and that acted along with them, led them to a potentially subversive insight: while many lauded the U.S. sanitary conquest of tropical Panama, entomological workers on the ground intimated that the ICC's real sanitary achievement was in overcoming environmental conditions that U.S. canal building had a strong hand in creating.

CONCEPTUALIZING THE TROPICS

When the United States entered the new nation of Panama in 1904, administrators faced a series of discrete engineering and public health challenges that they, like other Western imperial powers, tended to categorize as "tropical." U.S. canal builders, and American observers of that process, were by no means the first to think about the tropics as an environmental space that threatened the enterprises of white peoples from temperate zones. One can trace ideas about the medical and constitutional challenges of warm climates and torrid zones back to the ancient world. But a coherent discourse on the tropics emerged only during the early period of European exploration in the fifteenth century, and it assumed its modern form in the eighteenth and nineteenth centuries. As Nancy Leys Stepan has argued, modern notions of the tropics relied on a rigorous and geographically informed natural history, the development of modern human sciences such as anthropology, and a revival of medical geography. Together these disciplines made sense of the natural, human, and medical tropics, assembling them into a single geographical category at a moment when European powers had turned their imperial attention to these regions. Modern tropical thinking, then, was the product of several coalescing efforts at scientific exploration and categorization in an imperial age.[13]

Europeans led the way in crafting modern representations of the tropics, but, by the middle of the nineteenth century, expansionism of various sorts prompted North Americans to think about the tropics as well—and, influenced by the Prussian scientist and explorer Alexander von Humboldt, the American tropics in particular. Gold rush migrants and government explorers, travel writers and filibusterers, landscape painters and naturalists, capitalists and workers, diplomats and soldiers—all moved into and through the American tropics during the nineteenth century, writing about and representing the region to a substantial domestic audience. Americans also watched intently as the French floundered in their efforts to build a sea-level canal in Panama in the 1880s, losing more than twenty thousand workers to an environment that one prominent Frenchman described as "literally poisoned."[14] By the late nineteenth century, as a similarly diverse set of actors charted the

---

[13] On ancient notions of the tropics see Clarence J. Glacken, *Traces on the Rhodian Shore: Nature and Culture in Western Thought from Ancient Times to the End of the Eighteenth Century* (Berkeley: Univ. California Press, 1967); and May Berenbaum, *Bugs in the System: Insects and Their Impacts on Human Affairs* (Reading, Mass.: Helix, 1995), pp. 232–233. David Arnold argues for the fifteenth-century origins of tropical thinking in *Problem of Nature* (cit. n. 8), p. 143. On the formative importance of the eighteenth and nineteenth centuries see Stepan, *Picturing Tropical Nature* (cit. n. 8), pp. 16–17.

[14] McCullough, *Path between the Seas* (cit. n. 4), p. 80. For two very different interpretations of Humboldt

wilderness of the American West, parallel activities in Latin America fueled a growing American literature on and set of assumptions about the tropics that would serve U.S. expansion in the wake of the Spanish-American-Cuban War.

Historically, tropical thinking has been deeply ambivalent, with observers alternately portraying the tropics as edenic and hellish. Early European encounters tended to fall into the former category, with explorers depicting the tropics as bountiful and beautiful. But by the late eighteenth century, as it became clear that disease was a considerable obstacle to tropical colonization (and as colonial regimes transformed tropical environments in ways that made them more amenable to disease), what Philip Curtin has called the "terror" of the tropics emerged as an equal partner in the discourse.[15] The balance swung further in a negative direction during the nineteenth century, as Enlightenment ideas about human adaptability to environmental conditions yielded to notions about the impossibility of temperate peoples acclimatizing themselves to the tropics. Increasingly, temperate zone theorists argued, life in the tropics implied physical and moral degeneracy for northern peoples, and they looked on tropical denizens as physically and morally inferior in increasingly racialized ways. While an "affirmative tropicality" lived on in the legacy of Humboldt and other prominent naturalists who studied the tropics, Europeans and North Americans mostly came to see the tropics as a hostile environment and a vexing imperial problem.[16]

Concerns about race and labor informed the tropical thinking of temperate zone observers; their musings about exotic nature easily slid into questions about the attributes of tropical "natives," and vice versa. Why were tropical regions, where nature seemed so bountiful, so undeveloped by Western standards? Was the tropical climate so enervating that civilization could not flower there, or did the natural bounty of the tropics promote slothfulness among the residents? What would happen to the work ethic of white Americans, forged in supposedly edifying temperate climates, under tropical conditions? These were compelling questions for a nation about to undertake a vast and expensive public engineering enterprise in the tropics. And because racial distinctions appeared to mirror the tropical/temperate divide, many saw race as a "natural" way of organizing these ques-

and his influence see Pratt, *Imperial Eyes* (cit. n. 9); and Aaron Sachs, *The Humboldt Current: Nineteenth-Century Exploration and the Roots of American Environmentalism* (New York: Viking, 2006). On the Panama gold rush transit see Aims McGuiness, "In the Path of Empire: Labor, Land, and Liberty in Panama during the California Gold Rush" (Ph.D. diss., Univ. Michigan, 2001); and Brian Roberts, *American Alchemy: The California Gold Rush and Middle-Class Culture* (Chapel Hill: Univ. North Carolina Press, 2000). On American travel writing about Central America during this period see Larzer Ziff, *Return Passages: Great American Travel Writing, 1780–1910* (New Haven, Conn.: Yale Univ. Press, 2001). On American landscape painters in the tropics see Katherine Manthorne, *Tropical Renaissance: North American Artists Exploring Latin America, 1839–1879* (Washington, D.C.: Smithsonian Institution Press, 1989). On the development of tropical agriculture see Tucker, *Insatiable Appetite* (cit. n. 5); and Soluri, *Banana Cultures* (cit. n. 5). On the broader history of U.S. perceptions of Latin American nature and society see Frederick Pike, *United States and Latin America: Myths and Stereotypes of Civilization and Nature* (Austin: Univ. Texas Press, 1992).

[15] Arnold, *Problem of Nature* (cit. n. 8), pp. 144–145; Philip Curtin, *The Image of Africa: British Ideas and Action, 1780–1850* (Madison: Univ. Wisconsin Press, 1964), pp. 58–87; and John McNeill, "Epidemics, Environment, and Empire: Yellow Fever and Geopolitics in the American Tropics, 1620–1825," *Environment and History,* 1999, *5:*175–184.

[16] Arnold, *Problem of Nature,* p. 146; and Stepan, *Picturing Tropical Nature* (cit. n. 8), pp. 35–43. Philip Curtin charted the cost in lives of British attempts to settle the tropics in *Death by Migration: Europe's Encounter with the Tropical World in the Nineteenth Century* (New York: Cambridge Univ. Press, 1989). On the shift from acclimatization to degeneracy see the articles in the special section on "Race and Acclimatization in Colonial Medicine" in the *Bulletin of the History of Medicine,* 1996, *70*(1), including Warwick Anderson, "Disease, Race, and Empire" (pp. 62–67); Mark Harrison, "'The Tender Frame of Man': Disease, Climate, and Racial Difference in India and the West Indies, 1760–1860" (pp. 68–93); and Anderson, "Immunities of Empire: Race, Disease, and the New Tropical Medicine, 1900–1920" (pp. 94–118).

tions about labor efficiency and development. As the quotations with which I began this essay suggest, many saw the tropics as hostile to whites by nature.

Health also was integral to this tropical/temperate binary. White Americans brought with them to Panama deep anxieties about how the tropics would affect their bodies and constitutions.[17] They also wondered why tropical peoples seemed less affected by a climate and suite of diseases that struck white outsiders with a vengeance.[18] These fears were rooted in a miasmatic paradigm whose adherents postulated that emanations—bad air generated by swamps, other forms of putrefying vegetation, or the workings of the tropical climate on human filth—were the agents of tropical fevers such as malaria. While most writers on the subject, from diarists to medical experts, understood that such miasmas were not unique to the tropics, they assumed that tropical regions were uniquely productive of them. Swamps supposedly produced these emanations in abundance, particularly at night, and many feared that the turning over of tropical soils on the scale that would occur during the excavation of the Panama Canal would disinter these poisons and lay low those who were not immune. The miasmatic theory, though it sometimes encouraged careful attention to specific medical topographies, just as often led to a broad indictment of tropical nature and the tropical climate as constitutionally debilitating and disease producing. "When nineteenth-century Europeans wrote about the dangers of a tropical climate," Philip Curtin has written about the African setting, "they meant literally that temperature, humidity, [and] emanations from the soil were the sources of danger."[19] The same can be said for turn-of-the-century Americans, although, critically, they entered Panama with a more precise understanding of where these diseases came from and thus with several new options for understanding and attacking them.

MOSQUITO VECTORS AND NEW SANITARY CHOICES

Because modern Western notions about the tropics as a discrete region were forged within a miasmatic tradition, one might expect the mosquito vector discoveries for malaria and yellow fever to have thrown tropical thinking into flux. Surprisingly, they prompted only

---

[17] The tropics were in some senses one of the last redoubts of a once-prevalent American mode of thinking that connected the environment and bodily health, a mode that faded with the germ theory. On this mode of thinking see Conevery Bolton Valencius, *The Health of the Country: How American Settlers Understood Themselves and Their Land* (New York: Basic, 2002); Linda Nash, *Inescapable Ecologies: A History of Environment, Disease, and Knowledge* (Berkeley: Univ. California Press, 2007); and Gregg Mitman, *Breathing Spaces: How Allergies Shape Our Lives and Landscapes* (New Haven, Conn.: Yale Univ. Press, 2007). On these anxieties in another U.S. imperial setting see Warwick Anderson, *Colonial Pathologies: American Tropical Medicine, Race, and Hygiene in the Philippines* (Durham, N.C.: Duke Univ. Press, 2006).

[18] Tropical "native" status had long implied that one was nonwhite, disease resistant, and constitutionally suited to working under tropical conditions. These connections were made in the context of North American slavery. See, e.g., Peter Wood, *Black Majority: Negroes in Colonial South Carolina from 1670 to the Stono Rebellion* (New York: Knopf, 1974); and Mart Stewart, *"What Nature Suffers to Groe": Life, Labor, and Landscape on the Georgia Coast, 1680–1920* (Athens: Univ. Georgia Press, 1996). On disease exchanges and the physiology behind perceptions of disease resistance see Kenneth Kiple, *The Caribbean Slave: A Biological History* (New York: Cambridge Univ. Press, 1984); Kiple, ed., *The African Exchange: Toward a Biological History of Black People* (Durham, N.C.: Duke Univ. Press, 1988); Kiple and Virginia H. King, "Black Yellow Fever Immunities, Innate and Acquired, as Revealed in the American South," *Social Science History,* 1977, *1*:419–436; Richard B. Sheridan, *Doctors and Slaves: A Medical and Demographic History of Slavery in the West Indies* (New York: Cambridge, 1985); Philip D. Curtin, "Epidemiology and the Slave Trade," *Political Science Quarterly,* 1968, *82*(3):190–216; Curtin, "Disease Exchange across the Tropical Atlantic," *History and Philosophy of the Life Sciences,* 1993, *15*:329–356; and McNeill, "Epidemics, Environment, and Empire" (cit. n. 15).

[19] Philip Curtin, "Medical Knowledge and Urban Planning in Tropical Africa," *American Historical Review,* 1985, *90*:594–613, on p. 596.

a minor recalibration among most observers. Rather than seeing the tropics as miasmatic, many insisted that the region was uniquely productive of insect life. Others mixed miasmatic and mosquito theories, making little effort to appreciate the challenges these new discoveries posed to miasmatic logic. Still others began the slow process of teasing apart disease investigation—in this case figuring out the discrete etiologies of malaria and yellow fever—from questions about the physiological effects of the tropical climate, a project that would eventually challenge major assumptions about the capacity of white labor in the tropics.[20] But that process took time. While it might be tempting to assume that the vector discoveries, like a flash of light, replaced miasmatic superstition with scientific truth, the reality was more complex. The mosquito control work done in Panama was at the heart of that complexity.

While tropical nature remained definitive of the public health problem in the eyes of most Americans as the United States entered Panama, the mosquito vector facilitated a pernicious link that reframed tropical sanitary choices: for malaria and other vector-borne tropical diseases to survive, there had to be a reservoir of infected people from whom mosquitoes could obtain the parasite or virus before passing it on to others. That notion, a novel product of the vector theory, put the bodies and behaviors of "tropical" peoples under heavy scrutiny.[21] The miasmatic tradition, which largely blamed nature for disease, generally had justified imperial neglect of "native" public health, at least when it came to tropical fevers (contagious diseases were a different story); but the vector discoveries linked the health of natives and imported "tropical" laborers (mostly black West Indians in Panama) with that of white newcomers in important ways. Imperial public health authorities thus faced several options in controlling malaria as they adjusted to the vector theories: they could embrace racial segregation as a sanitary policy that separated vulnerable outsiders from diseased tropical peoples; they could broaden their public health efforts in an attempt to eliminate native disease reservoirs; or they could attempt to control the mosquito link in disease transmission.

Sanitary segregation had its supporters, in Panama as well as in other imperial settings. The approach emerged most clearly in portions of British West Africa, particularly in the work of S. R. Christophers and J. W. W. Stephens of the Liverpool School of Tropical Medicine. In his primer *Mosquito Brigades and How to Organize Them,* Ronald Ross, who had discovered the *Anopheles* vector for malaria, also suggested segregation as an approach in British India, though he was a stronger advocate of mosquito control.[22] Following these leads, several U.S. officials suggested sanitary segregation for Panama. Henry

[20] Weston Chamberlain, "Some Features of the Physiologic Activity of Americans in the Philippines," *American Journal of Tropical Diseases and Preventive Medicine,* 1913, *1*(1):12–32; and Chamberlain, "The Influence of Tropical Residence on the Blood," *ibid.,* 1914, *2*(1):41–55. On Chamberlain's work see Anderson, *Colonial Pathologies* (cit. n. 17), pp. 74–87.

[21] On these issues in other settings see the articles in the special section on "Race and Acclimatization in Colonial Medicine," *Bull. Hist. Med.,* 1996, *70*(1). The best treatments of yellow fever and malaria in the North American setting have been provided by Margaret Humphreys: *Yellow Fever and the South* (New Brunswick, N.J.: Rutgers Univ. Press, 1992); and *Malaria: Poverty, Race, and Public Health in the United States* (Baltimore: Johns Hopkins Univ. Press, 2001).

[22] S. R. Christophers and J. W. W. Stephens, "The Native as the Prime Agent in the Malarial Infection of Europeans," *Further Reports of the Malarial Committee of the Royal Society* (1900), pp. 3–19; Stephens and Christophers, *Practical Study of Malaria and Other Blood Parasites* (London: Univ. Press Liverpool, 1904); and Ronald Ross, *Mosquito Brigades and How to Organize Them* (London: George Philip, 1902). On segregation as a public health measure in India and West Africa see Curtin, "Medical Knowledge and Urban Planning in Tropical Africa" (cit. n. 19); and John Cell, "Anglo-Indian Medical Theory and the Origins of Segregation in West Africa," *Amer. Hist. Rev.,* 1986, *91*:307–335.

Rose Carter, director of hospitals in the Canal Zone, made such a suggestion before a meeting of the Canal Zone Medical Association in 1908:

> To prevent the infection of the mosquitoes which have access to the men we are protecting is simply to segregate the quarters of these men from those of the natives and colored laborers—a source of infection to the insects—a sufficient distance, which thing has been inculcated by the British writers for years. It is what is hardest to accomplish on the Isthmus of any of the measures that we attempt, yet it is fairly important and considerable sickness has been occasioned by failure to observe this principle.

And Charles F. Mason, who later became chief health officer of the Panama Canal, mentioned as a sanitary approach the "segregation of employees, that is, locating the dwellings of non-immunes far apart from the native villages, the inhabitants of which form reservoirs from which mosquitoes obtain their infection." These recommendations certainly fit with the segregationist policies the United States enacted in the Canal Zone, particularly the division of canal workers into "gold" and "silver" categories, with nonwhite "tropical" peoples largely confined to the second employment class.[23] And they certainly resonated with the strong connections between race, ethnicity, and public health then current within the United States. Indeed, as U.S. sanitarians turned their attention to the threat of diseased "natives" in places such as Panama, they had their own deepening reservoir of racialized domestic public health approaches from which to draw.[24] But sanitary records suggest that, during the construction era at least, effective sanitary segregation was difficult to achieve in a landscape of constantly shifting work sites and workers. In the Canal Zone, though the racial segregation of white Americans from the majority nonwhite workforce was everywhere a social and spatial reality, segregation was of secondary importance as a public health measure. Insofar as malaria control was concerned, sanitarians relied more heavily on other approaches.[25]

The ICC, again following other imperial examples, also considered mass quininization. Robert Koch, the pioneering German bacteriologist, was then the best-known advocate for quinine chemotherapy. On the basis of research he conducted in Africa, the Dutch East Indies, and New Guinea, Koch advocated a strict regimen of quinine both to protect work-

[23] Henry Rose Carter, "Malarial Fever Work on the Isthmus," *Proceedings of the Canal Zone Medical Association,* 1908, pp. 102–114, on p. 106; and Charles F. Mason, "Sanitation in the Panama Canal Zone," *Transactions of the International Engineering Congress, 1915,* rpt. in George Goethals, *The Panama Canal: An Engineering Treatise,* 2 vols. (New York: McGraw-Hill, 1916), Vol. 1, pp. 85–115, on p. 95. On segregation in the Canal Zone and the gold and silver rolls see Michael Conniff, *Black Labor on a White Canal: Panama, 1904–1981* (Pittsburgh: Univ. Pittsburgh Press, 1985).

[24] On these sorts of connections within the United States see Judith Walzer Leavitt, *Typhoid Mary: Captive to the Public's Health* (Boston: Beacon, 1996); Nayan Shah, *Contagious Divides: Epidemics and Race in San Francisco's Chinatown* (Berkeley: Univ. California Press, 2001); Natalie Molina, *Fit to Be Citizens? Public Health and Race in Los Angeles, 1879–1939* (Berkeley: Univ. California Press, 2006); and Tera Hunter, *To 'Joy My Freedom: Southern Black Women's Lives and Labors after the Civil War* (Cambridge, Mass.: Harvard Univ. Press, 1997). On the influence of U.S. medical practice in Panama on U.S. public health and eugenics see Alexandra Minna Stern, *Eugenic Nation: Faults and Frontiers of Better Breeding in Modern America* (Berkeley: Univ. California Press, 2005).

[25] As Charles Mason put it, sanitary segregation "has been carried out as far as is practicable, but it has not occupied as prominent a position as have the other measures referred to": Mason, "Sanitation in the Panama Canal Zone" (cit. n. 23), p. 104. In her study of malaria in the United States, Margaret Humphreys argues that sanitary segregation did not occur much in the U.S. South, for a variety of reasons; see Humphreys, *Malaria* (cit. n. 21), pp. 57–62, 73. Stephen Frenkel suggested in his dissertation that sanitary segregation was easier to achieve in Panama once canal construction was complete; see Frenkel, "Cultural Imperialism and the Development of the Panama Canal Zone, 1912–1960" (Ph.D. diss., Syracuse Univ., 1992).

ers from contracting malaria and to reduce the infectiousness of malarial patients. But American officials were skeptical of the effectiveness of quininization. Questioning Koch's advocacy specifically, Carter wrote in 1908, "I doubt if any sanitarian would be willing to attack the problem in the American tropics exclusively from the end of the human host." While the ICC distributed quinine throughout the construction period, to white and non-white workers alike, sanitarians never embraced it as a sole route to eliminating malaria. Consistently getting the correct doses to every worker, U.S. sanitary officials insisted, was too daunting a task for quinine to be much more than an individual prophylactic. Ultimately, American officials decided to attack the mosquito strand in the web of malarial transmission, following examples set by British sanitarians such as Ross and Malcolm Watson, who worked in British Malaya, as well as the pioneering work done by Gorgas in Havana.[26]

PURE AND APPLIED ENTOMOLOGY IN PANAMA

For mosquito control to be efficient and effective, U.S. sanitarians needed to understand vector ecology, and to figure out these details in Panama they turned to entomologists. Entomology had been a marginal, largely taxonomic science in the United States before the Civil War, but by serving postwar agricultural expansion entomologists assumed a central place within both the American scientific community and the growing federal agricultural bureaucracy. In the late nineteenth century, a series of high-profile insect threats to farms and forests—from Rocky Mountain locusts and citrus scales to Mexican boll weevils and gypsy moths—prompted a significant rise in federal sponsorship for applied or "economic" entomology, as the migratory nature of these pest species encouraged national and even international approaches to their control. As one historian of entomology put it, "in the latter half of the nineteenth century, it was as if the forces of nature were lobbying Washington on behalf of the entomologists."[27]

Entomologists built their expertise in the late nineteenth century with what might be called an ecological approach to agricultural pest problems. L. O. Howard, entomology's chief institution builder and a champion of its applied value, was at the vanguard of this effort. Howard received his B.S. and M.S. degrees from Cornell University, where he studied entomology with John Henry Comstock, one of the nation's premier entomologists.

[26] Carter, "Malarial Fever Work on the Isthmus" (cit. n. 23), p. 103. On Koch's quinine recommendations see Curtin, "Medical Knowledge and Urban Planning in Tropical Africa" (cit. n. 19), pp. 597–598; and Thomas D. Brock, *Robert Koch: A Life in Medicine and Bacteriology* (New York: Springer, 1988). Regarding the focus on mosquitoes see LePrince and Orenstein, *Mosquito Control in Panama;* and Ross, *Mosquito Brigades and How to Organize Them* (cit. n. 22). Malcolm Watson arrived in the Federated Malay States in 1901 to undertake mosquito control and wrote about it in two books: *The Prevention of Malaria in the Federated Malay States: A Record of Twenty Years' Progress* (London: John Murray, 1911); and *Rural Sanitation in the Tropics: Being Notes and Observations in the Malay Archipelago, Panama, and Other Lands* (London: John Murray, 1915).

[27] Berenbaum, *Bugs in the System* (cit. n. 13), p. 281. On the history of American entomology and its place in the American scientific establishment see Willis Connor Sorenson, *Brethren of the Net: American Entomology, 1840–1880* (Tuscaloosa: Univ. Alabama Press, 1995); A. Hunter DuPree, *Science in the Federal Government: A History of Policies and Activities* (1957; Baltimore: Johns Hopkins Univ. Press, 1985); Hae-Gyung Geong, "Exerting Control: Biology and Bureaucracy in the Development of American Entomology, 1870–1930" (Ph.D. diss., Univ. Wisconsin, 1999); L. O. Howard, "The Rise of Applied Entomology in the United States," *Agricultural History,* 1929, *3*(3):131–139; Jeffrey Lockwood, *Locust: The Devastating Rise and Mysterious Disappearance of the Insect That Shaped the American Frontier* (New York: Basic, 2004); Robert J. Spear, *The Great Gypsy Moth War: The History of the First Campaign in Massachusetts to Eradicate the Gypsy Moth, 1890–1901* (Amherst: Univ. Massachusetts Press, 2005); James Giesen, *The South's Greatest Enemy? The Cotton Boll Weevil and Its Lost Revolution, 1892–1930* (Chicago: Univ. Chicago Press, forthcoming); and Richard C. Sawyer, *To Make a Spotless Orange: Biological Control in California* (Lafayette, Ind.: Purdue Univ. Press, 2002).

Beginning in 1878, he served as an assistant to Charles Valentine Riley, the head of the U.S. Department of Agriculture's entomological work, and then, in 1894, succeeded him. Howard built the entomological division into the Bureau of Entomology, over which he presided until 1927. He insisted that an intricate knowledge of insect bionomics was crucial to controlling pests, as the simplification of modern agroecosystems opened them to invasions and as modern transportation technologies and networks made it easier for insects to move into previously isolated environments. "The enormous expansion of agriculture in North America" in the late nineteenth century, Howard recalled toward the end of his career, "resulted in types of agricultural practice peculiarly favorable to insect increase."[28] By the turn of the last century, under Howard's leadership, American entomologists were no strangers to serving as careful ecological observers of and problem solvers within a colonizing mission. Indeed, they were well prepared to bring their ecological approach to another round of American expansion.[29]

Federal entomologists gained a powerful new rationale for applied entomology with the mosquito vector discoveries.[30] And while malaria and yellow fever persisted in the United States during the early twentieth century, it was the commercial and military expansion of the United States into tropical Latin America and the Asian Pacific that most forcefully connected federal entomological expertise to public health campaigns. Indeed, these imperial campaigns helped to build federal public health capacity and to reframe disease control, traditionally a state responsibility, as a federal issue during the early twentieth century.[31] But to fulfill their roles as disease managers, entomologists first had to generate

---

[28] L. O. Howard, *A History of Applied Entomology (Somewhat Anecdotal)* (Washington, D.C.: Smithsonian Institution, 1930), p. 3. On Howard see Essig, *History of Entomology* (cit. n. 1), pp. 658–663; and Arnold Mallis, *American Entomologists* (New Brunswick, N.J.: Rutgers Univ. Press, 1971), pp. 79–86. Howard later became known for his martial and alarmist rhetoric about insects, rhetoric that presaged the postwar U.S. commitment to chemical control of insect pests. But for much of his early career he preached an applied ecological approach that was shared by many in the entomological community. Two of Howard's better-known books were *The Insect Menace* (New York: Century, 1931) and *Fighting the Insects: The Story of an Entomologist, Telling of the Life and Experiences of the Writer* (New York: Macmillan, 1933). See also Howard, *History of Applied Entomology (Somewhat Anecdotal);* Edmund Russell, *War and Nature: Fighting Humans and Insects with Chemicals from World War I to Silent Spring* (New York: Cambridge Univ. Press, 2001); and Russell, "L. O. Howard Promoted War Metaphors as a Rallying Cry for Economic Entomology," *American Entomologist,* Summer 1999, *45:*74–78.

[29] Sorenson makes this point in *Brethren of the Net* (cit. n. 27), pp. 258–259. Philip Pauly has argued that USDA scientists were deeply divided over the biological implications of the nation's increasingly global and imperial posture, with Howard and the Bureau of Entomology assuming a nativist or restrictionist position for fear of exotic insect invasions, while Beverly Galloway, David Fairchild, and others within the Bureau of Plant Industry took a more "cosmopolitan" approach to plant and other biotic introductions and exchanges. In the case of Panama, these two approaches seem to have coexisted without much conflict, perhaps because the applied entomology practiced there was medical rather than agricultural. See Philip J. Pauly, "The Beauty and Menace of Japanese Cherry Trees," *Isis,* 1996, *87:*51–73; and Pauly, *Biologists and the Promise of American Life: From Meriwether Lewis to Alfred Kinsey* (Princeton, N.J.: Princeton Univ. Press, 2000), Ch. 3.

[30] Early links between insects and disease transmission, both nationally and internationally, had already put a premium on applied entomological knowledge, spawning the fields of medical and veterinary entomology. In the imperial context, Sir Patrick Manson, the putative father of tropical medicine, figured out in the late 1870s that mosquitoes (*Culex fatigans,* to be precise) hosted and transmitted the filarial worms responsible for elephantiasis, a discovery that led medical officials to seek similar parasite-vector explanations for other diseases. See Roy Porter, *The Greatest Benefit to Mankind: A Medical History of Humanity* (New York: Norton, 1997), p. 467. And in the United States, the 1891 discovery by Theobald Smith that ticks (arachnids, not insects—but the problem was claimed by entomologists nonetheless) spread Texas fever among cattle in the South and West was a formative development. On the history of Texas fever see Claire Strom, "Texas Fever and the Dispossession of the Southern Yeoman Farmer," *Journal of Southern History,* 2000, *65:*49–74.

[31] Yellow fever outbreaks in 1905 in several southern cities were a critical turning point in this regard, as the U.S. Public Health Service (created in 1902 from the Marine Hospital Service) directly intervened with mosquito control efforts modeled on Gorgas's work in Havana and Panama, efforts that showed up ineffective state efforts. A similar if slightly less direct line of influence led from malaria control in Panama to subsequent federal efforts

some knowledge of mosquitoes, which to that point had not been extensively studied. As a first step, Howard put together a USDA Bulletin, *Notes on the Mosquitoes of the United States,* in 1900, and then a primer on the subject, *Mosquitoes: How They Live, How They Carry Disease, How They Are Classified, and How They May Be Destroyed,* in 1901. In the process, he corresponded extensively with leading entomologists and sanitarians throughout the world. Howard was not the first to engage in such transatlantic correspondence; nineteenth-century American entomologists had relied extensively on European collections and expertise. But his international correspondence was distinctive in the set of imperial connections it reflected, an underappreciated aspect of the Progressive Era's transatlantic conversation and one that was formative not just for entomologists but for many other experts within the nascent environmental management state.[32]

The adoption of mosquito control in Panama necessarily transformed the isthmus's mosquitoes from fauna incognita into subjects of close scientific study. Just prior to the U.S. entry into Panama, entomologists had begun the task of identifying and cataloguing the region's Culicidae (mosquito family). Recognizing "the imperfect character of our knowledge and especially the very great need for a competent monograph on the species of Culicidae of North and Central America and the West Indies, both from the biological and from the sanitary points of view," Howard and his colleagues, Harrison Dyar and Frederick Knab, wrote a grant to the newly created Carnegie Institution in 1902 seeking funding to undertake such a systematic study. Dyar was a lepidopterist with graduate degrees in entomology and bacteriology from Columbia, where he was teaching when Howard hired him to curate Lepidoptera for the National Museum of Natural History in 1897. Knab was a Bavarian-born artist and entomologist who would later become the chief curator of Diptera at the National Museum.[33] The three received the grant in 1903 and began assembling a network of collectors and consultants throughout the covered regions to assist in their efforts. Many of these collectors were associated with the nation's new imperial holdings, but the authors also came to rely on Latin American and Caribbean scientists and institutions, such as the Instituto Oswaldo Cruz in Rio de Janeiro. Howard not only tapped into the sort of transimperial circuits that have increasingly occupied the attention of colonial and postcolonial scholars; he and his colleagues also engaged in what

in the U.S. South, where the disease persisted until World War II. See Humphreys, *Yellow Fever and the South* (cit. n. 21), pp. 10–15; and Humphreys, *Malaria* (cit. n. 21), pp. 74–75.

[32] L. O. Howard, *Notes on the Mosquitoes of the United States: Giving Some Account of Their Structure and Biology, with Remarks on Remedies,* Bulletin No. 25, N.S., Division of Entomology, U.S. Department of Agriculture (Washington, D.C.: Government Printing Office, 1900); and Howard, *Mosquitoes: How They Live, How They Carry Disease, How They Are Classified, and How They May Be Destroyed* (New York: McClure, Philips, 1901). For Howard's correspondence during this period see National Archives and Records Administration, Record Group 7, Records of the Bureau of Entomology and Plant Quarantine (NARA RG 7), Washington, D.C., Entry 6, "General Records, Letters Sent, 1896–1908," and Entry 13, "Letters Received Regarding the Mosquito Investigations, 1900–1901." Howard had established himself as an expert on mosquitoes and their control in the 1890s, prior to the vector confirmations, and he carefully followed the vector discoveries and the various imperial sanitary efforts that came in their immediate wake. See Howard, "An Experiment against Mosquitoes," *Insect Life,* 1892, *5*(1):12–14; and Howard, "Another Experiment against Mosquitoes," *ibid.,* 1893, *6*(2):90–92. For the definitive treatment of the transatlantic Progressive Era conversation see Daniel Rodgers, *Atlantic Crossings: Social Politics in a Progressive Age* (Cambridge, Mass.: Belknap, 1998).

[33] L. J. Bruce-Chwatt notes that at the turn of the last century only twenty of the approximately four hundred species of *Anopheles* known today had been described; see Bruce-Chwatt, "Leland Ossian Howard" (cit. n. 1), p. 217. On Dyar see Marc E. Epstein and Pamela M. Henson, "Digging for Dyar: The Man Behind the Myth," *Amer. Entomol.,* Fall 1992, *38:*148–169; "Dyar, Harrison Gray," in Essig, *History of Entomology* (cit. n. 1), pp. 608–610; Mallis, *American Entomologists* (cit. n. 28), pp. 323–326; and Harrison Gray Dyar Papers, Record Unit (RU) 7101, Smithsonian Institution Archives (SIA), Washington, D.C. On Knab see Mallis, *American Entomologists,* pp. 393–394; and Frederick Knab Papers, RU 7108, SIA.

Stuart McCook has aptly called "creole science" by drawing on the expertise of Latin American and Caribbean scientists in ways that challenge stark notions of imperial science and medicine as hegemonic. The eventual result was a massive four-volume study, *The Mosquitoes of North and Central America and the West Indies,* coauthored by Howard, Dyar, and Knab, that appeared between 1912 and 1917. Building on the currency of the mosquito vector theories as well as the U.S. military, engineering, and agricultural expansion into the tropics, these federal entomologists produced a landmark study in mosquito systematics and ecology.[34]

Panama played an important if surprisingly belated role in the production of *The Mosquitoes of North and Central America and the West Indies.* When the Carnegie Institution money ran out after three years, the project was far from complete and in need of supplemental funding. The USDA provided some of that funding, but the authors also had to look elsewhere. Ironically, one of the notable gaps in their knowledge as of 1906 was Panama. During the first several years of the study, which coincided with the American entry into Panama, Howard had asked several amateur observers in the Canal Zone to send along mosquito information and specimens, but they proved unreliable. Sanitarians in Panama were focused on controlling the yellow fever mosquito, *Aedes aegypti,* which bred almost exclusively in artificial containers in and around human habitations. As a result, though broad mosquito control efforts were well under way, there had been little effort to survey the Canal Zone's mosquito fauna or to figure out which *Anopheles* species were responsible for malaria transmission. As it turned out, 1906 was a pivotal year in this regard: as a result of *Aedes aegypti* control, 1906 saw the last case of yellow fever of local origin during the construction era. But it was also a terrible year for malaria, with 21,934 hospitalizations and 195 deaths reported in the official statistics. The need to shift control efforts to *Anopheles* mosquitoes thus meshed neatly with the scientific goals of Howard, Dyar, and Knab.[35]

As the ICC'S Sanitary Department refocused on malaria, Howard and his coauthors arranged for a USDA entomologist named August Busck to visit the region and make a thorough survey of the mosquitoes there. Busck was born and educated in Demark, where he received his doctorate in 1893. He then relocated to the United States and, after a brief stint in the flower business, became an entomological assistant for the USDA. He also served as a specialist in Microlepidoptera for the National Museum, and he had traveled

[34] Leland O. Howard, Harrison G. Dyar, and Frederick Knab, *The Mosquitoes of North and Central America and the West Indies,* 4 vols. (Washington, D.C.: Carnegie Institution, 1912–1917). Information on the history of the project is from Vol. 1, pp. 2–6. Howard cites important Latin American figures in the introduction to his book *Mosquitoes* (cit. n. 32). See also his correspondence in NARA RG 7. On "creole science" see McCook, *States of Nature* (cit. n. 9); Peard, *Race, Place, and Medicine* (cit. n. 8); and Steven Palmer, *From Popular Medicine to Medical Populism: Doctors, Healers, and Public Power in Costa Rica, 1800–1940* (Durham, N.C.: Duke Univ. Press, 2003).

[35] LePrince and Orenstein, *Mosquito Control in Panama,* pp. 23–24. Howard was in touch with several people in Panama from the earliest days of the construction period. In early 1906, he scolded Joseph LePrince for not sending along mosquito samples, making clear that his interests went well beyond the applied: "But really, my dear Mr. LePrince, I have been so disappointed that you have not sent me a large collection of mosquitoes which you possibly might have captured, not in the towns but along the line of the Canal, in the woods or in the swamps. You see my interest is not only the medical and sanitary interest, but the zoological and entomological interest, and my monograph will cover all mosquitoes, whether they carry disease or whether they have not yet proven to have any such capacity. Can you not get some of them for me, and can you not interest other people in the search?" L. O. Howard to Joseph LePrince, 6 Jan. 1906, NARA RG 7, Box 33, Folder "L" [1 of 6]. Bureau of Entomology correspondence also contains letters between Howard and Dr. Arthur Kendall, a sanitary official with the ICC, that find Howard requesting the collection and shipment of Canal Zone mosquitoes as early as March 1905; see, e.g., Howard to Arthur Kendall, 7 Feb. 1906, NARA RG 7, Entry 6. Howard was also in touch with Major Louis LaGarde about mosquito collecting; see Howard to Major Louis A. LaGarde, Medical Department, Ancon, n.d., Box 32, Folder "L" [1 of 3].

extensively throughout Latin America and the Caribbean in the early twentieth century as a member of several scientific expeditions. Busck visited Panama from April to July 1907, which allowed him to collect during the first part of the rainy season. The ICC funded the trip, an indication that they appreciated its sanitary importance. Upon his arrival Busck reported immediately to Gorgas, who arranged for an inspection tour of several days. Then Busck settled into ICC accommodations at the town of Tabernilla, around which he did most of his own collecting. Busck received assistance from Joseph LePrince, the ICC's chief sanitary inspector, and LePrince's assistant, Herbert Canfield. The British-born LePrince had immigrated to the United States and received engineering degrees from Columbia in 1898 and 1899. He served as Gorgas's assistant during the Cuban sanitary campaign before following Gorgas to Panama, where he directed most of the mosquito control work. (After the canal's completion, LePrince went on to work for the U.S. Public Health Service, mostly on malaria-related mosquito control efforts in the U.S. South, where he was a key figure in bringing the mosquito control techniques of Panama back to the United States.)[36] LePrince instructed district sanitary inspectors to pass along samples of collected mosquito larvae to Busck, who then bred out adult mosquitoes. It was largely from these hatched larvae that Busck made his species identifications. He then sent his samples to Dyar and Knab for confirmation, and they made them a part of the Smithsonian's National Insect Collection.

Busck's expedition to Panama was a great success, particularly from a taxonomic standpoint. "Our knowledge of the mosquito fauna of the Isthmus of Panama has heretofore amounted to practically nothing," Dyar and Knab noted in an article about Busck's collecting trip; prior to Busck's survey, only seven mosquito species had been identified within the Canal Zone, but "Mr. Busck returned with more than 90 species, of which 30 were new to science." Busck also was able to recruit and train responsible collectors, including Allan Jennings of the Sanitary Department, who worked as Busck's assistant and went on to serve as an "entomologist" for the ICC.[37] Finally, physicians took note of Busck's presence and exhibited a hunger for entomological knowledge. For example, Dr. Perry Preston of the Empire District wrote the following note to Busck during his Panama stay: "Mr. Goodale, Sanitary Inspector of Empire, has been telling me about your collection of mosquitoes. If it is in no way an imposition on my part I would like very much to call on you in Tabernilla and see them. I have a number of circulars published by Prof. Howard and I would like to bring them with me and have you explain them to me." Beyond charting isthmian mosquito fauna, then, Busck's visit was a critical one for linking entomology and sanitation: it gave sanitarians a deeper appreciation of the entomological landscape; it gave entomologists a thorough look at early mosquito control efforts in Panama; and it trained sanitary workers in the careful arts of larvae collection and hatching, mosquito identification, and the location of breeding grounds.[38]

[36] On Busck see Mallis, *American Entomologists* (cit. n. 28), pp. 326–327; and August Busck Papers, RU 7129, SIA. On LePrince see F. E. Jackson and Son, comps. and eds., *The Makers of the Panama Canal* (Washington, D.C.: Division of the History of Technology, National Museum of American History, Smithsonian Institution, 1911); and Patricia M. LaPointe, "Joseph Augustin LePrince: His Battle against Mosquitoes and Malaria," *West Tennessee Historical Society Papers,* Dec. 1987, *41*:48–61.

[37] Harrison G. Dyar and Frederick Knab, "Descriptions of the New Mosquitoes from the Panama Canal Zone," *Journal of the New York Entomological Society,* Dec. 1907, *15*:197–212, on p. 197. See also Howard *et al., Mosquitoes of North and Central America and the West Indies* (cit. n. 34), Vol. 1, p. 5. Although Jennings became one of the most important entomological workers in the Canal Zone, particularly in relation to malaria control, I have not been able to find biographical information on him.

[38] Perry Preston to Busck, 11 May 1907, Busck Papers, Box 1. For details of Busck's 1907 trip see August

Panama had an extensive mosquito fauna, and to the entomologically uninformed all mosquitoes seemed a threat to public health. Indeed, given how little entomologists knew about mosquitoes, it is not surprising that the Americans failed to differentiate between species in their early control efforts. As Busck noted, aggressive mosquito control efforts were already under way in 1907, with brushing (clearing vegetation around settlements), drainage, oiling of potential breeding areas (covering standing water with a thin sheen of oil to asphyxiate larvae), and screening of quarters being the primary approaches taken. These tactics had been somewhat effective in reducing mosquito numbers, but Busck insisted that public health indictments be handed down only to the guilty species. There were pressures for fiscal efficiency, particularly after George Goethals took over as the ICC's chief engineer in the spring of 1907, and targeted mosquito control efforts promised to be most cost effective. As Busck put it:

> With all due credit to the truly excellent work and the undeniably brilliant results achieved, the work is nevertheless done more or less in the dark, at present, for lack of accurate knowledge of the enemy. It could undoubtedly be made both more effective in some ways and less expensive in others through a more intimate knowledge of the mosquitoes concerned, toward which the present investigation has made but a small beginning.

With his careful cataloguing of Canal Zone species and his attention to the "intimacies" of mosquito ecology, Busck moved the inquiry away from a broad assault on Panamanian nature and toward one that targeted specific species and their breeding grounds.[39]

In his final report Busck did note, as many other knowledgeable observers had and would, that the Panamanian climate and environment were conducive to mosquito life. Consistently high temperatures allowed for year-round breeding, and heavy rainfall during the eight-month rainy season meant that there was plenty of standing water in which mosquitoes might breed, even during the dry season. Moreover, Panama's soils often drained poorly, exacerbating problems with standing water. Panama's various natural bodies of water also supported the prolific growth of algae and aquatic plants, which provided mosquito larvae with food sources and some protection from predacious fish. Finally, tree holes and epiphytic plants in Panama's forests, and crab holes along the coastline, collected water and provided breeding habitat for certain mosquitoes, including several *Anopheles* species. There was much about Panamanian nature, in other words, that supported mosquito breeding.[40]

But Busck also noted that human-induced environmental change was critical to mosquito breeding success in Panama. He noted that "the ever-changing conditions due to the canal work are a continued source of trouble" and that "the progress of each steam shovel or of each of the extensive dumps produces new problems to be solved in the way of

Busck, "Report on a Trip for the Purpose of Studying the Mosquito Fauna of Panama," in *Smithsonian Miscellaneous Collections*, 1910, *52*(1921):49–77. See also Dyar and Knab, "Descriptions of the New Mosquitoes from the Panama Canal Zone"; and Harrison G. Dyar, "The Mosquitoes of Panama," *Insecutor Inscitiae Menstruus*, July–Sept. 1925, *12*(7–9):101–195.

[39] Busck, "Report on a Trip for the Purpose of Studying the Mosquito Fauna of Panama," p. 53. My focus on the term "intimacies" is meant to link this scientist–environment interaction with the sorts of social and sexual interactions upon which Ann Laura Stoler has focused in much of her work. For a discussion of her approach to empire see Stoler, "Empires and Intimacies: Lessons from (Post) Colonial Studies—A Round Table," *J. Amer. Hist.,* 2001, *88*:829–897.

[40] Busck, "Report on a Trip for the Purpose of Studying the Mosquito Fauna of Panama." See also LePrince and Orenstein, *Mosquito Control in Panama,* pp. 19, 22, 30–38.

drainage." More important, he intimated that mosquitoes breeding in these anthropogenic conditions were more likely to be malaria vectors than those breeding under natural conditions. Of the ninety mosquito species Busck identified, nine were Anophelines. He believed *Anopheles pseudopunctipennis* to be the most abundant, followed by *A. albimanus* (later observers would reverse this order). Busck did make a few anecdotal comments connecting particular species with malaria transmission. For instance, he noted that a July 1907 surge in *A. albimanus* around La Boca, the result of prolific breeding in "a temporarily dammed up swamp near the laborers' quarters," had corresponded directly with a spike in malaria among the workers quartered there. But entomologists and sanitarians still were not sure which *Anopheles* species were the critical vectors in Panama and which were less important.[41] Busck had done much to lay out the possibilities, but sanitarians awaited definitive experimental proof.

Samuel T. Darling, Chief of the Laboratory of the Board of Health in the Canal Zone, undertook that experimental work. Darling had received his M.D. in 1903 from the College of Physicians and Surgeons in Baltimore, and he arrived on the isthmus in 1905, where he quickly rose to a position of medical leadership. By 1907 he was not only in charge of the ICC lab but also head of the Canal Zone Medical Association, a group of physicians and sanitarians who met regularly to present tropical medical research; their *Proceedings* provide a thorough primary record of U.S. tropical medicine during the construction era. Darling's lab was part of a sophisticated medical infrastructure created in the Canal Zone that allowed doctors, like entomologists, to pursue both scientific and applied approaches to tropical medicine and public health. Indeed, the Canal Zone was an important incubator for American tropical medicine and public health expertise at a moment when tropical medicine was emerging globally as a distinctively imperial discipline.[42] In early 1908 Darling sought permission to engage in a series of experiments to determine the details of "the mechanical transmission of malarial parasites from man to man by mosquitoes." He was granted permission in early February of that year and proceeded to design a set of experiments using Spanish and West Indian malaria patients.[43]

As they had done for Busck, district sanitary inspectors collected *Anopheles* larvae and sent them to Darling, who bred them to adulthood. He then had these adult mosquitoes bite workers with transmissible cases of malaria. Once they did so the mosquitoes were

[41] Busck, "Report on a Trip for the Purpose of Studying the Mosquito Fauna of Panama," pp. 53, 57. Mosquito taxonomy is complex and fluid. Species names have changed over time with new interpretations of what constitutes a distinct species, and systematists are increasingly uncomfortable with rigid species definitions, now often talking about species complexes and lumping together all sorts of slight variations. Because of these complexities, I have chosen to use the species names assigned by the scientists I am examining, noting the changes over the time period I examine but not divergences with present understanding.

[42] On Darling see Enrique Chaves-Carballo, "Samuel T. Darling: Studies in Malaria and the Panama Canal," *Bull. Hist. Med.,* 1980, *54*:95–100. On tropical medicine's history see Roy Porter, "Tropical Medicine, World Diseases," in *Greatest Benefit to Mankind* (cit. n. 30), pp. 462–492. On the history of tropical medicine in Britain see Michael Worboys, "Manson, Ross, and Colonial Medical Policy: Tropical Medicine in London and Liverpool, 1899–1914," in *Disease, Medicine, and Empire: Perspectives on Western Medicine and the Experience of European Expansion,* ed. Roy MacLeod and Milton Lewis (London: Routledge, 1988), pp. 21–37.

[43] Samuel T. Darling to Superintendent, Ancon Hospital, 30 Jan. 1908, NARA RG 185, Entry 30, Box 253, Folder 37-H-7 (I). Specifically, Darling sought permission to "attempt the infection of certain black laborers by mosquitoes that have previously bitten patients with malarial fever," but his 1910 report suggests that he only went through with experiments that involved mosquitoes biting infected patients, after which the mosquitoes were examined to see if the plasmodia developed within them. Darling's study spoke volumes about the racial thinking and public health inequalities that marked the canal-building era, but for the purposes of this essay I want to focus on his entomological findings. See also William Gorgas to Major D. D. Gaillard, 7 Feb. 1908; Gaillard to Gorgas, 11 Feb. 1908; Executive Officer to Darling, 13 Feb. 1908; and Darling to Chief Sanitary Officer, 7 Oct. 1908: NARA RG 185, Entry 30, Box 253, Folder 37-H-7 (I).

removed and, after a necessary incubation period, dissected to see if malaria plasmodia were in their systems.[44] In March 1909 Darling reported his preliminary results, which he expanded on in a lengthy 1910 report, *Studies in Relation to Malaria.* Of the eleven species of *Anopheles* he studied (he worked with two species that Busck had not identified), he found that only three seemed capable of transmitting malaria. He dismissed the tree-hole and sylvan Anophelines, *A. cruzii* and *A. eiseni,* as malarial transmitters; found solely in forested landscapes remote from population centers, they had little contact with humans and seemed to prefer other blood sources, though he had not been able to disprove definitively that they were capable vectors. He also dismissed *A. malefactor,* a relatively abundant species, as a transmitter of malaria; none of the collected *A. malefactor* (despite their name) became "infected" with malaria. Sanitary inspectors and other entomological workers found several other species in such small numbers that, although Darling could not determine if they were capable vectors, he dismissed their importance. Of the three that did prove capable, Darling determined that *A. pseudopunctipennis* was marginally successful as a transmitter of malaria: only 12.9 percent of them became "infected" after biting malarial patients, and few adults were taken in workers' quarters. Darling did note that *A. tarsimaculata* had a high rate of infection acquisition (60 percent), but he said little about its role as a malaria transmitter beyond noting its capacity. Others would later point out that, because of its limited range (it was found mostly on the Atlantic side, in the vicinity of Colón), it was not particularly worrisome.[45]

By far the most important of Darling's findings was the central role in malarial transmission played by *A. albimanus,* which had an infection rate of 70.2 percent. He wrote: "Anopheles albimanus, the common white footed anopheles, is an extremely hardy, vigorous, rapidly developing adaptable mosquito and is the transmitter of Estivo Autumnal [Falciparum] and of Tertian [Vivax] malarial fever in the Canal Zone at this time. In the efforts at mosquito destruction, the extermination of A. albimanus is of paramount importance."[46] Although some of Darling's conclusions were later challenged, none of those challenges affected his essential public health message: when it came to malaria, *A. albimanus* was the major problem. Discerning the breeding and feeding habits of *A. albimanus* thus became a mosquito control priority.

Entomologists were coming to understand that the world's *Anopheles* species showed a wide variety of preferences when it came to breeding habitat. Some preferred clean water, others dirty water; some preferred water exposed to sunlight with lots of algae, others shady pure water, and still others brackish water; some favored natural bodies of water such as lakes and streams, though most avoided swiftly moving water, while others showed an uncanny attraction to artificial containers; and all larvae did best in habitats that provided protection from fish and other predators. As adults, *Anopheles* also showed varied habits, with the most important questions for sanitarians centering on their feeding preferences, their persistence at entering human habitations, and the distance they could fly. Some *Anopheles*

---

[44] Samuel T. Darling, *Studies in Relation to Malaria* (Washington, D.C.: Government Printing Office, 1910), pp. 13–23. See also James Stevens Simmons, *Malaria in Panama* (Baltimore: Johns Hopkins Press, 1939).

[45] Darling to Superintendent of Hospital, Ancon, 18 Mar. 1909, NARA RG 185, Entry 30, Box 253, Folder 37-H-7 (II). On *A. tarsimaculata*'s limited range see LePrince and Orenstein, *Mosquito Control in Panama,* p. 51.

[46] Darling to Superintendent of Hospital, Ancon, 18 Mar. 1909. There are four types of malarial parasites, two of which are important for the Panama case: *Plasmodium falciparum* and *Plasmodium vivax.* The former is a virulent strain of African origin that produces fairly high mortality rates; the latter is a milder strain, mostly of European origin.

seemed catholic in their tastes for mammalian blood, taking it where they could get it. But others showed decided preferences for either human or particular nonhuman blood-meal sources, and their choices in turn shaped disease ecology. Various species also exhibited distinct differences in their capacity to invade human habitations and thus pose a threat as disease transmitters. Finally, determining how far species could fly was critical in designing environmental control measures for areas around human habitations.

Where did *A. albimanus* fit in this range of breeding and adult habits, particularly in comparison with other *Anopheles* species? Busck, Darling, and others had already begun to sketch a portrait of *A. albimanus* as an aggressive, flexible species associated with malarial outbreaks, but it was left to other sanitarians and entomologists to fill out this portrait. Entomological workers already knew that *Anopheles* in general were country mosquitoes, not peridomestic ones like *Aedes aegypti.* Although some *Anopheles* inhabited sylvan or "wild" settings, most were found in and around villages in agricultural areas. But as entomological workers looked closely at the habits of *A. albimanus,* they found that it showed a marked affinity for landscapes of human disturbance. Allan Jennings, Busck's former assistant who came to occupy an important place among entomological workers in Panama, described its breeding habits in detail: "preference is shown by the mosquito for stagnant, fairly pure water, exposed to direct sunlight, with a growth of Spirogyra [algae]." "With the exception of foul or swift water," he continued, "they may occur in almost any collection of water, however small or seemingly unsuited to mosquito propagation. Hoof-prints, wheel-ruts, the smallest puddle or thinnest film of water seeping from the ground . . . are points of danger and must be included in the control work." Implicit in Jennings's description was an important point: human activity, even of the most minor sort, created many of the ideal breeding grounds for *A. albimanus.* In most natural bodies of water where such breeding conditions were found, including most swamps, fish predation controlled *Anopheles* larvae with varying levels of effectiveness. Tree holes, crab holes, and the water in bromeliads were important exceptions, though Darling's experiments had already shown that mosquitoes that bred in such natural conditions were not effective vectors in Panama. But fish infrequently found their way into human-produced depressions, and "mosquitoes instinctively avoid places that fish and other enemies reach easily."[47] Its larval preference for direct sunlight, moreover, meant that *A. albimanus* was not a forest species but one more likely to breed in cleared areas where water collected. There was no shortage of such areas in the Canal Zone.

A. albimanus's adult habits also suggested its troublesome nature. They were found to be aggressive biters with a particular taste for human blood, and they proved to be persistent entrants into human habitations. In surveys of *Anopheles* species captured in worker quarters, *A. albimanus* was by far the most prominent. It was also an efficient vector, achieving rates of infection in laboratory settings that were much higher than those seen in other vectors. *A. albimanus,* Jennings and others found, was a mosquito with an intimate connection to humans and their landscapes. Its preference for human blood, its persistence in entering human habitations, its capacity to breed in cleared landscapes, its absence from uninhabited forested areas, and its efficiency as a malaria vector all marked *A. albimanus* as a mosquito with a long coevolutionary history with humans and the malarial parasite.

---

[47] Allan H. Jennings, "Some Problems of Mosquito Control in the Tropics," *Journal of Economic Entomology,* Apr. 1912, *5:*131–141, on p. 133; and LePrince and Orenstein, *Mosquito Control in Panama,* p. 64. On the distribution of *Anopheles* see, e.g., J. A. LePrince, "Mosquito Destruction in the Tropics," *Journal of the American Medical Association,* 26 Dec. 1908, *51:*2203–2208.

Moreover, extensive human activities in the Canal Zone—the importation of tens of thousands of workers (lots of food) and the transformation of the Canal Zone environment (lots of breeding sites)—may well have favored its propagation over that of other mosquito species, including less dangerous Anophelines. To see this particular mosquito species as merely a pernicious product of tropical nature, Jennings intimated, was to miss both its evolutionary and ecological relationships to humans and the role of canal construction in facilitating its ubiquity. "While not domestic in the same sense as *Stegomyia calopus* [an earlier name for *Aedes aegypti*]," he wrote, "*Anopheles albimanus* is closely associated with man and finds its most congenial surroundings about his habitations and in conditions he creates in the course of agricultural, engineering and other work. This fact is correlated with the highly-developed blood-sucking habit and has been an active factor in its development and in establishing the economic importance of the species."[48] *A. albimanus* was a hybrid creature, difficult to explicate within an intellectual framework that separated tropical nature from U.S. efforts to transform the region. It was a nonhuman agent, to be sure, but its abundance in Panama is better seen as the product of a network of human and natural agencies and structures at work in the Canal Zone.

The most detailed case for seeing human contributions in the creation of malarial conditions in Panama came in *Mosquito Control in Panama,* the 1915 book cowritten by LePrince and his assistant, A. J. Orenstein. Orenstein, only in his mid-twenties when he first arrived in Panama in 1905 to work under Gorgas and LePrince, was an M.D. with degrees from the Jefferson Medical College in Philadelphia. Their mosquito control work led them to conclude that malaria "develops most rapidly when the soil is disturbed by large and extensive excavations and fills accompanied by the introduction of non-immune labor housed near the site of their work."[49] Here was not only a succinct summary of the potential for canal building to exacerbate malarial outbreaks but also a partial vindication of the discredited miasmatic theory's contention that turning over and disturbing soil would contribute to disease in Panama. Excavation had caused disease, but not because it exposed poisoned soil and let loose malarial emanations. Rather, it amplified the breeding success of the mosquitoes most productive of malarial transmission in Panama.

But LePrince and Orenstein went further. Building on the accumulated observations of district inspectors, they mapped out specific aspects of the Canal Zone's engineered landscape that had contributed to vector breeding.[50] For instance, ICC officials relied on a serpentine and ever-shifting network of rail lines for the rapid removal of materials excavated from sites such as the Culebra Cut. As a result, the construction landscape became riddled with corduroy-like rail-tie depressions that, when filled with rain, became ideal vector breeding places. Spoil dumping was also a problem, as it often blocked watercourses and drainage paths, creating new swamplands that entomological workers then identified as troublesome breeding sites (see Figure 1). District inspectors frequently complained of

[48] Jennings, "Some Problems of Mosquito Control in the Tropics," p. 133.

[49] LePrince and Orenstein, *Mosquito Control in Panama*, p. 39. After serving in Panama, Orenstein moved to South Africa, where he spent most of the rest of his long life (he died in 1972) working to improve public health, with a particular focus on the respiratory diseases that plagued the mining industry. See "Obituary: Major-General Alexander J. Orenstein," *Transactions of the Royal Society of Tropical Medicine and Hygiene,* 1972, *66*:815–817; and Randall Packard, "The Invention of the 'Tropical Worker': Medical Research and the Quest for Central African Labor on the South African Gold Mines, 1903–36," *Journal of African History,* 1993, *34*:271–292.

[50] The Canal Zone was divided into a series of sanitary districts, each with its own sanitary inspector and staff. These inspectors made regular reports to LePrince, who was the chief sanitary inspector, that were filled with accounts of mosquito breeding sites and conditions. These reports and related correspondence can be found in NARA RG 185, Entry 30, Boxes 242–244. There are individual folders for each sanitary district.

Result of blocking the Rio Cardenas. A change of species of *Anopheles* occurred here

Figure 1. *This was one of several photographs that LePrince and Orenstein included in their book to illustrate the ways in which the blocking of rivers and other watercourses as a result of spoil dumping and other construction activities could create new mosquito breeding habitat. In this case, they note a shift in* Anopheles *species as a result of the altered conditions. (From LePrince and Orenstein,* Mosquito Control in Panama, *opposite page 72.)*

spikes in *Anopheles* numbers and corresponding malarial outbreaks that coincided with the creation of these new wetlands. Extensive excavation also caused frequent landslides that similarly disturbed hydrological patterns, and as the canal cut went deeper it too began to collect water in ways that supported vector breeding (see Figure 2). One of the most vexing engineered breeding grounds, LePrince and Orenstein suggested, was a result of dredging. At various sites along the canal route, the ICC used hydraulic dredges that scoured the canal bottom and pumped tons of muddy slurry onto lands adjoining the canal. When the surface mud on these slurry fields began to dry it cracked, and water collected in the cracks. These watery crevices, exposed to sunlight but difficult for fish to reach, were ideal for *A. albimanus* breeding, and they were exceedingly difficult to mitigate, as workers who attempted to walk on the slurry fields would be quickly swallowed up by the mud (see Figure 3). Finally, the filling of Lake Gatun and Miraflores Lake, the two major reservoirs along the canal route, also proved troublesome from a mosquito-breeding standpoint, as these reservoirs created shallow edge environments, littered with flotsam, that favored vector breeding. All these various aspects of canal construction, then, created specific landscapes of breeding success.[51]

Even mosquito control efforts could, if not executed properly, be more of a problem

[51] These details are from LePrince and Orenstein, *Mosquito Control in Panama,* particularly pp. 43–78. Margaret Humphreys documents similar problems in the U.S. South as hydropower lakes filled and malarial rates increased there in the early twentieth century. See Humphreys, *Malaria* (cit. n. 21), pp. 88–92.

Constantly changing topography: near Miraflores

Figure 2. *The canal cut itself often held standing water and created new breeding sites for* Anopheles albimanus. *(From LePrince and Orenstein,* Mosquito Control in Panama, *opposite page 38.)*

than a solution. One of the most important mosquito control tactics was drainage. U.S. sanitarians cut ditches, laid tile, and sometimes even paved culverts in areas of persistent standing water. When well constructed and maintained, these drainage works could reduce mosquito breeding substantially. But when poorly constructed or maintained, they often exacerbated the problem. If, for instance, vegetation built up in the ditches, or if they were not of a sufficient grade, water could collect in pools, again providing favorable breeding conditions (see Figure 4). And such drainage problems were frequent; improper construction and maintenance of drainage systems was one of the biggest concerns voiced by the entomological workers who monitored mosquito breeding.[52]

LePrince and Orenstein concluded their lengthy summary by noting: "The topography, meteorological conditions, and constant changes due to the construction work, together with the character and constant moving of the population and their dwellings, and social conditions, were particularly unfavorable to the control undertaken."[53] This measured conclusion, implicating as it did the work of canal construction in exacerbating *Anopheles* breeding, stands in sharp contrast to the facile rhetoric of tropical conquest that many Americans engaged in. And while this nuanced perspective, gained by paying careful attention to the material environment, did not lead entomological workers to confront that rhetoric head on, it did create some revealing tensions among American administrators. Entomological workers constantly had to track down and apply control methods to new

[52] LePrince and Orenstein, *Mosquito Control in Panama,* pp. 71–72.
[53] *Ibid.,* p. 218.

Cracks in a hydraulic fill

Figure 3. *Hydraulic fill lands adjoining the canal route were among the most troublesome vector breeding grounds, as they were tremendously difficult for mosquito control workers to mitigate. (From LePrince and Orenstein,* Mosquito Control in Panama, *opposite page 120.)*

breeding grounds created by canal construction, and, as sanitary records reveal, they frequently scolded other ICC officials who took little heed of these anthropogenic hazards. LePrince and Orenstein also intimated that entomological workers considered "social conditions" and their relations to mosquito control, though they did so only in a shallow way. While entomological workers often resorted to racialized generalizations about the habits of nonwhite workers, whose alleged sanitary carelessness and moral failings threatened not only their own health but that of white Americans, thorough mosquito control also necessitated that they look carefully into mosquito breeding wherever it threatened worker health in the Canal Zone. In ways both expansive and limiting, the strategy of mosquito control privileged biological over social explanations, but it did not completely ignore the latter.

THE GREAT GRASS-CUTTING DEBATE

Tensions between entomological workers and other ICC administrators came to the fore in an unlikely debate over how short to cut the grass in the Canal Zone. Controlling adult mosquito populations was a critical aspect of the U.S. sanitary program, and the most important step taken here was brushing and grass cutting around work sites, homes, and barracks. Entomological workers had shown that such environments provided adult mosquitoes with daytime resting places and protection from wind and other elements and that, in some cases, long grass hid watery areas ideal for breeding. Eliminating such conditions in the immediate vicinity of work sites and human habitations reduced mosquito numbers substantially. Casual observers of this brushing and grass-cutting work often couched it in the language of conquest, as an effort to control the jungle's constant predatory advances

A ditch at Ancon: weeds retarding water flow

Figure 4. *Ditches of various sorts were critical to draining wet or swampy lands that bred mosquitoes in the Canal Zone. But when they were inadequately constructed or maintained, they could also themselves be troublesome breeding sites. This photo shows weeds retarding water flow, which could create pools of water ideal for vector mosquito breeding. (From LePrince and Orenstein,* Mosquito Control in Panama, *opposite page 20.)*

on human settlements. Indeed, sanitary clearing intersected in important ways with a larger landscaping effort in the Canal Zone, which sought to control and domesticate tropical vegetation as a symbol of American environmental mastery. But grass cutting also led to conflict between entomological workers, interested in the specifics of mosquito ecology, and those workers more concerned about the landscape aesthetics of environmental, and imperial, control.

The controversy began in 1908, during George Goethals's reorganization of the ICC and his efforts to streamline operations. During the first several years of canal construction, brushing and grass-cutting responsibilities were divided between the Department of Sanitation, which focused on mosquito control, and the Department of Labor, Quarters, and Subsistence, which cleared brush and cut grass around worker quarters for aesthetic reasons. When he became ICC chairman Goethals insisted, for the sake of efficiency, that all such work be brought under the control of the renamed Quartermaster's Department (QMD), with sanitary inspectors making specific requests to the QMD when sanitary cutting was necessary. Gorgas immediately raised several concerns about this new system.

He disliked the idea of putting mosquito control work in the hands of a department that did not appreciate its urgency or understand its specifics. He also questioned whether the QMD would carry out sanitary requests quickly and carefully, and he worried, presciently, that constant and often urgent requests from his department would become an irritant.[54]

Gorgas's sanitary grass-cutting goals were quite different from those that drove the QMD's work. Sanitarians did not worry until grass grew to a foot or more in height, the point at which it effectively harbored mosquitoes. This meant that they cut less often, focused their cutting in areas that served as critical mosquito habitat, and determined the areas to be cleared, usually a perimeter of 100–200 feet around housing and work sites, on the basis of observations of how far vector mosquitoes could fly. Under pressure to be cost efficient, entomological workers let their observations of mosquito behavior determine the extent of their brushing and cutting. Aesthetics were beside the point. Conversely, the QMD was more interested in neatly trimmed lawns in and around ICC facilities, particularly those that served and housed white American workers and their families, whether they were critical to mosquito control or not. While sanitarians trumpeted the need for precision mosquito control based on an intimate knowledge of Panama's malaria vectors, they feared that ignorant administrators were more interested in the aesthetic symbolism of unruly jungle giving way to finely cropped lawn.[55]

Gorgas's fears soon were realized. Within months of the reorganization, district sanitary inspectors complained that the QMD attended to requested sanitary cutting slowly, if at all; was sloppy in its execution—allowing, for instance, grass cut in and around ditches to block water flow; and gave excessive attention to cutting grass short near gold (white) married quarters while neglecting other important sanitary cutting in areas that bred and harbored vector mosquitoes. Indeed, whereas entomological workers seemed determined to provide thorough mosquito control, the QMD mostly manicured white American settlements, ignoring those areas where silver roll workers lived, which were often near disturbed and watery environments conducive to mosquito breeding. Moreover—most exasperatingly to Gorgas—the charges to the Sanitary Department for grass cutting had increased considerably since the reorganization, even as the QMD did the work less effectively. Indeed, Gorgas discovered that the QMD had taken to charging all grass cutting to his department under the assumption that its purpose was mosquito control, even though Gorgas thought that little of it was of any sanitary value.[56]

By 1910, two years after Goethals initiated his reorganization, Gorgas was ready to call the arrangement a failure. In a frank letter to Goethals, he explained: "Formerly grass cutting for sanitary purposes was done by the Sanitary Department, and that around the quarters, whose object was aesthetic, was done by the Quartermaster's Department. At present it is all done by the Quartermaster's Department from sanitary funds. The tendency of the local quartermaster, therefore, is to concentrate his attention upon grass cutting around quarters, and neglect that for sanitary purposes." Gorgas was quick to note that the QMD had "no especial knowledge of mosquito work," making it clear how tied he had become to seeing entomological expertise as proprietary, a part of what made entomolog-

---

[54] For correspondence outlining this controversy see NARA RG 185, Entry 30, Boxes 241–242, Folders 37-B-7 (I–V).

[55] The beginnings of this debate are laid out in Chief Sanitary Officer [Gorgas] to Chairman, Isthmian Canal Commission [George Goethals], 21 July 1908, NARA RG 185, Entry 30, Box 242, Folder 37-B-7 (I). Succeeding correspondence in this and following folders fills out the story.

[56] The billing situation is laid out in a series of letters and memos written in the fall and early winter of 1908: NARA RG 185, Entry 30, Box 241, Folder 37-B-7 (I).

ical workers both effective and necessary. "As I go about," he continued, "I am daily impressed, in all the details of execution, with the advisability of having the men in immediate charge of the work educated more or less in the habits of the mosquito so that they can carry out the measures understandingly." The QMD had few such men—and the results showed it, as official malaria rates for the critical months of June through August, in decline from 1906 to 1909, had risen again in 1910. Such recidivism, Gorgas believed, was the result of mosquito control work having been taken away from the entomologically informed.[57]

Goethals responded to Gorgas's concerns by creating a commission to study the issue. After hearings and extensive testimony, the commission recommended several changes to the system—including a better method to account for who ought to be billed for particular grass-cutting jobs. But the commission's final report nonetheless supported the reorganization, suggesting that there was not enough statistical evidence to prove that the shift in responsibility had been detrimental to public health in the Canal Zone—and particularly to the health of white American workers. Ultimately, Goethals refused to buy Gorgas's argument that, almost regardless of cost, sanitary work must be in the hands of those experts best equipped to limit mosquito breeding. That was insult enough. But then, in inquiring about the sanitary value of the QMD's obsessive grass cutting near gold quarters, Goethals revealed just how tenacious miasmatic perceptions could be: "Do you think the relief from mists gained by close grass cutting," Goethals asked Gorgas in a 1911 letter, "is of any sanitary or hygienic value?" "The only knowledge I have that short grass decreases mist," Gorgas responded with barely concealed exasperation, "is the statement in your letter to that effect. But accepting your statement that such is the case, I do not think that down here the presence or absence of mists is of any hygienic importance."[58]

This grass-cutting debate, however mundane, raised important tensions of empire, in the realms of discourse and practice. The battle over grass cutting provided a concrete example of how the dominant U.S. ideology of tropical conquest, manifest here as a landscape aesthetic, came into conflict with the perceived ecological dictates of mosquito control. For many American officials, closely manicured grounds not only reflected environmental mastery, the subjugation of a tropical jungle and its apparent threats to the health and enterprise of temperate white people; such a landscape *reproduced* the very distinction between hostile tropical nature and human mastery that informed its creation. As the canal neared completion, the journalist Albert Edwards celebrated this appearance of mastery: "On the sides of the hills," he wrote, "you see villages—clusters of homes, well-kept lawns where all that is beautiful in the jungle has been separated from what is noxious."[59] Here was the mainspring of the tropical ideology in the Canal Zone—a landscape of settlement that segregated the malignant tropics, environmentally and socially, from the benign and bountiful tropics (see Figure 5). The irony, of course, is that the white American landscape that Edwards celebrated—one that included well-built and well-

---

[57] Chief Sanitary Officer to Chairman, ICC, 21 Oct. 1910, NARA RG 185, Entry 30, Box 242, Folder 37-B-7 (I). On page 5 of this letter Gorgas cited yearly malarial hospitalization rates for the months of June–August, pointing out that the rate in 1909 had been 219 per 1,000 and that it had risen to 344 per 1,000 in 1910. But figures for the entire year showed a slight decline in malaria hospitalizations from 1909 to 1910. For a complete run of those statistics see W. C. Gorgas, *Report of the Department of Sanitation of the Isthmian Canal Commission for the Year 1913* (Washington, D.C.: Government Printing Office, 1914).

[58] Goethals to Gorgas, 9 Oct. 1911, NARA RG 185, Entry 30, Box 242, Folder 37-B-7 (III). On the larger controversy see NARA RG 185, Entry 30, Box 242, Folders 37-B-7 (III, V); and Gorgas to Goethals, 23 Oct. 1911, NARA RG 185, Entry 30, Box 242, Folder 37-B-7 (III).

[59] Albert Edwards, *Panama: The Canal, the Country, and the People* (New York: Macmillan, 1913), p. 512.

.Types of screened houses: Colon Hospital grounds

Figure 5. *This photograph of the Colón Hospital grounds illustrates the landscape that the QMD grass-cutting operations aimed to create around gold quarters and other critical U.S. facilities. (From LePrince and Orenstein,* Mosquito Control in Panama, *opposite page 204.)*

screened houses and clean, orderly streets as well as close-cropped lawns and other domesticated tropical plants—intersected only obliquely with the landscape of mosquito control. The grass cutting of the QMD was designed less to solve a discrete public health problem than to reinforce an environmental ideology of tropical conquest. Or, perhaps more accurately, many in the ICC assumed that pursuing their landscape aesthetic was an adequate approach to protecting a narrowly defined public health.

Entomological workers upset this logic at a couple of levels. Gorgas, their voice in this debate, argued that there were not only differences but real tensions between the QMD's landscape aesthetic and the work of mosquito control. There was no doubt that the work the QMD did around white quarters provided some protection against mosquitoes, but its expense and limited extent made it inadequate to the larger goal of eliminating vector mosquitoes wherever they threatened worker health. If the primary goal was improved public health, then much of the grass cutting accomplished by the QMD was either useless or inefficient, Gorgas argued. Moreover, entomological workers produced a landscape of mosquito control that was barely legible to those looking for an aesthetic of imperial environmental mastery. Their sanitary landscape was spotty and ragged; in it, the wild and tamed tropics could not be so easily separated and discerned. In this sense, it mirrored the ecology of *A. albimanus* breeding it was designed to mitigate. The unkempt sanitary landscape to which the QMD implicitly objected, then, was an apt symbol of how attending to vector ecology in Panama had made murky not only the aesthetic boundaries between tropical nature and its conquest through modernization, but also those between natural and human agency.

But that was not the only boundary that entomological workers, in their focus on mosquito control, softened and complicated. Their approach was also less beholden to another dichotomy critical to the logic of tropical triumphalism: that between the settlements of white gold employees—"the men we are protecting," to borrow Henry Rose Carter's telling phrase—and the areas where those on the silver rolls lived and worked. The QMD's maintenance of nice lawns around white American worker settlements functioned to create and perpetuate a landscape of racial as well as environmental differentiation, though the labor (much of it nonwhite) and expense that went into producing such manicured American enclaves was often obfuscated by a logic of differing racial capacities and drives. To many white observers, such lawns seemed like natural expressions of their racial superiority. Entomological workers' grass-cutting strategies worked at cross-purposes with this symbolic project. Critically, their attention to mosquito ecology did *not* lead entomological workers into a reasoned critique of the sharp racial and civilizational taxonomies so central to tropical thinking at the time. But their single-minded goal of eliminating vector mosquitoes did lead them to extend their work beyond white quarters to take account of breeding places wherever they threatened worker efficiency.

Because nonwhite workers were more likely to live near disturbed work sites in substandard housing where mosquitoes thrived, entomological workers had to pay some attention to the relationship between those "social conditions" and mosquito breeding. Though they often chalked up the persistence of a malarial reservoir among nonwhites to particular racial habits, sanitary inspectors and other entomological workers nonetheless worked to prevent and mitigate breeding conditions in and around silver quarters as well as in some "native towns," existing settlements in the Canal Zone that had swelled with the arrival of tens of thousands of workers and their dependents. By seeing mosquitoes as the problem, and by suggesting that landscape change had as much to do with malaria as tropical nature *per se,* entomological workers pursued a *relatively* color-blind course—not because they were critical of the racial logic behind the American tropical ideology or the medical and environmental inequalities everywhere evident in the Canal Zone, but because the taxonomies that interested them were less racial or civilizational than entomological. Entomological workers' relative color-blindness thus cut several ways. Compared to other U.S. officials, they were less likely to trace an effective sanitary perimeter atop the color line. But they were also largely blind to the broader set of power relations and networks of agency that tended to put nonwhite workers in harm's way.

Entomological workers' attention to mosquito ecology thus could be as reductive and compartmentalizing as it was challenging to tropical triumphalism. For them, the presence of particular *Anopheles* species *meant* the presence of malaria, and vice versa. Their attention to vector species in no way constituted a full or objective view of malaria and the forces that produced it. They came to know intimately only part of the malarial cycle, the part that occurred in mosquito bodies and the environments they required to breed and subsist. They paid less attention to what happened once malaria was in workers' bodies and to the socioeconomic determinants of malarial demography—low pay, crowded and substandard housing, and objectionable food among them. While their ecological focus encouraged entomological workers to attack mosquito breeding throughout the Canal Zone, not just in white areas, it also limited their capacity to look beyond the blinders of their ecological expertise to contend with the other social, economic, and racial dimensions of malarial parasitism and transmission.

Entomological workers' reductive focus on mosquito control also opened them to administrative critiques. While it might be tempting to portray the grass-cutting debate as one pitting heroic purveyors of nature's truths against ignorant senior administrators focused on expediency, it is important to recognize that entomological workers had a goal several layers removed from that of Goethals. For entomological workers, success was the elimination of vector mosquitoes; for Goethals, success was a completed canal, and mosquito control was only a means to that end. He was not going to pursue every last vector mosquito unless it was cost efficient to do so. Thus, while entomological workers gave voice to the material conditions that produced vector mosquitoes in Panama, they did so only by blocking out other voices and considerations. To know the "nature" of the problem was not to know it all.

CONCLUSION

The most important achievement of entomological workers in Panama was to control vector mosquitoes, which helped to eliminate yellow fever and greatly reduce malaria within the Canal Zone. The official malaria morbidity rate declined from a peak of 821 per 1,000 in 1906, just before Busck's trip and the beginnings of precision mosquito control, to 76 per 1,000 in 1913, as the canal neared completion.[60] Entomological workers, through careful observation and experimentation, came to understand that protecting workers against malaria necessitated not the control of tropical nature *per se* but, rather, the management of particular hybrid landscapes to which American enterprise substantially contributed. Where the logic of tropical triumphalism naturalized U.S. interventions in Panama and suggested that modernization was the surest path to unlocking the promise of the tropics, entomological workers noted that the environmental impacts of modernization-in-process were at the heart of the "tropical" threat to public health and that mosquito control in Panama involved conquering a nature that Americans helped to make.

Entomological workers were both critical *to* U.S. imperial success in Panama and, at times, implicitly critical *of* the environmental ideology that naturalized that success. Given that, how do we make sense of the positions and roles of these scientific workers as imperial actors? Were they agents of empire whose accomplishments served the goal of American imperial control, or were they critics of empire whose pursuit of truth challenged the fallacies of American tropical ideology? Neither characterization fits. Entomological workers had much to gain, in terms of professional status, by serving American expansion, a formula that had already built for them a central place in the federal scientific establishment. But entomological workers were not entirely captured by the instrumentality of the U.S. administration in Panama. Indeed, they insisted that their ecological expertise was built on a loyalty to nature and its workings, not solely to the U.S. mission or its animating environmental ideology. Just as important, entomological workers were loyal to the modernizing logic of science itself—the notion that science was a path to truth, administrative efficiency, and progress—and they increasingly identified with an international and trans-imperial community of scientists and sanitarians who saw themselves as crucial to controlling the tropics. While this could put them at odds with other national-imperial actors, as the grass-cutting debate suggests, it also meant that they were strongly inclined to an interventionist developmentalism that made their work to correct fallacies in tropical thinking its own kind of conquest. Finally, their loyalty to nature limited how far they could push into the world of the hybrid, for they still defined themselves as experts on nature

---

[60] Gorgas, *Report of the Department of Sanitation of the Isthmian Canal Commission for the Year 1913* (cit. n. 57), p. 6.

and thus avoided fully facing the "social conditions" that also contributed to malaria in Panama.

The scientific sensibility of entomological workers in Panama made them imperial agents of a particular kind, at once dutifully contributing to the goal of imperial control and developing an independent streak that made them agents for change. Their motives, like those of many imperial actors, emerged from competing identities, and their experiences on the ground challenged their ideological presuppositions. That men like Howard and Gorgas joined the triumphalist chorus when the canal was finished is not surprising. Claiming that mosquito control conquered the tropics for white northerners was a more satisfying and succinct boast than claiming that it resolved a set of problems created by a large-scale engineering project that relied on huge labor importations and massive disturbances under a particular set of environmental and social conditions. But to listen only to their triumphalist words would be to privilege their discourse over the material dimensions of imperial control to which they paid such close attention.

By carefully following entomological workers through the Panamanian environment that was their laboratory, we see environment intruding as a causal force, obliquely and in a networked form but also distinctly. The practitioners of American tropical triumphalism often misconstrued the nature that they sought to make culpable not only for disease but also for underdevelopment. That is something current science can certainly help us to explain, as we work with a more expansive knowledge base and a different set of cultural filters when we think and write about the tropics today. But it is more theoretically satisfying, if one's goal is coherently to link the material and the discursive in the imperial moment, to show how scientists contemporary to canal construction and engaged in an intimate reading of the environment themselves gave voice to a hybrid nature that talked back in confounding ways. In narrating the story of entomology in Panama in this way, I have tried to avoid the pitfalls of insisting that current science is the only route to a material environmental analysis. By attending to vector ecology, entomological workers learned that there was no simple nature to be conquered in Panama. Rather, they had to manage a series of environmental processes and entities so interlocked with human agency and action that they defied the dichotomies of tropical theorizing. American tropical triumphalism persisted, then, not in ignorance of a body of scientific knowledge yet to be constructed, but because its practitioners clung to useful binaries and ignored existing evidence to the contrary.

The Fire Ant Wars

Nature and Science in the Pesticide Controversies of the Late Twentieth Century

By Joshua Blu Buhs

ABSTRACT

This essay uses an approach borrowed from environmental history to investigate the interaction of science and nature in a late twentieth-century controversy. This debate, over the proper response to fire ants that had been imported into the American South accidentally and then spread across the region, pitted Rachel Carson and loosely federated groups of conservationists, scientists, and citizens against the U.S. Department of Agriculture. The analysis falls into three sections: an examination of the natural history of the ants; an examination of the views of the competing factions; and an examination of how those views, transformed into action, affected the natural world. Both sides saw the ants in terms of a constellation of beliefs about the relationship between nature, science, and democracy. As various ideas were put into play, they interacted with the natural history of the insects in unexpected ways—and with consequences for the cultural authority of the antagonists. Combining insights from the history of science and environmental history helps explain how scientists gain and lose cultural authority and, more fundamentally, allows for an examination of how nature can be integrated into the history of science.

I N HER 1962 BESTSELLER *SILENT SPRING,* Rachel Carson attacked the profligate use of pesticides, arguing that the chemicals did little to control insects but were deadly to wildlife, livestock, and humans. She pointed to the federal campaign to eradicate imported fire ants from the American South as evidence. The ants had arrived in Mobile, Alabama, in the late 1910s. By the 1950s they had spread across the South, and suddenly

Isis 93, no. 3 (September 2002): 377–400.

For their criticisms and encouragement I would like to thank J. Lloyd Abbot, Jr., Mark Adams, William Banks, Murray Blum, Peter Branum, Leonard Bruno, Eve Buckley, Alex Checkovich, Dwayne Cox, Pete Daniel, Nathan Ensmenger, John George, Becky Jordan, Robert Kohler, Henrika Kuklick, Linda Lear, Daniel Leedy, Susan Lindee, Clifford Lofgren, Kim McIlnay, Erin McLeary, Susan Miller, Beth Orton, Phil Pauly, Walter Rosene, Robert Rudd, Dan Speake, Jeffrey Stine, Elizabeth Toon, James Trager, Walter Tschinkel, Jeremy Vetter, Edward Osborne Wilson, and Audra Wolfe, as well as participants in the panel on invading organisms at the 2000 History of Science Society meeting and the 2001 graduate student conference on the Cold War at the University of California, Santa Barbara. This research was supported in part by grants from the National Science Foundation and the University of Pennsylvania. This essay won the History of Science Society's Schuman Prize for 2001.

there was a roar of complaint: the ants reportedly attacked crops, killed wildlife, worried livestock, built large earthen mounds that interfered with farm machinery, and stung painfully. Carson suspected that the U.S. Department of Agriculture (USDA) had fabricated these claims in order to justify a huge program to eradicate the insects and increase its bureaucratic strength. The measures taken killed quail and rabbits, cows and pigs, and threatened human health. Carson called the USDA program "an outstanding example of an ill-conceived, badly executed, and thoroughly detrimental experiment in the mass control of insects." Hers was not the only voice raised in protest. Concerned citizens, entomologists, hunters, nascent environmental groups, and wildlife biologists urged the department to end the eradication program. Complaints continued into the late 1970s, when the Environmental Protection Agency (EPA) finally banned the chemicals used to eradicate the fire ants. Carson's philippic was one salvo in a conflict that would last twenty years, a conflict so intense it was dubbed "the fire ant wars."[1]

This essay explores the interaction of nature and science in the fire ant wars by combining methodologies from the history of science and environmental history. Over the past quarter century, historians of science have staked out a constructivist approach that focuses on the way social processes are implicated in the manufacture of all natural knowledge. While not necessarily opposed to examining the role of the material world in the construction of scientific ideas, earlier constructivist analyses focused on the social machinery of science, ignoring nature. More recently, there have been a number of efforts to broaden constructivism by making the natural world an actor in historical narratives.[2] This essay attempts something similar by importing techniques from environmental history. Environmental historians take as their central topic the study of the interactions between humans and nature at different times and in different places. William Cronon has argued that to fulfill this agenda historians need to answer three intertwined questions: How does nature work in the time and place being studied? How do humans view the natural world and create ideas about nature? And how do those ideas, transformed into action, affect the

[1] Rachel Carson, *Silent Spring* (Boston: Houghton-Mifflin, 1962), p. 162. For histories of the fire ant wars see Harrison Wellford, *Sowing the Wind* (New York: Grossman, 1972), pp. 286–309; Phillip M. Boffey, *The Brain Bank of America: An Inquiry into the Politics of Science* (New York: McGraw-Hill, 1975), pp. 200–226; Thomas R. Dunlap, *DDT: Scientists, Citizens, and Public Policy* (Princeton, N.J.: Princeton Univ. Press, 1981), pp. 89–91; Christopher J. Bosso, *Pesticides and Politics: The Life Cycle of a Public Issue* (Pittsburgh: Univ. Pittsburgh Press, 1987), pp. 85–90; Pete Daniel, "A Rogue Bureaucracy: The USDA Fire Ant Campaign of the Late 1950s," *Agricultural History,* 1990, *64*:99–114; Elizabeth F. Shores, "The Red Imported Fire Ant: Mythology and Public Policy," *Arkansas Historical Quarterly,* 1994, *53*:320–339; Linda Lear, *Rachel Carson: Witness for Nature* (New York: Holt, 1997), pp. 305–306, 312–315, 332–336, 340–344; and Daniel, *Lost Revolutions: The South in the 1950s* (Chapel Hill: Univ. North Carolina Press, 2000), pp. 78–87.

[2] For overviews of the constructivist approach see David Bloor, *Knowledge and Social Imagery* (Boston: Routledge & Kegan Paul, 1976); Steven Shapin, "History of Science and Its Sociological Reconstructions," *History of Science,* 1982, *20*:157–211; Henrika Kuklick, "The Sociology of Knowledge: Retrospect and Prospect," *Annual Review of Sociology,* 1983, *9*:287–310; Shapin, "Here and Everywhere: Sociology of Scientific Knowledge," *ibid.,* 1995, *21*:289–321; and Jan Golinski, *Making Natural Knowledge: Constructivism and the History of Science* (New York: Cambridge Univ. Press, 1998). On the integration of nature into narratives in the history of science see John Law, ed., *Power, Action, and Belief: A New Sociology of Knowledge?* (Sociology Review Monographs, 32) (London: Routledge & Kegan Paul, 1986); Michael L. Smith, *Pacific Visions: California Scientists and the Environment, 1850–1915* (New Haven, Conn.: Yale Univ. Press, 1987); Bruno Latour, *The Pasteurization of France,* trans. A. Sheridan and J. Law (Cambridge, Mass.: Harvard Univ. Press, 1988); Robert Kohler, *Lords of the Fly:* Drosophila *Genetics and the Experimental Life* (Chicago: Univ. Chicago Press, 1994); Andrew Pickering, *The Mangle of Practice: Time, Agency, and Science* (Chicago: Univ. Chicago Press, 1995); Richard White, *The Organic Machine: The Remaking of the Columbia River* (New York: Hill & Wang, 1995); Adele Clark, *Disciplining Reproduction: Modernity, American Life Sciences, and "The Problem of Sex"* (Berkeley: Univ. California Press, 1998); and Angela N. H. Creager, *The Life of a Virus: Tobacco Mosaic Virus as an Experimental Model, 1930–1965* (Chicago: Univ. Chicago Press, 2002).

natural world?[3] The second question is that asked by constructivist historians of science; the first and the third investigate the role of the natural world. As these three questions are addressed, and connections are drawn between the answers, the integral place of the material world in histories of science is revealed.

Nature, in this essay, is embodied by the fire ants. The insects are opportunistic organisms, adapted to disruption, that exploited the changing ecology of the mid-twentieth-century American South. Neither the USDA entomologists nor their opponents focused on the cause of the ants' irruption, however. The federal employees, excited by the power of the new insecticides that had been introduced after World War II and worried that the ants threatened agricultural production, thought only of finding the most efficient methods to kill them. (See Frontispiece.) Carson, her allies, and her descendants, on the contrary, saw the insects as ecological innocents, not exploiters of the South's ecology but organisms that found a niche in North American ecology. The real threat, they said, was posed by bureaucrats who intervened in natural processes, disrupting nature and chipping away at personal liberties. Both sides expressed their views through vocabulary borrowed from debates over the structure of American democracy during the Cold War. Simultaneously reflecting alternative interpretations of the relationship between nature, science, and the state and offering a powerfully persuasive rhetorical tool, the Cold War imagery gave the fire ant wars their form and their urgency: for combatants on both sides, the imagined ends of the Cold War came to stand for the imagined ends of the fire ant wars, with both the response to the insects and the proper structuring of the American democratic system at stake.[4] The two sides alternated in seeing their views realized: first the USDA attempted to eradicate the ants; then, in the 1970s, the agency's opponents successfully banned the insecticides and the insects were allowed to integrate into the southern ecology. The ideas, however, were not simply imposed on a passive nature. The biology of the ants and the actions of the two groups interacted in unexpected ways, with repercussions for the insects and the world that they inhabited as well as for the humans who claimed to understand them.

THE NATURAL HISTORY OF THE IMPORTED FIRE ANTS

Fire ants belong to the subgenus *Solenopsis,* a diverse group of ants that originated in South America about sixty-five million years ago. All *Solenopsis* possess a stinger that gives the group its common name. The insects at the heart of the fire ant wars were actually two closely related species from this assemblage: *Solenopsis richteri,* a brown or black

---

[3] William Cronon, "The Uses of Environmental History," *Environmental History Review,* 1993, *17*:1–22. For reviews of the field more generally see Donald Worster, "History as Natural History: An Essay on Theory and Method," *Pacific History Review,* 1984, *53*:1–19; Kendall E. Bailes, ed., *Environmental History: Critical Issues in Comparative Perspective* (Lanham, Md.: Univ. Press America, 1985); Richard White, "American Environmental History: The Development of a New Historical Field," *Pacific Hist. Rev.,* 1985, *54*:297–335; Cronon, "Modes of Prophecy and Production: Placing Nature in History," *Journal of American History,* 1990, *76*:1122–1131; White, "Environmental History, Ecology, and Meaning," *ibid.,* pp. 1111–1116; Worster, "Transformations of the Earth: Toward an Agro-Ecological Perspective in History," *ibid.,* pp. 1087–1106; Cronon, "A Place for Stories: Nature, History, and Narrative," *ibid.,* 1992, *78*:1347–1376; I. G. Simmonds, *Environmental History: A Concise Introduction* (Oxford: Blackwell, 1993); Worster, "Nature and the Disorder of History," *Environ. Hist. Rev.,* 1994, *18*:1–15; and Mart A. Stewart, "Environmental History: Profile of a Developing Field," *History Teacher,* 1998, *31*:351–368.

[4] On the wider point see Steven Shapin and Simon Schaffer, *Leviathan and the Air-Pump: Hobbes, Boyle, and the Experimental Life* (Princeton, N.J.: Princeton Univ. Press, 1985), p. 15: "Solutions to the problem of knowledge are embedded within practical solutions to the problem of social order, and . . . different practical solutions to the social order encapsulate contrasting practical solutions to the problem of knowledge."

ant with a yellow stripe across its gaster; and *Solenopsis invicta,* a red species. Entomologists noted the color differences early on, but since the insects are otherwise hard to distinguish they lumped the two under the name *Solenopsis saevissima richteri,* "most savage fire ant." They were considered the same species that the English naturalist Henry Walter Bates had seen attacking the village of Aveyros in the Amazon River Basin. "A greater plague than all other [insects] put together," he had called them; they ate everything in sight and attacked people "out of sheer malice."[5]

The two species originally inhabited different parts of the world's largest wetland, an expanse of marshy land that follows the Río Paraguay and the Río Parana through Brazil, Paraguay, and northern Argentina to their confluence with the Río de la Plata and ultimately into the Atlantic Ocean. *Solenopsis richteri* lives on the periphery of this wetland, its range edging into the Pampas. *Solenopsis invicta* can be found throughout the marshy river basins. The area is characterized by frequent disturbances. During the dry season, thick grasses clog the riverbeds; when the rains come, the water is forced to cut new channels, eventually overflowing and flooding the landscape. The river's vagrancy has created a wealth of microclimates, and the area is dominated by a rich array of plants. In 1929 a geographer noted, "The most striking feature in the natural vegetation is its lack of uniformity." The ants have adapted to this situation by exploiting the disturbances. They are opportunistic—one entomologist calls them "weeds"—infiltrating disrupted areas, growing quickly—a single queen gives birth to 250,000 workers in three years—but are forced out when the ecology matures.[6]

Solenopsis richteri was the first of the two species to break from the wetland and travel north. In the late nineteenth and early twentieth century, Argentina's cattle industry flourished and international trade was brisk. The ants, which lived near major points of distribution, stowed away on ships, reaching Mobile, Alabama, around 1918. The new world the insects faced was climatically similar to South America but ecologically very different. "Extensive timberlands and swamps, almost quite impenetrable," surrounded Mobile. Approximately 80 percent of the land within a hundred-mile radius of the port city was thick forest. The rest of the southern coastal plain was equally uninviting, devoted to fields often

[5] Henry Walter Bates, *The Naturalist on the Rivers Amazon* (1863; London: Murray, 1892), p. 227. (The imported fire ants are no longer believed to be the same species as the ants Bates saw.) Fire ant biology is reviewed in Clifford S. Lofgren, William A. Banks, and B. Michael Glancey, "Biology and Control of Imported Fire Ants," *Annual Review of Entomology,* 1975, *20:*1–30; Edward O. Wilson, "The Defining Traits of Fire Ants and Leaf-Cutting Ants," in *Fire Ants and Leaf-Cutting Ants: Biology and Management,* ed. Lofgren and R. K. Vander Meer (Boulder, Col.: Westview, 1986), pp. 1–9; S. Bradley Vinson, "Invasion of the Red Imported Fire Ant (Hymenoptera: Formicidae): Spread, Biology, and Impact," *Bulletin of the Entomological Society of America,* 1997, *43:*23–39; and Stephen Welton Taber, *Fire Ants* (College Station: Texas A&M Press, 2000). I simplify a more complicated taxonomic history. In any case, the distinction between the imported fire ants and the fire ants of Aveyros was never that great and was largely erased by the late 1950s. For a fuller discussion see Joshua Blu Buhs, "The Fire Ant Wars: *Solenopsis* and the Nature of the American State, 1918–1982" (Ph.D. diss., Univ. Pennsylvania, 2001), Ch. 2.

[6] The ecology of the region is described in E. W. Shanahan, *South America: An Economic and Regional Geography with an Historical Chapter* (London: Methuen, 1927) (the quotation is from p. 90); and A. A. Bonettos and I. R. Wais, "Southern South American Streams and Rivers," in *River and Stream Ecosystems,* ed. C. E. Cushing, K. W. Cummins, and G. W. Minshall (Ecosystems of the World, 22) (Amsterdam: Elsevier, 1995), pp. 257–293. The comparison of the ants to weeds is from Walter Tschinkel, "History and Biology of Fire Ants," in *Proceedings of the Symposium on the Imported Fire Ant,* ed. S. L. Battenfield (Washington, D.C.: Government Printing Office, 1982), pp. 16–35; Tschinkel, "The Ecological Nature of the Fire Ant: Some Aspects of Colony Function and Some Unanswered Questions," in *Fire Ants and Leaf-Cutting Ants,* ed. Lofgren and Vander Meer, pp. 72–87; and Tschinkel, "Distribution of the Fire Ants *Solenopsis invicta* and *S. geminata* (Hymenoptera: Formicidae) in Northern Florida in Relation to Habitat and Disturbance," *Annals of the Entomological Society of America,* 1998, *81:*76–81.

left fallow by sharecroppers and groves of trees allowed to grow dense. With nowhere to go, the ants settled into the Government Street Loop, a rundown section of Mobile where the trolleys turned around.[7] Some two decades later, after the cattle industry reached deeper into the South American interior where *Solenopsis invicta* lived, the red ants also reached Mobile. As they arrived, the South was on the brink of a revolution that would alter the ecology of the region, opening vast new spaces for the ants to colonize.

Beginning in the 1930s the USDA began to modernize the South, making it more efficient, more like the Midwest. Tractors, harvesters, and combines replaced field hands and farms grew in size, doubling in Alabama and tripling in Georgia, Louisiana, and Mississippi between 1920 and 1969. Wastelands were plowed under, groves of trees felled, and fields seeded from fencepost to fencepost. New crops were cultivated, especially soybeans, and cattle. During World War II military contracts were sent south, absorbing the idle workforce and pulling rural citizens into urban areas. Cities sprawled "with little attention to urban planning and zoning," and suburbs suddenly appeared. The southern historian C. Vann Woodward called these interlocking changes "the bulldozer revolution." That revolution would last into the 1970s, transforming the South into the Sun Belt.[8]

The spread of *Solenopsis invicta* was an unexpected consequence of this modernization process. Thriving in disturbed areas, the ants were presented with a vast extent of disrupted habitats to exploit. Humans also unwittingly provided a means of transport out of Mobile and across the South. For decades southern nurseries had struggled against discriminatory railroad rates that favored the North. With the postwar economic boom road building increased, the trucking industry introduced cheap transport, and nurseries bloomed. Mobile became the nation's fifth largest horticultural center. The ants found their way into nursery stock and were shipped across the region and deposited in just the kinds of disrupted sites—suburban developments, highway rights of way—that they preferred. The late-

---

[7] On the Argentine cattle industry see Ysabel F. Rennie, *The Argentine Republic* (Westport, Conn.: Greenwood, 1945), pp. 142–150. For the migration of the ants see F. E. Lennartz, "Modes of Dispersal of *Solenopsis invicta* from Brazil into the Continental United States—A Study in Spatial Diffusion" (master's thesis, Univ. Florida, 1973); and James C. Trager, "A Revision of the Fire Ants *Solenopsis geminata* Group (Hymenoptera: Formicidae: Myrmicinae)," *Journal of the New York Entomological Society,* 1991, *99:*141–198. On the ecology of Mobile and the surrounding area see A. H. Howell, "Physiography, 1908," Box 1, Folder 17, Fish and Wildlife Service, U.S. Department of the Interior Field Reports, 1887–1961, Series 1, Record Group (RG) 7176, Smithsonian Institution Archives, Washington, D.C. (quotation); and Edward L. Ullman, "Mobile: Industrial Seaport and Trade Center" (Ph.D. diss., Univ. Chicago, 1942), p. 46. On the natural history of the area see Ecological Society of America, Committee on the Preservation of Natural Conditions, *Naturalist's Guide to the Americas,* ed. Victor E. Shelford (Baltimore: Williams & Wilkins, 1926); Havilah Babcock, *My Health Is Better in November* (Columbia: Univ. South Carolina Press, 1947); and Merle J. Prunty, "The Renaissance of the Southern Plantation," *Geographical Review,* 1955, *45:*459–491. The ants' settlement in the Government Street Loop is described in William Steel Creighton to Murray S. Blum, 22 Apr. 1968, William Steel Creighton Papers (hereafter cited as **Creighton Papers**), an unprocessed box of material incorporated in the uncatalogued E. O. Wilson Papers (hereafter cited as **Wilson Papers**), Library of Congress, Washington, D.C.

[8] Pete Daniel, *Standing at the Crossroads: Southern Life in the Twentieth Century* (New York: Hill & Wang, 1986), pp. 136–137; and C. Vann Woodward, *The Burden of Southern History,* 3rd ed. (Baton Rouge: Louisiana State Univ. Press, 1993), p. 6. The changes are documented in Gilbert C. Fite, "Southern Agriculture since the Civil War: An Overview," *Agr. Hist.,* 1979, *53:*3–21; Daniel, "The Transformation of the Rural South, 1930 to the Present," *ibid.,* 1981, *55:*231–248; Jack Temple Kirby, "The Transformation of Southern Plantations, ca. 1920–1960," *ibid.,* 1983, *57:*257–276; Fite, *Cotton Fields No More: Southern Agriculture, 1865–1980* (Lexington: Univ. Press Kentucky, 1984); Daniel, *Breaking the Land: The Transformation of Cotton, Tobacco, and Rice Cultures since 1880* (Urbana: Univ. Illinois Press, 1985); Kirby, *Rural Worlds Lost: The American South, 1920–1960* (Baton Rouge: Louisiana State Univ. Press, 1987); Bruce J. Schulman, *From Cotton Belt to Sunbelt: Federal Policy, Economic Development, and the Transformation of the South, 1938–1980* (New York: Oxford Univ. Press, 1991); Albert E. Cowdrey, *This Land, This South: An Environmental History* (Lexington: Univ. Press Kentucky, 1996); and U.S. Bureau of the Census, *Census of Agriculture* (Washington, D.C.: GPO), for various years.

arriving red ants were not as restricted in their distribution as their congeners and so were better prepared to take advantage of the changing southern ecology. *Solenopsis richteri* languished, reaching only parts of northern Alabama and northern Mississippi; *Solenopsis invicta* spread widely.[9]

By the mid 1950s, the red ants could be found in nine southern states. The ants are omnivorous: they prefer insects but take whatever food is available. As their population increased, they could not always find favored foods, and their turn to other caloric sources brought them to the attention of southerners. The ants ate seeds and crops and even young quail—which caused consternation, since the birds were the South's most important game animals. They colonized lawns and the open spaces of the newly built military bases, where they came into intimate contact with humans. In 1955 a boy in New Orleans died after being stung three times. Two years later ant stings sent three soldiers from Maxwell Air Force Base to the hospital. The imported fire ants especially favored cow pastures, fields that were open and constantly disrupted by the big beasts.[10]

It might have been possible to ignore many of these problems—the boy's death notwithstanding, the ants killed far fewer people than bees and wasps—except that the spread of the ants was so dramatic and so intense that ignoring the formicids was difficult. In traveling to North America, the ants had left behind predators and parasites; the bulldozer revolution diminished competition. Freed from constraints, the ants' population exploded—sometimes the insects built a hundred mounds on a single acre of land. The hundreds of thousands of workers in each nest scurried from the colony, looking for food and stinging gardeners and soldiers. Landowners could not use their tractors to mow pastures without breaking blades or knocking the angry insects onto their backs. (See Figure 1.) Laborers refused to harvest heavily infested fields.[11]

---

[9] Edward O. Wilson and William L. Brown, "Recent Changes in the Population of the Fire Ant *Solenopsis saevissima* (Fr. Smith)," *Evolution*, 1958, *12*:211–218; William F. Buren *et al.*, "Zoogeography of the Imported Fire Ants," *J. New York Entomolog. Soc.*, 1974, *82*:113–124; Geddes Douglas, ed., *The History of the Southern Nurserymen's Association, Inc., 1899–1974* (Nashville, Tenn.: Southern Nurserymen's Association, 1974); Roy A. Roe II, "A Biological Study of *Solenopsis invicta* Buren, the Red Imported Fire Ant, in Arkansas" (master's thesis, Univ. Arkansas, 1974); Richard P. White, *A Century of Service: A History of the Nursery Industry Associations of the United States* (Washington, D.C.: American Association of Nurserymen, 1975); and Henry W. Lawrence, "The Geography of the U.S. Nursery Industry: Locational Change and Regional Specialization in the Production of Woody Ornamental Plants" (Ph.D. diss., Univ. Oregon, 1985).

[10] Edmund P. Hill III, "Observations of Imported Fire Ant Predation on Nesting Cottontails," *Proceedings of the Annual Conference of the Southeastern Association of Game and Fish Commissioners*, 1969, *23*:171–181; Robert H. Mount *et al.*, "Predation by the Red Imported Fire Ant, *Solenopsis invicta* (Hymenoptera: Formicidae), on Eggs of the Lizard *Cnemidophorous sexlineatus* (Squamata: Teiidae)," *Journal of the Alabama Academy of Science*, 1981, *52*:66–70; Mount, "The Red Imported Fire Ant, *Solenopsis invicta* (Hymenoptera: Formicidae), as a Possible Serious Predator on Some Native Southeast Vertebrates: Direct Observations and Subjective Impressions," *ibid.*, pp. 71–78; James M. Mueller *et al.*, "Northern Bobwhite Chick Mortality Caused by Red Imported Fire Ants," *Journal of Wildlife Management*, 1999, *63*:1291–1298; and Craig R. Allen *et al.*, "Impact of Red Imported Fire Ant Infestation on Northern Bobwhite Quail Abundance Trends in Southeastern United States," *Journal of Agricultural and Urban Entomology*, 2000, *17*:43–51. On the importance of quail see Frank S. Arant, *The Status of Game Birds and Mammals in Alabama* (Wetumpka, Ala.: Wetumpka Printing, 1939); and Stuart A. Marks, *Southern Hunting in Black and White: Nature, History, and Ritual in a Carolina Community* (Princeton, N.J.: Princeton Univ. Press, 1991), pp. 169–199. The attacks are described in "Fire Ants Spread as Pastures Shrink," *Montgomery Advertiser*, 17 Feb. 1957; and "Maxwell Reports Plague of Fire Ants: Three in Hospital," 14 Mar. 1957, Box 62, Fire Ants—News Items, Plant Pest Control Division Papers (hereafter cited as **Plant Pest Control Papers**), RG 463, 86/6/3, National Archives and Records Administration, College Park, Maryland.

[11] Undated notes, and W. G. Bruce, H. T. Vanderford, and A. L. Smith, "Survey of the Imported Fire Ant *Solenopsis saevissima* var. *richteri* Forel," Special Report S-6, 19 Nov. 1948, Early Fire Ant Records File, Wilson Papers; Edward O. Wilson and James H. Eads, Jr., "A Preliminary Report on the Fire Ant Survey," Alabama Department of Conservation, SG 9977, and Leyburn F. Lewis, "Develop Methods of Control of the Imported

Figure 1. *Large fire ant mounds, hardened by the southern sun, presented a hazard for agricultural workers: they sometimes broke tractor blades, and the ants also stung laborers. In the early days of the invasion, as many as a hundred mounds covered a single acre. This formicary was unusually large. (From H. B. Green,* The Imported Fire Ant in Mississippi, *Mississippi State University, Agricultural Experiment Station, Bulletin 737, February 1967.)*

Many organisms introduced into new environments undergo similarly dramatic increases in their population and density. The increase is usually followed by an equally dramatic crash, as parasites and predators attack the newcomers, the ecology of the area stabilizes, and the imported organisms compete among themselves for increasingly scarce resources.[12] To a point the imported fire ants followed this pattern, as both their number and population density declined in the years after introduction, but there were also significant deviations. The bulldozer revolution accounts for some of the deviation, ensuring that there were always new areas for the ants to exploit as suburbs sprawled, old areas were razed, and new roads were built. Looking at the course of invasion over a single patch of ground—southern Alabama, say—reveals the familiar rise-and-fall pattern. On a regional scale, however, the irruption continued, as the ants found their way to more and more parts of the South. By 1957 they had spread from Mobile to cover twenty million acres.

The rest of the deviation from the expected pattern is explained by a biological quirk. In their homeland, *Solenopsis invicta* populations live in two social forms: monogynous

Fire Ant," Research Line 1—Investigations in the Biology of the Imported Fire Ant *Solenopsis saevissima* var. *richteri* Forel, Fire Ant Cooperative File, Alabama Department of Conservation Papers (hereafter cited as **Alabama Dept. Conservation Papers**), SG 17019, Alabama Department of Archives and History, Montgomery, Alabama; Sanford D. Porter *et al.,* "Fire Ant Mound Densities in the United States and Brazil (Hymenoptera: Formicidae)," *Journal of Economic Entomology,* 1992, *85:*1154–1161; Porter *et al.,* "Intercontinental Differences in the Abundance of Solenopsis Fire Ants (Hymenoptera: Formicidae): Escape from Natural Enemies?" *Environmental Entomology,* 1997, *26:*373–384; and Nicholas Gotelli and Amy Arnett, "Biogeographic Effects of Red Fire Ant Invasion," *Ecology Letters,* 2000, *3:*257–261.

[12] On the ecology of invasions see Charles S. Elton, *The Ecology of Invasions by Animals and Plants* (1958; London: Chapman & Hall, 1977); R. Groves and J. Burdon, eds., *Ecology of Biological Invasions* (Cambridge: Cambridge Univ. Press, 1984); A. Gray, M. Crawley, and P. Edwards, eds., *Colonization, Succession, and Stability* (Oxford: Blackwell Scientific, 1984); H. Mooney and J. Drake, eds., *Ecology of Biological Invasions of North America and Hawaii* (New York: Springer, 1986); and Geerat Vermeij, "When Biotas Meet: Understanding Biotic Interchange," *Science,* 1991, *253:*1099–1104.

and polygynous colonies. Monogynous colonies have a single queen that mates in aerial swarms and founds her colony independently; polygynous colonies contain several, sometimes several hundred, queens, most of which mate within the nest and form new colonies by adopting workers from the mother colony, walking to a likely nearby area, and building a new nest. Both forms arrived in America. Bigger and stronger, the monogynous queens initially predominated. They could spread widely and colonize many disturbed habitats quickly. But as the environment became saturated with ants, the polygynous colonies came to dominate. Young queens were protected in the nest and were subsidized by the mother colony when founding their own nests. The increasing prevalence of polygyny allowed more and more ants to be packed into the same area, softening the expected crash in the fire ant population. In the 1990s, some fields in Texas sagged under the weight of over four hundred imported fire ant mounds per acre.[13]

IDEAS ABOUT NATURE: THE FIRE ANTS IN A COLD WAR

Eradicating the Fire Ants

The spread of the fire ants occurred in a postwar America optimistic about the future and confident that it could use its natural resources, science and technology, and democratic institutions to solve any problem. The economy hummed, domestic problems seemingly obliterated by the power of mass consumption. Antibiotics had a death-grip on disease. The federal government was in the capable hands of the affable Dwight Eisenhower. But beneath this optimism was a dark layer of concern. Science brought not only antibiotics but also the bomb, radioactive fallout, and pollution. Conformity was the root not only of the good economy but also of totalitarianism. And democratic institutions, while guaranteeing that Americans were the freest people on earth, could also be perverted to squash the very individual liberties they purported to defend. How best to live in nature, how to use science and technology, and how to bring nature, science, and technology to bear on the maintenance of democratic traditions—these were all questions that would become intertwined with the fire ant wars.[14]

Control entomologists and their administrative allies within the USDA shared a vision of how nature, science, and democracy related. Nature, they thought, was imperfect: insects destroy crops, diseases kill livestock, and weather is foe as often as friend. Survival is a struggle, achieved only by the correct application of scientific ideas to hold the forces of nature at bay and protect civilization. One entomologist wrote, "To clothe and feed [the growing U.S.] population man must maintain his position of dominance, and our agricultural production must continue to increase even at the expense of the further displacement of native plants and animals."[15] By the time the fire ants had spread across the South, the

[13] Sanford D. Porter and D. A. Savignano, "Invasion of Polygyne Fire Ants Decimates Native Ants and Disrupts Arthropod Community," *Ecology,* 1990, 71:2095–2106; Rodger Lyle Brown, "Fire Ants: Buggy Battalions Beating Mere Mortals," *Atlanta Journal-Constitution,* 21 June 1992, Metro Sect., p. 2; and Kenneth G. Ross and Laurent Keller, "Ecology and Evolution of Social Organization: Insights from Fire Ants and Other Highly Eusocial Insects," *Annual Review of Ecology and Systematics,* 1995, 26:631–656.

[14] This reading of the 1950s is based on Paul A. Carter, *Another Part of the Fifties* (New York: Columbia Univ. Press, 1983); and Michael Sherry, *In the Shadow of War: The United States since the 1930s* (New Haven, Conn.: Yale Univ. Press, 1995).

[15] George C. Decker, "Pesticides' Relationship to Conservation Programs," *National Agricultural Chemicals Association News and Pesticide Review,* May 1959, pp. 4–7, 13–14, on p. 7. Compare David G. Hall, "Food, Wildlife, and Agricultural Chemicals," *Conservation News,* 1 July 1952, 17:1–4; W. L. Popham to Samuel H. Ordway, Jr., 26 Dec. 1957, Box 752, Regulatory Crops–1–Fire Ants File, and Ross E. Hutchins to Allen H.

USDA had adopted insecticides as a principal weapon in the struggle against nature. Control (or economic) entomologists who supported the use of the chemicals had elbowed aside other insect biologists (often called research entomologists) who supplemented the use of chemical insecticide with biological and cultural controls to manage insect outbreaks. They pointed out that DDT was safer than the arsenic and cyanide solutions used by earlier generations of entomologists and that it had been used in World War II to protect American soldiers from typhus and malaria. Now, they reasoned, that insecticide and its chemical relatives could be used to protect public health and agricultural production. Entomologists need not be limited to controlling insects—the mark of old-fashioned entomology—but could eradicate them completely. The chemicals, they admitted, might kill wildlife and other desirable animals as well, but the gain in farming efficiency was well worth the cost. One farmer made the calculus explicit, noting, "I believe I have been as much for conserving our wildlife as the next one and have spent a great deal of effort and money to see wildlife increase but if one of us should be hurt by treating the land for fire ants I do not believe it should be the man who owns the land and pays taxes on the same, especially when it means the survival of himself and family."[16]

Using science to control the ants did more than preserve public health and increase agricultural production, the federal economic entomologists contended. Insecticides helped the nation in its struggle with the Soviet Union. It was an article of faith at the USDA that American democracy grew from the soil: agriculture was a Cold War weapon. Byron Shaw, head of the USDA's Agricultural Research Service, for example, said in 1958, "I think the times were described rather aptly a year or so ago, when a Soviet premier told an American television audience that communism would win its contest with capitalism when the Soviets' per-capita production of meat, milk and butter surpassed that of the United States. He was really saying that a nation is as strong as its agriculture." Eradicating the ants would allow the United States to increase its productivity and win the Cold War. If, on the contrary, the ants were left to spread, agricultural output would plummet, dissatisfaction would increase, and the seeds of revolution would be sown.[17]

The Department of Agriculture had reason to believe that the imported fire ants represented an especially dire threat to the modernizing South. In the late 1940s the state of Alabama had tapped E. O. Wilson, then an undergraduate at the University of Alabama,

---

Morgan, 18 Feb. 1959, Box 113, Regulatory Crops–Fire Ants File, General Correspondence, 1954–1966, 1 (UD), Agricultural Research Service Papers (hereafter cited as **Agricultural Research Service Papers**), RG 310, National Archives and Records Administration, College Park, Maryland; C. H. Hoffman to Theodore Dashman, 23 Aug. 1957, Box 64, Insecticides, Fish, and Wildlife File, Entomology Research Division: General Correspondence, 1954–1958, 5 (UD), Agricultural Research Service Papers; John L. George, *The Program to Eradicate the Imported Fire Ant: Preliminary Observations* (New York: Conservation Foundation, 1958); Popham and Hall, "Insect Eradication Programs," *Annu. Rev. Entomol.,* 1958, *3:*335–354; "Conservation through Pest Control," 3 June 1958, Box 82, Information—Speeches File, and Woodrow O. Owen to J. F. Spears, 13 June 1960, Box 128, Fire Ant–Wildlife File, Plant Pest Control Papers; and U.S. Congress, Subcommittee on Agriculture, Committee on Appropriations, *USDA Appropriations FY 1960,* 86th Cong., 1st sess., 1959, p. 900.

[16] O. G. McBeath to E. D. Burgess, 17 Dec. 1958, Box 197, Fire Ant File, Plant Pest Control Papers. For more on the distinction between control and research entomologists and the rise of the eradication ideal see John H. Perkins, *Insects, Experts, and the Insecticide Crisis: The Quest for New Pest Management Strategies* (New York: Plenum, 1982).

[17] Byron T. Shaw, "Development of Research Facilities and the Role of Regional Laboratories in State and Federal Programs," 10 Nov. 1958, Box 34, Speeches File, Entomology Research Division: General Correspondence, 1954–1958, 5 (UD), Agricultural Research Service Papers. See also "Background on—Our Nation's Agriculture," Box 4007, Publications/Department, Jan. 1 to April 19 File, General Correspondence, 1906–1975, PI 191 and 1001 A-E (UD), Secretary of Agriculture Papers (hereafter cited as **Secretary of Agriculture Papers**), RG 16, National Archives and Records Administration, College Park, Maryland; and Bosso, *Pesticides and Politics* (cit. n. 1), p. 69.

to study the insects. Wilson and a classmate determined that the ants significantly damaged the state's agriculture. Worse, he concluded that the red ants were newly evolved mutants that were better adapted to life in North America and more aggressive than their black counterparts. The taxonomic decision explained the chronology of the ants' irruption without reference to the bulldozer revolution: the ants, Wilson argued, had remained unremarkable until the mutation appeared sometime around the end of World War II, allowing the insects to spread across the South. Wilson's ideas made the ants even more threatening and unpredictable—a constantly evolving pest. Thus C. C. Fancher, head of the USDA's fire ant program, could say without irony that the eradication of the insects would protect "the American way of life" and provide a "service for mankind."[18]

In popular culture more broadly, other connections were drawn between the ants and communist subversives. The insects lived in a hierarchically arranged social system that extinguished individualism—they were "regimented automatons, driven, dutiful in their prescribed pointless doing." The ants undermined the concept of private property, building their mounds without respect for property lines and making capitalist production difficult. They also perverted gender roles—the males reduced to mere bearers of sperm that died after mating and the females ruling the colony—just as communists promised to abolish sexual hierarchies. As the fire ants impinged on southern life, the connections between the insects and communists were made quickly and easily. One newspaper labeled the insects "the red peril" and "fifth columnists," while a hunting magazine noted, "This ferocious little ant . . . has carried communism to the ultimate, and its actions suggest a certain cold-blooded intelligence."[19]

Military metaphors permeate the history of insect control, and Americans have a long tradition of drawing parallels between insects and ostracized humans. But despite the triteness of the language (or maybe because of it), the analogy was useful to the USDA.[20]

---

[18] C. C. Fancher to Burgess, 27 June 1958, Box 198, Fire Ant—Wildlife Losses File, and Fancher to Spears, 7 June 1960, Fire Ant—1 File (1960), Plant Pest Control Papers. On Wilson's work see Field Notes File, SG 9977, and Wilson and Eads, "Preliminary Report on the Fire Ant Survey" (cit. n. 11); E. O. Wilson to Creighton, 3 Oct. 1949, Creighton Papers; Edward O. Wilson, "Invader of the South," *Natural History,* 1959, *68:*276–281; Wilson, "The Fire Ant," *Scientific American,* 1958, *198:*36–41, 160; Wilson, "In the Queendom of Ants: A Brief Autobiography," in *Leaders in the Study of Animal Behavior: Autobiographical Perspectives,* ed. Donald A. Dewsbury (Lewisburg, Pa.: Bucknell Univ. Press, 1985), pp. 465–486; and Wilson, *Naturalist* (Cambridge, Mass.: Harvard Univ. Press, 1994), pp. 107–115. On the importance of calling the insects mutants see portion of a letter to J. Lloyd Abbot, 27 June 1957, Box 62, Fire Ant File; Wilson to Robert L. Burlap, 11 Jan. 1959; Wilson to Clifford Lofgren, 11 Feb. 1959; and Wilson to Burgess, 11 Feb. 1959, Box 213, Fire Ant File; and Wilson to Philip Charam, 14 July 1964, Box 247, Imported Fire Ant—10 File: Plant Pest Control Papers.

[19] "Of Ants and Men," *Christian Century,* 13 Nov. 1957, *74:*1339 ("automatons"); "Fire Ants Sting Congress into Action," *Dayton Daily News,* Box 62, Fire Ant—News Items File, Plant Pest Control Papers ("red peril," "fifth columnists"); and John Foster, "Secrets of the Fire Ant," *Mississippi Game and Fish,* July 1957, *19:*4–5 ("ferocious little ant"). See also William Morton Wheeler, "Insect Parasitism and Its Peculiarities," *Popular Science Monthly,* 1911, *79:*431–449; Arpaud Ferenczy, *The Ants of Timothy Thümmel* (New York: Harcourt, Brace, 1924); "What Shall We Do about Ants? Efforts to Control New Invader," *Alabama Journal,* 9 Feb. 1957, p. 4-A; "Insect Saboteur on the March," *Senior Scholastic,* 4 May 1960, *76:*40; "Fire Ant a Hardy 'Soldier,' State's Poison War Gains," 22 May 1962, Box 128, Fire Ant—8 File, Plant Pest Control Papers; George Laycock, "The Determined Ant," *Field and Stream,* May 1963, *78:*61, 155–159; and Gregg Mitman, "Defining the Organism in the Welfare State: The Politics of Individuality in American Culture, 1890–1950," in *Biology as Society, Society as Biology,* ed. Sabine Maasen, Everett Mendelsohn, and Peter Weingart (Dordrecht: Kluwer, 1995), pp. 249–278.

[20] For a similar metaphorical linkage between a prominent problem and a less obvious one see Gary Alan Fine and Lazaros Christoforides, "Dirty Birds, Filthy Immigrants, and the English Sparrow War: Metaphorical Linkage in Constructing Social Problems," *Symbolic Interaction,* 1991, *14:*375–393. Edmund Russell has shown the connection between insects and America's enemies more generally; see Edmund P. Russell III, "War on Insects: Warfare, Insecticides, and Environmental Change in the United States, 1870–1945" (Ph.D. diss., Univ. Michigan,

To eradicate the ants, the federal agency needed to spread insecticides on all land, "without regard to location, land use, or ownership." If any land was left untreated, some ants might survive and spread, threatening American agriculture. Broad support was necessary to ensure that all land could be sprayed. Drawing on the shared Cold War imagery, the USDA worked to generate the needed mandate. A department press release, for example, noted, "Uncle Sam is ready to use a fleet of 60 planes to go to war against the dreaded fire ant. . . . Only the modern airplane, dropping insecticides on twenty million acres in the critical area, can hope to stop the menace." The word "menace," of course, had deeper connotations, evoking concerns over the Red Menace. Congressional testimony in favor of the eradication program drew on the same lexicon. One southerner testified, "The government should be building as big a defense against the fire ants as they are against the Russians. The ants have already invaded."[21] The rhetoric proved persuasive, and in late 1957 Congress gave the USDA $2.4 million to initiate the program. The agency took the money and transformed their ideas into action, spreading chemical insecticides onto one million acres in the South the first year and millions more before the fire ant wars ended. (See Figure 2.)

Naturalizing the Imported Fire Ants

As the USDA entomologists worked to put their ideas into practice, a loosely federated group of biologists, citizens, early environmentalists, and hunters offered an alternative vision of how nature, science, and democracy fit together and, consequently, an alternative response to the ants. To varying extents, the members of this group saw nature not as imperfect but as finely tuned and integrated. Over the course of the previous two decades, wildlife biologists had shown how animal populations kept each other in balance. For an even longer time, both research and control entomologists—before they were elbowed aside by the upstarts promising to use insecticides for eradication—had studied insects as part of an ecological community.[22] In their view, the job of the scientist was not to battle

---

1993); Russell, "'Speaking of Annihilation': Mobilizing for War against Human and Insect Enemies, 1914–1945," *J. Amer. Hist.*, 1996, *82*:1505–1529; and Russell, *War and Nature: Fighting Humans and Insects with Chemicals from World War I to* Silent Spring (Cambridge: Cambridge Univ. Press, 2001). Russell makes a very similar argument about the connection between pest control, the Cold War, and democracy but does so with a much broader array of evidence. Dunlap has also commented on the ubiquity of military metaphors in insect control: Dunlap, *DDT* (cit. n. 1), pp. 36–37.

[21] The phrase "without regard to location, land use, or ownership" was ubiquitous in the USDA; for one example see Popham to Lister Hill, 18 Jan. 1957, Box 752, Regulatory Crops—1—Fire Ants File, General Correspondence, 1954–1966, 1 (UD), Agricultural Research Service Papers. The story about Uncle Sam going to war against the ants was widely printed; for an example see Box 62, Fire Ant—News Clippings File, Plant Pest Control Papers. For the southerner's view see John Devlin, "Fire Ant Alarms South," *New York Times,* 19 Mar. 1957, p. 40. Illustrating how easy it was to make the connection between ants and threats to the social order, in 1954 the *New York Times,* reviewing the movie *Them!* about giant, mutated ants that threaten Los Angeles, found the "proceedings tense, absorbing and, surprisingly enough, somewhat convincing": rpt. in James Robert Parrish and Michael R. Potts, *The Great Science Fiction Pictures* (Metuchen, N.J.: Scarecrow, 1977), p. 318. Stories of invading ants were in vogue. In addition to *Them!* the story "Leiningen versus the Ants" was reported in numerous forms and articles about "army ants" proliferated: Carl Stephenson, "Leiningen versus the Ants," in *The Sixth New Year* (Chicago: Esquire, 1938), pp. 145–168; "Leiningen versus the Ants," *Senior Scholast.,* 5 Apr. 1948, *52*:25–26; "Go to the Army Ant," *Newsweek,* 12 Apr. 1948, *31*:54; Alan Devoe, "The World of Ants," *American Mercury,* Aug. 1949, *69*:225–229; B. Eddy, "Go to the Ants, and Be Warned," *New York Times Magazine,* 12 Dec. 1948, p. 22; T. C. Schnierla, "Army Ants," *Sci. Amer.,* 1948, *178*:16–23; "All about Army Ants," *Newsweek,* 28 July 1952, *40*:53; J. O'Reilly, "Swarming Killers of the Jungle: Army Ants," *Saturday Evening Post,* 16 May 1953, *225*:36; *The Naked Jungle* (Paramount Pictures, 1954); and "March of the Ants," *Newsweek,* 18 Mar. 1957, *49*:84.

[22] On the entomologists see W. Conner Sorenson, *Brethren of the Net: American Entomology, 1840–1880* (Tuscaloosa: Univ. Alabama Press, 1995); and Russell, "War on Insects" (cit. n. 20). More generally, see Susan

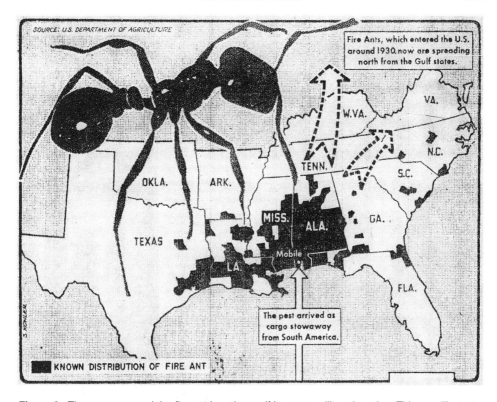

SOURCE: U.S. DEPARTMENT OF AGRICULTURE

Fire Ants, which entered the U.S. around 1930, now are spreading north from the Gulf states.

VA.

W.VA.

N.C.

TENN.

S.C.

OKLA.   ARK.

MISS.   ALA.   GA.

TEXAS

Mobile

LA.

FLA.

The pest arrived as cargo stowaway from South America.

S. KOHLER

■ KNOWN DISTRIBUTION OF FIRE ANT

Figure 2. *The press covered the fire ant irruption as if it were a military invasion. This map illustrates the spread of the insects using cartographic techniques reminiscent of World War II battle diagrams. Juxtaposing an enormously enlarged ant with the map also made the creature seem more monstrous, as well as capable of moving north, as the dotted arrows project. (From* Dayton Daily News, Box 62, Fire Ant News Items File [1957], Plant Pest Control Division Papers, RG 436, National Archives and Records Administration, College Park, Maryland.)

nature but to elucidate natural processes and find ways to accommodate human life to natural rhythms. This understanding of the relationship between science and nature was thought to serve democracy in several different ways. Some saw the protection of nature as the promotion of spiritual values above economic ones and thus a means for creating a better citizenry. Some felt that wildlife was one of the nation's most important natural resources and that its conservation was a way of maintaining U.S. strength. Others felt that a commitment to living in accord with nature proved the vitality of democratic institutions. If insecticides, say, were used without regulation, killing wildlife, it would mean that agricultural agencies had gained too much power, warping the political process and silencing those who voiced concern for wildlife. A rich, varied natural world was evidence of a strong democracy, in which policies were set to appease competing factions. The

Flader, *Thinking Like a Mountain: Aldo Leopold and the Evolution of an Ecological Attitude toward Deer, Wolves, and Forests* (Columbia: Univ. Missouri Press, 1974); Curt Meine, *Aldo Leopold: His Life and Work* (Madison: Univ. Wisconsin Press, 1987); Thomas R. Dunlap, *Saving America's Wildlife: Ecology and the American Mind, 1850–1990* (Princeton, N.J.: Princeton Univ. Press, 1988); and Donald Worster, *Nature's Economy: A History of Ecological Ideas,* 2nd ed. (New York: Cambridge Univ. Press, 1994).

USDA's policy of favoring agriculture over wildlife in the fire ant wars represented a threat to American democracy.[23]

Research by university entomologists provided evidence for these views. These studies did not examine the invasion as a regional phenomenon but instead studied the insects in limited locations that they had inhabited for some time—places where the irruption was ending and the damage done by the ants was less intense. For example, Kirby Hays, an entomologist at the Alabama Polytechnic Institute (renamed Auburn in 1960), visited South America in early 1957 and was told by local scientists that the ants were considered beneficial because they preyed on other insects. Research on the insects' behavior conducted in Alabama by Hays's brother Sydney, a graduate student at Auburn, substantiated these opinions. The younger Hays tested the feeding preference of laboratory-reared imported fire ants and found that they ate only insects, becoming cannibalistic rather than consuming plant material. Another Auburn graduate student showed that the ants consumed no more than 4 percent of young quail, and research in Louisiana determined that the ants were major predators of sugarcane borers and, perhaps, boll weevils. In 1958 a report by the Alabama state forester that had been written at the time of Wilson's survey reappeared. The ants, it claimed, were not pests.[24]

---

[23] For the spiritual view see Paul B. Sears, "The Road Ahead in Conservation," *Audubon,* Mar.–Apr. 1956, *58*:58–59, 80; Alfred G. Etter, "A Protest against Spraying," *ibid.,* July–Aug. 1959, *61*:153; Robert Rudd, "The Indirect Effects of Chemicals in Nature," in "The Effects of Toxic Pesticides on Wildlife," p. 16, Box 35, File 583, Rachel Carson Papers (hereafter **Carson Papers**), YCAL 46, Beinicke Rare Book Library, Yale Univ., New Haven, Connecticut; Rudd, "The Irresponsible Poisoners," *Nation,* 1959, *188*:496–497; and Samuel P. Hays, *Beauty, Health, and Permanence: Environmental Politics in the United States, 1955–1985* (Cambridge: Cambridge Univ. Press, 1987). On wildlife as a natural resource see Clarence Cottam, "Pesticides and Wildlife," 21–23 Feb. 1960, Box 35, File 581, Carson Papers; Ernest Swift to Ezra Taft Benson and National Wildlife Federation, press release, 21 Nov. 1957, Box 65, Fire Ant File, Entomology Research Division: General Correspondence, 1954–1958, 5 (UD), Agricultural Research Service Papers; Ross L. Leffler to Benson, 2 Dec. 1957, File 61, Fire Ant File, Plant Pest Control Papers; John L. George, *The Pesticide Problem* (New York: Conservation Foundation, 1957); "Peril in Attack on Fire Ant Seen," 18 Feb. 1958, Box 30, Fire Ant File, Entomology Research Division: General Correspondence, 1954–1958, 5 (UD), and Ira Gabrielson to Benson, 19 Feb. 1958, Box 944, Regulatory Crops—Fire Ants File, Agricultural Research Service Papers; Cottam, "A Conservationist's Views on the New Insecticides," in *Biological Problems in Water Pollution,* ed. C. M. Tarzwell (Washington, D.C.: U.S. Dept. Health, Education, and Welfare, 1960), pp. 42–45; and George, "The Pesticide Problem: Wildlife—The Community of Living Things," May–June 1960, Box 135, Folder 221 File, Paul B. Sears Papers (hereafter cited as **Sears Papers**), 663, Manuscripts and Archives, Yale Univ. On the connection between natural preservation and democratic institutions see Fairfield Osborn, "Conservation—The Core of Our Democracy," 10 Mar. 1948, Office of the President, Fairfield Osborn Papers, RG 2, New York Zoological Society Archives, New York; Olaus Murie to Lyle Watts, 7 Feb. 1952, Box 135, File 221, Sears Papers; Margaret K. Keath to John A. Blatnik, 29 July 1957, Box 2940, Insecticides June 1—File, Secretary of Agriculture Papers; Willhelmine Kirby Waller, "Poison on the Land," *Audubon,* Mar–Apr. 1958, *60*:68–70; Marjorie Spock to Rachel Carson, 6 June 1958, Box 105, Folder 20006, Carson Papers; Frances Martin to Department of Agriculture, n.d., Box 651, Publications—Rachel Carson Articles, 1963 File, General Correspondence, 1954–1966, 1 (UD), Agricultural Research Service Papers; Bill Ziebach, "Fire Ant Problem," *Mobile Press-Register,* 22 June 1958, p. 6B; "Farmers Protest Fire Ant Control Program 'Throat-Ramming,'" *Conserv. News,* 15 Aug. 1958, *23*:4–5; Charles H. Callison, "Pesticides and Wildlife," 21 Oct. 1959, Box 4, Insecticides—DDT—Wildlife File, Entomology Director's Correspondence, 1959–1965, 1055 (Al), Agricultural Research Service Papers; and "Experiments of USDA Violate Individual Rights," 11 Feb. 1961, Box 94, Fire Ant—8 File, Plant Pest Control Papers.

[24] Research at Auburn is covered in Frank S. Arant, Kirby L. Hays, and Dan W. Speake, "Facts about the Imported Fire Ant," *Highlights of Agricultural Research,* 1958, *5*:12–13; K. Hays, "The Present Status of the Imported Fire Ant in Argentina," *J. Econ. Entomol.,* 1958, *51*:111–112; Sydney B. Hays, "The Food Habits of the Imported Fire Ant, *Solenopsis saevissima richteri* Forel, and Poison Baits for Its Control" (master's thesis, Alabama Polytechnic Inst., 1958); "The Present Status of the Imported Fire Ant, *Solenopsis saevissima richteri* Forel, in Argentina," Report to the Governor, 16 June 1958, Gene Stevenson, press release, 21 Feb. 1958, and Kenneth B. Roy, press release, 24 Apr. 1958, Box 2, Ralph B. Draughon Papers (hereafter cited as **Draughon Papers**), RG 107, Auburn Univ. Archives, Auburn, Alabama; K. Hays, "Food Habits of *Solenopsis saevissima*

These reports became the basis of Carson's discussion of the ants in *Silent Spring*. While writing the book Carson had corresponded with Wilson and learned that he considered reports of the ants' beneficial traits exaggerated, but she ignored his conclusions and concentrated on the positive aspects of the ants' biology. Carson reinterpreted the meaning of the nests, for example, seeing them not as impediments to agricultural production but as necessary to the ecology of the earth. "Their mound-building activities," she wrote, "serve a useful purpose in aerating and draining the soil." Her book, she concluded, "thoroughly documented that the fire ant has never been a menace to agriculture and that the facts concerning it have been completely misrepresented." Ignoring the distinctions between the two species (or mutants) and the role that humans had played in the irruption, Carson argued that for most of the ants' time in North America they had been inconsequential. The sudden interest in them resulted from USDA propaganda, not biology; the ants were actually well-behaved parts of the ecosystem. In the late 1930s Carson had written an essay about the starling, a bird imported into America that many considered a pest but that she thought was becoming a necessary part of the American ecological order. She argued that it was time to give the starling citizen papers. Twenty years later, she was working to naturalize another immigrant, the imported fire ants.[25]

The insecticides, by contrast, remained outside the American ecological order, a true threat. Biologists monitoring the effect of the fire ant program found dead wildlife at every spot that they checked. The Alabama Department of Conservation, for example, found sixty-eight dead animals on a one-hundred-acre plot of land, while biologists with the U.S. Fish and Wildlife Service, scouting a two-acre sample plot in Georgia, found six dead quail, seven dead rabbits, twenty dead songbirds, three dead rodents, and one dead cat, all with enough insecticide in their bodies to account for their deaths; two months after the application, they could not find a single live bird. (See Figure 3.) Quail populations plum-

richteri Forel," *J. Econ. Entomol.* 1959, *52:*455–457; K. Hays, "Ecological Observations on the Imported Fire Ant, *Solenopsis saevissima richteri* Forel, in Alabama," *J. Alabama Acad. Sci.,* 1959, *30:*14–18; Albert S. Johnson, "Antagonistic Relationships between Ants and Wildlife with Special Reference to Imported Fire Ants and Bobwhite Quail in the Southeast," *Proc. Ann. Conf. Southeastern Assoc. Game and Fish Commissioners,* 1961, *15:*88–107; and Johnson, "Antagonistic Relationships between Ants and Wildlife" (master's thesis, Auburn Univ., 1962). Other research can be tracked in J. D. Long *et al.,* "Fire Ant Eradication Program Increases Damage by the Sugarcane Borer," *Insect Conditions in Louisiana,* 1958, *1:*10–11; Long *et al.,* "Fire Ant Eradication Program Increases Damage by the Sugarcane Borer," *Sugar Bulletin,* 1958, *37:*62–63; George Stromeyer, "Unraveling the Secrets of the Fire Ant," *Amer. Mercury,* July 1960, *91:*121–124; S. D. Hensley *et al.,* "Effects of Insecticides on the Predaceous Arthropod Fauna of Louisiana Sugarcane Fields," *J. Econ. Entomol.,* 1961, *54:*146–149; Leo Dale Newsom to Charam, 17 July 1964, Box 247, Imported Fire Ant—10 File, Plant Pest Control Papers; and Newsom, "Eradication of Plant Pests," *Bull. Entomolog. Soc. Amer.,* 1978, *24:*35–40. Mention of the forester's report is in John A. Fluno, office memo, 11 Aug. 1958, Box 84, Insects Affecting Man and Animals Branch Fire Ant File, Entomology Research Division: General Correspondence, 1954–1958, 5 (UD), Agricultural Research Service Papers.

[25] Wilson to Carson, 14 May 1959, 23 Oct. 1958, Box 44, Folder 841, Carson Papers; Carson, *Silent Spring* (cit. n. 1), p. 163 (on mound-building activities); Carson to Abbot, 6 Oct. 1961, Box 90, Folder 1586, Carson Papers (misrepresentation of facts); and Rachel Carson, "How about Citizenship Papers for the Starling?" *Nature Magazine,* June–July 1939, *32:*317–319. E. O. Wilson, then at Harvard University, and the USDA entomologists both dismissed the Hays brothers' work: the ants were mutants, they said, and their behavior in North America could not be predicted from their behavior south of the equator, nor could the actions of the insects in the wild be understood by studying them in a lab. They charged that the Auburn entomologists were also chronically confused by the taxonomy of the fire ants, often mistaking native fire ants for the imports, and so their conclusions could not be trusted. Creighton to Walter H. Grimes, 19 Nov. 1957, 23 Nov. 1957, and Smith to Creighton, 9 Dec. 1957, 19 Feb. 1969, Creighton Papers; portion of a letter to Abbot, 27 June 1957, Box 62, Fire Ant File; Wilson to Burlap, 11 Jan. 1959, Wilson to Lofgren, 11 Feb. 1959, and Wilson to Burgess, 11 Feb. 1959, Box 213, Fire Ant File; and Wilson to Charam, 14 July 1964, Box 247, Imported Fire Ant—10 File, Plant Pest Control Papers.

Figure 3. *In April 1958 the Alabama Department of Conservation surveyed a field in Autauga County where the USDA had recently sprayed heptachlor to eradicate the imported fire ants. The department released to the press a picture of some of the dead animals that its wildlife biologists had found, the image proof that the pesticides used to control the ants decimated wildlife. The department's director admitted that the survey was small but said, "Seeing them dead in the field following fire ant treatment is strong enough evidence for me." (From Ralph H. Allen, Jr., "The Fire Ant Eradication Program and Its Effect on Fish and Wildlife," Fire Ant File, Alabama Department of Conservation Papers, 1943–1951, SG 17018, Alabama Department of Archives and History.)*

meted by almost 90 percent. Livestock, as well, frequently died, and the chemicals were seeping into the milk supply and, possibly, into the bodies of children. Research entomologists working for the USDA added their voices to the chorus of complaint, advocating studies of biological and cultural control as a supplement to the use of chemical insecticides. The federal control entomologists were not working with nature, their opponents charged, but against it. The application of the insecticides, Carson wrote, "follow[ed] the impetuous, heedless pace of man, not the deliberate pace of nature."[26]

The eradication was more than a threat to wildlife. It was a threat to democracy, the

---

[26] Carson, *Silent Spring,* p. 7; Ralph H. Allen, Jr., "The Fire Ant Eradication Program and Its Effect on Fish and Wildlife," Fire Ant File, Alabama Department of Conservation Papers, 1943–1951, SG 17018, Alabama Dept. Conservation Papers; Daniel H. Janzen, "Effects of the Fire Ant Eradication Program upon Wildlife," 25 May 1958, Box 30, Fire Ant File, Entomology Research Division: General Correspondence, 1954–1958, 5 (UD), Agriculture Research Service Papers; Leslie L. Glasgow, "Wildlife and the Fire Ant Program," in *Transactions of the Twenty-fourth North American Wildlife Conference,* ed. James B. Trefethen (Washington, D.C.: Wildlife Management Inst., 1959), pp. 142–149; Daniel W. Lay, "Fire Ant Eradication and Wildlife," *Proc. Annu. Conf. Southeastern Assoc. Game and Fish Commissioners,* 1958, *12*:248–250; Walter Rosene, "Whistling-Cock Counts of Bobwhite Quail on Areas Treated with Insecticides and Untreated Areas, Decatur County, Georgia," *ibid.,* pp. 240–244; Otis L. Poitevint to Ray E. Tyner, 13 Oct. 1959, and Discussions of the Imported Fire Ant at Meetings of Georgia Sportsmen's League, 15–17 Oct. 1959, Box b-494, Peters File, National Audubon Society Papers (hereafter cited as **National Audubon Society Papers**), New York Public Library, New York; and Robert W. Murray, "A Synecological Study of the Effects of the Fire Ant Eradication Program in Florida," *Proc. Annu. Conf. Southeastern Assoc. Game and Fish Commissioners,* 1962, *16*:145–153. On the importance of milk contamination see Ralph H. Lutts, "Chemical Fallout: Rachel Carson's *Silent Spring,* Radioactive Fallout, and the Environmental Movement," *Environmental Review,* 1985, *9*:214–225.

USDA's opponents said, turning the federal agency's rhetoric on its head. The critique of the fire ant program raised a serious question: If the insects were not pests and the insecticides were so deadly, why would the USDA undertake to eradicate the ants? To answer these questions, the USDA's opponents called on a traditional American distrust of centralized governmental control. Antistatism has a long history in America, but it took on a particular intensity during the Cold War. As the historian Michael Hogan has shown, there was a widespread fear that in building a national security system against communism, the United States would take on the traits of its enemy. National defense required centralized control and secrecy and conformity, all characteristics of the Soviet Union. The Cold War, many feared, would transform the United States into a garrison state. The USDA had already defined itself as part of the national security system, and the agency's determination to pursue the eradication ideal demanded centralized control. Carson and others drew on these tropes to attack the USDA. Why would the USDA spread deadly chemicals against a pest that was not a pest? Because the agency was drunk on its own power and beyond democratic accountability.[27]

In *Silent Spring* Carson wrote, "Who has decided—who has the *right* to decide—for the countless legions of people who were not consulted that the supreme value is a world without insects, even though it be also a sterile world ungraced by the curving wing of a bird in flight? The decision is that of the authoritarian temporarily entrusted with power." Others followed the same line of reasoning. The wildlife biologist Clarence Cottam, for example, wrote, "I am convinced some of the philosophies expressed and actions taken by the pest control arm of our Federal Department of Agriculture strike at the very heart of American democracy. The problems, therefore, far transcend the control program or any entomological considerations." He urged others to agitate against eradication campaigns and avoid becoming "numbered pawns of the state." This form of critique was so powerful that it drew to wildlife groups some who opposed the growth of bureaucracies but had little interest in wildlife. The Mobile nursery owner J. Lloyd Abbot, for instance, joined the Alabama Wildlife Federation explicitly to stop the imported fire ant program not because he worried about the danger of insecticides but because "the threat to the continued existence of our Democracy, and whether or not we are going to be taken over by *internal* bureaucracies, could not be more clearly illustrated than it is by this whole reprehensible situation."[28]

[27] Michael Hogan, *A Cross of Iron: Harry S. Truman and the Origins of the National Security State, 1945– 1954* (New York: Cambridge Univ. Press, 1998). On antistatism more generally see Gary Wills, *A Necessary Evil: A History of American Distrust of Government* (New York: Simon & Schuster, 1999). For the South's particular blanching at centralized control see Twelve Southerners, *I'll Take My Stand: The South and the Agrarian Tradition* (New York: Harper, 1930); Paul Conklin, *The Southern Agrarians* (Knoxville: Univ. Tennessee Press, 1988); and Eugene Genovese, *The Southern Tradition: The Achievement and Limitations of an American Conservatism* (Cambridge, Mass.: Harvard Univ. Press, 1994). Some southern farmers did fear the government bureaucracy that undertook the fire ant program: W. A. Ruffin, "Tales about Insects," 3 Mar. 1958, Box 1, Ruffin File, Extension Entomology Papers, RG 842, Auburn Univ. Archives; E. V. Smith to Henry B. Gray, 1 Mar. 1957, and Smith to Draughon, 23 Mar. 1957, Box 2, Draughon Papers; and Joe P. Henderson to A. W. Todd, 10 Apr. 1958, Box 197, Fire Ant—Southern Region File, Plant Pest Control Papers.

[28] Carson, *Silent Spring* (cit. n. 1), p. 127. See also Rachel Carson, "A Sense of Values in Today's World," 17 Jan. 1963, Box 101, File 1895, Carson Papers; and U.S. Senate, Subcommittee on Reorganization and International Organizations, Committee on Government Operations, *Interagency Coordination in Environmental Hazards,* 88th Cong., 1st sess., 4 June 1963, pp. 206–216. Clarence Cottam, "Pesticides and Wildlife," 21–23 Feb. 1960, Box 35, File 581, Carson Papers; Cottam, "The Pesticide Problem," Series 1393, Carton 4, Florida State Archives, Tallahassee; Abbot to Members of the Board of Directors—Alabama Wildlife Federation, 24 June 1959, Box 1, Abbot File, Correspondence of Assistant Secretary for Fish and Wildlife Ross L. Leffler, 790 (A1), Secretary of Interior Papers (hereafter cited as **Secretary of Interior Papers**), RG 48, National Archives and

The USDA's opponents had more difficulties putting their vision of the relationship between nature, science, and democracy into practice than had the department. The agricultural agency was one of the most powerful federal bureaucracies, and control entomologists had embedded themselves deeply within a network of relationships with powerful allies. No attempts to stop pesticide use gained much political traction in the 1960s, but the protest against the imported fire ant program seemed the least likely to succeed. Just as *Silent Spring* was published in 1962, the USDA introduced a new insecticide for eradicating the ants, Mirex, that nullified the objections of the department's opponents. Billed as the perfect pesticide, Mirex was less harmful to vertebrates than its predecessors and was used at the incredibly low dose of one-seventh of an ounce per acre. With little left to object to and little power, the USDA's opponents turned their attention to other issues. When Carson died in 1965, the fire ant wars seemed to be over.[29]

"The Vietnam of Entomology"

By the 1970s, however, the situation had changed: Mirex was seen as dangerous and the environmentalists were on the ascendancy. Richard Nixon, wanting to coopt a Democratic constituency, created the Environmental Protection Agency and charged it with regulating the introduction of chemicals into the environment. An increasingly self-conscious environmental movement allied with the EPA to ban dozens of insecticides, most famously DDT, institutionalizing their ideas about the proper relationship between nature and science and endeavoring to create what they considered a more democratic nation.[30] In 1973,

Records Administration, College Park, Maryland. See also "Elements of Myth in the Fire Ant Peril," 11 Mar. 1957, Box 65, Fire Ant—News Clippings File, Entomology Research Division: General Correspondence, 1954–1958, 5 (UD), Agricultural Research Service Papers; Abbot to Byron T. Shaw, 25 Oct. 1957, Box 225, Quarantine File, and Abbot to Dwight Eisenhower, 31 May 1958, Box 3128, Insects, Aug. 8–Nov. 4 File, Secretary of Agriculture Papers; "Fire Ant Fight Intensified by State Group," 2 July 1958, Box 197, Fire Ant—News Clippings File, and Abbot to Marcus D. Byers, 14 July 1958, Box 225, Fire Ant File, Plant Pest Control Papers; Abbot to Smith, 24 Oct. 1958, Box 2, Draughon Papers; Abbot to Members of the Board of Directors—Alabama Wildlife Federation, 23 June 1959, 25 June 1959, Box 1, Abbot File, Correspondence of Assistant Secretary for Fish and Wildlife Ross L. Leffler, 790 (A1), Secretary of Interior Papers; Abbot to Paul B. Sears, 15 Aug. 1960, Box 135, Folder 221, Sears Papers; Abbot to Leffler, 17 Jan. 1961, Box 1, Abbot File, Correspondence of Assistant Secretary for Fish and Wildlife Ross L. Leffler, 790 (A1), Secretary of Interior Papers; and Abbot to Robert H. Michel, 2 Aug. 1961, Box 93, Imported Fire Ant—2 File, and "Fire Ant Spending," 16 Aug. 1961, Box 93, Fire Ant—8 File, Plant Pest Control Papers.

[29] Herbert L. Stoddard to Fancher, 30 June 1962, and Leo Iverson to Fancher, 5 Sept. 1962, Box 128, Fire Ant File; D. H. Janzen to Shaw, 31 Oct. 1962, Box 134, Insecticides—2—Toxic Effects File; Robert J. Anderson, "The Functions of the Federal Pest Control Review Board," 1 Nov. 1962, Box 134, Information—12—Talks, Speeches File; R. E. Lane to USDA, 3 Dec. 1962, Box 128, Fire Ant—9 File; Robert Anderson to Orville Freeman, 10 May 1963, and Maurice F. Baker, "New Fire Ant Bait," 1963, Box 165, Fire Ant File—5—Cooperation File; "Fire Ant Killer Tested at Refuge," n.d., Box 128, Fire Ant—8 File, Plant Pest Control Papers; "USDA Honors ARS Individuals, Groups . . . for Superior Service," *Agricultural Research,* June 1963, *11*:3–4; Jamie L. Whitten, *That We May Live* (Princeton, N.J.: Van Nostrand, 1966), p. 115; and Harrison Wellford, "Pesticides," in *Nixon and the Environment: The Politics of Devastation,* ed. James Rathelsberger (New York: Taurus, 1972), pp. 146–162. On the strength of the USDA see Grant McConnell, *The Decline of Agrarian Democracy* (Berkeley: Univ. California Press, 1953); McConnell, *Private Power and American Democracy* (New York: Knopf, 1966); Samuel R. Berger, *Dollar Harvest: The Story of the Farm Bureau* (Lexington, Mass.: Heath, 1971); Margaret W. Rossiter, "The Organization of Agricultural Sciences," in *The Organization of Knowledge in Modern America,* ed. John Voss and Alexandra Oleson (Baltimore: Johns Hopkins Univ. Press, 1979), pp. 211–248; Theda Skocpol and Kenneth Finegold, "State Capacity and Economic Intervention in the Early New Deal," *Political Science Quarterly,* 1982, *97*:255–274; and Gregory Hooks, "From an Autonomous to a Captured State Agency: The Decline of the New Deal in Agriculture," *American Sociological Review,* 1990, *55*:29–43. Bosso, *Pesticides and Politics* (cit. n. 1), elaborates the many connections between the USDA and southern congressmen who, because of the seniority system in the legislature and the South's *de facto* one-party system, controlled much of Washington's political machinery.

[30] "Toxic Substances: EPA and OSHA Are Reluctant Regulators," *Science,* 1979, *203*:28–32; and Dunlap,

prompted by the Environmental Defense Fund, the agency initiated a court case over the fate of the fire ant eradication program. In the years since it had been called the perfect pesticide, Mirex had been shown to accumulate in the fat of fish, kill shrimp and crabs, and, possibly, cause cancer.[31] The ants, on the contrary, were even more firmly established as nonthreatening. In the late 1960s some of E. O. Wilson's older contemporaries, feeling overshadowed by the young man's rapid ascent, revisited his taxonomic work—"*Someone* has to do the niddy-griddies to check out Wilson's theories while he continues onward and upward to still greater and greater glories"—and determined that the red and black ant forms were not mutants, but separate species, and that the red form was not as dangerous as Wilson had implied. Others suggested that the ants were a key part of the southern ecosystem that should not be removed. Even Wilson had changed his mind, calling the eradication program "the Vietnam of entomology": a battle with inchoate goals and no clear winners, only loss.[32]

The power of the EPA and the environmental movement slowly overwhelmed those promoting the eradication of the fire ants. The court case over the fire ant program and Mirex did not definitively settle the matter, but it made the cost of continuing too high for Allied Chemical, the maker of the insecticide. In 1976 Allied dropped out of the proceedings, selling its plant to the State of Mississippi for one dollar. Mississippi's waxing was

---

DDT (cit. n. 1). On the creation of the EPA see J. Brooks Flippen, "Pests, Pollution, and Politics: The Nixon Administration's Pesticide Policy," *Agr. Hist.,* 1997, *71*:442–456; and Flippen, *Nixon and the Environment* (Albuquerque: Univ. New Mexico Press, 2000).

[31] Regarding the court case see William A. Butler to Executive Committee and Executive Director, 11 May 1973, Box 68, Correspondence (from July 1, 1971) File; and Lee Rogers to Charlie Wurster, 29 May 1973, Box 71, Correspondence, Internal File, Environmental Defense Fund Papers (hereafter cited as **Environmental Defense Fund Papers**), MS 232, State Univ. New York, Stony Brook. On new findings about Mirex see C. C. Van Valin, A. K. Andrews, and L. L. Eller, "Some Effects of Mirex on Two Warm-Water Fishes," *Transactions of the American Fisheries Society,* 1968, *97*:185–196; Denzel Ferguson, "Fire Ant Eradication—Grandiose Boon-Doggle," 20 Mar. 1969, General Information to 1970 File, Series 2012: Fire Ant Correspondence, Mississippi Department of Agriculture and Commerce Papers (hereafter cited as **Mississippi Dept. Agriculture and Commerce Papers**), Mississippi Department of Archives and History, Jackson; J. R. Innes *et al.,* "Bioassay of Pesticides and Industrial Chemicals for Mutagenicity in Mice: A Preliminary Note," *Journal of the National Cancer Institute,* 1969, *42*:1101–1114; Lewis Nolan, "Further Study Suggested in Fire Ant Eradication Program," 5 Apr. 1970, General Information to 1970 File, Mississippi Dept. Agriculture and Commerce Papers; Jeff Nesmith, "Fire Ant Poison Kills Other Things," n.d. [1970?], Box 5227, Insects, Jan. 1 to Apr. 30 file [1 of 2], Secretary of Agriculture Papers; Deborah Shapley, "Mirex and the Fire Ant: Declines in the Fortune of a 'Perfect' Pesticide," *Science,* 1971, *172*:358–360; J. R. Gibson, G. W. Ivie, and H. W. Donough, "Fate of Mirex and Its Major Decomposition Products in Rats," *Journal of Agriculture and Food Chemistry,* 1972, *20*:1246–1248; T. E. Reagan, G. Coburn, and S. Hensley, "Effects of Mirex on the Arthropod Fauna of a Louisiana Sugarcane Field," *Environ. Entomol.,* 1972, *1*:588–591; E. G. Alley, "The Use of Mirex in Control of the Imported Fire Ant," *Journal of Environmental Quality,* 1973, *2*:52–61; K. M. Hyde *et al.,* "Accumulation of Mirex in Food Chains," *Louisiana Agricultural Experiment Station Bulletin,* 1973, *17*:10–11; and K. L. E. Kaiser, "The Rise and Fall of Mirex," *Environmental Science and Technology,* 1978, *12*:520–528. On the EPA's preoccupation with cancer see Edmund P. Russell, "Lost among the Parts per Billion: Ecological Protection at the United States Environmental Protection Agency, 1970–1993," *Environmental History,* 1997, *2*:29–51.

[32] William F. Buren to Creighton, 24 Apr. 1972, Creighton Papers (on the "niddy-griddies"); "Fire Ant War Lost by U.S.," 29 Sept. 1975, Box 73, Mirex—Press Releases File, Environmental Defense Fund Papers; and "Fire Ants: Vietnam of Entomology," news clippings, Mississippi Dept. Agriculture and Commerce Papers (for Wilson's new view). On the revision of Wilson's taxonomy see William F. Buren, "Revisionary Studies on the Taxonomy of the Imported Fire Ants," *Journal of the Georgia Entomological Society,* 1972, *7*:1–26; and Joshua Blu Buhs, "Building on Bedrock: William Steel Creighton and the Reformation of Ant Systematics, 1925–1970," *Journal of the History of Biology,* 2000, *33*:27–70. On the ants and the ecosystem see William L. Brown to W. Wallace Harrington, 7 May 1973, and "Environmental Defense Fund Response to 1972 Environmental Impact Statement," 17 Feb. 1973, both in *Miscellaneous Publications and Unpublished Documents Relating to the Use of Mirex to Control the Imported Fire Ant in the USA,* Comstock Library, Cornell Univ., Ithaca, New York. Cf. Brown to Rogers, 14 Dec. 1970, Box 68, Correspondence (up to July 1, 1971) File, Environmental Defense Fund Papers.

short-lived, however. Two years later, the EPA determined that Mirex and its by-products caused cancer and that the chemical could be found in the bodies of almost one in every two Mississippians. The use of Mirex was phased out.[33] In the years since it had been established, the EPA had also banned all other chemicals used to eradicate the imported fire ants. By the end of the 1970s, the fire ant wars were over. The ants were left to accommodate themselves to life in North America, safe from chemical attack.

IDEAS INTO ACTIONS: THE EFFECT OF THE FIRE ANTS WARS

The USDA and its opponents had focused on different aspects of the natural history of the imported fire ants, generalizing particular traits into the essence of the animal—like the five blind men of legend who touched different parts of an elephant and divined its essence from those parts, one touching the tail and deciding that an elephant was like a rope and another feeling the ear and deciding that the beast was like parchment. Committed to protecting American agriculture and confronting the ants on a regional scale, the federal entomologists focused on the negative aspects of the insects. (See Figure 4.) The dissenters, on the contrary, looking at the invasion on a smaller scale and wedded to the belief that nature was an integrated whole, studied the places where the irruption was dying and determined that the ants were no longer pests. They pointed to the way the opportunistic ants preyed on other insects as evidence of their beneficence and their acceptance into the American ecological order. Both sets of ideas, however, were simplifications of a more complicated natural history. When they were transformed into action and applied through American agricultural and environmental policy, the friction between the ideas and the reality created situations that no one expected.

The USDA entomologists believed that nature could be remade without consequence and so applied the insecticides fully expecting to eradicate the ants. The insecticides, however, were broad-spectrum poisons that killed huge numbers of insects and game animals. They disrupted the areas where they were applied, much as the vagaries of the rivers of South America did. Imported fire ants reinvaded the poisoned parcels of land, mocking the eradication ideal. In 1957, for example, Arkansas was declared ant-free after twelve thousand acres were sprayed. In 1958, however, the ants occupied ten thousand

[33] Shapley, "Mirex and the Fire Ant" (cit. n. 31), p. 359; Fancher to Sonny Montgomery, 17 Feb. 1972, Congressional Delegation, 1971–1972 File, Mississippi Dept. Agriculture and Commerce Papers; Maureen K. Hinkle to Judy Swiderski, 26 Apr. 1976, Box b194, Mirex File, National Audubon Society Papers; "Mississippi to Make Pesticide," 12 May 1976, Box 71, Correspondence, Internal File, Environmental Defense Fund Papers; "South Still Fights Its Longest War, Learns to Live with Surly Fire Ants," 28 June 1976, News Reports File, Mississippi Dept. Agriculture and Commerce Papers; William A. Banks *et al.,* "An Improved Mirex Bait Formulation for Control of Imported Fire Ants," *Environ. Entomol.,* 1973, 2:182–185; Banks *et al.,* "Imported Fire Ants: 10-5, an Alternate Formulation of Mirex Bait," *J. Econ. Entomol.,* 1975, 69:465–467; "Human Tissue Shows Mirex," 16 July 1976, and "Mirex Found in Bodies Fuels Pesticide Controversy," 8 July 1976, Box b194, Mirex File, National Audubon Society Papers; "Mirex Found in 44 per cent of Mississippians," 25 Aug. 1976, Box 73, Mirex—Press Releases File, Environmental Defense Fund Papers; "Mississippi Back in Mirex Business," n.d., "State Asks EPA to Ban Mirex Use in Two Years," n.d., and "Mirex's Phase Out," n.d., News Reports File, Mississippi Dept. Agriculture and Commerce Papers; "Mirex Cancellation Announced," *EDF Newsletter,* Sept. 1976, 7; "Spread of a Deadly Chemical—and the Ever-Widening Impact," *U.S. News and World Report,* 6 Sept. 1976, pp. 43–44; Jim Buck Ross to All News Media in Mississippi, 22 Sept. 1976, and "EPA Okays Phaseout Program for Mirex," 29 Oct. 1976, News Reports File, Mississippi Dept. Agriculture and Commerce Papers; "Mississippi Stymied by Lack of Fire Ant Poison," 23 Nov. 1976, Box 73, Mirex—Press Releases File, Environmental Defense Fund Papers; "Mirex Plant Shutdown Blocks War on Fire Ant," 27 Nov. 1977, News Clippings File, Mississippi Dept. Agriculture and Commerce Papers; and Ardith Maney, "New Influences on Agricultural Policy: Fire Ants, Environmentalists, and the Agricultural Research Service," Box 73, Mirex—Maney (Iowa State) Article File, Environmental Defense Fund Papers.

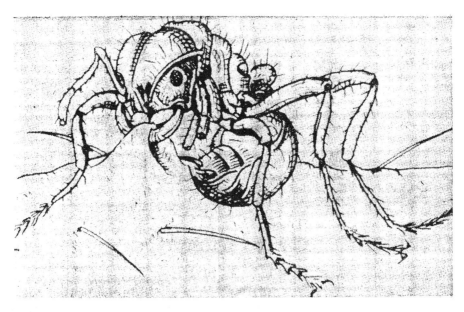

Figure 4. *A fire ant grasps human flesh with its mandibles and bends its gaster forward, prepared to sting. The pointed structures appearing here and there are human hairs. The USDA illustrated many of its publications with this picture of the imported fire ant; newspaper articles throughout the South reprinted the image. Some news publications labeled this picture "Monster," underlining the danger posed by the ants and the need to eradicate the insects. (From "$10 Million Sought to Fight Fire Ants," Montgomery Advertiser, 8 March 1957, Box 65, Fire Ant News Clippings File, Entomology Research Division: General Correspondence, 1954–1958, 5 (UD), Agricultural Research Service Papers, RG 310, National Archives and Records Administration, College Park, Maryland.)*

acres, many of them areas that had been treated previously. And in 1959 the ants could be found on twenty-five thousand acres. By 1960 the USDA noticed the reinvasion problem on a regional scale. One disappointed entomologist complained, "It is probable that the acreage being reinfested plus the expansion into new areas each year is as great as the acreage that can be treated with presently available funds. The program can hardly be termed an eradication program if there is no net gain in acreage free from ants." Over the course of the fire ant wars, the USDA tried a number of different insecticides and a number of different rates of application, but all failed.[34]

With the failure of eradication came a diminished belief in the professional competency

[34] Lamar J. Padget to Burgess, 11 Mar. 1959, and Padget to W. E. Blasingame, 29 June 1959, Box 213, Fire Ant File (on Arkansas); and "Appraisal Survey of the Cooperative Fire Ant Program, Sept. 1960, Summary and Comments," 12 Oct. 1960, Fire Ant File, 1960, Plant Pest Control Papers (quotation). For an overview of the eradication–reinfestation cycle see William F. Buren, "The Importance of Fire Ant Taxonomy," *Proceedings of the Tall Timbers Conference on Ecological Animal Control by Habitat Management,* 1978, 7:61–66; Buren, G. E., Allen, and R. N. Williams, "Approaches toward Possible Pest Management of the Imported Fire Ants," *Bull. Entomolog. Soc. Amer.,* 1978, 24:418–420; "Fire Ant Preventive Linked to Infestation," *Jackson [Mississippi] Clarion-Ledger,* 8 Mar. 1982, p. 8b; Tschinkel, "History and Biology of Fire Ants" (cit. n. 6); and Clifford S. Lofgren and David F. Williams, "Red Imported Fire Ants (Hymenoptera: Formicidae): Population Dynamics Following Treatment with Insecticidal Baits," *J. Econ. Entomol.,* 1985, 78:863–867. On the varied insecticide trials and failures see "Fire Ant Control," report no. 62 (Ames, Iowa: Council for Agricultural Science and Technology, 1976); and Anne-Marie A. Callcott and Homer L. Collins, "Invasion and Range Expansion of Imported Fire Ants (Hymenoptera: Formicidae) in North America from 1918–1995," *Florida Entomologist,* 1996, 79:238–251.

of control entomologists. Control entomologists had bet their scientific legitimacy on eradication: C. C. Fancher had announced at the beginning of the fire ant program, "It would
be a disgrace to entomologists of this country to permit the imported fire ant to become
established."[35] As the program progressed, though, the dream faded. The ants continued
to spread, as did other insects targeted by the USDA—bark beetles, gypsy moths, Japanese
beetles.

By the 1960s the embarrassment was acute, and it would only get worse. Even with the
introduction of the so-called perfect pesticide, many entomologists in the USDA doubted
whether eradication was possible.[36] Their backbone stiffened in 1969 when Nixon appointed J. Phil Campbell to the USDA. A former commissioner of agriculture in Georgia,
Campbell had built his career battling the ants and demanded that the agency renew its
commitment to eradication. Campbell's enthusiasm, though, could not stem the spread of
the insects. In 1978, deciding that the battle could not be won, the USDA shifted all funding
for the program away from control entomologists, awarding it instead to insect biologists
who hewed to older traditions in entomology, investigating insect natural history and employing biological and cultural control. Study of parasites and predators increased, and in
1995 the agency began releasing into the wild flies that decapitate the ants. The goal now
was control, not eradication. Indicating how far acceptance of the control entomologists'
terms had fallen, the Jackson, Mississippi, *Clarion-Ledger* reported, "'Eradication' is a
dirty word among the small corps of fire ant researchers."[37]

Environmental policy created similar, if less severe, professional problems, and for the
same reason: the unexpected outcomes of mixing human practices and the biology of the
ants. In her rehabilitation of the ants, Carson had excluded any consideration of the role
humans played in creating the irruption. This exclusion was part of Carson's more general
tendency to see humans as separate from the natural world, a habit of thought that was
taken up by later environmental groups and institutionalized by the EPA. This perspective
overlooked human responsibility for the fire ant problem and led to solutions that failed
to address the root cause of their spread. Insecticides were banned, but the bulldozer
revolution continued and the ants continued to spread, covering another hundred million
acres by the century's end. They reached western Texas, New Mexico, Arizona, Nevada,
and California. Wherever they appeared, their population exploded and they wreaked

---

[35] Fancher to Burgess, 20 May 1958, Box 197, Fire Ant—Southern Region File, Plant Pest Control Papers.

[36] Wellford, *Sowing the Wind* (cit. n. 1), pp. 302–303. D. R. Shepherd to F. J. Mulhern, 5 May 1969, Box 260,
File 8, Plant Pest Control Papers; John A. Schmittker to George Mehren *et al.,* 27 Oct. 1967, Box 4679, Insects
(October) File, Secretary of Agriculture Papers; Fancher to Ross, 20 Sept. 1971, Memos to Commissioner, 1970–
1971 File, Mississippi Dept. Agriculture and Commerce Papers; Joseph M. Robertson to Mehren, 7 Apr. 1967,
Doyle Conner to Spessard Holland, 10 Apr. 1967, and Charles L. Grant to Mehren and George Irving, 16 May
1967, Box 4680, Insects, April 1967–31 May 1967 File; Mehren to Gillespie F. Montgomery, 8 June 1967, and
Irving to the Secretary, 20 June 1967, Box 4679, Insects (June) File; and T. C., Byerly to Thomas T. Irvin, 20
Apr. 1970, Box 5227, Insects Jan. 1 to Apr. 30 File [1 of 2], Secretary of Agriculture Papers.

[37] "Fire Ants: The Experts Throw Up Their hands," *U.S. News World Rep.,* 17 Jan. 1977, p. 79; Alan Huffman,
"Environment May Pay the Price in Fire Ant War," *Jackson Clarion-Ledger,* 4 Nov. 1988; Sanford D. Porter *et
al.,* "Growth and Development of *Pseudacteon* Phorid Fly Maggots (Diptera: Phoridae) in the Heads of *Solenopsis* Fire Ant Workers (Hymenoptera: Formicidae)," *Environ. Entomol.,* 1995, *24:*475–479; L. W. Morrison
and L. E. Gilbert, "Oviposition, Behavior, and Development of *Pseudacteon* Flies (Diptera: Phoridae), Parasitoids
of *Solenopsis* Fire Ants (Hymenoptera: Formicidae)," *ibid.,* 1997, *26:*716–724; Porter, David F. Williams, and
R. S. Patterson, "Rearing the Decapitating Fly *Pseudacteon tricuspis* (Diptera: Phoridae) in Imported Fire Ants
(Hymenoptera: Formicidae) from the United States," *J. Econ. Entomol.,* 1997, *90:*135–138; Bruce Reid, "USDA
Launches Biological War on Fire Ants," *Jackson Clarion-Ledger,* 7 June 1998; Porter, "Host-Specific Attraction
of *Pseudacteon* Flies (Diptera: Phoridae) to Fire Ant Colonies in Brazil," *Florida Entomol.,* 1998, *81:*423–429;
and Porter and L. E. Alonso, "Host Specificity of Fire Ant Decapitating Flies (Diptera: Phoridae) in Laboratory
Oviposition Tests," *J. Econ. Entomol.,* 1999, *92:*110–114.

havoc.[38] Analyses that separated humans and nature obscured the critical importance of interaction between the two.

Many felt vulnerable without a way to stop the spread. The commercial insecticides that were available were for individual use, not widespread spraying, and they could not stop the introduction of the ants into a new area, only ameliorate the danger once they were ensconced. Meanwhile, alternative control techniques were slow in coming. It took the USDA twenty years to initiate a biological control program on a wide scale, and there remains no measure of its effectiveness. The EPA, for its part, blocked the introduction of other alternative techniques. In the early 1970s the biotechnology company Zoecon developed an analogue of a fire ant hormone, a chemical that would prevent ant larvae from developing into adults. This insect growth regulator, as it was called, was expensive to develop and had a smaller market than DDT and other such chemicals since the hormone analogue could be used only against the imported fire ants. But the EPA, following in Carson's tradition, was leery about introducing chemicals into the environment and required that the analogue undergo the same battery of tests as a broad-spectrum pesticide; when Zoecon's president pleaded for grants to help fund the necessary studies, the EPA refused. The cost was prohibitive and the insect growth regulator did not make it to market. Other species-specific chemicals, lacking support from the EPA, experienced similar fates.[39]

The EPA, then, had banned the chemicals perceived to control the ants, prevented the development of alternative forms of control, and left the ants to spread. Now it was the EPA—not the USDA—that looked like a bureaucracy unconcerned with civil liberties.[40] The president of Zoecon made this point when he noted that the Soviet Union had asked his company to develop new pest control methods: the most communist country on earth recognized the innovative possibilities, but an American bureaucracy could not. The USDA exploited the growing resentment toward the EPA in an attempt to win back some of the power it had lost to the agency. Here, too, was one of the roots of the Reagan-era backlash against environmentalism and the EPA, when the agency went into a decade-long

[38] William P. Mackay and Richard Fagerlund, "Range Expansion of the Red Imported Fire Ant, *Solenopsis invicta* Buren (Hymenoptera: Formicidae), into New Mexico and Extreme Western Texas," *Proceedings of the Entomological Society of Washington,* 1997, 99:757–758; Deborah Schoch, "Fire-Ant Fear on the March in Western States," *Los Angeles Times,* Orange County ed., 7 Feb. 1999; Schoch, "Invasion of Fire Ants Poses New Threat to Area Wildlife," *Los Angeles Times,* home ed., 1 Mar. 1999; "Imported Fire Ants a Red-Hot Nuisance," *Sacramento Bee,* 6 Mar. 1999, p. CL2; Schoch, "State Issues Its Marching Orders for Fire Ant War," *Los Angeles Times,* Orange County ed., 20 Mar. 1999; "State Turns up the Heat on Aggressive Fire Ants," *Sacramento Bee,* 22 Mar. 1999, p. A1; David Reyes, "Fire Ants Defeating Eradication Force," *Los Angeles Times,* home ed., 22 July 2000; "Fire Ants Have a Nasty Sting," *Sacramento Bee,* 23 Oct. 1999, p. CL12; "Is It 'Them!'?" *Sacramento Bee,* 3 May 2000, p. A4; Matthew Ebnet, "Fire-Ant Campaign Turns up the Heat," *Los Angeles Times,* Orange County ed., 6 Sept. 2000; and Kiley Russell, "Burned by Fire Ants," *Sacramento Bee,* 23 Feb. 2001, p. D1.

[39] Gene Bylinsky, "Zoecon Turns Bugs against Themselves," *Fortune,* Aug. 1973, 88:94–103; Carl Djerassi to Robert Long, 12 Nov. 1973, and Carl Djerassi, Christina Shih Coleman, and John Diekman, "Operational and Policy Aspects of Future Insect Control Methods," 12 Feb. 1974, Box 5855, Insects, April 19 to July 16 File, Secretary of Agriculture Papers; "John Diekman, Direct Testimony," 21 Feb. 1974, Box 68, Summaries File, Environmental Defense Fund Papers; and Djerassi to Paul A. Vander Myde, 14 May 1974, Box 5855, Insects, April 19 to July 16 File, Secretary of Agriculture Papers. For similar complaints against the EPA see Marc K. Landy, Marc J. Roberts, and Stephen R. Thomas, *The Environmental Protection Agency: Asking the Wrong Questions from Nixon to Clinton,* expanded ed. (New York: Oxford Univ. Press, 1994); and Mark Winston, *Nature Wars: People vs. Pests* (Cambridge, Mass.: Harvard Univ. Press, 1997), pp. 158–159.

[40] The EPA administrator John Quarles admitted the problem: "I am a federal bureaucrat. I am one of those bureaucrats everyone complains of," he began his 1976 book *Cleaning Up America.* "I am one of those whom George Wallace has called lazy, incompetent and 'pointy-headed.'" John Quarles, *Cleaning Up America: An Insider's View of the Environmental Protection Agency* (Boston: Houghton Mifflin, 1976), p. xi.

dormancy. Exterminator–cum–House Whip Tom DeLay, exercised at the ban on Mirex, charged the EPA with stifling necessary insect control operations under a pile of red tape. Texas Senator Phil Gramm also attacked the EPA's policy on the fire ants, looking—like DeLay—for an excuse to eviscerate the regulatory bureaucracy.[41]

The EPA's loss of status was not as stark as that of the control entomologists—the EPA remains the nation's largest regulatory body—nor was its failure as great—the EPA did remove carcinogenic chemicals from the market—but the decline in its legitimacy stemmed from the same cause. The biology of the ants and human activities interacted in unexpected ways. The ants were not ecological innocents, ready to become well-behaved citizens. They were opportunists, exploiting the disruption brought about by the bulldozer revolution and the constant building in the irrigated West. Failing to deal with the causes of the fire ant irruption, the EPA and the environmental community saw their predictions founder and their legitimacy weaken.

CONCLUSION

In 1976, when Mississippi purchased the Mirex manufacturing plant, environmentalists believed that their prophecies had come true. The ideology of eradication had led to American socialism: a state now owned a business. Journalists sympathetic to the anti-insecticide crusade lampooned "Magnolia-Scented Socialism." One newspaper wrote, "A losing effort to conquer a tiny, unconquerable insect has brought a strange and costly species of socialism to the Deep South. Today, the final citadel of this socialism is to be found, of all places, in Mississippi, whose politicians for decades urged the voters to resist any hint of socialism wherever it seemed to be nibbling at the woodwork of free enterprise."[42]

The situation, though, was not so cut and dried. The imagined ends of the Cold War—victory, capitulation, or transformation—that had come to stand for the imagined ends of the fire ant wars were too restrictive: magnolia-scented socialism soon collapsed. There was neither victory nor capitulation nor transformation: the ants continued to spread and entomologists continued to battle them in a country that continued to favor relatively weak bureaucratic governance. When it became clear that these imaginings were inadequate, a new view—one less dedicated to castigating or praising the imported fire ants—emerged. A Texas entomologist encapsulated this new view when he wrote in the late 1970s that imported fire ants wear neither a white nor a black hat, but a gray one. The insects do

[41] Nancy Mathis, "EPA under House Siege: DeLay Plans to Lead Assault for Changes in Environmental Policy, Agency Funding," *Houston Chronicle,* STAR ed., 8 Oct. 1995, p. 13A; and Scott Norvell, "Force of Nature," *Washington Post Magazine,* 23 June 1996, p. 14. For complaints about the EPA see Djerassi to Long, 14 Feb. 1974, Box 5855, Insects, April 19 to July 16 File, Secretary of Agriculture Papers; U.S. Congress, Subcommittee on Department Operations, Investigations and Oversights, Committee on Agriculture, *Fire Ant Eradication Program,* 94th Cong., 1st sess., 1975; and G. F. Engler to John Paul Hammerschmidt, 16 Aug. 1975, Louis Gregory to John C. Stennis, 19 Aug. 1975, and Earl McMunn, "Let's Reason Together," n.d., [1975?], Box 6012, Pesticides, July 1 to August 31 File, Secretary of Agriculture Papers. See also "Administrator's Letter," 26 Feb. 1976, and Shirley Schiebla, "Fighting the Fire Ant: Regulations Make It Almost Impossible to Do," 10 May 1976, Box 73, Mirex—Press Releases File, Environmental Defense Fund Papers; and "The Year of Insects," editorial, *Wall Street Journal,* 15 Aug. 1978. On seeds of the Reagan-era backlash see Hays, *Beauty, Health, and Permanence* (cit. n. 23), pp. 491–526.

[42] "Mississippi to Make Pesticide," *Washington Star,* 12 May 1976, p. A14, Box 71, Correspondence, Internal File, Environmental Defense Fund Papers; "'Magnolia-Scented Socialism' Invades Mississippi," *Gulfport Star Journal,* 2 Sept. 1976, News Reports File, Mississippi Dept. Agriculture and Commerce Papers; and "Solenopsis the Unconquerable," *Ecology Center Newsletter,* July–Aug. 1976, pp. 5–6, Box 73, Mirex—Press Releases File, Environmental Defense Fund Papers.

damage crops and they do attack young quail, but they also consume boll weevils.[43] This ambiguous ant is celebrated on the first Saturday of October each year in Marshall, Texas, when the city puts on a Fire Ant Festival. Participants are encouraged to vent their aggression toward the insects by tearing apart a stuffed ant. But they also choose a Miss Fire Ant in a beauty pageant, compete in a fire ant mating call contest, and judge a chili cook-off in which each pot must contain at least one fire ant. The environmentalist and singer-songwriter Bill Oliver has captured this new image in song:

> You who live in cities, you who live in neighborhoods
> Fire ants, it's understood, may come and take their stand
> The males that die in nuptials, the queens that come in multiples,
> The fire ant, combustible, is hard to understand.[44]

The combined tools of the history of science and environmental history, however, make the story of the ants less hard to understand. Environmental history teaches the need to pay attention to the natural world and the effects humans have on it. History of science teaches the need to look at the social processes that are constitutive parts of scientific thought and work. Put together, these techniques reveal intricate and important interconnections between nature and science. The biology of the fire ants mattered—and it was misunderstood by both the USDA and its opponents: the ants were neither specially adapted mutants nor invented bogeymen, but opportunists following the bulldozer across the South. The social history of the competing groups mattered, too, for the fire ant wars were not about Rachel Carson speaking truth to power but about the clash of two different visions of nature, science, and the state.

The interaction of the humans and the fire ants mattered as well. The legitimacy of the USDA control entomologists and their various opponents rested on this interaction, reliant on a nature that did not always behave as predicted but instead responded to a combination of its own rhythms and the actions of humans. Over the past three decades, historians of science have shown how scientists gain and maintain cultural authority through rhetorical techniques and political machinations. A focus on nature adds a new category of analysis to this familiar repertoire: cultural authority depends, as well, on interactions with nature, on making the material world perform in particular, specified ways. Historians of science need ways to integrate the natural world into their accounts to make sense of this—and other—aspects of science. Environmental history offers such tools, opening new paths in the field that build on the hard-won victories of past generations and provide for even fuller historical narratives.

---

[43] W. L. Sterling, "Imported Fire Ant . . . May Wear a Grey Hat," *Texas Agricultural Progress,* 1978, 24:19–20. See also Charles Seabrook, "Some Farmers Sing Praises of Fire Ant: South American Insect Preys on a Number of Crop Pests," *Atlanta Journal-Constitution,* 12 Jan. 1988, p. A1; and Joe Murray, "If You Can't Beat 'Em, Join 'Em: Raising Fire Ants for Fun and Profit?" *ibid.,* 8 Oct. 1996, p. A13.

[44] Bill Oliver and the Otter Space Band, "Queen Invicta (Fireant Invincible)," on *Have to Have a Habitat* (Texas Deck Music, BMI, 1995), lyrics available at "Bill Oliver's Environmental Songbook," *url: http:// www.geocities.com/mrhabitat/songbook.html,* accessed 2 Mar. 2002. See also Suzanne Gamboa, "Fire Ant Festival Offers a Chance for Revenge: 45,000 Expected at Marshall's Weekend Party," *Dallas Morning News,* 3 Oct. 1988; and Richard Conniff, "You Never Know What the Fire Ant Is Going to Do Next," *Smithsonian,* July 1990, pp. 48–57.

II

PATRONS, POLITICS, AND THE PHYSICAL SCIENCES

Patenting the Bomb

Nuclear Weapons, Intellectual Property, and Technological Control

By Alex Wellerstein

ABSTRACT

During the course of the Manhattan Project, the U.S. government secretly attempted to acquire a monopoly on the patent rights for inventions used in the production of nuclear weapons and nuclear energy. The use of patents as a system of control, while common for more mundane technologies, would seem at first glance to conflict with the regimes of secrecy that have traditionally been associated with nuclear weapons. In explaining the origins and operations of the Manhattan Project patent system, though, this essay argues that the utilization of patents was an *ad hoc* attempt at legal control of the atomic bomb by Manhattan Project administrators, focused on the monopolistic aspects of the patent system and preexisting patent secrecy legislation. From the present perspective, using patents as a method of control for such weapons seems inadequate, if not unnecessary; but at the time, when the bomb was a new and essentially unregulated technology, patents played an important role in the thinking of project administrators concerned with meaningful postwar control of the bomb.

N O TECHNOLOGY IN THE TWENTIETH CENTURY has been as intertwined with the policies of secrecy as nuclear weapons. In the United States, these weapons have always been manufactured by government monopoly, and the specific information about their manufacture has long been the target of policies meant to cloister that information within the highest levels of classification. The American patent system, on the other hand, has long been regarded as a tool of legal openness: the inventor is granted a temporary monopoly on the production of an invention in return for disclosing how it works.

As such, the patent system is not, from the present point of view, an obvious choice for

Isis 99, no. 1 (March 2008): 57–87.

I would like to thank Ellen Bales, Mario Biagioli, Cathryn Carson, Peter Galison, Michael Gordin, Hugh Gusterson, David Kaiser, and the anonymous *Isis* referees for their input. This essay was presented at the Con/texts of Invention Conference, hosted by the Society for Critical Exchange, Case Western Reserve University, 20–23 April 2006, and at the 2006 annual meeting of the History of Science Society in Vancouver, British Columbia, 2–5 November 2006.

dealing with nuclear secrets. But it is not well known that the United States government filed *thousands* of patent applications—in secret—on all aspects of the development and deployment of the first atomic bombs during the Manhattan Project. Was the U.S. government going to sue the Soviets for infringement if they developed their own bomb? Perhaps demand royalties? Would the government have let conflicting patent claims stand in the way of its nuclear ambitions? What did the patent system provide that could not be better, and more sensibly, achieved by other means? Why conceive of nuclear weapons within a system of intellectual property that now appears so ill adapted to them?

It is an approach that seems at first glance to run counter to every other aspect of security undertaken during the Manhattan Project. Instead of compartmentalizing information relating to bomb design, patenting would serve to centralize it. Instead of using code-names to refer to processes or materials, patents would all have to follow standard technical terminology. Instead of being secured within the barbed-wire fences that surrounded each of the major project production sites, the details of nuclear secrets would be kept in a safe in a government building visited daily by uncleared civilians, the U.S. Patent Office. And perhaps most obviously bizarre: patent laws were in theory supposed to encourage private innovation, and it hardly needs to be stated that this was not in the plans of the civilian and military administrators in charge of the country's atomic weapons program, itself created out of the fear that just such "private innovation" was taking place in Nazi Germany.

Why undertake such a policy? What did it mean to patent the bomb? The question was asked, indeed, when the program's existence was first made public six months after the end of the war. At a hearing before the Special Committee on Atomic Energy in February 1946, the officer in charge of the patent program, Captain Robert A. Lavender, informed the confused senators that the bomb itself had been patented:

> The CHAIRMAN: Now, are there any patent applications covering the making of the bomb?
> Captain LAVENDER: Yes.
> The CHAIRMAN: Are there any patent applications giving the bomb-making details in those patent applications?
> Captain LAVENDER: Well, I think that I had better give you that in executive session if you are going into any of the details of it.
> The CHAIRMAN: Not what details there were, but whether there were any of the details given in the patent application. You don't want to talk about that?
> Captain LAVENDER: Not any more than just to say that the bombs are covered by applications.
> The CHAIRMAN: I wonder what is the necessity for covering the bomb itself by applications for patents?

The perplexed chairman was none other than "Atomic" Senator Brien McMahon, the Connecticut sponsor of what would eventually become the Atomic Energy Act of 1946, the first postwar legislation defining the U.S. nuclear infrastructure. Captain Lavender explained that his office, on behalf of the U.S. government, had filed patent applications for all aspects of bomb manufacture in secret under the authority of the Commissioner of Patents because it had been feared that private inventors might file speculative patent applications and believed that the "first-to-file" status of the U.S. government would help in potential interference lawsuits. Rather than answering the question of why the Manhattan Project had turned to the patent system, Lavender's answer begged it. The senators were skeptical. "I didn't dream, frankly, up until this point," McMahon said, addressing a fellow senator and committee member, "that there was a patent application down there showing how the bomb was put together. Did you?" "No," the other senator replied. "Personally, I regret it."[1]

As the hearing went on, though, the senators stopped asking *why* it had been done and instead concentrated on *what* had been done and on what *should* be done in the future. The result was the legislative decree that "no patent shall hereafter be granted for any invention or discovery which is useful solely in the production of fissionable material or in the utilization of fissionable material or atomic energy for a military weapon"—a move that attempted to resolve the patent question by removing the issue altogether.[2] As "atomic energy" was then the only field of scientific and technological inquiry in which private patenting was explicitly prohibited by an act of Congress, it was an anomaly noted in many works on patent law and occasionally even in relatively recent court cases relating to patenting and patentability.[3]

The historiography of the atomic bomb is extensive: few topics in the history of science have received such thorough "historical" attention from their earliest inception (indeed, the first draft of the bomb's history was released only days after the Nagasaki bomb itself).[4] There are by now dozens of accounts of the Manhattan Project; most reproduce the sort of heroic, ideas-and-men narrative that has been present since the 1940s, with a heavy emphasis on physics, the wills and actions of individuals, and the idea that the atomic bomb "changed everything." New archival sources have been used in order to further the project of what Hugh Gusterson calls "nuclear salvage history," the attempt to rediscover the scientific authors whose names and activities were obscured or neglected altogether as a result of classification policies.[5] But works have also emerged that challenge this narrative, looking at aspects of the project that have been neglected by the traditional narrative and at the same time interrogating the construction of the narrative itself; they open the possibility that more work on the Manhattan Project can be done without resorting to, or relying on, further "salvage."[6]

The patenting program itself, like the rest of the Manhattan Project, was very large. It

[1] Statement of Capt. Robert A. Lavender, Atomic Energy Act of 1946 Hearings before the Special Committee on Atomic Energy, U.S. Senate, 79th Cong., 2nd sess., on S.1717, Pt. 3, 11 Feb. 1946, pp. 337–358, on p. 347.

[2] "An Act for the Development and Control of Atomic Energy," Public Law 585, 1 Aug. 1946, Sect. 11(a). The word "solely" is worth noting in particular here—technologies with multiple uses could still be patented.

[3] For the former see, e.g., the way it is discussed in Floyd L. Vaughan, *The United States Patent System: Legal and Economic Conflicts in American Patent History* (Norman: Univ. Oklahoma Press, 1956). A prominent example of the latter is its invocation by Chief Justice Warren E. Burger in *Diamond v. Chakrabarty*, 447 U.S. 303 (1980), to indicate that if Congress wanted to it could set biological organisms off-limits from patenting, as it had once done with atomic energy.

[4] Henry DeWolf Smyth, *Atomic Energy for Military Purposes: The Official Report on the Development of the Atomic Bomb under the Auspices of the United States Government, 1940–1945* (Princeton, N.J.: Princeton Univ. Press, 1945). For a critical view of the Smyth Report's influential role in shaping the narrative of the atomic bomb see Rebecca Press Schwartz, "The Making of the Historiography of the Atomic Bomb" (Ph.D. diss., Princeton Univ., in progress).

[5] Hugh Gusterson, "Death of the Authors of Death: Prestige and Creativity among Nuclear Weapons Scientists," in *Scientific Authorship: Credit and Intellectual Property in Science*, ed. Mario Biagioli and Peter Galison (New York: Routledge, 2003), pp. 281–307. For the "standard" narrative see, e.g., Richard G. Hewlett and Oscar E. Anderson, Jr., *The New World: A History of the U.S. Atomic Energy Commission, 1939–1946* (Berkeley: Univ. California Press, 1962); Stéphane Groueff, *Manhattan Project: The Untold Story of the Making of the Atomic Bomb* (Boston: Little, Brown, 1967); Richard Rhodes, *The Making of the Atomic Bomb* (New York: Simon & Schuster, 1986); and Lillian Hoddeson et al., *Critical Assembly: A Technical History of Los Alamos during the Oppenheimer Years, 1943–1945* (New York: Cambridge Univ. Press, 1993).

[6] One important but largely ignored work in this vein is Ruth H. Howes and Caroline L. Herzenberg, *Their Day in the Sun: Women of the Manhattan Project* (Philadelphia: Temple Univ. Press, 1999). Recent works of this sort include Michael D. Gordin, *Five Days in August: How World War II Became a Nuclear War* (Princeton, N.J.: Princeton Univ. Press, 2007); and Schwartz, "Making of the Historiography of the Atomic Bomb" (cit. n. 4).

enjoyed support from such top brass as Vannevar Bush, General Leslie Groves, and even President Franklin D. Roosevelt. It required direct intervention by the Commissioner of Patents, involved a platoon of lawyers spread across the country, and reached outside the confines of the government bureaucracy to grapple with the private sphere. All of this makes it particularly important to note that the existence of such a program is barely mentioned in secondary historical literature, though minor aspects of it occasionally surface as elements in other studies, and no comprehensive picture—or even an acknowledgment of a large, generalized patenting program—has yet emerged.[7] The reason for this lack of attention is, I suspect, at least partly related to the conceptual contradiction at the heart of the program: the deliberate placement of military-technical secrets into a system of legal openness. The traditional historical narrative of the atomic bomb, in which the "special" nature and "taboo" of the bomb are posited as having existed from the first moment, leaves little room for a technique like patent control as a feature of national security.[8]

When scholars talk about nuclear weapons in terms of intellectual property, they almost invariably conceptualize them within the model of trade secrecy. Nuclear weapon designs are conceived of as having a technical epistemology more similar to the formula for a soft drink than the design of an automobile: they are kept in a vault, the monopoly is not guaranteed by law, innovation (among national competitors) is discouraged, and the designs are thought to be at least somewhat concealable (though antisecrecy advocates have often challenged this). The paradigmatic nuclear secret, the Teller-Ulam design of the hydrogen bomb, is generally treated as a formulation of knowledge that could, theoretically, be kept from other parties, though they could—and did—come to the same knowledge independently.[9] Not surprisingly, the choice between two different models of intellectual property for a given technology depends as much on the way the technology will be deployed in the world as it does on the makeup of the technology itself.

This essay attempts to shed some light on the Manhattan Project's patenting program, to look at its motivations, its methods, its results, and, ultimately, its importance. I discuss the patent program in three discrete sections. The first section covers the evolution of patent clauses in research contracts as an attempt by government administrators to articulate a new relationship between federal funding and military-industrial-academic agencies just before and during World War II. The first attempts at government control of research technology by means of patents grew out of tensions that arose in that context, and it was this paradigm that was applied in full force to the patenting of the atomic bomb as the technology came to seem increasingly feasible. The second section examines the

---

[7] In general, the wartime patent issue surfaces in three ways: as a diplomatic dispute between the United States and the United Kingdom over the validity of French atomic patents; as a series of personal disputes between individual scientists and the federal government over patent rights; and in reference to the privatization of nuclear power as debated in the hearings for the Atomic Energy Act of 1946 and its 1954 amendments. All three of these topics will be discussed and referenced later in the essay, as appropriate.

[8] This omission is not primarily due to a lack of sources; all of the crucial primary sources used for this essay have not only been declassified for many decades but are available on microfilm at many universities worldwide.

[9] A recent article that looks at parallels between trade secrecy and national security secrets is Peter Galison, "Removing Knowledge," *Critical Inquiry*, 2004, *31*:229–243, esp. pp. 238–240. See also, e.g., the discussion of national and military secrecy in Sissela Bok, *Secrets: On the Ethics of Concealment and Revelation* (New York: Vintage, 1989). An interesting discussion of replication and independent discovery in relation to nuclear science is Spencer Weart, "Secrecy, Simultaneous Discovery, and the Theory of Nuclear Reactors," *American Journal of Physics*, 1977, *45*:1049–1060. Finally, there is a controversy over whether the Teller-Ulam design was *truly* independently developed (i.e., whether each country that developed it received some outside assistance in the form of collaboration or espionage), but this is beyond the scope of this essay.

ways in which patent policies that had been developed primarily for application *within* the Manhattan Project began to be used *outside* of the project, as unaffiliated scientists and inventors appeared to pose a threat to the government's patent monopoly. The third section then analyzes the ways in which the patenting program expanded and how it chafed against the ambitions of many Manhattan Project scientists as their research began to get closer to yielding results. In conclusion, I will articulate the reasons why the history of the patenting program, which touches on issues relating to government control, industry and educational contractors, and the autonomy of scientists, also demands a reconsideration of the historical relation of nuclear weapons and the legal mechanisms of their control.

THE SHORT FORM WITH A LONG REACH: THE EVOLUTION OF NDRC AND OSRD PATENT POLICY

The path to the patent policy of the Manhattan Project starts in its two most relevant predecessor organizations: the National Defense Research Committee (NDRC), founded by an executive order of President Roosevelt in 1940; and the Office of Scientific Research and Development (OSRD), which took over much of the NDRC's responsibilities in 1941, again by Roosevelt's executive order.[10] The influential scientist-administrator Vannevar Bush was the central figure in both organizations—he used his access to Roosevelt to negotiate their creation and his appointment as their head—and it is his individual approach to administering science that we find reflected in the creation of the patent policies that were inherited when the atomic bomb research was taken over by the Manhattan Project from the OSRD in 1943.

As Larry Owens has shown, Bush's unique approach to administering science featured a system of funding and control that used the tools of industry, primarily the contract, rather than the more open-ended tools associated with philanthropy, such as the grant.[11] NDRC patent policy was similarly rooted in contracts, and it was discussed at length by NDRC members even before the committee had been formally organized. Despite his adoption of the methods of big business and the tools of the marketplace, however, Bush was anxious to avoid charges of wartime profiteering, both for his agency and for his contractors, and the NDRC was from the beginning to be run on the principle of "no profit, no loss." When applied to patents, this principle came to be expressed in the "title taking" style of the first NDRC patent clause, adopted in August 1940 with the collaboration of the Commissioner of Patents. In brief, this first clause specified that for all inventions resulting—even in part—from NDRC-financed research, the NDRC retained the discretion to decide whether a patent application would be filed at all and whether the title of

---

[10] For the definitive administrative history of the NDRC and the OSRD see Irvin Stewart, *Organizing Scientific Research for War: The Administrative History of the Office of Scientific Research and Development* (Boston: Little, Brown, 1948). Stewart's book is also one of the few postwar sources that describes the OSRD patent program in any detail. Stewart was the OSRD Secretary and was often personally involved in patent administration, which helps to explain why he devotes an entire chapter to patent matters, though his account is colored largely by his vantage point and is not comprehensive or reflective.

[11] Larry Owens, "The Counterproductive Management of Science in the Second World War: Vannevar Bush and the Office of Scientific Research and Development," *Business History Review*, 1994, 68:515–576, esp. pp. 521–526. Also essential for understanding Bush's attitudes toward the wartime funding of research is Nathan Reingold, "Vannevar Bush's New Deal for Research; or, The Triumph of the Old Order," *Historical Studies in the Physical and Biological Sciences*, 1987, 17:299–344.

said application would go to the government or to the contractor who had developed the invention.[12]

The industrial contractors found this a little hard to swallow. A number of major contractors, including General Electric, RCA, Western Electric, and Westinghouse Electric, refused to sign any contract with these provisions and instead worked under "letters of intent" until the matter was settled to their liking. The objection of the contractors is understandable: the provision explicitly left all decisions relating to patent rights at the discretion of the NDRC. As NDRC Secretary Irving Stewart later characterized it, the NDRC patent provision

> was, in fact, somewhat anomalous. The United States was at peace and many people believed it would not become involved in the war being waged in Europe.. . .In effect NDRC was asking America's leading companies to take their best men off their own problems and put them (at cost) on problems selected by NDRC, and then leave it to NDRC to determine what rights, if any, the companies would get out of inventions made by their staff members.[13]

Furthermore, the War Department and the Navy Department rarely demanded similar concessions, allowing contractors to retain their commercial patenting rights while reserving royalty-free licenses for official military use. The policy was, in short, seen by the experienced industrial contractors as one that put them in an unpleasant and unfamiliar position, where their work for the government offered a relatively limited possibility of future benefits.

Five months later, a compromise was reached to end the contract stalemate. The new patent clause, forged out of lengthy negotiations with the contractors and modeled after Army and Navy patent practices, specified that the contractor's patent obligations to the government were only that

> ● the government received an irrevocable right to purchase, for a reasonable price, a license to any invention the contractor held title to that related to the subject matter of the contract;
> ● the contractor granted the government a royalty-free license to use any inventions created under the auspices of the contract;
> ● the contractor would, prior to final settlement of the contract, disclose all inventions made under the contract's auspices and indicate whether it intended to take out patents on them;
> ● and, if the contractor declined to file for a patent for an invention created under the contract, the government would have the right to file for the invention itself but would be required to grant the contractor a nonexclusive, royalty-free license to the invention.[14]

Because this clause contained more stipulations than the first one, it came to be known as the "long form," while the "title taking" clause was known as the "short form," even though it had the longer reach. The contrast between the "long" and "short" clauses is rather dramatic: with the new, "long" form the power relations have been reversed, with the power of discretion in patent arrangements vested in the contractor rather than the NDRC. At most, the NDRC retained the ability to take out a free license on NDRC-funded

---

[12] Owens, "Counterproductive Management of Science," p. 526 ("no profit, no loss"); and Stewart, *Organizing Scientific Research for War* (cit. n. 10), pp. 221–222.

[13] Stewart, *Organizing Scientific Research for War*, p. 222.

[14] This is my summary and distillation of Stewart, *Organizing Scientific Research for War*, p. 224; it is not a direct quotation, and I have simplified the legal aspects to their basic terms.

inventions and the ability to take full title in inventions only if the contractor decided not to bother.[15]

NDRC contracts did not, however, adopt the new form unilaterally. The "long form" was developed primarily to break the contract deadlock with major industrial contractors—which for the most part it did—and it continued to be used in cases where the contractor was well established in the field of investigation in question, had long-running vested commercial interests in the subject of research, and was using its previously developed research infrastructure for its NDRC work. The "short form" came to be used primarily in "central laboratory" contracts, the instances where the NDRC work was directly involved in establishing a research infrastructure and where commercial interests were less clear or, at the time, completely nonexistent; in most cases this meant research conducted at academic institutions and in fields where there were few if any preexisting groups of research specialists (including radar, rockets, antisubmarine warfare, and, eventually, atomic energy). Over the course of the war, the "long form" was used almost twice as often as the "short form," but the latter covered more of the most memorable wartime creations of the NDRC and the OSRD.[16]

It was in this contractual context, marked by concerns about profit, profiteering, and the obligations of the government, that the early atomic energy work was first situated under the NDRC as part of its S-1 Committee on Uranium. The atomic research was split between the two contract clauses, with those contractors that would qualify as "industrial contractors"—Westinghouse Electric and Standard Oil Development Company being the two largest—primarily using the new "long form" clauses, while the many sites that would qualify as "central laboratories"—mostly a mélange of academic research institutions— continued to use the original "short form" clauses. In practice, though, there were still many academic institutions that maintained "long form" clauses for the time being.

When the Office of Scientific Research and Development was created by Roosevelt in June 1941—at Bush's urging and with him as its head—it adopted the NDRC patent policies wholesale; and as its research establishment expanded beyond the original NDRC work, so did its patent program.[17] By October 1941 a dedicated staff working strictly on patent issues resulting from OSRD contracts had been assembled. After a request to the Secretary of the Navy, Bush received help from Commander (soon to be Captain) Robert A. Lavender, whom he appointed as the OSRD Advisor on Patent Matters (see Figure 1). Lavender, a graduate of the U.S. Naval Academy who also held an M.S. in electronics from Harvard University and a degree in legal studies from George Washington Law School, had over fifteen years of experience in negotiating different types of patent problems for the Navy by the time he arrived at the OSRD.[18] As the Advisor on Patent

---

[15] In the end there were a total of four variants of the "short" clause: the standard one (described in the text), two more that allowed contractors to retain certain commercial licensing and sublicensing abilities within their fields, and a fourth that helped exempt contractors from liability in the event of their infringement of other patents as a result of using "off the shelf" components in the course of their research. For the purposes of this study the differences are not important. See Statement of Capt. Robert A. Lavender, 11 Feb. 1946 (cit. n. 1), pp. 338–339.

[16] By 1946 the "long form" was used in 1,410 contracts, while the "short form" was used in 780. An additional type of contract clause, not discussed here, was created specifically for penicillin research, allowing commercial researchers to finance more of their research if they desired. See Stewart, *Organizing Scientific Research for War* (cit. n. 10), pp. 224–225.

[17] The NDRC was not dissolved, however; it became an advisory board to the OSRD, with James B. Conant as its head. See *ibid.*, Ch. 4: "NDRC of the OSRD," pp. 52–78.

[18] *Ibid.*, p. 226. Biographical information is from Statement of Robert A. Lavender, Economic Aspects of Government Patent Policies, Hearing before the Subcommittee on Monopoly of the Select Committee on Small

Figure 1. *Captain Robert A. Lavender, head of the OSRD and Manhattan Project patent programs, in 1948. Source: National Air and Space Archives, Fairchild Industries, Inc., Collection, Box 468, Folder 25.*

Matters, he was to serve as the personal representative to Bush on patent issues and became one of the key architects of OSRD patent administration.

Under Lavender the OSRD Patent Division hired patent lawyers—mostly commissioned officers with previous legal training; at its peak the division employed a dozen lawyers with offices in Washington, D.C., Boston, New York, and Chicago.[19] The patent program generally worked in a simple way: as part of the terms of their contracts, contractors were required to submit notice of any potentially patentable invention as an invention report to their OSRD division chiefs, who in turn would forward the material to the patent lawyers in Lavender's Patent Division. The reports were a distilled form of invention priority assignment—a brief description of the invention and a list detailing when the invention was first conceived, with whom it was discussed, whether or not it was

Business, United States Senate, 88th Cong., 14 Mar. 1963, pp. 274–281, on pp. 275–276. Lavender was, at this later date, serving as a witness in relation to U.S. Navy patenting practices.
    [19] Stewart, *Organizing Scientific Research for War*, p. 186.

described in an in-house report, and whether it was thought to be of much value—and were to accompany more detailed technical information and laboratory notebooks.[20]

After receiving the reports, the Patent Division would forward them to the branch of the armed services that had a predominating interest in the invention in order to determine whether the patent would be worth pursuing. Through the end of January 1946, the OSRD processed over 6,700 invention reports, of which about 2,600 were by that time definitely covered by patent applications. (The fate of an additional 2,200 had not, at that point, been decided; patent applications were not filed on the remaining reports.)[21]

In the spring of 1942, however, OSRD patent policy with regard to atomic energy research began to become a distinct and separate affair from that pertaining to the rest of the multi-million-dollar research the OSRD was contracting. Until this point, even though atomic energy research was situated somewhat differently within the OSRD hierarchy of research programs, it was not regarded differently in terms of contracting or patent clauses. The reason for the change appears simple enough on the surface—it had become much more realistic to think that the atomic research would yield military results applicable in wartime—but the bigger question remains: Why should atomic bomb research be controlled by different patent policies than, say, work on the proximity fuze, new explosive materials, or submarine research?[22] What makes atomic energy "special" here is that it was, indeed, considered "special" (the "S" in "S-1" stood for exactly that), but that in itself still falls short of answering the question of why *patent* policies would be a major locus of such change.

The specific change took shape as a coup of the "short form": Bush, in collusion with a number of other top S-1 project members—including James B. Conant, Arthur H. Compton, Ernest O. Lawrence, J. Robert Oppenheimer, and Harold Urey—decided that it would be prudent to try to convert all atomic energy–related contracts to "short form" clauses.[23] Bush knew, though, that the university contractors still on the "long form" would be recalcitrant—to say nothing of the industrial contractors, whose earlier refusal of the "short form" still stung. As Bush wrote to Conant in June 1942, it might take "a little pressure to do it"; that "pressure" would be nothing less than executive approval from Franklin Roosevelt himself. In a letter to Roosevelt in the early summer of 1942, Bush hammered out plans to expand the atomic research and also conveyed, as he later reported to Conant, that the OSRD "intended to have complete records of our experimental work so that patents could later be filed, and that I would attempt to see that as much patent control as possible resided in the hands of the government." Roosevelt provided Bush with exactly what he wanted, in the form of a brief memo on a small scrap of White House stationery:

---

[20] Two invention reports related to the plutonium work conducted at UC Berkeley are contained in *Bush-Conant File Relating the Development of the Atomic Bomb, 1940–1945*, Records of the Office of Scientific Research and Development, RG 227, microfilm publication M1392 (Washington, D.C.: National Archives and Records Administration, 1990?), Folder 6: "Patent Matters [1941–1945]," Roll 2, Target 1, Frames 71, 73. Items from this microfilm publication will hereafter be indicated by the abbreviation **Bush-Conant File.**

[21] Stewart, *Organizing Scientific Research for War* (cit. n. 10), pp. 227–228.

[22] The only other OSRD-financed research program to have program-specific patent issues was radar, but the reason in that instance was overlapping priority claims of MIT and Navy personnel, which owed more to the nature of the research program than to the substance of the technical work.

[23] A summary of these events is in Carroll L. Wilson to James B. Conant, interoffice memorandum, 29 Apr. 1942, Bush-Conant File, Folder 147: "Patents [1942–1944]," Roll 10, Target 5, Frame 298.

> I do not think I have replied to yours of June 19th in relation to the purchase of certain ore in Canada. I agree with you that we should encourage the Canadians to go ahead. Also, I wholly approve your patent control policy. I talked with Mr. Churchill in regard to this whole matter and we are in complete accord. F.D.R.

Bush was now completely unrestrained: this tiny memo gave him the confidence to declare that he had been given full support for total patent control. "The President," he wrote to the head of the U.K. Privy Council Office, "recognizing this aspect of the subject, has instructed me to acquire for this Office patent rights on this subject to as complete an extent as can be readily attained." "The President," he wrote, imploring University of California treasurer Robert M. Underhill to acquiesce to the "short form," "has fully grasped the significance of the project and the results of its solution and has stated that Government control should primarily be through the administration of patents. He has, therefore, directed me to arrange as far as possible for the vesting in the Government of the titles to the inventions and discoveries made and the patents that may be issued thereon that may be involved in this project." "In my capacity as an agent of the Government I must bear in mind," he wrote in a badgering letter to University of California president Robert Sproul, "that the President has directed me to obtain the assignment of patents in this field to the Federal Government."[24] These appeals to Roosevelt's authority, though not always as immediately effective as Bush had hoped, nonetheless played a key rhetorical role in convincing recalcitrant industry and university representatives to bend to his will.

Bush's aggressive approach in assigning atomic energy patents to the government is worth reflecting on, especially in light of the fact that he is generally remembered as a staunch advocate of allowing government-sponsored researchers to maintain their patent rights. This reputation comes primarily from his wartime congressional debate with Senator Harley M. Kilgore and from the 1945 report *Science—The Endless Frontier*, which outlined a proposal for postwar federal funding of scientific research. In both his showdown with Kilgore and his report, Bush took the position that as long as the government received a free license to use whatever patents contractors developed, it should not otherwise hinder their ability to control their inventions in the marketplace.[25] Bush understood the value of patents, and he took them seriously, long campaigning for patent law reform to prevent what he saw as abuses of the system by large corporations seeking to stifle commercial competition. There is some irony in the fact that, in the context of the Manhattan Project, Bush himself would use the patent system as a way of controlling technology, and this shift in roles was not lost on him. "I suppose that in the process," Bush later wrote of the wartime patent activities, "I personally destroyed more

---

[24] Vannevar Bush to Conant, memo, "Patent Aspects of S-1," 19 June 1942, Bush-Conant File, Folder 147: "Patents [1942–1944]," Roll 10, Target 5, Frame 293; Franklin D. Roosevelt to Bush, memo, 11 July 1942, Bush-Conant File, Folder 9: "S-1 British Relations Prior to Interim Committee [Folder] No. 1 [1942]," Roll 2, Target 4; Bush to Sir John Anderson, 1 Sept. 1942, Bush-Conant File, Folder 9: "S-1 British Relations Prior to Interim Committee [Folder] No. 1 [1942]," Roll 2, Target 4; Bush to Robert M. Underhill, n.d. [probably ca. late July 1942], Bush-Conant File, Folder 147: "Patents [1942–1944]," Roll 10, Target 5, Frame 288; and Bush to Robert G. Sproul, 13 Oct. 1943, Bush-Conant File, Folder 6: "Patent Matters [1941–1945]," Roll 2, Target 1, Frame 54.

[25] See Vannevar Bush, *Science—The Endless Frontier: A Report to the President by Vannevar Bush, Director of the Office of Scientific Research and Development, July 1945* (Washington, D.C.: Government Printing Office, 1945), esp. Sect. 5 of the proposed outline for the National Research Foundation. On the Kilgore debate and Bush's postwar patent stances see Daniel J. Kevles, "The National Science Foundation and the Debate over Postwar Research Policy, 1942–1945: A Political Interpretation of *Science—The Endless Frontier*," *Isis*, 1977, 68:4–26, esp. p. 14; and Jessica Wang, *American Science in an Age of Anxiety: Scientists, Anticommunism, and the Cold War* (Chapel Hill: Univ. North Carolina Press, 1999), pp. 25–37.

property in the form of patents than any other man living." Assigning a patent to the government, he later wrote, was the equivalent of destroying it completely from a commercial point of view, and he reflected that "it is paradoxical that I, who am a great believer in the system, should have been called upon to commit this particular sin."[26]

Paradoxical, perhaps, but not incomprehensible. Wartime inventions should not, in Bush's mind, be commercially successful—that would risk accusations of wartime profiteering. Moreover, in the case of atomic energy in particular Bush recognized the lingering question of whether even the aspect of the scientific development that was potentially commercially viable—civilian nuclear power generation—would ever be safe enough to be used by private industry. Writing to Roosevelt in December 1942, he noted, "It is clear that the utilization of atomic power must always be under close government control, not only because of the enormous hazards involved in such a process, but also because a super explosive appears as a possible by-product." The distinction between civilian and military uses of nuclear power, which are often blurred today in discussions of proliferation, was not at all clear in the early 1940s—the first reactors, after all, were developed for military purposes and produced heat energy only as an unwanted by-product. Bush also believed that patent control would facilitate international control of atomic energy. As he wrote to Sir John Anderson of the Privy Council Office in September 1942, explaining his attitude toward patent control:

> I have the strong feeling that much greater progress will be made if each government has in its hands a substantial part of the patent rights arising within the respective countries, for the problem of arriving at sound international relationships will then be much less likely to be complicated by reason of private interest in the outcome. . . . I am inclined to believe that this patent control in the hands of government will prove to be sufficiently strong so that this series of discoveries and inventions cannot be practiced at any point within our respective countries without government license based on the patent status. It would of course be entirely possible to superpose other controls, but the matter becomes somewhat simpler to handle if this is not necessary.[27]

The patent system, as Bush knew and approved, was about technological control. As a patent system advocate, Bush wanted to streamline and improve the system so that small inventors could properly protect their interests against big corporations; as a leader of wartime research, he was happy to use the strength of the government and of the patent system to control what he did not think should be encouraged to function independently in the marketplace. Bush was familiar with patents by the time he was head of the NDRC—his later autobiography describes in detail numerous patent battles in the early radio industry where he made his name—and when confronted with the new question of atomic energy, he understandably reached for the familiar.[28]

---

[26] In this context, Bush was speaking specifically of the MIT patent program, which also involved assigning inventor patents to the government. See Vannevar Bush, *Pieces of the Action* (New York: Morrow, 1970), p. 84.

[27] Bush to Roosevelt, memo, "Report on Present Status and Future Program on Atomic Fission Bombs," 16 Dec. 1942, in *Correspondence ("Top Secret") of the Manhattan Engineer District, 1942–1946*, microfilm publication M1109 (Washington, D.C.: National Archives and Records Administration, 1980), Folder 25: "Documents Removed from Gen. (L. R.) Groves' Locked Box, Plus Certain Documents of Historical Importance," Roll 3, Target 8; and Bush to Anderson, 1 Sept. 1942, *ibid.*, Folder 16A: "Summary of Facts Relating to Breach of Quebec Agreement," Roll 3.

[28] On the patent battles in the radio industry see esp. Bush, *Pieces of the Action* (cit. n. 26), pp. 197–200. This conclusion about Bush is consonant with Nathan Reingold's thesis that Bush observed a strict line between wartime OSRD policies and peacetime ones and that his reliance on existing legal mechanisms of control could

As his letter to Anderson indicates, Bush saw an intimate connection between patenting and the question of international control of atomic weapons. Before nations could try to work out agreements in relation to atomic energy, they would have to own the technology unambiguously. By doing this through patents, they could also control what was done within their own borders in regard to the technology. If these patents were owned outright by the government, there would be no threat of private industrial interests swaying policy for profit. Finally, accomplishing this control by means of the patent system would eliminate the need to worry about creating new, unprecedented forms of control. While it is easy in hindsight to see flaws with this scheme, from Bush's 1942 vantage point it made at least some sense, though it was far from being a necessary conclusion.

But Bush's attempt to take control of the technology behind atomic energy was not wholly successful. Contract clauses allowed control only over work that had been done under the official auspices of the OSRD; they did not cover situations relating to research done *outside* of the organization, and they did not necessarily apply to research that had been done *before* the inventor in question was part of the project. In both cases, the inventor would be largely beyond the reach of Bush's bureaucratic power and could seriously threaten his attempt at a complete government monopoly on atomic energy. The problem was neither trivial nor hypothetical, and it led to one of the stranger manifestations of wartime patent policy, a true fear of the "lone inventor."

FEAR OF THE LONE INVENTOR: PATENT SECRECY AND THE PRIVATE SPHERE

The "lone inventor" is a common trope used in discussions of patent law, in the same way that the "creative genius" is invoked in discussions of copyright. He is a character who represents the hypothetical beneficiary of a patent system, a legal fiction often trotted out as a rhetorical heuristic for comparing the effects of different interpretations of patent law. And even though it has been more than a century since the corporation has replaced the individual inventor as the primary beneficiary of patents (and copyrights), the "lone inventor" continues to be the "little guy" that politicians claim to care about and the reason to keep incentive-giving patent laws strong.[29] When the patentable subject matter is a nuclear bomb, however, the "lone inventor" turns into something else altogether.

The problem was first raised fairly early, when Bush became aware of the fact that in 1939 a team of French physicists at the Collège de France, led by Frédéric Joliot-Curie, had filed in France and the United Kingdom for a number of patents relating to atomic energy; these covered a wide variety of nuclear technologies, including the basic idea of a nuclear reactor and the means of controlling it (by early 1942, the French were negotiating for nearly a dozen patent applications filed in Britain).[30] In March 1942, around the same time that he was beginning his planning for the coup of the "short form," Bush met with Commissioner of Patents Conway Coe about what they called "the French problem"—one of the members of the Joliot-Curie team had registered their claims with

---

be interpreted as part of his desire not to disturb the "old order." See Reingold, "Vannevar Bush's New Deal for Research" (cit. n. 11).

[29] On corporations becoming "inventors" see David F. Noble, *America by Design: Science, Technology, and the Rise of Corporate Capitalism* (1977; New York: Oxford Univ. Press, 1979), Ch. 6.

[30] There was, additionally, a patent relating to an atomic explosive, but it was not filed with the British. The best overall coverage of the French patenting dilemma and British approaches to atomic patenting is Margaret Gowing, *Britain and Atomic Energy, 1939–1945* (New York: St. Martin's, 1964), pp. 201–215. For the point of view of the French scientists see Spencer R. Weart, *Scientists in Power* (Cambridge, Mass.: Harvard Univ. Press, 1979), pp. 93–102, 170–177.

the U.S. Patent Office while in Britain and was trying to use them as leverage for French science after the war. The problem was tricky: Bush and Coe did not want simply to give the French scientists patent control over such basic ideas of nuclear physics, but to argue about it publicly would disclose their own secret bomb program. The result the two administrators agreed on was to declare the French patent applications "secret."[31]

The ability to declare "secret" things not created within the confines of the government itself was not straightforward in the United States at this time. There existed, of course, examples of certain types of expression being suppressed during wartime, but with something like patents, where delays in filing or granting can have large economic consequences, such a use of executive power would have been a legal minefield. As such, Bush had requested an audience with no less an authority on patent law than the Commissioner of Patents, noting that the issue involved was "a matter of general policy of some difficulty . . . on which I certainly need your guidance."[32]

Fortunately for Bush, however, U.S. patent law had been specifically amended so as to permit patent applications to be ordered held in secret in extraordinary circumstances. The original legislation had been passed during World War I to allow patents with military implications to be declared "secret" during wartime, and in 1940 and 1941 the statute had been revised to apply during peacetime as well and to have stiffer penalties associated with the violation of secrecy orders (the original penalty having been simply loss of patent title).[33] The result of this legislative action was Public Law No. 700, a bill that allowed the Patent Office (via the authority of the Commissioner of Patents) to declare patent applications secret, preventing both their publication and access in the United States and also blocking their filing outside of the country. The question of whether the application would be granted was put on hold until the secrecy order had been lifted. If the patent was eventually granted, the inventor could then work out problems of interference with subsequently granted patents and could sue for compensation if the government had used the patent in the interim.

The statute and its later revisions each provoked substantial debate in Congress over their effect on the "lone inventor": on the one hand, an order of secrecy could drastically extend the life of the patent itself, since the patent's enforceability countdown did not begin until it was made public and granted; on the other hand, such an order could slow innovation in a field and tie the hands of the inventor, who was not eligible for compensation from the government unless the invention was actually used (inventions declared

[31] Bush to Conway Coe, 30 Apr. 1942, Bush-Conant File, Folder 6: "Patent Matters [1941–1945]," Roll 2, Target 1, Frame 9. Note that I have not found much evidence to indicate that "the French problem" was a motivation for the coup of the "short form" in the first place—in the historical record they appear as independent parallel developments, interconnecting on certain points but not strongly causally linked. It may be that Roosevelt was thinking of "the French problem" in his initial note to Bush mentioning Churchill, cited earlier, though I have not found direct evidence of that.

[32] Bush to Coe, 7 Mar. 1942, Bush-Conant File, Folder 6: "Patent Matters [1941–1945]," Roll 2, Target 1, Frame 4.

[33] On the World War I law see Hearings before the Committee on Patents, U.S. House of Representatives, 65th Cong., on H.R. 5269, 13 July 1917, pp. 3–10. On the 1940–1941 revisions see Unpublished Hearings before the Committee on Patents, U.S. House of Representatives, 76th Cong., 3rd sess., on H.R. 9928, 31 May 1940, 3 June 1940); and Hearings before the Committee on Patents, U.S. House of Representatives, 77th Congress, 1st sess., on H.R. 3359 and H.R. 3360, "Preventing Publication of Inventions and Prohibiting Injunctions on Patents," 20 Feb. 1941, 25–27 Feb. 1941, 11–12 Mar. 1941, 19–20 Mar. 1941, 22–23 Apr. 1941, pp. 1–376. Unfortunately, not very much has been written on the early history of secret patents in the United States; the only full-volume work on secret patents that I have come across is specifically on the situation in the United Kingdom, which seems to have developed in a considerably different historical and legal context than the U.S. case: T. H. O'Dell, *Inventions and Official Secrecy: A History of Secret Patents in the United Kingdom* (Oxford: Clarendon, 1994).

secret and not used would not be eligible for compensation). Putting the French patent applications under P.L. 700 removed them from sight and mind—for a while, at least. Even at its most influential, such patent secrecy could only forestall potential legal debates about priority claims pertaining to work on nuclear reactors and nuclear weapons—and only within the reach of U.S. patent law; patent secrecy statutes, even at their worst, could only wield economic threats (denying the granting of a patent in the United States) against foreign inventors. ("The French problem" would evolve into a rather trying diplomatic snafu later in the war, when it was discovered, to the horror of the Americans, that the British had made a secret agreement with the French scientists to share nuclear information in exchange for a guarantee to use the French patents in the postwar period.[34])

At one of their meetings regarding "the French problem," Coe had raised a far-reaching question: whether Bush should consider appointing someone within the OSRD to survey the existing field of patents (the "prior art") relating to fission research. The purpose would be to help the Patent Office examiners in their determination of which non-OSRD patents were worth paying attention to and which might need to be declared secret under P.L. 700. Bush thought it a good idea that applications "which have any significance" to the S-1 project "be withheld from issue," and he recognized that having a reviewer unconnected with other aspects of the wartime project was vital if they were to avoid accusations of conspiring against private inventors for the government's benefit.[35] Though he knew it would be hard to find someone with the appropriate scientific and legal classifications who would be free for the job, Bush thought he might have a candidate directly at hand—directly across the hall from his office, to be exact—who could be lent out for what Bush thought would be a small survey project lasting six weeks at most.[36]

William Asahel Shurcliff was a three-time Harvard graduate, having received his B.A. *cum laude* in 1930, a Ph.D. in physics in 1934, and a degree in business administration in 1935. Before the U.S. entry into World War II, he had been the head of the Spectrophotometric Laboratory at the Calco Chemical Division of the American Cyanamid Company, which was involved in using spectrometry to perform chemical analysis as well as in projects relating to electric amplifiers and camouflage (see Figure 2). He had been in charge of keeping patent records while at Calco and had filed a number of patents himself.

---

[34] This has been discussed in detail in a number of sources. See, e.g., Hewlett and Anderson, *New World* (cit. n. 5), pp. 284, 331–336; Barton J. Bernstein, "The Uneasy Alliance: Roosevelt, Churchill, and the Atomic Bomb, 1940–1945," *Western Political Quarterly*, 1976, 29:202–230, esp. pp. 227–228; and Weart, *Scientists in Power* (cit. n. 30), pp. 167, 171–174, 179–180, 205, 234. For more information on the earlier history of the British end of "the French problem" the best source is Gowing, *Britain and Atomic Energy* (cit. n. 30), pp. 201–215. As Gowing puts it, "The French patents run as a leitmotiv through the history of the United Kingdom atomic energy project" (p. 209). Gowing also discusses the tangled case of Anglo-American patent agreements in relation to atomic energy during the war.

[35] Bush to Coe, 23 Apr. 1942, 30 Apr. 1942, Bush-Conant File, Folder 6: "Patent Matters [1941–1945]," Roll 2, Target 1, Frames 14, 9. This latter point, though, would later raise a thorny problem: what if one of the applications in question did contain information that would be of use to OSRD scientists? "Problem as to how to get interesting patent application info not solved," the reviewer would later note in a memo to himself. In practice, though, the issue never came up: there is no evidence that anyone ever tried to patent anything that the reviewer thought the S-1 workers would find really useful. See William A. Shurcliff, memo, "Brief History of WAS S-1 Patent Work" [begun 2 June 1942, added to through at least 30 Sept. 1942], Bush-Conant File, Folder 14: "Material from Liaison Office Files—Primarily Shurcliff's Relations to S-1 Activities, Folder No. 2 [1942]," Roll 3, Target 1, Frame 154.

[36] Bush to Coe, 23 Apr. 1942, Bush-Conant File, Folder 6: "Patent Matters [1941–1945]," Roll 2, Target 1, Frame 13.

Figure 2. *A painting of William A. Shurcliff from 1948 by his father-in-law, the American artist Charles Hopkinson. Courtesy Arthur and Charles Shurcliff and Arthur Saltzman.*

In early 1942 a friend inside the OSRD suggested that he join the staff, an opportunity he leapt at, having feared that he might be drafted into the Army while at his civilian job.[37]

After working for a few months as a senior technical aide in the Liaison Office of the OSRD, Shurcliff was tapped by Bush in May 1942 to be brought into the S-1 program. His job was to review patents with possible implications for S-1 work, coming from both outside and inside the project, with the goal of helping the Patent Office learn when to apply secrecy orders. "I would be sworn to secrecy," Shurcliff wrote in a memo before beginning the work; and he hinted that even bigger ideas were being floated: "Taking over of [non-OSRD] patents or patent applications by the gov't is not *now* in view. An act of Congress might be required."[38]

By 1 July, as he wrote in a report to Bush, Shurcliff had found about thirty-five applications that were likely to require secrecy orders. He was keeping careful records, utilizing a system of six separate card indexes to keep track of patents, inventors, and subjects. He had also adopted a scientist-centric methodology, examining S-1 reports and "all relevant names" in *Physical Review* articles published between January 1939 and

---

[37] "Notes on the Training and Professional Experience of Dr. William A. Shurcliff," attached to *ibid.*; and William A. Shurcliff, "William A. Shurcliff: A Brief Autobiography," unpublished MS (Cambridge, Mass., 15 Dec. 1992), copy in Houghton Library, Harvard University, pp. 53–55, 187.

[38] William A. Shurcliff, memo, 14 May 1942, Bush-Conant File, Folder 14: "Material from Liaison Office Files—Primarily Shurcliff's Relations to S-1 Activities, Folder No. 2 [1942]," Roll 3, Target 1, Frame 172.

April 1942 to compile a list of six hundred individuals who he thought were doing work in the field of atomic energy (broadly defined). He would soon expand this approach to cover issues of the *Review of Modern Physics* and *Scientific Abstracts*, and he also requested that the National Academy of Sciences send him lists, culled from the Roster of Scientific Personnel, of those who had indicated expertise in nuclear physics and several other fields. His list grew to well over a thousand names by March 1943 and included many of those already employed by the OSRD on nuclear-related research—Enrico Fermi, Ernest O. Lawrence, Emilio Segrè, Leo Szilard, and Harold Urey are just a few of the most recognizable figures. Some on the list were doing wartime research that was not related to S-1 work, among them William Shockley and Robert Van De Graaff (both of whom had patent applications that attracted Shurcliff's attention), while numerous others were unaffiliated with government work—and of these others many were not even in the country (for example, all the members of the Collège de France team).[39] Though Shurcliff focused initially on physicists—no doubt owing to his own training—his scope would, over the course of his study, expand to include other disciplines (and rightly so, given that much of the development of fissile materials involved the work of chemists, metallurgists, and engineers, as well as physicists).[40]

What was initially to be a "survey of the art" became a full program to, as Shurcliff put it, "locate, examine, and make secret all non-gov't-controlled U.S. patent applications related to S-1."[41] Shurcliff would request patent applications from the Patent Office or would receive notice from contractors themselves about applications they were filing on behalf of their personnel.[42] He would then draw up a large list of application numbers on a notepad, penciling in the titles and inventors and labeling them "secrecy recommended" or "secrecy not recommended" (rubber stamps were later utilized for the purpose). If an application was, as he later put it, "hot"—that is, if it "had, or might have, an atomic-bomb connection"—he would designate that it be "put to sleep," which was accomplished by sending a brief letter to Captain Lavender with the specific level of secrecy recommended and an indication as to whether the government should attempt to acquire the title to the patent from the inventor. As the work continued, he and Lavender eventually worked out a standard form for requesting secrecy, on which Shurcliff could simply circle the specific responses ("Secrecy recommended: Yes / No").[43]

[39] Shurcliff, memo [forwarded to both Bush and Conant], "7/1/42 Progress Report on W.A.S. Secrecy Efforts Relating to Patent Applications Bearing on S-1 Subjects," 1 July 1942, Bush-Conant File, Folder 147: "Patents [1942–1944]," Roll 10, Target 5, Frame 291; Shurcliff, memo, "Manner of obtaining names for LAI cards" [2 June 1942, added to through 15 Aug. 1942], Bush-Conant File, Folder 14: "Material from Liaison Office Files—Primarily Shurcliff's Relations to S-1 Activities, Folder No. 2 [1942]," Roll 3, Target 1, Frame 150; Shurcliff to Joseph Morris, Roster of Scientific Personnel, National Academy of Sciences, 25 June 1942, Bush-Conant File, Folder 14: "Material from Liaison Office Files—Primarily Shurcliff's Relations to S-1 Activities, Folder No. 2 [1942]," Roll 3, Target 1, Frame 82.

[40] On this latter point see esp. Schwartz, "Making of the Historiography of the Atomic Bomb" (cit. n. 4).

[41] Shurcliff to Bush, 25 Sept. 1942, Bush-Conant File, Folder 14: "Material from Liaison Office Files—Primarily Shurcliff's Relations to S-1 Activities, Folder No. 2 [1942]," Roll 3, Target 1, Frame 38.

[42] In the beginning, though, Shurcliff had difficulty in getting the Patent Office to forward all of the applications he wanted, and he complained to Bush and the Patent Office about this. Bush stepped in at one point when a patent application Shurcliff had asked to examine was, in the meantime, granted and written about in the *New York Times*. See Shurcliff to A. E. Donnelly, 11 Aug. 1942, Bush-Conant File, Folder 14: "Material from Liaison Office Files—Primarily Shurcliff's Relations to S-1 Activities, Folder No. 2 [1942]," Roll 3, Target 1, Frame 45; Bush to Coe, 7 Oct. 1942, Bush-Conant File, Folder 14: "Material from Liaison Office Files—Primarily Shurcliff's Relations to S-1 Activities, Folder No. 2 [1942]," Roll 3, Target 1, Frame 27; and "Another Bomb Sight Is Patented: One Device Corrects Plane's Aim," *New York Times*, 4 Oct. 1942, p. A1.

[43] Shurcliff's notes are dispersed throughout Bush-Conant File, Folder 13: "Material from Liaison Office

Shurcliff was quite conscientious about his work, though; he did not impose secrecy where he did not think it reasonably called for, and in a number of cases he later rescinded secrecy orders he had issued when he changed his mind about the importance of the invention in question. In March 1943, for example, he rescinded the secrecy orders on seven patent applications for inventions relating to mass spectrometry because he decided that the applications "should be allowed to mature in the normal and unrestricted manner." The inventions "were at all times only of border-line interest," he wrote, and over the half-year since he had recommended secrecy his interest in the applications had "appreciably decreased." Furthermore, he was beginning to wonder what effect the secrecy program was having on private industry: "The damage done to industry by maintaining the secrecy orders must be increasing, especially in the petroleum industry and in the field of organic chemistry generally, all as attested by recent petitions filed by the individuals or assignees concerned with the cases listed above."[44] Industrial companies had contacted Shurcliff (via Lavender) a number of times, inquiring about the release of their patents, often so that they would be able to file them in other countries (filing in Canada in particular was a major concern). In a number of cases Shurcliff denied the requests flat out: the inventions were deemed too sensitive.[45]

From the point of view of the inventor, an order of patent secrecy could be irritating if not maddening. Contractors, of course, were well aware of why the patent secrecy was being ordered (at the very least, they knew it pertained to a specific secret war project—and they knew the rules of playing the OSRD's game), and their requests for secrecy to be lifted on their applications were primarily related to issues of filing abroad or for insuring that they gained a particular edge in their field. For those who were not associated with the bomb project, though, the orders could be an enigma, especially if the invention itself had no obvious conventional wartime application. The letters (emblazoned with the heading "SECRECY ORDER") came from the Patent Office, not the OSRD, and contained only an explanation of the basic provisions of patent secrecy law (do not publish, do not file an application abroad—or else).[46]

One inventor—a "lone inventor," apparently unaffiliated with any university or industrial

---

Files—Primarily Shurcliff's Relations to S-1 Activities, Folder No. 1 [1942–1944]," Roll 2, Target 8; and Bush-Conant File, Folder 14: "Material from Liaison Office Files—Primarily Shurcliff's Relations to S-1 Activities, Folder No. 2 [1942]," Roll 3, Target 1. Most are sparse; some contain notes about the specifics of the patent and even small sketches. For the quotation see Shurcliff, "William A. Shurcliff: Brief Autobiography" (cit. n. 37), pp. 59–60. For the standard form see Shurcliff to Robert A. Lavender, form letter, 27 Aug. 1944, Bush-Conant File, Folder 13: "Material from Liaison Office Files—Primarily Shurcliff's Relations to S-1 Activities, Folder No. 1 [1942–1944]," Roll 2, Target 8, Frame 814.

[44] Shurcliff to Lavender, 29 Mar. 1943, Bush-Conant File, Folder 13: "Material from Liaison Office Files—Primarily Shurcliff's Relations to S-1 Activities, Folder No. 1 [1942–1944]," Roll 2, Target 8, Frame 840. Shurcliff's concern with the petroleum industry and organic chemistry in general probably stems from his correspondence with representatives at Standard Oil Development (Eger V. Murphee and his patent attorney P. L. Young), which had a large contract for developing gas centrifuge enrichment technology (it was not, in the end, used during the war).

[45] E.g., a petition to remove secrecy from an application by Joseph Slepian at Westinghouse Electric for an "Ionic Centrifuge" (which was early on considered a route to uranium enrichment, though it was eventually abandoned as unlikely to produce rapid results) was refused by Shurcliff, who considered this particular application to be "one of the '*more important*' S-1 type applications." He elaborated on the terminology to Lavender: "I believe it is your policy to allow filing of 'important' S-1 type applications in Canada, but not 'more important' applications. Accordingly, I believe the present petition should be refused." Shurcliff to Lavender, 30 Sept. 1942, Bush-Conant File, Folder 14: "Material from Liaison Office Files—Primarily Shurcliff's Relations to S-1 Activities, Folder No. 2 [1942]," Roll 3, Target 1, Frame 36.

[46] E.g., Thomas Murphy, Assistant Commissioner of Patents, to Lawrence H. Johnston, 20 May 1953, regarding "Detonating Apparatus" (U.S. App. 165,171; later granted as U.S. Pat. 3,955,505); copy of letter courtesy of Lawrence H. Johnston.

research group—was in fact *encouraged* by the secrecy order, moved to declare his eagerness to lend his services if the government should desire them. On 14 May 1944 this man phoned Shurcliff at his office to inquire about an invention of his on which, he claimed, Shurcliff had recommended a secrecy order. Was the government using it? What action was the government taking with it? Was there any way he could help put it to use? Shurcliff bought some time by telling the caller that he would need to put his questions in writing; two days later he did so, addressing the letter to Shurcliff directly at his OSRD office.[47]

The inventor, Sol Wiczer, was then living in Washington, D.C., and had filed a patent titled "Separation of Isotopes" on 28 November 1942. Along with a large number of other patents on the same subject, it had been reviewed by Shurcliff in February 1943; on his notepad Shurcliff had written that he thought the application was "vague," but it commanded no particular attention apart from a secrecy order (as did practically all isotope separation patents, because the technology is key to uranium enrichment).[48]

The patent secrecy orders themselves were on Patent Office letterhead and were signed by the Commissioner of Patents, with the intent of giving no indication of where the secrecy order had actually originated or what concerns had led to its issuance. So how did this unaffiliated, uncleared, and unknown inventor find out not only what office had issued the secrecy order but precisely *who* had requested it, down to his office phone number? Something had gone horribly wrong. Shurcliff wrote Wiczer an official letter, oozing with governmental formality and faux ignorance, explaining that the Patent Office had issued the secrecy order and that therefore he could be of no assistance—and sent a copy to Lavender. Three days later he wrote a memo to Carroll L. Wilson, Liaison Office head and Bush's executive assistant, outlining in strict chronological order exactly what had happened—"A slightly suspicious incident is described"—and forwarding copies of his and Wiczer's correspondence. He explained that Wiczer's application was "moderately pertinent" to S-1 work and that the application had not indicated an institutional affiliation or any assignees.[49]

The day Wiczer's letter had arrived, Shurcliff's assistant (and eventual heir to the patent censoring job) David Z. Beckler had suggested that "enemy agents might file 'paper' applications on this subject to obtain leads as to U.S. secrecy policy and perhaps additional information." Shurcliff thought this worth considering: he told his superiors that he "could supply names" of all inventors on his lists without institutional affiliations and suggested that either Wilson or Bush himself "may wish to recommend to proper authorities that FBI or other investigations be made of the 10 or 20 'lone wolf' inventors who have filed applications in the 'S-1' field."[50] In the looking-glass world of Manhattan Project security, the much-lauded "lone inventor" was transformed into the much-feared "lone wolf."

Wilson forwarded Shurcliff's memo to Lieutenant Colonel John Lansdale, Jr., head of Manhattan Project security, who promised to look into the situation with Wiczer and expressed eager interest in the list of "lone wolves" and anything providing "factual basis

---

[47] Sol B. Wiczer to Shurcliff, 16 Mar. 1944, Bush-Conant File, Folder 13: "Material from Liaison Office Files—Primarily Shurcliff's Relations to S-1 Activities, Folder No. 1 [1942–1944]," Roll 2, Target 8, Frame 816.
[48] Shurcliff notepad, 1 Feb. 1943, Bush-Conant File, Folder 13: "Material from Liaison Office Files—Primarily Shurcliff's Relations to S-1 Activities, Folder No. 1 [1942–1944]," Roll 2, Target 8, Frame 752.
[49] Shurcliff to Wiczer, 17 Mar. 1944, Bush-Conant File, Folder 13: "Material from Liaison Office Files—Primarily Shurcliff's Relations to S-1 Activities, Folder No. 1 [1942–1944]," Roll 2, Target 8, Frame 815; and Shurcliff to Wilson, 20 Mar. 1944, Bush-Conant File, Folder 13: "Material from Liaison Office Files—Primarily Shurcliff's Relations to S-1 Activities, Folder No. 1 [1942–1944]," Roll 2, Target 8, Frame 811.
[50] Shurcliff to Wilson, 20 Mar. 1944.

for suspicion of a fishing expedition."[51] A special agent made two investigations into Wiczer's past—interviewing an employer and looking in personal records regarding his past employment and education—and determined what had happened: Wiczer had previously worked as a patent examiner and probably had connections within the Patent Office who would have been able to discover that Shurcliff had been his censor. As such, Wiczer was dismissed as a threat, but the larger specter of the "lone wolf" still loomed.

Once the list was actually compiled, Shurcliff's original estimate of "10 or 20" became only seven inventors who were totally unaffiliated, so far as Shurcliff knew, and he forwarded to Lansdale descriptions of their motley mix of inventions. Lansdale never wrote back to let Shurcliff know the results of the "lone wolf" investigations—they appear to have come to nothing, apart from provoking an angry letter from Captain Lavender to General Groves informing him that private patent applications should not be viewed by unauthorized project personnel.[52]

Shurcliff continued his patent censoring job through October 1944, when he was transferred to another office, and the patent watching job was then taken over by his former assistant. In his time as patent censor, Shurcliff "put to sleep" at least 131 patent applications (about half of the total number that he examined) from at least 95 separate inventors—a small percentage of all Manhattan Project patents filed, but still a considerable number in economic terms, since entire industries can rest on only a handful of patent claims.[53]

The "lone wolf" investigation was not simply an instance of wartime paranoia, though the idea that enemy agents would use mock patent applications to probe U.S. bomb development seems far-fetched—especially in retrospect, when we know that all significant wartime nuclear espionage was conducted by Allied forces (the Soviet Union) and in a much more direct fashion (with project participants volunteering information). Looked at one way, the wartime policy was a direct inversion of the traditional values of intellectual property legislation—the sacrosanct "lone inventor" became the "lone wolf," inherently suspect and requiring preemptive investigation on account of his "lone-ness." But it is worth remembering that the "lone inventor" himself is not only a legal and political fiction, but one often used to shroud the deliberately monopolistic aspects of intellectual property systems, where exclusivity is the name of the game. Though the long-term goals of a patent system are to encourage innovation, the methods of producing this all revolve around the short-term discouragement of competition, and there are even entire industries based on either "defensive" patenting (taking out a patent only for its use as a bargaining chip) or "offensive" patenting (taking out a patent only to use it to demand licenses from others).

[51] Wilson to Lt. Col. John Lansdale, Jr., 22 Mar. 1944, Bush-Conant File, Folder 6: "Patent Matters [1941–1945]," Roll 2, Target 1, Frame 77; and Lansdale to Wilson, 4 Apr. 1944, Bush-Conant File, Folder 13: "Material from Liaison Office Files—Primarily Shurcliff's Relations to S-1 Activities, Folder No. 1 [1942–1944]," Roll 2, Target 8, Frame 808. For more information on Lansdale see Gregg Herken, *Brotherhood of the Bomb: The Tangled Lives and Loyalties of Robert Oppenheimer, Ernest Lawrence, and Edward Teller* (New York: Holt, 2002), esp. pp. 58–59.

[52] Shurcliff to Wilson, 11 Apr. 1944, Bush-Conant File, Folder 13: "Material from Liaison Office Files— Primarily Shurcliff's Relations to S-1 Activities, Folder No. 1 [1942–1944]," Roll 2, Target 8, Frame 804; and Lavender to Gen. Leslie Groves, 13 Nov. 1944, Bush-Conant File, Folder 6: "Patent Matters [1941–1945]," Roll 2, Target 1, Frame 110.

[53] Shurcliff to David Z. Beckler, memo, "Remarks on Shurcliff's files on S-1-type patent application data and on secrecy recommendations thereon," 31 Oct. 1944, Bush-Conant File, Folder 6: "Patent Matters [1941–1945]," Roll 2, Target 1, Frame 107.

It is this aspect of the patent system that best defines the practices of the wartime OSRD patent policies, with the rather subversive twist that the agency that granted the exclusivity of the patents was in direct collusion with the agency that would benefit from them. Though the OSRD took pains to minimize this obvious conflict of interest by assigning to different people the roles of patent seeking and patent suppressing, in practice the collaboration between them was unmistakable.

But simply to characterize this collusion as an abuse of power would miss what it reveals about the *limits* of power. Despite Bush's belief that atomic research would be of dramatic wartime and postwar importance, when he made forays into matters of private property he did so carefully, within the structure of the existing law. As was previously noted, this is not entirely shocking from the point of view of bureaucratic considerations: after all, it is precisely the importance of the bomb that made Bush particularly inclined not to jeopardize postwar control through sloppy wartime legal handling of private inventors' patent rights. The possibility of real legal problems arising after the expediency of wartime had passed was not unknown to the participants: Lavender had spent much of his earlier work for the Navy handling messy suits resulting from U.S. infringement upon the claims of British inventors during World War I. It is only when we envision the Manhattan Project as a wartime power superior to all others that we risk losing sight of the legal constraints within which much of it took place, which had appreciable effects on policy and practice.

CONTROLLING SCIENTISTS, CONTROLLING THE ORGANIZATION

Bush had hoped that complete patent control could be accomplished through contract clauses and patent secrecy orders—the former would control developments internal to the project, while the latter would control those external to it. Apart from Sol Wiczer, there is no sign that any unaffiliated inventors came close to investigating—much less seriously contesting—the secrecy orders issued during the war, and so in that respect Bush's control scheme was successful. However, he soon found that controlling inventors *within* the organization—the brainy Nobel "prima donnas," as Groves characterized them—was considerably more difficult.

When the OSRD S-1 project was taken over by the Army Corps of Engineers and became the Manhattan Project—a process begun in the summer and fall of 1942 and officially completed by the beginning of 1943—the Manhattan Engineer District (MED) inherited the OSRD's bomb patent policy in its entirety. Patent administration became a vital and ever-enlarging aspect of the Manhattan Project: each of the many development sites had its own patent representative, invention reports streamed in by the thousands, and General Groves himself took a strong personal interest in the program.

Though the policies to discourage "lone inventors" can be seen as an inversion of the traditional values of the patent system, there are many aspects of implementation that were not inverted at all: patents were still about control, but the focus of this control was not just restricting external competition (as between corporations or between nations); it was also a response to a perceived threat from within the research organization itself. These policies in many ways mirror those undertaken in industry from the 1890s onward, as corporations began to replace individuals as the primary title-holders of patents, though in the case of the Manhattan Project the level of internal control is much more explicit, soaked in the rhetoric of national security.

Project scientists had their own patent concerns, separate from those of the Manhattan

Project bureaucracy. Nuclear physicists in particular had been taking out patents on their work in a rather systematic fashion since the 1920s, with people like Leo Szilard, Ernest Lawrence, and Enrico Fermi leading the way, although by the time the French scientists were thinking about taking out patents in 1939 the idea was still sufficiently foreign that they had to have a small discussion over its ethical merits first. In most cases the explicit goal seems to have been to try to guarantee that some of the profits made from a scientific invention would be reinvested in science, usually by means of assigning title to a third party such as the Research Corporation in the United States or the Centre National de la Recherche Scientifique in France.[54]

In at least two instances, in particular, impulses of this kind ran into trouble with the Manhattan Project patenting goals.[55] In both cases project scientists attempted to assert their patent ownership rights over the claims of the project patent program, and in both cases the difficulty hinged legally on the fact that the key scientific work involved in the invention happened *before* the scientists were on the OSRD payroll and under the sway of the OSRD "short form" clause. The first of these cases, that of the Hungarian physicist Leo Szilard, has been covered thoroughly elsewhere, and I will summarize it only by saying that Szilard attempted to use his patent claims to the first nuclear reactor as leverage to make more substantial demands for a voice in the growing project. He was, after some negotiation, given a choice between his fight for patent rights and the opportunity to participate in the project at all; in the end he chose the latter, as his goal of gaining more of a voice in the project would not be furthered by his exile from it, and so he more or less relinquished his patent claims.[56]

The second case, almost completely neglected in the secondary literature, is that of the plutonium researchers Glenn T. Seaborg, Emilio Segrè, Arthur C. Wahl, and Joseph W. Kennedy—a tangled three-body problem of scientists, university administrators, and the OSRD. The patents whose ownership was under dispute were extremely lucrative: they covered the production and basic chemistry of plutonium, as well as its basic use as a fissile material. Since the work had been performed before any of the scientists were under

[54] On the French scientists see Weart, *Scientists in Power* (cit. n. 30), pp. 93–96. Weart also offers a good summary of similar approaches by other nuclear physicists: *ibid.*, p. 97; another good source is J. L. Heilbron and Robert W. Seidel, *Lawrence and His Laboratory: A History of Lawrence Berkeley Laboratory* (Berkeley: Univ. California Press, 1989), esp. Ch. 3. Some of these patenting conflicts were in relation to the production of radioisotopes, an early market for nuclear physics. For specifics see Simone Turchetti, "The Invisible Businessman: Nuclear Physics, Patenting Practices, and Trading Activities in the 1930s," *Hist. Stud. Phys. Biol. Sci.*, 2006, *37*:153–172; more generally see Angela N. H. Creager, "Tracing the Politics of Changing Postwar Research Practices: The Export of 'American' Radioisotopes to European Biologists," *Studies in History and Philosophy of Biological and Biomedical Sciences*, 2002, *33*:367–388, esp. pp. 369–371.

[55] There are no doubt other cases as well; the finding aid to the Bush-Conant File indicates that Joseph Slepian was involved in a patent struggle over his centrifuge work (see note 45, above), and a recent essay by Simone Turchetti focuses on a patent struggle with administrators by Enrico Fermi. For the latter see Simone Turchetti, "'For Slow Neutrons, Slow Pay': Enrico Fermi's Patent and the U.S. Atomic Energy Program, 1938–1953," *Isis*, 2006, *97*:1–27; and Turchetti, "Invisible Businessman."

[56] This story was first well told in Carol S. Gruber, "Manhattan Project Maverick: The Case of Leo Szilard," *Prologue*, 1983, *15*(2):73–87; it was subsequently covered in Rhodes, *Making of the Atomic Bomb* (cit. n. 5), pp. 503–507. For a good overview of Szilard's patent history and his own thoughts on patents see the introduction by Julius Tabin to Pt. 5, "Patents, Patent Applications, and Disclosures (1923–1959)," in *The Collected Works of Leo Szilard*, Vol. 1: *Scientific Papers*, ed. Bernard T. Feld and Gertrud Weiss Szilard (Cambridge, Mass.: MIT Press, 1972), pp. 527–531. On the "more or less" aspect see Szilard's remarkable testimony before a frustrated House Committee on Military Affairs, in which he made it apparent that his cooperation was never quite complete: Statement of Dr. Leo Szilard, Hearings before the Committee on Military Affairs on an Act for the Development and Control of Atomic Energy, U.S. House of Representatives, 79th Cong., 1st sess., on H.R. 4280, 18 Oct. 1945, pp. 71–96.

OSRD contract, and while they were in the employ of the University of California, the assignment of the patent rights was more murky than Bush or Lavender would have liked.

Here patent control proved a difficult business, and the name of Roosevelt did not prove to be the universal balm Bush had hoped it would be. Outside of the OSRD, Bush's influence was far more limited than he liked; moreover, this was not an issue he could simply resolve by his standard contract system, and the University of California was a savvy and interested negotiator, well aware of the long-term benefits to be derived from patents developed by its scientists. All of the parties involved in the Seaborg dispute were looking to the future: the scientists wanted assurances of postwar research funding; the University of California saw future revenues in a hypothetical commercial nuclear power industry (and perhaps in radioisotope production); and Bush wanted to maintain his regime of technological control (he made no distinction between civilian and military patents during the war itself) and was not about to let any university administrators stand in his way. Because both the inventors and the Regents of the University of California were looking to a postwar world, the appeal to the requirements of wartime that worked so well with contractors had far less traction; and Bush, in Washington, D.C., was far away—literally, figuratively, and, in many ways, legally. Though he tried to impress Roosevelt's intentions on the other parties, in terms of the law the situation was far from clear cut and required months of negotiations.[57] Both the University of California and the scientists eventually relinquished their claims for the remainder of the war, though the scientists were later compensated $100,000 each for the patents by the Atomic Energy Commission (AEC)—a pittance in comparison to the worth of the patents, but much more than the $1 compensation they would have received under OSRD policy.[58]

This "internal" focus of the program is worth calling explicit attention to: all of the dealings with contracts and patent clauses were meant to strip contractors and inventors of any future claims of control, to avoid just the sorts of complicated and potentially compromising disputes that occurred in the Szilard and Seaborg cases (which, again, were problems only because they involved inventions created outside of contracts). This concern is, as already noted, different from the worry that external forces, such as interference suits, would attempt to control the government's nuclear ambitions, though as the handling of the "lone wolves" and "the French problem" shows this was clearly an issue as well. Patents became, for the project administrators, a convenient way to hedge their postwar bets in controlling this new technology to which they early on attributed twin auras of salvation and apocalypse.

By the time the Manhattan Project's authority was transferred to the AEC, on 1 January 1947, over 8,500 technical reports had been examined by the patent officers, over 6,300 technical notebooks had been scrutinized, and 5,600 different inventions in 493 different subject classes—covering everything "from the raw ore as mined to the atomic bomb"— were docketed by Lavender's office, resulting in some 2,100 separate patent applications being approved for filing, 1,250 of which had actually been filed with the U.S. Patent

---

[57] The only easily accessible version of this episode is in Glenn T. Seaborg, *The Plutonium Story: The Journals of Professor Glenn T. Seaborg, 1939–1946* (Columbus, Ohio: Battelle, 1994). There are many letters relating to the matter in Bush-Conant File, Folder 6: "Patent Matters [1941–1945]"; and it is briefly discussed in Robert M. Underhill, "Contract Negotiations for the University of California," oral history interview with Arthur Lawrence Norberg, 10 Feb. 1976, Bancroft Library, University of California, Berkeley. There are also many boxes of materials relating to this in the Glenn T. Seaborg Papers at the Library of Congress.

[58] Atomic Energy Commission, *Eighteenth Semi-Annual Report* (July 1955), p. 101. Note that there are some strong parallels to the story of Szilard and Fermi discussed in Turchetti, "'For Slow Neutrons, Slow Pay'" (cit. n. 55). There were a number of other cases of compensation claims reviewed by the AEC Patent Compensation Board in the 1950s; this is an aspect of the postwar history of atomic patents that could benefit from further study.

Office at that point. The magnitude of these numbers, if not immediately obvious, can perhaps be appreciated in light of the fact that the latter number would have been 1.5 percent of all the patent *applications* filed in 1946—more than one out of every hundred—or the fact that if all of the inventions docketed had been patented, they would have represented around 0.8 percent of all the patents *in force* at the time.[59] From an economic point of view, the program was massive: it was a deliberate and successful attempt to obtain the patents not just for key inventions but for the technological contents of numerous new industries in their entirety.

The exact number of applications that have been kept secret is not available, but a rough approximation of the scope can be extrapolated: of the approximately 85 patents, originating from Los Alamos research alone, filed between 1943 and the end of 1946, well over 60 had at least five years between their file and their award dates; well over 35 had at least a ten-year delay.[60] This long delay between filing and award is systemic for Los Alamos patents and is most likely caused by the fact that under P.L. 700 the patents that have been declared secret remain applications until they are declassified and awarded. The longest delay on any atomic patent that has since been issued—that is, on the patent that has thus far spent the longest time under a secrecy order—was nearly sixty years. This patent, filed in September 1945 but not granted until July 2004, was for a chemical process related to gaseous diffusion research done at Oak Ridge during the war.[61]

According to the Department of Energy lawyer who processed this long-delayed patent in its final stages, secret patents of this sort are reviewed annually and checked against changing classification guidance documents to see if they qualify for declassification. If a patent application is then determined to be declassified, the lawyers must decide whether it is a worthwhile expenditure of taxpayer funds to push the old application through to completion. A "small good-faith effort" is made to find the inventor, if he or she is still alive, and inform him or her of the patent's issuance, though this is only for purposes of credit (there is no monetary award of any sort, since the inventor has already signed the patent over to the government).[62]

Sometimes changes in classification guidelines can release what seem to be entire components of Manhattan Project weapons, suggesting an alternative way of viewing the atomic bomb: as a composite of many patents. For example, in May 1976 three secret patents were granted that had obvious implications for the war effort. The oldest was an

[59] For the totals coming out of Lavendar's office see "Manhattan District History: Book I—General, Volume 13—Patents," 31 Dec. 1946, in *Manhattan Project: Official History and Documents* [microform] (Washington, D.C.: Univ. Publications America, 1977), Sect. 5, pp. 1–4. The applications filed with the Patent Office were kept in a separate division reserved for military matters; see *ibid.*, Sect. 3, pp. 1–3. Overall patent figures were compiled from the U.S. Patent Office Web site by Bill Rankin, who generously shared his data. Depending on the years chosen, the figures can look even more impressive (patenting activity in general dropped dramatically during the war), but the most conservative version does enough work by itself.

[60] To obtain these numbers, patent information was extracted with a computer program from Los Alamos National Laboratory's library catalogues and tabulated. There is some evidence that the catalogue has undercounted, though probably not to a statistically significant degree for these dates. In any event, these should be taken as rough values only. Around 1,300 Los Alamos patents from many different years can be found through Los Alamos National Laboratory's online library search engine at http://library.lanl.gov.

[61] James P. Brusie, "Method of Determining the Extent to Which a Nickel Structure Has Been Attached by a Fluorine-Containing Gas," U.S. Patent 6,761,862 (awarded 13 July 2004). The majority of the few other patents that have endured comparable delays, according to the U.S. Patent Office's search engine and Google Patents, are either in the cryptographic field or are not in fact really "delayed"; they are, rather, typographical errors (i.e., the Patent Office has recorded the application date as 1938 when it was actually 1988).

[62] Paul A. Gottlieb, Assistant General Counsel for Technology Transfer and Intellectual Property, U.S. Department of Energy, communication with Alex Wellerstein, Aug. 2006.

application for a "Low Impedance Switch," filed in the name of Donald F. Hornig at Los Alamos in November 1945, which described a device that could close "a plurality" of electrical circuits within the space of 0.05 to 0.5 microseconds (the drawing shows sixteen circuits—a significant number when one knows that the "Trinity" gadget used thirty-two detonators). Then there was a "Detonating Apparatus" application, filed in May 1950 in the name of Lawrence H. Johnston, describing a device used "for detonation of high explosive in uniform timing" by means of a spark gap detonator. Last was a "High Explosive Compound" application, filed in January 1956 under the name of Theodore C. Crawford, describing one particular type of explosive with an alterable detonation velocity, useful for shaping the explosive wave in a very precise manner.[63]

The relevance of these three patents to nuclear weapons design should be fairly obvious—they describe key parts of an implosion nuclear weapon, in which the simultaneous detonation of carefully created explosive lenses is used to compress a plutonium core to supercriticality—though all are described in a generic technical language devoid of direct implications for bomb design, which serves to exempt them from the later ban on patents useful *only* for the detonation of nuclear weapons. Hornig's patent, the detonator, describes the invention in a manner that could hypothetically be applicable in many situations: "In certain types of ordnance and other equipment, it is necessary to energize a relatively large number of electrical circuits within periods of time of the order of 0.05 to 0.5 microseconds. . . . The principal object of the present invention therefore, is to provide improved switching apparatus for effecting the simultaneous closing of a plurality of electrical circuits of the type described above."[64] Nowhere, of course, does it note that among these "certain types of ordnance" is the atomic bomb.

These patents, in combination with Johnston's patent on the exploding bridge-wire detonator used in the "Trinity" and "Fat Man" devices (filed in 1944, granted in 1962) and the patent for the pressure switches used in the bombs dropped on Hiroshima and Nagasaki (filed in the name of Alan N. Ayers in 1946, granted in 1967), allow for a reasonable approximation of how an atomic bomb might look through the eyes of a patent lawyer. The pressure switches patent shows the same distinctive electrical harness that can be seen in pictures of the "Trinity" gadget (see Figures 3 and 4 and cover illustration). Many other Los Alamos patents, especially those awarded long after their applications were filed, are similarly suggestive, describing all aspects of bomb production, from explosive chemistry to reactor cooling, ion sources to electrical circuits, neutron sources to neutron detectors.

Two Manhattan Project scientists independently told me that they were asked to sign off their patent rights to the final bomb itself (both reported that they were promised a single dollar in compensation but never received it). An oral history conducted with another indicates that even something like last-minute changes to the design of the plutonium for

---

[63] Donald F. Hornig, "Low Impedance Switch," U.S. Patent 3,956,658 (awarded 11 May 1976); Lawrence H. Johnston, "Detonating Apparatus," U.S. Patent 3,955,505 (awarded 11 May 1976); Theodore C. Crawford, "High Explosive Compound," U.S. Patent 3,956,039 (awarded 11 May 1976); Lawrence H. Johnston, "Electric Initiator with Exploding Bridge Wire," U.S. Patent 3,040,660 (awarded 26 June 1962); and Alan N. Ayers, "Pressure Sensitive Switch," U.S. Patent 3,358,605 (awarded 19 Dec. 1967). A description of Hornig's development of the switch is in transcript, Donald F. Hornig Oral History Interview I, 4 Dec. 1968, by David G. McComb, Internet Copy, LBJ Library, Austin, Texas, pp. 3–6. Neither Hornig nor Johnston said that he could recall being made aware of his patent(s) being granted: Donald F. Hornig, communication with Wellerstein, May 2005; and Lawrence H. Johnston, communication with Wellerstein, May 2005.

[64] Hornig, "Low Impedance Switch," col. 1.

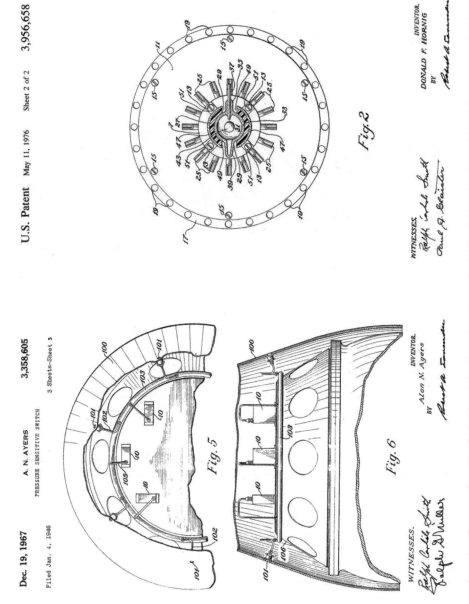

Figure 3. *Images from the "Pressure Sensitive Switch" and the "Low Impedance Switch" patents, showing a firing unit array very similar to that used in the X-Unit of the "Fat Man" bomb and "Trinity" gadget. Source: U.S. Patent Office.*

Figure 4. *The "Trinity" gadget detonated on 16 July 1945. The detonating switches are in the X-Unit firing device at the front of the bomb. Note the harness and the cables emanating from it. Source: Los Alamos National Laboratory, photo TR-297, part of collection LA-UR-06-1005.*

the "Fat Man" bomb required a secret patent application.[65] There are certainly patent applications from this period that have not yet been granted, and many (like the above) never will be. But this comes as no great surprise: of course, we reply, the government would never publish something that gave such sensitive information to the world—a response that assumes that a policy of institutionalized secrecy is the obvious approach. But if this is the model for dealing with nuclear weapons, why pursue the patents on such a vast scale? Why patent the bomb, why treat it like any other piece of technological intellectual property?

We have seen so far that these policies were made to control contractors and to control inventors both inside and outside the project; more to the point, the policies were meant to control atomic technology. Because patents did not become one of the major forms of proliferation control in the postwar era, it is easy to write off the patent program as a

---

[65] Oral communications with Philip Morrison, 9 Feb. 2005, and Robert F. Christy, 5 Dec. 2007. According to Robert Bacher, head of G-Division at Los Alamos during the war, the "Fat Man" core differed from the "Trinity" gadget in that the former "had three pieces instead of two . . . the funny part of it is when you look up the patents, you'll find . . . that little thing there has got a patent on it": Robert F. Bacher, oral history interview with Lillian Hoddeson, 3 Mar. 1986, transcript in the Robert F. Bacher Papers, 10105-MS, Caltech Archives, California Institute of Technology, Pasadena, Box 48, Folder 7, p. 8.

historical anomaly, but only by taking it seriously can we hope to understand its ultimate intent.

CONCLUSION: THE PLACE OF PATENTS IN WARTIME ATOMIC POLICY

In a report to Vice President Henry A. Wallace, Secretary of War Henry L. Stimson, and Chief of Staff General George C. Marshall dated 21 August 1943, Groves included an entire section on the patent program, introduced in powerful language: "If the possibility of world disaster through the development of this superexplosive and its possible military by-products is to be avoided and the enormous hazard involved in preparation minimized, the utilization of atomic power must always be under close control of governments interested in the welfare of mankind rather than in absolute domination and exploitation of other peoples." This rather dramatic statement was immediately followed by what today seems a non sequitur: "If the United States has a strong patent position, the achievement of the above will be facilitated." Lawyers were on the case, Groves explained, and all MED personnel ("both scientific and industrial") had for some time been required to sign over all patent rights to the government. "This program is sound for the war period, which now has first consideration," Groves concluded, "and will lay the groundwork for proper control thereafter." Though much of Groves's report had been cribbed from an earlier one written by Bush, the phrase "possibility of world disaster" was his own, and the level of concern expressed, for all its hyperbole, is commensurate with the scope of the patent-related policy undertaken by the MED, following in the footsteps of the OSRD.[66]

Groves's concern seems misplaced when viewed through modern eyes. The general's intentions to the contrary, patenting did not become a major way of controlling postwar nuclear development—civilian or military, national or international. Instead, U.S. nuclear weapons control during the Cold War took the form of the safeguarding of information (secrecy), attempted monopolies on raw materials, diplomatic agreements, international inspection agencies, and export controls on specific technologies and substances. Patents useful only for nuclear weapons were forbidden by the Atomic Energy Act of 1946. Civilian nuclear power controls took the form of centralized regulatory agencies (such as the Atomic Energy Commission and, later, the Nuclear Regulatory Commission) and operated within a legal framework quite different from patent law. So why did people like Bush and Groves think that patents were a central part of the "groundwork for proper control" of atomic energy during World War II?

Answering this question involves interrogating what exactly "control" means in relation to atomic weapons. Many different policies were undertaken to control atomic energy in the pre-Hiroshima period, and these involved a wide variety of practices and assumptions. Without attempting to provide a comprehensive taxonomy of "control," we can legiti-mately note that the practices of resource monopolies (entering into exclusive arrange-

---

[66] Groves and Military Policy Committee to Henry A. Wallace, Henry L. Stimson, and George C. Marshall, memo, "Present Status and Future Program on Atomic Fission Bombs," 21 Aug. 1943, in *Harrison-Bundy Files Relating to the Development of the Atomic Bomb, 1942–1946*, microfilm publication M1108 (Washington, D.C.: National Archives and Records Administration, 1980), Folder 6: "Military Policy Committee Papers—Minutes," Roll 1, Target 6. Cf. Bush to Roosevelt, memo, "Report on Present Status and Future Program on Atomic Fission Bombs," 16 Dec. 1942, in *Correspondence ("Top Secret") of the Manhattan Engineer District, 1942–1946*, microfilm publication M1109 (Washington, D.C.: National Archives and Records Administration, 1980), Folder 25: "Documents Removed from Gen. (L. R.) Groves' Locked Box, Plus Certain Documents of Historical Importance," Roll 3, Target 8.

ments with uranium-producing countries), information limitation policies (compartmen-talization, secrecy), personnel evaluation (background investigations and loyalty oaths), site security (barbed-wire fences and posted guards), and diplomacy (international agree-ments about cooperation and information sharing) constituted some of the basic and well-known categories of controlling the atomic bomb before it had ever been used (and when the most pressing security threat was German knowledge of the program).

In this schema, patents filled a gap between the policies of *individual* control, usually ascribed to "secrecy," and the policies of *international* control, usually described in terms of diplomacy, resource monopolies, and, later, export controls. Specifically, patent control was an attempt at *legal* control; that is, it was an attempt to remove any possible legal *problems* for U.S. government ownership of atomic technology during and after the war. The motivation for this ownership imperative shifted from an initial sense of civic responsibility—preventing corporations or individuals from profiting from government-funded research—to later concerns about the maintenance of technological sovereignty on the part of the government against the interests of foreign countries and individual inventors.

In itself, legal control would not be remarkable as a motivation: it becomes interesting only because we do not assume that there would ever be a legitimate legal challenge to the U.S. government's new atomic stockpile. The "special" nature of the bomb is usually implicitly taken to have given the government a blank check in regard to power, but in the period before the bombs were made public Manhattan Project administrators were forced for the most part to use existing legal structures in their efforts to enforce their own control over these new, "special" weapons. In some cases, of course, they were confident about what lines they could cross: in the area of internal security, for example, there seem to have been few quibbles about legality from project administrators. But with intellectual property—an area with a much stronger legal history, and one where wartime overstep-ping had historically been paid for in the postwar period—their concern is quite evident. The patenting program can be seen as one part of a multifaceted attempt to use ordinary laws to control extraordinary technology.

Though Bush and Groves clearly thought that atomic energy was something unique, they did not have our modern prejudice that all regulation of it would necessarily be unique. Though Bush had pushed for tight wartime secrecy restrictions on bomb infor-mation, he—like many project scientists—did not think that perpetual secrecy was likely to be an effective or efficient way of controlling the spread of nuclear weapons in the long run, and—again like many project scientists—he thought that ultimately the latter could be effectively achieved only through international control agreements.[67]

The patent program was, as Bush often reiterated, a form of control—technological control, an investment for the postwar era, a way to make sure that the individual scientists working on the project were not in a legal position to demand too much direct say over it, and a way to keep the government from having to answer to anyone, on a legal level,

---

[67] On this point see Vannevar Bush and James Conant to the Secretary of War, "Salient Points Concerning Future International Handling of Subject of Atomic Bombs," 30 Sept. 1944, in Record Group 77, Records of the Army Corps of Engineers, Manhattan Engineer District (MED), Harrison-Bundy Files (H-B Files), Folder 69, National Archives, available online through the National Security Archive at http://www.gwu.edu/~nsarchiv/NSAEBB/NSAEBB162/. There is a copious literature on postwar arms control schemes. See, e.g., Hewlett and Anderson, *New World* (cit. n. 5); Martin Sherwin, *A World Destroyed: Hiroshima and the Origins of the Arms Race* (New York: Vintage, 1987); and James Hershberg, *James B. Conant: Harvard to Hiroshima and the Making of the Nuclear Age* (Stanford, Calif.: Stanford Univ. Press, 1995).

in its goal of nuclear monopoly. If, as Bush and many others thought, atomic energy was going to be what separated the haves from the have-nots of the future, and if it could not—and should not—be kept secret indefinitely, then having all of the key patents—reactors, processes, emitters, receivers, switches, bombs, and all—secured in the name of the government of the United States would not only be a prudent use of $2 billion of the taxpayers' money; it would also be good for national security, economic and political. In this sense, the patent program is perhaps a glimpse at a postwar that never happened: a postwar where atomic energy would lead to a complete revitalization of international economies, where international control and shared information would eliminate the need for an arms race, and where the atomic bomb, while special, would be treated more like another technology than an apocalyptic symbol.

But it should not be misconstrued that there was a single bureaucratic impetus behind the program. As this narrative has attempted to show, in many ways the patenting program was an *ad hoc* adaptation of contract policy and intellectual property practices to address a changing menagerie of threats, whether they were accusations of profiteering, challenges by project scientists over ownership of their work (and a say in the project), or French scientists looking to secure postwar information-sharing deals. It is in part this *ad hoc* nature of the program that made it so difficult to understand in the postwar period: once it was assumed that atomic bombs would have to be regulated by legislation that codified their extraordinary status, all of the composite challenges—the ownership disputes, the French patents, the profiteering question—seemed comparatively easy to brush aside.

The patent program did other work for the Manhattan Project bureaucrats as well. For one thing, patent secrecy was, at the time, the only vehicle for achieving long-term, specifically technological secrecy available to administrators bound by a pre–atomic age legislative system: the 1917 patent secrecy statute was, in fact, the first U.S. statute to permit the government to impose secrecy restrictions on what could be entire categories of information developed *outside* of the government, foreshadowing the all-inclusive definitions of secrecy that would later come to fruition in the "born secret" clauses of the Atomic Energy Act of 1954.[68] Patent secrecy was, of course, considerably more limited in scope than the full system of perpetual nuclear secrecy that would materialize under the auspices of the Cold War atomic mentality: its threats were based primarily on the assumption that the denial of a potential patent in the United States would be enough of an economic inconvenience to deter infraction. But patent control allowed the Manhattan Project administrators to reach out into the private sphere, to declare things secret even if they had not been created within the project, and could serve as a stop-gap control measure until full legislation for atomic energy had been passed by Congress.[69]

The significance of the Manhattan Project patent program is, then, twofold. First, within the context of the history of the Manhattan Project itself, it represents one of the many paths to "control" that, in the end, did not become significant in the postwar era. In this, the definition of the bomb as a form of technology with specific authorship and intellectual property implications is one of the many competing definitions that existed in the 1940s,

[68] Arvin S. Quist, *Security Classification of Information*, Vol. 1: *Introduction, History, and Adverse Impacts* (Oak Ridge, Tenn.: Oak Ridge Classification Associates, 2002), p. 26; available online at http://www.fas.org/sgp/library/quist/.
[69] The evolving importance of the "secret" in the early Cold War is well covered in David Kaiser, "The Atomic Secret in Red Hands? American Suspicions of Theoretical Physicists during the Early Cold War," *Representations*, 2005, *90*:28–60.

before a more "Cold War" assessment of it became thoroughly cemented in the early 1950s with the spy trials and the Atomic Energy Act of 1954.

Second, in the context of nuclear historiography in general, the patent program can serve as an example of the ways in which historians, adopting those same visions of the bomb and its meaning that became so prevalent during the Cold War, have managed simply not to see large developments that did not sensibly fit into this model. While the Manhattan Project's massive patent program is emphasized in a number of accounts by participants, almost all subsequent histories omit any mention of it, and those few that do discuss it take notice only where it occasionally intersects with more "traditional" Manhattan Project narratives, as in the context of the struggle between scientists and bureaucrats or the places in which it served as a locus for diplomatic chafing. The patent program can serve as an important reminder of the need to revisit nuclear history, without reading events entirely through the lens of what *did* occur and, in the process, missing many of the other directions that history could have taken.[70] It is, in the end, a provocation to be skeptical of technological determinism, even in the case of an artifact as politically laden as the atomic bomb.

The patent program can be seen as something of a barometer of how administrative attitudes toward the atomic bomb changed over the course of its development. At first work on the bomb was regarded like any other wartime research venture—a speculative approach to an improved form of explosive whose success seemed possible but not necessarily probable. As it became a more realistic possibility, and a more massive development project, patent control became more important and at the same time more exceptional—the atomic bomb became something for which absolute patent control was required, and the act of patenting itself was contorted toward its most monopolistic extremes. Finally, when the bomb became a public reality, and its existence as a "special" technology was cemented after it was credited for Japan's surrender, the patent program— though in many ways an incredible attempt by the federal government to extend its ability legally to control technology—was seen as woefully inadequate, if not a liability. Its increasing omission from contemporary discussions and its almost total absence from the historical literature reinforce the idea that this was an insignificant effort, despite its immense scope and its importance to the leading project scientists and administrators.

In the absence of a clear model for how properly to control a technology that he thought would be revolutionary, and for whose realization he was dependent on expertise outside the government, it is not surprising that Bush looked to the patent system as a potential solution—and not so surprising that Roosevelt gave him the go-ahead. The patent system was a well-established and legally sound approach to technological control, both domestically and internationally, and, faced with a new—but still forming—conception of weaponry and power, Bush—himself an engineer with a rich history of involvement with the patent system—sought mastery first in preexisting systems of technological control while at the same time thinking about what systems would exist in the future.

After the war, all this would change. Attitudes about nuclear weapons policy, though never quite stable, coalesced into a far more rigid model. The notions of "secrecy"

---

[70] Michael Gordin's recent book on the bombing of Hiroshima and Nagasaki was in a way an inspiration for my thinking about competing visions for the bomb during World War II itself. It makes many salient historiographical points about reading what we know of the outcome back into the narratives of the war, arguing in part for a more epistemically flavored reevaluation of nuclear history, refusing to take for granted the technological uniqueness of the bomb itself and arguing that the construction of such a vision of the bomb must itself be analyzed. See Gordin, *Five Days in August* (cit. n. 6).

(information control) and "security" (physical and personnel control) used in the Manhattan Project seem to have collapsed into each other; and after the "nuclear club" was established the diplomatic and economic policies of international nonproliferation agreements became the new way of envisioning atomic control. Patents had very little to do with this: aside from their inherent limitations of scope, the Atomic Energy Acts effectively took the question of patenting the bomb out of the bureaucratic consciousness and made it a considerably less pressing issue from a security standpoint. Patenting policies within the nuclear complex continued, though it seems from a preliminary analysis that the patenting of inventions that clearly would never be declassified under the new regime of secrecy was more a matter of registering priority—and thus probably of establishing stepping stones for career advancement—than a question of legitimate short- or long-term technological control.[71] At the same time, postwar debates about atomic patent policy would refocus on economics rather than on security.[72] The legacy of the Manhattan Project patenting program seems to have persisted in its successor agencies, but the original impetus had waned.[73] As time went on, the belief that there was a strong connection between military security and patent control in relation to nuclear power faded into irrelevance and, because it defied an easy fit into the traditional narrative of the bomb, became increasingly incomprehensible.[74]

[71] Priority and secrecy relating to nuclear weapons have been provocatively discussed in Gusterson, "Death of the Authors of Death" (cit. n. 5). On behind-the-fence recreations of "open" institutions, such as "classified" journals and the like, much has been written; there is a good bibliography in Peter Westwick, "Secret Science: A Classified Community in the National Laboratories," *Minerva*, 2000, *38*:363–391, n. 1.

[72] Though there were extensive debates about patents in relation to the Atomic Energy Act of 1946 and the revisions of 1954, as the war faded into the background the debates focused almost exclusively on questions of private ownership for economic purposes—and not on security implications. For a discussion of these debates see Hewlett and Anderson, *New World* (cit. n. 5), pp. 495–498, 523–524, 527; James R. Newman and Byron S. Miller, *The Control of Atomic Energy: A Study of Its Social, Economic, and Political Implications* (New York: McGraw-Hill, 1948), Ch. 8; and Richard G. Hewlett and Jack M. Holl, *Atoms for Peace and War, 1953–1961: Eisenhower and the Atomic Energy Commission* (Berkeley: Univ. California Press, 1989), pp. 115–117. Note that Hewlett and Anderson, who are usually quite reliable, take their understanding of the patent issue from Newman and Miller, whose account in turn rests almost solely on Lavender's testimony and for this reason is extremely inadequate in terms of historical context (Newman's account of this matter should be read strictly as the point of view of one of the lawyers responsible for the first Atomic Energy Act, not as a synthetic historical presentation). One of the few instances in which the security question in relation to patents was raised in the postwar period was in the context of a House Un-American Activities Committee attack on David Lilienthal in 1947. See Jessica Wang, "Science, Security, and the Cold War: The Case of E. U. Condon," *Isis*, 1992, *83*:238–269, esp. pp. 244–248; and Creager, "Tracing the Politics of Changing Postwar Research Practices" (cit. n. 54).

[73] For a full discussion of Atomic Energy Commission and Department of Energy patent policies—and struggles—throughout the Cold War see Edward C. Walterscheid, "The Need for a Uniform Government Patent Policy: The D.O.E. Example," *Harvard Journal of Law and Technology*, 1990, *3*:103–166.

[74] A succinct example of precisely *how* incomprehensible it became can be found in Walterscheid, "Need for a Uniform Government Patent Policy." This former Deputy Laboratory Counsel for Los Alamos National Laboratory insists numerous times that the historical MED/AEC/DOE patent policies make little sense from a legal point of view, since "ownership of patent rights has very little to do with protection of national security" (p. 158). Such a point of view, though, had not yet solidified in the years of the Manhattan Project.

Nuclear Democracy

Political Engagement, Pedagogical Reform, and Particle Physics in Postwar America

By David Kaiser

ABSTRACT

The influential Berkeley theoretical physicist Geoffrey Chew renounced the reigning approach to the study of subatomic particles in the early 1960s. The standard approach relied on a rigid division between elementary and composite particles. Partly on the basis of his new interpretation of Feynman diagrams, Chew called instead for a "nuclear democracy" that would erase this division, treating all nuclear particles on an equal footing. In developing his rival approach, which came to dominate studies of the strong nuclear force throughout the 1960s, Chew drew on intellectual resources culled from his own political activities and his attempts to reform how graduate students in physics would be trained.

INTRODUCTION: MCCARTHYISM AND THE WORLD OF IDEAS

H istorians have studied several examples in which scientists framed details of their work in explicitly political language. German physiologists and physical scientists such as Rudolf Virchow, Ernst von Brücke, Emil Du Bois-Reymond, Hermann von Helmholtz, and others endeavored self-consciously to keep their dreams of political unity alive even after the crushing defeat of 1848 by pursuing methodological and epistemological unity within scientific knowledge. Just a few years later, the Swiss-French chemist Charles Fréd-éric Gerhardt suggested a "chemical democracy" in which all atoms within molecules

Isis 93, no. 2 (June 2002): 229–68.

My thanks to Stephen Adler, Louis Balázs, E. E. Bergmann, John Bronzan, Geoffrey Chew, Jerome Finkelstein, William Frazer, Carl Helmholz, Francis Low, Stanley Mandelstam, Howard Shugart, Henry Stapp, Kip Thorne, and Eyvind Wichmann for their interviews and correspondence with me. Thanks also to the Office for History of Science and Technology at the University of California, Berkeley, for its warm hospitality while most of the research for this paper was conducted, and to David Farrell of the Bancroft Library for his assistance with several collections. This paper has benefited from comments and suggestions from Cathryn Carson, Patrick Catt, Geoffrey Chew, James Cushing, Peter Galison, Tracy Gleason, Michael Gordin, Stephen Gordon, Loren Graham, Kristen Haring, Kenji Ito, Matt Jones, Alexei Kojevnikov, Mary Jo Nye, Elizabeth Paris, Sam Schweber, Jessica Wang, and five anonymous *Isis* referees.

would be treated as equals, a notion the German organic chemist Hermann Kolbe countered with a hierarchical, "autocratic" model of molecular structure. More recently, several Soviet theoretical physicists, including Yakov Frenkel, Igor Tamm, and Lev Landau, drew explicitly on their own life experiences under Stalin's rule—experiences that often included extended prison terms—when describing solid-state physics, referring, for example, to the "freedom" of electrons in a metal and particles' other "collectivist" behavior. The Indian astrophysicist Meghnad Saha emphasized during the 1920s that various chemical elements within the stars would respond to the same "stimulus" in varying ways, an analysis of atoms' agency deeply resonant with his own social and political struggles against caste hierarchies. Meanwhile, during the postwar period, the Japanese particle theorist Shoichi Sakata, thinking along explicitly Marxist lines, favored a strict hierarchy among subatomic particles, finding in such an arrangement appropriate base-superstructure relations.[1] Social metaphors abound within the physical sciences.

Like others in these earlier periods of tumult and turmoil, American scientists working after World War II experienced dramatic and fast-moving political currents. McCarthyism in America meant sweeping violations of civil liberties for thousands of citizens; blacklists and unfair firings from all manner of jobs affected people in and out of academia. But it also meant more than this: historians must supplement the tallies of such injustices with attention to the intellectual legacy of McCarthyism.[2] This essay explores the interplay during the early postwar decades between changing assumptions about political engagement, effective pedagogical approaches, and ideas about the behavior of subatomic particles. In particular, I will focus on the work of the prominent Berkeley particle theorist Geoffrey Chew and his concept of "nuclear democracy."

Beginning in the early 1960s, Chew railed against physicists' reigning approach to particle physics, quantum field theory, arguing that this framework offered no help for understanding the strong nuclear forces that kept atomic nuclei bound together. As he emphasized with great gusto at a June 1961 conference in La Jolla, California, quantum field theory was as "sterile" as "an old soldier" when it came to treating the strong interaction and hence was "destined not to die but just to fade away"—a memorable pronouncement that many of his peers repeated over the next several months.[3] In its place,

[1] On the German physiologists and physical scientists see Keith Anderton, "The Limits of Science: A Social, Political, and Moral Agenda for Epistemology in Nineteenth Century Germany" (Ph.D. diss., Harvard Univ., 1993), Ch. 2; cf. Timothy Lenoir, "Social Interests and the Organic Physics of 1847," in *Science in Reflection*, ed. Edna Ullmann-Margalit (Dordrecht: Kluwer, 1988), pp. 169–191. On Gerhardt's and Kolbe's positions see Alan Rocke, *The Quiet Revolution: Hermann Kolbe and the Science of Organic Chemistry* (Berkeley: Univ. California Press, 1993), pp. 208, 325 (my thanks to Michael Gordin for bringing this reference to my attention). On the Soviet physicists' outlook see Alexei Kojevnikov, "Freedom, Collectivism, and Quasiparticles: Social Metaphors in Quantum Physics," *Historical Studies in the Physical and Biological Sciences*, 1999, *29*:295–331; cf. Karl Hall, "Purely Practical Revolutionaries: A History of Stalinist Theoretical Physics" (Ph.D. diss., Harvard Univ., 1999). On Saha see Abha Sur, "Egalitarianism in a World of Difference: Identity and Ideology in the Science of Meghnad Saha," unpublished MS. On Sakata see "Philosophical and Methodological Problems in Physics," *Progress of Theoretical Physics*, 1971, *50*(Suppl.):1–248; Shunkichi Hirokawa and Shûzô Ogawa, "Shôichi Sakata—His Physics and Methodology," *Historia Scientiarum*, 1989, *36*:67–81; and Ziro Maki, "The Development of Elementary Particle Theory in Japan—Methodological Aspects of the Formation of the Sakata and Nagoya Models," *ibid.*, pp. 83–95 (my thanks to Masakatsu Yamazaki for discussions of Sakata's work).

[2] Cf. Loren Graham's suggestive work on the sometimes fruitful, generative appropriations of dialectical materialism by Soviet scientists: Loren Graham, *Science, Philosophy, and Human Behavior in the Soviet Union* (New York: Columbia Univ. Press, 1987); Graham, *What Have We Learned about Science and Technology from the Russian Experience?* (Stanford, Calif.: Stanford Univ. Press, 1998); and Graham, "Do Mathematical Equations Display Social Attributes?" *Mathematical Intelligencer,* 2000, *22*:31–36.

[3] A preprint of Chew's talk at the 1961 La Jolla conference is quoted in James Cushing, *Theory Construction and Selection in Modern Physics: The S Matrix* (New York: Cambridge Univ. Press, 1990) (hereafter cited as

Chew aimed to erect a new program based directly on the so-called scattering matrix, or *S* matrix, which encoded mathematical relations between incoming and outgoing particles while eschewing many of the specific assumptions and techniques of quantum field theory.

The single most novel conjecture of Chew's developing *S*-matrix program, and its most radical break from the field-theoretic approach, was that all nuclear particles should be treated "democratically"—Chew's word. The traditional field-theory approach, against which Chew now spoke out, posited a core set of "fundamental" or "elementary" particles that acted like building blocks, out of which more complex, composite particles could be made. As we will see in Section I, Chew and his young collaborators argued against this division into "elementary" and "composite" camps, instead picturing each particle as a kind of bound-state composite of all others; none was inherently any more "fundamental" or special than any other. Deuterons, for example, treated by field theorists as bound states of more "elementary" protons and neutrons, were to be analyzed within Chew's program in exactly the same way as protons and neutrons themselves; the "democracy" extended, in principle, all the way up to uranium nuclei. Chew described this notion by the colorful phrase "nuclear democracy."[4]

The larger physics community reacted swiftly to the string of early calculational successes that Chew and his Berkeley group produced under the "democratic" banner, one example of which I will examine in the next section. Ten months after his initial "call to arms" in La Jolla, Chew was elected a member of the National Academy of Sciences, an honor he attained before his thirty-eighth birthday. There followed a string of coveted invited papers at National Academy and American Physical Society meetings. "Such lectures invariably drew capacity crowds," Chew's department chair gloated to a dean in 1964, "since Chew is generally recognized as the outstanding exponent of a particular approach to the theory of elementary particles known as the S-matrix."[5] Earlier that year Murray Gell-Mann had introduced his "quark" hypothesis, whose proposed core, fundamental building blocks look to our eyes today like the very antithesis of Chew's nuclear democracy. Yet in his first papers on the quark hypothesis, Gell-Mann took pains to re-

Cushing, *Theory Construction*), p. 143. See also Murray Gell-Mann, "Particle Theory from *S*-Matrix to Quarks," in *Symmetries in Physics (1600–1980),* ed. M. G. Doncel, A. Hermann, L. Michel, and A. Pais (Barcelona: Bellaterra, 1987), pp. 479–497. This portion of Chew's unpublished talk was incorporated verbatim in the introduction to his 1961 textbook, *S-Matrix Theory of Strong Interactions* (New York: Benjamin, 1961), pp. 1–2. Several physicists cited Chew's unpublished La Jolla talk during the October 1961 Solvay conference in Brussels, as recorded in the Solvay conference proceedings: R. Stoops, ed., *The Quantum Theory of Fields* (New York: Interscience, 1961), pp. 88, 132, 142, 179–180, 192–195, 214–215, 222–224. In a recent interview, Chew likened his strongly worded talk at the 1961 La Jolla meeting to a "coming out of the closet" speech: Geoffrey Chew, interview with Stephen Gordon, Dec. 1997, quoted in Stephen Gordon, "Strong Interactions: Particles, Passion, and the Rise and Fall of Nuclear Democracy" (A.B. thesis, Harvard Univ., 1998) (hereafter cited as **Gordon, "Strong Interactions"**), p. 32.

[4] Geoffrey Chew, "Nuclear Democracy and Bootstrap Dynamics," in Maurice Jacob and Chew, *Strong-Interaction Physics: A Lecture Note Volume* (New York: Benjamin, 1964), pp. 103–152, on p. 105. For more on Chew's *S*-matrix program see esp. Cushing, *Theory Construction;* Tian Yu Cao, "The Reggeization Program, 1962–1982: Attempts at Reconciling Quantum Field Theory with *S*-Matrix Theory," *Archive for History of Exact Sciences,* 1991, *41:*239–282; and David Kaiser, "Do Feynman Diagrams Endorse a Particle Ontology? The Roles of Feynman Diagrams in *S*-Matrix Theory," in *Conceptual Foundations of Quantum Field Theory,* ed. Cao (New York: Cambridge Univ. Press, 1999), pp. 343–356.

[5] Burt Moyer to Dean W. B. Fretter, 30 Dec. 1964, quoted in Raymond Birge, "History of the Physics Department, University of California, Berkeley," 5 vols., ca. 1966–1970 (unpublished; copies are in the Bancroft and Physics Department Libraries, Berkeley) (hereafter cited as **Birge, "History"**), Vol. 5, Ch. 14, p. 50. On Chew's election to the NAS and his other invited lectures see *ibid.,* pp. 49–52; "Physics Professor Wins Prize," *Daily Californian* [Berkeley student newspaper], 3 Jan. 1963, p. 5; and Stanley Schmidt, "The 'Basic' Particle— It's Out of Date," *ibid.,* 23 Oct. 1963, pp. 1, 12.

assure his readers that quarks were fully compatible with Chew's "democratic" frame-work.[6] By the mid 1960s, Chew and his fast-growing group at Berkeley had changed the way most theoretical physicists approached the strong interaction.

Though the idea of nuclear democracy was put forth repeatedly throughout Chew's 1961 lectures, it took several more years before Chew could, by his own lights, escape "the conservative influence of Lagrangian field theory." His updated lectures, published in 1964, were distinguished from the older ones, Chew explained, by their "unequivocal adoption of nuclear democracy as a guiding principle." He began by contrasting at some length "the aristocratic structure of atomic physics as governed by quantum electrodynamics" with the "revolutionary character of nuclear particle democracy." Chew left his students and readers with little doubt: "My standpoint here . . . is that every nuclear particle should receive equal treatment under the law."[7]

All of this bluster about "conservative" field theory versus "revolutionary" nuclear de-mocracy might inspire a knee-jerk *Zeitgeist* interpretation: as Chew sat in Berkeley, with the 1964 "Free Speech Movement" taking flight all around him, a particular vision of "democracy" floated freely from Telegraph Avenue into the Radiation Laboratory. It is no coincidence, one might conclude, that "democracy" was "in the air" among increasingly radical Berkeley activists and students—and *therefore* also permeated the seemingly rar-efied project of Chew and his students. Like most knee-jerk reactions, however, this one is both too hasty and misplaced: whatever Chew was up to, his physics sprang from more than a vague "spirit of the times." Many roots of his new program extended back to theoretical developments from the mid 1950s. Chew's work on nuclear democracy cannot be read, in other words, simply as another instantiation of the strong Forman thesis, with Chew "capitulating" or "accommodating" his physics to a particular "hostile external en-vironment."[8]

Leaving aside claims of strong external determination, we are nonetheless left with a puzzle: Why did Geoffrey Chew combine these various theoretical developments, stretch-ing as they did over the better part of a decade, into his particular "democratic" interpre-tation? Why, moreover, did physicists working further and further from Chew's immediate group in Berkeley attribute different meanings to Chew's "democratic" physics? If we take a longer view of Chew's activities, intriguing questions and associations arise. In both his developing political activism (as discussed in Section II) and his unusual attempts to reform the training of graduate students (Section III), certain specific meanings of "democracy"

---

[6] Murray Gell-Mann, "A Schematic Model of Baryons and Mesons," *Physics Letters,* 1964, 8:214–215; and Gell-Mann, "The Symmetry Group of Vector and Axial Vector Currents," *Physics,* 1964, 1:63–75. See also George Johnson, *Strange Beauty: Murray Gell-Mann and the Revolution in Twentieth-Century Physics* (New York: Knopf, 1999), pp. 209–214, 225–227, 234. After the tide had shifted away from the *S*-matrix program and back to (gauge) quantum field theories, it was Chew who suggested that quarks and the nuclear-democratic formulation could be made compatible: Geoffrey Chew, "Impasse for the Elementary-Particle Concept," in *The Great Ideas Today,* ed. Robert Hutchins and Mortimer Adler (Chicago: Encyclopaedia Britannica, 1974), pp. 92–125, on pp. 124–125.

[7] Chew, "Nuclear Democracy and Bootstrap Dynamics" (cit. n. 4), pp. 104–106.

[8] Paul Forman's famous argument regarding the acceptance of acausal quantum mechanics in Weimar Germany may be found in Paul Forman, "Weimar Culture, Causality, and Quantum Theory, 1918–1927: Adaptation by German Physicists and Mathematicians to a Hostile Intellectual Environment," *Historical Studies in the Physical Sciences,* 1971, 3:1–115. James Cushing correctly dismisses an overly simplistic reading of Chew's "nuclear democracy" vis-à-vis the Free Speech Movement in *Theory Construction,* p. 217; cf. Gordon, "Strong Interac-tions," pp. 35–37. On the Free Speech Movement in Berkeley see esp. W. J. Rorabaugh, *Berkeley at War: The 1960s* (New York: Oxford Univ. Press, 1989), Ch. 1; and Todd Gitlin, *The Sixties: Years of Hope, Days of Rage,* rev. ed. (1987; New York: Bantam, 1993), pp. 162–166.

recurred: no one should be singled out as special, either for privileges or for penalties; all should be entitled to participate equally. In short, as he described his beloved nuclear particles in 1964, all should receive "equal treatment under the law." Weaving in and out of a series of specific contexts, these particular elements of "democracy" found explicit expression again and again. As evidenced by Chew's continuity of vocabulary when discussing his work in these various domains—a vocabulary that he took great care to hone—it appears that while developing his new program for particle physics, Chew drew on intellectual resources culled from his own efforts at political and pedagogical reform. Laboring to ensure a democracy among particles was only one way in which Chew's work took shape after the war.

Chew's work on particle theory during the 1950s and 1960s thus highlights how certain ideas about "democracy," brought to the surface by changing political and cultural conditions, could infuse a new vision both of how graduate students should be trained and of how nuclear particles interact. The story of "nuclear democracy" can thereby reveal certain connections—labyrinthine and indirect, to be sure, but connections nonetheless—between the Cold War national security state and changing ideas within theoretical physics. In the process, we learn why certain approaches and ideas within theoretical physics might have had special appeal or salience for Chew, as well as why these ideas found varying interpretations or associations outside of his immediate group.

It is important to realize, at the same time, that this is not a story about a well-formulated "ideology" or political philosophy steering the course of physical research. For one thing, Chew left no direct statements announcing that his interesting and influential ideas about particle physics were caused by his political convictions—nor would it make much sense to expect such pronouncements. American physicists after the war came to fashion themselves as eminently practical people, pragmatic tinkerers rather than philosopher-kings. The handful of theorists who did champion strong political or philosophical positions rarely acted on them with any consistency, as Sam Schweber's recent work on J. Robert Oppenheimer has shown. Moreover, those even rarer theorists who claimed to mingle such political commitments with their physical theorizing—such as David Bohm, dismissed from Princeton after pleading the Fifth Amendment before the House Un-American Activities Committee in 1949, or Sakata's Marxist group of particle theorists in Kyoto—were easily marginalized by mainstream American physicists, written off as "doctrinaire."[9]

Yet high-sounding principles and elaborate political philosophies are hardly prerequi-

[9] On Oppenheimer see Silvan Schweber, *In the Shadow of the Bomb: Oppenheimer, Bethe, and the Moral Responsibility of the Scientist* (Princeton, N.J.: Princeton Univ. Press, 2000). On Bohm see Russell Olwell, "Physical Isolation and Marginalization in Physics: David Bohm's Cold War Exile," *Isis,* 1999, *90:*738–756; Shawn Mullet, "Political Science: The Red Scare as the Hidden Variable in the Bohmian Interpretation of Quantum Theory" (B.A. thesis, Univ. Texas, Austin, 1999); and Alexei Kojevnikov, "David Bohm and Collective Movement," unpublished MS. On Sakata's school and their dismissal by American theorists see the sources in note 1, above; Robert Crease and Charles Mann, *The Second Creation: Makers of the Revolution in Twentieth-Century Physics* (New York: Macmillan, 1986), pp. 261–262, 295–296; and Johnson, *Strange Beauty* (cit. n. 6), pp. 202, 231–232. On postwar American physicists' views of themselves see Paul Forman, "Social Niche and Self-Image of the American Physicist," in *The Restructuring of Physical Sciences in Europe and the United States, 1945–1960,* ed. Michelangelo de Maria *et al.* (Singapore: World Scientific, 1989), pp. 96–104; David Kaiser, "The Postwar Suburbanization of American Physics," unpublished MS; and Schweber, "The Empiricist Temper Regnant: Theoretical Physics in the United States, 1920–1950," *Hist. Stud. Phys. Biol. Sci.,* 1986, *17:*55–98. American physicists' lack of explicit political philosophizing fit within broader trends during the American 1950s. See Daniel Bell, *The End of Ideology: On the Exhaustion of Political Ideas in the Fifties* (New York: Free Press, 1960); cf. Alan Brinkley, *Liberalism and Its Discontents* (Cambridge, Mass.: Harvard Univ. Press, 1998).

sites for political action. Rather than formulating an explicit political philosophy of democracy, Chew undertook a series of actions throughout the late 1940s and 1950s to fight for a (perhaps vague) notion of fair play and equal treatment. The fact that at the time these actions were sometimes interpreted quite differently by his peers, or that we might evaluate them today as something other than obviously or inherently "democratic," in no way diminishes the importance Chew himself invested in such experiences or the significance they carried for him. The absence of explicit pronouncements tying his political interests and physical research together in a neat and tidy package—the kind of "smoking gun" that is almost never to be found in historical investigations—hardly relieves us of the job of interrogating such episodes to tease apart the intellectual residue of McCarthyism in the world of ideas.

I. DIAGRAMMATIC BOOTSTRAPPING AND A DEMOCRACY OF PARTICLES

Much of Geoffrey Chew's work during the late 1950s and throughout the 1960s centered on new ways of interpreting and calculating with Feynman diagrams. Richard Feynman had first introduced the stick-figure line drawings that bear his name in the late 1940s. The diagrams were designed for service as a handy bookkeeping device when making lengthy calculations within quantum electrodynamics, the physicists' theory for how electrons interact with light. The full calculation of electron-electron scattering, for example, could be broken up into an infinite series of more and more complicated terms, each associated with more and more complicated ways in which the two incoming electrons could scatter. In the simplest contribution, one electron could emit a photon or quantum of light, which would then be absorbed by the other electron, with each electron thereby changing from its original momentum.[10] Feynman illustrated this process with the diagram in Figure 1. Uniquely associated with this diagram, Feynman continued, was a mathematical expression that yielded the probability for two electrons to scatter in this way.

The process illustrated in Figure 1, however, was only the start of the calculation; the two electrons could scatter in all kinds of more complicated ways, and these correction terms had to be included systematically as well. Feynman thus used his line drawings as a shorthand to keep track of these correction terms; examples of the diagrams that entered at the next round of approximation are shown in Figure 2. The key to the diagrams' use, as Feynman and his young collaborator Freeman Dyson emphasized, was the unique one-to-one relation between each element of the diagram and each mathematical term in the accompanying equation. Dyson, in particular, demonstrated that these calculational relations between diagram elements and mathematical expressions could all be derived rigorously from the foundations of quantum field theory. Soon after Feynman introduced the diagrams, however, theorists like Chew began to adopt them—and subtly adapt them—for applications well beyond the domain of electromagnetic interactions.[11]

---

[10] Richard Feynman, "Space-Time Approach to Quantum Electrodynamics," *Physical Review,* 1949, 76:769–789; F. J. Dyson, "The Radiation Theories of Tomonaga, Schwinger, and Feynman," *ibid.,* 1949, 75:486–502; and Dyson, "The *S* Matrix in Quantum Electrodynamics," *ibid.,* 1949, 75:1736–1755. "Virtual" particles, such as the photon exchanged in Figure 1, "borrow" excess energy and momentum for very short periods of time (as compared with the energy they would carry as free, noninteracting particles), as allowed by Heisenberg's uncertainty relation. According to quantum field theory, all interactions arise from the exchange of virtual particles.

[11] On the introduction and use of Feynman diagrams within quantum electrodynamics see Silvan Schweber, *QED and the Men Who Made It: Dyson, Feynman, Schwinger, and Tomonaga* (Princeton, N.J.: Princeton Univ. Press, 1994). On the dispersion of Feynman diagrams during the 1950s and 1960s for new and different types of calculations see David Kaiser, "Making Theory: Producing Physics and Physicists in Postwar America" (Ph.D. diss., Harvard Univ., 2000), pp. 271–470; and Kaiser, "Stick-Figure Realism: Conventions, Reification, and the Persistence of Feynman Diagrams, 1948–1964," *Representations,* Spring 2000, 70:49–86.

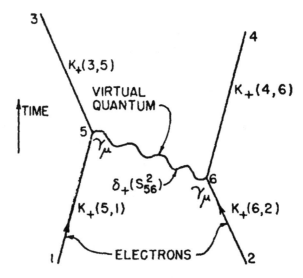

Figure 1. *The simplest Feynman diagram for electron-electron scattering. From R. P. Feynman, "Space-Time Approach to Quantum Electrodynamics," Physical Review, 1949, 76:769–789, on p. 772. Reprinted with kind permission of the American Physical Society.*

Chew pushed his new interpretation of the diagrams further than most of his peers by proclaiming that the diagrams' content, meaning, and calculational role could be severed entirely from the original field-theoretic framework in which they had been introduced. He emphasized in his published 1961 lectures, for example, that many of the new results of his developing program were "couched in the language of Feynman diagrams," even though, contrary to first appearances, they did not "rest heavily on field theory." "It appears to me," he further prophesied, "likely that the essence of the diagrammatic approach will eventually be divorced from field theory" altogether. Whereas field theorists had lectured to their students for over a decade that the lines within Feynman diagrams could only represent elementary particles—that is, the particles whose associated quantum fields appeared in the governing unit or basic interaction term—Chew countered that this distinction was not borne out by the diagrams themselves. As he expounded in his lecture notes, the diagrams themselves "contain not the slightest hint of a criterion for distinguishing elementary particles [from composite ones]. . . . If one can calculate the S matrix without distinguishing elementary particles, why introduce such a notion?"[12] He emphasized the next year, during his summer-school lectures at Cargèse, that the conjecture underlying

[12] Chew, *S-Matrix Theory of Strong Interactions* (cit. n. 3), pp. 3, 5. See also Chew's 1962 lectures at the Cargèse summer school: Geoffrey Chew, "Strong-Interaction *S*-Matrix Theory without Elementary Particles," in *1962 Cargèse Lectures in Theoretical Physics*, ed. Maurice Lévy (New York: Benjamin), Lecture 11, pp. 1–37, on p. 8. Chew was referring in particular to Lev Landau's 1959 modified rules for using the diagrams, which in turn had been based on Chew's 1958 work on the so-called particle-pole conjecture. The particle-pole conjecture stipulated that the only pole-like singularities within the scattering amplitude for a given process would occur uniquely at the values of mass and momentum corresponding to an exchanged particle. Landau used this speculation to derive general rules for isolating the singularities within generic scattering amplitudes. See Chew, "Proposal for Determining the Pion-Nucleon Coupling Constant from the Angular Distribution for Nucleon-Nucleon Scattering," *Phys. Rev.*, 1958, *112*:1380–1383; Lev Landau, "On Analytic Properties of Vertex Parts in Quantum Field Theory," *Nuclear Physics*, 1959, *13*:181–192; Cushing, *Theory Construction*, pp. 109–113, 129–131; and Kaiser, "Do Feynman Diagrams Endorse a Particle Ontology?" (cit. n. 4), pp. 345–349.

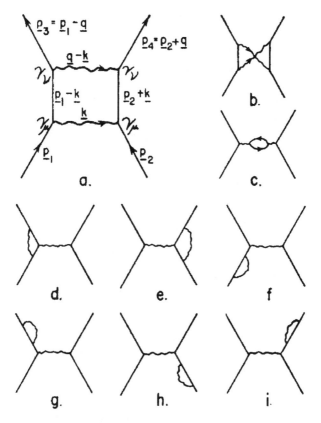

Figure 2. Feynman diagrams for electron-electron scattering correction terms. From R. P. Feynman, "Space-Time Approach to Quantum Electrodynamics," Physical Review, 1949, 76:769–789, on p. 787. Reprinted with kind permission of the American Physical Society.

this diagrammatic equality "grew out of field theory, particularly from Feynman graphs, but it is now believed that the principle can be formulated completely within the framework of the S matrix."[13]

Drawing liberally on a smattering of theoretical developments from the mid and late 1950s, Chew pointed to the diagrams when formulating his new "democratic" conclusion.[14]

[13] Chew, "Strong-Interaction S-Matrix Theory without Elementary Particles," pp. 6–7. Here Chew was again referring to his own 1958 particle-pole conjecture. He later emphasized the "decisive" role played by his new interpretation of Feynman diagrams, writing that by the end of the 1950s it had become clear to him "that graphs of the type invented by Richard Feynman for perturbative evaluation of a Lagrangian field theory are relevant to the analytic S matrix, independent of any approximation based on a small coupling constant." Geoffrey Chew, "Particles as S-Matrix Poles: Hadron Democracy," in *Pions to Quarks: Particle Physics in the 1950s,* ed. Laurie Brown, Max Dresden, and Lillian Hoddeson (New York: Cambridge Univ. Press, 1989), pp. 600–607, on p. 601.

[14] In addition to his particle-pole conjecture and Landau's rules for Feynman diagrams, Chew drew especially on Feynman diagrams' crossing symmetry (as established in 1954), single- and double-variable dispersion relations (which had been formulated during 1954–1958), and Tullio Regge's analysis of complex angular momenta in potential scattering (from 1959). On the dispersion relations work see Marvin Goldberger, "Introduction to the Theory and Application of Dispersion Relations," in *Relations de dispersion et particules élémentaires* [Proceedings of the 1960 Les Houches Summer School], ed. C. de Witt and R. Omnès (Paris: Hermann, 1960), pp. 15–157; and J. D. Jackson, "Introduction to Dispersion Relation Techniques," in *Dispersion Relations: Scottish Universities' Summer School, 1960,* ed. G. R. Screaton (New York: Interscience, 1961), pp. 1–63.

A particle that in one orientation of a Feynman diagram looked like the fundamental building-block constituent of a more complicated particle would appear, upon various rotations of the diagrams, as either the exchanged particle responsible for a force between other particles or as the end-state composite of other constituents. Two of Chew's young colleagues at Berkeley, Frederik Zachariasen and Charles Zemach, built directly on this work in their 1961–1962 treatment of the ρ meson, a heavy, unstable particle that decayed via the strong force relatively rapidly into pairs of pions. The ρ had been discovered by Berkeley experimentalists at the famous cyclotron in 1960, though its existence had actually been predicted earlier by a pair of Chew's graduate students.[15] Zachariasen and Zemach leaned on a series of simple Feynman diagrams, interpreted now along Chew's lines, to guide their calculations.[16] Figure 3a depicted a ρ meson being exchanged between two incoming pions, much as the two electrons in Feynman's Figure 1 exchanged a photon. The exchange of the ρ in this case would give rise to an attractive force between the pions. Figure 3b, on the other hand, showed two pions coming together to form a new bound-state composite particle, the ρ meson, which, being unstable, later decayed into a new pair of pions. If two pions could create a ρ meson, however, then interactions like that shown in Figure 3c had to be considered as well, in which a ρ meson acted just like an "elementary" particle, scattering with an incoming pion; in Figure 3c, meanwhile, a pion appeared as the exchanged force-carrying particle. The field theorists' labels of "elementary," "force-carrier," and "composite" swapped places with each rotation of a given Feynman diagram, Chew and his young colleagues charged. The only consistent theoretical framework, they therefore maintained, was a "democratic" one that made no distinctions between "elementary" and "composite" particles.

From this reinterpretation of Feynman diagrams, only a short step led Chew to the essential point of his autonomous S-matrix program: not only were composite particles to be treated as equivalent to elementary ones, but all (strongly interacting) particles could in fact be seen as composites of each other. Diagrammatic initiatives akin to those in Figure 3 were central to the new scheme, as Chew explained in his summer-school lectures during 1960:

> The forces producing a certain reaction are due to the intermediate states that occur in the two "crossed" reactions belonging *to the same diagram.* The range of a given part of the force is determined by the mass of the intermediate state producing it, and the strength of the force by the matrix elements connecting that state to the initial and final states of the crossed reaction. By considering all three channels [i.e., orientations of the Feynman diagram] on this basis we

Cushing treats the relation of these various ideas and approaches to Chew's S-matrix program in *Theory Construction,* Chs. 3–5.

[15] The experimental discovery of the ρ meson as a resonance in pion scattering was announced in A. R. Erwin, R. March, W. D. Walker, and E. West, "Evidence for a π-π Resonance in the $I = 1, J = 1$ State," *Physical Review Letters,* 1961, 6:628–630. It had been predicted earlier in William Frazer and José Fulco, "Effect of a Pion-Pion Scattering Resonance on Nucleon Structure," *ibid.,* 1959, 2:365–368; and Frazer and Fulco, "Partial-Wave Dispersion Relations for the Process $\pi + \pi \rightarrow N + \bar{N}$," *Phys. Rev.,* 1960, 117:1603–1608. Frazer completed his dissertation under Chew's direction in 1959, and Chew helped to advise Fulco's dissertation, completed in 1962 in Buenos Aires. Years later, Frazer recalled that "Geoff advised us every step of the way" with this work "but generously decided not to put his name on the paper": Frazer, "The Analytic and Unitary S-Matrix," in *A Passion for Physics: Essays in Honor of Geoffrey Chew,* ed. Carleton DeTar, J. Finklestein, and Chung-I Tan (Singapore: World Scientific, 1985) (hereafter cited as **DeTar *et al.,* eds., *Passion for Physics*)**, pp. 1–8, on p. 4.

[16] Frederik Zachariasen, "Self-Consistent Calculation of the Mass and Width of the $J = 1, T = 1$ ππ Resonance," *Phys. Rev. Lett.,* 1961, 7:112–113; erratum, *ibid.,* p. 268; and Zachariasen and Charles Zemach, "Pion Resonances," *Phys. Rev.,* 1962, 128:849–858.

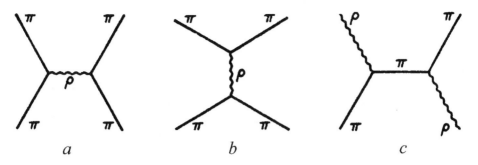

$$a \qquad\qquad b \qquad\qquad c$$

Figure 3. From Frederik Zachariasen and Charles Zemach, "Pion Resonances," Physical Review, 1962, 128:849–858, on pp. 850–851, 857. Reprinted with kind permission of the American Physical Society.

have a self-determining situation. One channel provides forces for the other two—which in turn generate the first.[17]

In this way, Chew explained his notion of a particle "bootstrap," saying simply that "each particle helps to generate other particles which in turn generate it."[18] Consider again the case of the ρ meson, one of the earliest successes for Chew's bootstrap model. In its force-carrying mode, as in Figure 3a, the exchange of the ρ would create an attractive force between the two incoming pions. Thus drawn together owing to this attractive force, the two pions could combine to produce a resonance—a new bound state or composite particle, as in Figure 3b. Next, the "self-determining" feature would come in: Chew and his young collaborators looked for self-consistent solutions such that the force-carrying process produced a resonance whose properties were precisely those of the force-carrying particle itself. If such a solution existed, then the ρ could, all by itself, bring the pions together so that they could produce a ρ. The ρ meson, in this series of diagrams, would

---

[17] Chew, *S-Matrix Theory of Strong Interactions* (cit. n. 3), p. 32 (emphasis added). The same paragraph appears in his summer-school lectures from 1960—Geoffrey Chew, "Double Dispersion Relations and Unitarity as the Basis of a Dynamical Theory of Strong Interactions," in *Dispersion Relations,* ed. Screaton (cit. n. 14), pp. 167–226, on p. 185—and in the proceedings of the 1960 Les Houches summer school—*Relations de dispersion et particules élémentaires,* ed. de Witt and Omnès (cit. n. 14), pp. 455–514. The duplication is not surprising since, as Chew explained, the 1961 lecture note volume "originated in lectures given at summer schools at Les Houches and Edinburgh in 1960": Chew, *S-Matrix Theory of Strong Interactions,* p. vi.

[18] Chew, "Nuclear Democracy and Bootstrap Dynamics" (cit. n. 4), p. 106. Chew's former postdoc and partner in the early bootstrap work, Steven Frautschi, explained simply in lectures from the 1961/1962 academic year that "bootstrap calculations lean heavily on 'crossing,'" that is, on the symmetries obeyed by scattering amplitudes as the associated Feynman diagrams underwent various rotations: Steven Frautschi, *Regge Poles and S-Matrix Theory* (New York: Benjamin, 1963), p. 176. The bootstrap notion was illustrated with the aid of crossed Feynman diagrams in Chew, "Nuclear Democracy and Bootstrap Dynamics," pp. 134, 136, and also in a text by one of Chew's former students: William Frazer, *Elementary Particles* (Englewood Cliffs, N.J.: Prentice Hall, 1966), p. 134. For more on Chew's bootstrap see James Cushing, "Is There Just One Possible World? Contingency vs. the Bootstrap," *Studies in History and Philosophy of Science,* 1985, *16:*31–48; Cushing, *Theory Construction,* Ch. 6; Cao, "Reggeization Program" (cit. n. 4); Kaiser, "Do Feynman Diagrams Endorse a Particle Ontology?" (cit. n. 4); and Yehudah Freundlich, "Theory Evaluation and the Bootstrap Hypothesis," *Stud. Hist. Phil. Sci.,* 1980, *11:*267–277. Chew expanded on his bootstrap idea in several popular pieces written after the idea had fallen from favor for most particle theorists. See Geoffrey Chew, "'Bootstrap': A Scientific Idea?" *Science,* 23 Aug. 1968, *161:*762–765; Chew, "Hadron Bootstrap: Triumph or Frustration?" *Physics Today,* Oct. 1970, *23:*23–28; and Chew, "Impasse for the Elementary-Particle Concept" (cit. n. 6). See also the interview of Chew by Fritjof Capra: "Bootstrap Physics: A Conversation with Geoffrey Chew," in *Passion for Physics,* ed. DeTar *et al.,* pp. 247–286.

have pulled itself up by its own bootstraps.[19] In Zachariasen and Zemach's numerical analysis, the self-consistent solutions for the mass and coupling constant of the ρ meson were surprisingly close to recent experimental results—this at a time when theorists working within more traditional field-theoretic traditions remained completely stymied in their attempts to analyze the reams of data pouring forth from the nation's accelerators.[20]

Chew and his postdoctoral student Steven Frautschi wondered whether every particle arose in this way: their bootstrap conjecture held that every strongly interacting particle was a composite particle, composed of just those other particles that were brought together by exchanging the first particle as a force. Chew elaborated on this point in his 1964 lecture notes:

> The bootstrap concept is tightly bound up with the notion of a democracy governed by dynamics. Each nuclear particle is conjectured to be a bound state of those S-matrix channels with which it communicates, arising from forces associated with the exchange of particles that communicate with "crossed" channels. . . . Each of these latter particles in turn owes *its* existence to a set of forces to which the original particle makes a contribution.

The bootstrap thus offered Chew the ultimate nuclear democracy: elementary particles deserved no special treatment separate from composite ones; in fact, there might not even exist any "aristocratic," elementary particles, standing above the composite fray. It was in this sense that Chew concluded that "every nuclear particle should receive equal treatment under the law."[21]

Feynman diagrams could be—and, indeed, were—deployed in a host of different ways throughout the 1950s and 1960s; the diagrams themselves did not dictate how physicists would use and interpret them. Chew built directly on his new and unprecedented interpretation of the diagrams to proclaim that all particles should be treated equally. Several of the ingredients for these new types of calculations had been forged over the previous decade (many of them by Chew himself); yet no one before Chew had put these particular elements of the calculation together in the same way. No one, moreover, had gleaned quite the same lesson about "democracy" from the structure of rotated Feynman diagrams. As we will see in Section IV, theorists working at Princeton at this time picked up a number of Chew's new calculational techniques. Yet they broke with him over the calculations' ultimate theoretical implications. Given the tremendous plasticity with which the diagrams had been appropriated ever since Feynman and Dyson originally introduced them, can we

---

[19] See G. F. Chew and S. C. Frautschi, "Unified Approach to High- and Low-Energy Strong Interactions on the Basis of the Mandelstam Representation," *Phys. Rev. Lett.,* 1960, *5:*580–583; Chew and Frautschi, "Principle of Equivalence for All Strongly-Interacting Particles within the S-Matrix Framework," *ibid.,* 1961, *7:*394–397; Chew and S. Mandelstam, "Theory of Low-Energy Pion-Pion Interaction," *Phys. Rev.,* 1960, *119:*467–477; Chew and Mandelstam, "Theory of Low-Energy Pion-Pion Interaction, II," *Nuovo Cimento,* 1961, *19:*752–776; and Chew, Frautschi, and Mandelstam, "Regge Poles in $\pi\pi$ Scattering," *Phys. Rev.,* 1962, *126:*1202–1208. See also the references cited in note 18, above, as well as Frederik Zachariasen, "Lectures on Bootstraps," in *Recent Developments in Particle Physics,* ed. Michael Moravcsik (New York: Gordon & Breach, 1966), pp. 86–151, and references therein.

[20] Zachariasen, "Self-Consistent Calculation" (cit. n. 16); and Zachariasen and Zemach, "Pion Resonances" (cit. n. 16). For further details on the steps within Zachariasen and Zemach's calculation see Kaiser, "Making Theory" (cit. n. 11), pp. 442–444. Zachariasen and Zemach improved on the closeness between their result and experimental data by including the exchange of three other particles in addition to the ρ in their calculation; Zachariasen's original calculation included only the pion-ρ interaction. Cushing treats some of the other early successes of Chew's S-matrix program in *Theory Construction,* pp. 145–151.

[21] Chew, "Nuclear Democracy and Bootstrap Dynamics" (cit. n. 4), p. 106.

understand what lay behind Chew's specific reading of them—let alone the fervor with which he turned the diagrams against their original field-theoretic birthplace?

Chew noted in his 1961 lectures that, more than anything else, "general philosophical convictions" (the details of which he left unstated) helped to guide him in his democratic reading of Feynman diagrams and in his conclusion that no particles were truly elementary. In these lecture notes, as elsewhere, Chew struggled to find a vocabulary that would support the conceptual breaks he aimed to make with quantum field theory. Finding the right terminology was no mean feat: "The language just didn't exist in physics," he later recalled. He remembers being "annoyed" by other physicists' sloppy terminology—it made him "gag"—"because the language wasn't there, there were no words, there was no way to say it."[22] The language he did produce to express his new physical concepts—his recurring incantations of "nuclear democracy," "equal treatment," and "equal participation"—therefore had to come from somewhere other than the stable repertoire of his fellow physicists. Guided in part by Chew's conscious choices of language, I will attempt in the next two sections to clarify what he called his "general philosophical convictions" and to study their evolution over time. In order to unpack Chew's nuclear democracy, we must begin not with his flipped Feynman diagrams of the early 1960s but, rather, with the battles fought in Berkeley over domestic anti-Communism, beginning in 1948.

II. GEOFFREY CHEW: A SCIENTIST'S POLITICS OF DEMOCRACY IN 1950s AMERICA

Geoffrey Chew, born in 1924, came of age in a generation of American theoretical physicists just after that of Richard Feynman and Julian Schwinger. Growing up in Washington, D.C., where his father worked in the U.S. Department of Agriculture, Chew graduated from high school at the age of sixteen. Four years later he completed his college education with a straight-A record from George Washington University. It being 1944, one of his undergraduate advisors, George Gamow, helped to arrange for Chew to head straight to Los Alamos, where he joined Edward Teller's special theoretical division, working on early ideas for a hydrogen "superbomb." Entering graduate school in February 1946 at the University of Chicago, Chew completed his doctorate in less than two and a half years, under the tutelage of Enrico Fermi. Ph.D. in hand, he and his fellow graduate student Marvin "Murph" Goldberger headed off to postdocs under Robert Serber at Berkeley's Radiation Laboratory. Before long, Berkeley's physics department took notice of Chew, who was already developing a reputation for both brilliance and clarity. The physics department appointed Chew as an assistant professor, to begin in the fall of 1949.[23] Chew's transit from the East Coast to the West Coast included some of the best stops along the way for young physicists at that time, and by the age of twenty-five he had arrived, this physicist's manifest destiny complete. (See Figure 4.)

---

[22] Chew, *S-Matrix Theory of Strong Interactions* (cit. n. 3), p. 4 ("general philosophical convictions"); and Chew, interview with Gordon, Dec. 1997, quoted in Gordon, "Strong Interactions," pp. 31–32.

[23] These biographical details are taken from Raymond Birge to Dean A. R. Davis, 27 Feb. 1949, in Raymond Thayer Birge Correspondence and Papers, call number 73/79c, Bancroft Library, University of California, Berkeley (hereafter cited as **Birge Papers**); and A. C. Helmholz to Dean Lincoln Constance, 25 Mar. 1957, Birge Papers, Box 40, Folder "Letters written by Birge, January–May 1957." Some clarification may be helpful here. Letters written by Birge are filed chronologically. The items cited in this essay are from Boxes 39 and 40; explicit folder titles will not be cited. Letters written to Birge (and other pertinent materials in this collection) will be cited with box number and folder titles. For further biographical information see also Birge, "History," Vol. 5, Ch. 19, pp. 43–51; and DeTar *et al.,* eds., *Passion for Physics.*

Figure 4. *Geoffrey Chew, circa 1960. Lawrence Berkeley National Laboratory, University of California, Berkeley. Courtesy of the Emilio Segrè Visual Archives, American Institute of Physics.*

This simple picture of professional progress grew complicated, however, nearly as soon as Chew was hired. Chew became increasingly engaged with political issues in the late 1940s, continuing with greater and greater intensity throughout the 1950s. His activities took him on an extended orbit that began in Berkeley and, seven years later, brought him back there again. His reasons for leaving Berkeley in 1950 can be understood only if we consider the political situation in which physicists found themselves soon after World War II and how Berkeley physicists, in particular, experienced the early years of the Cold War. The fast-moving descent into McCarthyism affected daily life in Berkeley's Radiation Laboratory and Department of Physics, shaping hallway discussions, straining old friendships, and altering many young physicists' career paths.

Politics and Physics at Berkeley, 1949–1954

Few American physics departments experienced the pains of transition to the postwar political scene more abruptly, or more publicly, than that at the Berkeley campus of the University of California. The House Un-American Activities Committee (HUAC) turned its sights on atomic espionage directly on the heels of its sensational probe of alleged Communists in the film industry.[24] Its first stop: Berkeley's Radiation Laboratory, built up to international prominence during the 1930s by Ernest O. Lawrence. During the war, Lawrence's famed laboratory on the hill had been staffed with teams endeavoring to separate the scarce, fissionable uranium-235 isotope from its more ubiquitous cousin, uranium-238. Some of this staff, HUAC began to insinuate nearly as soon as Chew arrived at the Rad Lab as a postdoc in 1948, had been "red."

Startling headlines greeted students and faculty returning to Berkeley in September 1948: a physicist who had worked at the Radiation Laboratory during the war, identified by HUAC only as "Scientist X," purportedly had leaked vital atomic secrets to the Soviets. Five physicists who had worked in the wartime Radiation Laboratory were singled out for intensive questioning. Though no evidence of espionage at the Rad Lab was ever uncovered, all were convicted of contempt of Congress and one of perjury, based on his testimony. Each lost his job immediately upon being indicted by Congress.[25]

By the time HUAC began its investigation, only one of these Rad Lab physicists remained in Berkeley. In September 1949 his case grabbed local attention when Berkeley's student newspaper, the *Daily Californian,* reported the front-page story, "T.A. Queried on Communist Ties." I. David Fox, the teaching assistant in question, had worked at the Radiation Laboratory during the war and was currently a graduate student in Berkeley's physics department. Fox refused to name names for his HUAC investigators, invoking the

[24] The literature on HUAC and McCarthyism is, of course, vast. On McCarthyism and American higher education in particular see esp. Ellen Schrecker, *No Ivory Tower: McCarthyism and the Universities* (New York: Oxford Univ. Press, 1986); Sigmund Diamond, *Compromised Campus: The Collaboration of Universities with the Intelligence Community, 1945–1955* (New York: Oxford Univ. Press, 1992); James Hershberg, *James B. Conant: Harvard to Hiroshima and the Making of the Nuclear Age* (Stanford, Calif.: Stanford Univ. Press, 1993), Chs. 19, 21–23, 31; Noam Chomsky *et al.,* eds., *The Cold War and the University: Toward an Intellectual History of the Postwar Years* (New York: New Press, 1997); Catharine M. Hornby, "Harvard Astronomy in the Age of McCarthyism" (A.B. thesis, Harvard Univ., 1997), esp. Ch. 2; Jessica Wang, *American Science in an Age of Anxiety: Scientists, Anticommunism, and the Cold War* (Chapel Hill: Univ. North Carolina Press, 1999); Lawrence Badash, "Science and McCarthyism," *Minerva,* 2000, *38*:53–80; Naomi Oreskes and Ronald Rainger, "Science and Security before the Atomic Bomb: The Loyalty Case of Harald U. Sverdrup," *Studies in History and Philosophy of Modern Physics,* 2000, *31*:309–369; and Schweber, *Shadow of the Bomb* (cit. n. 9). On the politicization of American scientists before World War II see Peter J. Kuznick, *Beyond the Laboratory: Scientists as Political Activists in 1930s America* (Chicago: Univ. Chicago Press, 1987).

[25] On the establishment of the Radiation Laboratory see John Heilbron and Robert Seidel, *Lawrence and His Laboratory: A History of the Lawrence Berkeley Laboratory,* Vol. 1 (Berkeley: Univ. California Press, 1990). On the HUAC investigation of the Rad Lab physicists see the San Francisco–area newspaper clippings in Department of Physics, University of California, Berkeley, Records, ca. 1920–1962, call number CU-68, Bancroft Library (hereafter cited as **Dept. Physics, Berkeley, Records, ca. 1920–1962**), Folder 4:12; "Thomas Committee Calls Ex-Instructor," *Daily Californian,* 22 Sept. 1948, p. 5; Louis Bell, "'No Great Surprise'; Identity of Scientist X Suspected Here," *ibid.,* 3 Oct. 1949, p. 1; "Ex-Physicist at U.C. Held for Perjury," *ibid.,* 26 May 1952, p. 1; and Schrecker, *No Ivory Tower,* pp. 126–148. The specifics of the legal case against the Rad Lab physicists are outlined in Carl Beck, *Contempt of Congress* (New Orleans, La.: Hauser, 1959), pp. 65–70. Two other cases involving Berkeley scientists also caught the attention of HUAC and the national media in 1948: those of Martin Kamen and E. U. Condon. See Martin Kamen, *Radiant Science, Dark Politics: A Memoir of the Nuclear Age* (Berkeley: Univ. California Press, 1985), Chs. 11, 12; and Jessica Wang, "Science, Security, and the Cold War: The Case of E. U. Condon," *Isis,* 1992, *83*:238–269. Condon was at the time the head of the National Bureau of Standards, having formerly been an undergraduate and graduate student at Berkeley and a consultant at the Berkeley Rad Lab during World War II. See also Schweber, *Shadow of the Bomb.*

Fifth Amendment twenty-five times. Three months later, the Board of Regents of the University of California dismissed Fox, without making any formal charges against him.[26] Even before Senator Joseph McCarthy had begun the anti-Communist activities with which his name forever will be associated, physicists in Berkeley were feeling the brunt of Mc-Carthyism.

The politics of domestic anti-Communism invaded physics departments far beyond Berkeley in the ensuing months and years. Physicists across the country debated the new proposal, in 1949, to require full background security checks for all recipients of Atomic Energy Commission graduate student fellowships. Thirty-four Berkeley physics graduate students wrote to the *San Francisco Chronicle* to protest what they saw as the proposed exclusion from training of "those among us who hold unpopular viewpoints," arguing that "education in a democracy must be available to everyone." The next year, after the final, contentious establishment of the National Science Foundation (NSF), prominent scientists such as Berkeley's iconic physics department chair Raymond Birge objected to congressional discussion that considered prohibiting NSF grants to members, past or present, of any organization listed on the attorney general's "subversive" list.[27]

With the outbreak of fighting in Korea that June, department chairs such as Birge found more and more of their time devoted to draft deferments for their students and young faculty and to problems involving personnel security clearances.[28] Berkeley physicists, like their colleagues across the country, found themselves routinely denied passports for foreign travel; meanwhile, foreign physicists experienced insulting delays or even rejections of their applications for visitors' visas. The Rad Lab, since 1947 under the auspices of the Atomic Energy Commission, was no longer permitted to have foreign scientists conduct any work there—classified or unclassified, paid or not.[29]

---

[26] "T.A. Queried on Communist Ties," *Daily Californian,* 28 Sept. 1949 [misprinted as 1948], p. 1 (see also the *Daily Californian* stories on 4 Jan. 1950, p. 1; 27 Mar. 1950, p. 7; and 31 Mar. 1950, pp. 1, 4); Schrecker, *No Ivory Tower,* pp. 126–127; and David Gardner, *The California Oath Controversy* (Berkeley: Univ. California Press, 1967), pp. 91–94.

[27] For the graduate students' protest see Letter to the Editor, *San Francisco Chronicle,* 27 May 1949; I. David Fox was among the signers. See also Birge to Lyman Spitzer, Jr., 26 May 1949, Birge Papers; E.R. [Eugene Rabinowitch], "The 'Cleansing' of AEC Fellowships," *Bulletin of the Atomic Scientists,* June–July 1949, *5:*161–162; "The Fellowship Program: Testimony before the Joint Committee," *ibid.,* pp. 166–178; "Loyalty Tests Cause Cut in AEC Fellowship Program," *ibid.,* Jan. 1950, *6:*32; "The Curtailment of the AEC Fellowship Program," *ibid.,* pp. 34, 62–63; "Loyalty Tests for Science Students?" *ibid.,* Apr. 1950, *6:*98; and Wang, *American Science in an Age of Anxiety* (cit. n. 24), Ch. 7. On the debates over the founding of the NSF see Daniel Kevles, "The National Science Foundation and the Debate over Postwar Research Policy, 1942–1945," *Isis,* 1977, *68:*5–26; Kevles, *The Physicists: The History of a Scientific Community in Modern America,* 2nd ed. (1978; Cambridge, Mass.: Harvard Univ. Press, 1987), Chs. 11–12; Nathan Reingold, "Vannevar Bush's New Deal for Research; or, The Triumph of the Old Order," *Hist. Stud. Phys. Biol. Sci.,* 1987, *17:*299–344; and Jessica Wang, "Liberals, the Progressive Left, and the Political Economy of Postwar American Science: The National Science Foundation Debate Revisited," *Hist. Stud. Phys. Sci.,* 1995, *26:*139–166. On the issue of the attorney general's list of "subversive" organizations and NSF grants see Birge to Robert G. Sproul, 14 Mar. 1950, Birge Papers.

[28] On draft deferments see Birge to R. C. Gibbs, 10 Aug. 1950, and Birge to Local Board No. 62, Santa Clara County, 8 June 1953, Birge Papers; and David Kaiser, "Putting the 'Big' in 'Big Science': Cold War Requisitions, Scientific Manpower, and the Production of American Physicists after World War II," unpublished MS. The concerns with draft deferments for physics students persisted well after fighting had ceased in Korea; see the correspondence from 1958 in the American Institute of Physics, Education and Manpower Division, Records, 1951–1973, Box 4, Folder "Scientific Manpower Commission, Washington, D.C." These records are held in the American Institute of Physics, Niels Bohr Library, College Park, Maryland, call number AR15. On security clearance troubles see Birge to K. K. Darrow, 11 Jan. 1955, Birge Papers; Wang, *American Science in an Age of Anxiety;* Ellen Schrecker, *The Age of McCarthyism: A Brief History with Documents* (Boston: Bedford, 1994), pp. 37–40, 150–164; and Adam Yarmolinsky, ed., *Case Studies in Personnel Security* (Washington, D.C.: Bureau of National Affairs, 1955).

[29] Regarding passport and visa problems see Birge to Darrow, 26 May 1955; Birge to Congressman Francis

Not all of the reactions to these fast-moving events were glum. Five years after the HUAC investigation of purported espionage, Berkeley student reporters concluded their five-part series on the Radiation Laboratory in December 1953 with the light-hearted story "Espionage at the Rad Lab—Naw!" After describing dozens of "scattered instruments painted a bright red with the white letters 'USSR' printed on them," the reporters explained that the letters stood for "United States surplus reserve," an old joke up at the laboratory. The gag, and visitors' stunned reactions to it, could still evoke "guffaws" from "the six foot, amiable, loose-jointed physicist" who gave these student reporters their tour.[30]

Still, the levity could only go so far. Birge remained more somber, hesitating before accepting his nomination as vice-president of the American Physical Society (APS). Writing in May 1953 to the society's secretary, Karl Darrow, he recalled a former "time when the American Physical Society was concerned only with physics. At the present time, however, I am afraid it is concerned almost as much with politics as it is with physics and I must say I do not like politics." Almost exactly one year later, just as Birge was preparing to assume the presidency of the APS, he and his Berkeley department were stunned over the decision by the Atomic Energy Commission to deprive J. Robert Oppenheimer, Berkeley's former star theorist and the world-famous "father of the atomic bomb," of his security clearance.[31]

These many developments shaped physicists' experiences across the country in the late 1940s and 1950s; reactions ranged from graduate student protests to the hand-wringing of the president of the American Physical Society. At Berkeley, however, all of these events took shape in the shadow of a local situation that dominated hallway discussions and faculty meetings for the better part of a decade: the "loyalty oath" controversy at the University of California. As we will see, the loyalty oath helped to prompt Geoffrey Chew's own political engagement as he struggled to define a working definition of "democracy" for scientists in Cold War America.

Walter, 15 July 1955; and Birge to Senator Harley Kilgore, 17 Nov. 1955: Birge Papers. The Federation of American Scientists focused on passport and visa problems; many of their efforts were reported in the *Bulletin of the Atomic Scientists,* a special issue of which (Oct. 1952, *8*) was dedicated especially to these problems. On the exclusion of foreign scientists from the Rad Lab see Birge to Walter Thirring, 8 Jan. 1952, Dept. Physics, Berkeley, Records, ca. 1920–1962, Folder 5:117. On the establishment of the AEC national laboratory system see Richard Hewlett and Francis Duncan, *A History of the United States Atomic Energy Commission,* Vol. 2: *Atomic Shield, 1947–1952* (University Park: Pennsylvania State Univ. Press, 1969), Ch. 8; Robert Seidel, "A Home for Big Science: The Atomic Energy Commission's Laboratory System," *Hist. Stud. Phys. Sci.,* 1986, *16:*135–175; and Peter Westwick, "The National Laboratory System in the U.S., 1947–1962" (Ph.D. diss., Univ. California, Berkeley, 1999).

[30] Sandra Littlewood and Skip Garretson, "Espionage at the Rad Lab—Naw!" *Daily Californian,* 11 Dec. 1953, p. 8. It is interesting to note that this was the only segment of the five-part series that Birge clipped and saved with his other newspaper clippings, which may be found in Dept. Physics, Berkeley, Records, ca. 1920–1962, Folder 4:12.

[31] Birge to Darrow, 22 May 1953, Birge Papers; Atomic Energy Commission, *In the Matter of J. Robert Oppenheimer: Transcript of Hearing before Personnel Security Board* (Washington, D.C.: Atomic Energy Commission, 1954); Philip Stern, *The Oppenheimer Case: Security on Trial* (New York: Harper & Row, 1969); John Major, *The Oppenheimer Hearing* (New York: Stein & Day, 1971); and Barton J. Bernstein, "'In the Matter of J. Robert Oppenheimer,'" *Hist. Stud. Phys. Sci.,* 1982, *12:*195–252. Of course, not all members of the Berkeley department were shocked by the news; some had even lobbied behind the scenes to ensure the outcome. On Berkeley involvement in and reactions to the hearing see "Oppenheimer Conflict: Former Professor Center of Dispute," *Daily Californian,* 15 Apr. 1954, p. 1; Birge to Edwin A. Uehling, 28 Mar. 1955, Birge Papers; Birge, "History," Vol. 5, Ch. 17; Nuel Pharr Davis, *Lawrence and Oppenheimer* (New York: Simon & Schuster, 1968), Chs. 8–10; Luis W. Alvarez, *Alvarez: Adventures of a Physicist* (New York: Basic, 1987), pp. 179–181; and A. Carl Helmholz with Graham Hale and Ann Lage, *Faculty Governance and Physics at the University of California, Berkeley, 1937–1990* (Berkeley: Regional Oral History Office, Bancroft Library, 1993), pp. 152–157, 276–279.

The California Loyalty Oath

Only a few months before David Fox was dismissed from his Berkeley teaching-assistant position, the Board of Regents of the University of California imposed a new "loyalty oath" on all university employees. This anti-Communist oath was drafted hastily during the lunch break of an otherwise routine monthly meeting of the regents on 25 March 1949, with only eleven of the twenty-four regents in attendance; these eleven adopted it unanimously. As historians such as David Gardner and Ellen Schrecker have emphasized, the new oath sprang more from questions regarding self-governance than from overriding fears on most of the regents' part about actual Communist infiltration of the university system. The California State Legislature was at that moment considering a proposal to wrest control over adjudicating the "loyalty" of the university faculty from the regents; enacting the new oath offered one way for the regents to show the legislature that it could manage such issues on its own. Soon the oath set off a long and bitter struggle between the regents and the faculty over this same question of self-governance: most members of the faculty claimed that the regents' act violated the faculty's traditional role of choosing its own members and, when necessary, keeping them in line.[32]

Adding to the furor, the regents failed to inform the faculty of the new requirement until mid-June 1949—over two months after the oath officially had been adopted and at just the time when many faculty were leaving Berkeley for the summer. The oath required all university employees—from janitorial staff to graduate student teaching assistants to tenured faculty—to swear that they were not members of the Communist Party; each signing had to be witnessed by a notary public. Letters to faculty members soon revealed that their "reappointment," and the payment of their salaries, was now conditional upon their signing the oath—even for those who believed that tenure had already removed all questions of reappointment.[33]

A vocal minority of the faculty, including several professors who had fled European dictatorships before and during World War II, immediately decried the oath as an infringement on academic freedom. Yet during the fall of 1949, in the midst of reports of the Soviets' detonation of their own atomic bomb and the "fall" of China to Communist leaders, the objections of most faculty members did not concern the ban on Communists per se. Instead, they opposed the fact that university employees had been singled out and held to an oath more stringent than that required of other state employees. Hundreds of faculty members at first refused to sign the oath in protest, a recalcitrance that, in turn, only strengthened the view of certain key regents that university faculties could not be trusted to govern themselves. Crowds of fifty to two hundred professors met each week at Berkeley's faculty club throughout the academic year 1949/1950 to discuss tactics and strategies.[34]

---

[32] Gardner, *California Oath Controversy* (cit. n. 26); and Schrecker, *No Ivory Tower* (cit. n. 24), pp. 116–125. See also George Stewart, *The Year of the Oath* (New York: Doubleday, 1950); Birge, "History," Vol. 5, Ch. 19; and Helmholz with Hale and Lage, *Faculty Governance*, pp. 96–97, 152–157. Gardner cautions against reading the regents' enactment of the oath too narrowly as a direct reaction to the state legislature's proposals, which, he concludes, provided only a clearly articulated measure of the "mood" of the times: Gardner, *California Oath Controversy*, p. 10.

[33] See the forms and notices in Dept. Physics, Berkeley, Records, ca. 1920–1962, Folder 3:41; and Gardner, *California Oath Controversy*, pp. 52–54.

[34] One of the early faculty leaders who spoke out against the oath because of what he perceived to be ties to fascist practices was Ernst Kantorowicz, a German-born medieval historian. See Grover Sales, Jr., "The Scholar and the Loyalty Oath," *San Francisco Chronicle*, 8 Dec. 1963, pp. 27–30, written soon after Kantorowicz's death. Robert Serber recalled that Gian Carlo Wick, a physicist at Berkeley who refused to sign the oath, said

Meanwhile, by February 1950 the regents' line had hardened: any university employee who had not signed the oath by the end of April would be dismissed. Alumni and the president of the university managed to pry a compromise from the regents, who agreed that the cases of all nonsigners would be reviewed on an individual basis by the faculty's Committee on Privilege and Tenure. Yet when North Korea invaded South Korea in late June 1950 and the United States entered the conflict, most of the remaining faculty holdouts simply signed the oath. Dismissing the faculty committee's recommendations, the regents then fired the remaining thirty-one "non-signers" on 25 August 1950. Even though none of those fired had ever been accused by the faculty or the regents of Communist Party membership or sympathy, their fates drew national attention to the question (and, as many proclaimed, the danger) of Communists in the classroom.[35]

The ensuing seven-year battle between the fired nonsigners and the regents, weaving in and out of the California State Supreme Court, left no department on the California campuses untouched. Yet few felt the full brunt of the controversy more, or in more ways, than Berkeley's Department of Physics. It is difficult to overestimate the effects of the oath controversy on daily life within the department. The department secretary had to rush off lists of graduate students who had signed the oath after an initial "oversight" to prevent their termination as course graders or teaching assistants. The examining committees for several students' dissertation defenses had to be rearranged at the last minute, since the regents ruled that faculty nonsigners could no longer serve on such committees. Birge circulated a memorandum to the department's faculty in early April 1950, cautioning them against putting their personal opinions regarding the oath in writing or even discussing them "at a meeting of a fairly large group." Writing to university president Robert Sproul in the midst of the controversy, Birge found himself wondering whether the "wave of hysteria now sweeping the country," as evidenced locally by the loyalty oath disaster, might even put "the entire democratic structure of this country . . . in some danger."[36] Whether interpreted ultimately as an anti-Communist witch-hunt, a principled fight over academic freedom, or a local power play pitting faculty against administration, the effects of the loyalty oath controversy on mundane daily life were palpable to students and faculty alike.

Several members of the physics department did more than just absorb the oath's after-

---

at the time that "he had been coerced into taking an oath once before in Italy, where he had to swear loyalty to Mussolini; he said he'd regretted it ever since and wasn't going to make the same mistake twice." Robert Serber with Robert Crease, *Peace and War: Reminiscences of a Life on the Frontiers of Science* (New York: Columbia Univ. Press, 1998), p. 171. Another Berkeley physicist, Emilio Segrè, on the other hand, later wrote that he had signed at least fifteen loyalty oaths while in Mussolini's Italy; thus he found them all meaningless, and therefore, as a practical matter, he saw little reason not to sign the California oath as well. In fact, as he put it, "I even remembered a pronouncement by Pope Pius XI, elicited by a Fascist oath, explicitly stating that under certain circumstances one could take such oaths with mental reservations that made them void. I dug the papal document out in the library and translated it, and some colleagues to whom I had sent it posted it in Los Alamos, which administratively depended on the Regents of the University of California. At Berkeley it circulated less openly." Emilio Segrè, *A Mind Always in Motion: The Autobiography of Emilio Segrè* (Berkeley: Univ. California Press, 1993), pp. 235–236. On the weekly meetings in the faculty club see Gardner, *California Oath Controversy*, p. 87.

[35] Gardner, *California Oath Controversy*, Chs. 5–6; and Schrecker, *No Ivory Tower* (cit. n. 24), pp. 120–122.

[36] O. Lundberg, University Controller, memo to "chairmen of departments, administrative officers, and others concerned," 27 Nov. 1950; RLY [Rebekah Young], Physics Department Secretary, to Lundberg, 30 Nov. 1950; M. A. Stewart, Associate Dean of the Graduate Division, memo to Physics Department Graduate Advisers, 14 Dec. 1950: Dept. Physics, Berkeley, Records, ca. 1920–1962, Folder 3:41. Rebekah Young to Robert Serber, 18 July 1951; and Serber to Young, 25 July, 1951: Dept. Physics, Berkeley, Records, ca. 1920–1962, Folder 3:4. Birge, "Memorandum to Members of the Physics Department Staff," 6 Apr. 1950; and Birge to Sproul, 14 Mar. 1950: Birge Papers.

effects. From the start, the department included active members on all sides of the controversy. The teaching assistant David Fox—who, ironically, had signed the oath—was nonetheless touted by certain regents as an all-too-visible reminder of the need to enforce loyalty among the faculty. Professor Francis Jenkins joined the "Operating Committee of Seven" to help coordinate faculty opposition to the oath and later served on the faculty's Committee on Privilege and Tenure. Robert Brode, also a senior professor in the department, served as the official custodian of the funds raised by the Operating Committee of Seven, which were intended in part to help pay the expenses of the nonsigners who had stopped receiving their salaries. And department chair Raymond Birge often played the part of back-room negotiator, pushing certain people (such as Wendell Stanley, later a Nobel laureate in biology) into visible positions on faculty committees, while presenting a series of prepared remarks against the regents' actions before Academic Senate and departmental meetings.[37]

On the other side, the physics department's Victor Lenzen, chair of the Committee on Privilege and Tenure for the northern section of the Academic Senate, helped to engineer the composition of this crucial committee by removing all nonsigners before agreeing to retire from the committee himself. In less explicit ways, Ernest Lawrence, Edwin McMillan, Luis Alvarez, and others, while remaining "aloof" from the campus-wide discussions of the controversy, created an "atmosphere" at the Radiation Laboratory that convinced several physicists that neither nonsigners nor their supporters would be welcome there. For young physicists at the Rad Lab, there was little room left for doubt: postdocs who had not signed the oath found notes on their desks on 30 June 1950, informing them that they had to turn in their badges and keys, clear off their desks, and leave by the end of that day.[38]

These many different types of participation in the controversy complicated daily life in the physics department. Even more concretely, Berkeley's physics department fell victim to the oath in losing six faculty members within one year. Two professors, Harold Lewis and Gian Carlo Wick, allowed themselves to be fired in August 1950, when the regents finally dismissed all remaining nonsigners. By June 1951 four other professors—Robert Serber, Wolfgang "Pief" Panofsky, Howard Wilcox, and Geoffrey Chew—had resigned in protest. Among those who left were all of the department's theoretical physicists. The very first to resign from the physics department over the issue—and perhaps the first in the entire university—was Geoffrey Chew.[39]

---

[37] Birge, "History," Vol. 5, Ch. 19, pp. 1–15; and Gardner, *California Oath Controversy* (cit. n. 26), pp. 123–124, 171–172. Some of Birge's speeches at Academic Senate and physics department faculty meetings are reprinted in Birge, "History," Vol. 5, Ch. 19, pp. 8–12.

[38] Gardner, *California Oath Controversy,* p. 248; reference to the "atmosphere" at the Rad Lab comes from Geoffrey Chew's letter of resignation, quoted below. See also Helmholz with Hale and Lage, *Faculty Governance* (cit. n. 31), pp. 96–97, 153; Serber with Crease, *Peace and War* (cit. n. 34), pp. 171–172; and Segrè, *Mind Always in Motion* (cit. n. 34), pp. 234–237. Jack Steinberger was one of the postdocs dismissed on 30 June 1950 for not signing the oath. See Jack Steinberger, "A Particular View of Particle Physics in the Fifties," in *Pions to Quarks,* ed. Brown *et al.* (cit. n. 13), pp. 307–330, esp. p. 311; and Steinberger, "Early Particles," *Annual Review of Nuclear and Particle Science,* 1997, *47*:xiii–xlii, esp. pp. xxxix–xl.

[39] Birge suggests that Chew was "apparently" the first professor to resign from all of the University of California over the oath controversy in Birge, "History," Vol. 5, Ch. 19, p. 45. Robert Serber recounts his own decision to leave in Serber with Crease, *Peace and War,* pp. 171–172. Serber had endured an extended, and at times hostile, personal security review in 1948, though perhaps because of his close affiliation with Ernest Lawrence his case did not stir the same media attention as the HUAC Rad Lab investigation did (*ibid.,* pp. 162–165). For more on Serber's continuing security woes see Barton Bernstein, "Interpreting the Elusive Robert Serber: What Serber Says and What Serber Does Not Explicitly Say," *Stud. Hist. Phil. Mod. Phys.,* 2001, *32*:443–486. Wolfgang Pauli kept abreast of the developments in Berkeley's physics department via his friend Erwin

Geoffrey Chew and the Politics of Democracy

Though no one on the faculty could have known it at the time, Chew's official letter of appointment to his assistant professorship arrived exactly one week after the regents secretly passed the new loyalty oath requirement.[40] Chew, who still maintained Q clearance to work on classified nuclear weapons projects, refused to sign what he called, in a letter to Oppenheimer, "the objectionable part of the new contract," which seemed to him to threaten "the right of privacy in political belief." He became further frustrated with what he saw as weak attempts by the rest of the faculty to fight the oath. At the end of his very first year of teaching Chew acted on his convictions, becoming the first person to resign from the physics department over the issue. As he explained to Birge in July 1950, one month before the regents finally dismissed the remaining nonsigners, Chew had decided "to get away from an intimidating and precarious situation."[41]

The firing of David Fox, Chew told Birge, had shown beyond doubt that the regents seemed bent on removing from the faculty the "right" to "maintain its own qualifications." The regents' actions with the oath, furthermore, aimed at nothing less than to "root out the last resistance" among the faculty. The few signs of "faculty solidarity" with the non-signers had all but vanished when fighting broke out in Korea in June 1950. As Chew pressed Birge one month later, "In a war-time situation, what security can a non-conformist have?" With the outbreak of fighting, the few remaining nonsigners on campus "have now become lepers who must keep out of sight." On top of this, Chew reported that the Radiation Laboratory, which was "the chief stimulus" of his scientific work, had made it clear that it "does not welcome non-signers. Even if I were allowed to maintain my affiliation [with the laboratory], the unsympathetic atmosphere would not be pleasant. This would be a more subtle form of intimidation." Though Chew found it a difficult decision, he left Berkeley in July 1950 and accepted a position at the University of Illinois in Urbana. He was promoted from assistant to associate professor within a year and became a full professor at Urbana in 1955, at the age of thirty-one.[42]

Panofsky, a senior art historian at the Institute for Advanced Study, whose son Wolfgang was one of the experimentalists to leave Berkeley's department because of the loyalty oath. In October 1950 Pauli forwarded to the elder Panofsky news that he had heard from the young theorist J. M. Luttinger. Pauli quoted from Luttinger in his letter to Panofsky: "Apart from Physics, the atmosphere is very unpleasant in Berkeley. Both [Gian Carlo] Wick and [Harold] Lewis have been fired for refusing to sign a Loyalty Oath, and both (so far as I know) are fighting the case in court. They have only a very slim chance of winning—on the whole it is a degrading business. *In addition to that the lab is full of secret work, and is overrun by petty officials and bureaucrats of all kinds.*" Wolfgang Pauli to Erwin Panofsky, 23 Oct. 1950, in Wolfgang Pauli, *Wissenschaftlicher Briefwechsel,* ed. Karl von Meyenn (New York: Springer, 1996), Vol. 4, Pt. 1, p. 179.

[40] Birge notes that Dean A. R. Davis sent the official letter of appointment to Chew on 1 Apr. 1949 (Birge, "History," Vol. 5, Ch. 19, p. 45); the regents enacted the new loyalty oath in their meeting on 25 Mar. 1949. While still a postdoc at the Rad Lab, Chew had delivered several talks on his research to the physics department, both formal and informal, so that Birge could introduce Chew as already "well known" at the Sept. 1949 departmental meeting. See Birge's handwritten notes, "First Dept. meeting, Wed., Sept. 28, 1949," Dept. Physics, Berkeley, Records, ca. 1920–1962, Folder 2:4.

[41] Geoffrey Chew to J. Robert Oppenheimer, 11 May 1950, quoted in Birge, "History," Vol. 5, Ch. 19, p. 45 (Birge also notes discussions with Chew over his frustration with Academic Senate resolutions regarding the oath); and Chew to Birge, 24 July 1950, Birge Papers, Box 5, Folder "Chew, Geoffrey Foucar, 1924–."

[42] Chew to Birge, 24 July 1950; and Birge, "History," Vol. 5, Ch. 19, p. 47. Chew's emphasis on the importance of the Korean war is echoed in several historians' recent studies of postwar American science policy. See Daniel Kevles, "Cold War and Hot Physics: Science, Security, and the American State, 1945–56," *His. Stud. Phys. Biol. Sci.,* 1990, *20:*239–264; Hershberg, *James B. Conant* (cit. n. 24), Chs. 27–28; Wang, *American Science in an Age of Anxiety* (cit. n. 24), Ch. 8; and Kaiser, "Putting the 'Big' in 'Big Science'" (cit. n. 28). The University of Illinois also had a standing loyalty oath requirement at the time Chew accepted his job there, but one that did not mention the Communist Party or any other group by name. Birge and University of California president

A few months after he left Berkeley, and after the regents dismissed the thirty-one remaining nonsigners, Chew reported on the struggle in the *Bulletin of the Atomic Scientists*. Like most of the faculty at the University of California, Chew objected that the faculty had been singled out and subjected to a more specific loyalty oath than was required for any other state employees. The fact that the regents then went beyond this, to threaten and eventually dismiss tenured faculty, constituted a further violation of "the cornerstone of academic freedom." He refrained from detailing any of his own experiences, reviewing instead the positions taken by the university president, the Academic Senate, and various factions within the Board of Regents. The controversy had been fanned in part, Chew explained, by what he called "fundamentalists," people who struck principled stands on questions like academic freedom even though they had lived uncomplainingly since 1942 with an official university policy excluding Communists from teaching there. The "moral of this very sad story," Chew concluded, was that more explicit procedures needed to be defined for tenure. The rights and roles of the faculty, the Academic Senate, and the Board of Regents needed similar attention and explication, to guarantee that faculty would not be singled out for special treatment again in the future. The procedures of due process might then guard against a repeat of "the present sad and humiliating situation."[43]

Over the course of the 1950s, while teaching in Illinois, Chew became more and more active in what has been called "the atomic scientists' movement." Soon after arriving on campus, Chew founded Urbana's local branch of the Federation of American Scientists (FAS), a national organization dedicated to moderate liberal causes. As Jessica Wang has detailed, by the late 1940s the FAS had become largely a bureaucratic organization, collecting information about some of the more severe abuses of McCarthyism and lobbying certain legislators for reform. In part because of pressure from HUAC and the FBI, the FAS had begun to refrain from its earlier pattern of public demonstrations by the time Chew joined the group, choosing instead the route of "quiet diplomacy."[44]

Chew participated directly in this FAS diplomacy, both on the Illinois campus and, soon, as a visible leader within the national organization. After founding the local FAS branch, Chew immediately began inducting friends and colleagues, such as his fellow physicist Francis Low. As Low recalls, Chew strode up to him soon after he arrived in Urbana and asked simply, "Okay, are you ready to join FAS?" "I was happy to do it," Low continues. "I thought it was a good organization. Geoff's position was very good, and I was happy to take part in it. It was a serious time." Under Chew's direction, the group organized monthly meetings on campus and hosted speakers on topics like the Fifth Amendment. On Chew's initiative, they also became a clearinghouse for campus-wide complaints about unfair treatment, such as problems in obtaining passports. Soon Chew's activities extended

---

Sproul found it ironic that Chew would agree to go to Illinois, but Chew explained that in Illinois this was "the same oath required of all state employees ... [and] no one feels it to be a restriction on his political activity. . . . The intent of the trustees, therefore, does not seem inimicable [*sic*] to academic freedom." In other words, as far as Chew was concerned, the Illinois oath did not single out faculty for special treatment or unfair scrutiny. Chew to Birge, 24 July 1950.

[43] Geoffrey Chew, "Academic Freedom on Trial at the University of California," *Bull. Atom. Sci.*, Nov. 1950, 6:333–336, on p. 336. The objection that faculty were singled out for closer scrutiny than other people was a common one among Berkeley faculty. Chew noted this in passing on p. 334 of his article; he gave it a more extended discussion in his letter to Birge of 24 July 1950. See also Birge, "History," Vol. 5, Ch. 19; Gardner, *California Oath Controversy* (cit. n. 26), Ch. 3; and Schrecker, *No Ivory Tower* (cit. n. 24), pp. 122–123.

[44] See esp. Wang, *American Science in an Age of Anxiety* (cit. n. 24). On the founding and early years of the FAS see also Alice Kimball Smith, *A Peril and a Hope: The Scientists' Movement in America, 1945–7* (Chicago: Univ. Chicago Press, 1965).

beyond the Urbana campus. In November 1955 he testified before a U.S. Senate subcommittee as chair of the FAS Passport Committee. The FAS objected to the State Department's unwritten policy of denying passports to scientists for political reasons and of further denying the applicants any rights of due process or means of appeal. Usually passports were denied with no reasons given, and appeals met delays lasting months and even years. One of the most famous cases at the time concerned Linus Pauling, who finally received a passport in 1954—after more than two years of attempts—when he applied to go to Sweden to receive his Nobel Prize in chemistry.[45]

In his testimony before the Senate Subcommittee on Constitutional Rights, Chew used several lesser-known cases to lobby for fairer treatment. A passport, he urged, must be "recognized as a right of the U.S. citizen, not merely a privilege." Due process must attend all dealings with passport applications, and only problems with demonstrated relevance to national security issues should result in denials. Applicants denied passports should be supplied with an explicit list of charges against them and given the opportunity for a prompt appeals hearing, at which "confrontation of witnesses and no concealment of evidence, should apply." All appeals hearings should be transcribed and copies made available to all parties. Most important of all, Chew argued, a channel outside of the State Department should be set up to handle further appeals: "We should like to see a well-defined channel" established, "so that applicants will have no uncertainty as to what to do." Both the loyalty oath and the passport situation convinced Chew that only "well-defined channels," operating under due process, could protect the equality and rights of academics. Just as he had explained in his report about the California loyalty oath, Chew labored to make it clear for the committee of senators during his 1955 testimony that academics, and scientists in particular, should neither be singled out for "special privileges" nor subject to special scrutiny or bias.[46] Clear and unambiguous procedures needed to be established so that disagreements would be settled fairly, providing equal treatment to all those affected. With these safeguards in place, Chew believed, scientists could participate in a democratic America as citizens, each equal under the law.

III. PEDAGOGICAL REFORMS: "SECRET SEMINARS" AND "WILD MERRYMAKING"

Chew's pedagogical efforts in the years following his congressional testimony resonated with his more explicitly political activities. Demonstrating the same attitudes as in his lobbying with the FAS, Chew endeavored to make certain that graduate students could work in such a way that none was singled out unfairly and all were encouraged to participate equally. His activities with his own graduate students shaped his approach to

---

[45] Francis Low, interview with the author, MIT, 11 Apr. 2001; David Kaiser, "Francis E. Low: Coming of Age as a Physicist in Postwar America," *Physics @ MIT,* 2001, *14*:24–31, 70–77, on pp. 71–72; and "Summary of Testimony of Linus Pauling," *Bull. Atom. Sci.,* Jan. 1956, *12*:28.

[46] Geoffrey Chew, "Passport Problems," *Bull. Atom. Sci.,* Jan. 1956, *12*:26–28, on p. 28. This article includes Chew's testimony from 15 Nov. 1955. See also "FAS Congressional Activity in 1955," *ibid.,* p. 45. The specific items Chew lobbied for were conspicuously absent in all kinds of hearings from this period, having been denied to witnesses in HUAC hearings, local security-clearance boards, and often even university committees. See Wang, *American Science in an Age of Anxiety* (cit. n. 24); Schrecker, *No Ivory Tower* (cit. n. 24); and Ellen Schrecker, *Many Are the Crimes: McCarthyism in America* (New York: Little, Brown, 1998). The FAS was quite active during the mid 1950s on the issue of passports and visas for scientists. The entire issue of the *Bulletin of the Atomic Scientists* for Oct. 1952 was dedicated to the topic. See also E.R. [Eugene Rabinowitch], "How to Lose Friends," *Bull. Atom. Sci.,* Jan. 1952, *8*:2–5; Victor Weisskopf, "Visas for Foreign Scientists," *ibid.,* Mar. 1954, *10*:68–69, 112; "American Visa Policy: A Report," *ibid.,* Dec. 1955, *11*:367–373; and John Toll, "Scientists Urge Lifting Travel Restrictions," *ibid.,* Oct. 1958, *14*:326–328.

enlisting collaborators for his autonomous S-matrix program. Years later, when his program lay largely abandoned by most particle physicists, Chew continued to assess the turnaround using the language of democratic participation.

Chew's "Little Red Schoolhouse" in Berkeley

While neglecting several large issues concerning tenure, academic freedom, and the legality of state-imposed loyalty oaths, the California Supreme Court ruled in favor of the dismissed nonsigners in October 1952, ordering that the regents reappoint them. This court decision, however, left unresolved the question of back pay, and so for several people the oath controversy lumbered on. This last issue was settled by the court, again in the nonsigners' favor, only in the spring of 1956. Yet as early as 1951, certain senior professors in Berkeley's physics department began to consider how best to lure Chew back to Berkeley. Soon after the first court decision was handed down, Birge tried to entice Chew to return. Reluctantly, Chew decided to stay in Illinois, which had made him a very generous counteroffer on hearing of Berkeley's actions. Still excited by the prospect of returning to Berkeley's stimulating campus, however, Chew spent the spring semester of 1957 there as a visiting professor. Eager to keep Chew in Berkeley, the new department chair, Carl Helmholz, performed some impressive financial gymnastics to convince the administration that it could afford to hire Chew as a full professor. Helmholz's schemes worked, and Chew accepted an appointment as a full professor, beginning in the 1957/1958 academic year.[47]

Immediately Chew began advising a large and growing group of graduate students within the department. Its size was especially notable in that Chew and all of his students were theoretical physicists, for whom working in large groups was still unusual. Often ten or more students would be under Chew's wing at a time, and Chew himself would be engaged in collaborative work with four or five of them; postdocs and research associates made the group even larger. A steady stream of Chew's students completed their Berkeley dissertations beginning in 1959, often with four or five finishing each year.[48] In choosing to train his students in this manner, Chew followed a pattern similar to that set by Oppen-

---

[47] With regard to Chew's return to Berkeley see the handwritten notes between Francis Jenkins, Robert Brode, and Raymond Birge, undated, ca. Apr. 1951, Dept. Physics, Berkeley, Records, ca. 1920–1962, Folder 5:25; on Chew's 1953 offer to return to Berkeley see Chew to Birge, 21 Apr. 1953, Birge Papers, Box 5, Folder "Chew, Geoffrey Foucar, 1924–"; and on his 1957/1958 appointment see Helmholz to Constance, 25 Mar. 1957, Birge Papers, Box 40, Folder "Letters written by Birge, January–May 1957." Helmholz's financial jockeying becomes clear in both *ibid.* and Helmholz to Chancellor Clark Kerr, 5 Mar. 1957, Dept. Physics, Berkeley, Records, ca. 1920–1962, Folder 1:26. The reappointment of the nonsigners was conditional on their signing a new statewide loyalty oath, the so-called Levering oath, which was even more explicitly anti-Communist than the original university oath had been. The key difference was that the Levering oath was imposed on all state employees, so that university faculty were no longer singled out for special treatment. See Gardner, *California Oath Controversy* (cit. n. 26), pp. 250, 253–254; and Schrecker, *No Ivory Tower,* pp. 123–125.

[48] Many of Chew's former colleagues and students recalled that his group was unusually large and that he still made time to work carefully with each of them. See Birge, "History," Vol. 5, Ch. 19, p. 51; Frazer, "Analytic and Unitary S Matrix" (cit. n. 15), p. 7; Georgella Perry, "My Years with Professor Chew," in *Passion for Physics,* ed. DeTar *et al.,* pp. 14–16, on p. 15; Steven Frautschi, "My Experiences with the S-Matrix Program," *ibid.,* pp. 44–48, on p. 44; Carleton DeTar, "What Are the Quark and Gluon Poles?" *ibid.,* pp. 71–78, on p. 77; David Gross, "On the Uniqueness of Physical Theories," *ibid.,* pp. 128–136, on p. 128; C. Edward Jones, "Deducing T, C, and P Invariance for Strong Interactions in Topological Particle Theory," *ibid.,* pp. 189–194, on p. 189; William Frazer, interview with the author, 7 July 1998; Jerome Finkelstein, interview with the author, 24 July 1998; Eyvind Wichmann, interview with the author, 13 Aug. 1998; and Henry Stapp, interview with the author, 21 Aug. 1998 (all interviews were conducted in Berkeley). A list of Chew's former graduate students, together with their years of graduation, appears in Frazer, "Analytic and Unitary S Matrix," pp. 7–8.

heimer at Berkeley in the 1930s: the students worked collectively, discussing their research projects regularly with the entire group. The large, close-knit group format contrasted starkly with the approach of Julian Schwinger, for example, who famously advised ten or more Harvard graduate students at a time during the 1950s and 1960s but met with any of them individually only rarely—and never with the whole group.[49]

Whereas Oppenheimer could intimidate students and colleagues alike with his notoriously sharp tongue, Chew's students uniformly recall a much more encouraging advisor, one who, in the words of a former student, "treat[ed] us as full partners in a common effort." In a further gesture of equality, Chew regularly joined the group for informal lunches in the Rad Lab cafeteria. Chew took Oppenheimer's pedagogical model a step further when he instituted what came to be known as the "secret seminar." His entire group of students met weekly to hear presentations from one another; often the meetings took place at Chew's house. The seminar sessions were "secret" because faculty members (other than Chew) were actively discouraged from attending: the goal was to make certain that no graduate students were too intimidated to participate equally with their peers. From deep within Lawrence's sprawling Radiation Laboratory, the original site of American "big science," Chew carved out what one of his former students described as a "little red schoolhouse."[50]

This "little red schoolhouse" approach also shaped how Chew and some of his Berkeley colleagues organized a special conference on the strong interactions, held in Berkeley in December 1960. As handwritten notes from an early planning meeting reveal, Chew, Carl Helmholz, Donald Glaser, and the other members of the committee wanted their conference to bring "new people up to date" on the status of strong-coupling particle physics. As emphasized in these notes, the conference was to be "non-exclusive." Minutes from this planning meeting likewise noted that graduate students' research, as part of the work of Berkeley's department, "should be strongly represented" at the conference.[51]

Meeting these goals would not be easy: physics conferences on special topics were rarely aimed at bringing nonspecialists up to speed, much less highlighting the contribu-

[49] On Oppenheimer's pedagogical approach see Robert Serber, "The Early Years," *Phys. Today,* Oct. 1967, *20*:35–39; Serber with Crease, *Peace and War* (cit. n. 34), Ch. 2; Alice Kimball Smith and Charles Weiner, eds., *Robert Oppenheimer: Letters and Recollections* (Cambridge, Mass.: Harvard Univ. Press, 1980), Ch. 3; and Kevles, *Physicists* (cit. n. 27), pp. 216–219. In 1958 Schwinger was technically advising sixteen Harvard graduate students; see "1958–59 Department Lists," in Department of Physics, Harvard University, Correspondence, 1958–60, Box A–P, Folder "1958–59 Department Lists," call number UAV 691.10, Harvard University Archives, Pusey Library, Cambridge, Massachusetts. Bryce DeWitt, who completed his dissertation under Schwinger in 1949, talked about Schwinger's style with me during several discussions; William Frazer raised the contrast between Chew and Schwinger during our interview.

[50] Gross, "Uniqueness of Physical Theories" (cit. n. 48), p. 128 ("full partners"). Both Carleton DeTar and Steven Frautschi recalled the lunches during their interviews with Stephen Gordon; see Gordon, "Strong Interactions," pp. 27–28. Details on Chew's "secret seminars" come from Frautschi, "My Experiences with the S-Matrix Program" (cit. n. 48), p. 44; A. Capella *et al.,* "The Pomeron Story," in *Passion for Physics,* ed. DeTar *et al.,* pp. 79–87, on pp. 86–87; and my interviews with Frazer, Finkelstein, Wichmann, and Mandelstam. The term "little red schoolhouse" comes from an interview between Carleton DeTar and Stephen Gordon, conducted May 1997, and quoted in Gordon, "Strong Interactions," p. 29. Several more of Chew's former students with whom Gordon spoke also recalled Chew's "secret seminar." In the 1930s Berkeley's physics department held informal weekly seminars, attended by faculty and graduate students alike, though this single department-wide meeting disappeared after World War II. Carl Helmholz and Howard Shugart discussed these older seminars during my interviews with them in Berkeley on 14 July 1998 and 29 July 1998, respectively.

[51] See the handwritten notes, dated Mar. 1960, on "Special Meeting APS," in Dept. Physics, Berkeley, Records, ca. 1920–1962, Folder 1:39, and the typed minutes from a planning meeting held on 4 Mar. 1960, in the same folder. The handwritten notes are probably by either Carl Helmholz, chair of the department at this time and head of the conference-planning committee, or Howard Shugart, who was a secretary to the conference-planning committee; the notes appear to match Helmholz's handwriting.

tions of graduate students. The Berkeley committee gained some help from the well-known MIT theorist Victor Weisskopf, who had worked in the FAS with Chew throughout the 1950s. As a member of the planning committee for the APS-sponsored special meeting in Berkeley, Weisskopf lobbied hard to obtain additional funding from the National Science Foundation so that the Berkeley conference could indeed include these younger participants. "You can well understand and I am sure you agree," Weisskopf urged, "that such conferences with open attendance are very important for the stimulation of young people or other people who are new in the field." Such openness was especially needed in particle physics, he continued: "The field of high-energy physics is, as you know, very strongly in the hands of a clique and it is hard for an outsider to enter. The Rochester conferences were the only conferences that dealt with that subject and they limited it to invited people only. The Berkeley conference is supposed to break this custom." "Open" meetings, intended for newcomers and students as much as for members of a "clique," were unusual in 1960. It took work on the part of Chew, Helmholz, Weisskopf, and the others to break the mold and keep their meeting "non-exclusive." The "open" meeting attracted about three hundred physicists.[52]

The theoretical portion of this special conference focused almost exclusively on recent work by Chew, Stanley Mandelstam, and Richard Cutkosky on a new framework for approaching particle physics. Just six months before Chew's more outspoken break with quantum field theory at the La Jolla meeting, the material discussed at the "open" Berkeley meeting helped to form the core of Chew's emerging S-matrix program. As Chew came to articulate more and more explicitly, the S-matrix approach relied on several general principles but eschewed much of the specific formalism of quantum field theory.[53]

The S Matrix and a Democracy of Practitioners

The independence of the S-matrix program from many of the esoteric niceties of field theory was given a doubly "democratic" spin by Chew as he championed the new approach and quickly spread its gospel far and wide. First, Chew began to argue for his concept of "nuclear democracy"—that all nuclear particles, subject to the strong nuclear force, should be treated equally, without dividing them into "elementary" and "composite" factions. As we have seen in Section I, Chew argued for this democratic treatment, and his related notion of the bootstrap, largely on the basis of his unusual interpretation of Feynman diagrams. In the rotated diagrams of Figure 3, for example, Chew and his students saw

---

[52] Victor Weisskopf to J. Howard McMillen, 14 Mar. 1960, Birge Papers, Box 29, Folder "Weisskopf, Victor Frederick, 1908–." As it turned out, the NSF refused any financial aid because the Berkeley meeting was under the auspices of the APS; additional funding for the meeting was provided by the AEC and United States Air Force. See the typed report "Conference on Strong Interactions," undated, Dept. Physics, Berkeley, Records, ca. 1920–1962, Folder 1:39; an unsigned, undated postconference report in this folder gives the attendance figure. The Rochester conferences on nuclear and particle physics began in 1950 at the University of Rochester; when they began to move to different venues during the mid 1950s, they still retained the name "Rochester conference." See, e.g., John Polkinghorne, *Rochester Roundabout: The Story of High-Energy Physics* (New York: Freeman, 1989). Like Chew, Weisskopf became quite active with the FAS during the 1950s; while Chew chaired the Passport Committee, Weisskopf headed the Visa Committee. See Victor Weisskopf, "Report on the Visa Situation," *Bull. Atom. Sci.*, Oct. 1952, *8*:221–222; and Weisskopf, "Visas for Foreign Scientists," *ibid.*, Mar. 1954, *10*:68–69, 112.

[53] Schedules and reports on the presentations at the 1960 Berkeley meeting may be found in Dept. Physics, Berkeley, Records, ca. 1920–1962, Folder 1:39. The general principles on which Chew and his collaborators hoped to build their non-field-theoretic S-matrix theory included analyticity, unitarity, Lorentz invariance, and crossing symmetry, not all of which are independent from each other. See esp. Chew, *S-Matrix Theory of Strong Interactions* (cit. n. 3); and Cushing, *Theory Construction*, Chs. 5–7.

the ρ meson move from force-carrier to bound-state composite to seemingly elementary particle. Arguing from the structure of these Feynman diagrams, Chew taught his many students to treat all nuclear particles the same way. But his "democratic" sentiment did not end with the new interpretation of Feynman diagrams; Chew fashioned his *S*-matrix work as "democratic" in a second sense as well. In addition to "democratic" diagrams, Chew championed a "nuclear democracy" among practitioners. Consider his remarks near the close of an ebullient invited lecture at the 1962 New York meeting of the American Physical Society: "I am convinced that a wild period of merrymaking lies before us. All the physicists who never learned field theory can get in the game, and experimenters are just as likely to come up with important ideas as are theorists. They may even have an advantage over us." Chew returned to this theme of nonexperts' advantage over field-theory experts in the *S*-matrix realm in a lecture at Cambridge University in 1963, reporting that "the less experienced physicists [have] an advantage in working with a new framework. (The inverse correlation of productivity with experience in a situation like this is remarkable.)" Meanwhile, Chew made good on his pledge to bring outsiders into the fold, delivering special lectures and seminars on the new material especially for experimentalists at Berkeley. (See Figure 5.) The 1963 *S*-matrix textbook by Chew's colleagues Roland Omnès and Marcel Froissart, *Mandelstam Theory and Regge Poles,* similarly carried the subtitle "An Introduction for Experimentalists."[54]

As Chew traveled around the country and beyond, feverishly working his campaign of "wild merrymaking," few could miss his obvious charisma and enthusiasm. As John Polkinghorne later recalled, "We used to call Geoff Chew 'the handsomest man in high energy physics.' I know of at least one senior secretary in a British physics department who kept a photograph of him near her desk. That frank and open face, with just a hint of his one-eighth Burmese ancestry, and his tall commanding figure, made him one of the few theorists in the pin-up class." Rumors spread far and wide that Chew had given up a potential career in professional baseball to work in particle physics. His personal charm and enthusiasm made Chew into an effective "salesman." Polkinghorne attests that "Geoff was definitely a man from whom one would be happy to buy a used car." His talks "were always eagerly awaited," given "their inspirational and encouraging tone."[55]

[54] Geoffrey Chew, "*S*-Matrix Theory of Strong Interactions without Elementary Particles," *Reviews of Modern Physics,* 1962, *34:*394–401, on p. 400 ("merrymaking"); Chew, "The Dubious Role of the Space-Time Continuum in Microscopic Physics," *Science Progress,* 1963, *51:*529–539, on p. 538 (nonexperts' advantage) (this article contains the text of Chew's 1963 Rouse Ball Lecture at Cambridge); and Roland Omnès and Marcel Froissart, *Mandelstam Theory and Regge Poles: An Introduction for Experimentalists* (New York: Benjamin, 1963) (Froissart spent time working with Chew's group in Berkeley during the early 1960s). Owen Chamberlain in particular recalled Chew's special lectures for experimentalists; see Owen Chamberlain, "Interactions with Geoff Chew," in *Passion for Physics,* ed. DeTar *et al.,* pp. 11–13, on p. 13. Chew's unusual ability and interest in instructing experimentalists was by this time long standing. As a visiting professor at Berkeley in 1957, Chew gave a seminar on the pion-nucleon interaction, which drew an unusual number of experimentalists from the Radiation Laboratory and from Livermore, in addition to the graduate students enrolled in the course. Carl Helmholz reported that the experimentalists "are getting considerable benefit from it even though the subject is quite abstract and mathematical": Helmholz to Constance, 25 Mar. 1957 (cit. n. 47). William Frazer also discussed Chew's informal seminar for experimentalists during our interview.

[55] John Polkinghorne, "Salesman of Ideas," in *Passion for Physics,* ed. DeTar *et al.,* pp. 23–25, on p. 23. Polkinghorne worked on *S*-matrix theory while based in Cambridge, England. For a similar analysis of the role of charisma in modern physics see Charles Thorpe and Steven Shapin, "Who Was J. Robert Oppenheimer? Charisma and Complex Organization," *Social Studies of Science,* 2000, *30:*545–590. Owen Chamberlain and Georgella Perry (Chew's former secretary) both mentioned Chew's baseball pretensions: Chamberlain, "Interactions with Geoff Chew," pp. 12–13; and Perry, "My Years with Professor Chew" (cit. n. 48), pp. 14–16. Several other physicists recalled these same rumors during interviews with Stephen Gordon: Gordon, "Strong Interactions," p. 15.

Figure 5. *Geoffrey Chew delivering an informal lecture at Berkeley, 1961. Reprinted with kind permission of the Lawrence Berkeley National Laboratory.*

His enthusiasm quickly suffused Berkeley's department and encouraged his graduate students and postdocs to participate as equals in the S-matrix campaign. Louis Balázs, who completed his Ph.D. under Chew's direction in the mid 1960s, reminisced recently that "it was an exciting experience being one of Chew's graduate students at UC-Berkeley in the early 1960s. New ideas were being discussed and developed continually and vigorously, particularly by the postdocs, and it seemed we were on the threshold of a new era in Physics." Those who were pushing the boundaries of the "new era" seemed to be drawn to Berkeley, if they weren't there already: "There was a constant stream of distinguished visitors," Balázs continued, "who seemed to be eager to learn about the new developments in Berkeley." And just as Chew had announced so exuberantly at the 1962 meeting, Balázs too recalled that "even graduate students found that they could make new independent contributions at that time." Another former graduate student, William Frazer, concurs, emphasizing that Chew worked hard to make sure that no students felt intimidated. "He really made a wonderful atmosphere for us to work in."[56] Both students who were still "innocent" of the elaborate quantum field theory formalism, as Chew put it, and experimentalists who had "never learned field theory" stood to contribute as equals in Chew's S-matrix program.

[56] Louis Balázs to David Kaiser, 6 Aug. 1998; and Frazer interview. See also Perry, "My Years with Professor Chew," pp. 15–16.

Chew reflected in several places on the best way to train these potential S-matrix contributors. As early as 1961, he noted that students who had not learned quantum field theory, and the usual ways of using and interpreting Feynman diagrams, seemed to fare best when approaching the new S-matrix material. He assured readers of his 1961 lecture note volume that "it is . . . unnecessary to be conversant with the subleties of field theory, and a certain innocence in this respect is perhaps even desirable. Experts in field theory seem to find current trends in S-matrix research more baffling than do nonexperts." The same sentiment appeared five years later, when Chew remarked in the preface to his 1966 textbook: "No background in quantum field theory is required. Indeed, as pointed out in the preface to my 1961 lecture notes, lengthy experience with Lagrangian field theory appears to constitute a disadvantage when attempting to learn S-matrix theory."[57]

Chew's own students largely followed this prescription. William Frazer, the first student to complete his dissertation under Chew after his return to Berkeley, worked through Eyvind Wichmann's "very rigorous field theory course" as a graduate student, though he found his work with Chew to be much more interesting. "For the student," Frazer recalled recently, "life was a bit schizophrenic. Either you read on your own the very dry mathematical structure of axiomatic field theory, or you tried to follow the more exciting material." Later students, such as Jerome Finkelstein, recalled that "there were quite a few of us in Chew's group who really did not spend a lot of time on field theory. I'd taken a course in it during my first or second year of graduate school, but that was all." Ramamurti Shankar, another of Chew's students from this time, recently put a humorous turn on this pedagogical approach:

> I had a choice: either struggle and learn field theory and run the risk of blurting out some four letter word like φ^4 in Geoff's presence or simply eliminate all risk by avoiding the subject altogether. Being a great believer in the principle of least action, I chose the latter route. Like Major Major's father in *Catch 22,* I woke up at the crack of noon and spent eight and even twelve hours a day not learning field theory and soon I had not learnt more field theory than anyone else in Geoff's group and was quickly moving to the top.

David Gross, who completed his dissertation under Chew's direction in 1966, paraphrased an often-heard refrain from colleagues who didn't realize that Gross, since the 1970s a preeminent field theorist, had come from Chew's group: "Funny—you don't look Chewish." One would search in vain to find much usage of specifically field-theoretic techniques in the sixty dissertations by Chew's graduate students.[58]

So much for Chew's immediate circle in Berkeley. With several well-known texts on quantum field theory already in print, the next question was how to spread the physics of "nuclear democracy" beyond Chew's large but geographically limited Berkeley group. Just as politicians debated a purported "missile gap" with the Soviets, Chew and his collaborators in the early 1960s faced a "textbook gap": given that so many physicists conceivably could participate in developing the "democratic" S-matrix program, the challenge was to reach these students, experimentalists, and other theorists and deliver the S-matrix message.

[57] Chew, *S-Matrix Theory of Strong Interactions* (cit. n. 3), pp. vii–viii; and Geoffrey Chew, *The Analytic S Matrix: A Basis for Nuclear Democracy* (New York: Benjamin, 1966), p. v.

[58] Frazer interview; Finkelstein interview; Ramamurti Shankar, "Effective Field Theory in Condensed Matter Physics," in *Conceptual Foundations of Quantum Field Theory,* ed. Cao (cit. n. 4), pp. 47–55, on p. 47; and Gross, "Uniqueness of Physical Theories" (cit. n. 48), p. 128. Dissertations by Chew's students from 1959 to 1983 may be found in the Berkeley Physics Department Library.

Toward this goal, Chew and many of his *S*-matrix students and collaborators delivered many sets of summer-school lectures—sometimes, like Chew in 1960, giving the same set of lectures at two different schools in the same summer.[59]

Chew and his postdocs also began to publish their lecture notes in inexpensive editions, nearly as soon as the lectures had been delivered. These books themselves reveal much about the quest to attract students to the *S*-matrix team. Most of the important *S*-matrix textbooks were part of the "Frontiers in Physics" series, which began to publish collections of lecture notes and reprints in 1961. Chew's 1961 *S-Matrix Theory of Strong Interactions,* based on his 1960 summer-school lectures, was one of the first books to be included in the new series. By 1964 *S*-matrix tracts constituted nearly one-third of the "Frontiers in Physics" books, even though the series was meant to treat all aspects of physics and not only particle theory.[60]

These books were rushed into print. *Regge Poles and S-Matrix Theory,* by Chew's postdoc Steven Frautschi, stemmed from lectures Frautschi had given once at Cornell University in 1961/1962 and then augmented for delivery at a June 1962 summer school. The lectures presented by Maurice Jacob and Chew in their 1964 *Strong-Interaction Physics* also had been delivered only once, during the academic year 1962/1963. As the series editor David Pines explained, "Frontiers in Physics" was intended to feature just this kind of "rough and informal" lecture notes rather than polished monographs. The very production of the books reflected their mission, as Pines explained: "Photo-offset printing is used throughout, and the books are paperbound, in order to speed publication and reduce costs. It is hoped that the books will thereby be within the financial reach of graduate students in this country and abroad."[61]

The progress of *S*-matrix theorists toward establishing an axiomatic foundation for their new work can also be read immediately from the material form of their books. Chew's early reports on the developing theory in 1961 and 1964 were printed in books that had not been carefully typeset but were printed on inexpensive paper and hurried into print. Chew's first textbook to treat the newly proposed axiomatic version of *S*-matrix theory, *The Analytic S Matrix* (1966), on the other hand, was not published as part of the "Frontiers in Physics" series. Its professional typesetting and glossy pages present a clear contrast with the earlier volumes.[62]

From summer-school lectures to "Frontiers in Physics" volumes, then, the task of cre-

---

[59] Chew's "Double Dispersion Relations and Unitarity as the Basis of a Dynamical Theory of Strong Interactions" (cit. n. 17) appeared in both *Relations de dispersion et particules élémentaires,* ed. de Witt and Omnès (cit. n. 14), pp. 455–514, and in *Dispersion Relations,* ed. Screaton (cit. n. 14), pp. 167–258. See also the "Editor's Note" *ibid.,* p. viii. One might compare Chew and his "missionaries" with Niels Bohr and the spread of the "Copenhagen spirit" as studied in John Heilbron, "The Earliest Missionaries of the Copenhagen Spirit," *Revue d'Histoire des Sciences,* 1985, *30*:195–230.

[60] Lists of the titles published in the "Frontiers of Physics" series were included in the front of each of the books within the series. The *S*-matrix books in the series published between 1961 and 1964 included Chew, *S-Matrix Theory of Strong Interactions* (1961); Omnès and Froissart, *Mandelstam Theory and Regge Poles* (1963); Frautschi, *Regge Poles and S-Matrix Theory* (1963); E. J. Squires, *Complex Angular Momenta and Particle Physics* (1963); and Jacob and Chew, *Strong-Interaction Physics* (1964). During this same period, only two books that focused on quantum field theory for particle physics were included in the series.

[61] See the "Editor's Foreword" by David Pines, dated Aug. 1961, which appears within each of the volumes in the series. Frautschi's book, for example, was reproduced from hand-typed originals. Chew mentioned his friendship with Pines, also a physicist at the University of Illinois at Urbana, as one of the reasons he decided to publish his *S*-matrix lecture notes in this series: Chew to Kaiser, 11 May 1998 (email).

[62] This reading of the material production of Chew's textbooks is inspired by the example of David Cressy, "Books as Totems in Seventeenth-Century England and New England," *Journal of Library History,* 1986, *21*:92–106.

ating and maintaining a community of S-matrix theorists involved far more than publishing research articles in the *Physical Review*. Chew was well aware that the dissociation of Feynman diagrams from their circumscribed meaning within quantum field theory would be difficult for students who had already mastered quantum field theory to grasp. His autonomous S-matrix program promised opportunities for those students, experimentalists, and other theorists who had not become overly attached to quantum field theory. Inexpensive pedagogical resources, such as lecture notes and reprint collections, could thus serve to broaden the base of "democratic" S-matrix practitioners. There is some evidence that this aggressive, democratizing textbook campaign worked: one physicist explained years later that, although he had not been a direct student of Chew's, he had learned particular calculational details of the S-matrix program, together with some sense of its special "philosophy," from reading Chew's 1961 *S-Matrix Theory of Strong Interactions*.[63]

The Language of Democracy for a Program in Decline

In addition to emphasizing his democratic team of contributors, Chew drew an increasing contrast during the 1960s and into the early 1970s between the kind of work his S-matrix program fostered as compared with that of the field theorists. The field theorists sought to explain the phenomena of particle physics with reference to a small set of basic or unit interactions taking place between a core set of "fundamental" or "elementary" particles. Chew mocked this approach of the "fundamentalists," arguing that such an "aristocratic" arrangement of fundamental particles could not provide an adequate framework for describing the strong interactions.[64] As in his 1950 article describing the loyalty oath controversy at the University of California, Chew reserved the term "fundamentalist" for colleagues who espoused a position at odds with his own "democratic" ideals.

By the late 1960s Chew's program appeared to many physicists to have lost its original focus, becoming mired in details and complexity. Treating the ρ meson within a democratic bootstrap framework was one thing, but moving beyond such a simple system to more complicated calculations had become much more frustrating. Quantum field theory, meanwhile, now augmented with a new emphasis on gauge symmetries and the quark hypothesis, was again attracting the attention of most particle physicists. Chew lamented this turn of events, arguing that it sprang more from physicists' "psychology" than from stubborn experimental data. The trouble, he wrote in the early 1970s, was that the "fundamentalists" dreamed "of the press conference that will announce to the world a dramatic resolution of their quest." Unlike the night thoughts of these fundamentalists, the autonomous S matrix was "the cumulative result of many steps stretching out over decades."[65]

Progress for the S-matrix camp, Chew explained, was necessarily a gradual game of

[63] Al Mueller, "Renormalons and Phenomenology in QCD," in *Passion for Physics*, ed. DeTar *et al.*, pp. 137–142, on p. 137.

[64] Chew used the language of "fundamentalists" throughout his articles "Hadron Bootstrap: Triumph or Frustration?" (cit. n. 18) and "Impasse for the Elementary-Particle Concept" (cit. n. 18). See also Chew, "'Bootstrap': A Scientific Idea?" (cit. n. 18).

[65] Chew, "Hadron Bootstrap," p. 24; and Chew, "Impasse for the Elementary-Particle Concept," p. 124. On the resurgence of quantum field theory see Cushing, *Theory Construction*, Chs. 6–7; Gordon, "Strong Interactions," Chs. 4–5; Andrew Pickering, *Constructing Quarks: A Sociological History of Particle Physics* (Chicago: Univ. Chicago Press, 1984), Chs. 4–8; Abraham Pais, *Inward Bound: Of Matter and Forces in the Physical World* (New York: Oxford Univ. Press, 1986), Ch. 21; Tian Yu Cao, *Conceptual Developments of Twentieth-Century Field Theories* (New York: Cambridge Univ. Press, 1997), Chs. 8–11; and Lillian Hoddeson, Laurie Brown, Michael Riordan, and Max Dresden, eds., *The Rise of the Standard Model: Particle Physics in the 1960s and 1970s* (New York: Cambridge Univ. Press, 1997).

constructing more and more partial models, incorporating the effects of more and more particle exchanges and interactions, all under the rubric of several reigning general principles (such as causality). Chew's program was therefore built on the assumption, he wrote in 1974, "that there will gradually develop a more and more dense coverage of the nuclear world by interlocking models no single model having preeminent status. Such a pattern might be characterized as a 'democracy of models.' " Chew reminded physicists in 1970 that "even though no press conference was called," the stepwise construction of these interlocking models within the *S*-matrix program had already scored several "breakthroughs," a fact he attributed to the "brilliant collective achievement of the high-energy physics community." Reinforcing this point with a history lesson, Chew recalled the "precedent of classical nuclear physics": "This model enjoyed an aristocratic status for almost thirty years, but eventually it was democratized."[66]

This notion of constructing interlocking models through a series of "collective achievements" fit well with Chew's "secret seminar" approach to training graduate students. His students' dissertations reveal this close-knit, mutually buttressing approach: Bipin Desai, completing his dissertation in April 1961, built directly on work by James Ball in his dissertation, filed almost exactly one year earlier; Yongduk Kim's dissertation from June 1961 in turn drew explicitly on the work by Desai and Ball. This pattern continued throughout the 1960s. The abstract from Shu-yuan Chu's May 1966 dissertation, for example, explained that "the method [employed in the dissertation] is an extension of [C. Edward] Jones' proof in the single-channel case, making use of an explicit expression of the determinant D constructed by [David] Gross." Jones completed his dissertation in 1964 with Chew, and Gross submitted his dissertation a few months after Chu.[67] Citations to work, both published and unpublished, by other graduate-student members of Chew's group, and acknowledgments of extended discussions with fellow students, fill nearly every one of Chew's students' dissertations. Chew's pedagogical ideal of equal participation melded seamlessly with his program for the piecewise construction of "interlocking models" by a "collective" of researchers.

By the 1970s, with his program all but abandoned by most physicists, Chew bemoaned the failure of his collective vision to attract those physicists who insisted instead on searching for a single "glamorous-sounding fundamental entity": "Few of the stars of the world of physics are content with the thought that their labors will constitute only a *fraction* of a *vast mosaic* that must be constructed before the complete picture becomes recognizable and understandable." These "stars" of physics, now nearly all within the "fundamentalist" camp, routinely embraced only models based on Lagrangian field theories, Chew noted. They accorded these models "special status," failing to "consider on an equal footing" models not derived from such a field-theoretical basis. The task of the collective-minded *S*-matrix theorist, on the other hand, remained "to view any number of different partially successful models without favoritism."[68]

[66] Chew, "Impasse for the Elementary-Particle Concept," p. 124; and Chew, "Hadron Bootstrap," p. 25.

[67] James S. Ball, "The Application of the Mandelstam Representation to Photoproduction of Pions from Nucleons" (Ph.D. diss., Univ. California, Berkeley, 1960); Bipin Desai, "Low-Energy Pion-Photon Interaction: The $(2\pi, 2\gamma)$ Vertex" (Ph.D. diss., Univ. California, Berkeley, 1961); Yongduk Kim, "Production of Pion Pairs by a Photon in the Coulomb Field of a Nucleus" (Ph.D. diss., Univ. California, Berkeley, 1961); and Shu-yuan Chu, "A Study of Multi-Channel Dynamics in the New Strip Approximation" (Ph.D. diss., Univ. California, Berkeley, 1966). Ball's, Desai's, and Kim's dissertations drew heavily on the work by William Frazer and José Fulco, in particular "Effect of a Pion-Pion Scattering Resonance on Nucleon Structure" (cit. n. 15) and "Partial-Wave Dispersion Relations for the Process $\pi + \pi \to N + \underline{N}$" (cit. n. 15).

[68] Chew, "Impasse for the Elementary-Particle Concept" (cit. n. 18), p. 124 (emphasis added); and Chew, "Hadron Bootstrap" (cit. n. 18), p. 27.

The late 1960s and early 1970s were a difficult time to be a young physicist in the United States, over and above Chew's frustrations with the fate of "nuclear democracy." With the onset of détente and dramatic cuts in defense spending, U.S. physicists rapidly slid into the worst job shortage the profession had ever witnessed. Enrollments in the placement service registries of the American Physical Society tell the grim tale: in 1968, nearly 1,000 applicants fought for 253 jobs; the next year, almost 1,300 competed for 234 jobs. Only 63 jobs were on offer at a 1970 American Physical Society meeting at which 1,010 young physicists were looking for work; 1,053 competed for 53 jobs at the 1971 meeting. Anecdotal evidence suggests that students who were steeped too heavily in S-matrix methods, to the exclusion of field-theoretic techniques, felt the crunch disproportionately. Although many of Chew's students from this later period have gone on to academic careers as theoretical physicists, several S-matrix students left the field after earning their Ph.D.'s to become medical doctors or lawyers; others were denied tenure at places like MIT. Although it is difficult to disentangle the root causes of these few theorists' difficulties from the overwhelming across-the-board cutbacks, several physicists to this day continue to associate S-matrix training with job-placement difficulties during the late 1960s and early 1970s.[69]

In frustration as in triumph, Chew spoke of his program in distinctly "democratic" terms. In the early 1960s it had seemed open to all: everyone could participate equally, and no one was singled out for special privileges. Later, when the program fell into neglect, the language of Chew's complaints was likewise laden with the tropes of democratic participation. Field theorists improperly elevated an "aristocracy" of particles and granted "special status" only to certain kinds of models. S-matrix theorists, on the other hand, strove for an equality of particles, models, and practitioners, all judged "without favoritism" as members of a collective. "Patience" should be the order of the day, Chew wrote, not the yearning for press conferences and the special privileges (such as increased government funding) such singular attention could foster.[70]

In between his many speeches and lectures, Chew built an approach to training students and colleagues that emphasized equal participation and the collective strivings of the group over the "cliquish" dreams of the "fundamentalists." In establishing his "secret seminar," designing new lecture note volumes and textbooks, and delivering special lectures and seminars for experimentalists, Chew sought to make theoretical particle physics a particularly democratic activity. Just as he lobbied for fair treatment of academics and scientists under a controlling state in the Cold War, Chew tried to produce within physics a community of peers, neither singled out for special treatment nor splintered between idea-producing theorists and fact-checking experimentalists. Each contributor to the "vast mosaic" of S-matrix theory was to be an equal partner under the law.

IV. THE VIEW FROM PRINCETON

Traces of Chew's outspoken stance on the failure of quantum field theory and on the need to treat all nuclear particles democratically may be found throughout his students' dissertations. Peter Cziffra, writing a year before Chew's famous La Jolla talk, echoed his ad-

---

[69] Several physicists drew these connections during their interviews with Stephen Gordon: Gordon, "Strong Interactions," pp. 50, 53–54. The statistics come from Kaiser, "Putting the 'Big' in 'Big Science'" (cit. n. 28), pp. 33–35.

[70] Chew, "Impasse for the Elementary-Particle Concept" (cit. n. 18), p. 125.

visor's attitude when he opened his dissertation by noting how "stymied" ordinary perturbative quantum field theory remained when treating the strong interactions. Well into
the campaign for nuclear democracy, Akbar Ahmadzadeh reminded readers of his dissertation that insisting with the field theorists on a strict division between "elementary" and
"composite" particles often leads to "absurd conclusions" when studying the strong interactions; instead, all particles should be treated as bound states, "on an equal footing."
Henry Stapp, a research associate at the Rad Lab when Chew returned there in the late
1950s, pursued an axiomatic foundation for S-matrix theory in the early 1960s. "Early
on," Stapp has recalled recently, "I was not really in close touch with Chew; I picked up
the S-matrix ideas by osmosis, since Chew's ideas were permeating the area."[71]

Still, despite the group's successes in spreading the word via summer-school lectures
and "Frontiers in Physics" volumes, their ideas did not "permeate" all departments of
physics in the same way. Princeton's department, in particular, provides a telling contrast
with Chew's Berkeley. By the early 1960s Princeton boasted a large and active group of
theorists working on many aspects of particle physics. In fact, the Princeton group, like
Chew's group in Berkeley, championed and extended many of the nonperturbative,
diagram-based tools with which Chew was tinkering. Though a large group of theorists at
Princeton pursued topics that fell under the S-matrix rubric, they did not share Chew's
zeal for a "nuclear democracy." Comparing the work by these Princeton theorists with that
of Chew's group may help, therefore, to highlight which elements of Chew's many-faceted
"democracy" remained particular to his Berkeley group.

One of the leaders of Princeton's group had shared many early stops with Chew along
a common trajectory. Marvin "Murph" Goldberger had been a graduate student with Chew
in Chicago immediately after the war. The two became fast friends, sharing office space,
arranging social outings together, completing their dissertations at the same time, and
moving together to Berkeley's Rad Lab as postdocs in 1948. Following his postdoc Goldberger took a job back at the University of Chicago, while Chew taught for one year at
Berkeley and then resigned over the loyalty oath. Once Chew landed in Urbana, he and
Goldberger were again in close proximity; they struck up an active collaboration during
the mid 1950s, together with Francis Low and Murray Gell-Mann, also both in the Midwest
at that time. Just when Chew left Urbana to return to Berkeley, Goldberger left Chicago
to take a position at Princeton, starting in February 1957. Though now separated by the
length of the continent, Goldberger and Chew continued to correspond.[72]

---

[71] Peter Cziffra, "The Two-Pion Exchange Contribution to the Higher Partial Waves of Nucleon-Nucleon
Scattering" (Ph.D. diss., Univ. California, Berkeley, 1960), p. 4; Akbar Ahmadzadeh, "A Numerical Study of
the Regge Parameters in Potential Scattering" (Ph.D. diss., Univ. California, Berkeley, 1963), pp. 2–3; and Stapp
interview. See Henry Stapp, "Derivation of the CPT Theorem and the Connection between Spin and Statistics
from Postulates of the S-Matrix Theory," *Phys. Rev.,* 1962, *125*:2139–2162; Stapp, "Axiomatic S-Matrix Theory," *Rev. Mod. Phys.,* 1962, *34*:390–394; Stapp, "Analytic S-Matrix Theory," in *High-Energy Physics and
Elementary Particles,* ed. Abdus Salam (Vienna: International Atomic Energy Agency, 1965), pp. 3–54; and
Stapp, "Space and Time in S-Matrix Theory," *Phys. Rev.,* 1965, *139*:B257–270. Stanley Mandelstam, a close
collaborator of Chew's and an architect of many of the S-matrix techniques, resisted following Chew and his
group in renouncing field theory. As early as the Dec. 1960 Berkeley conference, a conference report noted that
Chew's presentation based on Mandelstam's work "did not evoke Mandelstam's full assent": "Conference on
Strong Interactions," p. 10, Dept. Physics, Berkeley, Records, ca. 1920–1962, Folder 1:39. See also Cushing,
Theory Construction, pp. 131–132, 145.

[72] Annual Report 1956–1957, pp. 1–2, in Department of Physics, Princeton University, Annual Reports to the
[University] President, Seeley G. Mudd Manuscript Library, Princeton Univ., Princeton, N.J. (hereafter cited as
Dept. Physics, Princeton, Annual Reports); Marvin Goldberger, "Fifteen Years in the Life of Dispersion
Relations," in *Subnuclear Phenomena,* ed. A. Zichichi (New York: Academic, 1970), pp. 685–693; Goldberger,
"Francis E. Low — A Sixtieth Birthday Tribute," in *Asymptotic Realms of Physics,* ed. Alan Guth, Kerson Huang,

At Princeton Goldberger joined Sam Treiman, who had earned his Ph.D. from Chicago four years after Goldberger and Chew; a year and a half later, Richard Blankenbecler joined the group as a postdoc and later became a regular faculty member. Goldberger, Treiman, Blankenbecler, and their many graduate students spent much of their time during the late 1950s and 1960s on topics that Chew would have called S-matrix theory—just the sort of "interlocking models" that he hoped would bring clarity to the strong interaction. Goldberger and Treiman investigated the decay of unstable particles without resorting to field-theoretic Lagrangians; students completed dissertations on the analytic structure of scattering amplitudes and on how to incorporate unitarity, some of the key general principles from which Chew aimed to construct his autonomous S-matrix theory. On the surface, these projects all sound as if they could have been completed by Chew's students in Berkeley. Yet when Goldberger reported in 1961 on the Princeton group's many accomplishments, he categorized all of their work, with his characteristic sense of humor, as "the engineering applications of quantum field theory." "This work," Goldberger continued, "is complementary to the purer aspects of quantum field theory." Just at the time when Chew was announcing his clear and decisive break with quantum field theory in La Jolla, the Princeton group celebrated the close fit between their research and Chew's nemesis.[73]

The Princeton group's research was no closer to "engineering" than any of Chew's work; it only seemed more "applied" when compared with the work streaming out from Princeton's other group of theoretical particle physicists, headed by Arthur Wightman. Wightman championed an axiomatic approach to quantum field theory; his research remained at that time, unlike Chew's or Goldberger's, far removed from the details of recent experiments. No doubt with reference to Chew's standing-room-only invited lectures before the American Physical Society, the Princeton group reported in 1966 that "there are presently two approaches to relativistic quantum theory. These are axiomatic field theory and dispersion or S-matrix theory. Notwithstanding some passionate claims, there is not yet any evidence that the two are really different. . . . Both theoretical approaches have been and will continue to be pursued actively at Princeton."[74] Together with Wightman, then, Goldberger and Treiman erected a "big-tent" approach to quantum field theory: "engineering applications" based on S-matrix calculational techniques would coexist peacefully with the more "pure" investigations into the structure of quantum field theory. At Berkeley, meanwhile, there were no longer any senior field theorists in town to respond to Chew's challenge; the loyalty oath controversy had ensured that, when figures such as Gian Carlo Wick were fired as nonsigners. Years later, Chew mused on how he might have been more "intimi-

and Robert Jaffe (Cambridge, Mass.: MIT Press, 1983), pp. xi–xv; and Goldberger, "A Passion for Physics," in *Passion for Physics,* ed. DeTar *et al.,* pp. 241–245. Andrew Pickering has examined the active collaboration of Chew, Low, Goldberger, and Gell-Mann, emphasizing the importance of their geographical proximity: Andrew Pickering, "From Field to Phenomenology: The History of Dispersion Relations," in *Pions to Quarks,* ed. Brown *et al.* (cit. n. 13), pp. 579–599.

[73] M. L. Goldberger and S. B. Treiman, "Decay of the Pi Meson," *Phys. Rev.,* 1958, *110:*1178–1184; on the fit between the Princeton work and Chew's see Goldberger, "An Outline of Some Accomplishments in Theoretical Physics," Annual Report 1960–1961, pp. 12–13, Dept. Physics, Princeton, Annual Reports. On other work in the department see Annual Report 1958–1959, pp. 2, 7; Annual Report 1960–1961, pp. 12–13; Annual Report 1962–1963, p. 4; Annual Report 1963–1964, p. 66; Annual Report 1964–1965, pp. 79–81; Annual Report 1965–1966, pp. 4–5; and Annual Report 1967–1968, pp. 23–24: Dept. Physics, Princeton, Annual Reports. See also Treiman, "A Connection between the Strong and Weak Interactions," in *Pions to Quarks,* ed. Brown *et al.,* pp. 384–389; and Treiman, "A Life in Particle Physics," *Ann. Rev. Nuclear Particle Sci.,* 1996, *46:*1–30.

[74] Arthur Wightman, "The General Theory of Quantized Fields in the 1950s," in *Pions to Quarks,* ed. Brown *et al.,* pp. 608–629; and Annual Report 1965–1966, pp. 4–5, Dept. Physics, Princeton, Annual Reports (quotation).

dated," and less likely to dismiss quantum field theory outright, had Wick still been in Berkeley. A similar restraint might have been exercised by Francis Low, with whom Chew worked in Urbana in the mid 1950s. Although Low credits Chew with having provided most of the original ideas during their fruitful collaboration, it is likely that if they had remained together in Urbana, Chew's later flamboyant pronouncements about the death of quantum field theory would have been muted by Low's impressive and authoritative grasp of field theory's tenets.[75]

In keeping with the double-barreled approach to field theory at Princeton, and in clear contrast to Chew's group in Berkeley, Goldberger's and Treiman's graduate students actively studied quantum field theory as an essential part of their training. Stephen Adler, who completed his Ph.D. under Treiman's direction in 1964, remembers auditing Wightman's course, since Wightman's work "was seen as undergirding" the calculational techniques of dispersion relations and S-matrix theory. "We were doing an evasive physics," Adler continued: since no one knew how best to treat the strong interactions, "we used whatever methods we could. . . . Dispersion relations were a tool, but we also learned field theory methods because they were useful for treating symmetries." Other Princeton students from this period similarly recall an emphasis on learning quantum field theory.[76] This peaceful coexistence of S-matrix-style calculations with axiomatic quantum field theory also led to a different appraisal of Chew's S-matrix program than the one "permeating" the Berkeley area.

Adler recalls that when Chew came to Princeton to give a talk, "he sounded very messianic." Adler was hardly alone in comparing Chew's active campaigning to religious indoctrination. In fact, many of Chew's colleagues and former collaborators, now working at a distance from him, began to characterize his vigorous pronouncements as religious proselytizing. Goldberger wrote to Murray Gell-Mann in January 1962 that Chew had become "the Billy Graham of physics." He added playfully: "After his talk I nearly declared for Christ. Since I had already changed from Jewish to Regge-ish, it was the only thing I could think of." Years later, John Polkinghorne recalled that Chew proclaimed his ideas "with a fervour" like that "of the impassioned evangelist. There seemed to be a moral edge to the endeavour. It was not so much that it was expedient to be on the mass-shell of the S matrix as that it would have been sinful to be anywhere else." Rather than a ringing democratic political campaign, Chew's efforts often struck his former colleagues as runaway zealotry.[77]

[75] Geoffrey Chew, interview with Gordon, Dec. 1997, quoted in Gordon, "Strong Interactions," p. 33 (see also pp. 32–35). Low credited Chew with the main originality in their collaboration in his interview with me; see also Kaiser, "Francis E. Low" (cit. n. 45), pp. 71–72.

[76] Stephen Adler, telephone interview with the author, 16 Feb. 1999. This emphasis on semiphenomenological tools rather than overarching theory construction became a hallmark of Treiman's group and helped shape Adler's later work on current algebras. See Stephen Adler and Roger Dashen, *Current Algebras and Applications to Particle Physics* (New York: Benjamin, 1968); and Sam Treiman, Roman Jackiw, and David Gross, *Lectures on Current Algebra and Its Applications* (Princeton, N.J.: Princeton Univ. Press, 1972). See also Pickering, *Constructing Quarks* (cit. n. 65), pp. 108–114; and Cao, *Conceptual Developments of Twentieth-Century Field Theories* (cit. n. 65), pp. 229–246. The views of other former Princeton students are expressed in John Bronzan to Kaiser, 15 May 1997 (email), and E. E. Bergmann to Kaiser, 16 May 1997 (email). The same attitude is drawn out in the course notes taken by Kip Thorne when he was a graduate student at Princeton in the early 1960s. See, in particular, Thorne's notes from "Properties of Elementary Particles," a course given by Val Fitch (Spring 1963); "Elementary Particle Physics," taught by Sam Treiman (Spring 1963); "Intermediate Quantum Mechanics and Applications," taught by Goldberger (Fall 1963); and "Elementary Particle Theory," taught by Blankenbecler (Spring 1964). All notes in the possession of Professor Kip Thorne; my thanks to him for sharing copies of these notes.

[77] Adler interview; Marvin Goldberger to Murray Gell-Mann, 27 Jan. 1962, quoted in Johnson, *Strange Beauty*

As Adler recalled, Chew announced in his Princeton lecture "a great hope; but at the same time, Treiman was always a bit skeptical of any grand theory." Indeed, Treiman was skeptical. Whereas Chew's January 1962 lecture before the American Physical Society predicted a "wild period of merrymaking," Treiman's own lecture on S-matrix material, delivered ten months later, focused instead on how even the most promising-looking "partial results and insights" remained "all tangled up with approximations which have inevitably to be introduced and which vary in style and severity from one application to another, one author to another." It is important to note that Treiman was no critic of approximations, even "severe" ones, per se. His 1958 work with Goldberger on pion decay relied, in another reviewer's words, on "drastic assumptions" and "feeble arguments." Even in Treiman's own estimation, these were "hair-raising approximations" that he and Goldberger were "quite unable—apart from hand waving—to justify."[78] It wasn't the recourse to approximations that irked Treiman about Chew's program; it was the vehemence with which Chew pitted his work against field theory.

Treiman went on to dismiss the bootstrap hypothesis, trumpeted by Chew as the logical conclusion of nuclear democracy, as "amusing." As we have seen in Section I, the goal of the bootstrap work was to find a single self-consistent solution that would show that strongly interacting particles might each produce the very forces that led to their own production by other particles; each strongly interacting particle, then, might be said to "pull itself up by its own bootstraps." What had captured the imagination of the Berkeley group left Treiman unimpressed. Throughout his 1962 lectures, for example, Treiman emphasized the "conjectural" basis of the bootstrap work and "indulg[ed]" in what he characterized as "pessimistic remarks" regarding the bootstrap program.[79] When it came to relations between S-matrix theory and quantum field theory, Treiman was not Chew— and Princeton was not Berkeley.

These differences led to some subtle reinterpretations of the meaning of S-matrix work, not only of its relative importance. Princeton's L. F. Cook, another faculty member in the Goldberger-Treiman-Blankenbecler group, reported on his research on Chew's beloved bootstrap mechanism. Despite Treiman's deflating judgment, the bootstrap remained to Chew, Frautschi, and most other members of the Berkeley group the sought-for culmination of nuclear democracy, a kind of holy grail for the equal treatment of all nuclear particles: it *meant* that there were no "elementary" particles; each was a bound-state composite of others. Yet for Cook, surrounded by Princeton's field theorists, elementarity remained central. In his 1965 gloss, "The bootstrap embodies a philosophy which supposedly enables one to calculate" various parameters for "*elementary* particles." Cook's work went on to emphasize the lack of agreement between this favorite topic of Chew's

(cit. n. 6), p. 211; and Polkinghorne, "Salesman of Ideas" (cit. n. 55), pp. 24–25. Francis Low similarly remarked that Chew's later efforts seemed "religious" in character: Low interview.

[78] Adler interview; Sam Treiman, "Analyticity in Particle Physics," in *Proceedings of the Eastern Theoretical Physics Conference, October 26–27, 1962,* ed. M. E. Rose (New York: Gordon & Breach, 1963), pp. 127–174, on p. 149; Jackson, "Introduction to Dispersion Relation Techniques" (cit. n. 14), p. 50; and Treiman, "Life in Particle Physics" (cit. n. 73), p. 16. As Treiman himself later remarked of this calculation, "No one but Goldberger and I would have had the effrontery to do what Goldberger and I did": Treiman, "Connection between the Strong and Weak Interactions" (cit. n. 73), p. 388.

[79] Treiman, "Analyticity in Particle Physics," pp. 163, 143. Blankenbecler and Goldberger similarly characterized Chew's bootstrap work as a collection of "interesting speculations" lacking a "physical basis." Making explicit reference to Chew's 1961 La Jolla talk, they dismissed the entire discussion as having merely a "religious nature." R. Blankenbecler and M. L. Goldberger, "Behavior of Scattering Amplitudes at High Energies, Bound States, and Resonances," *Phys. Rev.,* 1962, *126:*766–786, on p. 784. This article was based on Blankenbecler and Goldberger's own talk at the 1961 La Jolla meeting, as indicated in a footnote on p. 766.

and existing experimental data.[80] Even when they turned to the topics most central to Chew's "democratic" campaign, then, Princeton's field theorists clung to the language of "elementary" particles rather than following Chew's group in their vocal break with the "fundamentalists." No matter how completely Chew's democratic vision might have permeated the Berkeley area, that vision did not command a single, unchanging interpretation from physicists further and further removed from Berkeley.

V. CONCLUSIONS: CONDITIONS OF DEMOCRATIC POSSIBILITIES

Perhaps it bears emphasizing that we needn't agree that Chew's many activities were inherently democratic. He maintained Q clearance for several years after his wartime Los Alamos work; such clearance already implied unequal access to research and resources.[81] His "secret seminar" actively discouraged other faculty members from "participating equally." The 1960 Berkeley conference served as a stepping stone for his own S-matrix work, featuring primarily the research of his collaborators. Several friends and former collaborators saw not a democratically minded enrollment campaign but, rather, quasi-religious zealotry in Chew's efforts to interest others in his "democratic" S-matrix program. It is not clear what his thoughts or actions were during the 1964 Free Speech Movement. And so on. Nor is it clear whether Chew was working with a fully articulated or consistent political ideology. While Chew was growing up in Washington, D.C., his father had worked in the Department of Agriculture, which was hardly a politically neutral bureaucracy during the New Deal; perhaps Chew had an ingrained interest, based on his early years, in political issues. Yet whether or not we agree on how "democratic" Chew's efforts ultimately were, or on whether they stemmed from a clear and consistent ideology, one thing remains crucial: Chew and many of his students and colleagues saw his program for strong-interaction particle physics as specifically "democratic"—and as special for that reason.

How, then, are we to interpret the self-proclaimed "democratic" work of Geoffrey Chew? Hard on the heels of his most overtly political activities, Chew began to teach graduate students in a manner consistent with his ideas about democracy. Springboarding from his "open," "noncliquish" conference at Berkeley in 1960, Chew proclaimed at every opportunity that his new physics invited all kinds of participation, from "innocent" students to experimentalists who had "never learned" the rival field-theory formalism. In the midst of these activities, which took Chew from Senate subcommittee hearings to "secret seminars," he began to articulate a new and unprecedented vision of how particles behave and how their interactions should be studied: a successful theory of the strong interactions—unlike quantum field theory, with its "aristocratic" elements—must make no distinctions between the many types of particles. None should be singled out, either for special privileges or for special neglect; all must receive, diagrammatically and mathematically, "equal treatment under the law."

---

[80] L. F. Cook, in Annual Report 1964–1965, pp. 79–80, Dept. Physics, Princeton, Annual Reports (emphasis added). Interestingly, Cook conducted this research with C. Edward Jones, a new Princeton postdoc who had just completed his dissertation under Chew's direction in Berkeley. Each of the courses on particle theory that Kip Thorne attended as a graduate student at Princeton during this period was labeled "elementary particle physics" or "elementary particle theory," further reinforcing this nonbootstrap view of the field. A second center on which it would be interesting to focus, to extend the analysis of the heterogeneity among S-matrix groups, would be Cambridge, England, which hosted a group centered on R. J. Eden, P. J. Landshoff, D. Olive, and J. Polkinghorne; see their textbook, *The Analytic S-Matrix* (New York: Cambridge Univ. Press, 1966), and references therein. Paul Matthews delivered his inaugural lecture at Imperial College in Nov. 1962 with the title "Some Particles Are More Elementary than Others" (London: Imperial College, Nov. 1962).

[81] On this point see esp. Oreskes and Rainger, "Science and Security before the Atomic Bomb" (cit. n. 24).

Were Geoffrey Chew's ideas about particle physics determined by these particular cultural and political ideas, dug up and exposed by the loyalty oath and McCarthyism? The unidirectional, causal story falls short here.[82] To begin with, not every physicist who experienced the rapid postwar political transitions in Berkeley went on to embrace a "democratic" physics in the way that Chew did. More important, Arthur Wightman—a quintessential "fundamentalist" in Geoffrey Chew's eyes for his insistence on retaining quantum field theory—became, with Chew, an active member of the Federation of American Scientists during the 1950s. No crude equations between voting behavior and theory choice will suffice here.

But the failure of such crude equations is just the beginning of our work as historians, not the disappointing end. Just as the straw-man story of sociopolitical determinism fails, so too does the contention that there was simply no connection between Chew's political engagement, pedagogical reforms, and particle physics. At the biographical level, perhaps Chew's frustration with unfair, anti-Communist practices at Berkeley helped to strengthen an iconoclastic resistance to unquestioned authority and a desire to follow out otherwise-unexplored options—he was, after all, already a self-proclaimed "non-conformist," as he wrote in his letter of resignation to Birge in July 1950. Speaking out against university regents and State Department officials, despite the prevailing orthodoxy, could well have prepped Chew for his similarly outspoken challenges to the field-theory orthodoxy some years later.[83]

Even so, iconoclasm alone cannot explain the specific details of Chew's nuclear democracy—its particular elements and the interpretation he gave to them. Substantive links—not just a consistent underdog stance—appear between the three realms of his postwar activities. In particular, the same language recurs again and again ("fundamentalists," "special status," "without favoritism," "equal partners"); models, particles, and collaborators were all "democratized." This was a vocabulary Chew had honed over a decade filled with angst and activism, in front of regents and senators, years before he began to apply it to Feynman diagrams and ρ mesons. Rather than asking "How much did politics affect Chew's physics?" or "What complicated admixture of politics and culture and society interacted in which complicated ways to produce Chew's ideas in physics?" we can

---

[82] Marcello Cini, who worked as a dispersion-relations theorist during the 1950s, has offered a somewhat strained argument that dispersion relations, which promised "utilitarian" correlations among the newly acquired reams of experimental data, took hold because of its fit with "the dominant ideology in the U.S." Marcello Cini, "The History and Ideology of Dispersion Relations: The Pattern of Internal and External Factors in a Paradigmatic Shift," *Fundamenta Scientiae,* 1980, *1*:157–172, on p. 157.

[83] On the question of iconoclasm, Chew's case warrants comparison with that of David Bohm. Just as Bohm was losing his Princeton job for not cooperating with HUAC, he published two long articles questioning the dominance of the standard interpretation of quantum mechanics and offering a new one in its place. Much as Chew would do in 1961, Bohm thus challenged the accepted physics orthodoxy in a manner consistent with his postwar political convictions. Unlike Chew, however, Bohm (at least later, starting ca. 1960) proclaimed that his ideas in physics were actually inspired by his political thinking. See David Bohm, "A Suggested Interpretation of the Quantum Theory in Terms of 'Hidden' Variables: I and II," *Phys. Rev.,* 1952, *85*:166–179, 180–193; Olwell, "Physical Isolation and Marginalization in Physics" (cit. n. 9); James Cushing, *Quantum Mechanics: Historical Contingency and the Copenhagen Hegemony* (Chicago: Univ. Chicago Press, 1994); F. David Peat, *Infinite Potential: The Life and Times of David Bohm* (Reading, Pa.: Addison-Wesley, 1997), Chs. 5, 6, 8; Mullet, "Political Science" (cit. n. 9); and Kojevnikov, "David Bohm and Collective Movement" (cit. n. 9). Wolfgang Pauli, for one, saw fit to merge Bohm's work in physics with his political troubles, describing Bohm's "younger fellow-travellers (mostly 'deterministic' fanaticists, more or less marxistically coloured)." Pauli believed that Bohm had blurred the line between political engagement and physical theorizing. See Pauli to Léon Rosenfeld, 16 Mar. 1952, in Pauli, *Wissenschaftlicher Briefwechsel,* ed. von Meyenn (cit. n. 39), Vol. 4, Pt. 1, pp. 582–583. Cf. Pauli to Abraham Pais, 7 Mar. 1952, in which he labels Bohm a *"Sektenpfaff,"* or, roughly, "cult leader" (*ibid.,* pp. 626–627).

thus build on Chew's curious continuity of language to turn the question around: Why did Chew's work emerge in the form that it did, at the time and in the place that it did? What were the conditions, in other words, that made "nuclear democracy" an intellectual possibility—and indeed not just a "possibility" but the dominant set of techniques for strong-interaction particle physics throughout the 1960s? Why, moreover, was the work subject to so many different, competing interpretations by physicists further and further removed from Chew's immediate group in Berkeley?

When we phrase the question this way, we are no longer driven to squabble over competing ledger sheets, trying to count up "how much political factors determined Chew's physics." Instead, when we ask, "Why then? Why there?" certain plausible connections relate Chew's choices of what to work on and what to lobby for. Feynman diagrams, those staple tools of quantum field theory, appealed to Chew with a usefulness and immediacy far beyond their narrow field-theoretic definitions—an appeal noted by scores of other theorists throughout the 1950s and 1960s as well. Deciding that the diagrams' "true" meaning was not dictated by field theory alone, Chew had some choice in how he would use and interpret them—just as scores of other theorists chose to read and interpret the diagrams in still different ways, toward different calculational ends. As Chew devoted more and more time to political and pedagogical alternatives to what he saw as infringements on equal treatment, perhaps a similarly democratic reading of Feynman diagrams, and of the particles they purported to describe, seemed particularly salient. In other words, perhaps this particular notion of "democracy" and "equal treatment" comprised what Chew mentioned in passing within his 1961 lecture notes as his "general philosophical convictions." That association, at least, would certainly help to explain why it was Chew who produced this fervently "democratic" reading of Feynman diagrams, amid the many other interpretations they received from other theorists. In this sense, "nuclear democracy" seems thoroughly enmeshed with Chew's time and place—it bears the marks of McCarthy-era Berkeley.[84]

These specific resonances, no doubt aided but not uniquely determined by the particular environment of late-1940s and 1960s Berkeley, further help to explain why many other physicists who worked on S-matrix theory at this time did different things with it. To Chew and his many postdocs and students, nuclear democracy and the bootstrap meant that quantum fields and virtual particles simply did not exist and that there was no such thing as an "elementary" particle. Yet many young theorists, such as Sam Treiman's graduate students at Princeton, completed dissertations that treated S-matrix theory neither as a self-consciously "democratic" pursuit nor, much less, as a raised-fist competitor to "aristocratic" quantum field theory—indeed, the bootstrap, at best, seemed to them to offer just one more technique for detailing the properties of the truly "elementary" particles. Pieces of Chew's new calculational machinery were picked up and taught at various places, often bundled with still different theoretical tools and deployed toward different calculational

---

[84] Regarding possible ties between Berkeley's political culture and his physics program, Chew responded recently that he "had never thought about it," though such connections are "a possibility worth considering": Chew, interview with Gordon, Dec. 1997, quoted in Gordon, "Strong Interactions," p. 37. Links between his specific pedagogical efforts and his theoretical approach to particle physics strike a similar chord these days from Chew, who recalls only that "I might have had some idea of that," though "it's hard to recapture the way one was thinking in an earlier period"—which, after all, occurred four decades ago: Geoffrey Chew interview with the author, Berkeley, 10 Feb. 1998. More recently, Chew wrote that an earlier draft of this essay, which made the case for substantive intellectual links between his politics, pedagogy, and physics, was "perceptive and accurate": Chew to Kaiser, 19 Aug. 1999.

ends. Just as Chew appropriated Feynman's diagrams, so too many theorists outside of Berkeley converted Chew's program into their own.

With this tale of democracy in postwar America, we may thus scrutinize and interrogate some complicated, attenuated connections between an intellectual legacy of McCarthyism and reactions to it, and certain ideas and practices within theoretical particle physics. Geoffrey Chew's repeated refrain of "equal participation" and "equal treatment under the law" marked his work—in its various guises and arenas—as a product of his specific time and place. Chew's postwar work thus provides a fertile test case to explore the politico-cultural changes of postwar America and physicists' changing place within it.

Amateur Scientists, the International Geophysical Year, and the Ambitions of Fred Whipple

By W. Patrick McCray

ABSTRACT

The contribution of amateur scientists to the International Geophysical Year (IGY) was substantial, especially in the arena of spotting artificial satellites. This article examines how Fred L. Whipple and his colleagues recruited satellite spotters for Moonwatch, a program for amateur scientists initiated by the Smithsonian Astrophysical Observatory (SAO) in 1956. At the same time, however, the administrators with responsibility for the IGY program closely monitored and managed—sometimes even contested—amateur participation. IGY programs like Moonwatch provided valuable scientific information and gave amateurs opportunities to contribute actively to the research of professional scientists. Moonwatch, which operated until 1975, eventually became the public face of a vast satellite-tracking network that expanded the SAO's global reach and helped further Whipple's professional goals. Understanding amateurs' interactions with the professional science community enables us better to understand the IGY as a phenomenon that enlisted broad participation and transcended traditional boundaries between professional and amateur scientists.

I N 2007 SCIENTISTS WILL MARK the fiftieth anniversary of the International Geophysical Year (IGY), arguably the most ambitious international science project of the twentieth century. Between July 1957 and December 1958, tens of thousands of professional scientists from sixty-seven nations manned hundreds of stations around the globe and researched topics in geodesy and geophysics, atmospheric sciences, oceanography,

Isis 97, no. 4 (December 2006): 634–58.

Research for this paper was supported by funding (grant no. 0323336) from the National Science Foundation—for which David DeVorkin and I are co-investigators—as well as by travel grants from the University of California. Janice Goldblum (National Academy of Sciences) and Pamela Henson (Smithsonian Institution) provided kind archival assistance. Teasel Muir-Harmony, now a graduate student at Notre Dame, deserves special recognition for the time she spent in the Smithsonian Archives helping me collect information. Thomas R. Williams and Jordan D. Marché graciously provided valuable research material from their own collections. Finally, Nicole A. Archambeau, Robert Smith, David DeVorkin, and the *Isis* referees gave my original drafts a careful reading and offered suggestions for improvement and clarification. Credit for resulting improvements are theirs; demerits for remaining errors are mine.

and other fields. Major achievements of the IGY include the detection of the Van Allen radiation belts around the earth, further exploration of Antarctica, and confirmation of a worldwide system of underwater mountains and ridges that helped further scientists' understanding of plate tectonics. Most stunning of all was the appearance of the first artificial satellites, beginning with the 4 October 1957 launch of *Sputnik* by the Soviet Union.[1]

Historians of science, international relations, and space policy have examined the establishment of the IGY, its political context, its role in fostering Big Science, and its legacy in terms of scientific results and international cooperation.[2] The focus of these studies has consistently been on the scientists and administrators who organized, managed, and did research during the IGY. Historians have not fully examined or appreciated the role of amateur scientists—a broad category that I will discuss in the next section—in IGY activities. Amateurs contributed substantially to the IGY, much to the surprise of detractors, especially in the area of satellite tracking. Indeed, the amateur satellite-spotting program of the Smithsonian Astrophysical Observatory (SAO) known as Moonwatch engaged the enthusiastic attention of thousands of amateurs for nearly two decades, continuing long after the IGY ended.

In 1975 the astronomer Fred L. Whipple, recently retired director of the Smithsonian Astrophysical Observatory (SAO), wrote to a worldwide corps of amateur astronomers, "THEY said it couldn't be done! THEY said it couldn't work!" But, as Whipple exulted, "THEY were dead wrong!" And who was the target of Whipple's glee? "They" were the "professional scientists, engineers, and administrators" who, according to Whipple, had claimed that amateurs were incapable of making useful and systematic contributions to the International Geophysical Year.[3]

By focusing most closely on Moonwatch, this essay explores the interaction between amateurs and professional scientists during the IGY and considers how scientific leaders negotiated this relationship. Against this backdrop, we can better understand the contri-

---

[1] Except where otherwise noted, I have followed the popular convention of referring to the first Soviet satellite simply as *Sputnik*. As many historians of science know, the first *Sputnik* involved at least two orbiting bodies— the 22.8-inch satellite itself and the much larger and more visible rocket body that accompanied it into orbit.

[2] The IGY has been treated in varying degrees and ways by historians, political scientists, journalists, and participants. General works on the history and accomplishments of the IGY include Walter Sullivan, *Assault on the Unknown: The International Geophysical Year* (New York: McGraw Hill, 1961); Sydney Chapman, *IGY: Year of Discovery* (Ann Arbor: Univ. Michigan Press, 1959); and J. Tuzo Wilson, *IGY: The Year of the New Moons* (New York: Knopf, 1961). Consideration of the political aspects of the IGY can be found in Harold Bullis, *The Political Legacy of the International Geophysical Year* (Washington, D.C.: U.S. Government Printing Office, 1973), while Allan A. Needell, *Science, Cold War, and the American State: Lloyd V. Berkner and the Balance of Professional Ideals* (Amsterdam: Harwood, 2000), Chs. 11, 12, treats the organization and management of the IGY and the launch of the first satellites. The support of the National Science Foundation for the IGY is discussed in J. Merton England, *A Patron for Pure Science: The National Science Foundation's Formative Years, 1945–57* (Washington, D.C.: National Science Foundation, 1983). The literature on *Sputnik* and the first satellites alone is voluminous. Book-length studies include Robert A. Divine, *The* Sputnik *Challenge: Eisenhower's Response to the Soviet Satellite* (New York: Oxford Univ. Press, 1993); Walter A. McDougall, *The Heavens and the Earth: A Political History of the Space Age* (Baltimore: Johns Hopkins Univ. Press, 1997); and Roger D. Launius, John M. Logsdon, and Robert W. Smith, eds., *Reconsidering* Sputnik: *Forty Years since the Soviet Satellite* (New York: Routledge, 2000). In the last book, Rip Bulkeley's contribution—"The Sputniks and the IGY" (pp. 125–159)—is especially useful, as is Michael J. Neufeld's "Orbiter, Overflight, and the First Satellite: New Light on the Vanguard Decision" (pp. 231–257). Jacob Darwin Hamblin's *Oceanographers and the Cold War: Disciples of Marine Science* (Seattle: Univ. Washington Press, 2005) considers the issue of international cooperation with reference to the IGY. Its influence on Big Science is examined in Robert W. Smith, "Large-Scale Scientific Enterprise," *Encyclopedia of the United States in the Twentieth Century*, ed. Stanley Kutler, Vol. 2 (New York: Scribner's, 1996), pp. 739–765.

[3] Fred Whipple to Moonwatch members, 15 June 1975, Folder "Test Alert #1," Box 45, Moonwatch Papers, RU 255, Smithsonian Institution Archives, Washington, D.C. (hereafter cited as **Moonwatch Papers**).

butions, meaningful as well as frivolous, of amateurs to the IGY. The essay furthers our understanding of the IGY by demonstrating how amateur scientists contributed to formal research programs and how professional scientists and administrators tried to set limits on amateur participation.

How Whipple mediated and organized the participation of amateurs to further his own goals is also explained. Moonwatch became the public face of a satellite-tracking network that expanded the SAO's global reach. Whipple used the satellite-tracking program as the primary means to secure funding for the observatory and to enlarge the SAO's staff in the late 1950s and early 1960s. These resources and the institutional connections they helped create provided a gateway for the observatory to participate in new research opportunities that arose in the early years of space exploration.

Finally, this essay demonstrates that the thousands of amateur scientists who contributed to the IGY were not merely passive collectors of data. Throughout the IGY, amateurs built and refined their equipment, developed new techniques, provided information to the public, and formed local and regional networks to communicate their work. While amateurs carried out research in a number of fields, ranging from aurora and variable star watching to seismology and oceanography, Moonwatch was the IGY's most successful amateur activity, and its impact persisted long after 1958. In fact, the SAO continued Moonwatch until 1975. Understanding the interactions of Moonwatchers with the professional science community enables us better to understand the IGY as a phenomenon that enlisted broad participation and transcended traditional boundaries between professional and amateur scientists.

WHO WAS AN AMATEUR SCIENTIST?

People who participate in amateur science activities like astronomy, birding, and archaeology fall into diverse categories: dabbler, hobbyist, recreation seeker, devotee, and serious amateur.[4] The same imprecise and flexible labels challenge our attempts strictly to separate amateur scientists from their professional counterparts. Historians have devoted considerable attention to the study of amateur scientists, their interaction with professional science communities, and the extent of amateur contributions to research. Amateur astronomers have received an especial amount of attention.[5] Most of this historiography focuses on the

---

[4] These categories appear in several publications by Robert A. Stebbins, including "Avocational Science: The Amateur Routine in Archaeology and Astronomy," *International Journal of Comparative Sociology*, 1980, *21*(1–2):34–48; "Amateur and Professional Astronomers: A Study of Their Interrelationships," *Urban Life*, 1982, *10*:433–454; and *Amateurs, Professionals, and Serious Leisure* (Montreal: McGill-Queen's Univ. Press, 1992). Thomas R. Williams uses a different set of criteria to separate amateur astronomers from what he terms "recreational observers," including "a serious intent to contribute to the advancement of astronomy" that is demonstrated over an extended period of time and involves communicating the results of one's work to others; see Williams, "Criteria for Identifying an Astronomer as an Amateur," in *Stargazers: The Contribution of Amateurs to Astronomy*, ed. S. Dunlop and M. Gerbaldi (Berlin: Springer, 1987), pp. 24–25.

[5] For a representative sample of this extensive literature see Sally Gregory Kohlstedt, "The Nineteenth-Century Amateur Tradition: The Case of the Boston Society of Natural History," in *Science and Its Public: The Changing Relationship*, ed. Gerald Holton and William Blanpied (Boston: Reidel, 1976), pp. 173–190; Susan Leigh Star and James Griesemer, "Institutional Ecology, Translations, and Boundary Objects: Amateurs and Professionals in Berkeley's Museum of Vertebrate Zoology," *Social Studies of Science*, 1989, *19*:387–420; Elizabeth Barnaby Keeney, *The Botanizers: Amateur Scientists in Nineteenth-Century America* (Chapel Hill: Univ. North Carolina Press, 1992); and Mark V. Barrow, *A Passion for Birds: American Ornithology after Audubon* (Princeton, N.J.: Princeton Univ. Press, 1998). Studies that look especially at amateur astronomy include John Lankford, "Amateurs versus Professionals: The Controversy over Telescope Size in Late Victorian Science," *Isis*, 1981, *72*:11–27; Marc Rothenberg, "Organization and Control: Professionals and Amateurs in American Astronomy, 1899–

nineteenth and early twentieth centuries. At that time, scientists were establishing their professional identities and delineating major research disciplines. By the time of the IGY, the traditional tensions historians have noted between professional scientists and amateurs were generally not an issue.

In the mid-twentieth century astronomy enthusiasts—the group most pertinent to this essay—formed a socially complex community. Prior to the start of the IGY, amateur astronomers made multiple attempts to organize themselves into local, regional, and national associations and met with varying success. At the same time, these amateurs engaged in activities that spanned a broad range of interests. Some oriented themselves primarily toward building their own telescopes, and, indeed, there was a renaissance in amateur telescope making in the United States beginning in the 1920s. Other amateur astronomy clubs had roots as civic organizations and encouraged recreational sky watching with an emphasis on entertainment and education. Finally, there were quasi-amateur groups like the American Association of Variable Star Observers and the Association of Lunar and Planetary Observers, which saw themselves as contributing to the research of professional scientists.[6]

Rather than trying strictly to distinguish the categories of "scientist" and "amateur" and to parse the latter group into more specific yet potentially confusing subgroups, in this essay I use the terms "professional scientist" and "amateur scientist" with the recognition that the boundaries between and the identities of these groups were indistinct and that they sometimes overlapped.

Consider, for instance, the case of Arthur S. Leonard, who organized and led a Moon-watch team near Sacramento, California, for years.[7] A professor of agricultural engineering at the University of California, Leonard had considerable professional training that, combined with an amateur's passion for astronomy, made him one of the most active and reliable Moonwatch volunteers. While nominally an "amateur" satellite spotter, Leonard's mathematical prowess and observational precision enabled him to make calculations of satellite orbits during the IGY that rivaled those of his "professional" counterparts for accuracy.

Many people with backgrounds similar to Leonard's took part in Moonwatch or other amateur IGY activities. Their involvement makes it clear that the identity of "amateur scientists" is more nuanced than one might first suspect. Credentials, institutional affiliation, and access to key equipment and other resources all serve as possible ways to separate amateur scientists from professional scientists engaged in the IGY. We may also inquire as to individuals' motives for taking part in the IGY and the degree of commitment they displayed. Moonwatch, especially after *Sputnik* appeared, naturally attracted people intrigued by its Space Age novelty. Their contributions were indeed often "amateurish"—and so perhaps not of much use to scientific research. Yet their participation in Moonwatch or other amateur science programs may have served civic or educational purposes that were valuable in other ways. What stimulated the interest and participation of amateur

1918," *Soc. Stud. Sci.*, 1981, *11*:305–325; Lankford, "Amateurs and Astrophysics: A Neglected Aspect in the Development of a Scientific Specialty," *ibid.*, pp. 275–303; and Ken Willcox, "The Golden Age of Amateur Astronomy," *Mercury*, 1996, *24*(1):32–34.

  6 See Thomas R. Williams, "Getting Organized: A History of Amateur Astronomy in the United States" (Ph.D. diss., Rice Univ., 2000); Williams, "Albert Ingalls and the ATM Movement," *Sky and Telescope*, Feb. 1991, pp. 140–143 (on amateur telescope making); and Leif J. Robinson, "Enterprise at Harvard College Observatory," *Journal for the History of Astronomy*, 1990, *21*:89–103 (on the quasi-amateur groups).

  7 Leonard's Moonwatch activities are recorded in Box 18, Moonwatch Papers.

scientists in the IGY? Did they hope to contribute to and further scientific knowledge, or were they simply caught up in the excitement the IGY stimulated? Did they set and follow established standards and practices? Were they members of a larger community that circulated news and technical tips? Did they meet with other groups of amateur scientists or interact with professional scientists in some fashion? These are key points to consider, and I will return to them in the final section of this essay.

EARLY PROPOSALS FOR AMATEUR PARTICIPATION

The degree to which amateurs contributed to the IGY was made possible, in part, by professional scientists like Whipple who cultivated their participation. Their contributions were, in turn, closely monitored and managed—sometimes even contested—by the administrators and scientists who had responsibility for the IGY program. In the United States, the IGY activities of professionals and amateurs alike were organized under the auspices of the National Academy of Sciences (NAS). Hugh Odishaw, formerly a scientist and administrator from the National Bureau of Standards, directed the IGY in the United States. He was assisted by dozens of scientists and administrators the academy asked to serve on the United States National Committee (USNC) or one of its working groups or technical panels.

In October 1955, only a few months after President Eisenhower announced that the United States would launch a series of satellites during the IGY, Clair L. Strong contacted Odishaw about whether the IGY had a "plan for any amateur participation." Strong, an electrical engineer for Westinghouse and expert tinkerer, wrote a popular column for *Scientific American* called "The Amateur Scientist." He argued that the enthusiasm of amateur scientists (people he defined as making an "avocation of one or another aspect of science") and their "fine grained network" would more than compensate for any shortcomings in their training and equipment.[8] These amateurs—he claimed there were at least a hundred thousand in the United States alone—included cooperative weather observers, aurora and variable star watchers, radio enthusiasts, and even amateur seismologists and particle physicists.

While Odishaw and the other leaders of the U.S. IGY program postponed making any decisions regarding the role of amateurs, Fred L. Whipple was already seriously considering the possibility of enlisting their services. By the conclusion of the IGY, Whipple and his colleagues at the SAO had done more than anyone else to engage the cooperation of amateur scientists and other enthusiasts worldwide.

---

[8] Clair L. Strong to Hugh Odishaw, 3 Oct. 1955, Folder "Volunteer Programs: Amateur Participation and Offers of Cooperation," Series 12.26, Papers of the International Geophysical Year, Archives of the National Academy of Sciences, Washington, D.C. (hereafter cited as **IGY Papers**, with appropriate series). Odishaw, immersed in the details of organizing funding and political support for the IGY program, responded perfunctorily, saying that he and his colleagues were aware of potential contributions from amateurs but had not yet decided how best to proceed: Odishaw to Strong, 10 Oct. 1955, Folder "Volunteer Programs: Amateur Participation and Offers of Cooperation," Series 12.26, IGY Papers. The recommendation that satellites be launched during the IGY was actually made as early as 4 Oct. 1954 by the Special Committee for the IGY (the Comité de l'Année Géophysique Internationale, or CSAGI). The National Security Council's "Draft Statement of Policy on U.S. Scientific Satellite Program," dated 20 May 1955, recommended the creation of a scientific satellite program as part of the IGY as well as the development of satellites for reconnaissance purposes; see John M. Logsdon *et al.*, eds., *Exploring the Unknown: Selected Documents in the History of the U.S. Civil Space Program*, Vol. 1: *Organizing for Exploration* (Washington, D.C.: NASA, 1995), doc. II-10. On the basis of this report, the NSC approved the U.S. satellite program for the IGY on 26 May 1955. However, it was not until 29 July that the Eisenhower administration made a public announcement.

Whipple was born in Red Oak, Iowa, in 1906. While not especially interested in astronomy as a child, he displayed great enthusiasm for investigating how things worked. Childhood hobbies taught Whipple the joy of building things, while chemistry experiments nurtured his interest in science—just as they would for later generations of children. As a graduate student at the University of California, Berkeley, Whipple studied under Armin O. Leuschner, one of the most adept campus political networkers in astronomy of his day.[9] Whipple brought his scientific and organizational expertise to the Harvard College Observatory in 1931, becoming the chair of Harvard's astronomy department eighteen years later. During World War II he was a member of several advisory panels that coordinated activities between the military and the scientific community. Whipple continued to advise the Office of Naval Research, the Air Force, and other agencies after the war, and he learned how to negotiate Washington's bureaucratic channels and to secure institutional funding and personal recognition.

In 1955 Whipple became the director of the Smithsonian Astrophysical Observatory when it moved from Washington, D.C., to Cambridge. Prior to Whipple's appointment, the SAO was a moribund institution with a small staff and a relatively obscure research program devoted almost solely to measuring solar radiation.[10] When Leonard Carmichael, the Smithsonian Institution's Secretary, chose Whipple to lead the SAO into the Space Age, he selected an ambitious and respected scientist who would transform the observatory into the world's largest astronomical institution in less than a decade. Whipple's decision to accept Carmichael's offer was motivated by his belief that Harvard's astronomy program had become "decadent" in the last years of Harlow Shapley's tenure. Whipple also clearly disagreed with Shapley's view that faculty should not receive military funds for research; moreover, Harvard itself forbade classified research on its campus. As he later recounted, "I took the job of directorship so that I could operate this photographic satellite observing program under the aegis of the Smithsonian, rather than Harvard."[11]

Whipple's tenure in Cambridge had served to acquaint him with the capabilities of amateur scientists, especially in the field of astronomy. For years, the Harvard College Observatory had hosted two of the principal organizations for amateur astronomy. The American Association of Variable Star Observers (AAVSO) was founded there in 1911. It became a model of amateur–professional interaction as its growing membership accumulated thousands of estimates of changing stellar magnitude. In addition, *Sky and Telescope*, the premier magazine for amateur sky watchers, was based at the Harvard College Observatory in the 1940s.[12]

Whipple also knew the problems that could result when amateur scientists and other enthusiasts interacted with professionals. For years he witnessed feuds between scientists

[9] My thanks to David H. DeVorkin for clarifying Leuschner's activities for me.

[10] On the SAO's move to Cambridge see Ronald E. Doel, "Redefining a Mission: The Smithsonian Astrophysical Observatory on the Move," *Journal for the History of Astronomy*, 1990, *21*:137–153. On its earlier research program see *ibid.*; and David H. DeVorkin, "Defending a Dream: Charles Greely Abbot's Years at the Smithsonian," *J. Hist. Astron.*, 1990, *21*:121–136.

[11] Fred L. Whipple, 29 Apr. 1977, oral history interview with David DeVorkin, Niels Bohr Library, American Institute of Physics, College Park, Maryland (hereafter cited as **Whipple oral history interview**). See also minutes of meetings of the Harvard College Observatory Council, Box 2, Fred L. Whipple Papers, RU 7431, Smithsonian Institution Archives, Washington, D.C. (hereafter cited as **Whipple Papers, SI**). By June 1960 the SAO employed some 265 people and its budget had risen from about $70,000 annually to well over $4 million. See "Reports on the Astrophysical Observatory," published each year in *Smithsonian Institution Annual Reports*.

[12] For Whipple's views on the capabilities of amateurs see, e.g., Whipple oral history interview, pp. 88–91. Both the AAVSO and *Sky and Telescope* later left the Harvard College Observatory, though they remained in Cambridge.

and other enthusiasts who collected and studied meteorites, and he complained that many of the amateurs were "quarrelsome" and prone to "bickering."[13] He was also on hand for the messy split between the AAVSO and the Harvard College Observatory in 1953, when Donald H. Menzel asked the amateur organization to leave as part of an overall reorganization of the observatory.[14] When it came to the IGY, however, Whipple believed that amateurs—properly managed—could make significant contributions to science.

Whipple's personal interest in satellite tracking stemmed from his enthusiasm for space exploration. His long-standing program in studying meteors as they entered the earth's atmosphere linked his research with that of geophysicists as well as with military interest in the upper atmosphere. For years, Whipple served on the Upper Atmosphere Rocket Research Panel, which guided research using rocket-borne instruments after World War II.[15] In 1954, when the military was considering plans—never realized—for what was known as Project Orbiter, he offered technical advice about how a small satellite could be tracked optically.[16] A year later, after the United States had formally committed itself to launching a satellite, the USNC convened the Technical Panel on the Earth Satellite Program (TPESP) to oversee the scientific and engineering aspects of the project, offer input on institutional relations, and inform the public.

Whipple envisioned a global network of specially designed instruments that could track and photograph satellites. This network, aided by a corps of volunteer satellite spotters and a computation bureau in Cambridge, would establish ephemerides—predictions of where a satellite would be at particular times.[17] The instruments at these stations were eventually designed by James G. Baker and Joseph Nunn and hence known as Baker-Nunn cameras (see Figure 1). Based on a series of super-Schmidt wide-angle telescopes and strategically placed at twelve locations around the globe, the innovative cameras could track rapidly moving targets while simultaneously viewing large swaths of the sky.

From the start, Whipple planned that the professionally manned Baker-Nunn stations would be complemented by teams of dedicated amateurs. Amateur satellite spotters would collect information that would be used to tell the Baker-Nunn stations where to look—an important task, given that scientists working on the U.S. satellite program likened finding a satellite in the sky to finding a golf ball tossed out of a jet plane.[18] Amateur teams would relay the information back to Cambridge, where professional scientists would use it to

---

[13] See, e.g., Whipple to Harrison Brown, 2 May 1949, and Brown to Whipple, 10 May 1949, Folder "B, 1940–1950," Box 3, Fred Whipple Papers, Harvard University Archives, Cambridge, Massachusetts (hereafter cited as **Whipple Papers, HUA**). Brown complained to Whipple of the "amazing hold that unadulterated scientific amateurism" had on the field, something with which Whipple agreed. See also Howard Plotkin, "The Henderson Network versus the Prairie Network: The Dispute between the Smithsonian's National Museum and the Smithsonian Astrophysical Observatory over the Acquisition and Control of Meteorites, 1960–1970," *Journal of the Royal Astronomical Society of Canada*, 1997, *91*(2):32–38.

[14] This episode is described in Robinson, "Enterprise at Harvard College Observatory" (cit. n. 6).

[15] Ronald E. Doel, *Solar System Astronomy in America: Communities, Patronage, and Interdisciplinary Science, 1920–1960* (Cambridge: Cambridge Univ. Press, 1996); and David DeVorkin, *Science with a Vengeance: How the Military Created the U.S. Space Sciences after World War II* (New York: Springer, 1992).

[16] "A Minimum Satellite Vehicle: Based on Components Available from Missile Developments," 15 Sept. 1954 [report by Wernher von Braun]: Logsden *et al.*, eds., *Exploring the Unknown* (cit. n. 8), Vol. 1, doc. II-7. See also Neufeld, "Orbiter, Overflight, and the First Satellite" (cit. n. 2), pp. 235–236.

[17] Similar instruments were employed by the Harvard Meteor Project, based in New Mexico, which Whipple directed. During the IGY, twelve Baker-Nunn cameras were eventually deployed in New Mexico, Florida, Hawaii, South Africa, Australia, Spain, Japan, India, Peru, Iran, Argentina, and Curaçao. The SAO's tracking program is described in E. Nelson Hayes, *Trackers of the Skies* (Cambridge, Mass.: Doyle, 1968); and Eloise Engle and Kenneth H. Drummond, *Sky Rangers: Satellite Tracking around the World* (New York: Day, 1965). Both volumes contain useful information yet lack contextual grounding and a critical perspective.

[18] Alton L. Blakeslee, "Volunteers Organized to Watch for Satellite," *Los Angeles Times*, 9 June 1957, p. A15.

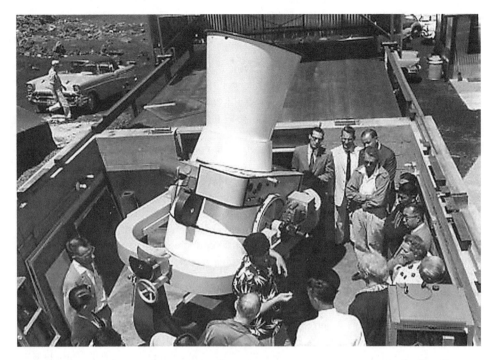

Figure 1. *One of the twelve Baker-Nunn satellite-tracking cameras operated by the Smithsonian Astrophysical Observatory. This one, shown at its dedication on 2 August 1958, was located on the mountain of Haleakala on the Hawaiian island of Maui. Photo courtesy of Dr. Walter Steiger, University of Hawaii.*

generate accurate satellite orbits. At this point, professionals at the Baker-Nunn stations would take over the task of photographing them.

In November 1955 Whipple presented his plan for enlisting amateurs to help track satellites to colleagues on the TPESP. Who did Whipple imagine would be the most likely respondents to his call? Amateur astronomy groups were obvious candidates, and these would in fact become the foundation of Moonwatch. Whipple explicitly identified groups like the American Association of Variable Star Observers and the Astronomical League, "many of whom have attained a high degree of proficiency in the observation of the skies."[19] As Whipple saw it, these amateur organizations had clearly demonstrated their value to professional scientists and were generally well regarded by them.

Participation was not limited to amateur astronomers, however. Whipple called attention to the "extensive network of the Ground Observer Corps," teams of volunteer aircraft spotters that monitored U.S. skies in the 1950s. Whipple's initial plans for Moonwatch bear considerable similarity to Operation Skywatch, a civil defense plan the Truman administration created in 1952 to scan the skies for hostile Soviet aircraft. While these Cold War–era aircraft spotters did not have "the extended sky watching experience of the bona fide amateur astronomer, the potential usefulness of [them] should not be overlooked."[20]

---

[19] "Proposal for the Initiation of an Optical Tracking and Scientific Analysis Program for the U.S. Earth Satellite Program," undated report (likely Nov. 1955), Folder "Project 30.3 Initial Development of Optical Tracking," Series 6.1, IGY Papers.
[20] *Ibid.*

Even the name Whipple chose for the amateur satellite-tracking program in the fall of 1955—the Visual Observer Corps—suggests the affinity between aircraft and satellite spotters in his mind at that point.

Moonwatch ultimately brought together a wide assortment of people, going well beyond vigilant aircraft spotters and amateur scientists conversant with astronomy and telescopes. All of these enthusiasts were organized under the aegis of the Smithsonian Institution, an entity "widely known for its activities in the dissemination of scientific information" and interest in encouraging amateur participation in science. Though he never directly acknowledged it, Whipple's proposal for a global network of satellite spotters harked back to the Smithsonian's first major project. In 1847, Secretary Joseph Henry had called for the collection of weather reports from a network of volunteer observers, a task the Weather Bureau took over when Congress established it in 1870.[21]

Amateur satellite spotters were not just to provide cheap labor for their professional counterparts. Their activities, Whipple's proposal argued, would "attract young people of scientific promise" and foster a "spirit of scientific cooperation." Even at this early stage, Whipple had considered the question of maintaining morale, something that had bedeviled leaders of the Ground Observer Corps. While bonuses like a rewards and recognition program could boost morale, Whipple believed that the "satisfaction of participation in a significant scientific program" would provide sufficient motivation for most amateurs.[22]

In late 1955 the NAS awarded the SAO $50,000 to initiate a program of optical satellite tracking. In the spring of 1956 the SAO received some $3.4 million more to carry out all of the optical tracking of satellites during the IGY.[23] These funds provided Whipple with resources to expand the scope and visibility of the SAO's activities and represented one of the largest grants made to a civilian institution during the IGY.

Spurred in part by growing media attention, amateur scientists sent scores of letters to the National Academy of Sciences.[24] Some correspondence came from high school teachers wanting to encourage student participation in the IGY, while other letters were from individuals offering their companies' services. By January 1956 the NAS had compiled a lengthy list of individuals, professional organizations, and industries that wanted to contribute to the IGY in some fashion. Most of the correspondence came from people interested in some aspect of astronomy or the satellite program.[25]

Odishaw and those managing the IGY were soon challenged with reconciling often conflicting tasks. They needed to respond to the public's mushrooming interest in the IGY while controlling both the participation of amateurs and the enthusiasm of people like Whipple who were eager to enlist and employ them. At the same time, they needed to

[21] *Ibid.* See also Daniel Goldstein, "'Yours for Science': The Smithsonian Institution's Correspondents and the Shape of Scientific Community in Nineteenth-Century America," *Isis*, 1994, *84*:573–599. The Smithsonian officially transferred its volunteer corps to the Weather Bureau in early 1874.

[22] "Proposal for the Initiation of an Optical Tracking and Scientific Analysis Program for the U.S. Earth Satellite Program" (cit. n. 19).

[23] Only $89,500 was initially allocated for the amateur observing program: "Summary of Fiscal Information," 23 Jan. 1957, Folder "Project 30.9 Administration of Visual Observing Programs," Series 6.1, IGY Papers.

[24] In 1954 and 1955 alone, the *New York Times* mentioned the IGY in over a hundred articles. Popular magazines like *Scientific American* and *National Geographic* regularly featured articles on the IGY. See, e.g., Hugh L. Dryden, "The International Geophysical Year: Man's Most Ambitious Study of His Environment," *National Geographic*, Mar. 1956, pp. 285–298; Heinz Haber, "Space Satellites: Tools of Earth Research," *ibid.*, Apr. 1956, pp. 487–509; and James A. Van Allen, "The Artificial Satellite as a Research Instrument," *Scientific American*, Nov. 1956, pp. 41–46.

[25] See "Memorandum for the Files," 19 Jan. 1956; and memo, 14 Mar. 1956: Folder "Volunteer Programs: Amateur Participation and Offers of Cooperation," Series 12.22, IGY Papers.

ensure that professional scientists benefited from amateurs' involvement. Nowhere were these tensions more evident than in debates and discussion about Moonwatch.

WHIPPLE PROMOTES PLORB

As soon as the task of tracking satellites optically had been assigned to the SAO, Whipple began to promote amateur participation. Believing that "the importance of amateur astronomers in the project can hardly be overestimated," he gave a talk to the Institute of Aeronautical Sciences in New York City in January 1956. He explained that the IGY satellite program offered "sky watchers an unparalleled opportunity to be of significant service to science." Amateur scientists were especially needed to spot satellites after their launch and transmit their location, via the SAO, to the Baker-Nunn stations around the world. Amateur participation was also vital for monitoring the death throes of satellites in the moments before they reentered the atmosphere. Finally, in the event that a satellite's radio transmitter failed—not unlikely, in the brand new era of microelectronics—amateur sky watchers might be "the only means of locating it."[26]

Whipple assumed that groups of amateur astronomers would form the initial nuclei around which Moonwatch would coalesce. After publicizing Moonwatch to the amateur astronomy community, he reached out to broader audiences in an effort to alert ordinary citizens that they too could play a part in the IGY. In September 1956 the *Saturday Review* published his article with the catchy title "Moontracking: The New Global Science-Sport." During the IGY, Whipple told readers nationwide, "thousands of men and women of all hues, creeds, and ideas" will work together on a project "so new it does not even have a name." To remedy this deficiency, Whipple suggested "PLORB: the *pl*acing of artificial moons in *orb*its in space." He went on to describe how amateurs everywhere could contribute to what might be "the biggest scientific venture ever shared by the common man."[27] Whipple's promotional campaign paid off, and the SAO received hundreds of letters from people around the world eager to be satellite spotters.

In January 1956 Whipple recruited J. Allen Hynek, a professor of astronomy at Ohio State University, to direct the SAO's entire optical tracking program, including Moonwatch.[28] Born in 1910, Hynek received his bachelor's degree from the University of Chicago before earning his doctorate while working at Yerkes Observatory. During World War II he took leave from his university position to help develop proximity fuse technologies at the Applied Physics Laboratory run by the Johns Hopkins University. After the war Hynek returned to Ohio State, but he remained involved with the lab's research projects using instruments carried by V-2 rockets, work that introduced him to Whipple.

The SAO also asked Armand N. Spitz to generate grassroots enthusiasm among amateur scientists. Spitz was an enthusiastic amateur astronomer and active science popularizer whose Philadelphia company manufactured planetaria for schools and science museums.

---

[26] "Amateurs to Observe Satellites," *Sky Telesc.*, Mar. 1956, p. 203. Whipple's speech was accompanied by an SAO press release on the same date: Folder "Earth Satellite Program, 1956," Box 13, Papers of the Office of the Secretary, RU 50, Smithsonian Institution Archives, Washington, D.C. (hereafter cited as **Office of the Secretary Papers, SI**).

[27] Fred L. Whipple, "Moontracking: The New Global Science-Sport," *Saturday Review*, 1 Sept. 1956, pp. 37–39, on pp. 37, 38. A similar article is Whipple, "Wanted: Spotters for Satellites," *Science Digest*, Dec. 1956, pp. 33–37. His appellation never took hold with the public. One reader complained in verse: "To have this brand new fun celestial/ruined by a word so bestial/PLORB! . . . We need a brand new Keats/To name with grace such stellar feats." Persis Smith, "Watcher of the Skies," *Saturday Rev.*, 5 Jan. 1957, p. 41.

[28] Folder "Hynek, J. Allen," Box 2, Whipple Papers, SI.

He would embark on a series of trips around the United States to help enlist the support of "amateur astronomers and other observers."[29] Whipple, Hynek, and Spitz still needed a name for their program. One idea they considered briefly was SEESAW, as in "I see it . . . I saw it." By the time Spitz began his recruitment travels, however, the three men agreed that "Moonwatch" was the best moniker for a program in which people would be looking for what were, in fact, new moons. By May 1957 Spitz's efforts had paid off: more than sixty teams were registered in the United States, while others were forming in Chile, Japan, South Africa, and over a dozen other countries.

MARKING TERRITORY

While Hynek and Spitz organized teams of amateur satellite spotters around the world, Whipple fended off criticism of the SAO's plans to enlist the participation of amateurs. One skeptic was Homer E. Newell, the coordinator of science programs for the Naval Research Laboratory (NRL), which managed Project Vanguard, the U.S. IGY satellite program effort. The NRL had responsibility for tracking satellites electronically using Minitrack, a global network of specially designed radio facilities. Newell recognized the importance of an optical tracking program but argued that "amateur astronomers cannot be expected to serve this purpose satisfactorily."[30] In an attempt to undercut Whipple's initiative, Newell proposed that the NRL establish its own network of stations employing salaried observers.[31]

The turf war between Whipple and naysayers at the Naval Research Laboratory continued throughout 1956. In sticking to his guns, Whipple ran a certain risk. While the Air Force had taken over support of his meteor research in 1954, the Navy had generously funded his work on meteors and astroballistics for years. In 1956, Whipple expected to receive some $200,000 in contracts and grants from the Department of Defense, and he was obviously concerned about jeopardizing future funding.[32]

In December 1956 the TPESP finally ended the debate when it declined the NRL's proposal to establish its own visual tracking network manned by paid professionals. William H. Pickering, director of the Jet Propulsion Laboratory and a key member of the TPESP, supported this decision, noting that "the use of paid observers might adversely affect the morale of the unpaid volunteer observers of Moonwatch." Whipple concurred—

[29] J. Allen Hynek to Armand N. Spitz, 1 May 1956, Folder "Spitz, Armand," Box 5, Whipple Papers, SI. For information on Spitz see Brent Abbatantuono, "Armand Spitz: Seller of Stars," *Planetarian*, 1995, *24*(1):14–22; and Jordan D. Marché, *Theatres of Time and Space: American Planetaria, 1930–1970* (New Brunswick, N.J.: Rutgers Univ. Press, 2005), Ch. 5.

[30] Homer E. Newell to TPESP, 25 Jan. 1956, Folder "3rd Meeting of TPESP," Series 4.10, IGY Papers. Constance McLaughlin Green and Milton Lomask, *Vanguard: A History* (Washington, D.C.: NASA History Office, 1969), Ch. 8, discusses the various tracking programs that were part of Project Vanguard.

[31] TPESP members predicted that the NRL's radio tracking program had a roughly fifty-fifty chance of working properly the first time: "January 28, 1956 minutes of the 3rd meeting of the TPESP," Folder "3rd Meeting of TPESP," Series 4.10, IGY Papers. The minutes are unclear as to who pointed out the potential unreliability of NRL's Minitrack system, but Whipple is an obvious candidate. The NRL objected to Moonwatch on several likely grounds—its belief that amateurs were not up to the task, its desire to compete for a share of the substantial funding allocated to the SAO for IGY satellite spotting, and perhaps a wish to centralize acquisition and tracking activities in one agency. The fact that Moonwatch was generating favorable publicity for the SAO also should not be overlooked. Green and Lomask, *Vanguard*, notes that Moonwatch received as much if not more publicity than Vanguard. The wide coverage given to Moonwatch in major newspapers throughout 1956 and 1957 supports this assertion.

[32] Whipple to Leonard Carmichael, 7 Oct. 1955, Folder "Whipple, Fred, 1954–55," Box 24, Office of the Secretary Papers, SI.

fostering the interest of amateurs was a "delicate matter of personal sensitivities [for] people who are doing research for nothing. It's rather important to make them feel that it's all worthwhile and their efforts are appreciated." Whipple assured his colleagues that "amateurs will help dependably at least in the early stages [of the IGY] when the glamour is new" and invited staff at the NRL to cooperate with Moonwatch.[33]

By the end of 1956, Whipple had achieved his initial goals of establishing the SAO's claim on the optical tracking of IGY satellites. This was a major coup for the observatory, which was still establishing itself after its move from Washington. The SAO's expansion into satellite tracking also provided a significant window of opportunity for amateur scientists who wished to participate in the IGY.

LIMITS ON AMATEUR PARTICIPATION

Throughout the summer and fall of 1956, the USNC and the SAO attempted to manage amateurs' participation. They preferred that professional scientists participate directly with amateurs, that the amateurs have technical experience, and that their activities be coordinated through official IGY channels, especially overseas. At the same time, the USNC became equally concerned with controlling how the SAO was organizing and promoting amateur participation.

The organizers of the IGY were obliged to consider professional scientists' reactions to amateur participation. Odishaw and other USNC members wanted professional scientists to become involved in amateur programs, believing that this might reinforce the public's image of the IGY as a professional scientific undertaking. Odishaw reminded Hynek, for example, that professional scientists were a valuable resource to be tapped and that "to restrict the [satellite tracking] program to the high precision approach [i.e., the Baker-Nunn network] and 'amateurs' may well kill professional interest abroad." He encouraged Hynek to "stimulate scientists" to help improve the performance of amateur groups that lacked formal scientific training or experience. Odishaw also asked that the SAO's publicity campaign for Moonwatch convey the amateurs' appropriate place. As he told Whipple, "The visual observer has an important, a significant role, but I don't think he should feel his role is bigger than it is." Whipple promised Odishaw that the SAO would carefully articulate how amateur satellite spotters could participate in the IGY.[34]

Whipple had managerial reasons to limit the participation of amateurs in Moonwatch. He recognized that solitary professional astronomers probably had the background and equipment to track satellites with "quite surprisingly good results." However, he was less sanguine about encouraging such activity among "isolated observers who are non-professional. . . . How would we know who was good and who was not?" The program, he insisted, would function best using teams of amateurs; it was not for the "lone wolf" observer.[35]

As amateurs' interest in the IGY grew, the USNC staff worried that they would be

[33] "December 3–4, 1956 minutes of the 9th meeting of the TPESP," Folder "9th Meeting of TPESP," Series 4.10, IGY Papers; and "December 3, 1956 minutes of the Tracking and Computation Working Group," Folder "4th Meeting of Tracking and Computation," Series 4.11, IGY Papers.

[34] Odishaw to Hynek, 26 July 1956; Odishaw to Whipple, 19 May 1956; and Whipple to Odishaw, 24 May 1956: Folder "IGY Office of Information, Volunteer Programs, Moonwatch," Series 12.26, IGY Papers.

[35] Whipple to Hynek, 6 Apr. 1956, Folder "Hynek, Allen," Box 2, Whipple Papers, SI; and *Bulletin for the Visual Observers of Satellites*, no. 1, July 1956, p. 2. The full run of the *Bulletin*—which appeared periodically in *Sky and Telescope*—may be found in Folder 1, Box 61, of the Moonwatch Papers. See also Whipple, "Moontracking" (cit. n. 27), p. 38.

besieged by citizens not only seeking news and information but wanting actually to participate. Odishaw and others at the NAS tried to restrict participation to those with advanced skills. When a writer from *Popular Mechanics* contacted Odishaw for a story on
Moonwatch, his staff's reply emphasized that while contributions from "amateur-
professionals" were encouraged, the scientific programs did not want to be "overwhelmed
with masses of uncalibrated data of various levels of professional worth." Odishaw's office
preferred to steer the reporter away from Hynek to "someone with restraint. . . . In that
way, we might better slant the story . . . instead of inviting all amateurs to come batter
down our doors and overwhelm us with unnecessary data."[36]

Odishaw and his assistant, S. Paul Kramer, were also concerned about the international
ambitions Whipple had for Moonwatch. Whipple had long promoted the possibility of
amateur participation in satellite tracking beyond the borders of the United States. Like
Odishaw, Kramer, who had a background in military intelligence, was keenly sensitive to
public relations. Before and during the IGY, for example, he worked with the U.S. Information Agency to spread the message that it was a civilian, scientific enterprise rather than
a militaristic, engineering endeavor. Kramer feared that the SAO was promoting Moonwatch too quickly. He was especially concerned that the observatory's plans would be too
developed before the Comité de l'Année Géophysique Internationale, the international
group that directed the IGY worldwide, made an official announcement. Premature attempts by the SAO to mobilize amateurs, especially in the United States, might "give the
[U.S.] satellite a nationalistic tone."[37] Kramer insisted that the SAO postpone further public
announcements about the international character of Moonwatch until IGY officials made
a formal statement at their plenary meeting in Barcelona in September 1956.[38] The delay,
however, placed a greater administrative burden on the SAO, leaving it less time to recruit
and coordinate Moonwatchers overseas.

Odishaw and his colleagues also suspected that the SAO was unprepared to manage
Moonwatch. Media gaffes by Spitz did not bolster their confidence. For example, in late
April 1956, while visiting Montreal, Spitz compared the relative scientific merit of the
U.S. satellite program with what the Soviets were doing and predicted that the Soviets
would beat the Americans into space. His statements were picked up by the Associated
Press wire service just as he was about to travel around the United States to promote local
interest in Moonwatch. Joseph Kaplan, the USNC chair, called Spitz's statements "disturbing," while an internal memo labeled Spitz as "one of the two-bit publicity hitchhikers"
who had to be carefully rebuked lest he go to the press and strike a "martyr pose."[39]

[36] Memo on phone conversation originating in Odishaw's office, 22 May 1956, Folder "IGY Office of Information, Volunteer Programs, Moonwatch," Series 12.26, IGY Papers.

[37] S. Paul Kramer to Odishaw, memo, 5 June 1956, Folder "IGY Office of Information, Volunteer Programs,
Moonwatch," Series 12.26, IGY Papers. For expressions of concern over perceptions of military influence on
the U.S. IGY program see Kramer to Odishaw, 18 June 1956, Folder "IGY Office of Information Chron. File
Apr–Jul 1956," Series 12.1, IGY Papers; and Needell, *Science, Cold War, and the American State* (cit. n. 2), pp.
333–336. On Kramer's background and intent see Fae L. Korsmo, "Shaping Up Planet Earth: The International
Geophysical Year (1957–1958) and Communicating Science through Print and Film Media," *Science Communication*, 2004, *261*(2):162–187, esp. p. 170.

[38] Correspondence in the IGY Papers indicates that Whipple and others at the SAO backed off on their
organizing plans somewhat, at least with regard to amateurs outside the United States. As Whipple told Kramer,
"We have not made steps to set up international relations because we have received a firm 'No' until very recently
from your central committee": Whipple to Kramer, 18 July 1956, Folder "IGY Office of Information, Volunteer
Programs, Moonwatch," Series 12.26, IGY Papers.

[39] Joseph Kaplan to Spitz, 9 May 1956 (with attached memo, carbon copies to Whipple and Carmichael),
Folder "IGY Office of Information, Volunteer Programs, Moonwatch," Series 12.26, IGY Papers.

The *Bulletin for Visual Observers of Satellites*, first available in July 1956 and included regularly in issues of *Sky and Telescope*, reflected efforts to control amateur participation in Moonwatch. The first *Bulletin* introduced Moonwatch with no explicit mention of international participation. Instead, it focused on the basic facts of satellite orbits and the U.S. satellite program. Interested parties were directed to contact members of a National Advisory Committee the SAO had formed, not the USNC. As published, the *Bulletin* took a professional tone and employed considerable technical jargon. It attempted to draw a boundary between amateurs with science backgrounds and curiosity seekers stimulated by IGY-related media attention. It emphasized that participants in Moonwatch needed to meet established qualifications, take part in practice sessions, be "interested in science," and serve as part of an organized team of "completely dependable" amateurs.[40] It emphasized, in short, that while Moonwatchers might be amateurs, they were also fortunate participants in an important scientific enterprise.

AMATEURS MOBILIZE FOR ACTION

In September 1956 the *New York Times* ran a series of articles describing how professional and amateur scientists would spot and track IGY satellites. Besides highlighting the valuable role that amateur observers could play in the IGY program, articles such as these provided the public with information about how a typical Moonwatch team would function. Each member would observe with a specially designed "satellite spotter," a short aluminum tube with optics that combined a wide-angle view with modest magnification. As the magazine *Natural History* noted, "Any amateur who has ever made a telescope should be able to assemble the device."[41] Those lacking such experience could order a "Satellite Scope" from the Edmund Scientific Company for $49.50. Because using the telescope for extended periods could give satellite spotters neck pains, amateurs' telescopes were often designed so that users actually looked down and aimed their telescope at a fixed mirror (see Figure 2).

So equipped, groups of observers were to organize themselves into an "optical fence" by positioning themselves along a north–south meridian, in the center of which was a tall pole with a crossbar at the top. Team members would mount their telescopes along this meridian and align them on the pole. Each observer's view, therefore, would take in a different part of the sky, while slightly overlapping those of his neighbors on either side. The cover illustration shows one such group, a Moonwatch team based at a private school for boys in New York, posing in front of the observatory they built themselves.

Critical times for observing satellites were dusk and dawn, when the objects would reflect sunlight and be most visible in the semidark sky. Observers would record the exact moments when the satellite entered their field of view, when it passed the crossbar, and, finally, when it left their field of view. In popular articles on satellite tracking, the SAO recommended that Moonwatch teams use a tape recorder that registered short-wave time signals broadcast by the National Bureau of Standards in Maryland. The machine would also record the voices of spotters calling out prearranged signals. After a satellite was spotted, the team leader would transmit the information to the SAO by telegram or collect

---

[40] *Bulletin for Visual Observers of Satellites*, no. 1, July 1956, insert in *Sky Telesc.*, July 1956.

[41] John T. Kane, "Operation Moonwatch," *Natural History*, Mar. 1957, pp. 126–129. The *New York Times* articles—all authored by staff writer Walter Sullivan—appeared 1–3 Sept. 1956. The 2 Sept. piece was devoted solely to Moonwatch.

Figure 2. Donald Charles, a member of a Moonwatch team based in Walnut Creek, California, with a homemade Moonwatch telescope. Note the school protractor integrated into the mount, as well as the flat mirror in front of the telescope positioned to enable an observer to look down rather than up. The astronomer's hand holds a pushbutton device to mark passage of a satellite through his field of view. Photograph courtesy of Jack Borde.

phone call. Using electronic computers and Moonwatch data, SAO staff would calculate a predicted orbit for the satellite and send it to the Baker-Nunn sites, which would then take over tracking and photographing duties.

Descriptions of Moonwatch in the popular press followed the SAO's lead. They emphasized the need for team-based observing and described Moonwatch teams as well-organized groups primed for action. These articles also stressed that Moonwatch members, who had taken the time to practice and refine their skills, formed a global data-gathering fraternity whose contributions would be a valuable part of the IGY program. As Hynek told one writer, "The amateur astronomy teams of Moonwatch may well be the backbone of the visual tracking assignment."[42]

The SAO was besieged with inquiries from interested people around the world who wanted to take part. In October 1956, to help manage the Moonwatch program more

[42] Kane, "Operation Moonwatch," p. 129. Similar articles appeared throughout 1956 and 1957 in *Popular Science*, *Popular Mechanics*, and other such magazines. The article in *Popular Mechanics* even described how readers could build their own spotting scopes: Richard F. Dempewolff, "How You Can Spot the Satellites," *Popular Mechanics*, Aug. 1957, pp. 72–76, 234, with directions on p. 170.

effectively, the SAO hired Leon Campbell, Jr. Like Whipple, Campbell was well acquainted with the American amateur astronomy community and its potential to contribute to scientific research. His father, Leon Campbell, Sr., had been a staff member of the Harvard College Observatory and an active participant in the AAVSO from 1915 until his death in 1951. The younger Campbell, in addition to being exposed to amateur and professional astronomy communities throughout his life, had professional experience in public relations and journalism.[43] As the SAO's coordinator for Moonwatch stations around the world, Campbell fielded the dozens of requests for IGY and satellite information the SAO received each week and served as a liaison between the professionals at SAO and the amateur teams.

With Campbell's oversight, the SAO began promoting Moonwatch abroad. To assist these efforts, Whipple enlisted the help of Teofilo Tabanera, a scientist in Argentina, to organize teams in South America. Tabanera also translated the *Bulletin for Visual Observers of Satellites* into Spanish and helped distribute it to local teams.[44] By the time *Sputnik* was launched, Moonwatch teams were in place throughout Argentina, Chile, Peru, South Africa, and Australia.

Outside the United States, Japan showed the most enthusiasm for Moonwatch. Newspapers and other companies stimulated amateurs' interest through team sponsorships. By October 1957 Japan fielded over seventy Moonwatch teams, initially coordinated by Masasi Miyadi, a professor at Tokyo Astronomical Observatory. In fact, Miyadi faced such a surge of interest that he lamented that there were "so many amateurs proposing to participate in the work that it is rather difficult to us to qualify them [all]."[45]

Back in Cambridge, the SAO was experiencing Miyadi's organizational dilemmas on a grander scale. The scores of Moonwatch groups that sprang up around the world presented the SAO with a management challenge that brought it into conflict with the Smithsonian Institution's Washington-based administration. Leonard Carmichael, the Smithsonian Secretary, and his staff were especially alarmed that the SAO's "crash program" in satellite tracking had a budget that rivaled the outlays of the entire Smithsonian operation. Moreover, Smithsonian staff in Washington questioned the SAO's management of the globally dispersed Moonwatch program and Baker-Nunn stations.[46] By early 1957, however, the SAO's satellite tracking program had acquired enough funding, institutional momentum, and international publicity that Smithsonian officials could not easily cancel or curtail these efforts.

As more teams joined Moonwatch, the SAO struggled to cultivate adequate public relations while maintaining morale among its amateur groups. Hynek warned Whipple that it was "preposterous" to expect Campbell to "handle such a far-flung operation" without

[43] "Resume of Leon Campbell, Jr.," Folder "Campbell, Leon," Box 1, Whipple Papers, SI.

[44] Folder "Argentina Teofilo Tabanera Correspondence," Box 25, Moonwatch Papers. Tabanera was vice president of the International Astronautical Federation. The Moonwatch papers at the Smithsonian are full of requests from and correspondence with participants and team leaders throughout the world, including Cuba, Egypt, the Philippines, and Iran.

[45] Masasi Miyadi to Whipple, 7 Feb. 1957, Folder "Japan, Correspondence with the Coordinator, 1956–70," Box 34, Moonwatch Papers.

[46] John L. Keddy, Assistant Secretary of the Smithsonian Institution, note for file, 14 Mar. 1956, Folder "Earth Satellite Program, 1956," Box 13, Office of the Secretary Papers, SI. Keddy claimed that he would not be happy until the SAO had established some "centralized administrative control" and charged Whipple and Hynek with "abdicating their jobs." See Carmichael's notes, Folder "August 23, 1957 Meeting," Box 19, Office of the Secretary Papers, SI. Whipple, on the other hand, resented what he perceived as micromanagement from Washington and, after he retired, reflected that at least one of the administrators he had to deal with in this period was "an S.O.B. of the first order": Whipple oral history interview.

additional staff or funding, given the extraordinary amount of public and administrative attention the program was receiving.[47] Nevertheless, the Smithsonian administration continued to exert pressure on the SAO as Whipple's estimates for when the professionally manned Baker-Nunn cameras would be operational proved overly optimistic. This scrutiny only increased after the launch of *Sputnik* and the media blitz that followed.

At the local level, amateur scientists struggled to organize and train their teams before the IGY began. Raising funds to start and equip a team was a fundamental challenge. Sponsorship varied from town to town, as leaders tried with varying degrees of success to solicit donations from businesses to support community teams. In a few cases, national companies got involved. The soft drink company Seven-Up sponsored several teams around the United States, while Beech Aircraft contributed handsomely to erect a permanent station for a team in Wichita, Kansas.

Meanwhile, Moonwatch's five thousand members were a relatively diverse lot, at least in terms of their occupations. As the caption of a cartoon in the *Great Plains Observer*, a newsletter for Midwestern amateur astronomers, noted, "The thing about Moonwatch that intrigues us is the considerable range of society it draws from. Old men shiver, junior high kids romp, local bankers rub elbows with the guy who sweeps out the bank." While a demographic study of Moonwatch would be extremely difficult to do and certainly incomplete, photographic and documentary evidence suggests that Moonwatch primarily engaged the interest of white, middle- to upper-class Americans. This was perhaps not surprising, given that expected costs for a Moonwatch station were about $2,000, equal to several months' salary for most Americans.[48]

Moonwatch appears primarily to have attracted men. Nevertheless, some women were active participants. Two women served on the national advisory committee, women led Moonwatch teams, and many girls participated with high school groups. One Texas team tried actively to recruit Girl Scouts, encouraging their participation with the jingle "Boop de-boop-boop, boop de-boop-boop. We're the girls from the Moonwatch group. We don't squint and we don't blink and we don't close our eyes to think—Our team saw the satellite!"[49]

Some communities used Moonwatch to expand existing amateur science and education programs and connect to programs in other cities. For example, in 1933 Dr. Eldred R. Harrington, a science teacher at Albuquerque High School, started a "Dawn Patrol" program for promising students and science buffs. Vioalle Hefferan, director of the school's

[47] Hynek to Whipple, memo, 8 Feb. 1957, Folder "Earth Satellite Program, 1957," Box 13, Office of the Secretary Papers, SI. *Time, Life,* Disney, and local and national radio and television stations had all contacted the SAO. See Hynek to Whipple, 23 Sept. 1957, Folder "Moonwatch," Box 33, Papers of the Smithsonian Astrophysical Observatory, 1954–1966, RU 188, Smithsonian Institution Archives, Washington, D.C. (hereafter cited as **SAO Papers**). As Hynek pointed out, Campbell was close to the breaking point in his "attempt to turn back the tide of work."

[48] The cartoon was reprinted in the *Bulletin for the Visual Observers of Satellites,* no. 8, Mar. 1958, p. 8. The information on cost comes from "The Moonwatch Program," undated (likely late 1957), Folder "Moonwatch," Box 33, SAO Papers. Information on members' occupations is available for some groups via the membership rolls they provided to the SAO. The Moonwatch team in Alamogordo, New Mexico, e.g., reported some 150 observers, listing 23 senior Boy Scouts, 20 senior Girl Scouts, 30 airmen and officers, 40 "adult engineers, technicians, and scientists," 10 high school science students, 10 "Church, Lions Club, similar adults," and 25 "townsmen and housewives." See E. P. Martz to Leon Campbell, 21 Sept. 1957, Folder "Alamogordo, NM," Box 4, Moonwatch Papers.

[49] Martz to Campbell, 27 Apr. 1958, Folder "Alamogordo, NM," Box 4, Moonwatch Papers. Some of the tropes identified in Kristen Haring, "The 'Freer Men' of Ham Radio: How a Technical Hobby Provided Social and Spatial Distance," *Technology and Culture,* 2003, *44:*734–761, can also be seen in the correspondence pertaining to Moonwatch, although the latter appears to have been more inclusive in terms of gender.

Astronomy Club, expanded the Dawn Patrol's offerings by forming a Moonwatch team in late 1956 with sponsorship from a local bank. During the IGY, Hefferan was one of the most active Moonwatch leaders, and her team of high school students established an impressive record for accurate satellite spotting.[50]

Moonwatch was not the only opportunity for amateur scientists to contribute to the IGY. Despite its expressed pessimism about the abilities of amateurs, the Naval Research Laboratory organized a parallel effort called Project Moonbeam that enlisted ham radio operators to record the passing of satellites.[51] In the Soviet Union, scientists and government officials established similar amateur tracking efforts using both radio and optical means. The Soviets encountered many of the same challenges the SAO faced, such as organizing volunteers, obtaining proper equipment, and training teams.[52] While satellite spotting was the primary way in which amateurs chose to take part in the IGY, the USNC eventually identified other activities in which volunteers could participate. Throughout the IGY, Clair Strong's monthly column in *Scientific American* described a whole range of projects amateurs were carrying out.[53]

Articles in popular American magazines, combined with the SAO's proselytizing, helped generate the belief among amateurs that they could usefully participate in the IGY. As the start of the IGY neared, ordinary citizens and curiosity seekers joined hundreds of amateurs with experience in astronomy and other fields who wished to make a genuine contribution to IGY research and perhaps gain recognition for their activities in the process. The abilities of Moonwatch teams and the SAO's management acumen would be tested in the hurly-burly weeks following the launch of *Sputnik*.

THE MONTH OF NEW MOONS

The first test arrived early for Vioalle Hefferan and her Albuquerque team. Hefferan returned to her apartment in the afternoon of 4 October 1957 to find the phone ringing. It

---

[50] Albert Q. Maisel, "'Doc' Harrington's Dawn Patrol of Young Scientists," *Reader's Digest*, Nov. 1956, pp. 142–146. Correspondence and other materials pertaining to Hefferan's team are located in Box 4, Moonwatch Papers.

[51] "Ham Participation in IGY," *Radio and TV News*, Jan. 1958, pp. 8, 142. Project Moonbeam was nowhere near as well organized or successful as Moonwatch, in part owing to the much greater cost of establishing an adequate radio receiving setup. Smaller amateur tracking efforts appeared during the IGY—e.g., Phototrack, a program organized by the Society of Photographic Scientists and Engineers and based in Washington, D.C. Phototrack was a modestly funded operation whose goal was to track as well as photograph passing satellites; it never competed with Moonwatch. My thanks to Dr. Victor Slabinski of the United States Naval Observatory for sharing documents in his possession regarding Phototrack; see also "Operation Phototrack," *Sky Telesc.*, June 1958, p. 387.

[52] See, e.g., A. G. Massevitch, "Optical and Radio Tracking of Satellites (and Interplanetary Probes)," in *Space Age Astronomy: Proceedings of an International Symposium Held August 7–9, 1961, at the California Institute of Technology in Conjunction with the 11th General Assembly of the International Astronomical Union*, ed. Armin J. Deutsch and Wolfgang Klemperer (New York: Academic, 1962), p. 127. The USSR also organized its large network of amateur radio operators through its Ministry of Communication, an effort described in Rip Bulkeley, "Harbingers of *Sputnik*: The Amateur Radio Preparations in the Soviet Union," *History and Technology*, 1999, *16*:67–102. Comparatively little is known about the involvement of Soviet amateurs in satellite tracking, although Whipple once quipped that he suspected "there was more compulsion on the [Soviet] observers to get out early in the morning and late at night and make observations [than in the U.S.]": "Introductory Remarks," 18 June 1960, Folder "STP-Moonwatch," Box 117, SAO Papers.

[53] Similar pieces appeared in other magazines like *Sky and Telescope*. While most of these articles were written for adults, the USNC also provided information for teenagers in the 25 Oct. 1957 issue of *Senior Scholastic*, the entire issue of which was devoted to the IGY. Moonwatch, Project Moonbeam, and aurora watching were activities where "the teenager may make his greatest contribution" (p. 21). See also Kramer to Strong, 8 Feb. 1957, Folder "IGY Office of Information, Volunteer Programs, Moonwatch," Series 12.26, IGY Papers.

was less than an hour after the SAO had received word that *Sputnik* was in orbit, and the voice from Cambridge asked if she could have her Moonwatch team ready to observe at twilight. Hefferan quickly called her students—many of whom had to cancel dates for the homecoming game—and convened them for an evening of sky scanning. Their prompt response resulted, in part, from the extensive training they had done with Hefferan. This included spotting pebbles tossed over the crossbar of their mast, registering the flight of moths, and participating in national Moonwatch alerts carried out with the cooperation of the Civil Air Patrol. Despite their preparations, that night they spotted neither the satellite nor the more visible rocket body that had boosted it into orbit. Unbeknownst to them at the time, these objects were not yet visible from their location in New Mexico, and Hefferan's group did not spot one of the orbiting bodies until 19 October. The sight elated her team of high school students, who "swaggered a bit" in the school hallways afterward.[54]

Whipple and Campbell were both in Washington, D.C., on 4 October, the former attending a meeting of the USNC. Members of the committee grilled Whipple about the lagging schedule for the construction of the Baker-Nunn cameras needed to equip the SAO's network of satellite-tracking stations. By *Sputnik*'s launch, none had yet been deployed, and the first camera system was still being tested at the Boller & Chivens factory in Pasadena.[55]

Because of production delays, when *Sputnik* went into orbit the Moonwatch teams were the only means available to track satellites optically. Rather than supplementing the professionally manned stations, as planned, amateur teams quickly became an essential stopgap, fulfilling a task the SAO never intended for them. By all accounts, the amateur scientists in the Moonwatch program acquitted themselves admirably. Teams in Australia made the first confirmed *Sputnik* observations on 8 October, eleven days before the first Baker-Nunn camera—hastily assembled and mounted in Pasadena—took photographs of the orbiting rocket body. A Moonwatch team in New Haven, Connecticut, made the first confirmed sighting in the United States on 10 October.

The launch of *Sputnik* put extraordinary demands on Moonwatchers, who found themselves not only watching for satellites but also responding to a flood of inquiries from citizens in their communities. Not surprisingly, even more people wrote the SAO with requests to join Moonwatch. The observatory's staff was further taxed as reporters from around the world descended on Cambridge, eager for information. Whipple and Hynek found themselves on the front lines of a media blitz, obliged to answer questions—serious and silly—about *Sputnik*'s significance. When a major national newspaper printed off-the-cuff remarks from SAO staff members predicting that the Soviets would soon land a rocket on the moon, Leonard Carmichael insisted that the SAO route all communications regarding satellites through the main Smithsonian office in Washington. Throughout the IGY, Smithsonian administrators monitored SAO staff for any comments that might provoke similar "grave misunderstandings" in the politically charged atmosphere.[56]

The success of Moonwatch, ironically, somewhat embarrassed the professional science community. It drew attention to the fact that the IGY satellite program depended on the contributions of amateurs because of the failure to have professionally manned optical and

[54] "Story of the Albuquerque High School Moonwatch Team," Aug. 1958, Folder "Team Histories, A–F," Box 43, Moonwatch Papers. On 26 Nov. Hefferan's team spotted the *Sputnik* satellite itself.

[55] "Minutes of the Twelfth Meeting of the TPESP," 3 Oct. 1957, Series 4.10, IGY Papers.

[56] Carmichael to Whipple, telegram, 11 Nov. 1957, Folder "Earth Satellite Program 1957," Box 13, Office of the Secretary Papers, SI. Whipple and Hynek were featured in the 21 Oct. 1957 issue of *Life*.

radio tracking stations operating at full effectiveness when *Sputnik* went into orbit.[57] As pressure for news and data mounted, the USNC and the Smithsonian scrutinized the SAO's management of the entire satellite-tracking program, calling it a "very un-professional looking operation" and a "confused circus."[58]

At hastily convened meetings in October and November, TPESP scientists like William Pickering challenged Whipple's request for an additional $200,000 to support Moonwatch, a plea necessitated by the sudden spike in the program's expenses. Hynek reported that two dozen new teams were waiting to be registered in the United States, while "99 percent of the inquiries we get at SAO have to do with Moonwatch." It would be "a most inauspicious move" to ignore these people or cancel Moonwatch, especially "as this is the public's chance to get in on the act." Athelstan Spilhaus, a geophysicist from the University of Minnesota and USNC member, came to Moonwatch's defense: "This is the one program," he said, "where for comparatively little money you can get the ordinary person to play a part in IGY. . . . The stimulus that this thing gives at a fairly small cost is very considerable."[59] The USNC approved Whipple's request for additional funds, with the caveat that it would send a representative to Cambridge to offer suggestions on improving operations there.

Despite criticism about the SAO's management of its satellite-tracking program, Whipple and his colleagues not only kept the amateur program going but expanded it throughout the IGY. Moreover, in the emotionally and politically charged opening weeks of the Space Age, amateurs had a clear advantage over the community of professional scientists. The network of amateur scientists, after all, was the only global system capable of providing crucial visual tracking information regarding the world's first satellites. Moonwatchers around the world, despite the low expectations many professional scientists initially had, found themselves an essential component of the IGY's professional research program.

WAS HISTORY "MADE AT A SMALL TELESCOPE"?

In the summer of 1957, Fred Whipple helped draft an article for *National Geographic* about Moonwatch and the opportunities for amateur participation. While the article was eclipsed by rapidly unfolding events in the autumn of 1957 and never appeared, it predicted that amateurs would take part in "history being made at a small telescope."[60] Was Whipple's boast accurate? What did amateur scientists contribute to the work of their professional colleagues?

The number of Moonwatch teams peaked during 1958, when some 230 teams around the world were formally registered with the SAO. Of the 128 teams registered in the United

---

[57] In fairness to the SAO, it should be noted that the radio tracking program run by the Naval Research Laboratory was also incomplete, while its amateur-based Project Moonbeam was slow to yield useful information.

[58] "Minutes of the Fourteenth Meeting of the TPESP," 6 Nov. 1957, pp. 149, 151, Series 4.10, IGY Papers. This concern is reflected in the fact that, by April 1958, the Smithsonian Institution, acting on a recommendation from the National Academy of Sciences, commissioned an "Administrative Analysis of the Optical Satellite Tracking Program." While primarily directed toward the implementation of the Baker-Nunn system, the analysis by a Boston management firm referred to the SAO's "unstable administrative situation," which was "in danger of developing a 'crash psychosis'": Folder "Earth Satellite Program 1958," Box 14, Office of the Secretary Papers, SI.

[59] "Minutes of the Thirteenth Meeting of the TPESP," 22 Oct. 1957 (Spilhaus's defense of Moonwatch is on p. 92); and "Minutes of the Fourteenth Meeting of the TPESP," 6 Nov. 1957, pp. 137 ("99 percent of inquiries"), 138 ("most inauspicious move"): Series 4.10, IGY Papers.

[60] Draft article, n.d. (likely Aug. 1957), unpublished, Folder "Whipple, Fred 1957," Box 24, Office of the Secretary Papers, SI.

States, over 80 percent held productive observing sessions (1,659 in all) that yielded some 3,500 "scientifically useful observations."[61] The contributions made by amateurs during the IGY cannot be measured solely in terms of numbers. The SAO's Baker-Nunn network did not become fully operational until July 1958; Moonwatch's amateurs provided a valuable and cost-effective optical tracking system in the interim. Amateur observations helped professional scientists predict the passage of satellites launched during the IGY, an especially useful function when the batteries on *Sputnik I* and *Sputnik II* failed and radio transmissions ceased. Moonwatchers made the first sightings of *Explorer I*, the first U.S. satellite, and participated in the "death watch" of *Sputnik II* as it reentered the atmosphere in April 1958. The plethora of Moonwatch observations helped the SAO refine models of the upper atmosphere and the shape of the earth. Even more important, Moonwatch teams in the United States and Australia located "lost satellites" when professional predictions were flawed and other tracking systems could not find them.

Many professional scientists were initially skeptical about enlisting the help of amateurs for the IGY. Their reluctance at first stemmed from the perceived need to maintain boundaries between the research efforts of amateurs and professionals. Once amateur participation had proved useful, attention focused on how such a "polyglot assortment of some thousands of men and women" could best be managed and coordinated. While President Eisenhower thanked "the hard-working Moonwatch teams" for their contributions, some amateurs were perhaps even more pleased when the National Academy of Sciences formally acknowledged the "importance of volunteer programs" and began to report their activities in the *IGY Bulletin*.[62]

Just as substantial as amateurs' contributions to IGY research is the influence Moonwatch had on communities of amateur scientists. During the IGY, amateur scientists did not perform solely as passive observers, watching the night skies and reporting their findings to the SAO and other institutions. Instead, they actively and enthusiastically pursued a range of scientific programs. Besides tracking satellites, dedicated amateurs developed innovative ways of making seismic observations, counted meteors using radio-wave reflection, and detected solar flares.[63] The IGY presented amateurs with opportunities to strengthen their social networks, enlist new members, and embark on new areas of study. In the broadest sense, Moonwatch helped unite thousands of amateur scientists from all around the globe for a mission that had important scientific, cultural, and political dimensions.

Moonwatchers were especially creative in taking the initiative to modify and improve their equipment and observing techniques.[64] Amateur satellite spotters communicated the results of their work to other groups as well as to professional scientists. Many Moonwatch

[61] *Bulletin for the Visual Observers of Satellites*, no. 9, July 1958, p. 8.

[62] Daniel Lang, "Earth Satellite No. 1," *New Yorker*, 11 May 1957, p. 116; Dwight D. Eisenhower to Carmichael, 1 Feb. 1958, telegram (with copies to the SAO), Folder "Earth Satellite Program, 1958," Box 14, Office of the Secretary Papers, SI; and *IGY Bulletin*, no. 6, Dec. 1957. The full run of the *Bulletin* is in the IGY Papers.

[63] These ideas appeared in C. L. Strong's "The Amateur Scientist" column in *Scientific American* during and after the IGY; see, e.g., articles in July 1957, January 1958, July 1958, October 1958, September 1960, and January 1961.

[64] Some Moonwatchers, e.g., decided to abandon the use of the vertical mast with crossbar and opted instead for the more challenging task of locating the satellite's position in the sky using star maps. Those Moonwatch groups with access to local sponsors built elaborate observing stations so their members could make observations in greater comfort. Many teams also used local planetaria to get observers acquainted with the night sky or took advantage of "satellite simulators" that the SAO sent around the country to help train sky watchers. These examples are drawn from various issues of the *Bulletin for Visual Observers of Satellites*.

teams in the United States organized regional meetings to share tips for observing satellites more effectively and published newsletters describing their activities.[65] With help from the SAO, they established accepted community standards of work, recognized those groups that performed especially well, and encouraged others to try harder. Some Moonwatchers also established connections with other amateur groups more interested in activities like ham radio tracking and satellite photography.

When the IGY ended, interest in satellite spotting among the general public lapsed. For the overworked SAO staff, this was something of a relief.[66] As less dedicated amateurs—"joy riders," as one team leader referred to them—dropped out, the program's character changed. The SAO refashioned Moonwatch to make use of fewer teams composed of better trained and more committed amateurs who contributed increasingly precise data for satellite tracking. The SAO adapted to the needs and wishes of the "hard cores" who remained and gave Moonwatch teams more challenging assignments, such as locating extremely faint satellites and improving the precision of their observations.[67]

Did programs like Moonwatch have any lasting effect on communities of amateur scientists after the IGY, or were they simply curiosities, a fad of the early Space Age? After *Sputnik*, amateur science clubs experienced a surge in membership and there was a renewed interest in "do-it-yourself" science activities such as telescope making. Meanwhile, companies that made telescopes and science kits reported growth in sales.[68] The serious observers who remained in Moonwatch after 1958 continued to develop ever more sophisticated techniques; eventually their work rivaled that of the professionally manned Baker-Nunn stations. Throughout the 1960s, these two branches of the SAO's tracking network shared data and, in a few cases, personnel. These collaborations further blurred the divide between amateurs and professionals.[69]

Some teenagers parlayed the experience they gained as amateur participants in the IGY and Moonwatch into long-term careers in science, especially astronomy.[70] One young man from the Albuquerque Moonwatch team credited Moonwatch for his prizewinning entry in a national science fair, admittance into MIT, and eventual career as a professional astronomer and university professor.[71] When the SAO discontinued Moonwatch in 1975, another participant eulogized that "[it was] my greatest single contribution to life in my 42 years."[72]

---

[65] Series 10 of Record Unit 255 at the Smithsonian Institution Archives has a large collection of these materials.

[66] As early as the spring of 1958, the SAO had stopped trying to recruit new members. Campbell noted that his office was already "simply swamped by the detail work and cannot cope": Campbell to Whipple, memo, 1 Apr. 1958, Folder "Moonwatch Program General," Box 35, SAO Papers.

[67] "Summary of Consultation with Arthur S. Leonard and Dr. Armand N. Spitz, January 19–21, 1959," Folder "Moonwatch Correspondence 1959," Box 125, SAO Papers.

[68] Donald G. Cooley, "Astronomical Number of Sky Watchers," *New York Times Magazine*, 26 Jan. 1958, p. 21; and Williams, "Getting Organized" (cit. n. 6), Ch. 12.

[69] One satellite observer, Russell Eberst of Scotland, produced extremely accurate and efficient observations—the 31 July 1972 *Moonwatch Newsletter* (the full run of newsletters can be found in Boxes 61–63 of the Moonwatch Papers) reported that Eberst once made 127 observations of some 72 transits of 40 different satellites in a single night. He also recorded over 40,000 observations during his time as a Moonwatch member.

[70] Several of these people were identified during the course of this research, in part through a query David DeVorkin and I circulated in the *Bulletin of the American Astronomical Society* in 2004. Among those who were active in Moonwatch as youngsters is Stephen P. Maran, an astronomer and longtime press officer for the AAS, who was a member of the Junior Astronomy Club of New York during the IGY. The late James P. Westphal, a noted astronomer and instrument designer at Caltech, is another example.

[71] Personal communication between Joel Weisberg and W. Patrick McCray, 3 Nov. 2005. Numerous other stories and examples exist; they will be presented in more detail in the book-length study of Moonwatch I am preparing.

[72] William Griffin to Albert Werner, 15 June 1975, Folder "G," Box 2, Moonwatch Papers.

The recognition of what amateurs could contribute to professional scientists extended to other fields of IGY-related research. As noted, Clair Strong's articles for *Scientific American* described a whole range of amateur efforts in seismology, meteorology, and amateur rocketry. Relatively little attention has been paid to amateur contributions in these areas, and an exploration of how and to what extent these groups of amateurs interacted with professional scientists would be worthwhile.

Amateur scientists' participation in the IGY also suggests that there are still issues to consider with regard to the historiography of amateur–professional relations. As noted earlier, much of the work by historians has focused on the nineteenth and early twentieth centuries, when the various disciplines of science were still professionalizing. In astronomy, at least, the roles of amateurs and professionals were not firmly fixed. Indeed, some tensions remained between the two groups as they negotiated their relations and determined how they could derive benefit from each other. Moreover, the story of Moonwatch told here focuses largely on the United States. Amateurs also formed scores of teams overseas. Japan alone had more than seventy at the peak of activity, many of which were more closely connected to professional organizations than those in the United States, while the Soviet Union operated a program similar to Moonwatch. A broader global perspective could yield additional insights into amateur–professional relations in other countries as well as, perhaps, international relations between scientists. Professional scientists and journalists hailed the IGY as a model of international cooperation in science.[73] Did this ideal also extend to amateur scientists? If so, how was such cooperation achieved?

Moonwatch stands out among IGY activities because of its scope, scale of organization, and relative standardization. All around the world, amateurs built or bought similar equipment, developed standard observing techniques, and mobilized for the common purpose of spotting satellites. This standardization of purpose and practice was, of course, enabled by the very nature of artificial satellites. Unlike a seismic event or a meteor shower, which only professionals and amateurs in a particular locale could witness, satellites that circled the entire earth were objects that everyone could watch for and study.

Are there any historical analogues to Moonwatch in fields other than amateur astronomy? Since May 1999, millions of people have participated in SETI@home. This international effort harnesses their inactive personal computers to analyze radio telescope data for signs of extraterrestrial communications. While the venture is an intriguing social experiment and quite possibly (according to the *Guinness Book of World Records*) the largest single computation to date, it is hard to interpret SETI@home, which runs as a background program on participants' computers and requires little specialized skill or knowledge, as an example of active participation in amateur science.[74]

Another historical analogue might be the annual Christmas Bird Count (CBC), an event initiated in 1900 and carried out today under the supervision of the National Audubon Society. This annual survey provides useful information about avian species status and distribution for ornithologists and wildlife managers. Over fifty thousand people participate in this "citizen science" event each year, providing systematic information about bird

---

[73] Lloyd Viel Berkner, "The International Geophysical Year, 1957–1958: A Pattern for International Cooperation in Research," *Proceedings of the American Philosophical Society*, 1957, *101*(2):159–163; and Walter Sullivan, "The IGY—Scientific Alliance in a Divided World," *Bulletin of the Atomic Scientists*, 1958, 2:68–72.

[74] SETI@home's mission is described on the organization's web site: http://setiathome.ssl.berkeley.edu/ (accessed 30 Mar. 2006). Within astronomy itself, the best analogue obviously is the nearly century-old American Association of Variable Star Observers.

populations across wide swaths of North and South America.[75] The CBC, however, lasts only a few weeks each year, while people in each locale record data for only one day; Moonwatchers stayed active for months and years at a time.

Perhaps the most salient comparison is the extensive record-keeping activities of amateur meteorologists. In 1970, for example, amateurs formed the Climatological Observers Link, the "enthusiasts' weather observer network for the United Kingdom," which shares data with professional scientists.[76] But, again, the scale and international membership of groups like these do not compare with the scope of Moonwatch.

One conclusion is that the IGY presented a unique opportunity for amateurs to interact with the professional science community. Emerging as it did in the context and with the support of a massive global scientific enterprise, the IGY provided amateur scientists with a mission—one that, to the satisfaction of Fred Whipple, outlasted the IGY itself. Years after Moonwatch ended, a representative of the SAO cited the program as a model of amateur–professional collaboration, indicating that it helped change the perception of what amateurs could contribute to professional science.[77]

During the IGY, amateur scientists were not the only individuals who benefited. Fred Whipple, for example, used IGY funding of the SAO's satellite-tracking program to initiate other large-scale projects. Satellite tracking also enabled the SAO to de-emphasize its long-running but moribund program of solar constant research while simultaneously connecting the observatory to new federal and military patrons. All of this was done while fulfilling the Smithsonian Institution's traditional mission of "aiding the increase and diffusion of knowledge among men."[78]

Once the IGY ended, Whipple capitalized on Moonwatch's popularity and the eventual success of the SAO's Baker-Nunn stations to secure years of support for satellite tracking from the newly formed National Aeronautics and Space Administration (NASA). NASA and military contracts for satellite tracking, worth millions of dollars annually, were the SAO's largest source of revenue in the early 1960s. They enabled Whipple to oversee a rapid expansion of the SAO into promising new areas of research such as space studies, planetary science, and astrophysics while strengthening its expertise in meteoritic and cometary studies.[79] These endeavors were linked to Whipple's own interests in the upper atmosphere and meteorites and greatly expanded his observatory's research networks.[80] In the 1960s, the SAO adopted the Moonwatch model for other programs, such as the Center for Short-Lived Phenomena, which monitored the earth for new natural events.[81]

[75] Barrow, *Passion for Birds* (cit. n. 5), pp. 167–169, discusses the role of the ornithologist Frank Chapman in establishing the Christmas Bird Count. Statistical and historical information about the CBC is available on the National Audubon Society's web site: http://www.audubon.org/bird/cbc/index.html (accessed 28 Mar. 2006).

[76] Information comes from the COL's web site: http://www.met.rdg.ac.uk/~brugge/col.html (accessed 30 Mar. 2006).

[77] James Cornell, "The Moonwatch Program: A Model for Amateur Contributions to the ISY," in *Stargazers*, ed. Dunlop and Gerbaldi (cit. n. 4), pp. 181–182. Articles and conferences about the resurgence of amateur–professional collaboration in astronomy support this observation. See, e.g., Jeffrey Kluger and David Bjerklie, "Calling All Amateurs," *Time*, 11 Aug. 1997, p. 68.

[78] See the Smithsonian Institution's history page: http://www.si.edu/about/history.htm.

[79] E.g., in terms of personnel, the SAO hired over 120 new staff between 1959 and 1961, according to "Reports on the Astrophysical Observatory" in *Smithsonian Institution Annual Reports*.

[80] On Project Celescope, another program Whipple and the SAO initiated to expand the observatory's research and funding profile, see David H. DeVorkin, "SAO during the Whipple Years: The Origins of Project Celescope," in *The New Astronomy: Opening the Electromagnetic Window and Expanding Our View of Planet Earth*, ed. Wayne Orchiston (New York: Springer, 2005), pp. 1–20. My thanks to DeVorkin for providing me with an advance draft.

[81] The Center for Short-Lived Phenomena was established and based at the SAO in 1968 and run by former

During the IGY, amateurs demonstrated that they could make a meaningful contribution to one of the largest science enterprises in history. For several months, their participation proved essential to the success of the IGY's satellite program. Leon Campbell, Jr.'s, assessment of Moonwatch—that "probably no organization of laymen in all history has contributed so valuably to a scientific program"—may be an overstatement that reflected his fondness for amateur science programs.[82] Nevertheless, amateur participation created favorable publicity for the Smithsonian and the IGY, while Moonwatch was the public manifestation of Whipple's emerging empire of research and data collection stations. Just as important, the IGY and programs like Moonwatch provided opportunities for amateurs to earn respect from their professional counterparts and contribute both to a prominent Big Science endeavor and to the opening of the Space Age.

Baker-Nunn station manager Robert Citron. Another example is the Prairie Network, an SAO sky-monitoring program that amateurs were encouraged to contribute to.

[82] "Summary of Moonwatch Operations for 1960," 1 Mar. 1961, Folder "Moonwatch," Box 33, SAO Papers.

"Industrial Versailles"

Eero Saarinen's Corporate Campuses for GM, IBM, and AT&T

By Scott G. Knowles and Stuart W. Leslie

ABSTRACT

Eero Saarinen may be a familiar name to architectural historians for his designs for Dulles Airport, the St. Louis Arch, and other late modernist landmarks. Yet his biggest commissions were for corporate research laboratories for General Motors, IBM, and Bell Laboratories. In 1951 *Fortune* sent a photographer to document GM's sprawling "research campus," just beginning to take shape in suburban Detroit. The photographs capture what the editors called "a new and serene integration" of modern architecture and modern science and engineering. The GM Technical Center (1956), the IBM Thomas Watson Research Center (1961), and Bell Laboratories at Holmdel (1962) symbolized a postwar ideology of corporate research that emphasized basic research and took the university as the appropriate model for organizing science. But as the people who worked in and managed these laboratories over the following decades would learn the hard way, R&D, in the sense of turning scientific inquiry into product and profit, does not necessarily thrive in an "Industrial Versailles."

Since the war, there has been a phenomenal uprising of research laboratories. Nearly every major corporation, and many a minor one, has launched into building a shining new sanctum of centralized research. Nearly all have followed a pattern of ultramodern decor on park like acres, far from the pressures and distractions of production departments or corporate offices.
—Fortune *(1951)*

The Finnish-born architect Eero Saarinen designed the most architecturally distinguished corporate laboratories for the most discriminating corporate clients—GM, IBM, and

Isis 92, no. 1 (March 2001): 1–33.

We would like to thank Mark Coir and Ryan Wieber at the Cranbrook Archives, Sheldon Hochheiser at the AT&T Archives, Laura Mancini at the General Motors Media Archives, and Robert Godfrey at the IBM Archives for their invaluable assistance in locating documents and photographs. We would also like to thank Emily Thompson for sharing her recollections of Holmdel, as well as her broader insights into laboratory architecture and design; and Lindy Biggs, Tom Carroll, Ross Bassett, and Tom Gieryn for so warmly welcoming a couple of new workers into the vineyard. Comments on earlier versions of this essay given at the Johns Hopkins History of Science, Medicine, and Technology Colloquium and at conferences organized by Catherine Westfall and Robert Crease at SUNY Stony Brook and by Louis Galambos at Johns Hopkins raised more good questions than we can hope to answer, though we'll keep trying.

Entrance to IBM Watson Research Center. As in the GM Tech Center, Saarinen designed a signature staircase for the lobby, visually and literally connecting the slate floor and the local fieldstone column. Note the ever-present clocks, a legacy of Thomas Watson, Sr., plus the back-to-back chairs more appropriate to one of Saarinen's airport terminals. (Courtesy of IBM Archives.)

AT&T. In three stunning commissions, he reinterpreted the laboratory for American corporations in research centers that dwarfed any university. The GM Technical Center (1956), at $125 million, was worth more than most university endowments. Saarinen's dramatic glass facades and modular interiors for IBM's Thomas Watson Research Center (1961) and for Bell Laboratories at Holmdel, New Jersey (1962), followed the GM center in setting new standards for laboratory architecture.[1]

Saarinen's laboratories provided what corporate executives considered essential outposts on the "endless frontier" of postwar science. They had seen for themselves how science had helped win the "physicists' war" and had come away from that experience convinced that science would be their not-so-secret weapon for capturing postwar markets. They shared a belief in what the historian David Hounshell has called the "linear model" of industrial research, with fundamental science at one end driving profitable products out the other. Universities had proven themselves to be essential partners in creating wartime wonder weapons. So business leaders understandably concluded that they needed their own version of the university campus, a place with the academic appeal of a conventional campus, to lure the best scientists and engineers away from academic and government positions, but without the inconvenience of students, a place where science could be put to work without the corporate distractions of deadlines and bottom lines. "The shibboleths of this new age were that basic science and well-funded scientists knew better than generals, engineers, or industrialists what new science to pursue, which new technologies to develop, and how best to deploy those new technologies," Hounshell argues. "Seldom have the lessons of war been more fundamentally misunderstood. Seldom have such misunderstandings been more important, for they governed the course of national policy and the direction of U.S. industrial R&D until the 1960s."[2] No one put those misinterpretations into more beguiling form than Saarinen and his corporate patrons.

Saarinen shared with his corporate clients the convictions that the isolated campus was the ideal model for R&D and that basic research required a new spatial and symbolic identity and an elaborate public stage. His laboratories, like the new corporate headquarters simultaneously redefining the American skyline, projected an appropriately modern image but rarely contributed to the corporate bottom line in any direct way. Saarinen's laboratories were the kind of public relations only monopolies could afford and, conveniently enough, an investment that could pass muster with antitrust watchdogs, who had long interpreted R&D, especially fundamental science, as a legally acceptable strategy for securing future markets.[3]

Surprisingly, Saarinen's laboratories, and their many imitators, have attracted scant attention from historians of science. Recent studies of the "architecture of science" have

[1] "Portraits in Architecture: A Review of the Most Recent Buildings of the Late Eero Saarinen," *Industrial Design*, May 1963, *10*:62–71; for the epigraph see *Fortune*, Dec. 1951, *44*:82. For a contemporary appraisal of Saarinen's career see Allen Temko, *Eero Saarinen* (New York: Braziller, 1962). Carter Wiseman, *Shaping a Nation: Twentieth-Century American Architecture and Its Makers* (New York: Norton, 1998), perceptively places Saarinen's career in the context of late modernism, between Mies van der Rohe and the International School and Robert Venturi and postmodernism. Venturi, who also designed important academic laboratories, worked for Saarinen and for Louis Kahn, another master builder of campus laboratories, before opening his own architectural firm.

[2] David A. Hounshell, "The Evolution of Industrial Research in the United States," in *Engines of Innovation: U.S. Industrial Research at the End of an Era*, ed. Richard S. Rosenbloom and William J. Spencer (Boston: Harvard Business School Press, 1996), pp. 13–85, on p. 41.

[3] *Ibid.*, p. 46.

considered virtually every space and place of experiment—private homes, museums, hospitals, universities, national laboratories, even gardens—except corporate laboratories.[4] Business and economic historians have provided detailed accounts of "science and corporate strategy," but again with little sense of place or of how the distinctive spatial arrangements and geographical locations of these laboratories shaped the kind of science being done there, as if innovation somehow flowed directly from corporate memoranda.[5]

Just as careful studies of other kinds of laboratories can, as Peter Galison suggests, "help us position the scientist in cultural space," so the architecture and design of industrial laboratories can reveal changing corporate attitudes toward science and scientists. Thomas Gieryn has done more than anyone else to demonstrate how to read the laboratory as text, highlighting, for instance, the distinction between public and private space and how each defines and so constrains the other. Elsewhere, he has shown how seemingly incidental architectural details—an otherwise-mundane all-weather connecting corridor, for instance—can provide powerful clues for interpreting the political intentions and compromises literally built into a modern biotechnology laboratory. His larger point is that laboratory design, internal and external, gives meaning and identity to both the building and its occupants. Going beyond architectural semiotics to questions about how design and layout structure scientific practice requires attention to matters Saarinen considered crucial in theory but, curiously, neglected in his actual laboratory designs. Despite a serious interest in watching how scientists moved about their laboratories, despite a willingness to spend hundreds of thousands of dollars on scale models and mock-ups, and despite a remarkable sensitivity to how the built environment could alter patterns of behavior, Saarinen's laboratories always ended up being more about corporate image making than about making scientific breakthroughs. His laboratories, like corporate pavilions at the World's Fairs, reassured corporate executives and the public alike that their money was being well spent, even when it wasn't.[6]

In a fascinating study of Saarinen's corporate headquarters building for John Deere and Company in Moline, Illinois, commissioned the same year as the Holmdel and Watson Research Center laboratories, the anthropologists Mildred and Edward Hall asked some

[4] Peter Galison and Emily Thompson, eds., *The Architecture of Science* (Cambridge, Mass.: MIT Press, 1999), Crosbie Smith and Jon Agar, eds., *Making Space for Science: Territorial Themes in the Shaping of Knowledge* (New York: St. Martin's, 1998), and Frank A. J. L. James, ed., *The Development of the Laboratory: Essays on the Place of Experiment in Industrial Civilization* (Basingstoke, Hampshire: Macmillan, 1989), include not a single industrial laboratory among their dozens of case studies, though they do provide some excellent examples of how to read laboratory architecture.

[5] Even the best of the literature—including Leonard Reich, *The Making of American Industrial Research: Science and Business at GE and Bell, 1876–1926* (New York: Cambridge Univ. Press, 1985); George Wise, *Willis R. Whitney and the Origins of American Industrial Research* (New York: Columbia Univ. Press, 1985); David Hounshell and John Kenly Smith, *Science and Corporate Strategy: DuPont R&D, 1902–1980* (New York: Cambridge Univ. Press, 1988); Margaret B. W. Graham, *The Business of Research: RCA and the VideoDisc* (New York: Cambridge Univ. Press, 1986); and David F. Noble, *America by Design: Science, Technology, and the Rise of Corporate Capitalism* (New York: Knopf, 1979)—rarely places its subject in a physical setting, so that it is often difficult to figure out where the scientists and engineers actually were.

[6] Peter Galison, "Buildings and the Subject of Science," in *Architecture of Science*, ed. Galison and Thompson (cit. n. 4), pp. 1–25, on pp. 2–3. On the laboratory as an architectural "text" see Thomas Gieryn, "Biotechnology's Public Parts (and Some Private Ones)," in *Making Space for Science*, ed. Smith and Agar (cit. n. 4), pp. 281–312; for better photographs see the same piece in Arnold Thackray, ed., *Private Science: Biotechnology and the Rise of the Molecular Sciences* (Philadelphia: Univ. Pennsylvania Press, 1999), pp. 219–253. On architectural details see Gieryn, "Two Faces on Science: Building Identities for Molecular Biology and Biotechnology," in *Architecture of Science*, ed. Galison and Thompson, pp. 423–455. On architecture as "reassurance" see Roland Marchand and Michael L. Smith, "Corporate Science on Display," in *Scientific Authority in Twentieth-Century America*, ed. Ronald G. Walters (Baltimore: Johns Hopkins Univ. Press, 1997), pp. 148–182.

probing questions about how architectural design reinforced or discouraged particular patterns of behavior. Though almost fawning in their praise of Saarinen's design, they noticed that the building was, in their words, "a direct reflection of William A. Hewitt, Chairman and chief executive," and that "one of Saarinen's main attractions was his desire to capture the spirit of the company and the personality of its Chairman."[7] Saarinen's laboratories likewise captured the spirit of research, as understood by top management, but missed its essence.

Beyond location, what distinguished postwar corporate laboratories from their prewar predecessors was a style and an attitude reflected in their architecture and in their leadership. Where the pioneering industrial laboratories of General Electric, DuPont, General Motors, AT&T, and Westinghouse had the look and feel of factories and were often housed in former manufacturing facilities near production divisions, their postwar counterparts were deliberately isolated from the rest of the corporation. Whereas the emphasis had once been on the "D" (development) in R&D, the new watchword would be *basic* research," with attention to fundamental science rather than its direct application. And where earlier research directors had generally come up through the ranks, often without the benefit of advanced degrees (the founders of Bell Laboratories being conspicuous exceptions), their successors nearly always came from universities or government agencies and boasted Ph.D.'s, most often in physics, the new king of the disciplinary hill.

The physical isolation that Saarinen exploited so masterfully in his architectural designs—the sweeping vistas at the GM Tech Center, the great glass arch at Watson Research Center, the giant mirrored facade and reflecting pool at Holmdel—deliberately distanced these laboratories geographically, organizationally, and intellectually from the rest of the corporation, but at the cost of losing a vital link to products and production. More and more, the crucial task of linking research to production fell to branch laboratories located near manufacturing facilities, such as the IBM laboratory at San Jose, where disk storage was invented, Bell Laboratories' branch at Allentown, Pennsylvania, where Western Electric produced state-of-the-art vacuum tubes and, later, semiconductors and integrated circuits, and GM's Delco Radio facility in Kokomo, Indiana, where automobile radios gave rise to semiconductor research and missile guidance systems.

While GM, IBM, and AT&T still dominated their respective industries to the point of risking antitrust intervention, they could afford corporate campuses as showplaces and showcases for research. Saarinen's laboratories remain architectural masterpieces, but they have had to try to reinvent themselves as scientific workplaces in a corporate environment where the balance sheet puts far less emphasis on scientific achievement and far more on return on investment. Paradoxically, Bell Laboratories' old Murray Hill facility, the preeminent corporate laboratory of its day, won no architectural awards—only Nobel Prizes and markets. Its defiant functionality continues to challenge conventional thinking about whether imaginative science demands equally imaginative architecture. Murray Hill's apparent flaws—cramped offices, long corridors, distant public spaces—more often than not turned out to be secret strengths its more architecturally sophisticated cousins could not match.

Saarinen's laboratories are now largely monuments to a golden age of American science, when asking whether something was worth doing was primarily an intellectual rather than

[7] Mildred Hall and Edward Hall, *The Fourth Dimension in Architecture: The Impact of Building on Man's Behavior: Eero Saarinen's Administrative Center for Deere and Company, Moline, Illinois* (Santa Fe, N.M.: Sunstone, 1974), pp. 10–11.

a financial, question. *Life* correctly judged the GM Tech Center a "Versailles of Industry," not simply for its extravagance and architectural coherence but as a tangible expression of an era, a distinction it came to share with Watson Research Center and Holmdel.[8] Yet like their illustrious predecessor, these Industrial Versailles symbolized an increasingly dangerous isolation from the outside world that would eventually threaten to undermine the very legitimacy of the regimes they so powerfully expressed.

THE GENERAL MOTORS TECHNICAL CENTER

General Motors is a metal-working industry; it is a precision industry; it is a mass-production industry. All these things should . . . be expressed in the architecture of its Technical Center. Thus the design is based on steel—the metal of the automobile. Like the automobile itself, the buildings are essentially put together, as on an assembly line, out of mass-produced units. And, down to the smallest detail, we tried to give the architecture the precise . . . look . . . characteristic of industrial America. . . . Some sort of campus plan seemed right. . . .
—*Eero Saarinen (1956)*

For three decades, from 1919 until his retirement in 1947, the legendary inventor Charles Kettering defined research, GM-style. Kettering was a self-proclaimed "screwdriver and pliers man" whose aphoristic dismissals of academic science—"You can't do a good job with educated people. They want to do it the way they are educated"—were nearly as famous as his electric self-starter or leaded gasoline. In 1925 GM moved Kettering's laboratory from its original home in Dayton, Ohio, to the center of the GM empire and hired the architect Albert Kahn to add a wing to the back of the recently completed General Motors Building in downtown Detroit.[9]

Kahn imagined the GM laboratory just as he had imagined his famous automobile factories such as Highland Park and the River Rouge, built for Henry Ford, as form accommodating function. Informality and practicality set the tone. One visitor reported finding "250 skilled mechanics, 150 technical researchers, 100 students, clerks, and others. . . . There are few house rules, not much nickel-plated apparatus, no Hollywood-size retorts or atom smashers. A new x-ray machine is apt to be rigged on boards from the packing case in which it arrived, and the researchers, without their glasses, would be hard to distinguish from the greaseballs who work in the garage downstairs."[10] Surviving photographs of the laboratory suggest how fully the machine-shop culture of Delco, where Kettering and a small staff built electric self-starters by hand, survived in the Detroit laboratory. Metal lathes, grinders, milling machines, and drill presses competed for floor space with spectrometers and chemical apparatus (see Figure 1). This was blue-collar science, where a bruised knuckle was as common as a notebook entry.

GM chairman Alfred Sloan, however, had decided that if the company wanted to meet the challenges of the postwar world, and to compete with other firms and with universities for the best and brightest scientific and engineering talent, it would have to make a new

---

[8] "Architecture for the Future Constructs a Versailles of Industry," *Life*, 21 May 1956, *40*:102–103.

[9] Stuart W. Leslie, *Boss Kettering: Wizard of General Motors* (New York: Columbia Univ. Press, 1983), p. 337; for the epigraph see Aline Saarinen, ed., *Eero Saarinen and His Work* (New Haven, Conn.: Yale Univ. Press, 1968), p. 30. On Kahn's career see Federico Bucci, *Albert Kahn: Architect of Ford* (New York: Princeton Architectural Press, 1993); Lindy Biggs, *Designing the Rational Factory* (Baltimore: Johns Hopkins Univ. Press, 1996); and W. Hawkins Ferry, "Foreword," in Detroit Institute of Arts, *The Legacy of Albert Kahn* (Detroit, Mich.: Gaylord, 1970).

[10] Leslie Velie, "Kettering of GM," *Coronet*, Sept. 1945, pp. 123–124; and "General Motors IV: A Unit in Society," *Fortune*, Mar. 1939, p. 46 (quotation).

Figure 1. Kettering's "boys" turn out the products that "kept the customer dissatisfied" in Albert Kahn's General Motors Research Corporation laboratories—which looked more like a machine shop—in the 1930s. (Courtesy of General Motors Media Archives.)

place for science. When Kettering retired in 1947 Sloan replaced him with the nuclear physicist and former Atomic Energy Commissioner Lawrence Hafstad, who seemed to embody everything that was modern about modern science. Hafstad was anything but a "screwdriver and pliers" scientist. Like the best physicists of his generation, he looked to Los Alamos and Oak Ridge—not automobile factories—as the appropriate models for a great laboratory.

If this new breed of scientist worked best in an academic environment, GM executives reasoned, then they required a laboratory that would mimic, or at least approximate, the functions of a real campus. Sloan envisioned a research center that could "house almost five-thousand scientists, engineers, designers, and technicians, and their supporting personnel [in a] . . . campus-like atmosphere . . . so that these surroundings will stimulate creative thinking and excellent work." He wanted enough distance between the laboratory and the rest of the corporation, geographically and organizationally, to assure his new scientists sufficient independence.[11]

But who should design such a laboratory? Some top executives thought GM's own

[11] Albert Christ-Janer, *Eliel Saarinen: Finnish-American Architect and Educator* (Chicago: Univ. Chicago Press, 1948), p. 116 (quotation); and *Architectural Record*, May 1956, *119*:151. For the GM "promotional" view on the Tech Center idea see General Motors, "Where Today Meets Tomorrow," n.d., Saarinen, Eero, and Associates File, Cranbrook Academy Archives, Bloomfield Hills, Michigan.

engineers and draftsmen were equal to the task. They argued that "any emphasis on high aesthetic standards might be detrimental to the practical operations of the center."[12] Master stylist Harley Earl, however, insisted that the company should look outside its own ranks for an architect of sufficient stature to take on what promised to be perhaps the most expensive commission of modern times. Everyone GM talked to seemed to recommend the Saarinen firm. Sloan was impressed enough with the Saarinen design for the Ethyl Corporation laboratories in Detroit to settle the issue immediately and began negotiations with the Saarinens in July 1945.

Designed, redesigned, and constructed over the next eleven years, the General Motors Technical Center marked Eero Saarinen's first major statement on the relationship between modernist industrial architecture and modern research. His father, Eliel, had made his name largely in Finland, but a lack of work after World War I convinced him to enter a competition for the design of the Chicago Tribune Building in 1922. Winning the $20,000 second-place award gave Saarinen the opportunity to move his family to America. In 1925 he constructed the Cranbrook School in Bloomfield Hills, Michigan, for the newspaper tycoon George C. Booth. Arguably the most influential work of his career, Cranbrook recreated certain elements of Saarinen's own home (Hvittrask) in Finland, as well as Frank Lloyd Wright's Taliesin—places where collaborative artistic work called for a flexible, decentralized architecture.

At Cranbrook, the Bauhaus-influenced arts and crafts ethos flexed a sculpting hand on Eliel's son, young Eero Saarinen, who, following the Cranbrook philosophy, trained in a variety of media. At nineteen Eero designed the tables and chairs for the school's cafeteria, and for a time sculpture was his chosen art, though he continued drawing and designing at his father's side. Surrounded as he was by experts in ceramics, sculpture, weaving, furniture design, and the associated visual arts, the younger Saarinen came to see the architect's task as conceiving of the whole and blending the parts. He was learning to think of the architect's role as that of a "symphony conductor."[13] The final physical shape of the Tech Center, with its wide variety of expertly designed craft elements, reflected this attitude. In a sense the Tech Center represented a first and not so tentative meeting between corporate America and the late modernist aesthetics Saarinen imbibed at Cranbrook.

Graduating from Yale in 1934, Eero Saarinen toured Europe to study architecture, then returned to Bloomfield Hills in 1939 to take up a teaching post at Cranbrook. By 1939 he was helping Charles Eames design plans for the new Smithsonian art gallery (see Figure 2); he would win a furniture competition at New York's Museum of Modern Art the next year. Saarinen's chosen architectural language was still evolving in these years, tending toward Bauhaus-style internationalism at times but most closely related to the industrial styles of Albert Kahn and the spare glass and metallic modernism of Ludwig Mies van der Rohe's Illinois Institute of Technology campus. When the elder Saarinen died early in the planning stages of the Tech Center, he left to his accomplished but not yet famous son the task of completing the final stage of modeling and sketches. With the Tech Center Saarinen would find his own architectural voice, a modernism with a distinctive American industrial accent. As he wrote during the construction of the Tech Center: "America has a very different background; very little tradition plus a fast industrialization has made America lose whatever it had of taste and tradition of the former, therefore the American

[12] Alfred Sloan, *My Years with General Motors* (Garden City, N.Y.: Doubleday, 1963), p. 304.
[13] Joy Hakanson, "A Walk in the Shadow of Genius," *Detroit News Pictorial Magazine*, 28 June 1966, p. 20.

Figure 2. *The pride of Cranbrook—Charles Eames (left) and Eero Saarinen—together at the drawing board, around the time of their Smithsonian collaboration. (Courtesy of Cranbrook Academy Archives.)*

designer has to look wholly towards forms sympathetic to and derived from the machine era."[14]

To Saarinen this meant delivering a complex of buildings that would bring to mind the image of the central focus in the GM universe: the automobile. The complex, then, needed to be big enough that it could be properly appreciated only from behind the wheel. To this end, and to provide the geographical separation of the Tech Center from top management, GM acquired a 900-acre site in Warren Township, several miles northwest of Detroit. When completed in 1956, the General Motors Technical Center would provide an entirely new kind of space where GM scientists and engineers could invent the world of tomorrow so vividly brought to life in "Futurama," the company's popular exhibit at the 1939 World's Fair. Like "Futurama," the Technical Center would be a master stroke of corporate relations, a public stage for a more select audience. Appropriately enough, the center would house Harley Earl's styling staff along with the research laboratory, engineering, and manufacturing. In the postwar world, mastering "science, the endless frontier," would be considered as important as mastering styling, marketing, manufacturing, and production engineering.

In line with Miesian theory and his own ideas about an architecture appropriate for industry, Saarinen chose mass-produced, modular steel panels as his building blocks and had them made by one of the GM divisions. Making use of the Detroit laboratory itself in the fabrication process, Saarinen and GM researchers then developed neoprene gasket material to hold glass in place, a process much like sealing a windshield into place on a Cadillac. Saarinen put GM's chemists to work formulating paints appropriate for his massive, brightly colored ceramic walls.

Saarinen's design included separate buildings for different research functions—styling, engineering, research, administration, and process development—arranged around a rectangular lake, with a large tower providing a central visual focus. Though the design derived from his father's original sketches, Saarinen gave the Tech Center his distinctive stamp, especially in the research building, where Eero rejected his father's peculiar aerofoil-shaped design in favor of a "more classic interpretation of the formal needs of the Technical Center." In designing for his client's specific industrial needs, Saarinen clearly echoed Albert Kahn. None of the Tech Center buildings rose higher than three stories. Glass curtain walls allowed for generous doses of natural light. Saarinen surrounded the site with woods and used greenery and landscaping to help bring steel and concrete stretches to life. More important, he made most of the interior lab offices and bench spaces flexible through the use of modular, movable wall panels, with hallways and cafeterias designed to encourage casual meetings. This rendered the plan responsive to Saarinen's sense that researchers liked to meet informally and that they tended to change their lab arrangements from time to time. Saarinen himself remarked: "It has been said that in these buildings I was very much influenced by Mies. But this architecture really carries forward the tradition of American factory buildings which had its roots in the Middle West in the early automobile factories of Albert Kahn."[15]

The sheer size and expense of the Tech Center project pushed Saarinen into areas left

[14] Astrid Sampe Collection of Eero Saarinen Correspondence, 1/6, Letter 9, n.d. (ca. 1948–1949), Cranbrook Academy Archives.

[15] Rupert Spade, *Eero Saarinen* (New York: Simon & Schuster, 1971), p. 12; and A. Saarinen, ed., *Eero Saarinen and His Work* (cit. n. 9), p. 30. Kahn had designed rural laboratories for Henry Ford in Northville, Waterford, and Wanki, near Detroit, around 1918, bowing to Ford's insistence that industry and natural beauty were not incompatible.

unexplored by Kahn. In style and tone the Tech Center was a world apart from Kahn's 1922 GM laboratory. According to the architecture critic Allen Temko: "Saarinen . . . was the first architect fortunate enough to work on a titanic scale without serious budgetary restrictions, just as he was the first of the younger Modernists to enjoy the chance . . . to apply the principles of their pioneer predecessors to a problem of this magnitude." The entire complex was scaled to a 5-foot module, allowing researchers to put together and take apart office and lab space as needed. This deceptively simple flexibility, invisible to the casual visitor, hid the finest of Saarinen's innovations. Air conditioning, power delivery, and water delivery systems all coursed through the buildings virtually unseen yet could be routed and rerouted according to a scientist's whims in a matter of hours.[16] Former GM research director Robert Frosch recalls that even in the 1980s researchers were still amazed at the ease of converting one type of lab into another—say, a mechanical engineering lab into a chemistry lab—in no time at all. While the Tech Center would win acclaim for its artistic achievements, it was this commitment to a modular and completely flexible design that would be Saarinen's legacy to the next generation of laboratory architects.

This flexibility apparently pleased GM researchers, but executives took greater pride in the ornamentation of the center—a sort of "research opulence" that fit with GM's new high-style image. Drawing on the talents of his Cranbrook colleagues, Saarinen commissioned Marianne Strengell to design carpets and tapestries intended to "soften and humanize the great expanses of glass, the pre-fabricated units of the walls," while respecting "the desire of the client for strong, practical . . . interiors." Harley Earl, the sharpest dresser at GM, got a glass office "curtained in sheer white, with a ceiling in grey and silver and rough-textured upholstery in brilliant colors—orange, red, blue, and snow white." Charles Eames supplied the furniture, and a handful of other Cranbrook notables contributed work in ceramics and sculpture. Alexander Calder designed a tripod water tower and "water wall" for the center of the lake, which reportedly pumped more water than all of the original Versailles' fountains combined.[17]

Above the heads of the researchers sat the regal quarters of the research manager. It was from here that the entire conception of the Industrial Versailles made the most sense— the vice president sat at his hexagonal teakwood table, back to the purple and blue tapestries, eyes fixed outside the glass curtain wall on the Japanese tea garden or the parklike acres of Warren stretching out into the distance, or hosted Eleanor Roosevelt, Gary Cooper, John Glenn, the king and queen of Belgium, or that other grand purveyor of images, Walt Disney. It was in just this sort of language that GM managers described their new Tech Center—a campus that might attract and keep top-quality scientists and engineers and shield them from the "practical" needs of the divisions while at the same time showing off the sort of wealth that a company such as GM could flaunt in the 1950s.[18]

The opening of the Tech Center in 1956 transformed the look of research at GM. Widely praised by architects, scientists, research directors, and the business community, the Tech Center seemed to embody modernity and originality. Here scientists and engineers could work unfettered, free to pursue their interests without the deadlines and restricted budgets of applied research labs. Predictably, Kettering hated it. He showed up at the ceremonial

[16] Temko, *Eero Saarinen* (cit. n. 1), p. 21; and "GM Nears Completion," *Architectural Forum*, 1954, *101*:100–119.

[17] Marion Holden Bemis, "Marianne Strengell: Textile Consultants to Architects," *Handweaver and Craftsman*, Winter 1956–1957, pp. 6–7; and Spade, *Eero Saarinen* (cit. n. 15), p. 12.

[18] "Architectural History of Design Staff," n.d., GM Technical Center Design Library, Detroit.

Figure 3. *The laboratory of the future arrives at General Motors. Gone are the rolled-up shirt sleeves and the metal lathes. Now the laboratory coat and the analytical instrument are standard issue. (Courtesy of General Motors Media Archives.)*

groundbreaking, where he posed at the controls of a diesel shovel, in coveralls and rarely visited the Tech Center once it opened. Too many white lab coats, too many Ph.D.'s, and too much scientific equipment for his taste, he said (see Figure 3). In a touch of corporate one-upmanship, GM commissioned the artist Charles Sheeler, whose photographs and paintings had made the River Rouge factory an American icon, to capture the "spirit of research" on canvas. Simply titled *General Motors Research*, Sheeler's mural portrayed the apparatus of a physics experiment designed to measure the gyromagnetic effect of the atom against the backdrop of Saarinen's signature spiral staircase, perfectly capturing the Tech Center's blend of basic science and modernist impulse.[19]

IBM'S THOMAS WATSON RESEARCH CENTER

The architectural challenge in designing this research center seemed to be that of reconciling two apparently contradictory requirements. We wanted to make the laboratories and research offices the most efficient and flexible of twentieth-century research centers. We also wanted a building that would be appropriate to the personality of the users and the site and the environment. . . . It has always seemed to me that many scientists in the research field are like university professors—tweedy, pipe-smoking men. We wanted to provide them with a relaxed,

---

[19] Peter Galison, *How Experiments End* (Chicago: Univ. Chicago Press, 1987), includes a thoughtful reading of Sheeler's painting, which is reproduced as its title page illustration.

"tweedy" outdoor environment as a deliberate contrast to the efficient, precise laboratories and offices.

—*Eero Saarinen (1961)*

Under founder Thomas Watson, IBM's inventions departments in Endicott, New York, were headed by graduates of the machine-shop culture of hard knocks, men who turned out ingenious variations of punch-card calculators and tabulators on lathes, grinders, and drill presses. Watson emblazoned his corporate motto, "Think," on the entrance of every IBM facility, next to a clock, forceful reminders of the Watson philosophy that getting the jump on the competition demanded imagination as well as timing. IBM's North Street Laboratory, established in 1932, looked more like a vocational high school than a corporate laboratory. It housed the experienced mechanics and metalworkers Watson liked to call "his" inventors. Virtual freelancers in the Edisonian tradition, IBM's inventors had their own shop space, competed with one another for resources, patents, and the boss's ear, and gained a reputation for secrecy bordering on paranoia. Nonetheless, they consistently delivered the electromechanical tabulating machines that kept IBM on top, with 90 percent of the punch-card office machine market. By any measure, IBM's R&D effort lagged far behind those of comparably sized companies; in 1942 its research staff numbered only 350, not a single Ph.D. among them.[20]

Though no engineer himself, Watson understood that electronics would ultimately revolutionize the office machine industry. To prepare for that day, he established the Watson Scientific Computing Laboratory at Columbia University in 1945 to "serve as a world center for the treatment of problems in various fields of science whose solution depends on the effective use of applied mathematics and mechanical calculation." The Watson Laboratory's small staff could never be more than an in-house consultant to the rest of IBM. And the real challenge facing IBM was incorporating modern electronics, including the recently invented transistor, into its standard line of office machines. Here the initiative came from Thomas Watson, Jr., executive vice president and heir apparent, who firmly believed that conventional punch-card machines were history. To underscore the message, he chose the 1949 introduction of the IBM 407, the very latest in electromechanical accounting machines, to tell the IBM sales force that by the end of the 1950s he expected all IBM products to be entirely electronic.[21]

To wrest control of electronics from what he considered "a bunch of monkey-wrench engineers," Watson, Jr., centralized IBM's electronics research at Poughkeepsie, New York, the company's largest wartime factory, and quadrupled the overall research staff from one thousand to four thousand. Where research had previously been about perfecting and updating old products, it would now be about thinking up entirely new ones. Following the lead of companies such as GM, RCA, and GE, IBM decided in 1955 to create a centralized research laboratory reporting directly to corporate headquarters and hired Emanuel Piore to head it.[22] Like Hafstad at GM, Piore was trained as a physicist, then

---

[20] For the early history of IBM research see Emerson Pugh, *Building IBM: Shaping an Industry and Its Technology* (Cambridge, Mass.: MIT Press, 1995), pp. 77–87. On IBM's business strategy see Rowena Olegario, "IBM and the Two Thomas J. Watsons," in *Creating Modern Capitalism*, ed. Thomas K. McCraw (Cambridge, Mass.: Harvard Univ. Press, 1997), pp. 356–360. For the epigraph see A. Saarinen, ed., *Eero Saarinen and His Work* (cit. n. 9), p. 70.

[21] Jean Ford Brennan, *The IBM Watson Laboratory: A History* (New York: IBM, 1971), p. 1 (quotation); and Steven W. Usselman, "IBM: Making Waves in the Computer Business," *Understanding Innovation*, 7 June 1997, p. 15 (move to electronic products).

[22] Pugh, *Building IBM* (cit. n. 20), pp. 237–238.

made his reputation in government service as chief scientist of the Office of Naval Research (ONR). At ONR Piore had been a strong advocate of supporting academic research, in the conviction that today's basic research would become tomorrow's advanced weapons systems. He brought the same philosophy to IBM: support fundamental science and future applications would take care of themselves.

Watson, Jr., who had been named chief executive officer in the spring of 1956, just weeks before his father's death, finally had a free hand to run IBM as a *computer* company. To Watson, Jr., that meant a world-class image and a world-class laboratory to go with world-class sales, marketing, service, and products. For years, Watson, Jr., had chafed under what he considered a stodgy corporate personality inherited from his father. What IBM needed, he decided, was a complete corporate makeover. "I thought you ought to be able to look at an I.B.M. factory, at an I.B.M. product, even at an I.B.M. curtain, and say it's I.B.M.," he recalled. He hired the industrial designers Eliot Noyes and Paul Rand with a mandate to make IBM's image as modern as its products—starting with a face-lift of Watson, Jr.'s own office on the sixteenth floor of IBM World Headquarters in New York City.[23]

Watson also launched the biggest building binge in corporate history. Over the next fifteen years IBM would construct an astounding 150 plants, laboratories, and office buildings throughout the world. Watson wanted these buildings to have style but not uniformity, and he asked Noyes for names of the best modern architects. At the top of Noyes's list was Eero Saarinen, whom IBM immediately hired to design a factory complex for the Data Processing Division in Rochester, Minnesota. Noyes admired the GM Tech Center, and his own product development laboratory for Poughkeepsie frankly emulated it.

"When I recommended Saarinen for the job," Noyes said shortly after the completion of the Rochester plant, "I was not thinking about what appearance his building would have. I was thinking that if he does the job, I will not have to worry about its integrity or its modernity, and these are certainly the qualities that IBM should represent."[24] Saarinen's Rochester facility, though a manufacturing plant, had a distinct campus feel, with low, sprawling buildings separated by gardens and set off against the Minnesota prairie with stunning blue porcelain–enameled panels and glass.

With the GM Tech Center to his credit, Saarinen seemed the ideal choice for IBM's own research center. IBM had selected a site in Yorktown Heights, about halfway between Poughkeepsie and corporate headquarters in midtown Manhattan, on a bluff overlooking 240 acres of wooded Westchester County. Saarinen's original plan for the Watson Research Center looked nothing like the final version. Instead of a single structure, he proposed a campus complex with dozens of small buildings connected by pathways across courtyards and gardens. He intended the complex to grow organically from the center as demands for space increased. Robert Gunther-Mohr recalled meeting Saarinen as a member of a Research Advisory Committee set up to assist in planning the new laboratory:

> He first got us together and asked us for our concepts of research because he wanted to have a kind of general idea of what went on in research such that he could create a building responsive

[23] Thomas J. Watson, Jr., "Good Design Is Good Business," in *The Art of Design Management*, Tiffany/ Wharton Lectures, n.d., IBM Archives, Yorktown, New York, p. 58 (complete corporate makeover); "I.B.M. Banishes Dowdiness," *Fortune*, June 1959, p. 129 (quotation); and Hugh B. Johnston, "From Old IBM to New IBM: The Story of a Company's Increasing Sense of Design," *Indust. Design*, Mar. 1958, pp. 48–53, esp. p. 51 (office face-lift).

[24] Watson, "Good Design Is Good Business," p. 59; and Johnston, "From Old IBM to New IBM," p. 51.

to the needs of the people doing research. At the time it was believed that research was conducted in the universities, with a group of people who had certain academic leanings and who clustered together to carry out these activities. So there were centers in electrical engineering, and there were centers in process science, and so on. He visualized these centers being all independent buildings, and that these buildings would then be connected together with walkways so the philosopher-like people would be strolling on these walkways, discussing things, but they would have homes that would be focused on their disciplines. So he was kind of thinking in terms of universities with departments. . . . I well remember that he came to us, with a whole set of mock-ups showing how this would look on this hilltop, and then we got asking him questions about how all of this all would go. One of the questions we asked him, "Well you know, in winter here, it blows a little hard, and how do you visualize people getting from one building to the next since they are all separate?" And he said blithely, "The thing to do is to just have tunnels."[25]

This particular vision faltered when Saarinen explained to Watson, Jr., just how much the tunnels were going to cost. Fortunately, the architect had an ace up his sleeve: plans for a comparable facility that Bell Laboratories had just commissioned for Holmdel. Virtually overnight, Saarinen returned with a strikingly bold blueprint for an entirely different kind of campus. *Architectural Forum* succinctly captured the iconoclasm of Saarinen's masterpiece: "his new IBM Research Center suggests, first, that the best laboratory or office space may conceivably be a windowless room; it suggests, secondly, that the shortest line between two points may be a curve; and it suggests, finally, that the handsomest facade in a rolling landscape may be a 1,000 foot ribbon of near-black metal and glass." After scrutinizing research behavior, Saarinen had decided that conventional laboratory design, where everyone of importance got a window, no longer made sense in an era of air conditioning and fluorescent lighting. "Windows are like fireplaces, nowadays," he remarked. "They are nice to have but rarely used for their original purpose."[26] Indeed, most researchers seemed to close the blinds or draw the shades if they had the opportunity, or simply sat with their backs to the windows.

Saarinen's radical alternative was an enormous arc, 1,000 feet long but a mere 150 feet wide, three stories high, with alternating corridors of offices and laboratories and dramatic glass-enclosed promenades along the front and rear. Placing offices and labs back-to-back, with a utility spine running between the labs, encouraged local traffic flow across the aisles separating the offices and labs and provided convenient access to the exterior walkways and their unobstructed views of the countryside. No office or laboratory was more than 60 feet from one of the promenades. The great arc kept sight lines along the walkways to less than 100 feet, avoiding the vertigo induced by a 1,000-foot corridor. In the not-so-distant future, Saarinen predicted, "the growth of IBM would lead to a complete circling of the hill-crest and the fortress of technology would be complete."[27] That might have provided a symbolic echo of the emerging System 360, a corporate vision as closed-minded and self-contained as the Watson Reseach Center itself.

IBM's only firm guidelines were "(1) that the interior would be completely flexible, and (2) that the building be capable of easy expansion to meet the growing program of IBM Research." To meet those demands, Saarinen designed the offices and laboratories on a 4- by 6-foot modular grid, with movable partitions of various combinations of walls and

[25] Johnston, "From Old IBM to New IBM," p. 110 (intended growth); and E. W. Pugh interview with Robert Gunther-Mohr, 4 Oct. 1983, IBM Technical History Project Interview, IBM's Early Computers (copy courtesy of Ross Bassett).

[26] "Research in the Round," *Architect. Forum*, June 1961, *114*:80–85, on p. 80.

[27] Spade, *Eero Saarinen* (cit. n. 15), p. 15.

doors that could be locked into place in minutes. Typically, the offices were 12 feet deep and the laboratories 24 feet deep, with a 6-foot aisle between them. Offices could range in size from 8 by 12 feet up to 8 by 24 feet, while laboratories could be as large as 12 by 128 feet. In exceptional cases, the aisles could be bridged to create very large laboratories or conference rooms up to 42 by 128 feet without fundamentally altering the basic design. The service core carried water, power, compressed air, vacuum, gases, and other essential wet-laboratory services, which could then be connected to any particular laboratory as needed. In practice, 457 (64.3 percent) of the original offices were 8 by 12 feet, another 242 offices (34.1 percent) were 12 by 12 feet, and just 11 offices (1.5 percent), for top executives, were 16 by 12 feet. Likewise, three-quarters of the laboratories ranged from 12 by 24 feet to 20 by 24 feet, with a few others larger or smaller.[28] But not even the best office, originally reserved for Watson, Jr., himself, had exterior windows; in this regard he was first among equals.

Saarinen built mock-ups of the offices at Poughkeepsie and spent several weeks watching how researchers moved about, what they used, where they sat, what they liked, and what they did not. Most of them "detested the idea of not having windows" except onto the interior aisles, but Saarinen convinced them that he had observed enough research scientists to know that "emotionally you may think you want windows, but actually you never use them." He did agree, reluctantly, to use frosted rather than clear glass along the interior aisles. Since no one wanted to live in a fishbowl, the first thing the scientists did after moving into the new research center was to paper over the glass, for privacy and to cut down on noise.[29]

Saarinen believed strongly in the advantages of the modular offices, whose rear walls could accommodate any arrangement or rearrangement of file drawers, bookshelves, and cubbies in 4-foot units. Unfortunately, the back-to-back office design made storage space a zero-sum game. A long file drawer on one side left only enough room on the other for a small bookshelf. Understandably, the first occupants staked their claims as quickly as possible, leaving later arrivals little choice but to store their files in cardboard boxes on the floor.

Saarinen placed a wall of local fieldstone along the interior of the outer corridors, with matching end walls, to provide a visual cap to the glass wall and a connection between the structure and the surrounding landscape. In the cafeteria, too, he stressed natural materials—a black slate floor, teak and white maple tables, stone walls and planting boxes— again blurring the boundaries of inside and out. The cafeteria also housed IBM's collection of scale models of Leonardo da Vinci inventions, presumably as an inspiration to ingenuity, and supplied a notepad at each table (the original IBM Thinkpad?) so that researchers with an idea could jot it down immediately. The library, with its distinctive wine-glass chairs and slate-topped reading tables, continued the same theme, creating a "tweedy" environment considered appropriate for deep thought and reflection.[30] For the lobby (see Frontispiece), Saarinen designed another of his signature stairways as a dramatic opening gesture.

[28] "The New IBM Research Center at Yorktown," pp. 6 (guidelines), 34 (actual configurations), A 1602–3, IBM Archives; and "Unique Cross-Curve Plan for IBM Research Center," *Architect. Rec.*, June 1961, pp. 137–146, on p. 140 (possible configurations).

[29] Mike Girsdansky, "Yorktown: Headquarters for Innovation," *Research*, Nov./Dec. 1981, p. 21 (observations); Stuart Leslie interview with Emerson Pugh and Phillip Summers, 13 Oct. 1998 (quoting Pugh); and Watson, "Good Design Is Good Business" (cit. n. 23), p. 58 (papering over glass, quoting Pugh).

[30] "Unique Cross-Curve Plan for IBM Research Center" (cit. n. 28), p. 145 (cafeteria); and "New IBM Research Center at Yorktown" (cit. n. 28), p. 19 (library).

The back-to-back circles of chairs in the waiting area, on the other hand, seemed better suited for his TWA terminal or Dulles Airport than for visiting scientists.

In the offices Saarinen insisted on an elaborate color-coding scheme that irked many researchers. He painted the aisle walls stark white. But on the inside walls of the offices he placed baked enamel panels in ninety-three hues arranged so that the colors faded, panel to panel, from pale in the offices nearest the rear parking lot to dark at the front of the arc. Each aisle represented a particular season: red to brown for autumn, blue to green for winter. One amused visitor compared the effect to a "high fashion kitchen" and mused that "an unlucky stenographer who fancies herself in red may find herself scheduled to work against a bright orange backdrop." Saarinen turned Henry Ford's famous dictum about color on its head: you could have whatever color you liked, as long as it was already on the wall. Most researchers painted over the panels as soon as possible with one of the approved versions of beige. Like their counterparts at Saarinen's John Deere headquarters, another commission intended to update a stodgy corporate image, IBM researchers found themselves constantly on display. Surely they would have appreciated the lament of one Deere employee—"We're inmates of the edifice."[31]

Saarinen gave serious consideration to encouraging informal conversation among different groups of researchers. One point of the promenades was to get people out of their offices and talking with one another (see Figure 4). He placed benches in strategic areas along the front corridor, but they were uncomfortable and didn't orient people the right way; the alcoves never got much use, even after management replaced the benches with leather chairs and chalkboards. Instead, people met in offices or conference rooms or rendezvoused at the coffee cart as it made its rounds. In later years, researchers could track its progress simply by typing the command "coffee" into their computers, which then gave back the current location, say, 17-3-21!

Saarinen could never have anticipated how quickly conventional wet labs would become obsolete, at least in the computer industry, or what new demands software design would put on his laboratory. The original offices had small desks, since none of the researchers did their own typing, leaving the next generation to scramble for desktop space and electrical outlets. Most of the laboratories have long since been turned into offices and the gas lines shut off. Saarinen may have been right about the windows. When IBM made the intended additions in 1977 it gave researchers cubicles instead of offices, with surprisingly few complaints. The new "wizards of Yorktown" didn't seem to mind what a visiting reporter described as a cramped and airless beehive.[32] Their window on the world was a computer screen.

When Piore took over as research director, he promised his troops "an opportunity to build a laboratory with the proper research-campus environment, and to create an intellectual center of the world in all areas of interest to IBM." Saarinen delivered it. On 25 April 1961 Watson, Jr., officially dedicated the Thomas Watson Research Center to the memory of his father at a gala open house for employees, stockholders, and visiting dignitaries, including New York governor Nelson Rockefeller. Saarinen, of course, attended. The highlight of the day was burying a time capsule (or "information capsule," as IBM insisted on calling it) containing samples of what IBM considered its most important recent

---

[31] Betsy Brown, "Scientists Designing Machines for Space Age Housed in Giant New IBM Research Center," *Mount Kisco Patent Trader*, 2 Mar. 1961, p. 14 ("high fashion kitchen"); Leslie interview with Pugh and Summers (switch to beige); and Hall and Hall, *Fourth Dimension in Architecture* (cit. n. 7), p. 24 (quoting Deere employee).

[32] Linda Ashear, "Wizards of Yorktown," *New York Times Magazine*, 22 Feb. 1981, p. G8.

Figure 4. *Outer walkway of the IBM Watson Research Center. Obviously posed, this photograph captures the kind of face-to-face interaction Saarinen intended his design to inspire. In fact, most researchers avoided the exterior walkways in favor of shorter traffic routes along the interior corridors. Saarinen's benches and alcoves were generally empty. (Courtesy of IBM Archives.)*

innovations: a FORTRAN manual, instruction books for several IBM computers, magnetic heads for a disk drive, IBM-designed semiconductor components, and examples of machine translation from English into Russian and French. Piore underscored his vision for IBM research: "Here, we want our scientists to become the top creators—and the nucleus of an intellectual community. We hope that some of them will become Nobel laureates."[33] Five IBM scientists have—though not, ironically enough, anyone from the Watson Research Center.

Piore and his successors built the world's largest campus for computer science research, with some 1,200 scientists and engineers and world-class departments in mathematics, solid-state physics, lasers, linguistics, thin films, and cryogenics. But the center would end up contributing disappointingly little to IBM's direct bottom line. Indeed, the center's very success as a campus attracted scientists and engineers with relatively little interest in the commercial applications of their work. As one longtime IBM researcher conceded, "As we got people here from universities who were not experienced in working with manu-

---

[33] *IBM Poughkeepsie Newsletter*, Oct. 1956, p. 2 (Piore's promise); Girsdansky, "Yorktown" (cit. n. 29), p. 21 (time capsule); and "Research Center Opens Its Doors," *Business Machines*, May/June 1961, p. 10 (Piore's vision).

facturing groups and development groups, they tended to isolate themselves from such groups because there was no obvious contact. And so it's been very difficult, in fact, to bring the type of contact that's desirable between basic research groups, applied groups, and development groups. And I'd say we never really completely solved that problem." Those scientists who decided not to spend their entire careers with IBM generally returned to universities. "If you didn't stay here for your entire career, if you ever did leave, you'd leave and head towards academia. That was just the culture."[34]

Watson Research Center too often fulfilled the first half of the mission enunciated by Ralph Gomery, head of IBM's research division from 1970 to 1986—"famous for its science"—at the expense of the second—"to be vital to IBM."[35] In compiling a list of its "Top 12" innovations for its fiftieth anniversary (counting from the founding of the original Watson Scientific Computing Laboratory in 1945), the Watson Research Center staff mentioned high-temperature superconductivity and the scanning tunneling microscope (which won four Nobel Prizes but no markets); fractals and speech recognition (of great academic but little industrial interest); and RAMAC, thin-film magnetic recording heads, and the relational database (all from San Jose). The rest—RISC, one-device memory cells, FORTRAN, scalable parallel systems, and the token ring (an early version of a local area network)—more often brought big payoffs to other companies. For years, nobody really cared because IBM had a lock on the market. "Up until 1960, people who were hired into IBM stayed with IBM, so you could have a researcher and he could work for twenty years for IBM and produce great things and somewhere along the line that could be picked up in the development laboratories and it was of value," explains one longtime Watson Research Center veteran, a slight edge in his voice. "But in today's times, he can be bought off by somebody and he's gone, and so it's increasingly important that the results of research get picked up by your own development people and your own manufacturing people immediately, before the [research] people are scooped off by some other company that's going to pay them a few more bucks for what they've learned here at IBM."[36] In the increasingly competitive computer market of the 1980s, simply asserting that basic research would eventually pay off was no longer sufficient. Top management demanded tangible results and insisted on written agreements of understanding between Watson Research Center and development laboratories.

Architectural Forum praised the Watson Research Center as not only a dramatic architectural statement but, "at only $23 per square foot," a genuine bargain. The Architectural League of New York awarded it a Gold Medal. Yet just two years after the dedication, in 1963, IBM assigned all construction projects to a Corporate Facilities Planning Department. Its "package buildings" earned contempt from architects and architectural critics alike; they believed that "IBM has become less a patron than a hard-nosed client out to build fast and cheap."[37]

Looking back over IBM's decade-long experiment with late modernism, *Architectural Forum* concluded that "IBM turned to the architects not so much for quality or efficiency but for an image . . . and the danger for the architect as image-maker is clear: the day may

[34] Leslie interview with Pugh and Summers (quoting Pugh).

[35] Mason Southworth, "Helping IBM Scientists Pursue Their Goals," *Think*, 1989, *3*:17; and "50 Years of Research: The Science Behind the Solutions," press release, 1995, IBM Archives.

[36] "Fifty Years of Research," *IBM Research*, 1995, *3–4*:9–10; and Leslie interview with Pugh and Summers (quoting Summers).

[37] "Research in the Round" (cit. n. 26), p. 80 (bargain); and John Morris Dixon, "IBM Thinks Twice," *Architect. Forum*, 1966, *124*:33–39, on p. 33 (image).

come when the client no longer wants that image."[38] Saarinen's research campus, the architectural expression of IBM's commitment to basic research on the university model, would seem increasingly outdated in an era when success would be measured by time-to-market rather than Nobel Prizes, a corporate extravagance even IBM could no longer comfortably afford.

BELL LABORATORIES, HOLMDEL

The challenge for us in the new research center for Bell Telephone Laboratories at Holmdel, N.J., was to make today the same advance in the science of planning research laboratories for our time as the Bell Laboratories at Murray Hill, built twenty years ago, did for their time. To that end, we tried to benefit from all the experience, technology and ideas of amenities in research centers arrived at during the past two decades. As a result, we have achieved an entirely new kind of plan and a new technical development in building.

—*Eero Saarinen (1959)*

From its founding in 1925, Bell Laboratories has been the ultimate yardstick against which other industrial research laboratories have measured themselves, internationally recognized as not merely the biggest but the best. Bell Laboratories generated the patents (more than twenty thousand before divestiture in 1984) and the products that consolidated AT&T's control over telephony and made the idea of "One Policy, One System, Universal Service" a commercial reality.[39]

While AT&T certainly appreciated the importance of the grand architectural gesture, a prominent example being its sumptuous Greek Revival headquarters at 195 Broadway, it hid Bell Laboratories away in a former Western Electric manufacturing plant at 463 West Street in Manhattan. Almost immediately researchers complained about overcrowding and about the vibration, dust, noise, and electrical interference that played havoc with sensitive acoustic and radio experiments. When the city decided to run a railroad line within two blocks of the laboratory, research director Harold Arnold realized that he needed "an acreage laboratory," somewhere with room to grow far from the urban hubbub. At his urging, Bell Laboratories president Frank Jewett acquired some 240 acres in suburban Murray Hill, New Jersey, twenty-five miles west of New York City, and in August 1930 announced plans for building there a branch laboratory to be designed by Vorhees, Walker.[40]

Arnold died unexpectedly in 1933. Murray Hill might have died with him except for his successor, Oliver Buckley, who firmly believed that it could be far more than what Jewett called "a satellite laboratory" of 463 West Street, that it might indeed become a model for an entirely new kind of corporate laboratory. Instead of a conventional campus, Buckley imagined "a single building that retained most of the advantages of separate buildings but assured more intimate contact among departments and discouraged departmental 'ownership' of space."[41]

Later architectural critics would lambaste the Vorhees, Walker design as a classic ex-

[38] Dixon, "IBM Thinks Twice," p. 38.

[39] See Reich, *Making of American Industrial Research* (cit. n. 5), pp. 129–184, on p. 153; for the epigraph see Eero Saarinen, "Statement about Bell Telephone Laboratories for Holmdel, N.J.," Saarinen Family Papers, Cranbrook Academy Archives.

[40] H. D. Arnold, "An Acreage Laboratory," 19 July 1929, AT&T Archives, Warren, N.J., AT&T 464-06-02-11; and Bell Telephone Laboratories press release, 1 Aug. 1930, AT&T 464-06-02-11.

[41] Oliver E. Buckley, "Reminiscences," p. 26, AT&T 55-10-01-16.

Figure 5. *Bell Laboratories, Murray Hill, New Jersey, 1941. Designed by Vorhees, Walker, Murray Hill was Bell Laboratories' first research campus. The long corridor of the main building, though inconvenient for scientists, engineers, and managers, provided unexpected opportunities for chance encounters among different research groups. (Courtesy of AT&T Archives.)*

ample of AT&T's willingness "to hide its technology behind facades of studied mediocrity . . . in buildings made very well of brick but immersed in colonial design as conservative as the creosote bath given telephone poles" (see Figure 5). But Murray Hill's plain-Jane buff brick disguised some imaginative rethinking of laboratory space. Buckley was looking for the kind of flexibility an old Western Electric factory like 463 West Street could never provide. He also wanted to give the researchers themselves a significant role in designing the new laboratory and appointed a "Murray Hill Project Department" to give them a voice. Indeed, contemporary architectural critics considered Murray Hill a successful example of "designing from the inside out" and an attractive alternative to the "insensitive starkness or the cold austerity of much Modern Architecture."[42]

Bell Labs researchers–turned–architectural consultants started with what they knew best, the laboratory bench, and worked outward from there. They drew up a list of do's and don'ts drawn from their own experience and from what they had seen in other laboratories. Their cornerstone was a 6-foot module, equipped with the standard wet-chemical services of the day such as compressed air, hydrogen, nitrogen, oxygen, steam, vacuum,

---

[42] "The Telephone Company Dials the Moon," *Architect. Forum*, Oct. 1962, *117*:88–97, on p. 88; and "The Murray Hill Unit of Bell Telephone Laboratories," *New Pencil Points*, Aug. 1942, pp. 34–66, on p. 34.

three-phase current, and, of course, a telephone line.[43] These modules could then be expanded to any appropriate size up to a 24-foot width. To assess the idea, Vorhees, Walker
built a full-scale "test house" with five of the standard modules, including working laboratory services. This allowed both researchers and architects to check the appropriateness
of the basic dimensions, to modify the design to place benches, hoods, drains, and outlets
as needed, and to show subcontractors exactly what they would be constructing.

Perhaps the most original architectural feature of Murray Hill was its metal partitions.
These double steel panels—4 feet wide, 10 feet, 8 inches high, and 3 inches thick, including 2 inches of rock wool insulation—came either as a door and transom or as a full
panel, were completely interchangeable with one another, and could be snapped into special fittings in the floor and ceiling in a matter of minutes. In plan, Murray Hill resembled
a truncated H: there were two parallel corridors, one 340 feet long at the front of the
complex, and the other, 675 feet long, centered behind it; each was 58 feet wide, and they
were connected by a 144-foot transverse section. Projecting wings, each 55 feet long,
jutted out from the main buildings at regular intervals so researchers could get an office
window with a view—if only, for many of them, of an adjoining wing. The entire complex,
by Buckley's estimate, cost just over $3 million to build and another $2 million to equip
and furnish.[44]

Compared with the best of the International Style, Murray Hill may have seemed
bland—or worse. But at the time it attracted considerable attention from laboratory directors of other companies, more intrigued by its flexible frame than its plain face. As Buckley
later remarked, "It has been so successful a model that scarcely any large industrial laboratory has subsequently been built without taking ideas from it and some laboratories are
fairly close copies of it."[45]

Bell Labs moved the first group of 140 researchers from 463 West Street to Murray Hill
on 17 November 1941, just in time for the massive wartime buildup that would swell the
Bell Laboratories staff from 4,600 in 1941 to 8,000 by war's end. With Bell Laboratories
as the R&D arm and Western Electric as the manufacturing arm, AT&T designed and
manufactured about half of all American radar systems, far more than any other company,
and become a major supplier of sonar, electronic fire control, and communications systems
of all sorts.[46] After the war AT&T transferred most of its military projects, by then a huge
and specialized business that included the Nike air defense system and its successors, to
its laboratory at Whippany, New Jersey, ten miles north of Murray Hill.

Murray Hill, meanwhile, became the center for Bell Laboratories' basic research effort,
particularly in electronics and materials science research, as more and more departments
moved out of West Street into more modern quarters. Building 2, a five-story structure by
Vorhees, Walker built on the same "flexible space" modular plan as the original laboratory,
opened in the fall of 1949, more than doubling the total floor space at Murray Hill and
boosting the total staff there to 2,300. A decade later Bell Labs added Building 3, an office
complex designed to free up laboratory space in the other buildings. Designed in a slightly

[43] "Murray Hill Unit," p. 36.

[44] On the partitions see Franklin L. Hunt, "New Laboratories of Bell Telephone Laboratories at Murray Hill,
N.J.," June 1943, pp. 8–10, AT&T 464-06-02-11. On the overall shape see *ibid.*, p. 4; and "Murray Hill Unit,"
pp. 44–45. The cost estimates are from O. E. Buckley to C. M. A. Stine, 28 Sept. 1944, AT&T 74-05-03.

[45] Buckley, "Reminiscences" (cit. n. 41), p. 26.

[46] M. D. Fagen, ed., *A History of Engineering and Science in the Bell System: National Service in War and
Peace* (Murray Hill, N.J.: Bell Telephone Laboratories, 1978), provides a comprehensive survey of Bell Labs
military projects.

more modern style, with alternating bands of brick and glass instead of conventional brick piers, Building 3 housed another 900 employees, bringing Murray Hill's total staff to more than 3,000. William Baker, a polymer chemist and future Bell Laboratories president who came to Murray Hill in the first wave, fondly remembered just how much influence he and other researchers had on the layout of the laboratory spaces: "By adding new equipment and designing labs, generations of scientists and engineers at Murray Hill have controlled their own surroundings" (see Figure 6).[47]

Recognizing that the intellectual center of gravity had gradually shifted to Murray Hill, Bell Laboratories relocated its headquarters there in 1959. The complex got a final face-lift in 1972, a sloping futuristic facade of bronze-toned steel and tinted glass bestriding the main east–west axis of the older buildings. The design, by Vincent Kling, looked more like a World's Fair pavilion than an entrance to a corporate laboratory. It bears an uncanny resemblance to GE's "Horizons" exhibit entrance, built at Walt Disney's EPCOT at about the same time. If the original entrance had been inconspicuous, this one gave visitors the disconcerting feeling that they were checking into a John Portman hotel. With all the additions, the total population of Murray Hill topped 4,000 by the 1980s, including some 2,000 scientists and engineers.[48]

Always a leader, "The Lab," as locals dubbed it, set the pattern for New Jersey's "Research Row." Celanese, Ciba, Allied Chemical, Air Reduction, and Standard Oil all followed Bell into the suburbs, building strikingly similar, though much smaller, laboratories. As one reporter commented, "To the passerby, Bell's 225-acre retreat looks more like a college campus than some campuses. The Air Reduction Laboratory, with its roof-top restaurant, has been compared to a country club in appearance." In townships such as New Providence, a quarter of the new residents might be scientists and engineers employed at the laboratories. Their families put some pressure on local schools and municipal services, but the boroughs welcomed them, as they did their employers, as just the sort of neighbors they were looking for—well educated, well groomed, and prosperous. After all, the laboratories paid taxes—and wages. Murray Hill alone would end up paying some $3 million a year in local taxes and had a total annual payroll of $139 million.[49]

In scale, Murray Hill was closer to a national laboratory than to other industrial laboratories, and in international recognition, closer to Harvard, MIT, and Princeton combined. During the 1930s Mervin J. Kelly, as general research director, had built up a small but extremely good group in solid-state physics that included such future Nobel Prize winners as William Shockley, Charles Townes, and John Bardeen. Their mission, from the laboratories' perspective, was studying solid-state alternatives to vacuum tubes, essential for the kind of electronic switching systems already on the drawing board. They beat other industrial and academic groups to the transistor largely because of the interdisciplinary collaboration and engineering resources unique to Murray Hill. Even cramped cubicles had their advantages. Bardeen had to share an office with the semiconductor experts Walter Brattain and Gerald Pearson, and in trying to make sense of their experiments he worked out the theoretical foundations for solid-state electronics. The transistor, announced in

[47] "Second Murray Hill Building Unit Opened by Bell Laboratories," *195 Bulletin*, Nov. 1948, p. 3; "Murray Hill Building 3 Opens," *Reporter*, Mar. 1958, pp. 1–4, AT&T 464-06-02-11; and Loren Osterman, "Murray Hill Turns 50," 11 Nov. 1991, AT&T 464-06-02-04 (quoting Baker).

[48] "AT&T Bell Laboratories—Murray Hill," ca. 1983, AT&T 464-06-02-12.

[49] Lawrence G. Foster, "Research Row," *Newark Sunday News*, 13 Nov. 1949, p. 12 (quotation); Robert J. Klein, "Jersey Boomtowns," *ibid.*, 2 Dec. 1951, pp. 6–8 (welcome neighbors); and "AT&T Bell Laboratories—Murray Hill" (taxes and payroll).

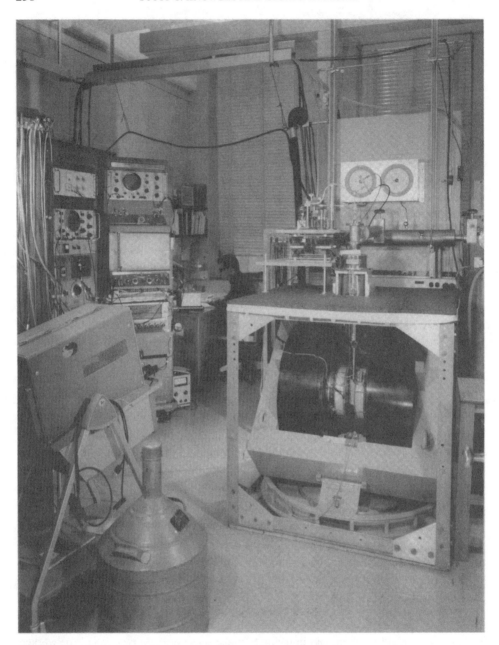

Figure 6. A typical laboratory layout at Murray Hill in the 1940s. The modular interior set new standards for flexibility. With common service cores set into the exterior walls and floor, and interchangeable wall partitions, such laboratories could be rearranged within a matter of hours. Note the drawn shades. Perhaps Saarinen knew the habits of corporate researchers better than they cared to admit. (Courtesy of AT&T Archives.)

1948, turned out to be Murray Hill's most visible triumph, winning Nobel Prizes for Shockley, Bardeen, and Brattain and creating a completely new industry.[50]

Another of Murray Hill's apparent architectural weaknesses, its seemingly endless hallways, proved to be one of its secret strengths. Former Bell Laboratories president Ian Ross, who worked at Murray Hill for some years before moving over to Holmdel, explained: "Look, the Murray Hill building was stupidly designed. It's a long building, so you put the main conference hall at one end, you then move the library from the middle and put it at the other end. I one time ordered a bicycle to get to the library. [He didn't get it!] You had to walk a quarter mile to get to the library or a quarter mile to the conference hall."[51] Yet all that walking had its rewards, because researchers could not help but run into one another, walk past each other's offices and laboratories, and find out what was going on in different departments.

By 1955, despite one major addition and another on the way, Murray Hill was overcrowded. Western Electric had a policy of never putting more than 5,000 employees in any one plant, and that seemed like a sensible idea for Bell Laboratories as well. To keep pace with the technical challenges of modern electronics communication, especially allelectronic switching, Bell decided it needed a facility as big as or bigger than Murray Hill and chose to build it at Holmdel, twenty-seven miles south of Murray Hill and about equally distant from New York City.

Since 1929 Bell Laboratories had maintained a research facility at Holmdel, where its then-rural isolation provided sufficient radio silence to study "noise." Karl Jansky, assigned to identify the sources of shortwave radio static on AT&T's new transoceanic radio telephone service, noticed a disturbance on his shortwave antenna that seemed to be coming from somewhere beyond the solar system; thus he inadvertently became the world's first radio astronomer, though he soon moved on to other projects. Holmdel also became a pioneering center in microwave electronics. After the war Holmdel turned its attention to radio relay repeaters and other circuit components for AT&T's growing nationwide radiotelephony system.[52] Its TD-2 transcontinental microwave relay system went into operation in 1950.

Resident engineers dubbed Holmdel the "Turkey Farm," and the plain-framed structure did indeed resemble a poultry coop more closely than a research laboratory. However primitive architecturally, Holmdel had what Bell Laboratories needed most: 450 vacant acres within a reasonable distance of its facilities at Murray Hill and Whippany. In selecting an architect, Bell Laboratories was looking for someone who could ensure "that Holmdel would point the way in the Sixties, as Murray Hill had in the Forties," but who was at the same time willing to work closely with its researchers and with Western Electric's construction engineers in designing the new laboratory. In signing up Saarinen in 1957, Bell Laboratories was deliberately shedding its conservative image in favor of something more in keeping with telecommunications in the space age. As one architectural critic perceptively commented after getting a look at the finished design: "It is clear at last that the

[50] Jeremy Bernstein, *Three Degrees above Zero* (New York: Scribner's, 1984), pp. 97–114, includes an accessible section on transistor development at Bell Labs; Michael Riordan and Lillian Hoddeson, *Crystal Fire: The Birth of the Information Age* (New York: Norton, 1997), is the best account of the invention and legacy of the transistor.

[51] Stuart Leslie interview with Ian Ross, 18 Aug. 1998.

[52] On Jansky see Bernstein, *Three Degrees above Zero* (cit. n. 50), pp. 201–203; and H. T. Friis, "The Holmdel Laboratory," *Bell Laboratories Record*, Dec. 1934, *13*:117–121. On postwar directions see "Holmdel," *Reporter*, Dec. 1952, pp. 9–11; and "Bell Laboratories at Holmdel Has Long, Distinguished History," *Bell Telephone Laboratories News*, 24 Sept. 1962, AT&T 464-06-02-07.

good old telephone company has stepped out of its village-uncle role and accepted a frank new characterization, that of the futuristic scientist."[53]

Saarinen believed that the "futuristic scientist" needed more peace and quiet and less of the hustle and bustle so characteristic of Murray Hill. In place of the warren of Murray Hill, he proposed a "compact building, with short communication lines and utmost privacy for groups of scientists." It would consist of four identical rectangular blocks, 700 by 135 feet each, six stories high (counting the basement, which would house most of the building's service facilities), separated by a vast interior courtyard and covered by a common roof. Each block was intended to accommodate about a thousand researchers, what Bell Laboratories considered a "university-sized" unit.[54] The main corridors, enclosed by a glass curtain wall, would run around the outside of the blocks, with smaller aisles running through the blocks to offices and laboratories, six cross-aisles in each section, and twelve sections on each floor. As in the Watson Research Center, all of the offices and laboratories would be placed in interior space, arranged back-to-back, with service or storage corridors in between, though on a rectangular grid rather than an arc. Researchers would only have to walk across the hall to reach their main work spaces and would not be disturbed by foot traffic from the main corridors.

Holmdel, like Murray Hill, was designed on a 6- by 6-foot modular pattern that could be quickly reconfigured into any multiple, though at Holmdel the standard-issue office was a more generous 12 feet deep and the standard laboratory 24 feet deep, separated by 6-foot-wide aisles. As he had at IBM, Saarinen put token glass panels on the upper portion of the side and aisle partitions and interchangeable shelves, file cabinets, and bookcases on the rear walls, with blank panels in leather-textured vinyl. Where the Watson Research Center had been a kaleidoscope of bold reds, blues, and yellows, Holmdel would be all muted grays, blacks, whites, and off-whites. In place of the local fieldstone that distinguished Watson, Holmdel got bush-hammered concrete in the corridors, the elevator towers, and even the lobby.[55]

Holmdel's signature would be its glass facade, five stories of 6- by 3-foot panes (6,800 of them) held in aluminum mullions by black neoprene gaskets: "the biggest mirror ever"—or the perfect Miesian glass box (see Figure 7). From the inside, the panes—a unique sandwich of glass and aluminum-chromium film that reflected 70 percent of the sun's heat while letting in 25 percent of its light—looked like conventional tinted glass. From the outside, they looked like a low-brightness mirror, either blinding the viewer or dissolving into the sky and the landscape, depending on the angle. At night the glass panels disappeared entirely, revealing the stunning building within. Unfortunately for Saarinen, the contractor could deliver only enough reflecting panels to cover the south section of the original building. The rest got conventional gray-tinted glass. "Instead of offices and laboratories with small windows, covered with Venetian blinds or other sun-screening devices," Saarinen declared, "there can be a continuous floor-to-ceiling glass wall for the corridors around each building block. The individual, emerging from concentration in laboratory or office, will come upon magnificent, uninterrupted views of the surrounding countryside and the winter-garden interior courts as he walks, in moments of relaxation, down these periphery main corridors."[56]

---

[53] "Telephone Company Dials the Moon" (cit. n. 42), p. 88.

[54] *Ibid.*, p. 93.

[55] *Bell Laboratories News*, 26 Sept. 1962, p. 7, AT&T 464-06-02-06.

[56] "The Biggest Mirror Ever," *Architect. Forum*, Apr. 1967, *126*:33–41; and Saarinen, "Statement about Bell Telephone Laboratories for Holmdel, N.J." (cit. n. 39), p. 2.

Figure 7. Bell Laboratories, Holmdel, New Jersey, 1966. Saarinen's last and most monumental laboratory commission, Holmdel combined the scale of the GM Tech Center with the alternating corridors of laboratories and offices of the IBM Watson Research Center. Essentially four blocks connected by a giant atrium, Holmdel's glass facade perfectly captured the research ethos of late modernism; it was the perfect Miesian box. (Courtesy of AT&T Archives.)

Dominating that view would be a 6-acre reflecting lagoon (and cooling pond) with a fountain display worthy of Versailles itself, all encircled by an enormous traffic oval. The architectural critic Rupert Spade judged it the best of Saarinen's "Royal Hunting Lodge[s]" and "an even more grandiloquent indulgence in the landscape world of Louis XIV." Indeed, the grounds would be planted with some 7,900 maples, willows, pines, junipers, yews, and other trees and scrubs, giving Holmdel the "isolated grandeur worthy of a Renaissance chateau."[57]

Saarinen designed the laboratory so that it could be built in stages. Phase 1, begun in 1959 and completed in September 1962, just one year after his death, cost $21 million and provided space for 1,300 researchers and staff. Phase 2, begun in 1964 and completed in 1966, doubled the size of the original building and cost $17.5 million, providing space for 4,500 employees. Phase 2 allowed Saarinen's protégé Kevin Roche to complete the mirrored facade and add the stunning garden courtyard—700 feet long, 100 feet wide, and 70 feet high—that Saarinen had in mind from the beginning. The atrium, with some 3,600 plants, shrubs, and trees, from houseplants to exotic tropicals, resembled a modern Hanging Gardens of Babylon, with overlooking walkways and bridges between the buildings at every floor, a reception area at one end, and an employee lounge at the other.[58]

---

[57] Spade, *Eero Saarinen* (cit. n. 15), p. 15; and Temko, *Eero Saarinen* (cit. n. 1), p. 37.
[58] "Biggest Mirror Ever" (cit. n. 56), p. 38.

Where Murray Hill was about as far toward the "R" on the R&D spectrum as an industrial laboratory could get, Holmdel was closer to the "D." Its primary mission was electronic switching—everything from massive central station equipment and top-secret military systems to consumer products such as the push-button phone and Touch-Tone dialing. Holmdel engineers often considered Murray Hill rather "blue sky." They worked much more closely with their counterparts in Western Electric manufacturing plants who made switching systems for local telephone companies or cordless telephones than with the scientists at Murray Hill. Holmdel's big project, at least in the early years, was developing an electronic switching system to replace the relays, switches, and wired-in logic of conventional electromechanical switching systems. Holmdel's staff reached 1,700 by the end of 1961, 3,300 by 1965, and 4,200 by 1967, overtaking Murray Hill as Bell's largest laboratory.[59]

Paradoxically enough, while communication was Holmdel's main mission, encouraging communication within the new laboratory turned out to be one of its most perplexing challenges. As one former researcher joked, in sly parody of AT&T's best-known advertising campaign, the motto for Holmdel might well have been "Reach out and touch someone; just not here." In practice, Saarinen's interior aisles ended up isolating the researchers from one another. Without any natural means of circulating from one floor to another, researchers rarely strayed from their own aisle. Ian Ross noticed the difference immediately after arriving from Murray Hill: "I had to take steps to get them to communicate with one another, because they didn't have that [long] corridor. So it had its drawbacks. It's nice and peaceful and quiet but you have to force people to meet one another."[60] The laboratories were flexible, and groups could be moved from one building to another in a few days and with little disruption, but the offices and laboratories themselves seemed strangely sterile. Predictably, people chose the shortest distance between two points, generally the walkways, bridges, and elevators, and ignored the exterior corridors altogether, despite the view. Almost no one met there, since there was no place to sit, not even uncomfortable wooden benches like those at the Watson Research Center. "All you could do was go and lean against the railing and look out the window," one engineer recalled. Unlike Watson Research Center, where no one ever got a corner office, Ross insisted on this traditional management perquisite: "I built an office on the corner there and I had my window, and it was nice, too."[61]

Like the GM Tech Center and the Watson Research Center, Holmdel won its share of accolades and awards. Watching the first stage reach completion, the architectural critic Allan Temko said it just might prove to "be one of the influential buildings of the age. . . . These softly gleaming laboratories, everywhere organized with quiet perfection, will be a civilized beacon of human knowledge. . . . Bell is a work of fully matured genius." *Industrial Research*, the trade journal for corporate research managers, named Holmdel its "Laboratory of the Year" over such contenders as the National Bureau of Standards campus in Gaithersburg, Maryland, the Corning Glass Works Sullivan Park Science Complex, and I. M. Pei's striking National Center for Atmospheric Research, set against the backdrop of the Rocky Mountains in Boulder. The prize committee, headed by RCA Laboratories vice president James Hillier (whose son would later become one of America's premier

[59] "Bell Laboratories at Holmdel Has Long, Distinguished History" (cit. n. 52), pp. 5–6 (primary mission); and Guy Accettura, "A Neighbor Looks at Monmouth County," 13 Dec. 1967, p. 12, AT&T 464-06-02-04 (staff size).

[60] Stuart Leslie interview with Emily Thompson, 6 July 1999; and Leslie interview with Ross.

[61] Leslie interview with Thompson; and Leslie interview with Ross.

laboratory architects), praised Holmdel's flexibility, functionalism, and "masculine design," though it did criticize the inevitable "loss of communication in a floor-over-floor configuration." A corporate spokesman told the committee, "We are trying to maintain the campus environment. The type of people we hire here naturally are drawn to campus life, and in many ways we are in competition."[62]

In the end, Holmdel may have turned out to be a perfect architectural expression of Bell Laboratories and its parent company—monumental, impersonal, inward looking, and self-contained. Summing up the international race for commercial fiber-optic cables, the industry analyst Jeff Hecht writes: "In many ways, Bell Labs was a surprising also-ran, a world-class laboratory that scored few conceptual breakthroughs and spent untold millions chasing dead ends. Bell suffered from being fat and happy and too quick to reject outside ideas in favor of its own. . . . Had the prestigious lab become stuck in the mud?" Temko had correctly predicted that "later architects will find it difficult . . . to surpass the splendor of Bell's palatial baroque park which, in this country at least, is unrivaled as a formal setting for a technological building."[63] This modern Versailles may have matched the original in cost, in architectural assurance, and as an expression of prosperity and power. But its "hall of mirrors" faced outward, resisting closer scrutiny and suggesting, if only symbolically, that everything worth knowing was already inside.

CENTERS AND PERIPHERIES

There was a time—and it was not so very long ago—when the architectural profession was led by people whose work was at once popular and profound, at once of interest to the public at large and to their most serious colleagues. The work of a few crucial practitioners seemed to provide the key to understanding an entire time. In the 1950s, for example, Eero Saarinen's buildings seemed to sum up everything that decade was about—a sense of technological daring, a sense of bold geometric forms, and yet also an underlying respect for the status quo.
—*Paul Goldberger (1983)*

Do Saarinen's masterpieces have a future in a corporate world so obsessively focused on the bottom line and the quarterly report? Can they survive, much less prosper, in an environment where high technology companies like Intel and Compaq openly and successfully flout conventional wisdom about corporate support for basic research? Saarinen's corporate laboratories, like the national laboratories they so much resemble, must now search for meaning and mission in a world turned upside down. The business journalist Robert Buderi may be right in calling today's corporate laboratories "engines of tomorrow," but they are as different from yesterday's engines as a Honda CVCC with lean-burn technology is from an Oldsmobile 455 Rocket.[64]

---

[62] Temko, *Eero Saarinen* (cit. n. 1), pp. 40–41; and Graeme C. Keeping, "I.R. Laboratory of the Year," *Industrial Research*, May 1967, pp. 63–68, on p. 68.

[63] Jeff Hecht, *City of Light: The Story of Fiber Optics* (New York: Oxford Univ. Press, 1999), p. 231; and Temko, *Eero Saarinen*, p. 40.

[64] Robert Buderi, *Engines of Tomorrow: How the World's Best Companies Are Using Their Research Labs to Win the Future* (New York: Simon & Schuster, 2000); for the epigraph see Paul Goldberger, *On the Rise: Architecture and Design in a Postmodern Age* (New York: Times Books, 1983), p. 19. On the dilemmas of national laboratories at the dawn of the new millennium see Robert Crease, *Making Physics: A Biography of Brookhaven National Laboratory* (Chicago: Univ. Chicago Press, 1999); Jack Holl, *Argonne National Laboratory, 1946–96* (Champaign: Univ. Illinois Press, 1997); Michele Gerber, *On the Home Front: The Cold War Legacy of the Hanford Nuclear Site* (Lincoln: Univ. Nebraska Press, 1992); and Bruce Hevly and John Findlay, eds., *The Atomic West* (Seattle: Univ. Washington Press, 1998).

Farsighted Bell executives had long understood that "one system . . . and universal service" could be achieved only if the ideas brought to life at Murray Hill and, later, Holmdel eventually made their way out of the laboratory and into the Bell System. In a remarkably prescient memorandum written in 1943, future Bell Laboratories president Mervin J. Kelly tackled the issue of coordinating R&D with manufacturing. During what Kelly called the "non-competitive period" of the 1920s and 1930s, AT&T had essentially owned long-distance telephony through its mastery of (and patents for) vacuum thermionic triodes. But with wartime advances in radar and microwave electronics, most of them funded by the government, Bell suddenly found itself facing unexpected competition from companies like RCA, which had nearly as much experience in high-frequency communications. "We have been a conservative and non-competitive organization," Kelly stressed.

> We engineer for high quality service with long life, low maintenance costs, high factor of reliability as basic element [sic] in our philosophy of design and manufacture. But our basic technology is becoming increasingly similar to that of a high volume, annual model, highly competitive, young, vigorous and growing industry. . . . Our greatest element of weakness in the development-design and design sectors stems from the separation in space and organization of this work from that of engineering for manufacture and the manufacturing plants.[65]

By combining the unique research resources of Bell Laboratories with the unmatched manufacturing know-how of Western Electric, branch laboratories would give AT&T an important edge in the struggle for the control of postwar electronics. Even at the high noon of science as the endless frontier, research never represented more than a small fraction of AT&T's overall R&D budget. In 1954, for instance, all but $23 million of the Bell Labs total budget of $110 million was under contract to Western Electric for development projects or to the Department of Defense for military projects. That $23 million, just 0.5 percent of the Bell System's gross operating revenues, supported virtually everything that could be called research at Murray Hill.[66]

Kelly got the opportunity to put his branch laboratory idea into practice with the building of Western Electric's Allentown electronics manufacturing plant in 1947, first for klystrons, magnetrons, and other state-of-the-art microwave electronics components, and later for the transistor and its descendants—the integrated circuit, the microprocessor, and VLSI. From the original team of 30, Bell's Allentown group grew to 180 by 1958; in addition, 175 Western Electric engineers were assigned to the branch laboratory. To put its two-million-component megachip into production in the 1980s, Bell Laboratories increased the staff at Allentown to 1,000. In 1988 Ian Ross formally dedicated an $85 million Solid State Technology Center (SSTC) on a 200-acre site ten miles southwest of Allentown. In the spirit of Silicon Valley, the Solid State Technology Center promised not only "a campus-like environment that inspires employees and fosters creativity" (with no reserved parking!) but also to anchor "AT&T's Photonics Valley" and make Bell Laboratories a leader in light-wave technology, including optical relays, optical data links, lasers, photonic switching microchips, and other components for Bell System customers.[67]

[65] M. J. Kelly, "A First Record of Thoughts Concerning an Important Postwar Problem of the Bell Telephone Laboratories and Western Electric Company," May 1943, AT&T 464-06-02-04.

[66] Ibid.; for the figures see Francis Bello, "Industrial Research: Geniuses Now Welcome," Fortune, Jan. 1956, 53:142.

[67] "Allentown Marks Its First Decade," Reporter, Apr. 1958, p. 12; "40th Anniversary Marks AT&T's History of Providing Service," Allentown Morning Call, 22 Feb. 1987, AT&T 464-06-02-04; and Russ Glover, "Quest for World-Changing Technologies," AT&T Focus, 22 Dec. 1988, pp. 26–27 (quotations).

Throughout the postwar era, branch labs played a vital role in coupling Murray Hill and, later, Holmdel to AT&T's operating divisions. Their projects may have been less visible within the scientific community, and certainly never won any Nobel Prizes, but they gave Western Electric the competence and the confidence to move into the solid-state era, a bold move considering the standards of reliability expected from the Bell System. Saarinen's glass box got the headlines, but the branch labs and Western Electric made the money.

At General Motors, too, the very opulence and visibility of the Tech Center tended to overshadow just how much the divisions continued to contribute to GM innovation. The divisions, which had always maintained an uneasy relationship with the Kettering laboratory, never entirely surrendered their relative autonomy to the college boys in the Tech Center. If anything, the divisions became more suspicious of the Research Division after 1956 and worked harder than before to shore up the capabilities of their own labs. A look through the pages of the *General Motors Engineering Journal* in the years just before and for the first decade after the Tech Center opened gives some sense of the relative proportions of research output generated by the Tech Center and the divisions. Divisional engineers held their own against Hafstad and his physicists. Collectively, they outpublished the research laboratory staff five to one and filed more patents. Many divisions refurbished older research centers or built new ones in the postwar years. To be sure, none of these labs fit the campus model, but neither did they limit themselves to Kettering-style applied research. Arguably, divisional engineers contributed more to GM's products and bottom line than did their more academically inclined counterparts at the Tech Center. Delco Radio, for instance, did much more than design car radios. In 1954 it established its own Semiconductor Research and Engineering Department in Kokomo, Indiana, where researchers learned to grow high-purity silicon crystals competitive with those from the best academic and industrial labs and won independent defense contracts for the servo circuits, preamplifiers, and power controls for the ballistic missile guidance systems being designed by MIT's Instrumentation Laboratory. Rarely did the Tech Center simply invent and the divisions apply, as the linear model might imply. In fact, GM's own engineers had a tough time recalling who had contributed what to the Toronado, GM's mid-1960s experiment with front-wheel drive. The Tech Center certainly played a critical role in early thinking about a front-wheel-drive automobile, but Oldsmobile, with the help of other divisions like Buick and Saginaw, seems to deserve the lion's share of the credit (or blame) for getting the Toronado off the drawing board and onto the assembly line.[68]

Even by the crassest measure—corporate budgets—the GM Tech Center came in a distant second to the divisions. Geniuses may have been welcome at the new central laboratories of the 1950s, as *Fortune* put it, but they did not necessarily get much money to play with. The Tech Center cost $125 million, but Hafstad's research budget in 1954 was a paltry $12 million a year, or about 0.1 percent of sales. The divisions, by contrast, spent perhaps $200 million on their own research and development.[69]

The IBM innovations that mattered most often came out of smaller development laboratories attached to manufacturing facilities, whether in San Jose or Poughkeepsie. Some of the Watson Research Center's best work, such as RISC architecture for small computers

[68] Henry Barr, "Product Engineering in General Motors," *General Motors Engineering Journal*, 1966, *13*(1):9–10 (division strengthening and the Toronado); and James H. Guyton and Frank E. Jaumot, Jr., "The Organization of Research and Engineering Activities at Delco Radio Division," *ibid.*, Jan./Feb./Mar. 1961, pp. 24–25.

[69] Bello, "Industrial Research: Geniuses Now Welcome" (cit. n. 66), p. 142.

and its router technology, ended up being orphaned by IBM, at least temporarily, and perfected by upstarts like Sun Microsystems and Cisco.[70] The invention of disk memory by IBM's San Jose Research and Engineering Laboratory provides an instructive contrast with the later false starts and failed follow-ups of the Watson Research Center. Founded in 1952 adjacent to IBM's West Coast punch-card factory (IBM produced sixteen billion cards a year in those days), the San Jose laboratory had two primary goals: to attract West Coast electronics engineers reluctant to relocate to scenic Endicott, New York, and to think about alternatives to punch-card storage and retrieval systems. As a replacement for the cumbersome card, San Jose's engineers came up with the first magnetic disk memory, a stack of rotating magnetic disks, much like a stack of record albums, with magnetic heads to read and record information on the concentric tracks of the disks, much like a phonograph needle. They dubbed it "jukebox memory."[71] IBM San Jose followed up the RAMAC success with a series of cheaper and denser storage files, culminating in the innovation that defined the personal computer industry, the disk drive.

In the popular imagination, IBM San Jose may have seemed on the margins of Silicon Valley, but technically it was as central as the more famous Xerox PARC (see Figure 8). Despite its buttoned-down, East Coast origins, IBM San Jose fit perfectly into the emerging Silicon Valley culture. A reporter covering the plant dedication in 1958 captured the new spirit of West Coast electronics:

> Its tremendous success may be attributed to the simplest little formula—accumulate a lot of brains, give them the facilities, turn 'em loose—and they'll come up with the answers. . . . It's so different out there in those campus surroundings, so different from the usual pell-mell workday experiences, that it's like being in another world. . . . The whole place is peopled by young brains, no long hairs, and they all give the impression that they're exceedingly proud and gladly would die for old IBM.[72]

Not all of them, apparently. Maverick Al Shugart, who led IBM's major advances in magnetic storage, defected to Memorex in 1969 to head its storage products development group, eventually taking more than a hundred former IBM engineers with him. Shugart quit Memorex in 1973 to found Shugart Associates, which pioneered the hard disk for personal computers, and later founded disk master Seagate Technologies, the other major manufacturer of hard drives. Utterly unconventional by East Coast IBM standards, Shugart's career path virtually defined Silicon Valley. Of the hundred IBMers who followed Shugart to Memorex, fully a third ended up heading their own companies. Meanwhile, back at the Watson Research Center, IBM's research director mused that getting the most out of his 900 Ph.D.'s was tougher than trying to conduct a symphony of soloists: "It's an orchestra full of composers."[73]

Saarinen, of course, considered himself a master composer, but his architectural symphonies, as visually arresting as they were, rarely achieved the sustained harmony he intended. At their best, his laboratories froze the worldview of one era in a way that the

[70] Pugh, *Building IBM* (cit. n. 20), p. 240.

[71] David W. Kean, *IBM San Jose: A Quarter Century of Innovation* (San Jose, Calif.: IBM, 1977), pp. 5, 27–32.

[72] *Ibid.*, p. 48.

[73] Hubert Kay, "Harnessing the R. and D. Monster," *Fortune*, Jan. 1965, *71*:163. On Shugart see "Award for Achievement," *Electronics Week*, 14 Jan. 1985, *58*:40–44; Tom Diederich, "Al Shugart Has Just One Question: Why?" *Computerworld*, On-line News, 23 July 1998; and "Al Shugart," in *Jones Telecommunications and Multimedia Encyclopedia*, www.digitalcentury.com.

Figure 8. IBM San Jose Research Laboratory. Though less architecturally distinguished than its big brother in Yorktown Heights, the San Jose laboratory's close interaction with development and manufacturing gave it an important edge in getting products out of the laboratory and into the market. It spawned entrepreneurs along with innovations, including thirty future presidents of Silicon Valley start-ups. (Courtesy of IBM Archives.)

worldview of the next has only begun to thaw. As R&D centers for monopolies, regulated or otherwise, they could pursue fundamental research in the expectation of eventual payoff, since no competitor could capture more than a small share of the economic return. Under those conditions, a visible research laboratory could even be valuable evidence that market control rested on technological leadership rather than raw economic power.

Neither the Justice Department nor the companies themselves could have guessed how quickly global competition in the 1970s and 1980s would transform their respective industries. By 1980 Japan would be manufacturing more automobiles than the United States, while GM's share of the U.S. market was shrinking from half to less than a third. IBM would remain the world's largest computer maker, but its main competition would be from U.S. firms entering the business from unexpected directions, such as Hewlett-Packard and Compaq, and from Japanese companies such as Fujitsu, NEC, and Hitachi, which by 1994 would have higher combined sales than IBM. While IBM would continue to dominate the large-system market, it would face intense competition in mid-range systems, workstations, software, peripherals, and of course personal computers.[74] Even Bell Labs suddenly found

[74] Olegario, "IBM and the Two Thomas J. Watsons" (cit. n. 20), pp. 377–378, 394–395.

itself playing the innovation game under entirely new rules, where the score was kept in terms of patents, products, and profits rather than scientific papers, prestige, and prizes.

In response, all three corporations cut back research expenditures substantially and re-oriented remaining projects toward more short-term objectives. IBM cut its R&D budget by a fifth in the wake of its near-collapse in 1993, laid off half its physicists, and dropped "Famous for Its Science and Technology" from its motto. Deep Blue, the famous chess-playing computer, got the headlines, but less glamorous workstations, backed by a new branch laboratory in Austin, Texas, and magnetoresistive heads for laptops, from Almaden (the successor to San Jose), grabbed the markets. Isamu Noguchi's famous 1964 *Garden of the Future*, commissioned for IBM's corporate headquarters in Armonk, New York, with its emphasis on how fundamental science was charting the world of tomorrow, seems as quaint a relic as a Selectric typewriter. Nowadays, joint programs with development divisions account for 25 percent of IBM's total research budget (estimated at $550 million), and research targeted at ongoing development efforts accounts for a sizable share of the rest.[75]

Even at Bell Labs, Nobel laureate Arno Penzias, as research director, has had to cut back the very programs that led to his prize for detecting the background radiation from the Big Bang; he admits that "a lot of today's basic research goes far beyond any commercially interesting objectives. . . . The change in research priorities is not a nice message for fundamental science. But the opportunities in applied science are greater than ever."[76]

General Motors still claims one of the biggest R&D budgets of any corporation, yet all but a few percent goes for engineering work in the divisions. In May 2000 it placed the Tech Center on the National Registry of Historic Places. Though the purpose was to claim a tax credit for renovating Saarinen's campus, there could hardly be a more fitting symbol of changing times.

As Paul Goldberger astutely observed, Saarinen's work perfectly captured the spirit of corporate America in the 1950s because it never fundamentally challenged its assumptions. Saarinen certainly appreciated the idea—or, perhaps more correctly, the ideal—of research, at least as it was understood by corporate executives. Yet he demonstrated surprisingly little appreciation for its practice or its practitioners. His laboratories celebrated a classical interpretation—the linear model—that would be obsolete almost before his masterpieces were finished. One of their toughest challenges in the years ahead will be overcoming an ideology of research embodied in their very architecture.

---

[75] On IBM's problems see Hounshell, "Evolution of Industrial Research in the United States" (cit. n. 2), p. 56; and Buderi, *Engines of Tomorrow* (cit. n. 64), pp. 171, 129. On Noguchi see Isamu Noguchi, *A Sculptor's World* (New York: Harper & Row, 1968).

[76] Penzias is quoted in Malcolm W. Browne, "Prized Lab Shifts to More Mundane Tasks," *New York Times*, 10 June 1995, p. C1.

Rethinking Big Science

Modest, Mezzo, Grand Science and the Development of the Bevalac, 1971–1993

By Catherine Westfall

ABSTRACT

Historians of science have tended to focus exclusively on scale in investigations of large-scale research, perhaps because it has been easy to assume that comprehending a phenomenon dubbed "Big Science" hinges on an understanding of bigness. A close look at Lawrence Berkeley Laboratory's Bevalac, a medium-scale "mezzo science" project formed by uniting two preexisting machines—the modest SuperHILAC and the grand Bevatron—shows what can be gained by overcoming this preoccupation with bigness. The Bevalac story reveals how interconnections, connections, and disconnections ultimately led to the development of a new kind of science that transformed the landscape of large-scale research in the United States. Important lessons in historiography also emerge: the value of framing discussions in terms of networks, the necessity of constantly expanding and refining methodology, and the importance of avoiding the rhetoric of participants and instead finding words to tell our own stories.

I F WE WANT TO UNDERSTAND THE DYNAMICS of large-scale research—that is, team research employing complicated equipment or procedures—should we focus only on the burgeoning development of the very largest projects? It has been easy to answer in the affirmative, especially for those of us who have studied the history of research performed with U.S. particle accelerators. Our choice has seemed clear cut: high-energy physics has simply appeared more grand in every aspect than other types of accelerator-based science, from the size and expense of high-energy physics equipment to the expansive personalities of high-energy physicists and their intense determination to conquer enor-

Isis 94, no. 1 (March 2003): 30–56.

This work has been authored by a contractor of the U.S. government under contract no. DE-AC03-76SF00098. Accordingly, the U.S. government retains a nonexclusive royalty-free license to publish or reproduce the published form of this contribution, or allow others to do so, for U.S. government purposes. I conducted all the interviews cited, and all interviews and documents are in my possession unless otherwise noted. I am indebted to Peter Galison, Robert W. Smith, Karen Rader, Margaret Rossiter, and the unnamed *Isis* referees for many useful comments and suggestions and to Lee Schroeder, Reinhard Stock, Hans Gutbrod, and especially Gary Westfall for technical help and advice.

mous technological complexity for the sake of dramatic scientific simplicity—exploration of the fundamental nature of matter. One consequence has been that the recorded history of large U.S. accelerators (which is sometimes referred to as a type of Big Science or Big Physics) is really little more than the history of high-energy physics.[1]

Thanks to this tendency, there are many untold tales of large-scale research, including the story of the Bevalac, an accelerator built in the 1970s at the California laboratory made famous by Ernest Lawrence in the 1930s, the Lawrence Berkeley Laboratory (LBL).[2] The Bevalac was not a high-energy physics accelerator, nor did it require unprecedented funding. This research tool was designed not simply to accelerate protons but to accelerate nuclei—and relatively heavy nuclei at that—for studies of nuclei and nuclear matter. This heavy-ion accelerator for nuclear physics was formed in 1974 by joining two preexisting accelerators: the world-renowned Bevatron, which had outlived its usefulness after its high-energy proton beam yielded many particle discoveries, and the somewhat less famous SuperHILAC, a heavy-ion accelerator of more modest scale. After accelerator experts built a beam-transport line to connect the two machines and made some upgrades to both accelerators—at a cost of less than $10 million over a period of eight years—heavy ions could be produced in the SuperHILAC, transported down the beam line, and then accelerated to relativistic energies, that is, close to the speed of light.

The Bevalac's modest price tag and the kind of science it made possible are not the only features that set it apart from the usual high-energy physics project. Bevalac users, unlike their counterparts in high-energy physics, were neither expansive nor renowned. In fact, one of the puzzling (and particularly revealing) turns of the Bevalac story is that when the accelerator first began operating, Berkeley's fine crew of high-energy physicists generally refused to join the research effort. Indeed, at this point the machine attracted few users at all. Its initial unpopularity could be understood as merely the result of limited research potential if the cobbled-together Bevalac had begun with unpromising prospects.[3] Instead, the machine offered glittering research prospects from the start—and indeed it ushered in an entirely new subfield of physics and eventually inspired construction of the $500 million Relativistic Heavy Ion Collider (RHIC), built at Brookhaven National Laboratory in New York in the late 1990s.[4]

Clearly the untold tale of the $10 million Bevalac warrants our attention, even if this project was dwarfed by the largest high-energy physics projects of the time, such as Fermilab, the $250 million facility built in Illinois. Indeed, a close look at this "mezzo science"

[1] The study of non-U.S. particle accelerators has been less restrictive. Two examples of studies of smaller, non-high-energy physics accelerators are Gregers Hansen, "The SC: Isolde and Nuclear Structure," in *History of CERN*, Vol. 3, ed. John Krige (Amsterdam: Elsevier, 1996), pp. 327–413; and Siegfried Buchhaupt, *Die Gesellschaft für Schwerionenforschung: Geschichte einer Großforschungseinrichtung für Grundlagenforschung* (Frankfurt: Campus, 1995).

[2] For a comprehensive history of the beginnings of Lawrence's laboratory see J. L. Heilbron and Robert Seidel, *Lawrence and His Laboratory: A History of the Lawrence Berkeley Laboratory,* Vol. 1 (Berkeley: Univ. California Press, 1989). After the 1970s the Lawrence Berkeley Laboratory was renamed the E. O. Lawrence Berkeley National Laboratory (Berkeley Lab). Since the older name was in use as the Bevalac developed, it will be used here except in notes that give the current address of the laboratory.

[3] The Intersecting Storage Ring built at the European high-energy physics laboratory CERN provides an example of a machine that fit the preferences of managers (in this case because it provided technological challenges) but not the research interests of prospective users. See Arturo Russo, "The Intersecting Storage Rings: The Construction and Operation of CERN's Second Large Machine and a Survey of Its Experimental Programme," in *History of CERN*, Vol. 3, ed. Krige (cit. n. 1), pp. 97–170.

[4] The relativistic heavy-ion project at Brookhaven was facilitated by the existence of a tunnel constructed for a high-energy physics accelerator, ISABELLE, that was never built.

project—that is, a medium-sized large-scale project—reveals that by transcending the preoccupation with bigness historians of science can gain a glorious new view that reveals how disconnections and connections, as well as the interconnections of research enterprises of different types and scales, formed a new kind of science and thereby transformed the landscape of large-scale research in the United States.

For help in setting a course through this landscape, I first examine how historians and scientists framed initial discussions about U.S. large-scale research in terms of "Big Science," then explore more recent refinements of the historiography of large-scale research. Drawing on insights gained from this historiography, I next unravel the mystery of the Bevalac. How were plans for the project set into motion? Why was the machine initially so unpopular? How was an unpopular machine transformed into a trendsetter, and what were the consequences of its success? And, in particular, what was the connecting tissue that eventually formed the new science of relativistic heavy-ion physics? I will end by assessing how the Bevalac story helps to expand and refine our understanding of large-scale research in the United States.

THE STUDY OF "BIG SCIENCE" AND THE PREOCCUPATION WITH BIGNESS

For more than thirty years, scholars of many stripes have discussed "Big Science." The terms of this discussion were framed by the popular writings of the physicist Alvin Weinberg and the physicist/historian Derek de Solla Price in the 1960s. Both authors focused on the rapidly increasing size and expense of scientific projects, both identified this increase in scale as a distinctive feature of science in the post–World War II era, and both worried about the consequences of the "disease" of Big Science.[5]

Ever since, scientists have tended to shape the discussion of Big Science along the lines of professional self-interest. For example, Weinberg, an administrator in the 1960s at the national laboratory in Oak Ridge, Tennessee, suggested that the funding for large-scale research be confined to national laboratories to "prevent the contagion" of Big Science from spreading. Big Science bashing was still popular decades later, as Daniel Kevles has noted: those who coveted money allocated for the Human Genome Project in the 1990s labeled that project "Big Science," invoking the pathological connotations of the term in the hopes of derailing its funding.[6]

Other scientists found a way to turn criticism of Big Science inside out. For example, those who promoted the construction of new, larger high-energy physics accelerators used the size and expense of their enterprise to advocate even larger funding allocations. Exploiting the American predilection for the grand, such promoters argued that the rapidly increasing federal expenditures for high-energy physics equipment signaled the superior value of this type of research. The next step was to argue that this superior science, which

[5] Alvin M. Weinberg, "Impact of Large-Scale Science," *Science,* 1961, *134*:161–164; Weinberg, *Reflections on Big Science* (Cambridge, Mass.: MIT Press, 1967); Derek J. de Solla Price, *Science since Babylon* (New Haven, Conn.: Yale Univ. Press, 1961), Ch. 5; and Price, *Little Science, Big Science . . . and Beyond* (New York: Oxford Univ. Press, 1963). Weinberg spoke of the "triple diseases" of Big Science—"journalitis, moneyitis and administratitis": Weinberg, "Impact of Large-Scale Science," p. 162; Price described the dangers of large-scale research in "Diseases of Science," in *Little Science, Big Science . . . and Beyond,* pp. xix–xx. They are quoted in James H. Capshew and Karen A. Rader, "Big Science: Price to the Present," *Osiris,* 1992, *7*:3–25, on pp. 5, 6. This source gives a more thorough explanation and critique of the views of Weinberg and Price.

[6] Capshew and Rader, "Big Science," p. 6; and Daniel J. Kevles and Leroy Hood, "Reflections," in *The Code of Codes: Scientific and Social Issues in the Human Genome Project,* ed. Kevles and Hood (Cambridge, Mass.: Harvard Univ. Press, 1992), pp. 300–328, esp. pp. 301–303.

was intrinsically expensive, required more funding in order to continue to grow and prosper and thus meet its grand destiny.[7] Big Science promotion would continue, alongside Big Science bashing, through the 1980s.[8]

Historians have also proceeded down the trail blazed by Weinberg and Price. Initial studies mapped the effects of growth in a variety of fields—high-energy physics research, space science, laser research, and large-scale biological science—focusing primarily on the post–World War II period.[9] Spurred by the insights gained in such studies, historians in the last few years have begun to pay increasing attention to methodology in the hopes of devising ways to gain an even more incisive understanding of the phenomenon.

As part of this effort, historians have reassessed various parameters of large-scale research. One issue has been appropriate periodization. The standard notion is that Big Science began in the radar and atomic bomb projects in World War II, prospered during the Cold War, and showed signs of decline at the end of that era, as signaled by the 1993 cancellation of the multibillion-dollar high-energy physics project the Superconducting Supercollider (SSC). In 1992 Peter Galison, along with coauthors Bruce Hevly and Rebecca Lowen, used the story of research at Stanford University to challenge that standard creation story: they pointed out that Stanford actually had ties with industry well before World War II and that the research changes in Palo Alto resulting from federal sponsorship occurred in the early war years, not in the immediate postwar era.[10]

There were also other challenges to the standard periodization. Price noted as an aside

[7] These arguments can be seen, e.g., in the Atomic Energy Commission's "Policy for National Action in the Field of High Energy Physics," written in 1965 to lay out an ambitious program for funding high-energy physics accelerators and other aspects of high-energy physics research. The report first notes that "the success of the U.S. program has been based on the exceptional talent and enthusiasm of the individuals involved and on the willingness of the Federal Government to support a broad range of high energy physics activities," including accelerator projects. It then asserts that "during the past decade the rate of expenditures of high energy physics has increased markedly. . . . Continued strong Federal support is required to maintain a vigorous high energy physics research program." U.S. Congress, Joint Committee on Atomic Energy, *High Energy Physics Program: Report on National Policy and Background Information* (Washington, D.C.: Government Printing Office, 1965), p. 9. The long-standing American preference for large-scale projects has been discussed, e.g., in Charles C. Gillispie, *The Professionalism of Science: France, 1770–1830, Compared to the United States, 1910–1970* (Kyoto: Doshisha Univ. Press, 1983), p. 13; and Daniel J. Kevles, *The Physicists: The History of a Scientific Community* (New York: Knopf, 1978), pp. 82–83.

[8] A report by the High Energy Physics Advisory Panel (HEPAP) urging construction of the multibillion-dollar Superconducting Super Collider provides a good example of recent Big Science promotion arguments. E.g., the report states: "In summary, we are excited at the prospects for high energy physics in the U.S. during the near future. Our present frontier position is largely due to our past investment in the development of new technologies and in resulting new facilities. The continued future health of high energy physics in the U.S. will be insured by timely construction and exploitation of the SSC." HEPAP Subpanel, *Report of the HEPAP Subpanel on High Energy Physics and the SSC over the Next Decade* (Washington, D.C.: Government Printing Office, 1989), p. 4.

[9] References to works in this vein can be found in Capshew and Rader, "Big Science" (cit. n. 5), and examples can be found in Peter Galison and Bruce Hevly, eds., *Big Science: The Growth of Large-Scale Research* (Stanford, Calif.: Stanford Univ. Press, 1992).

[10] The role of World War II and the Cold War in the rise of large-scale research is discussed in Kevles, *Physicists* (cit. n. 7), pp. 117–154; and Robert Seidel, "The National Laboratories of the Atomic Energy Commission in the Early Cold War," *Historical Studies in the Physical and Biological Sciences,* 2001, *32*:145–162. The declining prospects for large-scale research at the end of the Cold War are discussed in Peter Galison, *Image and Logic: A Material Culture of Microphysics* (Chicago: Univ. Chicago Press, 1979), p. 671; and Catherine Westfall, "Panel Session: Science Policy and the Social Structure of Big Laboratories," in *The Rise of the Standard Model: Particle Physics in the 1960s and 1970s,* ed. Lillian Hoddeson *et al.* (Cambridge: Cambridge Univ. Press, 1997), pp. 364–383. The demise of the SSC is discussed in Daniel J. Kevles, "Big Science and Big Politics in the United States: Reflections on the Death of the SSC and the Life of the Human Genome Project," *Hist. Stud. Phys. Biol. Sci.,* 1997, *27*:269–297; and Michael Riordan, "A Tale of Two Cultures: Building the Superconducting Super Collider, 1988–1993," *ibid.,* 2001, *32*:125–144. On research at Stanford see Galison, Bruce Hevly, and Rebecca Lowen, "Controlling the Monster: Stanford and the Growth of Physics Research, 1935–1962," in *Big Science,* ed. Galison and Hevly, pp. 46–77.

that team research using complicated devices and procedures did exist long before World War II—for example, in astronomical observatories. In the early 1990s Robert Smith and others built on this idea and began distinguishing between recent and older large-scale research.[11] In the midst of this work on periodization, James Capshew and Karen Rader inaugurated a new focus on the typology of large-scale research. In their view there is "Big Science," the rhetorical construction popularized by Alvin Weinberg in 1961 and subsequently used to describe various post–World War II projects such as the Human Genome Project. In addition, there is "big science"—for example, observatories—"a generic label for the forces of growth that have propelled the scientific enterprise since the seventeenth century." Smith continued the effort to refine periodization and typology in a 1996 encyclopedia article describing large-scale research, and in the next several years Galison and Kevles came up with their own typologies, all based on how work was conducted and controlled.[12]

The focus on methodology has also led to a more analytical approach to the development of large-scale research. Capshew and Rader complain that "although the litany of money, manpower, machines, media, and the military provides a convenient mnemonic for the chief features of Big Science, it also tends to obscure the process of growth by focusing on its end results." They go on to argue that if we stop thinking of Big Science à la Weinberg, assuming that large-scale research is "almost entirely" a post–World War II phenomenon, and drop the "exclusive concern with the attributes of 'bigness,'" then we can employ "an evolutionary approach" that looks "at how science," including science in previous centuries, "*becomes* big(ger)." In addition to recasting "issues of scientific growth within a framework sensitive to relative changes in scale, scope, and significance," such an approach places contemporary large-scale research alongside earlier examples of the phenomenon and thus can "be used to historicize Big Science." Smith has made further attempts to historicize "Big Science" by finding many early examples of its use, dispelling the commonly held belief that Weinberg coined the term.[13]

Capshew and Rader's attempts are part of a larger effort to forge a new interpretive framework for understanding large-scale research. Indeed, their emphasis on process and growth—and their de-emphasis on the attributes of individual end results—is part of a trend to view large-scale research in a less particular and more global perspective. Smith provides an early example of this trend. In 1989, drawing on models from political science and the sociology of science, he described the development of the Hubble Space Telescope

---

[11] Price felt that the growth of large-scale research was less explosive in the past and therefore of less note: Price, *Little Science, Big Science . . . and Beyond* (cit. n. 5), p. 3. Smith pointed out that as early as 1903 an observer had noted that the nineteenth-century Greenwich observatory "introduced production on a large scale into astronomy": Robert W. Smith, "A National Observatory Transformed: Greenwich in the Nineteenth Century," *Journal for the History of Astronomy,* 1991, 22:5–20, on p. 13.

[12] Capshew and Rader, "Big Science" (cit. n. 5), p. 22. Smith's categories are "machine centered," "expedition driven," "coordinated"—which includes numerous work sites—and "mixed." See Robert W. Smith, "Large-Scale Scientific Enterprise," in *Encyclopedia of the United States in the Twentieth Century,* ed. Stanley Kutler et al. (New York: Scribner's, 1996), p. 739. Galison identified various "axes" in Big Science: "geographic," "economic," "multidisciplinary," and "multinational." He also distinguished between the "microintegrated big science" of the 1950s and the "macrointegrated big science" of the 1970s in which physics and engineering "were thoroughly intercalated." See Peter Galison, "The Many Faces of Big Science," in *Big Science,* ed. Galison and Hevly (cit. n. 9), pp. 1–17, on p. 2; and Galison, *Image and Logic* (cit. n. 10), p. 559. Kevles identifies three types of U.S. Big Science—"centralized," "federal," and "mixed"—on the basis of the form of control used in establishing and conducting research. See Kevles and Hood, "Reflections" (cit. n. 6), p. 307.

[13] Capshew and Rader, "Big Science," p. 4. As Capshew and Rader point out (*ibid.,* p. 5), Smith has noted that "Hans A. Bethe used the term in his review of Robert Jungk, *Brighter Than a Thousand Suns,* in *Bulletin of the Atomic Scientists,* 1958, *14*:426–428, on p. 428."

in terms of the interplay of technical and nontechnical forces and spoke of the utility of viewing large-scale research as a series of systems and networks rather than as stand-alone projects. In a later study Smith and Joseph Tatarewicz built on this insight by examining the development of the space telescope's sophisticated camera in terms of "a 'heterogeneous network' that included technologies, institutions, and social networks."[14]

A similar point of view has penetrated studies of the institutions surrounding large instruments. Recent studies by the American Institute of Physics focus on the network of multi-institutional collaborations in high-energy physics and other fields. Robert Seidel, and later Peter Westwick, chose the *system* of national laboratories as a research subject.[15] The network/systems point of view has also been applied to the study of individual laboratories: Robert Crease, the historian at Brookhaven National Laboratory, explained that one of the key challenges for the laboratory historian is to take into account that an institution's history is "entangled in different networks or events, institutions, instruments, politics and people." In a similar vein, German scholars formulated a definition of "Big Science" (*"Großforschung"*) that emphasizes the scientific, technical, social, and political connections necessary for large-scale enterprises.[16]

A particularly explicit example of the network/systems approach comes from Galison's study of a series of devices used to detect elementary particles. Galison, whose analysis is suffused with metaphors of connection, disconnection, and interrelation, casts his study in terms of the "intercalation of diverse sets of practices (instrument making, experimenting, and theorizing)" as a means of examining the "borders in transition" that drove the evolution of twentieth-century high-energy physics.[17]

[14] Robert W. Smith, *The Space Telescope: A Study of NASA, Science, Technology, and Politics* (Cambridge: Cambridge Univ. Press, 1989), p. 432; and Smith and Joseph Tatarewicz, "Counting on Invention: Devices and Black Boxes in Very Big Science," *Osiris,* 1994, *9:*101–123, on p. 101. Historians of large-scale science also benefit from the work of Thomas Hughes and Bruno Latour, who in different ways show the possibilities of thinking of the scientific and technological workplace in terms of networks and systems. See Thomas Hughes, *Networks of Power: Electric Supply in the U.S., England, and Germany, 1880–1920* (Baltimore: Johns Hopkins Univ. Press, 1983); and Bruno Latour, *Science in Action* (Cambridge, Mass.: Harvard Univ. Press, 1987).

[15] Joan Warnow-Blewett and Spencer R. Weart, eds., *AIP Study of Multi-Institutional Collaborations, Phase I: High-Energy Physics* (New York: American Institute of Physics, 1992), *Phase II: Space Sciences and Geophysics* (College Park, Md.: American Institute of Physics, 1995), *Phase III: Ground-Based Astronomy, Materials Science, Heavy Ion and Nuclear Physics, and Computer-Mediated Collaborations* (College Park, Md.: American Institute of Physics, 1999); Robert Seidel, "A Home for Big Science: The Atomic Energy Commission's Laboratory System," *Historical Studies in the Physical Sciences,* 1986, *16:*135–175; and Peter Westwick, *The National Labs: Science in an American System, 1947–1974* (Cambridge, Mass.: Harvard Univ. Press, forthcoming). A partial summary of the American Institute of Physics research can be found in Joel Genuth *et al.,* "How Experiments Begin: The Formation of Scientific Collaborations," *Minerva,* 2000, *38:*311–348.

[16] Robert Crease, "Labs as Crucibles of Uncertainty," *Physics World,* July 2001, p. 16; and Crease, *Making Physics: A Biography of Brookhaven National Laboratory, 1946–1972* (Chicago: Univ. Chicago Press, 1999). The German scholars' definition is "(1) The binding together of different scientific-technical disciplines . . . in one project, with a large apparatus often standing at the center; (2) binding together extensive resources in manpower and finances . . .; (3) predominant financing by the State . . .; (4) orientation towards concrete, middle- to long-term projects . . .; (5) connecting basic and applied research in an industrial context; (6) orientation towards goals which are considered politically and socially especially relevant . . .; (7) dualism of a clear political goal and far-reaching autonomy of scientists in the setting of concrete work goals." Burghard Ciesla and Helmuth Trischler, "Legitimation through Use: Rocket and Aeronautical Research in the 'Third Reich' and USA," in *Science and Ideology: A Comparative History,* ed. Mark Walker (London: Routledge, 2002).

[17] Galison, *Image and Logic* (cit. n. 10), pp. 19, 839. Galison produces a sophisticated and thought-provoking analysis of ways in which the scientific workplace can be studied in terms of a "trading zone" where professionals from many different disciplines interact in order to create and use large-scale equipment and find meaning from the research it produces (see pp. 46–63 and Ch. 9). He draws on many sources. Particularly useful for historians of large-scale research is the work of Susan Star and James R. Griesemer, who look at the interactions of various professionals in the study of early twentieth-century ecology: Susan Star and James R. Griesemer, "Institutional

Although this work in periodizing, typing, historicizing, and interpretive recasting allows a much more penetrating view of large-scale research, these methodological discussions also betray an elemental concern about the very use of the term "Big Science." Galison, for example, begins his analysis of modern high-energy physics detectors with the following disclaimer: "My view is this: as an analytic term, 'big physics' is about as helpful to the historian of science as 'big building' would be to a historian of architecture."[18]

Although Galison acknowledges that the "big" label does little to help us understand large-scale research, he continues to use it. And despite their statement that an exclusive concern with the nature of bigness obscures our view of such research, Capshew and Rader merely shift focus from "big" to the more fluid "bigger." Indeed, although the methodology for studying large-scale research has become more sophisticated in many ways, most authors have continued to be preoccupied with growth or some other attribute of bigness, perhaps because it has been easy to assume that comprehending a phenomenon dubbed "Big Science" hinges on an understanding of bigness.

There are a few exceptions to this preoccupation with bigness. For example, Galison's analysis examines various types of nongrowth. In contrast to the oft-told story that emphasizes the triumphs achieved through ever-increasing scale, he notes that the detector collaboration of one of the largest high-energy physics detectors of the 1980s was so large that it had to be primarily concerned "with the simple task of remaining intact against the forces of disaggregation." In a similar vein, he identifies the detectors of the failed SSC of the 1990s as the "limiting case of [high-energy] physics energy, collaboration size, budgetary expenditure, political commitment and brute square mileage." Another exception to the familiar focus on bigness comes in a study by Lillian Hoddeson and myself on the building of Fermilab in the 1960s. This work focuses on the pressure to reduce scale, explaining that the constraints of building an expensive research facility in a stringent funding environment forced builders to abandon the standard of reliability for the sake of reducing the size and, hence, the expense of individual accelerator components.[19] Even these exceptions, however, go only so far in examining how increases in bigness ultimately lead to some sort of limit on growth.

The Bevalac story offers new vistas for the examination of large-scale research. Because this mezzo science project was formed by joining a very large accelerator and a smaller machine, its story showcases the mechanisms, connections, and interconnections of large-scale research at a variety of scales. In addition, the composite nature of the Bevalac highlights the importance of connections in large-scale research—just as constructing the Bevalac required the interconnection of accelerator components, enabling research with the new machine required making connections to gather financial support for the project, assemble a cadre of users and equipment, and forge the intellectual underpinnings for a new line of scientific inquiry. In short, the Bevalac story provides the opportunity to extend efforts to examine the development of large-scale research in terms of connections and shows the more refined understanding that emerges when we consider more than bigness and growth when assessing large-scale research.

Ecology, 'Translations,' and Boundary Conditions: Amateurs and Professionals in Berkeley's Museum of Vertebrate Zoology, 1907–39," *Social Studies of Science,* 1989, *19*:337–419.

[18] Galison, *Image and Logic,* p. 553.

[19] *Ibid.,* pp. 688, 671; and Catherine Westfall and Lillian Hoddeson, "Thinking Small in Big Science: The Founding of Fermilab, 1960–1972," *Technology and Culture,* 1996, *37*:457–492.

A VOID OPENS AND THE BEVALAC EMERGES

An explanation of the circumstances that led to the Bevalac—including the conditions that gave rise to the machine's initial unpopularity—must start with an understanding of the two groups of large-scale researchers at the Berkeley laboratory and the failed projects they sponsored in the 1960s: a 200 GeV proton synchrotron and a multipurpose heavy-ion accelerator. Edward Lofgren and others at the Berkeley laboratory's Physics Division initiated plans for the proton synchrotron in 1960 in hopes that the machine, which would be the largest of its kind and the most expensive accelerator of its time, would continue the Bevatron's glorious chain of particle discoveries.[20]

However, Lofgren and his team made the mistake of creating a design that ensured reliability but ignored expense, an unpopular move at a time when the federal budget was tightening. Ironically, Berkeley's position was further weakened by the laboratory's high-powered reputation: other physicists worried that their famous Berkeley colleagues would monopolize the expensive, one-of-a-kind accelerator. When Berkeley laboratory director Edwin McMillan refused to bend to the resulting pressure to assure equitable access to outside users, Atomic Energy Commission (AEC) and National Academy of Sciences officials took over the planning, fearing that divisiveness within the physics community would undermine efforts to fund the costly project. By the time Lofgren submitted the completed design to the AEC, in 1965, the project was managed by a nationwide consortium of universities and the AEC had organized a formal competition to decide where the accelerator would be located.[21]

While the fate of the 200 GeV proposal was still being debated, Albert Ghiorso of Berkeley's Nuclear Chemistry Division (NCD) proposed another large machine for Berkeley. The Omnitron, which would be used for both nuclear science and biomedical research, was to be an accelerator with "two concentric alternating-gradient rings, a rapid-cycling synchrotron and a storage ring." The machine would provide "ion beams of all elements" with higher energy or intensity than other accelerators.[22]

The dual 200 GeV and Omnitron proposals continued a pattern: since the 1957 construction of the HILAC, the Berkeley laboratory had supported NCD-operated accelerators devoted to heavy-ion research alongside the Bevatron and other high-energy physics machines operated by the Physics Division. NCD founder Glenn Seaborg was a nuclear chemist, and he and others in the NCD had joint appointments with the UC Berkeley Chemistry Department, while Luis Alvarez, McMillan, and others in the Physics Division had joint appointments with the UC Berkeley Physics Department. NCD researchers were known, in particular, for transuranic element discoveries (elements above uranium, element 92 in the periodic table). For example, in 1961 Ghiorso and others at the HILAC first produced element 103, which was later named Lawrencium to honor the Berkeley lab's founder.[23]

---

[20] An electron volt (eV) is a unit of energy describing the energy obtained by an electron accelerated through 1 volt of potential difference. "GeV" stands for "billion electron volts"; "MeV" stands for "million electron volts." At that time in the United States, 200 billion electron volts was referred to as "200 BeV"—a source of the Bevatron's and the Bevalac's names. Since this energy is now referred to as "200 GeV," this is the term I use in the text. This episode is described in detail in Westfall and Hoddeson, "Thinking Small in Big Science."

[21] For more information on the site competition see Catherine Westfall, "The Site Contest for Fermilab," *Physics Today*, 1989, *42*:44–52.

[22] Lawrence Radiation Laboratory, "The Omnitron: A Multipurpose Accelerator" (Lawrence Radiation Laboratory, 1966).

[23] Darleane Hoffman, "The Heaviest Elements," *Chemical and Engineering News*, May 1994, pp. 24–34,

Although the two types of accelerators operated side by side at the same laboratory, they supported different types of large-scale research: since cutting-edge NCD science could be done less expensively, research was done at a more modest scale in the NCD than in the Physics Division. One measure of the difference between the modest large-scale research of the NCD and the grand large-scale research of the Physics Division can be taken from the relative costs of their proposed accelerator projects. Whereas Ghiorso estimated in 1966 that he would need $24 million to construct the Omnitron, Lofgren had estimated the year before that he would need more than ten times that, $348 million, to build the 200 GeV machine.[24]

In line with its grander equipment, the Physics Division in the late 1960s had a more impressive array of luminaries on its staff. By that time the NCD had lost its one Nobel Prize winner—Seaborg, who shared the 1951 chemistry prize with McMillan for the discovery of plutonium, left Berkeley in the late 1950s to become the chairman of the AEC. The Physics Division boasted three Nobel Prize winners, all of whom had gained the honor for particle discoveries at the Bevatron.[25]

Along with the differences in scale and staff came a difference in style. While Lofgren's team, and others in the Physics Division, were famous for meticulous, carefully executed projects and tightly organized, hierarchical management, Ghiorso and his NCD colleagues had a more informal, loosely organized research environment with a family-like team that was led, after Seaborg's departure, by Ghiorso, who lacked formal graduate training in science.[26] Because the Physics Division hosted the largest accelerators and had more famous scientists, it gained more publicity and had more clout within the laboratory—and its grander science had more national and international visibility.

In the late 1960s Berkeley became the scene of compounded loss. In early 1967 the Physics Division designers sustained a stunning blow: the AEC announced that the 200

---

esp. p. 25. The arsenal of NCD accelerators also included the sector-focused 88-inch Cyclotron, which was built in 1961. Lawrence Radiation Laboratory, "Nuclear Chemistry" (Lawrence Radiation Laboratory, 1968), p. 23. The HILAC, which was originally proposed by Luis Alvarez, was part of a worldwide network of heavy-ion accelerators that by the early 1970s included major centers at Dubna in the Soviet Union, Orsay in France, and Brookhaven, as well as Yale and Berkeley, and a host of smaller accelerators, mostly tandem electrostatic accelerators. Although most high-energy physics accelerators were supported through the AEC Division of Physical Research, funding for heavy-ion accelerators, which were used for applied as well as basic research, was more diverse. The twenty-seven heavy-ion facilities supported by the AEC in 1972 received funding through five AEC divisions: Military Application, Biomedical and Environmental Research, Reactor Development and Technology, Controlled Thermonuclear Research, and Physical Research. See Atomic Energy Commission, *Heavy-Ion Science: Research, Facilities, and Outlook: A Report to the Joint Committee on Atomic Energy* (Washington, D.C.: Atomic Energy Commission, 1972), p. iv; D. Allan Bromley, "The Development of Heavy-Ion Nuclear Physics," in *Treatise on Heavy-Ion Science*, ed. Bromley (New York: Plenum, 1984), pp. 3–132; and Committee on Nuclear Science, "Report of the Ad Hoc Panel of Heavy Ion Facilities" (National Academy of Science, National Research Council, 1974), pp. 28, 67.

[24] The estimates for the respective machines reflect Ghiorso's tendency to rough out costs and Lofgren's tendency to calculate expense carefully; nonetheless, the difference in scale is apparent. Lawrence Radiation Laboratory, "Omnitron" (cit. n. 22), p. I-4; and Lawrence Radiation Laboratory, "200 BeV Design Study," Vol. 1 (Lawrence Radiation Laboratory, 1965), p. I-8.

[25] Emilio Segrè and Owen Chamberlain won the Nobel Prize in Physics in 1959 for the discovery of the antiproton; Luis Alvarez won the Nobel Prize in Physics in 1968 for the study of subatomic particles. See Borgna Bunner, ed., *Time Almanac 2000* (Boston: Time, 2000), p. 257.

[26] Interview with Robert Stevenson, 9 Mar. 1994; interview with Richard Gough, 14 Apr. 1994; and interview with Albert Ghiorso, 1 Mar. 1994. Sharon Traweek points out that high-energy physicists in Japan tend to develop more family-like experimental teams than those found in the United States. The differing cultural styles that shaped research at Berkeley show that such differences can arise in different subfields as well as in different countries. See Sharon Traweek, *Beamtimes and Lifetimes: The World of High Energy Physicists* (Cambridge, Mass.: Harvard Univ. Press, 1988), pp. 148–149.

GeV accelerator would not be built in California. Instead, a new laboratory (which would later be named Fermilab) would be built in Illinois to house the machine. The decision meant more than the loss of a single project: in the competitive world of high-energy physics, the Physics Division had lost its chance to dominate the search for new particles and the Berkeley laboratory had lost its position as host of the world's largest accelerator projects.

In the wake of this devastating news, the mood of McMillan and his Physics Division colleagues turned ugly. Lofgren declined the offer to construct the Illinois laboratory, and when the eventual appointee to the job visited Berkeley in February he described the 200 GeV design team as "completely and ostentatiously negative and unfriendly." Berkeley's eminent high-energy physicists were also nettled that, as AEC chairman, former NCD member Seaborg had helped choose the Illinois site. It was easy for anger to spill over onto the Omnitron, a project that Seaborg championed. One Berkeley researcher went so far as to write a letter telling Seaborg and an AEC colleague of the "gnawing fear" among Berkeley high-energy physicists that the Omnitron would be given to Berkeley as a "sop" to appease the loss of the 200 GeV. McMillan soon dealt another blow to the project: when a member of the powerful Joint Committee on Atomic Energy asked him whether the Omnitron was essential, he replied, "No, it's senseless, it's useless, it's not essential."[27]

In the prevailing climate of tight budgets, those promoting the Omnitron could scarcely afford McMillan's vote of no confidence. Although the proposed machine was relatively modest and inexpensive compared with other planned accelerators, the proposal emerged just as the AEC was struggling to adjust to an abrupt decline in federal research funding and build two other accelerators of unprecedented expense: the $100 million Stanford Linear Accelerator Center (SLAC), which had just begun operation, and the 200 GeV accelerator, which cost $250 million. Although Seaborg twice succeeded in restoring the Omnitron to the AEC budget, the executive branch watchdog, the Bureau of the Budget, eventually deleted the proposal in 1967.[28]

Although the NCD could have requested funding for the proposal in subsequent years, Ghiorso decided that the struggle was not worth the effort. Realizing that a more modest request promised a surer way to satisfy his own immediate research needs, he abandoned the fight for the Omnitron and instead submitted a proposal for upgrading the HILAC so that it would accelerate a slightly wider range of heavy ions but not increase beam energy. By reducing his expectations, Ghiorso had dramatically decreased costs: in contrast to the $24 million requested for the Omnitron, the upgrade to the HILAC—which would subsequently be called the SuperHILAC—cost $2.6 million.[29]

[27] Robert Wilson, tape recording, "1967 Berkeley Meeting"; Edward Lofgren to Norman Ramsey, 12 Jan. 1967, Fermilab History Collection, Batavia, Illinois; Dave Judd to Glenn Seaborg and Gerald Tape, 10 Feb. 1967, Box 5, Files of Edwin McMillan, Berkeley Lab, Berkeley, California; and interview with Edwin McMillan, 16 May 1984.

[28] Spencer Weart, "The Physics Business in America, 1919–1940: A Statistical Reconnaissance," in *The Sciences in the American Context: New Perspectives,* ed. Nathan Reingold (Washington, D.C.: Smithsonian Institution Press, 1979), pp. 295–358, esp. p. 328; and Albert Ghiorso, "On the Origin of the BevaLac," paper presented at "Bevatron/Bevalac: Its People and Science," Berkeley Lab, 1993. For more information on the planning of the SLAC see Galison *et al.,* "Controlling the Monster" (cit. n. 10); Stuart W. Leslie, *The Cold War and American Science: The Military-Industrial-Academic Complex at MIT and Stanford* (New York: Columbia Univ. Press, 1993), pp. 160–187; and Rebecca Lowen, *Creating the Cold War University* (Berkeley: Univ. California Press, 1997), pp. 177–186. For information on the founding of Fermilab see Westfall and Hoddeson, "Thinking Small in Big Science" (cit. n. 19).

[29] Interview with Ghiorso, 1 Mar. 1994; and Joint Committee on Atomic Energy, *AEC Authorizing Legislation, Fiscal Year 1970,* 99th Cong., 1st sess., 1969, p. 499.

Thus while NCD researchers had a machine that suited them in the works in the early 1970s, for the first time in the Berkeley laboratory's history the Physics Division's high-energy physicists did not. Since Berkeley accelerator experts had not become deeply involved in the effort to build Fermilab, they too were without a major project. The situation became even more worrisome for Berkeley managers when rumors began to circulate that since the aging Bevatron was of limited research utility, the AEC might shut it down, leaving the Bevatron crew out of work. The loss of the 200 GeV project had clearly left a void damaging to the Physics Division.[30]

The idea for filling this void did not come, however, from McMillan or the Physics Division, nor did it come in response to Physics Division needs. Instead, Ghiorso stepped into the breach because, as he later explained, he felt guilty about abandoning the Omnitron and thereby leaving Berkeley's biomedical researchers without the higher-energy heavy-ion machine they needed. On its face, his idea sounded quirky: because the SuperHILAC was located on a hill above the Bevatron, he proposed building a beam line to connect the two machines so that heavy ions could first be produced in the SuperHILAC, transported down the hilly slope separating the machines, and then accelerated to relativistic energies in the Bevatron. The combined machine, which he called "the BevaLac," would "furnish the biomedical groups with ample beams of fast heavy particles at a very moderate cost" while allowing the SuperHILAC to continue operation as before. As an added bonus, the combined machine would open "a whole new field of theoretical and experimental science . . . at minimal cost." Instead of considering the idea quirky, McMillan and Lofgren, who was then head of the Bevatron, were immediately enthusiastic and began casting more concrete plans.[31] Although the failed plans for the 200 GeV and the Omnitron project had led only to disconnection and disappointment, the resulting void created conditions ripe for the interconnection of Berkeley's two styles of large-scale research and the creation of research of a different scale and type.

SOME EASY CONNECTIONS, BUT UNPOPULARITY FORCES A DISCONNECTION

Many of the connections and interconnections necessary for creating the Bevalac were easy to make. The project seemed to fill a worrisome void and therefore had the strong support of Berkeley managers. Since it was composed mostly of existing equipment, the accelerator was inexpensive to build; and given its low price tag and high scientific promise, enthusiastic Berkeley managers, who had Seaborg's support, were readily able to convince government officials to open a conduit for funding to flow to the Bevalac. In stark contrast to the usual wrangling at the beginning of new accelerator projects (the 200 GeV being a case in point), the Bevalac got started without controversy or oversight from federal advisory committees. Indeed, funding for the project proceeded at lightning speed, with little time taken for discussion outside Berkeley.

In April 1971, only a month after Ghiorso first announced his idea for the Bevalac,

---

[30] According to Ghiorso, by 1971 laboratory rumor had it that "the Bevatron would have been shut down about six months" later: interview with Ghiorso, 1 Mar. 1994. See also Joint Committee on Atomic Energy, *AEC Authorizing Legislation; Fiscal Year 1971,* 91st Cong., 2nd sess., 1970, p. 511; and interview with Edward Lofgren, 3 May 1984.

[31] Interview with Ghiorso, 1 Mar. 1994; and Albert Ghiorso, "Proposal: The 'Bevalac'—A Versatile Accelerator Concept," 22 Mar. 1971, "Bevalac Reports and Specifications," Hermann Grunder, Deputy Director's Bevalac Files, 1970–1983 (hereafter cited as **Grunder Files**) Box 1 of 2, Archives and Records Office, Berkeley Lab.

Lofgren had a conceptual design report in hand thanks to the efforts of Hermann Grunder, an accelerator physicist who worked under his direction, and others. In a few more months Lofgren had obtained $200,000 from the AEC Division of Biology and Medicine to get the project started. After receiving the initial design report, which spoke glowingly of the high-energy physics, nuclear physics, and biomedical experiments to be made possible by the machine, in early 1972, the division granted $2 million more to start construction, raising the project's budget to $2.2 million. Eventually, the Department of Energy provided $5 million more to finish modifications allowing the acceleration of elements as heavy as uranium, bringing the total construction cost to $7 million.[32]

Making the necessary connections and modifications so that the SuperHILAC and the Bevatron could operate as an interconnected accelerator system was also not a terribly difficult task. After a few months of delay caused by a rash of last-minute problems, including a mud slide, Grunder and his team completed construction of the transfer line to bring heavy ions from the SuperHILAC to the Bevatron, and SuperHILAC accelerator experts readied that machine for combined operation. On 1 August 1974 the Bevalac operated for the first time, accelerating carbon beams. By the end of the year heavier beams—neon and argon—were available for research.[33]

Physics research also got off to a quick start. In 1971, in the first months after the Bevalac was proposed and three years before the transfer line was completed, Grunder figured out a way to obtain heavy ions without the line: after he fashioned a specially designed ion source, nitrogen beams were accelerated to about 2 GeV per nucleon, and a research group led by the Berkeley physicist Harry Heckman modified a cosmic-ray detector and used it to collect the very first Berkeley relativistic heavy-ion data. (See Figure 1.)[34]

Despite this apparently seamless beginning, a problem soon arose. As John Teem, director of the AEC Division of Physical Research, noted in 1973, "none" of the "well established HEP-people" were "considering heavy ions." SuperHILAC users were similarly uninterested. Bevalac leader Lee Schroeder later joked that recruitment prospects seemed so dim that he felt like walking into bars in downtown Berkeley to plead with the

---

[32] "Briefing on High Energy Heavy Ion Facility and Program," 29 June 1971, "HEHIF-Meeting, June 29, 1971," Grunder Files, Box 1 of 2; Edward Lofgren, "Status and Prospects of the Berkeley High Energy Heavy Ion Program," 5 Oct. 1971, Edward Lofgren Papers (hereafter cited as **Lofgren Papers**), Berkeley Lab; Lawrence Berkeley Laboratory, "Design Study Report: High Energy Heavy Ion Facility (BEVALAC)," LBL-786 (Lawrence Berkeley Laboratory, Jan. 1972); and "High Intensity Uranium Beams from the SuperHILAC and the Bevalac," Proposal-32 (Lawrence Berkeley Laboratory, 1975), p. ii.

[33] Accelerator Division, "Annual Reports, July 1, 1972–December 31, 1974," LBL-3835 (Berkeley, 1973), pp. 3–5. Over the next several years accelerator experts worked to make further modifications—including the installation of a new injector for the SuperHILAC—so that the Bevalac could reach its ultimate design goal: the acceleration of elements as heavy as uranium to relativistic energies. The most difficult obstacle at this stage was finding a way to improve the vacuum of the Bevatron, an elderly machine designed for the easier task of accelerating protons. This obstacle was eventually overcome by a vacuum liner scheme that used circuit-board material cooled to very low temperatures to obtain the necessary cryogenic properties. Construction then proceeded quickly: in March 1982 the new injector and liner were brought into operation, in May the Bevalac accelerated its first uranium beam, and in September the machine began accelerating uranium at relativistic energies. See Jose Alonso, "Thirty Years at the Forefront—A Perspective on the Bevatron/Bevalac," LBL-18546 (Berkeley, 1984), p. 5; Accelerator and Fusion Research Division, "Annual Report, 1980," LBL-12263 (Berkeley, 1981), p. 4; and Accelerator and Fusion Research Division, "Annual Report, 1982," LBL-16125 (Berkeley, 1982), p. 6.

[34] Hermann Grunder, Walter Hartsough, and Edward Lofgren, "Acceleration of Heavy Ions at the Bevatron," *Science,* 1971, *174:*1129; and Harry Heckman *et al.,* "Fragmentation of Nitrogen-14 Nuclei at 2.1 GeV per Nucleon," *ibid.,* p. 1130. A program was also developed that used relativistic heavy ions for cancer diagnosis and therapy, but the story of that program is outside the scope of this essay.

Figure 1. *Press conference in 1971 to announce that heavy ions were accelerated at the Bevalac for the first time. To the far left at the table is the Berkeley physicist Harry Heckman, who is sitting next to Edwin McMillan. Edward Lofgren sits on the right in a white shirt. Also pictured is Cornelius Tobias, who is talking about the use of the Bevalac for cancer treatment with a diagram showing his technique on the blackboard behind him. On either side of Lofgren are accelerator experts Tom Budinger and Walter Hartsough. (Photo courtesy of Lawrence Berkeley National Laboratory.)*

customers: "Would anybody, anybody at all, like to do a Bevalac experiment?"[35] This was a new experience at Berkeley. For the first time an accelerator was taking shape that lacked a sizable cadre of enthusiastic would-be users within the laboratory to help stimulate worldwide interest. As Bevalac leaders stepped up recruitment efforts, they worried that they were building an unpopular machine that would sit idle.

Why was the Bevalac initially so unpopular? After all, Bevalac research was, at least on the face of it, a logical extension of the scientific work of both Bevatron and SuperHILAC scientists—for Bevatron users it meant using heavier particles at familiar high energies, while for SuperHILAC users it meant using familiar particles at higher energies. In addition, the Bevalac's relativistic heavy-ion collisions opened the opportunity to explore, for the first time, nuclear matter at high temperatures and densities, an explo-

---

[35] "Meeting with Thorne, John Teem, Lofgren, Wheeler, WAE, WDH, and HAG," 8 Mar. 1973, "Bevalac Early Notes," Grunder Files, Box 1 of 2; and interview with Lee Schroeder, 7 Mar. 1994.

ration that would provide information about nuclear matter itself as well as other, related phenomena, such as the behavior of supernovas, which compress to nuclear densities during their lifetimes.

One explanation for the Bevalac's unpopularity is that the capabilities of the new machine fit poorly with the existing research agendas of Bevatron and SuperHILAC users. A prime research objective for SuperHILAC users was to accelerate heavy ions at an energy one hundred times *lower* than that of the Bevalac to hit a heavy-ion nucleus so that protons were added to it, creating a new element that could then be identified and studied. (See Figure 2.) Accelerating heavy ions to higher, Bevalac energies would not facilitate such work because in the violent relativistic collisions nuclei would not undergo the fusion necessary for element production. Compared with the lower-energy, modest scale of the SuperHILAC, the Bevalac was too big.

The Bevalac was also a poor tool for achieving the prime research goal of Physics Division physicists. They wanted to accelerate light particles to strike a light target using the highest possible energies to produce very rare, individual events and thereby discover and study new particles. (See Figure 3.) The Bevalac was useless for this type of research because its collisions produced a messy complex of debris rather than new particles that could be detected individually with available equipment and because Bevalac energies

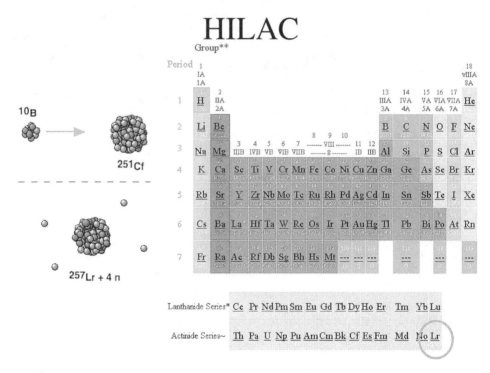

Figure 2. *An example of heavy-ion research. On the left is a pictorial representation of the collision of a boron-10 ion with a target made of nuclei of element 98, Californium. The boron-10 has five protons and five neutrons. These fuse with the target and four neutrons are emitted. This was the sort of collision used to discover Lawrencium at the HILAC. On the right is a picture of the periodic table of the elements showing the placement of the newly discovered element Lawrencium. (This drawing was made by Gary Westfall from information gained from Darleane Hoffman, "The Heaviest Elements,"* Chemical and Engineering News, *May 1994, pp. 24–34.)*

Bevatron

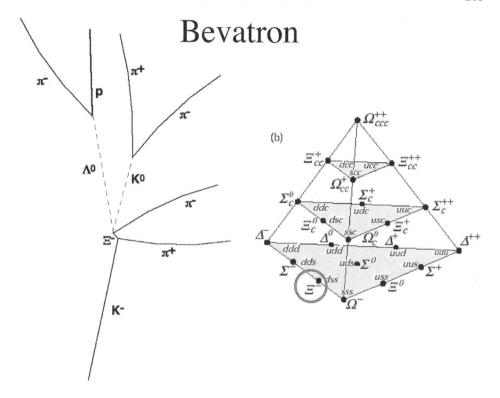

Figure 3. *An example of particle physics research. On the left is the sort of collision that would have been produced and studied in the 1950s at the Bevatron. A secondary beam of negative kaons produced from the Bevatron's proton beam collides in a hydrogen bubble chamber, hitting a hydrogen nucleus and producing one of the new particles discovered in the 1950s, the Ξ. On the right is a particle map, which classified particles according to quark structure. This portion of the map shows the position of the Ξ. The map is based on the Standard Model, which was developed in the late 1970s. (This drawing was made by Gary Westfall with information gained from the Particle Data Group and M. Russell Wehr and James Richards, eds.,* Physics of the Atom *[Reading, Mass.: Addison-Wesley, 1967], p. 420.)*

were a hundred times *lower* than those used by cutting-edge accelerators, such as the 200 GeV machine being built in Illinois. By the grand standards of high-energy physics, the Bevalac was too small.

Of course, Berkeley researchers could simply have abandoned their research agendas for the sake of using a new, local tool to explore new scientific territory. It is perhaps not terribly surprising that SuperHILAC users refrained from making such a move—after all, they had a newly built machine at their disposal for ongoing research. But why did rank-and-file Physics Division physicists, who were suffering from the lack of a machine, not embrace the idea of the Bevalac, just as Lofgren and McMillan had? The attitude even caught some high-energy physicists by surprise. Oreste Piccioni, for example, agreed to rally support for the Bevalac, only to find that "the overwhelming vote of experimental physicists" was "in favor of the biggest machines," even if that meant "the discontinuance" of accelerator-based physics at the Berkeley laboratory.[36]

[36] Oreste Piccioni to Andrew Sessler, 21 Dec. 1973, "Bevatron Users Mtg. January 1974," Andrew M. Sessler Files (hereafter cited as **Sessler Files**), Director's Office Records, 1973–1980, Archives and Records Office, Berkeley Lab, Box 2 of 26.

For his part, Andrew Sessler, who succeeded McMillan as LBL director in 1973, was "disappointed . . . but not surprised." Sessler, who was best known for his work in accelerator development, had long been surrounded by high-energy physicists and had observed that "they are not interested in other things, they are single minded."[37] After assuming the directorship, Sessler sought to find an administrative solution to the problem that the two parts of the Bevalac were in separate divisions and in the process do his part to encourage interdivision cooperation in the development of the Bevalac research program. As he floated the possibility that the combined machine might be put in the Physics Division, it became clear that the single-mindedness of high-energy physicists—at least at Berkeley— arose, in part, from a distaste for the mezzo science of the Bevalac and the desire to affiliate exclusively with a sort of physics they considered more grand, precise, and prestigious.

For example, the high-energy physicist Dave Nygren noted that if the Bevalac was housed in the Physics Division but used by others, then high-energy physicists would face the "psychological and political" difficulty "of having to compete for 'Nuclear Science' money outside [their] own division," given that Bevalac research would be funded from the Department of Energy nuclear science allocation. NCD head Bernard Harvey fumed that such comments showed that "the words 'Chemistry' and 'Nuclear' turn people off in high energy physics." In a letter to Sessler, the high-energy theorist David Jackson pointed to the underlying difficulty. Noting that "more interaction" among Berkeley scientists in the administration of the Bevalac and in preparations for its research program "would lead to better science," he nonetheless warned that "shot gun marriages are not usually successful. Collaborations arise from shared attitudes and interests and mutual respect, things that seem to be lacking in some measure here."[38]

The suggestion that the Bevalac be placed in the NCD also met with resistance. Users of the SuperHILAC and the even more diminutive 88-inch Cyclotron worried that if the machine were put in their division it would divert resources from their research efforts. In fact, many NCD researchers resented the advent of the Bevalac and the mezzo science it brought to their doorsteps. Ghiorso, for example, would later regret conceiving the Bevalac because once built the larger machine became "the tail that wags the dog," at the expense of SuperHILAC research efforts. One of the few Bevalac users to come from the Physics Division, the Japanese researcher Shoji Nagamiya, later summarized the criticisms: high-energy physicists said that Bevalac physics was "unclear physics," while those working with heavy ions at lower energies said that the Bevalac physicists were "very expensive, not useful physicists."[39]

By the time of these discussions, Nygren was already working on the Time Projection Chamber (TPC), an enormous detector project that Physics Division physicists would build with others and use at SLAC. Work at SLAC and other large high-energy physics accelerators was greatly stimulated by the 1974 discovery of the J/φ particle and the consequent development of the Standard Model; and in the end only a handful of physicists in the Physics Division joined the Bevalac effort. For their part, most SuperHILAC users eschewed work with the combined machine in favor of using the lower-energy heavy-ion beam produced by the SuperHILAC, where further new elements were identified and

[37] Interview with Andrew Sessler, 3 Mar. 1994.

[38] David Nygren to Sessler, 28 July 1975; B. G. Harvey to file, 21 Aug. 1975; and J. D. Jackson to Sessler and E. K. Hyde, 13 Aug. 1975: "Nuclear Chemistry Divisional Reorganization," Sessler Files, Box 23 of 26.

[39] Interview with Ghiorso, 1 Mar. 1994; and Shoji Nagamiya, "Banquet Speech Presented at the 4th International Conference on Nucleus-Nucleus Collisions, Kanazawa, Japan, June 10–14, 1991." For more information on heavy-ion physics at this time see Bromley, "Development of Heavy-Ion Nuclear Physics" (cit. n. 23).

discovered.[40] Indeed, the single-minded attitude toward research held by both Physics Division and NCD scientists, reinforced by the prevailing pecking order and long-held rivalries, conspired to make the Bevalac a decidedly unpopular machine, despite its research promise.

Berkeley managers found it easy to make the necessary connections so that Bevalac funding flowed from Washington to Berkeley, and accelerator experts found it easy to interconnect the two machines so that heavy ions could travel from the SuperHILAC to the Bevatron. Nonetheless, in the mid 1970s the Bevalac effort was thrown into crisis because of a crucial disconnection: too few users had agreed to join the research effort.

ORPHANS NO MORE: FORMING A NEW SCIENCE

Just when the Bevalac's unpopularity seemed insurmountable, those struggling to recruit users enjoyed several strokes of good fortune. Perhaps most important was the background fact that in other corners of the world the mezzo science of the Bevalac was seen as neither too modest nor too grand. In fact, Bevalac science seemed just right to researchers in Japan, where nuclear science enjoyed higher status than high-energy physics, and in Germany, where other factors facilitated interest.[41]

By the 1970s the German government had built separate laboratories, the Deutsches Elektronen-Synchrotron (DESY) for high-energy physics and the Gesellschaft für Schwerionenforschung (GSI) for heavy-ion nuclear physics. Hans Gutbrod, a GSI researcher who had done experiments at the SuperHILAC, noted that German heavy-ion physicists found it easier to promote new projects since they didn't have to play the role of "stepchildren" to high-energy physicists within their own laboratory, "which was always a problem at Berkeley."[42] One such project was the UNILAC, a SuperHILAC-like machine for the study of nuclear structure that would be completed in 1976. Unlike their Berkeley counterparts, however, GSI lower-energy heavy-ion researchers did not spurn work with relativistic heavy ions. One reason was that the Frankfurt theorist Walter Greiner had made the intriguing suggestion that "shock waves" could compress into nuclear matter after a nuclear collision, a finding that stimulated a great interest in the reaction mechanisms that would emerge from relativistic heavy-ion collisions. In both Japan and Germany, in fact, scientists were promoting the construction, in the next decade, of Bevalac-like machines; they anticipated gaining experience in the short term at the Bevalac that would they would use in the long term at home.[43]

Japanese interest in the sort of work the machine made possible led to the participation

[40] For a history of the development of the TPC see Galison, *Image and Logic* (cit. n. 10), pp. 553–688. Galison also presents a detailed discussion of the visual and electronic detector traditions that are combined in TPC detectors. On the SuperHILAC work see Hoffman, "Heaviest Elements" (cit. n. 23). For a history of high-energy physics during this period see Hoddeson et al., eds., *Rise of the Standard Model* (cit. n. 10).

[41] For information on attitudes about subatomic science in Japan see Lillian Hoddeson, "Establishing KEK in Japan and Fermilab in the U.S.: Internationalism, Nationalism, and High Energy Accelerators," *Soc. Stud. Sci.,* 1983, *13:*1–48.

[42] Private communication with Hans Gutbrod, 15 Jan. 2002. For information on the founding of GSI see Buchhaupt, *Gesellschaft für Schwerionenforschung* (cit. n. 1); for information on DESY see Claus Habfast, *Großforschung mit kleinen Teilchen: Das Deutsche Elektronen-Synchrotron: DESY 1956–1970* (Berlin: Springer, 1989).

[43] "How Darmstadt Became the Center of Heavy Ion Physics," GSI website: www-aix.gsi.de. For a review of the early German theoretical work on nuclear compressibility see Walter Greiner and Horst Stöcker, eds., *The Nuclear Equation of State, Part A* (New York: Plenum, 1984). On expectations as to how Bevalac research would later be useful in their home countries see interview with Reinhard Stock, 15 Apr. 1994; and interview with Isao Tanihata, 28 Oct. 1993.

of Nagamiya, an essential player in the early years of Bevalac research, Japanese govern-ment support for some equipment, and limited funding for other Japanese experimenters. German support was considerably more substantial. At the urging of one of the founders of the GSI, Rudolf Bock, GSI director Christoph Schmelzer agreed to send a team of German researchers to Berkeley with a sizable equipment budget. Reinhard Stock, who had done relativistic heavy-ion experiments at the Princeton-Pennsylvania Accelerator (PPA) and lower-energy heavy-ion research at Berkeley, brokered the arrangements in 1974, striking a deal to join Arthur Poskanzer, one of the few NCD members planning work at the Bevalac. Over the next ten years the resulting German-American collaboration, which would eventually split into two separate lines of inquiry, would create what D. Allan Bromley has called "the most extensive" of "all the Bevalac programs." (See Figure 4.)[44] While the funding and personnel that came from Japan and Germany were crucial to the long-term success of the Bevalac, the enthusiasm of Japanese and German scientists in the mid 1970s was even more important, for without this foreign enthusiasm, which provided the necessary momentum to generate further interest and recruit more experimenters, it is unclear whether the local effort would have gotten off the ground.

At just this juncture momentum for Bevalac physics received another boost. In 1974 the Nobel laureate T. D. Lee visited Berkeley and presented a talk in which he suggested the possibility, based on calculations made with his colleague Giancarlo Wick, that a new form of matter would be created. One year later a German group led by Erwin Schopper from the University of Frankfurt reported seeing "shock wave" peaks that seemed to confirm the prediction by Greiner.[45] Although Bevalac researchers would ultimately fail to confirm the existence of either Lee-Wick matter or shock waves, the possibility that exotic matter could be created in relativistic heavy-ion collisions generated considerable excitement.

New administrative arrangements allowed Berkeley to capitalize on this and other ex-citing new results. Starting in 1973, Schroeder organized a series of summer studies at Berkeley, complete with side trips to the nearby wine country, to entice potential Bevalac users. In the words of one young Japanese researcher, thanks to these congenial occasions by 1974 "Berkeley had become the center" for discussions of relativistic heavy-ion col-lisions. Sessler did his part by creating a new Accelerator Division, headed by Lofgren, to administer both the Bevalac and the SuperHILAC, abolishing the NCD, and forming the Nuclear Science Division (NSD), which joined the Bevalac groups that had been spread between the Physics Division and the NCD. This division was headed by former NCD head Harvey, an 88-inch Cyclotron researcher who happily accepted the new arrangement once he got assurance from the AEC that budgets for research at the smaller machines "would not be pushed out of shape." In addition to creating a more sensible scheme for managing Bevalac research, this change made the laboratory's organizational structure fit

[44] Nagamiya, "Banquet Speech Presented at the 4th International Conference on Nucleus-Nucleus Collisions" (cit. n. 39); Bromley, Development of Heavy-Ion Nuclear Physics" (cit. n. 23), p. 99; Arthur M. Poskanzer, "A History of Central Collisions at the Bevalac," in *Nuclear Equation of State, Part A,* ed. Greiner and Stöcker, pp. 447–462, esp. p. 452; and interview with Stock, 15 Apr. 1994. The very first relativistic heavy-ion experi-ments were in fact performed at the Princeton-Pennsylvania Accelerator (PPA), which was originally built to accelerate protons but was modified to run heavy ions. The PPA work started at about the time the Bevatron was modified to accelerate nitrogen to 2 GeV in 1971. This PPA work was part of a heavy-ion run that terminated after one year.

[45] T. D. Lee, "A Possible New Form of Matter at High Density," May 1974, Lofgren Papers; and H. G. Baumgardt et al., "Shock Waves and Mach Cones in Fast Nucleus-Nucleus Collisions," *Zeitschrift für Physik,* 1975, *A273:359–371.*

Figure 4. *Arthur Poskanzer and Reinhard Stock in front of a spherical scattering chamber they built in the late 1970s.*

more snugly with that of the AEC, which had recently reorganized to consolidate "all research which is concerned with studying nuclear matter and nuclear phenomena."[46]

Schroeder found that the advent of the NSD was a big boon for the developing research program, since the division had new positions to fill, hosted visitors, and created "a hospitable environment" for researchers. Recruitment efforts were also aided by the surplus of new Ph.D.'s due to the surge of baby boomers through the U.S. educational system and federal belt tightening in the wake of the Vietnam War which made jobs tight elsewhere. And the mezzo science of the Bevalac was tailor-made for the young—they were at no

---

[46] Interview with Tanihata, 28 Oct. 1993; Bernard Harvey to Sessler, 12 Apr. 1973, "Bevalac Funding," Sessler Files, Box 18 of 26; Sessler to the LBL Staff, 30 Sept. 1975, "Nuclear Chemistry 1975 Divisional Reorganization," Sessler Files, Box 23 of 26; and Joint Committee on Atomic Energy, *AEC Authorization Legislation, Fiscal Year 1975,* 93rd Cong., 2nd sess., 1974, p. 1359.

disadvantage since no one had experience and the subfield's novelty was itself attractive. Steven Koonin, who visited Berkeley from Caltech as a newly minted theorist, later explained: "The thing about Bevalac physics was that it wasn't high energy physics, it wasn't nuclear physics, it was new physics—and that's what made it so exciting."[47]

Thanks to the contributions of German and Japanese scientists, the excitement of new results, and the influx of new hires and visitors, the Bevalac became a decidedly popular machine as the 1970s wore on. However, the growing cadre of Bevalac researchers faced yet another crucial task: they needed to forge the connections necessary to shape a new kind of science.

Since relativistic heavy ions were already available, an obvious first step was to accumulate data. Acquiring the necessary equipment was the easy part. When seeking equipment to use on the floor of the former Bevatron, the recycled high-energy physics accelerator, Bevalac researchers again turned to the recycle bin of high-energy physics. As Nagamiya explained, he visited many Berkeley high-energy physics groups "to beg" for "phototubes, cables, electronics," and other used items. As one of Nagamiya's Japanese collaborators concluded: "Everything we used for the first experiment [was] all junk."[48]

Fashioning a framework for understanding the data that emerged from these early experiments was much more difficult because no quick-and-dirty recycling would do the job. Instead, Bevalac science was *recombinant*—that is, it required that researchers combine insights and expertise from various subfields in new ways to create a brand new outlook. Because an ordinary heavy nucleus is used to start a heavy-ion collision, Bevalac researchers required a detailed understanding of what the nucleus is and what it is like under ordinary conditions as a point of departure for understanding what happens after collision. Nuclear physicists who had worked with lower-energy heavy ions had this insight. For their part, high-energy physicists were experienced in using accelerators to produce and detect collisions at the Bevalac's relativistic energy range, though they had used lighter particles. Cosmic ray physicists were the ones who had already worked with relativistic heavy ions, though they had not used accelerators. Since most of those who permanently joined the Bevalac research effort were nuclear physicists, summer study opportunities—which attracted high-energy theorists as well as a wide range of other experts—were crucial in bringing together the range of specialists necessary to forge the new understanding.

A key early challenge was to find the appropriate scientific language and concepts to understand and describe Bevalac collisions. A turning point came when Bevalac researchers realized that they could use specialized methods for dealing with high-energy interactions (relativistic kinematics) from high-energy physics. From this kinematics, the researchers borrowed a new set of variables. Relativistic variables enabled them to plot a large range of data in a comprehensible manner. Another turning point came when cosmic ray researchers suggested classifying relativistic events in terms of central collisions, in which nuclei collide head-on, and peripheral collisions, in which one nucleus glances off another.[49]

To get a further handle on relativistic heavy ions, researchers needed theoretical guid-

[47] Interview with Schroeder, 7 Mar. 1994; interview with Steven Koonin, 27 Oct. 1994; and Weart, "Physics Business in America, 1919–1940" (cit. n. 28), p. 329.

[48] Nagamiya, "Banquet Speech Presented at the 4th International Conference on Nucleus-Nucleus Collisions" (cit. n. 39); and interview with Tanihata, 28 Oct. 1993.

[49] Interview with Harry Heckman, 19 Jan. 1994; and interview with Schroeder, 7 Mar. 1994. Relativistic kinematics includes reference-frame transformation and the treatment of energy, momentum, and mass at high energies. An example of the switch to relativistic variables is the use of rapidity instead of parallel velocity.

ance. Since they knew from the outset that Bevalac collisions created a highly complex reaction system, theorists started at the beginning, first devising very simple schemes to model collision processes: cascade models, which viewed nucleons interacting as "billiard balls"; hydrodynamic models, which saw the nuclear system as a fluid; and thermodynamic models, which considered nuclear matter as a quantum interacting gas. However, such theoretical explanations were so basic that they all seemed to work equally well, offering researchers little help in sorting out what was important in the flood of new data. The theorist Miklos Gyulassy derisively noted that in fact they were faced with a "zoo of models" to explain relativistic heavy-ion collisions.[50]

By this time, however, theoretical efforts were starting to pay off. A couple of years earlier a group of theorists had suggested viewing collision products in terms of "spectators" and "participants." This point of view took hold when it became apparent that it allowed researchers to focus on particularly telling aspects of the data and led to the realization, for example, that it was important to determine how central a collision is— that is, how great the overlap was between colliding nuclei.[51]

The growing theoretical understanding guided experimental efforts, which spanned a wide range of research topics. As part of this program the German-American collaboration constructed a series of detectors designed to zero in on central collisions. The data emerging from these detectors allowed Gary Westfall, a postdoctoral fellow involved in the collaboration, to combine the participant-spectator model with thermal measurements to create the Nuclear Fireball Model (and a later refinement, the Firestreak Model), which described the "fireball" created in a relativistic heavy-ion collision in terms of the spectator-participant point of view. (See Figure 5.) In Poskanzer's words, with the advent of the Fireball Model, Bevalac researchers "were in the exciting position that a rough description of the data was at hand."[52]

Having developed a basic framework for understanding their data, researchers' next major goal was to gain a more comprehensive view of relativistic heavy-ion collisions. The first important detector for this purpose was the Streamer Chamber, a device originally used by high-energy physicists that had been modified by Stock and others for use with heavy ions. Unlike previous electronic detectors, which detected only some particles from each collision, this device captured an image of most collision products. To increase their research bounty even further, Poskanzer and Gutbrod set their sights on devising a 4π detector that would complement the Streamer Chamber by detecting a large fraction of the particles emitted from the collisions, measuring their energy, identifying them electronically, and reading out the information electronically rather than by visual scanning. This powerful new tool would provide more specific information and take much more data.[53]

Gutbrod hit on the idea of designing a detector based on the geometry of a state-of-the-art 4π high-energy physics detector, the Crystal Ball. After obtaining both advice and

[50] Miklos Gyulassy, "Comparison of Models of High Energy Nuclear Collisions," Proceedings of the Special Bevalac Research Meetings, 1–3 Nov. 1977, LBL-7141 (Lawrence Berkeley Laboratory, 1978).

[51] J. D. Bowman, Wladyslaw J. Swiatecki, and Chin Fu Tsang, "Abrasion and Ablation of Heavy Ions," LBL-2908 (Berkeley, 1973); and S. Das Gupta and Gary Westfall, "Probing Dense Nuclear Matter in the Laboratory," *Phys. Today,* 1993, *46:*34–40.

[52] Gary Westfall *et al.,* "Nuclear Fireball Model for Proton Inclusive Spectra from Relativistic Heavy-Ion Collisions," *Physical Review Letters,* 1976, *37:*1202–1205; J. Gosset *et al.,* "Calculation with the Nuclear Fire-streak Model," *Physical Review C,* 1978, *18:*844–855; W. D. Myers, "A Model for High-Energy Heavy Ion Collisions," *Nuclear Physics A,* 1978, *296:*177–188; and Poskanzer, "History of Central Collisions at the Bevalac" (cit. n. 44), pp. 451–452.

[53] Such devices are called 4π detectors because they cover the full (4π) steradian solid angle.

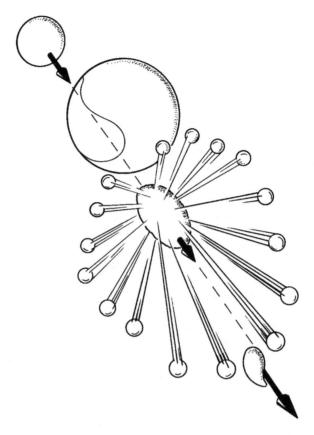

Figure 5. *The Fireball Model describes a heavy-ion projectile, which hits a stationary-target heavy-ion nucleus. After collision most of the target nucleus remains a "spectator" to the event, as does a piece of the speeding projectile, which continues on its original path, as shown with the arrows and the dotted line. The "participants" in the event are the overlapping portions of the target and projectile, which merge in a central collision to create a "fireball" of hot, dense nuclear matter. (Photo courtesy of Lawrence Berkeley National Laboratory.)*

drawings from SLAC researchers, he specially fashioned a device capable of identifying the large number of particles emitted in a relativistic heavy-ion collision. He dubbed this mezzo science version of the detector the "Plastic Ball" because it was constructed using plastic scintillators rather than lead glass.[54] Although smaller than the largest high-energy physics detectors of the time, the Bevalac's multimillion-dollar Plastic Ball was large for a nuclear physics detector and much larger than early Bevalac devices, which were only a few feet long.

The Plastic Ball was finished in 1982, in time for use with the heavier beams being produced by the Bevalac. (See Figure 6.) As before, better data led to better theory and

[54] Interview with Gutbrod, 26 Oct. 1993. The necessary revisions included, e.g., the development of an unorthodox method of optically coupling two scintillators to a single phototube, which allowed researchers to pack many detector components into a small area so that more particles could be detected. Although the Plastic Ball researchers came up with the idea independently, they later learned that this notion had been presented twenty-seven years earlier by Sir Denys Wilkinson, who coined the term "phosphor sandwich," or "phoswich." See Poskanzer, "History of Central Collisions at the Bevalac" (cit. n. 44), p. 452.

Figure 6. *The completed Plastic Ball. Hans Gutbrod and his colleague Hans-Georg Ritter are in front; Arthur Poskanzer is in back. (Photo courtesy of Lawrence Berkeley National Laboratory.)*

increased understanding. In the words of one reviewer, thanks to "a mini-renaissance" in the development of relativistic heavy-ion theories inspired by the advent of the Plastic Ball, systematic studies were made of collective effects in both Streamer Chamber and Plastic Ball data.[55]

In place of the early crude models, Bevalac researchers at this stage were working on what they called the nuclear equation of state, a coherent, detailed explanation of nuclear matter at high temperatures and densities. Owing to the ability of the Plastic Ball to take

[55] Poskanzer, "History of Central Collisions at the Bevalac," p. 459; and Accelerator and Fusion Research Division, "Annual Report, 1985," LBL-22750 (Berkeley, 1986), pp. 5–6. On the work to make the acceleration of heavy elements possible see note 33, above.

and analyze data quickly, over the next several years the Plastic Ball group was able to provide concrete answers to clues provided by early Streamer Chamber results; thus Bevalac researchers had achieved, in Poskanzer's opinion, "a quantitative understanding of the nuclear equation of state."[56]

Research aimed at gaining greater understanding of the nuclear equation of state would continue at the Bevalac through the early 1990s—for example, with the Equation of State (EOS) Time Projection Chamber, yet another device built from a high-energy physics model.[57] In the mid 1980s, however, the Bevalac again became unpopular, this time as a consequence of its own success. Ironically, the very lure that drew researchers to Berkeley in the mid 1970s—the possibility that exotic phenomena would be produced in the high temperatures and densities of relativistic heavy-ion collisions—at this point pulled them elsewhere. Thanks to the Plastic Ball and Streamer Chamber data, theorists now predicted that with energies much greater than those produced in the Bevalac researchers could hope to find exotic phenomena, perhaps even the long-sought free or deconfined quarks—that is, quarks existing by themselves instead of being bound in clusters, like all other known examples. Yet again, high-energy physics equipment was recycled for the cause: researchers at Brookhaven National Laboratory in New York and the European laboratory CERN in Geneva retooled the Alternating Gradient Synchrotron and the Super-Proton Synchrotron to produce the desired higher-energy heavy ions. The Plastic Ball eventually joined the parade it had helped to start: in 1986 the device was put into a crate and shipped to CERN.[58]

By the late 1980s, LBL researchers joined others to cast longer-range plans, which were built on the experiences at Berkeley, Brookhaven, and CERN, for a larger, even higher-energy heavy-ion machine, the Relativistic Heavy Ion Collider, that was meant to carry the field into the twenty-first century. As researchers drifted away to forge detector plans and form new collaborations, the nuclear science community reassessed its funding priorities in light of the RHIC's $500 million price tag and the drain in resources that came in the wake of planning for it. With its network of human and hardware connections grievously damaged, the Bevalac lost its constituency, paving the way for the decision to terminate operation of the machine in 1993. For the Bevalac, the ultimate consequence of its success as a trendsetter was its own demise.[59]

Before the 1970s, accelerator-produced relativistic heavy-ion physics did not exist. By the time the Bevalac was shut down in 1993, the subfield was a thriving enterprise, with eighty publications in U.S. journals that year. By 2001, with RHIC in operation, the number had risen to ninety-six. And along the way researchers had gained much new understanding. By the last years of the Bevalac, earlier views of relativistic heavy-ion collisions and the nuclear equation of state had been sharpened and extended, as represented

[56] Poskanzer, "History of Central Collisions at the Bevalac," p. 460.

[57] One of the originators of EOS, the Berkeley researcher Howell Pugh, noted that the detector was aptly named because Eos was the goddess of the dawn and had long been associated with the creation of the universe. See Howell G. Pugh, Grazyna Odyniec, Gulshan Rai, and Peter Seidl, "EOS: A Time Projection Chamber for the Study of Nucleus-Nucleus Collisions at the Bevalac," LBL-22314 (Berkeley, 1986).

[58] Gordon Baym, "Major Facilities for Nuclear Physics," *Phys. Today,* 1985, *38:*40–48, esp. p. 43 (the expected exotic phenomena); and Poskanzer, "History of Central Collisions at the Bevalac" (cit. n. 44), p. 460 (move of the Plastic Ball). Many of the German and Japanese researchers who had worked at the Bevalac were among those who shifted their effort to the new accelerators, in these cases at the expense of the Bevalac-like accelerators at their home institutions.

[59] Baym, "Major Facilities for Nuclear Physics," pp. 43–48. As Smith notes, large-scale projects "demand continued justification": Smith, *Space Telescope* (cit. n. 14), p. 424.

in Figure 7. According to this view, researchers had new insights not only into relativistic heavy-ion collisions, but also into the conditions to be found in neutron stars and in the early universe. In addition, they had evidence for the existence of the Quark Gluon Plasma, a unique new state of matter with deconfined quarks.[60]

Although the machine of recycled parts died, the recombinant science of the Bevalac thrived in the decade after its death. By forging links to interconnect machinery, provide money and workers, and rewind separate scientific strands, Bevalac researchers set into motion a new line of inquiry that inspired ever more refined efforts to penetrate the fabric of subatomic and cosmic matter.

THE BEVALAC AND THE HISTORIOGRAPHY OF LARGE-SCALE RESEARCH

Growth is certainly part of the Bevalac story. In line with the often-described pattern, Bevatron, HILAC, and Bevalac researchers sought larger accelerators to continue fruitful

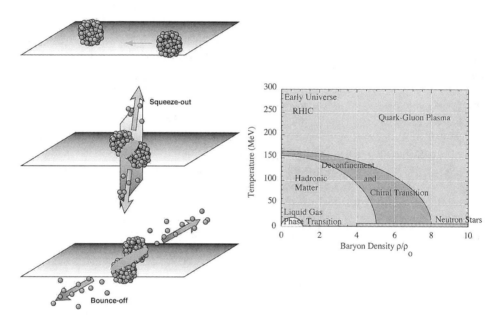

Figure 7. *The left side of the figure shows the view of relativistic heavy-ion collisions in 2001. According to this view, when two heavy nuclei collided at relativistic energies, a hot compression zone formed, squeezing out hot, dense nuclear matter—which perhaps included quarks and gluons—at right angles. After this squeeze-out, the two nuclei bounce off each other. The right side of the figure shows a representation of the nuclear equation of state. This description of nuclear matter in terms of compression and temperature shows that at high density and low temperature nuclear matter is like neutrons stars, whereas at low density and high temperature nuclear matter is like the early universe in the first few microseconds after the big bang. One particularly compelling phenomenon of the early universe is the Quark Gluon Plasma, a phenomenon whose existence is not yet confirmed in 2002. The Quark Gluon Plasma would be a new state of matter, the only type with deconfined quarks. (Gary Westfall made this drawing with information from Hans Gutbrod and Horst Stöcker, "The Nuclear Equation of State,"* Scientific American, *Nov. 1991, pp. 58–66.)*

[60] Hans Gutbrod and Horst Stöcker, "The Nuclear Equation of State," *Scientific American,* Nov. 1991, pp. 58–66. The publications were counted in the following U.S. journals: *Physical Review Letters, Physical Review C* and *D,* and *Reviews of Modern Physics.* I thank Gary Westfall for compiling this information.

lines of scientific inquiry. Although some researchers did not get everything they asked for, in each case larger accelerators were built (even if not at Berkeley), providing more proof that growth is a distinctive feature of large-scale research. But growth is hardly the only significant feature of the Bevalac tale. In fact, the Bevalac story expands and refines every aspect of ongoing discussions of the historiography of large-scale research.

Periodization is a case in point. Since the Bevalac paved the way for RHIC, a new $500 million U.S. accelerator project that came into operation in the late 1990s, it is clear that the 1993 demise of the SSC did not signal an end to large research accelerators, much less large-scale research in the United States. Just as it is necessary to push back the beginning point of large-scale research, it is necessary to push forward its ending point. The obituaries were premature: at the beginning of the twenty-first century large-scale research is alive and well in the United States.[61]

Similarly, the pivotal roles played by specialists working at a variety of scales in the development of a new subfield suggest an expansion of the typology of large-scale research. Alongside considerations of time period and the organization and control of work, we need to note relative scales, for large-scale research comes in a variety of sizes— modest, mezzo, and grand. If we let researchers working at a large (but not the largest) scale get lost in the shadows cast by grander colleagues, we will miss the different varieties and differing cultures of large-scale research.

The story of the Bevalac, with its tangled web of connections, interconnections, and disconnections, also underlines the value of framing discussions of the development of large-scale research in terms of networks. In line with the cases examined by Smith, Galison, and others, the Bevalac story shows that a research project is built piecemeal from the painstaking connection of a diverse collection of parts—in this case conduits for funding, pieces of equipment, practitioners, theoretical ideas, experimental methods, and a supporting organizational framework.

The Bevalac story lends the additional insight that the relevant network for a project can be quite wide reaching, uniting different kinds of science practiced at different scales by researchers from different countries. In fact, the Bevalac's transformation from an unpopular machine to the harbinger of a new type of science shows the extent to which successful knowledge production can *hinge* on the ability of researchers to draw on such a wide-reaching network so as to circumvent disconnections (like the lack of local Bevalac users) and bridge a variety of gaps (for example, by intertwining machinery as well as concepts previously used for very different purposes).

The Bevalac story also reveals two features that result from the network of large-scale research. One is the recycling of equipment and personnel, which typically proceeds from the larger to the smaller types of research. Another is recombinant research, that is, the combining of insights and experience from various parts of the network in novel ways to create a new type of science. Both features make a striking contribution to the breadth and longevity of large-scale research.

---

[61] Further evidence of the health of large-scale research in the United States was the development in the 1980s and 1990s of the $800 million Advanced Photon Source at Argonne National Laboratory in Illinois and the $500 million Thomas Jefferson National Accelerator Facility in Newport News, Virginia. For information on the development of these facilities see Jack Holl, *Argonne National Laboratory, 1946–96* (Urbana: Univ. Illinois Press, 1997), pp. 470–473; and Catherine Westfall, "A Tale of Two More Laboratories: Readying for Research at Fermilab and Jefferson Laboratory," *Hist. Stud. Phys. Biol. Sci.*, 2002, *32*:369–407. The Bevalac also started radioactive beam research, which has prompted the proposal for an $800 million Radioactive Beam Accelerator that, if approved, would be built in the latter part of the first decade of the twenty-first century.

The Bevalac story contains other lessons. Particularly important is the demonstration of what is lost if historians rely on the rhetoric of participants. The problem has been that along with the term "Big Science" we pick up values that may not suit our purposes. For example, for high-energy physicists talk of Big Science underscores the superiority of *their* science, with its focus on probing the fundamental nature of matter with the grandest, most expensive equipment. For them, the Bevalac, like all smaller forms of large-scale research, does not figure into the discussion because it is puny and useless in the pursuit of superior science. Although belief in the peerless grandeur of their subfield may help high-energy physicists compete for resources, the preoccupation with bigness merely restricts and obstructs our view. If we continue to frame discussions in terms of Big Science, in fact, there will continue to be many untold tales and therefore many missed lessons.

To remedy the problem, we need to abandon talk of Big Science and find rhetoric that fits our own agendas. As a step in that direction, this discussion of the formation of the Bevalac and relativistic heavy-ion physics from various types and sizes of science speaks of modest, mezzo, and grand science. This is only one example of finding words to tell our own story. Thinking in terms of "expensive science" might allow us to focus more precisely on the important issue of how much large-scale science of various kinds costs taxpayers, not only in construction costs but in continuing yearly expenses. We might also consider "multidisciplinary large-scale science." Argonne National Laboratory's Advanced Photon Source, for example, serves physicists, material scientists, biologists, and others. How are the networks of large-scale research extended by researchers from such disparate fields, and what new features of large-scale research arise from the disconnections, connections, and interconnections of these networks?

Finally, the almost untold saga of the Bevalac should serve as a cautionary tale. We must continue self-critical analysis as we further expand and refine our methodology. Only by constantly searching for more incisive analyses can we uncover new parameters and dimensions to reveal the full, vivid panorama of large-scale research.

III

SOCIAL POLICIES, SCIENTIFIC PRACTICE, AND THE LAW

Families Made by Science

Arnold Gesell and the Technologies of Modern Child Adoption

By Ellen Herman

ABSTRACT

This essay considers the effort to transform child adoption into a modern scientific enterprise during the first half of the twentieth century via a case study of Arnold Gesell (1880–1961), a Yale developmentalist well known for his studies of child growth and the applied technologies that emerged from them: normative scales promising to measure and predict development. Scientific adoption was a central aspiration for many human scientists, helping professionals, and state regulators. They aimed to reduce the numerous hazards presumed to be inherent in adopting children, especially infants, who were not one's "own." By importing insights and techniques drawn from the world of science into the practical world of family formation, scientific adoption stood for kinship by design. This case study explores one point of intersection between the history of science and the history of social welfare and social policy, simultaneously illustrating the cultural progress and power of scientific authority and the numerous obstacles to its practical realization.

D URING THE TWENTIETH CENTURY, child adoption was reimagined as a scientific enterprise. Research on nature and nurture that utilized adoption data, along with numerous studies of adoption practices and outcomes, attempted to join the production of knowledge to the promotion of child and family welfare. In a transaction that turned biological strangers into kin, scientific professionals saw an unusual experimental opportunity, a challenging social laboratory, and a series of intricate operations that promised intellectual discovery as well as hope for children in need.[1] Adoption was subjected to empirical inquiry and technical innovation by geneticists probing the relationship between heredity and environment, psychologists charting the developmental course through childhood and into adult life, sociologists seeking to unearth the logic of social roles, and social

Isis 92, no. 4 (December 2001): 684–715.

[1] The term "adoption" is used here to refer to adoption by nonrelatives (adoption by "biological strangers") because scientific professionals, including Arnold Gesell, have been almost exclusively interested in this type of adoption. Where adoption by blood relatives was presumed to be natural and easy, the difficulties associated with adoption by strangers justified public regulation and professional oversight. Most children adopted during the twentieth century were probably placed with nonrelatives, but a large number were adopted by natal relatives and stepparents. Adoptions by relatives have increased substantially since 1970.

workers refining procedures for placing children who needed parents with adults willing to take them.[2]

This article explores one aspect of the modern science of family making by considering Arnold Gesell's contribution to "scientific adoption." Scientific adoption promised to reduce the numerous hazards presumed to be inherent in adopting children, especially infants, not one's "own." "Science has been uncovering facts that mean for the family, and particularly for the adoptive parent and child, a welcome emancipation from fear," two observers noted in 1939, equating science with safety. In addition to preventing disastrous placements by detecting genetic "lemons" wholly unsuited to family life and identifying children destined to disappoint the expectations of adopters, advocates of scientific adoption aimed to improve children's lives by importing insights and techniques from the world of science into the practical world of family formation. More than a token of humanitarian concern for vulnerable and dependent children, the science of home emerged as a developmental imperative at a time when the ideology of institutional care went into precipitous decline and family life was championed as "the highest and finest product of civilization."[3] Scientific adoption was a central aspiration for many human scientists, helping professionals, and state regulators during the first half of the twentieth century. Professional elites turned to engineering methods and metaphors in their quest to modernize adoption precisely because they believed that science was both antithetical and superior to sentimental, commercial, accidental, and other idiosyncratic ways of making up families. Scientific adoption stood for making families methodically, on a foundation of empirical knowledge supported by the protective machinery of bureaucratic regulation, disinterested professionalism, and technical competence. We might term this enterprise "kinship by design."

Scientific adoption is a worthwhile topic for scholars interested in how science works as well as how science matters. Although not widely realized in practice before 1945, scientific adoption was increasingly visible from the 1910s to the 1930s—in mass-circulation magazines and newspapers, in government pamphlets, on the radio, and in professional literature of many kinds—as a benchmark against which all adoptions should be measured. The first popular book-length manual for would-be adopters, *The Adopted*

---

[2] A few examples of adoption as an opportunity to explore issues about nature and nurture are David Howe, *Patterns of Adoption: Nature, Nurture, and Psychosocial Development* (Oxford: Blackwell, 1998); Joyce Maguire Pavao, *The Family of Adoption* (Boston: Beacon, 1998); Robert Plomin and John C. DeFries, *Origins of Individual Differences in Infancy: The Colorado Adoption Project* (Orlando, Fla.: Academic, 1985); and Marie Skodak and Harold M. Skeels, "A Final Follow-Up Study of One Hundred Adopted Children," *Journal of Genetic Psychology,* Sept. 1949, 75:85–125. On social roles see H. David Kirk, *Shared Fate: A Theory of Adoption and Mental Health* (New York: Free Press of Glencoe, 1964). On adoption practices and outcomes see Donald Brieland, *An Experimental Study of the Selection of Adoptive Parents at Intake* (New York: Child Welfare League of America, May 1959); Child Welfare League of America, ed., *Quantitative Approaches to Parent Selection* (New York: Child Welfare League of America, 1962); David Fanshel, *Far from the Reservation: The Transracial Adoption of American Indian Children* (Metuchen, N.J.: Scarecrow, 1972); Michael Schapiro, *A Study of Adoption Practice,* 3 vols. (New York: Child Welfare League of America, 1956); Rita James Simon and Howard Alstein, *Transracial Adoption* (New York: Wiley, 1977); and Sophie van Senden Theis, *How Foster Children Turn Out,* Publication No. 165 (New York: New York State Charities Aid Association, 1924).

[3] Lee M. Brooks and Evelynn C. Brooks, *Adventuring in Adoption* (Chapel Hill: Univ. North Carolina Press, 1939), p. 12 ("welcome emancipation from fear"). The often-quoted declaration about family life by the first White House Conference on Children in 1909 sometimes obscures the fact that the theory of institutional care was repudiated long before its practice ended. Not until the 1950s did the number of children living in temporary foster families exceed the number of children living in institutions, and it was not until the 1960s that the number of adoptive placements surpassed the number of institutional placements. See Bernadine Barr, "Spare Children, 1900–1945: Inmates of Orphanages as Subjects of Research in Medicine and in the Social Sciences in America" (Ph.D. diss., Stanford Univ., 1992), p. 32, Fig. 2.2.

Child (1936), for instance, contained chapters on "Heredity and Environment" and "Intelligence Tests in Adoption." These detailed the views of "acknowledged leaders" in various scientific fields, even though the author, Eleanor Gallagher, worried that overweening professionals would squeeze common sense out of adoption. Gallagher noted that publicity about scientific adoption had already changed the behavior of many adults in search of children. Prospective adopters from small towns, where professional agencies were scarce, were more likely than in the past to travel to cities, where their demands for the scientific selection of children could be satisfied.[4]

Scientific adoption exhibited the worth of science talk even as it demonstrated persistent obstacles to the exercise of comprehensive scientific authority in child and family policy. As a symbolic ideal, scientific adoption of the kind advocated by Gesell arguably shaped more and more Americans' expectations and experiences of adoption over time. Warnings that only scientific practices could prevent adoption catastrophes and produce positive outcomes eventually sank in, if only because such practices were becoming habitual. After World War I the mainstreaming of testing techniques in schools, the military, and workplaces made classification technologies routine features of public administration and popular consciousness. The practice of classifying children so as to meet adopters' desires or promote children's interests opened the private sphere to a regulatory style that clashed with the ideology of familial love and autonomy, but many educated adopters desired the safety and knowledge that science offered. They learned to associate them with professional practices, such as developmental and intelligence testing.

One couple who told their Depression-era adoption story to a broad reading audience freely admitted that "we wanted the best material available." Though "fatalistic" about the hereditary credentials of adoptable children, their agency obtained "a baby who answered most of our requirements," including a high I.Q. and birth parents who had graduated from college. The adoptive parents attributed the eventual success of the adoption—a deep love for their adopted daughter in spite of initial worries that they would not love her as they did their "own" son—to this "cold-blooded" approach. At midcentury, a choosy New York couple applied for a pretty girl "with violet-colored eyes, of pure Nordic background, and an I.Q. of a hundred and thirty-five or above." Roundly criticized as inflexible by agency workers, such detailed orders for children illustrate that promises of scientific certification had not fallen on deaf ears. More than a simple hedge against the unknown, scientific adoption boldly ventured to compensate participants for membership in families labeled artificial, unnatural, and less "real" than the biological norm. Compromised by inauthenticity, adoption could at least be privileged by a measure of predictive certainty unavailable to people who acquired children in the ordinary way. "Consider the interesting fact that the foster parent does not appear on the scene until someone else has run the early, major hazards of childbearing!" pointed out two observers of the adoption scene in the late 1930s. "He can choose, he does not have to take what comes along." At the same time, anecdotal evidence suggested that "social class position is inversely related to the capacity of couples to accept deviations from normal in adoptive children."[5] Adopt-

[4] Eleanor Garrigue Gallagher, *The Adopted Child* (New York: Reynal & Hitchcock, 1936), p. 96.

[5] Anonymous, "We Adopt a Child," *Atlantic Monthly*, Mar. 1940, *165*:318, 322 ("cold-blooded" approach); Katherine Kinkead, "Our Son," *New Yorker*, 4 Mar. 1950, 26:32–57, on p. 38 (New York couple); Brooks and Brooks, *Adventuring in Adoption* (cit. n. 3), p. 17 (advantages of choice); and Edmund V. Mech, "Trends in Adoption Research," in *Perspectives on Adoption Research*, ed. Child Welfare League of America (New York: Child Welfare League of America, 1965), pp. 7–32, on p. 9. Note that the term "foster" was used to denote both temporary and permanent family placements until well after World War II.

Figure 1. *Arnold Gesell. Reprinted by permission of the Gesell Institute.*

ers at the bottom of the socioeconomic ladder were less discriminating, more likely to discount scientific methods of quality assurance and welcome "special needs" adoptions than either wealthier adopters or professionals.

"Diversity" has been a keyword in the recent ideological war over the family, a strategy for celebrating different ways of forming families and acquiring children in the face of conservative mobilization around "family values." For much of the twentieth century, however, science offered some of the most persuasive answers to what was widely perceived to be the stigma of adoption, a fact that illustrates how technological optimism shaped a social policy debate far from the conventional world of science. That scientific approaches largely aimed to make adoption conform to a unitary model of "natural" kinship—opting for sameness as a vision more likely to succeed than difference—is one indicator of how dramatically both scientific and cultural conversations about family life and nature have changed course in recent decades.

MAKING ADOPTION MODERN BY MAKING ADOPTION SCIENTIFIC

Arnold Gesell (1880–1961) articulated the central premise of scientific adoption when he asserted that "nothing in the field of social welfare needs more deliberate and conscious regulation than child adoption." (See Figure 1.) It is worth quoting him at some length to emphasize the logical premise of scientific adoption, the causal link between professionalism and positive outcomes for deserving children and families.

> [Adoption] can not be intrusted altogether to good will or to intuitive impulse, or even to unaided common sense. There are too many opportunities for error and miscarriage. The combined critical judgment of the social investigator, the court, the physician, and the mental examiner should enter into the regulation of adoption. . . . Systematic psychoclinical examinations not only will reduce the wastes of error and miscarriage but will serve to reveal children of normal and superior endowment beneath the concealment of neglect, of poverty, or of poor repute.

Clinical safeguards can not solve all the problems of child adoption but they can steadily improve its methods and make them both more scientific and humane.[6]

Families could and should be made up scientifically because science offered the safest way to adopt and the surest way to ensure that adoption turned out well. At a time when numerous scientists and many social welfare professionals voiced eugenic doubts about the value of adoption as a social institution, Gesell championed its potential for children and for society. His scientific and technological ingenuity contributed to a new era of confidence in adoption, anticipating the decline of hereditarianism and the unprecedented rise of adoption's cultural visibility and statistical popularity over the course of the twentieth century.[7] Few things were more important to the transformation of adoption's reputation during the twentieth century than the promise made in the name of science: the risks of acquiring children born to others could be known and predicted in advance, hence reduced to a point of relative insignificance.

Gesell was a nationally renowned developmental psychologist and physician. He was best known for devising scales of normative development that were widely utilized by clinicians working in medical and educational fields during the interwar period. With his Ph.D. in psychology from Clark University in 1906 and a M.D. from Yale in 1915, Gesell's career coincided with the formative years of a self-conscious developmental science. The experimental ethic in American psychology had roots in the late nineteenth century, when laboratories were first established in American universities on the model pioneered by Wilhelm Wundt at the University of Leipzig in 1883. It was not until the first decades of the twentieth century, however, that American psychologists forged a consensus that psychological research required a sharp division of labor between scientific subjects and sci-

---

[6] Arnold Gesell, "The Psychological Welfare of Adopted and Foster Children," 26 Oct. 1937, p. 3, Box 45, Folder: "Subject File: Adoption," Arnold Gesell Papers, Library of Congress Manuscript Division, Washington, D.C. (hereafter cited as **Gesell Papers**); and Arnold Gesell, "Psychoclinical Guidance in Child Adoption," in U.S. Children's Bureau, *Foster-Home Care for Dependent Children*, Publication No. 136 (Washington, D.C.: Government Printing Office, 1926), pp. 193–204, on p. 204.

[7] Statistics on twentieth-century domestic adoptions are unreliable because a national reporting system existed only between 1945 and 1975, when the U.S. Children's Bureau and the National Center for Social Statistics collected data voluntarily supplied by states and territories. It is clear that legal adoption was still a rare and exotic procedure in 1900. Recent studies conservatively estimate that one million adopted children are currently living with adoptive parents, that five million adoptees of all ages are alive in the United States, and that 2–4 percent of all American families have adopted. Approximately 125,000 adoptions have taken place annually in recent years. The numerical high point for twentieth-century adoption occurred around 1970, when adoptions reached 175,000. International placements, almost all nonrelative adoptions, became more numerous after World War II and have increased dramatically in recent years; in 1999 there were 16,396, more than double the 1990 figure. Like international placements, transracial adoptions have received a great deal of attention, but they are far less significant statistically than symbolically. The largest number of transracial adoptions occurred in the years around 1970, when there were perhaps a few thousand annually. One of the only national surveys of black children adopted documented 4,336 adoptions in 1969, of which almost one-third were transracial placements. See "Adoption of Black Children in 1969," Box 26, Folder: "Oregon—Adoption,"International Social Service/ American Branch Papers, Social Welfare History Archives, University of Minnesota, Minneapolis. See also the statistical profile compiled by the Evan B. Donaldson Adoption Institute: http://www.adoptioninstitute.org/research/ressta.html; Anjani Chandra *et al.*, "Adoption, Adoption Seeking, and Relinquishment for Adoption in the United States," *Advance Data from Vital and Health Statistics of the Centers for Disease Control and Prevention/National Center for Health Statistics* , no. 306, 11 May 1999; and Kathy S. Stolley, "Statistics on Adoption in the United States," *Future of Children,* Spring 1993, 3:26–42. On eugenics see Hamilton Cravens, *The Triumph of Evolution: American Scientists and the Heredity-Environment Controversy, 1900–1941* (Philadelphia: Univ. Pennsylvania Press, 1978); Carl N. Degler, *In Search of Human Nature: The Decline and Revival of Darwinism in American Social Thought* (New York: Oxford Univ. Press, 1991); and Daniel J. Kevles, *In the Name of Eugenics: Genetics and the Uses of Human Heredity* (Cambridge, Mass.: Harvard Univ. Press, 1995).

entific observers, that normal psychology was at least as significant a focus of inquiry as abnormal psychology, and that the point of psychological science was to generalize in quantitative terms about groups rather than explore the qualitative uniqueness of individuality. "The triumph of the aggregate," as one insightful historian has called it, was the necessary prerequisite for developmental perspectives based on the study of large numbers of ordinary children and oriented toward the efficient, hygienic administration of institutions, from schools and orphanages to juvenile courts and families. This normalizing, managerial shift within a new and still small scientific community facilitated psychology's imperial expansion and enormous cultural success later in the century.[8]

Gesell's career also paralleled the formative stages of modern adoption. Although exchanging children for reasons of love, labor, and property is a practice as ancient and varied as human culture, adoption took a historically unprecedented turn beginning early in the twentieth century. State legislators, who had passed special-purpose adoption statutes since the mid-nineteenth century, devised legal regulations after 1900 that subjected adoptive kinship to new forms of public inspection and professional regulation, from the certification of child placers to mandatory home studies and lengthy periods of postplacement supervision. By midcentury, lawmakers everywhere in the country endorsed the idea that adoptive families ought to be made up in public and on purpose, with as much professional help as possible. A few states eventually banned all nonprofessional placements, but these were largely symbolic acts, ineffective in preventing adults from crossing state lines in search of the children they desired. State adoption laws retained stubborn particularities, and no uniform adoption code was ever passed.[9] Loopholes persisted even as the formal network of expert control over adoptive family formation expanded.

Throughout his long academic career, Gesell championed professional child welfare and adoption practices. He worked with the most important national and local child welfare organizations of his day, including the U.S. Children's Bureau (a federal bureaucracy established by Congress in 1912) and the Child Welfare League of America (a federation of public and private service organizations still in existence today). Such public and private entities were major players in the campaign to encourage and enforce kinship by design. Gesell often advised these and other groups on policies related to placement age, preplacement testing, and clinical supervision in adoption.[10] The advantage of scientific adoption was one of many topics on which he spoke and wrote, helping to establish its credibility with popular audiences. Gesell's public influence as a psychological expert was a hallmark of his career.

Kinship by design amounted to a legal and bureaucratic revolution in the adoption process, but this revolution was geographically and culturally lopsided, effectively realized

[8] "The triumph of the aggregate" is from Kurt Danziger, *Constructing the Subject: Historical Origins of Psychological Research* (New York: Cambridge Univ. Press, 1990), esp. Ch. 5. On the study of large numbers of children see Hamilton Cravens, *Before Head Start: The Iowa Station and America's Children* (Chapel Hill: Univ. North Carolina Press, 1993). On the expansion and success of psychology see James H. Capshew, *Psychologists on the March: Science, Practice, and Professional Identity in America, 1929–1969* (New York: Cambridge Univ. Press, 1999); and Ellen Herman, *The Romance of American Psychology: Political Culture in the Age of Experts* (Berkeley: Univ. California Press, 1995).

[9] For one midcentury summary of state-by-state variations in adoption laws see Frances Lockridge and Sophie van S. Theis, *Adopting a Child* (New York: Greenberg, 1947), Ch. 12.

[10] E.g., see "Adoption," Memorandum re: Child Welfare League of America Program, 7 June 1939, Box 45, Folder: "Subject File: Adoption [Law]"; Maud Morlock to Arnold Gesell, 22 July 1944, and Gesell to Morlock, 27 July 1944, Box 45, Folder: "Subject File: Adoption, 1939–56"; Memorandum re: Child Adoption, 30 Jan. 1939, Box 45, Folder: "Subject File: Adoption [Memoranda]"; and Memorandum re: meeting of Committee, CWLA, 9 June 1939, Box 45, Folder: "Subject File: Adoption, 1939 [CWL of America]": Gesell Papers.

in those parts of the country where the new expertise was most heavily concentrated: cities in the east and north. Professionally staffed agencies were rare or nonexistent in many parts of the country during the first half of the century, and a majority of adoptions were independently arranged by relatives, doctors, midwives, lawyers, orphanage staff, and other baby brokers. Scandalous publicity about the dangers of the black and grey adoption markets, along with zealous declarations about the indispensable virtues of agency adoption, probably convinced some would-be parents that professionals knew best. But for all those who agreed that agencies were uniquely equipped to safeguard adoption, many others avoided expert supervision. Perhaps they could not satisfy the growing number of agency requirements, or would not tolerate humiliating submission to professional judgment, or sought types of children—newborns or infants—that were rarely placed by agencies in the early decades of the century.[11]

After World War II professional agencies made substantial gains, in part because their competitive strategy against independent child placers encouraged them to offer what many applicants wanted: healthy white infants. The definitive survey of professional adoption practice and philosophy at midcentury, conducted by the Child Welfare League of America in 1953, described adoption as "a laboratory" for testing a variety of theories from the human sciences. Many agencies had moved decisively toward earlier placements, reflecting more faith in environmental influences, less worry about hereditary taint, and new sensitivity to the power of early attachment as well as to adopters' wishes to "start from scratch." The survey illustrated that most agencies used psychological testing services "in order to reduce risks to a minimum." While 55 percent of agencies used standardized mental testing to help them determine children's eligibility for adoption, a full 75 percent used it as a routine way of placing children in families suited to their intellectual potential; such selective placement practices were considered even more important than matching by race or religion, not only because mentally similar children and parents naturally belonged together but because of the emotional comfort that similarity of intellect allegedly engendered.[12] By midcentury, even the vast majority of Catholic agencies considered "level of intelligence" the most important factor in selecting adoptive homes and routinely employed psychological testers and consultants, clear evidence that scientific adoption had spread to some child placers whose religious motivations had previously led them to advocate families made by God rather than by science.[13] The appeal of early placements,

---

[11] Until at least 1930, for example, most agency-mediated adoptions in New York involved children over age two. Even in the late 1930s, when agencies began making more infant placements, they considered a four- to six-month period of preadoptive placement an absolute minimum: "A Study of the Adoption Situation of New York City as It Relates to Protestant Children," Jan.–Apr. 1938, CWLA Papers, microfilm reel 3, Social Welfare History Archives. By 1940 age at placement in New York was decreasing, but very few babies were placed for adoption before six months of age and many were still placed between the ages of one and four years. See Lockridge and Theis, *Adopting a Child* (cit. n. 9), p. 12.

[12] For the survey, which included 270 agencies in all forty-eight states, see Schapiro, *Study of Adoption Practice* (cit. n. 2), Vol. 1, pp. 32 (quotation), 54–58 (on matching); see the table of "matching factors" on p. 84. The study's second volume, *Selected Scientific Papers Presented at the National Conference on Adoption, January, 1955*, includes a great deal of additional material on mental measurement in adoption. One 1956 how-to manual suggested that newborn placements traded the benefits of developmental knowledge gained with time for the emotional security of permanent placement during the first months of life; such placements remained controversial among professionals even in the mid 1950s. See Ernest Cady and Frances Cady, *How to Adopt a Child* (New York: Whiteside/Morrow, 1956), p. 69.

[13] National Conference of Catholic Charities, *Adoption Practices in Catholic Agencies* (Washington, D.C.: National Conference of Catholic Charities, 1957), p. 96, Table A-24, p. 100, Table A- 29. See also "Draft of Adoption Study," Box 89, Folder: "Adoption Study Drafts (1)," National Conference of Catholic Charities/ Catholic Charities USA Records, Archives of the Catholic University of America, Washington, D.C. The NCCC

the promises of mental measurement, and vigorous public relations efforts to correct the "lack of public understanding" that professionals blamed for the persistence of independent operators all helped adoption professionals. By the early 1970s, the high point of professional authority in adoption, only 21 percent of nonrelative adoptions were privately arranged.[14]

The "matching" mandate, a paradigmatic feature of modern adoption, presented unique opportunities for scientific approaches claiming to mirror nature, but matching was not an edict favored only by elites. It combined adopters' clear preference for children who looked like them with professional conviction that resemblance portended love whereas difference spelled trouble. Matching required that adults who took in children be married heterosexuals who looked, felt, and behaved as if they had themselves conceived children born to others. Matching was an optimistic and arrogant objective that suggested that adoption could and should reenact reproductive nature. Popular resistance to professional authority in adoption, which remained tenacious throughout the century, occasionally took the form of resistence to matching.[15] But in general matching was an overwhelmingly popular policy, especially among infertile couples who sought to adopt after failing to conceive children of their "own." By midcentury, "sterility" had become an unquestioned qualification for adoption, reinforcing the notion that matching practices compensated for failures of reproductive nature.

Today, the early opponents of matching—including the jurist Justine Wise Polier, the novelist Pearl S. Buck, and Helen Doss, whose 1954 memoir, *The Family Nobody Wanted*, described a large family forged transracially and transnationally—are conspicuous not only because their critique was prophetic but because it departed so dramatically from majority opinion at the time. According to the matching ethos, sturdy families were founded on continuities of appearance and intelligence, a vision of reproductive nature characterized by emphatic sameness and singularity. "There need be no question of superiority or inferiority raised in a rule to limit placements generally to similar personal, racial, or national types," pointed out one early proponent of both matching and professionalism. "No good can come from, and much harm may be done by, wilful violations of customs and comity in the placement of children."[16] The ubiquity of variation in families made naturally was simply ignored.

The work of constructing complete and transparent substitutes for natal kin relied on visible characteristics—eye, hair, and skin color—as well as on more elusive qualities associated with mentality and intelligence. The latter were considered especially important because adoption frequently involved dramatic upward mobility, moving children of im-

---

conducted this national survey in part because of its conflict with the CWLA over how absolute religious matching should be. The following statement appears on page 4 of the adoption study draft but not in its final published form: "Although it seldom happens, we would rather keep a child in a Catholic foster home, or even in a Catholic institution, over a long period of time than to have him placed in a home of a different religious faith."

[14] Stolley, "Statistics on Adoption in the United States" (cit. n. 7), pp. 30–31, Fig. 3; and Penelope L. Maza, *Adoption Trends: 1944–1975* (Child Welfare Research Notes, 9) (U.S. Children's Bureau, Aug. 1984), Table 3, Fig. 2. Since the early 1970s the percentage of independent adoptions has risen steadily.

[15] See Judith S. Modell, *Kinship with Strangers: Adoption and Interpretations of Kinship in American Culture* (Berkeley: Univ. California Press, 1994); for resistance, see Julie Berebitsky, *Like Our Very Own: Adoption and the Changing Culture of Motherhood, 1851–1950* (Lawrence: Univ. Press Kansas, 2000), esp. Ch. 5; and Barbara Melosh, *Strangers and Kin: A History of Adoption in the United States* (Cambridge, Mass.: Harvard Univ. Press, forthcoming).

[16] Helen Doss, *The Family Nobody Wanted* (1954; Evanston, Ill.: Northwestern Univ. Press, 2001); and W. H. Slingerland, *Child-Placing in Families: A Manual for Students and Social Workers* (New York: Russell Sage Foundation, 1919), p. 125.

poverished and uneducated parents into working-, middle-, and upper-class families. In addition to shielding adoptees from the lasting shame of their (often illegitimate) birth and ensuring their sense of belonging, the desire to institutionalize adopters' exclusive entitlement was surely one factor that led many states, beginning with Minnesota in 1917, to keep adoption proceedings confidential. Confidentiality was an utterly novel policy, intended to shield adoptive families from the stigmatizing gaze of outsiders. For decades curious adoptees had access to records, including original birth certificates, but after 1950 these were increasingly shrouded in secrecy.[17] Prior to the twentieth century, formal adoptions were few in number, rarely severed ties between individual children and their natal relatives permanently, and were legally obligated to match children and adults solely on the basis of religion.[18] In contrast, modern adoption was reimagined as a series of concrete steps resulting in wholesale kinship replacement. What could be a clearer case of social engineering? One family was substituted for another so carefully, systematically, and completely that the old family was rendered invisible and unnecessary. The new family—and the identity and sense of belonging it offered the child—did not appear to be "made up" at all.

Gesell was a vigorous advocate for this paradoxically designed-but-not-designed adoption process. For him—and for countless others in human science and human service— modernizing adoption meant making families by using procedures that relied heavily on clinical control, standardization, specialization, quantification, mountains of paperwork, and other telltale signs of rationalization. "Standards" and "safeguards" were for decades virtually interchangeable terms, at once embodying the tight fit between uniformity and child protection and the deficiency of fickle forms of child exchange. Charity, for example, was a woefully inadequate reason to arrange adoptions, but pecuniary motivations had an especially bad reputation. In the early twentieth century, scandals still surfaced regularly in connection with for-profit baby farms, adoption advertisements in newspapers, and instances of benevolent but amateur adoption gone tragically wrong. In a child-placing landscape that was open to just about anyone, scientific professionals imagined themselves as the only sincere defenders of children's interests against a host of hazards and abuses. Proponents of scientific adoption, like Gesell, did not much care whether alternative methods of making up families were compassionate or corrupt. Sentimental and commercial adoption were equally inferior because they were equally unscientific.[19]

ARNOLD GESELL: BACKGROUND AND CAREER

Gesell's life and historical location help make sense of his lifelong commitments to both science and social welfare. Like other educated Americans with visions of constructive

---

[17] E. Wayne Carp, *Family Matters: Secrecy and Disclosure in the History of Adoption* (Cambridge, Mass.: Harvard Univ. Press, 1998); and Katarina Wegar, *Adoption, Identity, and Kinship: The Debate over Sealed Birth Records* (New Haven, Conn.: Yale Univ. Press, 1997). Minnesota was also the first state to require expert endorsement before an adoption could be finalized in court.

[18] Brian Paul Gill, "The Jurisprudence of Good Parenting: The Selection of Adoptive Parents, 1894–1964" (Ph.D. diss., Univ. California, Berkeley, 1997), esp. Ch. 3; and Joan H. Hollinger, "Introduction to Adoption Law and Practice," in *Adoption Law and Practice*, ed. Hollinger (New York: Bender, 1994), 1-24–1-47. Religious matching mandates were frequently violated, sometimes trumped by racial and ethnic factors, and resulted in much controversy well into the twentieth century. See Nina Bernstein, *The Lost Children of Wilder* (New York: Pantheon, 2001); Linda Gordon, *The Great Arizona Orphan Abduction* (Cambridge, Mass.: Harvard Univ. Press, 1999); and Ellen Herman, "The Difference Difference Makes: Justine Wise Polier and Religious Matching in Twentieth-Century Child Adoption," *Religion and American Culture*, Winter 2000, *10*:57–98.

[19] Gesell disdained commercial child care and adoption arrangements. See the description of Mrs. B's baby farm in Arnold Gesell, *The Pre-School Child from the Standpoint of Public Hygiene and Education* (Boston: Houghton Mifflin, 1923), pp. 139–141.

social change in the early decades of the century, Gesell believed that rationality and reform were compatible rather than contradictory goals. Born in Alma, Wisconsin, to a photographer and a teacher steeped in the small-producer republican ethic that was fast disappearing in modern America, Gesell attended the University of Wisconsin and appeared headed for a career in education. The commencement address he delivered in 1903 was devoted to the problem of child labor, an emblematic cause for reformers trying to come to terms with the human costs of industrial capitalism as well as with the new emotional and psychological (rather than economic) value of children.[20] Gesell went on to work as principal of a high school before enrolling at Clark University to pursue doctoral studies in psychology with G. Stanley Hall, a founding figure in American psychology and an authority on development.

After completing his Ph.D. Gesell moved to New York, where he taught elementary school and lived in the East Side Settlement House. The settlement movement, a characteristic expression of Progressivism closely identified with figures like Chicago's Jane Addams, set out to organize the immigrant-dominated urban neighborhoods in which reformers lived. "Settling" appealed to elite Americans who longed to make a social difference and also served as a key field site for pioneering scholars. The extra-academic intellectual arenas created by settlement houses have recently been reclaimed as crucial incubators of American social science, especially nourishing for early studies of urbanism, industrial labor, immigration, ethnicity, poverty, crime, and youth.[21] In them the relationship between activists and intellectuals could be organic and reciprocal rather than automatically hierarchical and tension filled. For Gesell and many of his contemporaries, the settlement model offered persuasive evidence that social action and scientific work were two sides of the same coin.

The piles of data gathered by settlement workers may have been crucial to establishing the legitimacy of social scientific work, but the distance between "applied" and "basic" research in this field was associated from the start with the fact that the world of Progressive-era reform—especially the settlement house network and anything related to children and families—belonged to women, whereas the world of the research university certainly did not. The gendered locations and dimensions of knowledge taught men like Gesell important lessons about the hazards of associating too closely with the female world of intellect and politics.[22]

---

[20] Viviana A. Zelizer, *Pricing the Priceless Child: The Changing Social Value of Children* (New York: Basic, 1985). On Gesell see Benjamin Harris, "Arnold Lucius Gesell," in *American National Biography*, ed. John A. Garraty and Marc C. Carnes (New York: Oxford Univ. Press, 1999), pp. 877–878; and Boyd R. McCandless, "Arnold L. Gesell," in *International Encyclopedia of the Social Sciences*, ed. David L. Sills (New York: Macmillan, 1968), pp. 157–158. For Gesell's own autobiographical reflections see "Arnold Gesell," in *A History of Psychology in Autobiography*, Vol. 4, ed. Edwin G. Boring *et al.* (Worcester, Mass.: Clark Univ. Press, 1952), pp. 123–142.

[21] Kathryn Kish Sklar, "Hull-House Maps and Papers: Social Science as Women's Work in the 1890s," in *Gender and American Social Science: The Formative Years*, ed. Helene Silverberg (Princeton, N.J.: Princeton Univ. Press, 1998), pp. 127–155; Mary Jo Deegan, *Jane Addams and the Men of the Chicago School, 1892–1918* (New Brunswick, N.J.: Transaction, 1988); and Ellen Fitzpatrick, *Endless Crusade: Women Social Scientists and Progressive Reform* (New York: Oxford Univ. Press, 1990).

[22] Dorothy Ross, "Gendered Social Knowledge: Domestic Discourse, Jane Addams, and the Possibilities of Social Science," in *Gender and American Social Science*, ed. Silverberg, pp. 235–264; Kevin P. Murphy, "Socrates in the Slums: Homoerotics, Gender, and Settlement House Reform," in *A Shared Experience: Men, Women, and the History of Gender*, ed. Laura McCall and Donald Yacovone (New York: New York Univ. Press, 1998), pp. 273–296; and Laurel Furumoto, "Shared Knowledge: The Experimentalists, 1904–1929," in *The Rise of Experimentation in American Psychology*, ed. Jill G. Morawski (New Haven, Conn.: Yale Univ. Press, 1988), pp. 94–113.

Gesell never relinquished the substance of his reform commitments or his interest in children, but he did pursue a highly controlled, laboratory-based, methodologically self-conscious, and theoretically grounded science directly at odds with the domain of female reform. His medical degree may have helped to imbue his work with the desired masculine aura, and perhaps this was one of the reasons he sought to supplement his Ph.D. with an M.D. For Gesell, making developmental science men's work also meant snubbing important female developmentalists—Myrtle McGraw was one—whose research was closely related to his own but whom he never publicly credited.[23] Gesell's scientific sexism helped to solidify his reputation as an objective scientist in spite of his enduring belief in the constructive partnership between (basic) child science and (applied) child welfare. In this sense, at least, Gesell exemplified the direction of psychology at large. With "efficiency" and "control" as Progressive keywords, pioneers in a range of practical fields, from intelligence testing to industrial and public relations, sensibly bet on engineering metaphors as the quickest route to authority in the new age of scientific management.[24] Twentieth-century psychology was a can-do profession whose supporters pointed with pride to tangible results far outside of the ivory tower, but they never failed to advertise its credibility as an experimentally based science and its legitimacy in the male-dominated academic world.

Because Gesell worked in a milieu that exalted scientific over social values and made applied work (like adoption itself) appear second best, the fact that he spent virtually his entire career ensconced at Yale University, the very model of the new research institution, buttressed the credibility of his work. Gesell was the head of the Juvenile Psycho-Clinic, founded in 1911 and subsequently known as the Clinic of Child Development. (Upon Gesell's retirement in 1948, Yale abandoned its sponsorship and a group of his colleagues founded an independent organization, the Gesell Institute of Child Development.) Gesell was best known for the normative scales of child development that he devised on the basis of longitudinal studies of hundreds of New Haven children from 1919 through the 1930s. Although retrospective critics have reproached Gesell for his obliviousness in failing to incorporate children of diverse class and cultural backgrounds into his experimental population, Gesell was not oblivious at all. He believed that diversity would be an obstacle to, rather than a precondition for, progress in developmental research. Gesell deliberately constructed "a homogeneous group from the standpoint of socio-economic status and

---

[23] Thomas C. Dalton and Victor W. Bergenn, "Reconsidering McGraw's Contribution to Developmental Psychology," in *Beyond Heredity and Environment: Mrytle McGraw and the Maturation Controversy,* ed. Dalton and Bergenn (Boulder, Colo.: Westview, 1995), pp. 1–36, esp. pp. 11–12. On the female reform tradition see Fitzpatrick, *Endless Crusade* (cit. n. 21); Estelle B. Freedman, *Maternal Justice: Miriam Van Waters and the Female Reform Tradition, 1887–1974* (Chicago: Univ. Chicago Press, 1996); and Robyn Muncy, *Creating a Female Dominion in American Reform, 1890–1935* (New York: Oxford Univ. Press, 1991). In his history of the Iowa Child Welfare Station Hamilton Cravens similarly emphasizes that legitimizing developmental science required male scientists and university officials to distance themselves from the women and reform organizations that had helped to establish the station and secure initial funding for it: Cravens, *Before Head Start* (cit. n. 8), Chs. 1–3.

[24] On intelligence testing see JoAnne Brown, *The Definition of a Profession: The Authority of Metaphor in the History of Intelligence Testing, 1890–1930* (Princeton, N.J.: Princeton Univ. Press, 1992); Michael M. Sokal, ed., *Psychological Testing and American Society, 1890–1930* (New Brunswick, N.J.: Rutgers Univ. Press, 1987); and Leila Zenderland, *Measuring Minds: Henry Herbert Goddard and the Origins of American Intelligence Testing* (New York: Cambridge Univ. Press, 1998). On industrial management see Richard Gillespie, *Manufacturing Knowledge: A History of the Hawthorne Experiments* (New York: Cambridge Univ. Press, 1991). On public relations see Larry Tye, *The Father of Spin: Edward L. Bernays and the Birth of Public Relations* (New York: Crown, 1998). On scientific management see Robert Kanigel, *The One Best Way: Frederick Winslow Taylor and the Enigma of Efficiency* (New York: Viking, 1997).

educational background of parents and race.''[25] All of the children Gesell studied were born in the United States. All had grandparents of Northern European (German and British) ancestry. All had parents located in the middle class with respect not only to employment (as measured by the Barr Scale of Occupational Intelligence) and schooling but with regard to cultural indicators such as avocational interests and even household furnishings. The vast majority of children Gesell studied came from families who lived in rented apartments, owned their own cars, read at least one daily newspaper, and attended church. From Gesell's perspective, comparability rather than representativeness was the true measure of developmental science. Norms manifested in uniform populations were accurate and valid. They were also generalizable.

GESELL'S SCIENCE AND TECHNOLOGY OF DEVELOPMENTAL MEASUREMENT

Gesell's research project proceeded roughly as follows. After placing children in a carefully designed laboratory environment called the observation dome and presenting them with challenges ranging from flashing lights, ringing bells, and bouncing balls to strangers, spoons, and stairs, Gesell scrupulously documented every aspect of their behavioral and social responses at four-week intervals: how their faces looked and bodies moved, what sounds or words they uttered, how they interacted with mothers and test-givers. (See Figures 2 and 3.) The result was an age-graded atlas of normal development beginning at birth. Because the children were alike and his observations were meticulous, Gesell claimed that his norms were biologically and sociologically typical: "They express general developmental trends and sequences.''[26]

For instance, at four weeks of age only 30 percent of babies could be relied upon to place a rattle in their mouths, whereas by sixteen weeks 59 percent did. Since Gesell defined "normal" as whatever more than half the children he studied did consistently, mouthing a rattle became a developmental "norm" at sixteen weeks. Similarly, playing pat-a-cake was normal at one year, while peekaboo was not normal until considerably later. Climbing up two stairs was normal at forty-eight weeks, but climbing up three required fifty-two weeks and ascending to the fourth was normal only at fifty-six weeks of age. With permanent behaviors the evidence of growth tended to follow a pattern of increasing frequency, but some behaviors were temporary, advancing to higher levels or even disappearing over time. For instance, when presented with a small pellet, a normal twenty-eight-week-old infant would place an open hand over it, while a normal forty-week-old showed no sign of doing any such thing. The forty-week-old grasped the pellet instead, with thumb and index fingers flexed; by forty-four weeks normal children held tightly to the pellets placed in front of them. Other reactions appeared for the first time as development unfolded. Before twelve weeks, for example, 100 percent of Gesell's subjects were indifferent to strangers, making it perfectly normal to be that way. Between thirty-two and fifty-two weeks, however, the percentage dropped precipitously. Stranger accep-

---

[25] Arnold Gesell and Helen Thompson, *Infant Behavior: Its Genesis and Growth* (New York: McGraw-Hill, 1934), p. 5. Several scholars have noted the central institutional role Yale played in advancing psychology's scientific credentials during the first half of the century, but they have ignored Gesell. See James H. Capshew, "The Yale Connection in American Psychology: Philanthropy, War, and the Emergence of an Academic Elite," in *The Development of the Social Sciences in the United States and Canada: The Role of Philanthropy*, ed. Theresa Richardson and Donald Fisher (Stamford, Conn.: Ablex, 1999), pp. 143–154; and J. G. Morawski, "Organizing Knowledge and Behavior at Yale's Institute of Human Relations," *Isis*, 1986, 77:219–242.

[26] Gesell and Thompson, *Infant Behavior*, p. 7.

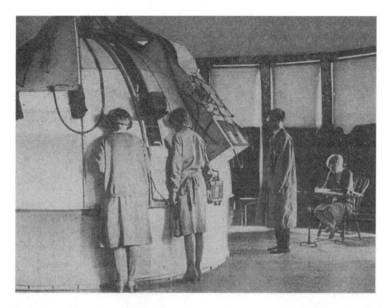

Figure 2. *Observers and recorder stationed outside the photographic dome. From Arnold Gesell,* Infancy and Human Growth *(New York: Macmillan, 1928), page 67, figure 25.* Reprinted by permission of the Gesell Institute.

tance consequently changed from normal to abnormal and stranger rejection from abnormal to normal as the first year of life progressed.[27]

The technology that resulted from this scientific effort was a scale. The scale meticulously translated developmental data into a concrete problem-solving tool that gauged children's proximity to or distance from developmental norms at a series of key ages. Gesell defined norms as "a standardized tool for discriminative characterization," the closest possible thing to a scientific unit of developmental measurement. "Growth" could not be seen directly, Gesell freely admitted, but it could be approximated through a comparative method that reduced similarities and differences between children to precise mathematical terms. He called his technology "developmental diagnosis" and envisioned it at the center of an ambitious new clinical endeavor extending public health regulation beyond bodies to minds, beyond preventing disease to promoting optimal growth. The principle involved was easy to understand, even intuitive: in development, age and growth were linked. But Gesell cautioned that developmental diagnosis was a job for skilled professionals, not amateurs. He wanted clinicians to think of his technological achievement not only as "a measuring rod or a calibrated scale," but also as an experimental situation in which ordinary cubes and bells were *"controlled devices . . . which excite the infant's nervous system to symptom responses."*[28] To the uninitiated, toys might appear to be merely toys. Scientists knew that toys were technical.

Managing the geography in which children were observed and tested was a crucial step

[27] *Ibid.*, pp. 70 (stair-climbing behavior), 110 (rattle behavior), 174 (pellet behavior), 258 (pat-a-cake and peekaboo; stranger behavior).

[28] *Ibid.*, p. 327 ("standardized tool"); and Arnold Gesell and Catherine S. Amatruda, *Developmental Diagnosis: Normal and Abnormal Child Development, Clinical Methods, and Practical Applications* (New York: Hoeber, 1941), pp. 6, 17 (emphasis in original).

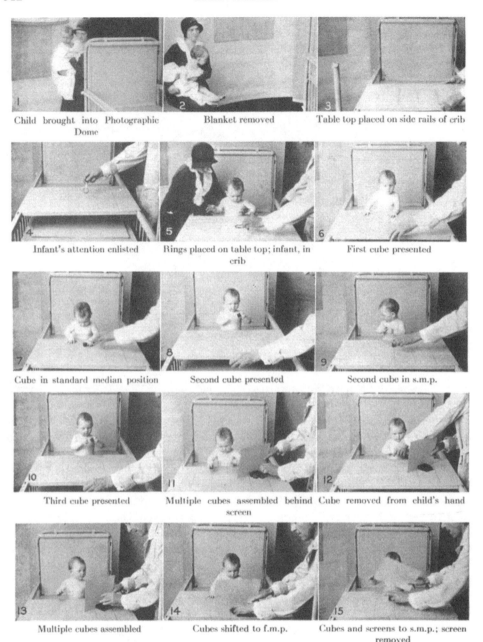

Child brought into Photographic Dome

Blanket removed

Table top placed on side rails of crib

Infant's attention enlisted

Rings placed on table top; infant, in crib

First cube presented

Cube in standard median position

Second cube presented

Second cube in s.m.p.

Third cube presented

Multiple cubes assembled behind screen

Cube removed from child's hand

Multiple cubes assembled

Cubes shifted to f.m.p.

Cubes and screens to s.m.p.; screen removed

Figure 3. *Normative examination procedures. From Arnold Gesell and Helen Thompson,* The Psychology of Early Growth Including Norms of Infant Behavior and a Method of Genetic Analysis *(New York: MacMillan, 1938), figure 13.* Reprinted by permission of the Gesell Institute.

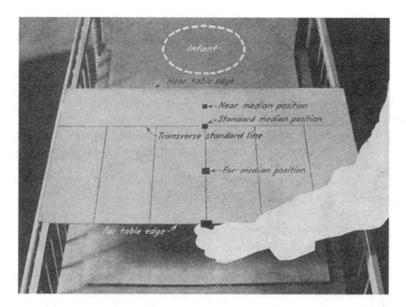

Figure 4. *Location points of examination table. From Arnold Gesell and Helen Thompson,* Infant Behavior: Its Genesis and Growth *(New York: McGraw-Hill, 1934), page 32, figure 9.* Reprinted by permission of the Gesell Institute.

for developmental science. As experimenters, the clinic personnel administering behavioral items had painstakingly to erase their presence from the experimental field, or at least neutralize it. Except in very unusual circumstances, observers were supposed to initiate test items with ruthless consistency—always placing balls and cubes in exactly the same position on the examination table, for instance—then remove themselves from the child's line of sight and observe circumspectly. (See Figure 4.) Mothers who accompanied their babies to the clinic were not allowed to be present during the developmental exam unless the infants became positively frantic. In that case the mother was permitted inside the observation dome but was positioned as inconspicuously as possible, so that the child would know that she was there but still could not see her. Whatever independent stimulus was associated with the human factors in the testing situation was deliberately marginalized. Gesell's goal was to witness the behavioral manifestations of growth in pure form, as a natural process uncontaminated by social interference.

The highly controlled situations that might yield significant developmental insight were therefore very different from the unpredictable and variable situations in which babies and children actually lived from day to day and year to year. Gesell did not wish to replicate home or school environments. He aimed instead to achieve invariable experimental conditions. Testers were given detailed instructions about how to administer the battery of challenges, observe the reactions of their subjects, record the results on an array of pre-printed schedules, and tally the scores with shorthand that utilized + (item adequately performed at age level), − (item not adequately performed at age level), and + + (item not adequately performed because a more mature pattern was exhibited instead). The scales standardized developmental norms, beginning at birth, in four behavioral fields: motor, linguistic, adaptive, and personal-social. Each test item was designated by letter (M, L, A, or P) as belonging to one of these fields. Inconsistent ratings between these developmental spheres were theoretically possible, but Gesell rarely found them. He attributed the coordination between M, L, A, and P to an overarching "developmental complex" that channeled overall growth patterns into retarded, average, or accelerated directions. Administering the scales took fifteen minutes at four weeks but expanded to thirty minutes at twenty weeks and to forty-five minutes after that. They could be administered once or repeated several times before obtaining a "final developmental index" that efficiently summarized the child's current developmental level and future potential.[29]

Gesell pioneered the use of photography in developmental science. The camera was a literal expression of his desire to see the evolution of a larger developmental logic, or what he sometimes called the "structuralization" of behavior.[30] His specially designed photographic dome, along with various one-way screens, allowed Gesell and his Yale colleagues to capture literally thousands of images of behavior in still and moving pictures without being seen. All infants were observed in the nude, or just in diapers, to enhance the possibility of an all-encompassing photographic gaze.

The camera was a "servant of science." Although it could not capture growth as such, or reveal the developmental mechanisms lurking below the behavioral surface, it was nevertheless indispensable. Because photographs were behavioral cross-sections, they yielded a picture of the developmental course when taken at appropriate intervals. Photographic evidence was also objective and complete: "it sees everything with instantaneous vision; and it remembers infallibly." Gesell sometimes used medical metaphors to describe his pictures as a type of "biopsy which requires no removal of body tissue" or as "seriated optical records—records which do not fade with time nor warp with prejudice, but which perpetuate with impartial fidelity the configuration of the original event."[31]

The Yale clinic amassed an enormous photographic research library, and most of Gesell's publications included a selection of illustrations drawn from it. Several of his books amounted to pictorial encyclopedias exhibiting optical records in sequence and showing that development was a process that unfolded in comfortingly visible order, from birth onward no less than from conception to birth. Development conceived as a journey in which normal children moved at the same speed and on the same path merged description with prescription. How normal children developed in the aggregate became the yardstick for how any particular child should develop. This notion—that normal development displays unmistakably natural regularities against which deviations can be measured—is so commonplace today, so pervasive in the popular literature on parenting and childrearing, as to be unremarkable.[32] Gesell was an originator of this novel way of thinking about development and its management.

SCIENTIFIC APPLICATIONS IN ADOPTION

As applied to adoption, the Gesell scale was a major advance over earlier testing technologies such as the Stanford-Binet. Similarly founded on the doctrine that age governed

[29] For a full description of the testing, recording, and scoring techniques see Gesell and Amatruda, *Developmental Diagnosis*, esp. Ch. 4 and App. A.

[30] For an example of the use of the term "structuralization" see Gesell and Thompson, *Infant Behavior* (cit. n. 25), p. 22.

[31] Arnold Gesell, *Infancy and Human Growth* (New York: Macmillan, 1928), p. 56 ("servant of science"); Gesell and Thompson, *Infant Behavior*, pp. 21 (on photographic evidence), 22 ("biopsy"); and Gesell, *Infancy and Human Growth*, p. 57 ("seriated optical records").

[32] For example, the popular *What to Expect* series by Arlene Eisenberg is organized around developmental milestones. Penelope Leach's baby and child care manuals similarly emphasize one normal developmental step at a time.

development, that test focused solely on intelligence, relied exclusively on language, and therefore could not be administered to infants or toddlers under the age of three without considerable improvisation. (Although items that three-year-olds could answer were included on the Stanford-Binet, that test, like the original Binet scale, was devised for use with school-aged children. The Kuhlman-Binet was the adaptation most often used with children under three.) There were creative mental testers who took liberties with available tests, of course. Gesell himself administered a great many Stanford-Binet tests prior to 1920. So did a number of psychologists employed by major professional agencies, including the New England Home for Little Wanderers and the New York State Charities Aid Association, where intelligence testing was a standard procedure that distinguished the professional placement process and marked it as superior to independently arranged adoptions that lacked the benefit of mental measurement.[33]

By 1920, as Gesell was launching his long-term developmental study, the testing ethos had penetrated the professional child-placing world. Elites considered it little short of malpractice to make placements without mental measurement, and they fumed about the many families formed without the benefits of such technology. Their goal was not simply to distinguish between normal, subnormal, and supernormal children but to make refined discriminations among the various subnormal types—idiots, imbeciles, morons, and dullards—whose promiscuous placement in family homes had, they were convinced, contributed to adoption's negative reputation. In his classic 1919 text on family placement, W. H. Slingerland called on all child-placing agencies to utilize professional psychologists to conduct "a scientific study of mentality and personality." This study would have two purposes: to identify those who were so "constitutionally defective" as to be unadoptable and in need of institutional care, and to pair adoptable children with parents who were just like them. "To put a low grade mental defective in a family home where a normal child was expected is a social crime, once to be condoned because of ignorance, but now inexcusable in a well-ordered and progressive child-placing agency," Slingerland chided his colleagues in the rapidly professionalizing field of social work. Like testers intent on detecting feeblemindedness among immigrants, soldiers, and schoolchildren, child placers embraced intelligence as a proxy for social status that could withstand the egalitarianism of American democracy. "You must bear in mind that there are first-class, second-class, and third-class children," Slingerland added, "and there are first-class, second-class, and third-class homes."[34]

The earliest mention of adoption in Gesell's published work came in 1923, when he

---

[33] For Gesell's Stanford-Binet tests see Gesell Papers, Box 58, which contains clinical case records for the years before 1920. At the NEHLW Rose Hardwick instituted a comprehensive mental testing program in September 1915 and eventually wrote a dissertation about her experience as a mental tester working with hundreds of young children between June 1918 and October 1923: Rose S. Hardwick, "The Stanford-Binet Intelligence Examination Re-Interpreted with Special Reference to Qualitative Differences" (Ph.D. diss., Radcliffe, 1924). At the NYSCAA intelligence testing by a psychologist was as fundamental to placement as a complete physical exam: Sophie van Senden Theis and Constance Goodrich, *The Child in the Foster Home*, Pt. 1: *The Placement and Supervision of Children in Free Foster Homes: A Study Based on the Work of the Child-Placing Agency of the New York State Charities Aid Association* (New York: School of Social Work, 1921), p. 14. For comparative purposes see the study of 852 adoptions finalized in Suffolk and Norfolk counties in Massachusetts between 1922 and 1925. Few involved mental examinations prior to placement and over 70 percent were arranged independently. These were deplorable facts, according to the author, demonstrating the urgency of legal and social reform: Ida R. Parker, *Fit and Proper? A Study of Legal Adoption in Massachusetts* (1927), rpt. in David J. Rothman and Sheila M. Rothman, eds., *Women and Children First: Social Reform Movements to Protect America's Vulnerable, 1830–1940*, Vol. 8: *The Origins of Adoption: Two Reports* (New York: Garland, 1987), pp. 29, 96.

[34] Slingerland, *Child-Placing in Families* (cit. n. 16), pp. 73, 69, 118.

reported that his assistant, Margaret E. Cobb, had evaluated the potential of 198 candidates for adoption at the Yale Clinic. The results, based on the Stanford-Binet, equated adoptability with educability. Cobb concluded that only 2 percent of the children had college potential, 7 percent could be expected to finish high school, 17 percent could do some high school work, 35 percent might benefit from vocational training after completing elementary school, 21 percent might finish the fifth or sixth grade, and 18 percent were unsuited for any kind of regular education but would benefit from special training. Gesell's 1926 discussion of the "clinical phases of child adoption" concluded on a similarly somber note: "Although it is a grave responsibility to prejudice in any way the opportunities for adoption, we ought within judicious limits to attempt to forestall all the pangs and aggravations which may come from ill-considered adoption."[35]

In 1926 the U.S. Children's Bureau published a widely distributed pamphlet authored by Gesell, "Psychoclinical Guidance in Child Adoption." It argued that "purely impulsive adoption should be discouraged and the whole procedure should be surrounded with clinical and supervisory safeguards." Gesell spoke regularly on radio programs, gave public lectures, attended conferences, and otherwise publicized the advantages of scientific adoption, warning that only expert guidance could protect against the "the intense suffering" and parental heartbreak caused by "bungled" adoptions.[36] His devotion to standards, established by systematic tests carried out by duly trained and certified professionals, was typical of the pioneering generation of adoption reformers. Their commitment to children required that humanitarianism be put to the rugged test of rationalization. Benevolence was admirable, but the desire to help was not enough. It produced sturdy families only by accident, never by design; often it resulted in tragedy. Good will, sentimentality, impulse, intuition, tradition, shame, desires for secrecy . . . all of these motives would have to give way to exacting scientific standards if adoption were ever truly to benefit children.

Gesell never doubted that adoption was risky, even inappropriate for certain children, but he believed that the risks could be measured and hence known in advance. By 1939 his Yale clinic had studied at least 1,500 adoption candidates; Gesell estimated that one in every ten was grossly defective, therefore unadoptable. Without assessment, he estimated, many average and even superior children who should be placed would be passed over, and those who did find parents were likely to be placed where they did not belong. By 1939 the errors of "underplacement" and "overplacement"—which gave bright children to dull parents and dull children to bright parents—had been lamented for at least two decades. One close study of fifty-six New England adoptions in the late 1920s and 1930s concluded that testing had become a more common and reliable adoption practice over time but also noted that "the saddest tragedies to be found in the history of past adoptions have resulted primarily from the error which psychological testing is devised to prevent—i.e., the under-placement or over-placement of children in relation to their capacity." Advances in testing by the 1930s made failures to match children and parents mentally more galling than ever. "From the test results in frequent examinations, the child

---

[35] Gesell, *Pre-School Child from the Standpoint of Public Hygiene and Education* (cit. n. 19), p. 137 (for Cobb's original research report see Margaret Evertson Cobb, "The Mentality of Dependent Children," *Journal of Delinquency*, May 1922, 7:132–140); and Arnold Gesell, *The Mental Growth of the Pre-School Child: A Psychological Outline of Normal Development from Birth to the Sixth Year, Including a System of Developmental Diagnosis* (New York: Macmillan, 1926), p. 428 (Ch. 36 is titled "Clinical Phases of Child Adoption").

[36] Gesell, "Psychoclinical Guidance in Child Adoption" (cit. n. 6), p. 193; and Arnold Gesell, "Child Adoption in Connecticut," p. 3, address delivered to the quarterly meeting of the probate judges of Connecticut, 17 May 1939, Box 45, Folder: "Subject File: Adoption," Gesell Papers.

may be found fit for adoption if a particular type of home is chosen."[37] Average children belonged with average parents, but superior parents and children belonged together too. Mental matching was believed to maximize children's future happiness by helping them take advantage of adoption opportunities, but it was also used to deliver children who would not disappoint their parents by failing to "measure up."

Developmental scales like Gesell's were not the first technologies utilized for placement purposes, but they allowed clinicians in medicine and social work who worked with children to move beyond the limitations of the first generation of mental tests. Gesell's scales were not immediately or universally employed, and adoption based on commercial and sentimental considerations persisted, much to the chagrin of professionals. But developmental technologies were a hallmark of scientific adoption, and not only at Yale. Publications aimed at both professional and popular audiences described Gesell's work to illustrate the metamorphosis of psychology from "the theoretical and the experimental to the dependable and the practical," proof positive that "science helps the adoptive family to approximate the biological family."[38]

Adoption professionals wrote to Gesell from around the country to inquire about the application of his scales in practice. One psychologist employed by a statewide agency in Kansas began, in the early 1930s, to administer the Gesell scales to the babies of mentally retarded mothers. By the end of the decade the agency was routinely testing all babies in need of placement, confident that Gesell's technology had saved normal babies from wrongful institutionalization, accurately predicted children's future development, and guaranteed parents that children would "fit" their families and live up to their expectations.[39] Gesell's scales were also used for many professionally arranged adoptions in New York State in the 1920s and 1930s. The State of Iowa, concerned that retarded children might be adopted unwittingly, began in 1934 to test all children placed as babies prior to the issuance of adoption decrees. These tests could be "a highly emotional experience for the parents, who understood that the psychologist's word was final in approving or disapproving completion of the adoption."[40] The Menninger Clinic, one of the best-known psychiatric institutions in the country, established an infant testing service in 1944. Used mainly by adoption agencies, it routinely administered the Gesell scales. Sophie van Senden Theis, a pioneering figure in adoption modernization, claimed by the late 1940s that psychological testing was routinely utilized in agency practice around the country to determine children's developmental baselines, to ascertain developmental progress over time in borderline cases, to predict future developmental capabilities, and to place children with

[37] Iris Ruggles Macrae, "An Analysis of Adoption Practices at the New England Home for Little Wanderers" (M.S. thesis, Simmons College, School of Social Work, 1937), p. 67; and Marie Wilson Peters, "What the Social Worker Expects from the Psychologist," *Family*, Oct. 1934, *15*:179–183, on p. 181. For a detailed description of one case of overplacement see Ira S. Wile, *The Challenge of Childhood: Studies in Personality and Behavior* (New York: Seltzer, 1925), pp. 254–259 (case of Paul).

[38] Brooks and Brooks, *Adventuring in Adoption* (cit. n. 3), Ch. 7 ("Scientific Aids"), pp. 84, 89. See also Gallagher, *Adopted Child* (cit. n. 4), Ch. 6 ("Intelligence Tests").

[39] Avis Carlson, "To Test a Baby," *Atlantic*, June 1940, pp. 829–832. For one example of how the Gesell scales were used in the placement process see Louise B. Heathers to Gesell, 17 Sept. 1946, and attached MS, "Psychologists Look at Adoption," Box 45, Folder: "Subject File: Adoption," Gesell Papers.

[40] On New York practices see Edith F. Symmes, "An Infant Testing Service as an Integral Part of a Child Guidance Clinic," *American Journal of Orthopsychiatry*, 1933, *3*:409–430. Symmes reported on two hundred children referred to her clinic by New York agencies prior to adoption during a short period in the late 1920s. The Gesell scales were administered to them all, along with supplementary exams including the Buhler Tests, the Merrill-Palmer Performance Scale, and the Kuhlman-Binet. On practices in Iowa see Skodak and Skeels, "Final Follow-Up Study of One Hundred Adopted Children" (cit. n. 2), p. 94. In this case the Stanford-Binet was used for the small number of children over age three and the Kuhlman-Binet was administered to the rest.

PRE-TESTING

FOR

NORMALCY

Figure 5. *Pretesting for normalcy. From Frances Lockridge and Sophie van Senden Theis,* Adopting a Child, *condensed ed. (New York: Reader Service, 1947), page 9.*

parents who mirrored their mentality and would expect neither too little nor too much of them. By 1947 the Gesell scales were the most widely used developmental tests in the United States.[41] Employed for many purposes, the scales helped to carve out new roles for testers and demarcated the enterprise of making up families as fresh scientific terrain. (See Figure 5.)

Promises to predict the developmental future for children whose background was either unknown or unsavory addressed a central dilemma for child placers: the perception that blood was thicker than water and that adoption was therefore a dangerous arrangement, prone by definition to difficulty and disaster. Gesell's theory of infant and child development held maturation to be one of the fundamental problems in the biological sciences and suggested that patient, careful scientific effort could expose its logic. "Growth *is* lawful and in no sense whimsical, fortuitous, or even wholly unpredictable in its nature," Gesell asserted. No miracle or mystery, it was a process governed "by laws and forces just as real as those which apply to an internal combustion engine." Children about whom little was known, or about whom what was known was bad, could be comprehended more clearly and convincingly than ever before. People concerned about adoption could take heart from the knowledge that "because it [development] is lawful it is within certain limits predictable."[42]

"Within certain limits" was a crucial qualification. Though lawful, children's growth was neither automatic nor invulnerable to external influences. Gesell admitted that growth was more encompassing than mental development and more difficult to conceptualize than ability or achievement. He defined it as a sort of elastic potential for change, an urgent tendency to mature, then mature again. The fact that all growth was based on past growth was perhaps its most important characteristic. This historical dimension illustrated that development was reflexive, that it incorporated the experience of growth into the growth

    [41] Sibylle Escalona, "The Use of Infant Tests for Predictive Purposes," *Bulletin of the Menninger Clinic*, July 1950, *14*:117–128; Lockridge and Theis, *Adopting a Child* (cit. n. 9), Ch. 6; and Simon H. Tulchin, "Psychological Testing in Social Welfare," in *Social Work Year Book*, ed. Russell H. Kurtz (New York: Russell Sage Foundation, 1947), pp. 366–374, on p. 368.
    [42] Gesell, *Infancy and Human Growth* (cit. n. 31), p. 21 (emphasis in original); Arnold Gesell, "The Social Significance of a Science of Child Development," 15 May 1943, p. 7, Box 146, Folder: "Speech, Article, Book File: Address, The Social Significance of a Science of Child Development, 1943," Gesell Papers; and Gesell, "Reducing the Risks of Child Adoption," *Child Welfare League of America Bulletin*, 15 May 1927, 6:1–2, on p. 2.

curve and made even the youngest baby a "growing action system." Technically, individuals could never fail developmentally, since even the most seriously disabled and retarded children were actively engaged in growing. But it was obvious that developmental patterns varied widely, with deviations traceable to prematurity, birth defects, accidents, disease, and malnutrition, among other factors. Growth potential, Gesell concluded, "probably resides in the inherent protoplasmic plasticity of the individual."[43] But he never doubted that it was also conditioned by social institutions—families, schools, communities—that provided (or withheld) opportunities and inspired (or stifled) motivation. Development was remarkably stable in aggregate terms, but individual differences proliferated within the limits of lawfulness.

THE NATURE AND NURTURE OF CHILDHOOD: GESELL IN CONTEXT

Gesell's confidence in the knowability of children and the visibility of development had been nourished by the child study movement, initiated in the late nineteenth century by such figures as G. Stanley Hall, a pioneer in developmental science. While engaged in doctoral studies with Hall at Clark University, Gesell imbibed his commitment to theoretical knowledge that would transform the practices of teachers, medical personnel, and parents. Fired by equal measures of scientific zeal and passion for social reform, members of the child study movement transcribed interviews, gathered documents, took photographs, and compiled mountains of statistics all over the country. Convinced that facts were the key to unraveling the secrets of human growth, they founded organizations, including the Federation for the Study of Child Nature, dedicated to spreading the child study gospel. Like Henry Herbert Goddard, a lifelong friend and occasional collaborator who was one of the first American scientists to conduct sustained research on children, Gesell was profoundly influenced by the goals of child study enthusiasts: making childhood a resource for science and science a resource for children. During the first half of the century movements for parent education and child guidance, although more professionalized and less collaborative than child study, helped to keep these commitments alive.[44]

Equally important in shaping Gesell's approach to developmental science was the triumph of experimentalism in psychology and the behavioral sciences early in the twentieth century. This paradigm made quantitative methods, segregated laboratory environments, and controlled observation the *sine qua non* of objective knowledge. Just as, in adoption, matching imitated the presumed superiority of biogenetic nature, so too did psychological inquiry strive to achieve equivalence with the physical and biological sciences by adopting their methods. In significant ways, the experimental paradigm marginalized the partnerships of amateurs and professionals that child study had promoted, along with the naturalistic field approach that characterized the movement's work. By placing a premium on extensive scientific training, technical apparatus design, and sophisticated statistics, experimentalism changed the philosophy and location of developmental science as radically as it reoriented other fields of psychological inquiry.[45] Instead of going where children

---

[43] Gesell and Amatruda, *Developmental Diagnosis* (cit. n. 28), p. 3; and Gesell, *Infancy and Human Growth*, p. 13.

[44] Julia Grant, *Raising Baby by the Book: The Education of American Mothers* (New Haven, Conn.: Yale Univ. Press, 1998); and Kathleen W. Jones, *Taming the Troublesome Child: American Families, Child Guidance, and the Limits of Psychiatric Authority* (Cambridge, Mass.: Harvard Univ. Press, 1999).

[45] Morawski, ed., *Rise of Experimentation in American Psychology* (cit. n. 22); and Danziger, *Constructing the Subject* (cit. n. 8). For dissent from the experimental paradigm see Katherine Pandora, *Rebels within the Ranks: Psychologists' Critique of Scientific Authority and Democratic Realities in New Deal America* (New York: Cambridge Univ. Press, 1997).

were, in schools and homes, Gesell had all children past four weeks of age brought to his Yale clinic. His many detailed descriptions of his specially designed and equipped nursery, observation alcove, and photographic dome make it abundantly clear that he considered the laboratory environments in which experiments took place a major element in scientific discovery. Observing children move and talk and interact was no less rigorous experimentally than watching cells divide or chemicals react. The scientific process was ideally invisible in each case. Nature itself—and not cultural conventions for learning about nature—was on display.

There were moments of tension between Gesell's science and his social commitment to child welfare. Gesell crusaded for new laws and policies that he believed would improve children's lives and often announced that heredity and environment were reciprocal rather than antithetical forces in human growth. But his scientific work betrayed a strong belief in the constitutional and genetic factors governing most aspects of individuality. The developmental trajectory was "for the most part hereditary in nature," its outside limits determined by "original equipment," especially the maturing nervous system. While Gesell allowed that personality could be molded through training and socialization, he denied that the same was true of temperament or the potential for growth. Both were fundamentally rooted in the embryological phase and "in no sense derived from the external environment."[46]

Committed both to exploring the forces that determined growth and to countering social fatalism, Gesell tried mightily to resolve the contradiction between nature and nurture. He never succeeded. Technologies like developmental diagnosis made it possible to exert a revolutionary new level of social control over childhood. But what exactly was the point if social conditions had little or no effect on outcomes "primarily determined by heredity and constitutional factors which undergo their basic organization in the uterine period"?[47] Gesell qualified statements like these by pointing out that "determination" was not synonymous with "destiny," but the distinction was contrived and confusing. The maturational process he described was almost as much like fate as fate itself. It unfolded inexorably, universally, with an awesome and predictable kind of beauty. In contrast to Gesell's thick descriptions of orderly developmental cycles unfolding beneath watchful eyes in his dome at Yale, the cultural texture of child life outside the clinic's walls was thin and uninspiring.

That Gesell's thought contained a distinctly hereditarian strain alongside his advocacy of environmental intervention hardly made him unusual. Gesell never suggested that all children were adoptable. Like his contemporaries, he believed that only some children made good candidates for family life, and he was more concerned about bright children likely to be overlooked than dull children likely to be overplaced. "The more superior a child is, the more urgently does he demand placement in a home with optimum opportunity. The more defective a child is, the less he is harmed by institutional care." Just as they would not consider adopting children with heart disease or tuberculosis, Gesell told would-be parents that they "should of course investigate the inheritance and mental status of the child; to make sure that capacity is at least normal." Gesell obviously assumed that most potential parents would not want children who deviated from the norm by falling below

---

[46] Gesell, *Infancy and Human Growth* (cit. n. 31), pp. 362, 373; and Gesell and Thompson, *Infant Behavior* (cit. n. 25), p. 308.

[47] Gesell and Thompson, *Infant Behavior*, p. 326. See also Esther Thelen and Karen E. Adolph, "Arnold Gesell: The Paradox of Nature and Nurture," in *A Century of Developmental Psychology*, ed. Ross D. Parke *et al.* (Washington, D.C.: American Psychological Association, 1994), pp. 357–387.

it. Was he correct? Cases were certainly reported in which parents indignantly returned "defective" children, but there were others in which agency staff had to insist that parents give up children whom they had already grown to love, sometimes by arguing that it was selfish of them to keep subnormal children and deprive children "of good mental endowment" of homes. In those rare instances where parents insisted on adopting despite Gesell's recommendation against it, he encouraged child welfare authorities not to block the adoption if the parents would promise to prevent the child from marrying. Announcing that normal children were adoptable and subnormal children were not was typical at the time, as was the eugenic anxiety that adoption would result in "defective germ-plasm . . . mated with normal stock, thereby passing on the defect and causing much preventable misery." In contrast, discerning the boundaries of normality and establishing with certainty "the degree of normality required" for adoption were substantial scientific and technological challenges.[48]

Gesell was somewhat unusual in his willingness to grant that normality might live in places that many of his colleagues considered exceedingly unlikely or even off limits. For example, he believed that many unmarried birth mothers ought to surrender their children for adoption and that women who were unable to make clear decisions about placing their babies should be given a two-year time limit. A technology that probed the growth process scientifically, Gesell further believed, would open a door for some children whose out-of-wedlock births threatened to place them forever outside the protective circle of normal family life in which they belonged by virtue of their normal (or better) developmental prospects.

Placement of such children ran counter to the eugenic sensibilities of many professionals, reformers' commitment to keeping birth families together, and the public health view that breast-feeding was the key to reducing infant mortality, a view institutionalized in new state laws criminalizing the placement of infants away from their birth mothers. Maryland, for instance, in 1916 banned the removal and placement of children under six months of age. Gesell knew, as everyone did, that many candidates for adoption were "illegitimate" children and that these represented the vast majority of nonrelative adoptions (his estimate was 80 percent).[49] In the case of children born to feebleminded birth mothers (who were considered disproportionately likely to have out-of-wedlock children of subnormal mentality), Gesell considered the tendency to keep mother and child together "a sentimental idea" and urged that adoption be considered in order to "give a child a good home rather than a bad one." Being born to a feebleminded parent was a burden and a liability; living with such a parent was acceptable to Gesell only "if it does not cost

[48] Gesell, *Pre-School Child from the Standpoint of Public Hygiene and Education* (cit. n. 19), p. 138 (on superior and defective children); Arnold Gesell, "Adoption" radio talk, 3 Nov. 1930, p. 2, Box 45, Folder: "Subject File: Adoption [Law]," Gesell Papers (investigate mental capacity); Symmes, "Infant Testing Service" (cit. n. 40), p. 427 (case II, Jack; on giving up a "defective" child); Supplementary Memorandum for Child Welfare Association, Re: Rebecca Nearing, 16 Nov. 1923, Box 45, Folder: "Subject File: Adoption, 1923 [Cases]," Gesell Papers (preventing marriage) ("Rebecca Nearing" is a pseudonym; confidentiality of client names was a condition of access to Gesell's papers); Amey Eaton Watson, "The Illegitimate Family," *Annals of the Amerian Academy of Political and Social Science,* May 1918, 77:103–116, on p. 113 (eugenic anxiety); and Gesell, untitled draft, p. 1, Box 45, Folder: "Subject File: Adoption [Law]," Gesell Papers (discerning boundaries of normality).

[49] U.S. Children's Bureau, *The Welfare of Infants of Illegitimate Birth in Baltimore,* Publication No. 144 (Washington, D.C.: Government Printing Office, 1925); and Gesell, "Psychological Welfare of Adopted and Foster Children" (cit. n. 6), p. 2.

the child too dear.''[50] Growing up in an institution and belonging to no one, he believed, was the most damaging possibility of all.

Gesell's nature-centered maturational narrative was roomy enough to accommodate a carefully engineered social environment. He advocated placement even for some children on the border between mental normality and subnormality since for these children, even more than others, social factors could tip the developmental balance into the normal range. "Mental hygiene, like charity, begins at home.''[51] His emphasis on the salience of social arrangements anticipated environmentally centered work on institutional retardation in the 1930s, the positive significance of attachment in the 1940s, mother-infant bonding in the 1970s, and even the articulation of "psychological parenthood" in the 1970s and 1980s. He argued that "children need to take root, to attach themselves to someone, to have someone they can count on, someone they are sure of." "We do not sufficiently serve the neglected child if we simply put a roof over his head," he noted. "There must be affection and understanding beneath that roof." Three-year-old Sarah seemed to bear this out. Mentally pegged as "low average" by Gesell's diagnostic staff when she was placed in an adoptive home at age three, her test results jumped to "high average" in no time. Gesell interpreted Sarah's progress as an "example of response to placing in *home of her own!''*[52] Meeting children's psychological needs for permanence and belonging might not guarantee normality or developmental improvement, but in some adoptions it made all the difference. The difficulty of recognizing such cases was another rationale for clinical judgment based on science and skill.

Eugenic science, conventional sexual morality, and higher regard for custodial institutions led many of Gesell's colleagues, including his close friend Henry Herbert Goddard, to different conclusions. Goddard, a famous figure in the history of mental testing and mental retardation, pronounced adoption "a crime against those yet unborn" and suggested that institutionalizing feebleminded mothers and children alike offered them the best possible lives and society its only effective means of controlling their reproductive sexuality and the quality of the gene pool. Most Progressive-era child welfare officials expressed skepticism about the adoption of illegitimate children for reasons that mixed

---

[50] Gesell, untitled draft, pp. 3, 1, Box 45, Folder: "Subject File: Adoption [Law]," Gesell Papers. Gesell believed in sealing the original birth records of young adoptees but also thought that they should be given identifying information about birth parents on reaching adulthood. For the view that feebleminded birth mothers tended to have illegitimate children of subnormal mentality see Charlotte Lowe, "Intelligence and Social Background of the Unmarried Mother," *Mental Hygiene*, 1927, *11*:783–794; W. E. McClure and Bronett Goldberg, "Intelligence of Unmarried Mothers," *Psychological Clinic*, May–June 1929, *18*:119–127; McClure, "Intelligence of Unmarried Mother, II," *ibid.*, Oct. 1931, *20*:154–157; and Lilian Ripple, "Social Work Studies of Unmarried Parenthood as Affected by Contemporary Treatment Formulations: 1920–1940" (Ph.D. diss., Univ. Chicago, 1953).

[51] Gesell, *Pre-School Child from the Standpoint of Public Hygiene and Education* (cit. n. 19), p. 138; and Arnold Gesell, "The Nursery School Movement," *School and Society*, 1924, *20*:130–150, on p. 148 (quotation). Characteristically, Gesell qualified this testimony about the power of the family over development; his next sentence read: "But it does not end there." I thank Ben Harris for sharing this reference with me.

[52] Arnold Gesell, draft manuscript, p. 2, Box 45, Folder: "Subject File: Adoption [Law]," Gesell Papers; Gesell, "Psychological Welfare of Adopted and Foster Children" (cit. n. 6), p. 1; and index card describing Sarah's case (emphasis in original), Box 45, Folder: "Subject File: Adoption, 1957," Gesell Papers ("Sarah" is a pseudonym). On institutional retardation see Barr, "Spare Children" (cit. n. 3). On attachment theory see Inge Bretherton, "The Origins of Attachment Theory: John Bowlby and Mary Ainsworth," in *Century of Developmental Psychology*, ed. Parke *et al.* (cit. n. 47), pp. 431–471. On bonding see John Bowlby, *Maternal Care and Mental Health* (Geneva: World Health Organization, 1952); and Diane Eyer, *Mother-Infant Bonding: A Scientific Fiction* (New Haven, Conn.: Yale Univ. Press, 1992). On the notion of "psycholoogical parenthood" see Joseph Goldstein, Anna Freud, and Albert J. Solnit, *Beyond the Best Interests of the Child* (New York: Free Press, 1973).

eugenic beliefs like Goddard's with a variety of noneugenic considerations. They objected to separating unmarried mothers and babies just because the mothers were poor. They suspected that adoption would encourage more illegitimacy. Or they steadfastly insisted that the moral redemption of unwed birth mothers depended on their willingness to keep and rear their children. The most extreme of these original "family preservationists" went so far as to call adoption "abortion after birth."[53]

One premise of scientific adoption was that the disinterested scientific gaze was the only realistic antidote to the adoption abuses and errors publicized in the early part of the century. Gesell certainly held nonscientific methods responsible for many unregulated placements that were too early, too late, or just plain bad because of "haphazard, under-cover, boot-leg, hit-or-miss adoptions."[54] Adults could adopt and intermediaries could place children for frivolous reasons or worse. Women fancied babies to dress in pretty clothes or decided to adopt children rather than pets on a whim. Couples on the verge of divorce hoped that adopting would save their marriages. Grieving parents wished to replace dead sons and daughters. And elderly (by which they generally meant middle-aged) single women acquired children simply in order to indulge eccentric childrearing theories.

The fact that unregulated adoptions remained common at the end of Gesell's scientific career, as they had been at the beginning, did not convince Gesell that adoption itself was flawed, but simply that it needed to be brought under an even stricter regime of scientific and legal authority. Adoptions arranged with "intelligent social control," he contended, were "a rich addition to human happiness." They were good for children and good for adults who wanted children, and Gesell never hesitated to recommend adoption to people in his own family. "Every well-conceived child adoption is an asset to society."[55]

IMPROVING OUTCOMES

"The proof of the pudding" in adoptions was how children turned out. Longitudinal data associated with developmental studies like his own made it more likely that they would turn out well, becoming "sober, honest, useful citizens."[56] Gesell was sure that he had reduced the risks of adopting a child by testing that child's developmental status and predicting its future course. Gesell's scales assessed whether any given child was lagging behind the norm, keeping pace with it, or outstripping it. This was important not only because it helped professionals to make crude, if important, distinctions between those children who were and were not good candidates for family life, but because it facilitated the far more refined process of intellectual matching and emulated the similarity of intellect that was presumed to occur when families were made naturally. Technologies like the Gesell scale turned matching from a clumsy moral theory that based love on simple likeness

[53] Henry H. Goddard, "Wanted: A Child to Adopt," *Survey*, 14 Oct. 1911, *27*:1006; and J. Prentice Murphy, "Mothers and—Mothers," *ibid.*, 3 May 1919, *42*:171–176, on p. 176. Although many obstetricians, gynecologists, nurses, and midwives arranged adoptions, hostility to adoption was common in medicine, frequently for eugenic reasons. See R. L. Jenkins, "On Adopting a Baby: Rules for Prospective Adoptive Parents," *Hygeia*, 1935, *13*:1066–1068.

[54] Gesell, "Psychological Welfare of Adopted and Foster Children" (cit. n. 6), p. 3.

[55] Arnold Gesell, "Is It Safe to Adopt an Infant?" n.d., pp. 2, 11, Box 45, Folder: "Subject File: Adoption [Law]," Gesell Papers. Gesell recommended agency adoption to his infertile young nephew: Raymond Stephens to Gesell, 16 Jan. 1958, and Gesell to Stephens, 21 Jan. 1958, Box 45, Folder: "Subject File: Adoption," Gesell Papers.

[56] Gesell, untitled draft; and Gesell, draft manuscript: Box 45, Folder: "Subject File: Adoption [Law]," Gesell Papers.

into a technical operation whose success could be calculated. Showing that kinship could be socially designed so as to approximate the results of biogenetic nature helped to mitigate adoption's reputation as second best and strengthened the case for scientific adoption.

Gesell resisted pressures from parents and professionals to pinpoint an ideal age for adoptive placements. Because individual children displayed broad growth patterns in unique ways, he believed that placement age should vary with the individual under consideration. Sound adoption policies and practices had to be founded simultaneously on very general and very individualized knowledge. Gesell considered placements before six months of age unwise in general, an opinion that reflected widespread professional reluctance to endorse perilous "cradle adoptions."[57] But adopters overwhelmingly desired the youngest possible babies. Gesell was willing to allow "early" placements, at four months, if detailed information about the infant's heredity and development suggested that it was absolutely safe. Such early placements, he admitted, had the advantage of maximizing affection between parents and child and minimizing public expense. In theory Gesell's scales were equally accurate predictors at four months and beyond, but in practice he recommended repeated tests supplemented by a variety of other diagnostic techniques so that consistency of results would shape a reasoned prognosis. When a life hung in the balance, patience and probability were more trustworthy than a judgment based on a single exam.

Caution was almost always more important than speed in making up families well. Older children, whose qualities and characters were already well established and known, had distinct advantages as adoption candidates. For example, Gesell advised parents who wished to know whether their adopted children would be intelligent enough to attend college to adopt children no younger than three.[58] No infant under six months should ever be legally adopted, he warned, because the chances were too great that the adopters would unwittingly take in defectives or even that white parents might adopt "colored" children. Nor should babies be placed permanently after six months if any concern remained about the child's fundamental normality. "Hair-raisingly bad" family histories were all too common among children available for adoption, and they offered a good pretext to hold off, even if children themselves tested in the normal range.[59]

For Gesell, smart placements were patient wait-and-see placements. Delay was the right course of action until the child's normality was clearly documented and unambiguously known. Sometimes this could take years. One child, described by Gesell as "the most extreme case in our files," was a baby with a dizzying number of problems, from premature birth and syphilis to prolonged hospitalization for respiratory difficulties and fifteen months

[57] Albert H. Stoneman, "Adoption of Illegitimate Children: The Peril of Ignorance," *Child Welfare League Amer. Bull.*, 15 Feb. 1926, 5:8 ("cradle adoptions"). Adoption timing was a problem often featured in the popular press. See Honoré Willsie, "When Is a Child Adoptable?" *Delineator*, Dec. 1919, p. 35.

[58] E. J. Mandeville to Yale Psycho-Clinic, 11 July 1940, Box 45, Folder: "Subject File: Adoption, 1923–43 [cases, with individuals concerning]," Gesell Papers. The first major "outcome" study, Theis's *How Foster Children Turn Out* (1924) (cit. n. 2), found that children placed at younger ages were more likely to be treated by parents as their "own" children, more likely to be adopted in court than left in a perpetual state of legal limbo, and more likely to turn out well according to straightforward criteria including school success, economic self-sufficiency, and observance of the law. Although this study was widely discussed at the time, I have been unable to discover whether Gesell read it or what he thought of it.

[59] Gesell's papers include a number of descriptions of the difficulty of making racial determinations and the risks of transracial adoptions. E.g., see Arnold Gesell, "Child Adoption," 29 June 1937, Box 45, Folder: "Subject File: Adoption"; and fragment: "case 6," n.d., Box 45, Folder: "Subject File: Adoption [Memoranda]": Gesell Papers. See also list of children recommended and not recommended for adoption, 1934–35, Box 45, Folder: "Subject File: Adoption [Law]," Gesell Papers.

of institutional living. Her first exam indicated "a serious degree of retardation," but four years, several foster homes, and at least five developmental tests later, "she was an alert, attractive child, reaching almost a complete average performance on the developmental schedules—a remarkable realization of latent normality. . . . She was now in every sense adoptable."[60]

Gesell never failed to stress the advantages of developmental diagnosis to successful adoptions, on the one hand, and the costs of unscientific adoption, on the other. His articles and speeches catalogued the bad things that could happen if parents relied on superficial impressions or emotional reactions rather than disinterested knowledge. One telling case involved a "cute" baby girl who "was just the kind of child who would smite the heart of questing adoptive parents." But Gesell suspected that she would never even complete high school and predicted that "there may be genuine pangs of regret" in store for any parents foolish enough to adopt her as an infant, before her true developmental potential could be known with certainty. Another child illustrated how easily physical appearance could mislead. Four-year-old Rose, brought to Yale for adoption assessment in 1923, "is physically attractive, of cheerful disposition, obeys readily and plays in a lively manner with the objects at the table of the clinic. Actual examination and analysis, however, clearly show that her deportment is much more in harmony with the 3 year level than with the 4 year level of development. . . . We feel warranted in classifying this child as definitely subaverage in spite of the general normality of the picture."[61]

The most worrisome instances of mistaken development were children like Rose. They appeared normal but were not, and Gesell made object lessons out of cases in which parental disappointments might have been avoided with a bit of judicious technological intervention. Occasionally, however, the reverse occurred. Adopters who took in children, young and old, sometimes demanded the reassurance of experts when they suspected that the agencies promising to give them normal children had actually given them something less. In 1917, before Gesell had started work on his scales, ten-year-old Caroline was placed with a family by the New York State Charities Aid Association. Although the agency administered the Stanford-Binet as a placement aid and Caroline performed normally—her score was 103—she was rejected by two separate families because, according to her records, "they feel that Caroline is not developing as she should . . . there is something wrong with the child's mentality. . . . They would not want to keep a child who would not amount to something."[62] Caroline did find a permanent home, more than a year and several tests later, when a third family finally agreed that she was normal. In this and other cases, the habit of administering repeated intelligence tests to calm parental jitters suggests that scientific claims about mental measurement were becoming persuasive and meaningful to ordinary people.

[60] Arnold Gesell, "Clinical Aspects of Child Adoption," in Gesell and Catherine S. Amatruda, *Developmental Diagnosis*, 2nd ed., rev. (New York: Harper & Row, 1947), pp. 328–345, on p. 341. One Kansas agency that utilized Gesell's scales also followed his instructions to wait patiently. In one case, a baby whose mother was mentally retarded was kept in a temporary placement for two and one-half years before being placed with adoptive parents, just to be sure that repeated test results would confirm his normality. See Carlson, "To Test a Baby" (cit. n. 39), p. 830.

[61] Gesell, "Psychoclinical Guidance in Child Adoption" (cit. n. 6), pp. 200–201; and Clinical Memorandum in Regard to Rose Golden, 13 Nov. 1923, Box 45, Folder: "Subject File: Adoption, 1923 [Cases]," Gesell Papers ("Rose Golden" is a pseudonym). For a personal testimony from an adopter who was informed by a psychologist that the "gorgeous looking boy of ten months" she wanted "would be dull when he reached school age" see Anonymous, "A Baby in Your Arms," *Child Welfare League Amer. Bull.*, Dec. 1937, p. 2.

[62] Theis and Goodrich, *Child in the Foster Home* (cit. n. 33), pp. 129–146.

In 1930 Alice Taylor, a psychologist for the Children's Aid Society of Pennsylvania, reported to Gesell that the new parents of thirty-five-week-old Matilda worried constantly "because her tongue hangs out a great deal of the time." Taylor, who had put Matilda to the test of the Gesell scales on several occasions, assured the adopters that the baby was perfectly normal. In fact, her response on every testing item placed her well ahead of the normal growth curve. Cases like those of Caroline and Matilda strengthened Gesell's view that developmental norms were the soul of fair play as well as the essence of scientific practice. In addition to offering parents and professionals security through accurate prognosis, developmental assessment "will serve to reveal children of normal and superior endowment beneath the concealment of neglect, of poverty, or of poor repute." In popular media coverage, fairness to the individual and scientific progress were frequently equated.[63]

The democratic credentials of developmental testing cannot be taken at face value, however. Technologies that mapped children's current growth status and future potential were understandably perceived by some would-be parents as a means not only of knowing what they were getting but of ensuring that they got what they wanted. Hadn't they been told repeatedly that "children may be successfully selected with scientific precision for most worthy and particular families"?[64] Would-be adopters frequently expressed preferences for children with certain racial and ethnic backgrounds (light rather than dark), conventionally attractive features (blue eyes, blonde curls), educational potential (college material), and excellent prospects for physical and mental health. The rhetoric of systematic selection not only endorsed such preferences but surely led some adopters to view technologies like Gesell's as scientific warranties.

Highly educated would-be parents, worried about whether adoptees could meet their personal and community standards, were most inclined to credit the promises of scientific adoption. "I want very much to be unselfish and charitable in planning for the welfare of a child who needs help," one college-educated physician's wife wrote to Gesell defensively, but she reported that the absence of information about available children made her "mentally panicky." She wanted a child who would fit into her upper-middle-class milieu, where children were expected to be bright. "I feel that I need impersonal advice from a properly trained person who knows what may and may not be expected of children. Will you try to help me? . . . I feel it is only wise to try to be sure that I am not being led by sympathy and sentimentality into a situation which is essentially unworkable." One indelicate Yale graduate announced to Gesell that, even during the Depression, he had a very substantial income and was willing to "do everything possible to secure a child that will have the capabilities of making the most of a college education and all that goes with it." "We can give a child a great many advantages," pleaded yet another questing parent. "Out of fairness to ourselves as well as the child, we desire to avail ourselves of the latest scientific achievements, to insure a happy outcome." "We have understood from our reading on the subject that you are able to judge mental capacity of a child with fair accuracy even at such an early age," a Williams College physicist wrote in hopes of securing a highly intelligent infant boy with Gesell's aid. "We feel that adopting a baby

[63] Alice Taylor to Gesell, Sept. 1939 (including test schedules and scoring sheets for Matilda), Box 60, Folder: "Subject File: Clinical Records, Matilda Harris, [Adoption case—Penn.], 1938," Gesell Papers ("Matilda Harris" is a pseudonym); Arnold Gesell, *The Guidance of Mental Growth in Infant and Child* (New York: Macmillan, 1930), p. 217; and Carlson, "To Test a Baby" (cit. n. 39), pp. 829–832 (equating fairness and scientific progress).

[64] C. V. Williams, "Before You Adopt a Child," *Hygeia*, 1924, pp. 423–427, on p. 424.

is less hazardous if this is true.''[65] The applied developmental science that Gesell trusted to enhance child welfare helped to legitimize kinship by design. But it is difficult not to conclude that it also functioned as a handy method of product certification and class maintenance in the very adoption market whose abuses and idiosyncracies had provoked such fervent efforts to replace commerce and consumption with science and technology.[66]

Gesell and other adoption professionals were understandably uncomfortable with the expectation that they should infallibly predict how children would turn out. "Perfection of pedigree is a rarity," Gesell warned. "There is some danger that adoptive parents will go to extremes in seeking such perfection." Requests for conclusive developmental forecasting were apparently common enough that Gesell regularly had to remind would-be adopters that biogenetic kinship had its own risks. "Normal parentage does not insure normal development of natural born children," Gesell patiently wrote to one Ohio man seeking iron-clad identification of a child from fine stock and with great potential. "A normal amount of risk and faith must enter into every adoption." Picky parents who were "unreasonably detailed and exacting in their specifications" paradoxically believed both too much and too little in the power of science to deliver what they wanted.[67] Letters like those quoted here exaggerated the power of developmental knowledge and simultaneously discounted its value, reducing it to one more service to be purchased in the course of the adoption process.

REFLECTING ON THE HISTORY OF FAMILIES MADE BY SCIENCE

However well intentioned Arnold Gesell and other advocates of scientific adoption were, their promises to disqualify "unadoptables" systematically and smoothly match children with parents fueled an unrealistic vision of adoption as a mirror of biogenetic kinship subject to scientific control. Later generations of developmentalists and their allies in social work retreated from the idea that adoption was a risk-free arrangement and criticized would-be adopters whose specifications indicated that they wanted guarantees that nature itself did not offer. "During the scientific-minded twenties and thirties," one advice giver noted in the 1950s, agencies took so much time testing and scrutinizing children in their care that "a baby was halfway out of his babyhood before an agency was ready to risk a placement."[68] The earlier wave of confidence in the efficacy of technologies that promised to predict development by measuring normality waned. In part, this shift represented disagreement between Gesell and other developmentalists engaged in longitudinal research, such as Nancy Bayley. Bayley, who directed the Berkeley Growth Study (which began in

[65] Elizabeth Comeau to Gesell, 29 June 1950, Box 45, Folder: "Subject File: Adoption" (physician's wife); Thurston Blodgett to Yale Psycho-Clinic, n.d. but probably late 1920s, Box 45, Folder: "Subject File: Adoption, 1923, 1932 [Agencies]" (Yale graduate); Mandeville to Yale Psycho-Clinic, 11 July 1940, Box 45, Folder: "Subject File: Adoption, 1923–43 [cases, with individuals concerning]" ("insure a happy outcome"); and Ralph P. Winch to Department of Human Relations, Yale Medical School, 29 Mar. 1939, Box 45, Folder: "Subject File: Adoption, 1923–43 [cases, with individuals concerning]" (Williams physicist): Gesell Papers.

[66] Early in the century, some physicians also claimed that they could "certify babies" on the basis of basic pediatric examination and developmental observation: Thyra Samter Winslow, "Certified Babies," *Technical World Magazine*, Feb. 1915, pp. 829–832.

[67] Gesell, "Is It Safe to Adopt an Infant?" (cit. n. 55), p. 8; Gesell to D. Oberteuffer, 25 Sept. 1937, Box 45, Folder: "Subject File: Adoption, 1923–43 [cases, with individuals concerning]," Gesell Papers; and Gesell, "Psychoclinical Guidance in Child Adoption" (cit. n. 6), p. 199.

[68] Carl Doss and Helen Doss, *If You Adopt a Child: A Complete Handbook for Childless Couples* (New York: Holt, 1957), p. 77. For an example of a writer critical of demands for guarantees see Louise Raymond, *Adoption . . . and After* (New York: Harper, 1955), p. 32.

1928), thought that making predictions on the basis of infant testing was useless and irresponsible because preschool children displayed considerably more variability in their scores, upon retesting, than did school-age children. She was nevertheless as dedicated as Gesell to precise measurement and produced a number of developmental technologies herself. The best known was the Bayley Scale of Infant Development.[69]

Others were more reluctant to admit that their previous faith in prediction had been misplaced. A follow-up study of all of the infants Gesell had examined in the process of adoption between 1942 and 1947 showed that babies' original scores had nothing what-soever to do with their subsequent performance on the Stanford-Binet. "All our results are obviously contrary to the implications of Gesell's claims," lamented researcher J. Richard Wittenborn, who was motivated to undertake the study by "the hope of finding some indication that the infant examination has a practically useful predictive validity."[70] His hopes were dashed.

Broad social and political forces also buffeted those members of the scientific com-munity concerned with developmental applications in adoption during the 1930s and 1940s. The specter of Nazism cast a sinister pall over all things eugenic; and as adoption gained in popularity and the demand for children grew, the scientific tables turned. After 1945 the adult participants in the adoption transaction came under heightened scrutiny, and prospective parents were as likely to be ruled ineligible as children. A new consensus emerged that would-be adopters who could not tolerate the risks of the process were probably not emotionally mature enough to be good parents. And good parents, more than any quality inherent in children, were the decisive determinants of good developmental outcomes, according to this new emphasis on the psychological motivation to adopt. After professionals did their investigations and made their choices, after all, parents had to be trusted to create the crucial emotional environments that would surround children and shape their personalities during childhood and throughout their entire lives.[71]

The environmentalist ethos of post-1945 science and culture inaugurated a new and emphatically psychologized stage in the ideology of adoption professionals, who repudi-ated the hereditarian rhetoric of their Progressive- and interwar-era counterparts. Nothing illustrated this more clearly than the expanding notion of adoptability itself. The very psychologists and social workers who had formerly promised to warranty children against damage and defect became the proud authors of an inclusive adoption creed: "Adoption is appropriate for any child without family ties who is in need of a family and for whom a family can be found to meet his need." This was the credo of professional adoption after midcentury, and it has been the basis of "special needs" adoption ever since. There is every reason to believe that the power of blood and biology endured as the central cultural tributaries feeding adoption history during the second half of the twentieth century. But the rhetorical emphasis on (children's bad) blood declined, the trend toward infant place-ments accelerated, and new reproductive technologies intensified the vogue of scientific matching and design previously visible chiefly in adoption.[72]

[69] Nancy Bayley, "Value and Limitations of Infant Testing," *Children*, Aug. 1958, 5:129–133. See also Judy F. Rosenblith, "A Singular Career: Nancy Bayley," in *Century of Developmental Psychology*, ed. Parke *et al.* (cit. n. 47), pp. 499–525.

[70] J. Richard Wittenborn, *The Placement of Adoptive Children* (Springfield, Ill: Thomas, 1957), pp. 58, 178.

[71] Ellen Herman, "Rules for Realness: Child Adoption in a Therapeutic Culture," *Society* 39(January–Feb-ruary 2002): 11–18.

[72] Schapiro, *Study of Adoption Practice* (cit. n. 2), Vol. 1, p. 9. A "new biologism" has accompanied the renaissance of genetic research since 1970 and reinvigorated the belief that blood is much thicker than water.

It is not surprising, however, that during the 1920s and 1930s Gesell championed developmental diagnosis as a means of child protection and quality control and that he took credit for devising new technologies that could protect good-hearted adopters and guarantee that they would not be "cheated" out of the normal children they deserved.[73] Developmental technologies like those Arnold Gesell pioneered were definitive steps in the elaboration of scientific adoption, which became a key ingredient in adoption modernization, which in turn recast an ancient social institution in starkly new terms.

The hope that knowledge and procedures deemed "scientific" could improve the way adopted children and families turned out marks twentieth-century adoption as a distinctive, if incomplete, example of social engineering. The modern adoption process never replaced the messy value conflicts of family formation with neat efficiencies and confident certainties, but it did introduce scientific actors and practices as significant contenders for authority in an important social institution never previously associated with science. Modern adoption enlisted science as a means of advancing particular social, cultural, and moral ends: providing new parents and families for children who needed them, making them as "real" as the biogenetic ideal, and building into them the prerequisites for secure identity and belonging. Bold, even utopian, claims to match children and parents so flawlessly that families made artificially would appear wholly authentic, and therefore undetectable, are without any historical precedent. For their trouble, Gesell and his colleagues achieved the sort of power that human scientists bent on social relevance often achieved in the twentieth century: "a space in which to be heard, and the frustration of listening to others challenge what they had to say."[74]

The campaign to design kinship to resemble an undesigned reproductive nature was inspired by benevolent desires to dignify and equalize adoption and endow participants in it with every opportunity to form lasting, loving bonds. Yet, ironically, by conceding that kinship made naturally was superior to kinship made socially, it ensured the persistence of the blood bias that stigmatized adoption as less real—as lesser, period. The campaign to construct families rationally also supposed that "the natural" was hermetically sealed off from "the social" but that scientifically trained experts could transcend the contaminating influences and corrupting values that necessarily permeated unscientific means of making adoptive families. This was the dream—with its characteristic exaltation of science and devaluation of society—that Gesell's developmental technology nurtured and sought to realize.

In retrospect, the effort to reenact nature through testing and matching in adoption can

---

Even the search and reunion movement is premised on the view that the essence of human identity and connection is genealogical rather than social, that our truest selves reside in our genes. There has been a great deal of commentary on the new genetics/eugenics. For a sample see Richard Lewontin, *It Ain't Necessarily So: The Dream of the Human Genome and Other Illusions* (New York: New York Review of Books, 2000); Dorothy Roberts, *Killing the Black Body: Race, Reproduction, and the Meaning of Liberty* (New York: Pantheon, 1997); and Arlene Skolnick, "Solomon's Children: The New Biologism, Psychological Parenthood, Attachment Theory, and the Best Interests Standard," in *All Our Families: New Policies for a New Century*, ed. Mary Ann Mason, Skolnick, and Stephen D. Sugarman (New York: Oxford Univ. Press, 1998), pp. 236–255. The profound impact of the new biologism on reproductive decision making related to adoption is poignantly illustrated in Jill Bialosky and Helen Schulman, eds., *Wanting a Child: Twenty-Two Writers on Their Difficult but Mostly Successful Quests for Parenthood in a High-Tech Age* (New York: Farrar, Straus & Giroux, 1998).

[73] Gesell used the word "cheated" to describe a mother who had unwittingly adopted a feebleminded baby: Gesell, "Clinical Aspects of Child Adoption" (cit. n. 60), p. 331.

[74] John Carson, "The Science of Merit and the Merit of Science: Mental Order and Social Order in Early Twentieth-Century America," in *States of Knowledge: Science, Power, and Political Culture*, ed. Sheila Jasanoff (forthcoming). Typescript in author's possession; the quotation is from p. 41.

be interpreted as a distinctive social artifact, located in the temporal geography of modernity during the first two-thirds of the twentieth century. Since the 1960s, Kuhnian scholarship has bluntly revealed the particularism lurking beneath the surface of universalist knowledge claims, including the claims of human science to probe and predict the developmental course. Today, psychological testing—including Gesell's technology—remains a routine practice in many adoptions, and the "naturalness" of matching still has ardent defenders, especially with regard to race. But adoptions that deliberately violate the matching paradigm—by crossing racial, ethnic, and national lines—also indicate how far matching has gone into eclipse since 1970. Formerly conventional notions about human nature, the nature of family, and the naturalness of homogeneity to kinship, love, and belonging have correspondingly eroded. Nature itself is not what it used to be.[75]

Matching may consequently appear as one among many ways to make up families—and a culturally and temporally peculiar one at that—rather than as a method founded on universal truths about human development, rational recipes for emotional intimacy, or the best way to ensure that either growth or love corresponds to nature, as Gesell and its other supporters claimed. That kinship by design seems in retrospect so imperious an enterprise, so bent on replicating an exclusive definition of family, and so sadly out of touch with contemporary understandings of authenticity and pluralism in personal life is an instructive lesson in the rapidly changing, curiously overlapping cultural histories of science and family.

[75] Susan R. Harris, "Race, Search, and My Baby-Self: Reflections of a Transracial Adoptee," *Yale Journal of Law and Feminism*, 1997, 9:5–16. In this personal narrative Harris recovers the records of her first fourteen months of life, prior to adoption, in 1963–1964. They include the administration of the Gesell scales. The most famous condemnation of transracial placements and defense of racial matching in adoption is National Association of Black Social Workers, "Position Statement on Trans-Racial Adoption" (1972), in *Children and Youth in America: A Documentary History*, ed. Robert H. Bremner, Vol. 3: *1933–1972*, Pts. 1–4 (Cambridge, Mass.: Harvard Univ. Press, 1974), pp. 777–780. On the particularism that haunts universalist knowledge claims see David A. Hollinger, *Postethnic America: Beyond Multiculturalism* (New York: BasicBooks, 1995), esp. Ch. 3. On "nature" nowadays see Marilyn Strathern, *After Nature: English Kinship in the Late Twentieth Century* (Cambridge: Cambridge Univ. Press, 1992).

Blind Law and Powerless Science

The American Jewish Congress, the NAACP, and the Scientific Case against Discrimination, 1945–1950

By John P. Jackson, Jr.

ABSTRACT

This essay examines how the American Jewish Congress (AJC) designed a legal attack on discrimination based on social science. This campaign led to the creation in 1945 of two new AJC commissions, the Commission on Community Interrelations and the Commission on Law and Social Action. The AJC's attack on discrimination highlights the difficulties and potentialities of using social science research in the law. On the one hand, the relationship between the communities of law and social science was inherently unstable, with constant conflicts over their mission, work styles, timetables, audiences, and standards for success. On the other hand, the AJC's commissions generated new types of arguments and new types of evidence that were powerful tools against racial discrimination and eventually the key to toppling legal segregation. The essay concludes that while the processes by which lawyers and scientists produce their arguments are very different, the products of science can be very useful in the courtroom.

O N 17 MAY 1954, in one of the most important civil rights cases of the century, the U.S. Supreme Court declared segregation unconstitutional in *Brown v. Board of Education*. Social science received prominent mention in a footnote in Chief Justice Earl Warren's opinion. Soon after the decision, Warren's citation of social science became a source of controversy, as scholars tried to sort out its importance. New York University professor of jurisprudence Edmond Cahn expressed concern that Warren's opinion rested on the "flimsy foundation" of social science rather than on solid legal reasoning. Cahn believed that in the *Brown* case social scientists had overstepped the bounds of proper

Isis 91, no. 1 (March 2000): 89–116.

An earlier version of this essay was delivered at the 1996 meeting of the Southern Conference on Afro-American Studies, Tallahassee, Florida. This research was supported by the National Science Foundation's Program in Law and Social Science (NSF Award Number SBR-9421729). I am grateful to Arthur L. Norberg, Leila Zenderland, Benjamin Harris, Stephen Berger, Will Maslow, Margaret Rossiter, and three anonymous referees for their patience and expert editorial advice.

science and ventured into the realm of advocacy. Given the undeveloped state of social psychology, Cahn thought it unable to provide truly objective scientific evidence. Hence social psychology posed a danger to the law. It was imperative, Cahn argued, for judges to "learn where objective science ends and advocacy begins." At present, he insisted, it was still possible for the social psychologist to "hoodwink a judge who is not over wise."[1]

Kenneth B. Clark, a psychologist who had served as the chief liaison between the National Association for the Advancement of Colored People (NAACP) and the social science community during the *Brown* litigation, was quick to defend his work against Cahn's allegation that he had been an "advocate" rather than an objective scientific advisor. How could this be, Clark asked, when the "primary research studies were conducted ten years before these cases were heard on the trial level"? Clark concluded that "one would have to be gifted with the power of a seer in order to prepare himself for the role of advocate in these specific cases ten years in advance."[2]

Clark was certainly correct that the social scientific community had no crystal ball to predict the course of the *Brown* litigation. Yet it is also true that social scientists had been working to use their expertise to dismantle legal segregation in the years immediately following the end of World War II. The most famous social scientific evidence used in the *Brown* litigation was Kenneth and Mamie Clark's projective tests using black and white dolls to measure psychological damage caused by segregation. The "doll tests" have become so well known that commentators on the case often focus on them to the exclusion of any other social science testimony. Because of his work in *Brown*, Kenneth Clark became the exemplar of the socially conscious social scientist in the postwar era, but his doll tests were not the only social scientific evidence used by the NAACP, nor even the most important. A group of social scientists centered at the American Jewish Congress (AJC) had mounted a social scientific attack on discrimination that was expressly designed to be integrated with a legal attack. Although Clark worked briefly for the AJC in the 1940s, this group carried out its research program without his involvement. The AJC social scientists exemplified what Sheila Jasanoff has called "scientific subcultures that . . . co-alesced in and around the processes of adjudication."[3]

[1] *Brown v. Board of Education of Topeka,* 347 U.S. 483 (1954); and Edmond Cahn, "Jurisprudence," *New York University Law Review,* 1955, *30*:150–169, on pp. 157–158.

[2] Kenneth B. Clark, *Prejudice and Your Child* (Boston: Beacon, 1963), p. 191.

[3] Sheila Jasanoff, *Science at the Bar* (Cambridge, Mass.: Harvard Univ. Press, 1995), p. 8. The perception that the Clarks' doll tests were central to *Brown* is incorrect. Many authors, however, treat them as the only empirical evidence in *Brown*. A sampling of this work includes Hadley Arkes, "The Problem of Kenneth Clark," *Commentary,* Nov. 1974, *58*:37–46; Harold Cruse, *Plural But Equal: Blacks and Minorities in America's Plural Society* (New York: Morrow, 1987), pp. 72–74; William H. Tucker, *The Science and Politics of Racial Research* (Urbana: Univ. Illinois Press, 1994), pp. 144–146; Roy L. Brooks, *Integration or Separation? A Strategy for Racial Equality* (Cambridge, Mass.: Harvard Univ. Press, 1996), pp. 12–16; and Ernest van den Haag, "Social Science Testimony in the Desegregation Cases: A Reply to Professor Kenneth Clark," *Villanova Law Review,* Sept. 1960, *60*:69–79. In the doll tests, Kenneth and Mamie Clark presented young children with two dolls identical in every way except skin color. The children were asked to identify the dolls as white or black. Then the children were asked to identify one doll with themselves and to identify one doll as "good" or "bad." The Clarks argued that the identification of a black doll as "bad" by African American children was a sign of psychological damage. A complete description of the doll tests can be found in William E. Cross, *Shades of Black: Diversity in African-American Identity* (Philadelphia: Temple Univ. Press, 1991). On the Clarks' work more generally see *ibid.;* Ben Keppel, *The Work of Democracy: Ralph Bunche, Kenneth B. Clark, Lorraine Hansberry, and the Cultural Politics of Race* (Cambridge, Mass.: Harvard Univ. Press, 1995); and Darryl Michael Scott, *Contempt and Pity: Social Policy and the Image of the Damaged Black Psyche, 1880–1996* (Chapel Hill: Univ. North Carolina Press, 1997). A work that situates Clark as a "public psychological expert" is Ellen Herman, *The Romance of American Psychology: Political Culture in the Age of Experts* (Berkeley: Univ. California Press, 1995). On Clark's brief tenure at the AJC see Gerald Markowitz and David Rosner, *Children, Race, and Power: Kenneth and Mamie Clark's Northside Center* (Charlottesville: Univ. Virginia Press, 1996), p. 81.

This essay will examine how the AJC designed an attack on discrimination that was founded on the assumption that law and social science could be merged. This assumption led to the creation of two new AJC commissions, the Commission on Community Inter-relations (CCI) and the Commission on Law and Social Action (CLSA). The AJC's attack on discrimination highlights the difficulties and potentialities of combining social science research and the law. On the one hand, the relationship between the communities of law and social science was inherently unstable, with constant conflicts over their mission, work styles, timetables, audiences, and standards for success. On the other hand, the AJC's merger generated new types of arguments and new types of evidence that were powerful tools against racial discrimination and eventually the key to toppling legal segregation. In short, while the AJC failed to institutionalize any long-term working relationship between lawyers and social scientists, it did succeed in creating a powerful attack on racial dis-crimination.

The first section of the essay looks at the postwar context of social science research on race prejudice and at the AJC's place in that context. The second section considers the design of each commission within the AJC. Next, I examine the work each commission undertook in order to carry out the attack on discrimination. Of particular importance here is how AJC attorneys developed new legal arguments that required new kinds of social scientific evidence and how social scientists stepped up to provide that evidence. The fourth section outlines how the NAACP began to use social scientific data in ongoing litigation against segregated schools and how it began to turn to the AJC to supply that data. The final section of the essay examines the end of the collaboration of social scientists and lawyers at the American Jewish Congress and considers the ramifications for our under-standing of the interplay between the law and science.

THE POSTWAR FIGHT AGAINST RACE PREJUDICE

World War II had a profound effect on minority/majority group relationships in the United States. After the war was won, many Americans turned their attention to the elimination of racism and prejudice at home. Social science was one tool that could be enrolled against race prejudice. (See Figure 1.) Even before the war, most social scientists had abandoned notions of innate racial traits and had begun to study intergroup relations and race preju-dice.[4] But the war transformed the way social scientists approached their work, giving a new urgency to the social scientific study of race prejudice.

There had always been two strands of social scientific thought in the United States. On the one hand, the cultural authority of science depends on the image of scientists as de-tached, impartial experts who are immune from political or moral concerns. On the other hand, social science makes a claim to social utility by offering to solve numerous political and moral problems.[5] The second strand of social scientific thought was exemplified by

[4] Philip Gleason, "Americans All: World War II and the Shaping of American Identity," *Review of Politics,* 1981, *43*:483–518; David Southern, *Gunnar Myrdal and Black-White Relations: The Use and Abuse of* An American Dilemma, *1944–1969* (Baton Rouge: Louisiana State Univ. Press, 1987); Elazar Barkan, *The Retreat of Scientific Racism* (Cambridge: Cambridge Univ. Press, 1992), pp. 343–346; Carl N. Degler, *In Search of Human Nature* (New York: Oxford Univ. Press, 1991), pp. 187–211; and Franz Samelson, "From 'Race Psy-chology' to 'Studies in Prejudice,' " *Journal of the History of the Behavioral Sciences,* 1978, *14*:265–278.

[5] There is an extensive literature on this "paradox" of social science in the United States. Major monographs include Joanne Brown, *Definition of a Profession: The Authority Metaphor in the History of Psychological Testing, 1890–1930* (Princeton, N.J.: Princeton Univ. Press, 1992); Mary O. Furner, *Advocacy and Objectivity: A Crisis in the Professionalization of American Social Science, 1865–1905* (Lexington: Univ. Kentucky Press, 1975); John M. O'Donnell, *The Origins of Behaviorism: American Psychology, 1870–1920* (New York: New York Univ. Press, 1985); and Mark Smith, *Social Science in the Crucible: The American Debate over Objectivity and Purpose, 1918–1941* (Durham, N.C.: Duke Univ. Press, 1994).

RACE EXPERTS SPEAK ON COMPARATIVE INTELLIGENCE

"No one has been able to demonstrate that ability is correlated with skin color or head shape or any of the anatomical characteristics used to classify races." -- Dr. Otto Klineberg
Race Differences

"Actually there is no scientific evidence for the theory that there are fewer Negroes than whites possessing high intelligence."   -- Prof. Gerhart Saenger
The Social Psychology of
Prejudice

"None of these tests (devised to measure the relative intelligence of human groups) demonstrates that racial differences in intelligence exist."   -- Dr. Ralph L. Beals &
& Dr. Harry Hoijer
An Introduction to Anthropology

"We may use the term 'race' to call attention to groups of people who are more or less alike among themselves and more or less different from others, but just as soon as we proceed a step further and make 'race' mean differences in mental characteristics and moral quality we have gone beyond the facts and have entered the region of unjustified theories and assumptions."
-- Dr. Edmund D. Soper
Racism: A World Issue

Figure 1. *Social science arguments as anti-racist propaganda in the postwar United States. (From the Collections of the Manuscript Division, Library of Congress, Washington, D.C.)*

the Society for the Psychological Study of Social Issues (SPSSI), a group of activist social psychologists that came together in 1936.

SPSSI was born in the Great Depression, a time of profound social and economic distress. The social psychologists of SPSSI hoped to create a new form of social psychology, one that could empower their research subjects and be used for positive social change. Members championed a nonreductivist social psychology that moved out of the laboratory and into the lives of real Americans. They rejected what they saw as positivist, experimentalist psychology and sought to bring forth an unapologetic political psychology that would make American society more democratic.[6]

SPSSI's project was not without contradictions. While professing an ideology that could be used to empower workers and create a true socialist democracy, SPSSI was also a tool affording its members professional advancement within the psychological discipline. Within a year after its founding, SPSSI was a force to be reckoned with, at least within the world of psychology, as one of every six American Psychological Association members was also a member of SPSSI. Soon SPSSI began publishing a "SPSSI Bulletin" as part of the *Journal of Social Psychology* and produced *Industrial Conflict*, a book on labor conflict. World War II gave many of these social psychologists unprecedented opportunities to work closely with the government; they were involved in strategic bombing surveying, propaganda analysis, psychological warfare, studies of civilian and enemy morale, and surveys of public opinion.[7] In all of these efforts, psychologists were convinced that their special expertise was necessary to direct governmental power in the most efficient manner. War work led social psychologists to reconceptualize their mission as social engineers in two important ways.

First, social scientists began to focus on the government as an agent of change in society. In the 1930s many SPSSI members had considered themselves "radicals" and "outsiders" and believed that society was best remade by the people, or the workers, rather than by a powerful elite represented by government or state action. After World War II, however, many SPSSI members began to emphasize the role of government, and of legal change, in altering society. Their experiences in World War II linked social psychologists to government to such an extent that even dedicated leftists began writing about how the government could undertake social engineering for the benefit of the people.[8]

Second, after the war social scientists began to focus intently on race relations and race prejudice. A common theme in much of the research done on race prejudice during the war was that social science needed not only to understand race prejudice but to work to eliminate that prejudice. Hitler's rise to power, the struggle against Nazi ideology, and the perceived need to unify the nation behind the war effort transformed the study of prejudice into a struggle against totalitarianism. This viewpoint was epitomized by Gunnar Myrdal's *An American Dilemma,* which laid the blueprint for two decades of social engineering focused on the elimination of race prejudice. At the close of World War II two social psychologists who would soon be active within the AJC program, Ronald Lippitt and Marian Radke, declared that "the need for an understanding of the dynamics of prejudice has no equivalent in importance in the social sciences. In no other aspect of interpersonal and intergroup relationships is there a more urgent need for social sciences to 'get out and do something.' "[9] For Lippitt and Radke, as for many others, race prejudice was the chief

---

[6] Lorenz Finison, "Unemployment, Politics, and the History of Organized Psychology," *American Psychologist,* 1976, *31*:747–755; and Finison, "Unemployment, Politics, and the History of Organized Psychology, II: The Psychologists League, the WPA, and the National Health Program," *ibid.,* 1978, *33*:471–477.

[7] On SPSSI see Benjamin Harris, "Reviewing Fifty Years of the Psychology of Social Issues," *Journal of Social Issues,* 1986, *42*:1–20. On the membership see Finison, "Unemployment, Politics, and the History of Organized Psychology," p. 753. SPSSI's publishing ventures are discussed in Lorenz Finison, "The Psychological Insurgency: 1936–1945," *J. Soc. Issues,* 1986, *42*:21–34, on pp. 28–29. On the war work of social psychologists see James Capshew, *Psychology on the March* (Cambridge: Cambridge Univ. Press, 1998); and Blair T. Johnson and Diana R. Nichols, "Social Psychologists' Expertise in the Public Interest: Civilian Morale Research during World War II," *J. Soc. Issues,* 1998, *54*:53–78.

[8] Finison, "Psychological Insurgency," pp. 31–33.

[9] Ronald Lippitt and Marian Radke, "New Trends in the Investigation of Prejudice," *Annals of the American*

threat to the democratic way of life. Such a threat demanded more than dispassionate study; it demanded eradication. This understanding contrasted sharply with prewar views that saw race prejudice as irrational but seldom portrayed it as a danger to society.

The change in how social scientists viewed race prejudice coincided with a change in Jewish leadership. After World War II a host of middle-class American Jews of Eastern European descent with socialist-labor backgrounds began to take over Jewish leadership roles. In contrast to the older generation of leaders, these men were professional civil rights workers who were not interested in placating the WASP elite but were eager and willing to join public battle for Jewish rights.[10]

The new militancy in the Jewish leadership was owed in part to the horrors of Nazi Germany. The Holocaust united the American Jewish community against anti-Semitism with the goal of eliminating it forever. New funds poured into the coffers of Jewish organizations dedicated to fighting anti-Semitism, including the decades-old American Jewish Congress, the American Jewish Committee, and the Anti-Defamation League of B'nai B'rith. Unlike other organizations, however, the American Jewish Congress was not willing to rely on moral pleading and the process of education to eliminate anti-Semitism. It preferred a more direct approach through litigation and lobbying for legal change.[11]

The AJC's belief that legal changes were necessary to eliminate prejudice dovetailed with the interests of activist social scientists who wanted to use governmental power to create a more just and fair society. The next section examines how lawyers and social scientists came together at the AJC.

TWO BLUEPRINTS FOR MERGING SCIENCE AND THE LAW

Founded in 1918, the American Jewish Congress was one of the more militant and confrontational Jewish organizations. From the beginning—and in contrast to the older and more staid American Jewish Committee—the American Jewish Congress protested, confronted, and publicized anti-Semitism. After World War II, when the full horror of the Nazi regime became public, and fearing rampant anti-Semitism in the United States, the AJC redoubled its efforts. Drawing on the insights of two innovative émigré scholars, Kurt Lewin and Alexander Pekelis, it launched a "comprehensive program of legal, legislative and social action which would protect and safeguard the rights of Americans . . . by outlawing every form of discrimination on grounds of race, creed, color or national origin."[12] At the heart of the program was the belief that education and moral exhortation against

Academy of Political and Social Science, 1946, *244*:167–176, on p. 167. On the transformation to a struggle against totalitarianism see Herman, *Romance of American Psychology* (cit. n. 3), pp. 174–207. For Myrdal's book see Gunnar Myrdal, *An American Dilemma* (New York: Harper, 1944); on the immense impact of Myrdal's work in creating a "liberal orthodoxy" concerning race relations see Walter Jackson, *Gunnar Myrdal and America's Conscience* (Chapel Hill: Univ. North Carolina Press, 1990).

[10] On this change in Jewish leadership generally see Lenora E. Berson, *The Negroes and the Jews* (New York: Random House, 1971), pp. 98–107.

[11] Edward S. Shapiro, *A Time for Healing: American Jewry since World War II* (Baltimore: Johns Hopkins Univ. Press, 1992), pp. 16–17; and Stuart Svonkin, *Jews against Prejudice: American Jews and the Fight for Civil Liberties* (New York: Columbia Univ. Press, 1997), pp. 79–112.

[12] David Petegorsky, "Report of the Executive Director to the Biennial National Convention of the American Jewish Congress," 31 Mar.–5 Apr. 1948, p. 10, American Jewish Congress Papers, Box 19, American Jewish Historical Society, Waltham, Massachusetts (hereafter cited as **AJC Papers**). Accounts of the formation of the AJC can be found in Morris Frommer, "The American Jewish Congress: A History, 1914–1950" (Ph.D. diss., Ohio State Univ., 1978); and Henry L. Feingold, *A Time for Searching: Entering the Mainstream, 1920–1945* (Baltimore: Johns Hopkins Univ. Press, 1992).

prejudice would always fail if official discrimination continued. Official discrimination—whether in the form of quotas determining the number of Jews allowed into medical school or the more blatant segregation statutes of the South—would have to be eliminated to eliminate prejudice. Only then could education to combat prejudice have some hope of success. In short: attacking discrimination was the key to attacking prejudice.

To attack official discrimination the AJC created two new divisions, both concerned with using both the law and social science. The Commission on Community Interrelations (CCI) hoped to translate social science research into social action in order to lead a "scientific attack on anti-Semitism and other minority problems in the United States."[13] CCI recognized that the law could play a valuable role in its work. The Commission on Law and Social Action (CLSA) was to do for Jewish Americans what the NAACP had been doing for African Americans—litigate to protect their rights. CLSA attorneys viewed social science as a valuable resource for the creation of new laws against discrimination and segregation.

Commission on Community Interrelations

The Commission on Community Interrelations was the brainchild of one of the most influential social psychologists of the time, Kurt Lewin. Educated at the University of Berlin, Lewin made several fundamental contributions to social psychology in Europe, concentrating on worker education and job satisfaction. In 1934 he fled Europe for the United States. Finding himself in a strange new country, Lewin turned his attention to issues surrounding group identification, prejudice, aggression, and Jewish identity. After two years at Cornell University, he spent 1935–1945 at the Child Welfare Research Station at the University of Iowa.[14]

In 1944 some of Lewin's publications came to the attention of AJC president Rabbi Stephen Wise. The AJC had earmarked $1 million for the creation of a research center for the study of intergroup relations, and Wise thought Lewin should lead it. Lewin saw an opportunity to carry out scientifically informed social engineering. He held that social engineering was parallel to industrial engineering: "as industrial plants have found out that physical research pays, social organization will soon find out that social research pays."[15] The AJC initiative seemed tailor-made for Lewin's vision of a social engineering organization, and he quickly agreed to undertake the creation of the new commission.

In July 1944 Lewin submitted a plan for the new organization, "Memorandum for the Commission on Anti-Semitism," to the American Jewish Congress. It outlined a vision of lawyers and social scientists working side by side to fight for democracy. Lewin argued that many existing programs were based on an insufficient understanding of the causes of and cures for anti-Semitism. His program would be based on two criteria. First, "it has to be objective, i.e. it has to uncover the essential facts in an unbiased scientific manner."

[13] "New Scientific Attack on Anti-Semitism Launched by American Jewish Congress," 28 June 1945 (press release), Alfred J. Marrow Papers, Box 20, Folder: "CCI, Pamphlets, Public Relations," Archive for the History of American Psychology, Akron, Ohio (hereafter cited as **Marrow Papers**).

[14] Mitchell Ash, "Cultural Contexts and Scientific Change in Psychology: Kurt Lewin in Iowa," *Amer. Psychol.*, 1992, *47*:198–207.

[15] Kurt Lewin, "The Place of the Commission on Community Interrelations within the Work of Jewish Organizations," address to the National Community Relations Advisory Council, 16 Nov. 1944, Kurt Lewin Papers, Folder 19: "American Jewish Congress, CCI," Archive for the History of American Psychology (hereafter cited as **Lewin Papers**). See also Alfred J. Marrow, *The Practical Theorist: The Life and Work of Kurt Lewin* (New York: Basic, 1969), pp. 161–164.

Second, the program "has to be practical, i.e. it should lead to coordinated significant actions." Lewin envisioned an organization with two main divisions: a research division that would consist of community sociologists, opinion analysts, group psychologists, individual psychologists, and statisticians; and an "operational division" that would combine two existing AJC commissions that included numerous lawyers: the Commission on Economic Discrimination and the Commission on Law and Legislation. In the new commission, research workers and operational personnel would be equal participants. As Lewin wrote, "The need for action determines the content of the research; scientific requirements determine its technique." Lawyers and other legal personnel, he believed, should work hand in hand with social science researchers. Fundamental to CCI's views on prejudice was the notion that "if we can break down the social segregation and discrimination which defines a racial or religious group as a sanctioned target for prejudice and scape-goating, time will take care of the individual prejudice."[16]

CCI's focus on official discrimination rather than the attitude of prejudice is reflected in one of its first projects: a general survey of the state of knowledge about prejudice and discrimination undertaken by Goodwin Watson. As one of the founding members of SPSSI, Watson had long been concerned with using psychology to create a more democratic and just world. Watson had made one of the first attempts to measure race prejudice in his Ph.D. dissertation at Columbia's Teacher's College in 1925. His survey for CCI was part of his struggle to become, as he put it, a "social engineer" in the postwar world and also reflects the postwar emphasis on government action in the creation of a just society.[17]

In his report for CCI Watson surveyed different techniques for fighting prejudice, including education, moral exhortation, and other methods. He concluded that "*it is more constructive to attack segregation than it is to attack prejudice.*" Segregation was amenable to public control, unlike people's private prejudices. Moreover, education would fail to end prejudice unless official "caste barriers" were torn down because "habits built around those barriers will silently undo anything we accomplish."[18]

CCI believed that law could do what education could not: break through the irrational attitude of race prejudice. Thus the law became an essential component in fighting discrimination and prejudice. Stuart W. Cook, who would later play a pivotal role in the *Brown* litigation, came to CCI in 1948 to serve as co-director with Lewin. A University of Minnesota Ph.D. (1938), Cook had long been interested in the effects of intergroup contact on attitude change. Soon after joining the staff of CCI, he set forth the commission's philosophy regarding the relationship between education and legal change:

> Educational programs aimed at reducing discrimination are likely to make slow headway when there are no anti-discrimination laws. But passage of a law changes the atmosphere in which education is carried on. Once legislation exists, an educational program can draw support from the law-abiding tradition of most citizens. After a law is passed, an educational campaign designed to explain the law's purpose and to encourage compliance with it is no longer an

[16] Kurt Lewin, "Memorandum on Program for the Commission on Anti-Semitism of the American Jewish Congress," pp. 1, 9, Lewin Papers, Box M946, Folder 22: "CCI"; and "Some Basic Issues," 12 Dec. 1944, Marrow Papers, Box M1938, Folder 21: "CCI Papers."

[17] Watson laid out his blueprint for psychology's role in the postwar world in Goodwin Watson, "How Social Engineers Came to Be," *Journal of Social Psychology,* 1945, *21*:135–141. For details of Watson's life and work see Ian A. M. Nicholson, "The Politics of Scientific Social Reform, 1936–1960: Goodwin Watson and the Society for the Psychological Study of Social Issues," *J. Hist. Behav. Sci.,* 1997, *33*:39–60; and Nicholson, " 'The Approved Bureaucratic Torpor': Goodwin Watson, Critical Psychology, and the Dilemmas of Expertise, 1930–1945," *J. Soc. Issues,* 1998, *54*:29–52.

[18] Goodwin Watson, *Action for Unity* (New York: Harper, 1947), p. 64.

inefficient technique but is in a position to produce a great return for a relatively small investment.[19]

Cook was presenting the key idea for these researchers: that the law is not an alternative
to education against group prejudice but is best used in conjunction with such education.

Isidor Chein, CCI's research director, made a similar point. Chein, like Kenneth B.
Clark, had received his Ph.D. in psychology (1939) from Columbia, one of the centers of
activist social psychology in the 1930s. Before joining CCI Chein worked with one of the
first organizations in New York City dedicated to the study of intergroup relations, the
Committee on Unity established in the wake of the 1943 Harlem riots. In 1946 he wrote
that legal changes "constitute virtually the only means of breaking into the vicious circle
[of prejudice and discrimination]; legislation, in all areas to which it may be applied,
against discriminatory practices in employment, education, housing, and so on." Hence,
Chein was quite receptive to the AJC's program of direct legal action against discrimination. He joined CCI as a research associate and became director of research in 1950. At
CCI Chein continued to maintain that "education and the law ... go hand in hand; each
approach helps to bring out the best potentialities of the other. Either, alone, is apt to be
fruitless."[20]

The particular method Lewin developed for CCI was "action research." Lewin envisioned social scientists researching a problem and "actionists" implementing a reform
program. "Actionists" included not just lawyers but also community workers, local leaders,
religious personnel, and others who lived and worked in the community under study. From
the beginning, CCI personnel anticipated difficulties with these two groups of people
working side by side. Ronald Lippitt, a student of Lewin's from the Iowa days, made the
issues clear in an internal memorandum. Social scientists, he wrote, were "likely to feel
that action personnel have no appreciation of the problems and requirements of data collection" and were "likely to be put in situations where [their] previous sources of satisfaction—recognition for competence in technical publication, theorizing, etc., are not relevant." By contrast, action personnel were "likely to feel that social research isn't practical
enough yet to make improvements on the great mass of experience which has led to certain
more or less institutionalized practices" and that "data collection procedures are an un-
warranted nuisance when they begin to call for certain modifications of action plans."[21]
As we will see, Lippitt's argument would prove prescient, for while CCI pursued valuable
research that would later become important in the *Brown* litigation, social scientists and
lawyers would coexist only uneasily within the institutionalized context of the AJC.

[19] Stuart Cook to Will Maslow, 5 Mar. 1947, Marrow Papers, Box M1938, Folder 14: "1946–1947, CCI
Correspondence."

[20] Isidor Chein, "Some Considerations in Combating Intergroup Prejudice," *Journal of Educational Sociology,*
1946, *19*:412–419, on p. 416; and Isidor Chein to Garner Roney, 9 Feb. 1949, Marrow Papers, Box M1938,
Folder 15: "1948–1949, CCI Correspondence." A brief biography of Chein can be found in "Awards for Dis-
tinguished Contributions to Psychology in the Public Interest: 1980," *Amer. Psychol.,* 1981, *36*:67–70. On the
Committee on Unity see Gerald Benjamin, *Race Relations and the New York City Commission on Human Rights*
(Ithaca, N.Y.: Cornell Univ. Press, 1972), pp. 38–70. On activist social psychology in the 1930s see Katherine
Pandora, *Rebels within the Ranks: Psychologists' Critique of Scientific Authority and Democratic Realities in
New Deal America* (Cambridge: Cambridge Univ. Press, 1997).

[21] Ronald Lippitt, "Action Research—Idea and Method," 17 July 1945, Marrow Papers, Box M1938, Folder
13: "1943–1945, CCI Correspondence." For a more complete description of CCI's program of "action research"
see Frances Cherry and Catherine Borshuk, "Social Action Research and the Commission on Community Interrelations," *J. Soc. Issues,* 1998, *54*:119–142.

In 1945, however, despite reservations, most CCI social scientists felt that the problems linked with fusing research and action could and would be worked out. At a February 1945 meeting, Stephen Wise announced that the AJC would fund Lewin's "Commission of Community Interrelations" for five years and that its purpose was to investigate scientifically the causes and cures of anti-Semitism and race prejudice. The expansion of CCI's mission to include forms of race prejudice beyond anti-Semitism was in keeping with the larger agenda of the AJC. Shad Polier, an attorney and Wise's son-in-law, put the matter this way:

> We make no claim that our activities are compounded wholly of altruism, or that they are entirely divorced from our more partisan and proximate objective of enhancing the security of Jews in America. On the contrary, since we believe Jewish interests to be inseparable from the interests of justice we have always contended that for the Jewish community there is an unfailing advantage to be derived from performance of the principled act.[22]

The focus on all forms of discrimination rather than on anti-Semitism alone was mirrored in other branches of the AJC. As the work of CCI was getting under way, the AJC was also forming the Commission on Law and Social Action; it too began with great plans for merging social scientific research and legal action.

Commission on Law and Social Action

Although Lewin had hoped that CCI would incorporate the existing Commission on Law and Legislation and Commission on Economic Discrimination, in November the AJC instead merged them into a new Commission on Law and Social Action under the leadership of Will Maslow.

Maslow had come to the United States when he was three years old. He had studied economics at Cornell but switched to law at Columbia when he discovered that universities hesitated to hire Jewish professors. Unfortunately, most large law firms were equally hesitant to hire Jewish attorneys. The newly formed bureaucracies of the New Deal, however, had no such strictures, and Maslow worked as a trial attorney for the National Labor Relations Board and later supervised the Fair Employment Practices Commission.[23]

In 1945, when Maslow was hired to head the newly formed CLSA, he found Alexander Pekelis waiting for him. Pekelis had been the director of the Commission on Law and Legislation until it was collapsed into CLSA. Born in Russia, Pekelis spent most of his adult life in Italy and France as a professor of jurisprudence. He had fled to the United States in 1941, just before the Nazis entered France. He took a position on the Graduate Faculty of the New School for Social Research and in 1942 entered the Columbia School of Law, where he became editor of the *Columbia Law Review*. Because he was not yet a member of the bar, Pekelis could not serve as a staff attorney for CLSA, but Maslow created a special position for him—"Chief Consultant."[24] Pekelis's first job for CLSA was

---

[22] H. Epstein, " 'Forward,' Accent on Action: A New Approach to Minority Group Problems in America," 1945, Stuart W. Cook Papers, Box M2337, Folder 2: "Commission on Community Interrelations," Archive for the History of American Psychology (hereafter cited as **Cook Papers**); and Shad Polier, "Why Jews Must Fight for Minorities," 4 Nov. 1949, pp. 2–3, Shad Polier Papers, Box 16, Folder: "Polier/AJC on Civil Liberties," American Jewish Historical Society.

[23] Murray Friedman, *What Went Wrong? The Creation and Collapse of the Black–Jewish Alliance* (New York: Free Press, 1995), pp. 133–134.

[24] M. R. Konvitz, "Introduction," in *Law and Social Action: Selected Essays of Alexander H. Pekelis*, ed. Konvitz (Ithaca, N.Y.: Cornell Univ. Press, 1950), pp. v–vii, on p. vii.

to prepare a memorandum that would set out the commission's organization and philosophy, just as Kurt Lewin was doing for CCI.

Pekelis based his understanding of the law on an older American legal tradition, "legal realism." There are as many definitions of legal realism as there were legal realists. Generally the term referred to "that body of legal thought produced for the most part by law professors at Columbia and Yale Law Schools during the 1920s and 1930s."[25] These law professors were following the lead of progressive theorists who rejected nineteenth-century legal reasoning.

At the end of the nineteenth century American jurisprudence was dominated by the belief that judges discovered, rather than made, law. According to this view, law was a process of deductive reasoning that took legal rules and case precedents as its major premises and the facts of a particular case as its minor premises. The judge could make a proper ruling in each case by following a formal procedure that inevitably led to the correct decision. The case method of legal education, pioneered by Christopher C. Langdell at Harvard, perpetuated this formalistic system by teaching that legal precedent, combined with proper reasoning, would lead to uniform law.[26]

In the twentieth century the formal legal system came under increasing attack, as legal theorists and practicing lawyers began to demand that the law pay more attention to how the world actually worked. The Columbia law professor Karl Llewellyn was the first to attempt a definition of the new jurisprudence in a 1930 law review article, "A Realistic Jurisprudence: The Next Step." Llewellyn called for the findings of the behavioral sciences to be used by lawyers and judges to bring jurisprudence more into line with the way the world truly operated. The legal realists, as they came to be known, attacked two major concepts. First, they denied that judges "discovered" law. Rather, argued the realists, judges *made* law, and the law either helped or hindered certain social policies. Second, the realists argued for new sources of information to replace the abstractions of nineteenth-century jurisprudence. New sources of information, including social science, began flooding the legal system with facts about how the law operated in society.[27]

At least some of those that could be called legal realists were interested in using social science in legal proceedings. Progressive lawyers such as Charles E. Clark and William O. Douglas attempted to use social science research to further their reforms. Unfortunately, social research initiated for use in litigation could not be completed in time to be useful in the courtroom. In addition, if the results of the research did not suit the needs of the lawyers it was ignored.[28] The same issues would soon haunt CCI researchers and CLSA attorneys in their attempt to integrate social science and the law.

Pekelis had entered Columbia as a mature scholar; law school for him was almost a formality, a credential he needed to gain access to the American bar. The central notions of realism—that judges make rather than discover law and that information about the "real" world should play a central role in the creation of law—clearly resonated with him. "Similar theories had been developed in Europe long before legal realism became popular

[25] Morton J. Horowitz, *The Transformation of American Law, 1870–1960: The Crisis of Legal Orthodoxy* (New York: Oxford Univ. Press, 1992), p. 169.

[26] Edward A. Purcell, *The Crisis of Democratic Theory: Scientific Naturalism and the Problem of Value* (Lexington: Univ. Kentucky Press, 1973), pp. 74–75.

[27] Karl Llewellyn, "A Realistic Jurisprudence: The Next Step," *Columbia Law Review,* 1930, *30*:431–465. A detailed treatment of how social science made inroads into the legal culture is John W. Johnson, *American Legal Culture, 1908–1940* (Westport, Conn.: Greenwood, 1981).

[28] See, e.g., John Henry Schlegel, *American Legal Realism and Empirical Social Science* (Chapel Hill: Univ. North Carolina Press, 1995).

here," he wrote in 1943.[29] Pekelis wrote the central doctrines of legal realism into his memorandum for the AJC leadership.

At the heart of Pekelis's proposal was the unique nature of anti-Semitism in the United States. In contrast to European anti-Semitism, which was a function of official government action, American anti-Semitism "comes from the forces of society itself. Anti-Semitism here is private or communal, not public or governmental in nature." These social forces—in the form of unofficial quotas on Jews in professional schools, for instance—would be much harder to detect than anything so blatant as a law. Hence, to combat American anti-Semitism, Jews needed the sorts of data that could only be provided by social science. "Contemporary experience," Pekelis wrote, "has shown that no political, social, administrative, or legal action can be conducted efficiently unless means are found to narrow the gap between those who devote themselves to the study of social reality and those who, in legislative communities and courts, shape the law of the community. . . . Law without a knowledge of society is blind; sociology without a knowledge of law, powerless." Pekelis wanted to insure successful cooperation between social scientists and lawyers by having them work in the "same functional unit." In a line that Lewin might have written, Pekelis urged "a close-knit integration of projects. . . . The same type of integration must be achieved between social and legal action and between research and operational activities."[30]

Although Pekelis did not realize his dream of social scientists working within the legal department, his ideas resonated with the working attorneys of CLSA. Just as Pekelis predicted, CLSA attorneys found social science materials necessary because of the special problems of American anti-Semitism. Maslow and his colleagues recognized that social science data would be necessary for much of their litigation to be successful. A February 1946 memorandum argued that "legal skills, social science training and the capacity for social action must be joined if specific tasks are to be defined intelligently and pursued successfully." In 1949, contrasting the sorts of problems confronted by CLSA, which wanted to eliminate anti-Semitism, with those confronted by the NAACP, which wanted to eliminate discrimination against African Americans, a "Note" in the *Yale Law Journal* observed that "discrimination against Jews in the United States is usually non-governmental, non-violent, and extremely subtle" and that "an organization concerned with Jewish problems must employ sociological research to expose the more subtle discrimination to which Jews are subjected."[31]

Pekelis, like Lewin, had set out a blueprint for collaboration between social scientists and lawyers in the fight against discrimination. Both argued not only that merging the law and social science would be advantageous in the battle against discrimination but that it was vital for lawyers and social scientists actually to plan and carry out their projects together. What they envisioned went beyond the citation of a few research results in a legal brief, involving lawyers' partnership with social scientists in the creation of that brief. Moreover, social scientists would not merely study the effects of particular legal changes but would be active agents in creating those changes. Looking beyond mere "cooperation,"

[29] Alexander H. Pekelis, "The Case for a Jurisprudence of Welfare" (1943), in *Law and Social Action*, ed. Konvitz (cit. n. 24), pp. 1–41, on p. 3.

[30] Alexander H. Pekelis, "Full Equality in a Free Society: A Program for Jewish Action" (1945), in *Law and Social Action*, ed. Konvitz, pp. 218–259, on pp. 256–257.

[31] CLSA Memorandum, Feb. 1946, AJC Papers, Box 33, Folder: "CLSA Memorandum"; and "Private Attorneys-General: Group Action in the Fight for Civil Liberties" (note), *Yale Law Journal*, 1949, 58:574–598, on pp. 589, 594.

Lewin and Pekelis wanted lawyers and social scientists to work in an almost symbiotic relationship. In reality, however, the lawyers and social scientists would have trouble working as closely as Lewin and Pekelis had hoped. While social scientists did try to design research that would be of use in the legal arena, and while lawyers designed arguments that required social scientific data, the close collaboration Lewin and Pekelis anticipated never emerged.

GENERATING RESEARCH FOR THE LEGAL ARENA

The research sponsored by CCI demonstrated social scientists' belief that education and the law could be partners. CCI researchers attempted to show that a change in the actual physical circumstances of society could precede a change in attitudes. In other words, the law did not have to wait for a change in social climate to be effective; it could itself change attitudes. Two lines of research illustrate how CCI attempted to merge social science with an attack on discrimination: research conducted on the separation of attitudes from behavior and research on the effects of interracial contact.

On the separation of attitudes from behavior, CCI's social scientists built on the work of Richard T. Lapiere. In the 1930s Lapiere traveled through the United States with a Chinese couple. They stayed in hotels or auto camps and ate in a total of 184 restaurants. Except in one hotel, they were served without incident. Six months later, Lapiere sent a questionnaire to these establishments asking if they served "members of the Chinese race." Over 90 percent of those responding indicated they would not, despite the fact that they had done just that six months earlier.[32]

At CCI Bernard Kutner successfully duplicated Lapiere's research when he sent two white women into New York restaurants. They were later joined by an African American woman, who was seated without incident. When Kutner inquired as to the policies of the restaurants, however, he was informed that they did not serve African Americans. Gerhardt Saenger conducted a similar study on the integration of sales personnel, discovering that customers who, moments earlier, had been assisted by African American salesclerks at a large New York department store would tell a pollster that they would never trade at a store that employed African Americans.[33]

Of more lasting influence than the research on attitudes and behavior was that on interracial contact. CCI social scientists envisioned a vicious circle of discrimination and prejudice: because discrimination seemed to teach people that minority groups were inferior, it led to prejudicial attitudes; these attitudes, in turn, led to the erection of more discriminatory barriers preventing minorities from fully entering society. CCI saw interracial contact as the point at which the circle of discrimination and prejudice could be broken.

As a program of scientific study, CCI researchers attempted to discover the circumstances under which contact between individuals of different races tended to decrease prejudice. The "contact hypothesis"—that contact between different races decreased prejudice—was an open question in the late 1940s. While some unsystematic evidence indicated that interracial contact decreased prejudice, there was no firm consensus, and, indeed, some evidence suggested that contact increased prejudice. In a 1948 survey of various

---

[32] Richard T. Lapiere, "Attitudes vs. Actions," *Social Forces*, 1934, *13*:230–237.

[33] Bernard Kutner, C. Wilkins, and P. R. Yarrow, "Verbal Attitudes and Overt Behavior Involving Racial Prejudice," *Journal of Abnormal and Social Psychology*, 1952, *47*:649–652; and Gerhardt Saenger and E. Golbert, "Customer Reactions to the Integration of Negro Sales Personnel," *International Journal of Opinion and Attitude Research*, 1950, *4*:57–76.

programs designed to decrease prejudice, the Cornell social scientist Robin M. Williams called for further research into interracial contact. Noting that "establishment of the effects of segregation per se will be an extraordinarily difficult task," Williams argued that social scientists should study the effects of segregated and nonsegregated situations in order to answer questions such as, "Where is friction greatest? Where are the areas of high and low intensity and incidence of verbal prejudice? How do stable areas of intermingling compare with shifting areas and 'invasion' points?"[34]

Soon CCI researchers were attempting to answer the sorts of questions posed by Williams. In 1948 CCI research director Stuart Cook wrote that "insofar as successful action against discriminatory practices brings about a decrease in segregation this will mean increased contact between persons from different backgrounds that will, under favorable circumstances, create a reduction in prejudice. Such a reduction in prejudice should, of course, have the consequences of further reduction of discriminatory practices and hostile behavior."[35]

As soon as he arrived at CCI in 1947, Cook was testifying before city commissions that contact between the races served to decrease racial tensions. Before the City Commission of Jersey City Cook stated that "joint occupancy of the same housing community by Negroes and whites has consistently worked out: Initial, mistaken ideas are soon corrected through day-to-day contact and, in many places, members of the different races have come to share identical responsibilities in a completely democratic way."[36]

Building on earlier work on the desegregation of the armed forces, CCI's researchers concentrated on interracial public housing and employment. These situations grew out of the migration of African Americans into New York and the newly desegregated employment and public housing opportunities for them. CCI seized the research opportunities offered by these real-life situations to study interracial contact in a series of what can best be described as field studies.

In one of the first studies of interracial housing, for instance, two CCI staffers conducted interviews with families living in two desegregated and two segregated housing projects. The researchers found that white prejudice was much higher in the segregated projects. They posited that the contact possible in integrated neighborhoods gave individuals the opportunity to realize that their prejudices had no basis in reality. By contrast, in the segregated neighborhoods no opportunity existed for prejudiced individuals to overcome their stereotypes about African Americans. A parallel set of studies explored the effects of interracial workplaces. John Harding and Russell Hogrefe polled the white workers on a newly integrated sales floor. They found that while the basic attitude of the white workers toward their African American co-workers may not have changed significantly, they could nonetheless work peacefully side by side. What CCI and other researchers on interracial contact attempted to discover were the specific conditions under which interracial contact would decrease prejudice. In the early 1950s the Harvard psychologist Gordon Allport summarized what social science had learned about conditions necessary for contact to reduce prejudice: "first, that the contact must be one of equal status; and second, that the members must have objective interests in common." In order to achieve these conditions,

[34] Robin M. Williams, Jr., *The Reduction of Intergroup Tensions* (New York: Social Science Research Council, 1948), p. 91.

[35] Stuart W. Cook, "The Program of CCI," 17 Dec. 1948, Marrow Papers, Box M1938, Folder 15: "1948–1949, CCI Correspondence."

[36] Stuart W. Cook, "Some Psychological and Sociological Considerations Related to Interracial Housing," Nov. 1947, p. 2, AJC Papers, Box 21, Folder: "CCI, 5/20/49."

Allport argued, "artificial segregation should be abolished. Until it is abolished equal status contacts cannot take place. And until they take place cooperative projects of joint concern cannot arise. And until this condition is fulfilled we may not expect widespread resolution of intergroup tensions. Hence, nearly all the investigators agree that the attack on segregation must continue."[37] This would be a key point for the social scientists when they became involved in the litigation campaign: that the abolition of segregation was a *necessary* rather than a *sufficient* step toward bettering race relations. In terms of the "contact hypothesis," for example, social scientists were arguing not that all that was required to reduce prejudice was to eliminate legal segregation but, rather, that *nothing* could be done to reduce prejudice until legal segregation was eliminated. The elimination of segregation was not the end of the journey, but the beginning.

The work on the contact hypothesis, like the work on the separation of attitudes and behavior, was designed to show that social change could be enacted through legislation or court order. What CCI social scientists were arguing was a reversal of William Graham Sumner's famous dictum, "Stateways cannot change folkways." On the contrary, the CCI social scientists insisted, stateways could indeed change folkways. Eliminating discrimination—for example, through a Fair Employment Practices Act or through unsegregated housing projects—would go a long way toward reducing prejudicial attitudes. Pointing to the body of research that CCI had amassed, the sociologist Arnold Rose, Gunnar Myrdal's collaborator on *An American Dilemma*, argued that segregation statutes had always had the effect of increasing prejudice and that the latest research demonstrated that the law could similarly decrease prejudice. Rose declared:

> It has been thus demonstrable for a long time that law and power could create or increase attitudes of prejudice, it should not be surprising from newly available evidence that law and power would also decrease prejudice. Yet the latter conclusion has been contrary to most experts' opinions for a long while. Perhaps the sociologists have misled us with their notions of "mores," "folkways," and the "inevitable" slowness of social change. . . . Now we know that law and authority can reduce prejudice.[38]

THE LEGAL USES OF SOCIAL SCIENCE

While CCI was building the scientific case for legal change, CLSA was taking action against discrimination, drafting model Fair Employment Practices statutes and campaigning heavily for a federal statute. It brought four test cases before the New York State Commission against Discrimination, sued Columbia University Medical School for its "unofficial" quota on Jewish applicants, filed six cases in three states against discrimination

[37] Morton Deutsch and Mary Evans Collins, "Intergroup Relations in Interracial Public Housing: Occupancy Patterns and Racial Attitudes," *Journal of Housing,* 1950, 7:127–129; Deutsch and Collins, *Interracial Housing: A Psychological Evaluation of a Social Experiment* (Minneapolis: Univ. Minnesota Press, 1951); Daniel M. Wilner, R. P. Walkley, and S. W. Cook, "Residential Proximity and Intergroup Relations in Public Housing Projects," *J. Soc. Issues,* 1952, 8:45–69; Wilner, Walkley, and Cook, *Human Relations in Interracial Housing: A Study of the Contact Hypothesis* (Minneapolis: Univ. Minnesota Press, 1955); John Harding and Russell Hogrefe, "Attitudes of White Department Store Employees toward Negro Co-workers," *J. Soc. Issues,* 1952, 8:18–28; and Gordon Allport, *The Resolution of Intergroup Tensions* (New York: National Conference of Christians and Jews, 1952), pp. 22–23.

[38] Arnold Rose, "The Influence of Legislation on Prejudice" (1949), rpt. in *Race Prejudice and Discrimination: Readings in Intergroup Relations in the United States,* ed. Rose (New York: Knopf, 1951), pp. 545–555, on p. 554.

in housing, and fought the granting of a radio operator's license to the *New York Daily News* on the grounds that it was prejudiced against Jews and African Americans.[39]

Social science played an important role in many of these cases. In the cases before the New York State Commission against Discrimination, for example, CLSA established statistical data as proof of discrimination. In the case against the *Daily News,* CLSA established content analysis as a technique with probative value. However, while these cases made use of social science evidence, that evidence came from research conducted by CLSA itself rather than by CCI. For example, in the *Daily News* case, Pekelis and his staff performed the content analysis themselves rather than having CCI undertake the task.[40] There would be one case, however, that would lead to a close relationship between the arguments generated by the CLSA attorneys and research conducted by CCI researchers.

Westminster v. Mendez

A case against segregated education provided the first opportunity for a true collaboration between CCI and CLSA. The case was *Westminster v. Mendez,* a California school segregation case involving Mexican Americans. Segregation of Mexican American schoolchildren had been a common practice in California since the 1920s, when immigration made Mexicans the state's largest minority. Although the practice was not sanctioned by California law, the white or "Anglo" populations of many communities insisted that their local school boards create separate schools for Mexican American children, ostensibly because of language problems.

On 2 March 1945 five Mexican American parents filed a suit in Federal District Court to enjoin the segregation of their children on the grounds that such segregation constituted a violation of Fifth and Fourteenth Amendment equal protection guarantees. The school districts argued that such segregation was not based on race but served sound educational purposes involving bilingual education. Moreover, they argued that control of the schools was a local matter and hence outside the jurisdiction of the federal courts. Finally, they claimed that the separate facilities were in any case equal and therefore constitutional under the "separate but equal" doctrine of *Plessy v. Ferguson,* the 1896 U.S. Supreme Court case that had entrenched segregated facilities in constitutional law.

The Federal District Court's ruling came nearly a year later. Because California had no segregation statutes, the court ruled that there was indeed a violation of the equal protection clauses of the Constitution. The court found that, in the absence of such segregation statutes, the "separate but equal" doctrine did not apply. Finally, it found no sound educational reason for the segregation of Spanish-speaking pupils; in fact, the court argued, assimilation would proceed more rapidly in integrated schools. The school district quickly appealed the case to the Ninth Circuit Court of Appeals.[41] During this appeal, the case came to the attention of the American Jewish Congress.

The *Westminster* case, though not originated by CLSA, provided a unique opportunity. The case was "low risk"—that is, very few AJC resources were expended in the litigation

---

[39] Petegorsky, "Report of the Executive Director" (cit. n. 12).

[40] "Content Analysis—A New Evidentiary Technique" (note), *University of Chicago Law Review,* 1948, 15:910–924; and *Memorandum in the Nature of Proposed Findings Submitted at the Direction of the Federal Communications Commission by the American Jewish Congress,* AJC Papers, Box 189, Folder: "Brief, Nov. 12, 1946, Federal Communications Commission."

[41] This account of the case is taken from Charles Wollenberg, *All Deliberate Speed: Segregation and Exclusion in California Schools, 1855–1975* (Berkeley: Univ. California Press, 1976), pp. 110–121.

process. An *amicus* brief provided an opportunity for a direct attack on the "separate but equal" doctrine—one of the first chances for CLSA to present its views on racial segregation and discrimination.

Although not a member of the bar, Alexander Pekelis wrote the CLSA brief for the *Westminster* case. In his introduction he noted that CLSA firmly agreed with the central point that the NAACP had made in its own *amicus* brief: "If facilities were really duplicated, financial ruin of the local bodies of the states would ensue. If financial disaster is to be avoided, the facilities granted to minorities are bound to be physically inferior." However, so as not to duplicate the argument made by the NAACP, Pekelis predicated his arguments on the assumption that facilities were "identical" rather than attempting to show that those provided to the minority students were inferior. To prove that even identical facilities were inherently unequal, he relied heavily on sociological and psychological data.[42]

As all attorneys arguing against segregation were obliged to do, Pekelis had to deal with the *Plessy* precedent. Pekelis's strategy was to accept the legal doctrine, propounded in *Plessy,* that separate but equal facilities were, in fact, constitutional. However, he analyzed the findings of the 1896 case as firmly anchored in "factual" rather than "legal" grounds. He noted that *Plessy* found segregated railroad cars constitutional because "it proceeded on the factual and sociological assumption that such segregation did '*not necessarily imply the inferiority of either race to the other.*' " Pekelis then argued that the legal "fiction" of *Plessy*—that segregation does not connote the inferiority of one race to another—was contradicted by nearly all social scientific knowledge available. Pekelis asked starkly: "Will any court today, in the light of the sociological and psychological findings made in the last fifty years, prove so lacking in candor and so blind to realities as to subscribe to the fiction of benevolent segregation on which *Plessy v. Ferguson* relies? That is the issue. Not the legal doctrine of *Plessy v. Ferguson* is in question but the factual fallacy on which it rests." To prove his point that *Plessy* was grounded in a "factual fallacy," Pekelis argued that

> whenever a group, considered "inferior" by the prevailing standards of a community, is segregated by official action from the socially dominant group, the very fact of official segregation, whether or not "equal" physical facilities are being furnished to both groups, is a humiliating and discriminatory denial of equality to the group considered "inferior" and a violation of the Constitution of the United States and of treaties duly entered into under its authority.

Pekelis first noted that all parties agreed that separate facilities had to be equal if they were to pass the constitutional test of *Plessy.* Second, he maintained that "equality" was defined not by "mere identity of physical facilities" but by "identity of substantial similarity of their *values.*" Pekelis then defined "values" by their "social significance and psychological context, or in short, on the community judgment attached to them." For example, Pekelis maintained that a probate court would not hold two physically identical houses equal if one were in a slum and the other in a fashionable neighborhood.[43]

Having established that the law recognized the reality of what he called "social inequality," Pekelis then showed how the segregation of a group previously deemed socially

---

[42] *Westminster School District v. Mendez,* Brief for the American Jewish Congress, *amicus curiae,* 1946, p. 3, Case Files, Ninth Circuit Court of Appeals, Record Group 276, Box 4464, Folder 11310, National Archives—Pacific Sierra Region, San Bruno, California (hereafter cited as **CLSA Brief**).

[43] CLSA Brief, pp. 21, 22, 4, 5. See also *Plessy v. Ferguson,* 163 U.S. 537 (1896).

inferior in and of itself constituted a legal inequality. The act of segregation, he insisted, was tantamount to an official declaration of the inferiority of the segregated group. Once the "legal inferiority" of segregation is adopted, it reinforces and intensifies the "social inequality," leading to a vicious circle of discrimination and prejudice.

Such problems were especially acute in segregated education, according to Pekelis: "The official imposition of a segregated pattern based on notions of inferiority and superiority produces its deepest and most lasting social and psychological evil results when applied to children." He argued:

> Since segregation reinforces group isolation and social distance it helps to create conditions in which unhealthy racial attitudes may flourish. By giving official sanction to group separation based upon the assumption of inferiority it helps to perpetuate racial prejudice and contributes to the degradation and humiliation of the minority child. The crippling psychological effects of such segregation are in essence a denial of equality of treatment. In this sense segregation is burdensome and oppressive and comes within the constitutional prohibition.[44]

Pekelis argued that the internal psychology of schoolchildren should be a constitutional issue, regardless of the effect of segregation on the larger society. This is the "damage" argument that would underpin the *Brown* decision eight years later. Pekelis supported the argument with a wide variety of social science materials on segregation and discrimination.

On 14 April 1947 the Ninth Circuit Court of Appeals ruled unanimously that the segregation of Mexican American schoolchildren in California schools violated the Fourteenth Amendment. The basis of the opinion was not the far-reaching legal case made by CLSA but a narrow legal ground: segregation was unconstitutional because California law had no provisions for the segregation of Mexican American schoolchildren. Governor Earl Warren made any further litigation moot when, on 14 June 1947, he signed a repeal of California's education segregation statutes that ended *de jure* segregation in the state.[45]

This case showed that the existing social science literature was not adequate to support Pekelis's argument. Given the dearth of psychological studies specifically on segregated education, Pekelis had relied on studies of general childhood development and general studies of segregation. Except for an article by Howard Hale Long, these studies did not directly address segregated education.[46] Neither did the social science literature isolate segregation as required by law as a separate variable from segregation that arose by custom. Finally, there was no social science that could distinguish the psychological effects of segregation in general from the psychological effects of being segregated to inferior facilities. CCI researchers could have undertaken research projects to fill these holes in the social science literature. Such projects, however, would have been tremendously difficult to design, and CLSA might have had to wait months or years for the results to be reported. In order to make a more authoritative pronouncement, CLSA turned to CCI's social scientists, who found a way to enroll scientific authority without undertaking additional studies.

Survey of Social Science Opinion

Isidor Chein, after consultation with CLSA, decided to poll a wide range of social scientists on the issue of the psychological damage of segregation.[47] The results of this survey,

---

[44] CLSA Brief, pp. 14–15.

[45] Wollenberg, *All Deliberate Speed* (cit. n. 41), p. 132.

[46] The article was Howard Hale Long, "Some Psychogenic Hazards of Segregated Education of Negroes," *Journal of Negro Education*, 1935, 4:336–350.

[47] Tracy S. Kendler, "Contributions of the Psychologist to Constitutional Law," *Amer. Psychol.*, 1950, 10:505–510.

designed explicitly for use in legal proceedings, would become one of the most cited articles of social science in subsequent briefs by CLSA and by the NAACP in cases against segregation.

The survey was mailed to 849 social scientists in May 1947. Respondents were asked if they agreed that "enforced segregation has detrimental psychological effects on members of racial and religious groups which are segregated, even if equal facilities were provided." A parallel series of questions asked about psychological effects on "groups which enforce the segregation." The respondents were also asked to state the basis of their opinions; they could choose four options: "My own research findings. Research findings of other social scientists. My own professional experience. Professional experiences of other social scientists which have been made available to me." The cover letter was explicit as to the motivations for the survey: "For the purpose of providing legislative bodies, courts and the general public with a consensus of responsible scientific opinion, we are asking social scientists to indicate their position in this issue."[48]

Chein presented the results at the 1948 Eastern Psychological Association meeting. He reported that 90 percent of the respondents believed segregation psychologically damaging to the segregated group and 83 percent believed it also damaged the enforcing group. All but 10 percent of the respondents checked one of the four alternatives offered as the basis for their opinion. "All in all, then," Chein concluded, "we may not only say that there is widespread agreement among social scientists that enforced segregation is psychologically detrimental despite equal facilities, but we may add that the majority of these social scientists believe that there is a factual basis for such agreement." Chein noted that there were problems with the existing research regarding the questions he posed in his survey. "Since equal facilities are in fact not provided," he explained, "the proposition that enforced segregation does have detrimental psychological effects, even under conditions of equal facilities, seems impossible to prove." Indeed, Chein found "virtually nothing in the published literature that is explicitly devoted to this problem." That did not mean that his respondents had no basis for their opinions, however, for there was a large literature on segregation "from which one may cull a great deal of information pertinent to [the problem]."[49] Chein called for social scientists to turn their attention to research designed to prove psychological damage even given equal facilities.

Chein's survey was a curious piece of social scientific research. As a document for a legal trial, however, it would be tremendously useful. Attorneys fighting segregation could now make part of the legal record the fact that the vast majority of social scientists found segregation to be psychologically damaging. That there was no study that specifically isolated *de jure* segregated education with equal facilities would not be immediately relevant in a court of law. An expert witness is often called to provide an opinion; the reasons for the opinion are seldom queried. Hence the opinions of social scientists, as discovered

[48] The cover letter and the survey are reprinted in Max Deutscher and Isidor Chein, "The Psychological Effects of Enforced Segregation: A Survey of Social Science Opinion," *Journal of Psychology,* 1948, 26:286–287. Deutscher was a statistician who helped Chein crunch the numbers in interpreting his survey.

[49] Isidor Chein, "What Are the Psychological Effects of Segregation under Conditions of Equal Facilities?" paper read at the Nineteenth Annual Meeting of the Eastern Psychological Association, 16–17 Apr. 1948, Kenneth B. Clark Papers, Box 22, Folder: "American Jewish Congress Reports, 1946–49," Manuscript Division, Library of Congress, Washington, D.C. (hereafter cited as **Clark Papers**). A version of this paper, which includes more commentary than the essay cited in note 48, above, was published: Chein, "What Are the Psychological Effects of Segregation under Conditions of Equal Facilities?" *Int. J. Opinion Attitude Res.,* 1949, 3:229–234; the quotations are from pp. 232, 259.

by Chein's survey, would be relevant. How they came to those opinions was simply not that important for the lawyers.[50]

The results of Chein's survey were published in May 1948. Chein, with his colleague Max Deutscher, explained that the survey was inspired by the recent Ninth Circuit Court of Appeals decision in *Westminster*. "According to the Court," they announced, "the basic evidence for this decision consisted of studies in race relations made by anthropologists, psychologists, and sociologists, demonstrating that legal segregation does in reality imply the inferiority of one 'race' to the other and implements such status." The characterization of the Ninth Circuit decision as primarily based on social science material was, in fact, completely inaccurate. As noted earlier, *Westminster* was decided on a relatively narrow point of law. It did not address either the larger question of the constitutionality of separate but equal facilities or the sociological and psychological questions raised in CLSA's *amicus* brief. Nevertheless, Deutscher and Chein announced that the present study was designed for use in the legal arena. They continued: "Final decision on the legality of enforced segregation, regardless of equal facilities, has not yet been rendered by the Supreme Court. Here too, social science may be a significant if not crucial factor. For social scientists interested in 'social engineering' this represents a concrete opportunity to apply the relevant findings of social science data."[51]

More than any other single work, Chein's survey represented a confluence of interests between social scientists and attorneys. For CLSA attorneys, the survey represented a terse, nontechnical presentation of social science opinion that could add authority to legal briefs. For the CCI social scientists, it represented an opportunity for research to be meaningfully translated into legal action. The social science survey was particularly well timed for use in the battle against legalized segregation. By 1948, CCI social scientists were beginning to attract the attention of the NAACP, just as it was entering a new phase of litigation against segregated education. To understand the NAACP's use for social science, it is necessary to understand the shape of its campaign against segregated education.

THE NAACP AND THE USE OF SOCIAL SCIENCE

The roots of the NAACP campaign against segregated education can be traced to 1930, when the association hired a young lawyer, Nathan Margold, to draft a legal strategy that could be used to secure adequate educational facilities for African American children. Margold counseled against a futile campaign of litigation aimed at the equalization of facilities in the "separate but equal" world of the South. Such a campaign would entail a separate lawsuit for each of the thousands of southern school districts to prove that facilities in each of them were unequal, quickly depleting the NAACP's meager resources. Margold argued, however, that if the NAACP "boldly challenge[d] the constitutional validity of segregation if and when accompanied irremediably by discrimination," it could eliminate segregated schooling in one stroke and thus guarantee African American children adequate educational facilities.[52]

---

[50] See, e.g., "Rogers on Expert Testimony," NAACP Papers, Series IIB, Box 138, Folder: "Schools, Kansas, Topeka, Brown v. Board of Education, Expert Witnesses, 1951," Manuscript Division, Library of Congress (hereafter cited as **NAACP Papers**).

[51] Deutscher and Chein, "Psychological Effects of Enforced Segregation" (cit. n. 48), p. 259.

[52] Mark V. Tushnet, *The NAACP's Legal Strategy against Segregated Education, 1925–1950* (Chapel Hill: Univ. North Carolina Press, 1987), p. 27.

The Margold report was the blueprint for the NAACP's challenge to segregated education, providing a simple but powerful strategy for the elimination of Jim Crow schools. Margold's legal strategy was well timed. It arrived just as the NAACP's legal staff was undergoing significant changes, the most important of which was the arrival of Charles Hamilton Houston, dean of the Howard University School of Law. Houston would be responsible for the planning and execution of the litigation envisioned by Margold.

Houston was a 1922 graduate of the Harvard University School of Law, where he was an exceptional student who earned a doctorate in juridical science rather than the more conventional J.D. He came to the attention of two of the most famous and demanding members of Harvard's faculty, Felix Frankfurter and Roscoe Pound. Frankfurter, one of Harvard's legal realist professors, became Houston's doctoral advisor. Frankfurter taught Houston that the law was a tool of social engineering and that an attorney needed an understanding of the law's social setting in order to be successful. Pound, one of the founders of the earlier school of "sociological jurisprudence," believed that the law must include an understanding of the social sciences in order to operate effectively. As dean of Howard Law School, Houston brought the realist emphasis on the use of social science to an entire generation of African American attorneys who would lead the fight for civil rights.[53] The NAACP's later reliance on social science during the school segregation cases might have been expected, given that much of the legal staff received its training at Howard, an important center of legal realist thinking and its concomitant reliance on "real world" factual data.

In 1933 Houston began the NAACP litigation campaign against segregation. The NAACP focused on suits to eliminate segregated graduate and professional schools, because Houston felt that these were most vulnerable to a direct attack on segregation.[54] Segregated states seldom made a pretense of offering equal opportunities for African Americans seeking a graduate education. Seventeen of the nineteen states that required segregated education had no graduate schools whatsoever for African Americans. Only three states—West Virginia, Missouri, and Maryland—offered out-of-state scholarships that would pay the tuition for African American students who wished to pursue graduate education. Houston felt that this near-absolute absence of opportunities for graduate education would enable him to demonstrate easily the inequality of segregation.

Five years after the Margold report, Houston was joined in his effort by a former pupil, Thurgood Marshall. Marshall was born in Baltimore, which he described as the "most segregated city in the United States," and raised in the heart of its African American middle-class community. He went to Lincoln University in Pennsylvania in 1925, and then to the newly-accredited Howard Law School. Marshall excelled under the strict hand of Houston at Howard and, after graduation, moved into what turned out to be a brief private practice. In 1936 he joined the NAACP staff, with an initial commitment from the organization to pay his salary for six months. Marshall quickly moved up in the hierarchy, however. When Houston left the NAACP in 1938 to return to private practice Marshall took charge of the association's legal activities.[55]

Under Marshall's direction, the NAACP won a major victory in 1938 when the Supreme Court decided that the state of Missouri had a constitutional obligation to provide a legal

[53] Genna Rae McNeil, *Groundwork: Charles Hamilton Houston and the Struggle for Civil Rights* (Philadelphia: Univ. Pennsylvania Press, 1983), pp. 49–56, 76–85; and Tushnet, *NAACP's Legal Strategy*, p. 118.

[54] Tushnet, *NAACP's Legal Strategy*, p. 34; and McNeil, *Groundwork*, pp. 116–117.

[55] Mark V. Tushnet, *Making Civil Rights Law: Thurgood Marshall and the Supreme Court, 1936–1961* (New York: Oxford Univ. Press, 1994), pp. 28–29.

education to Lloyd Gaines, a young African American represented by Marshall. However, the Court left open the question whether Missouri could have fulfilled its obligation if it had a separate law school for African Americans.[56]

World War II interrupted the litigation campaign, and further suits were suspended "for the duration." Thus, as the war drew to a close in 1945, the NAACP had one Supreme Court victory to its credit: *Gaines.* That was a limited victory, however, since the Court left open the possibility that separate graduate education programs for African Americans would be constitutionally permissible. Southern states quickly went about setting up such programs. After the war, the task of the NAACP would be to prove that these hastily assembled programs could not possibly be equal to their long-established counterparts for white students, a task that the organization assumed would be relatively simple. In fact, proving any two schools unequal was extraordinarily difficult.[57] One reason the NAACP turned to social science was to prove that separate systems of education were necessarily unequal and, hence, discriminatory.

In a 1947 case to desegregate the University of Texas Law School, *Sweatt v. Painter*, the NAACP sought social science data and social scientists who could serve as expert witnesses. The NAACP turned to CLSA for the names of social scientists who could testify at the *Sweatt* trial. Will Maslow sent a copy of the social science survey to Thurgood Marshall, even though it had not yet been published, and asked if "it can be used in any form in the University of Texas suit." The survey would be of some use, but what the NAACP really needed was a prominent social scientist willing to testify to the evils of segregation. If possible the social scientist should be at a southern institution, minimizing the appearance of northern hostility toward the South. Unfortunately, Will Maslow had to inform Marshall, "CCI believes there are no nationally known psychologists in the South."[58]

The NAACP did have the University of Chicago anthropologist Robert Redfield testify in Texas. Redfield was a rare resource for the NAACP. The son of a prominent Chicago attorney, he had received his J.D. from the University of Chicago Law School in 1921. After a brief practice, Redfield found himself dissatisfied with the law and returned to the university, receiving his Ph.D. in anthropology in 1928. His conversion from the law to social science may have been encouraged by his wife, Margaret Lucy Park, the daughter of the founder of the Chicago school of sociology, Robert Park. In any event, Redfield was much happier as an anthropologist than as a lawyer and published several influential books on Latin American folk culture.[59]

If his almost unique credentials of degrees in both the law and anthropology were not enough to bring Redfield to the attention of the NAACP, then his concern for racial justice would have been. During World War II Redfield proclaimed that African Americans were the victims of a society that treated them as "half-citizen[s]." The problems of African Americans, he declared, were in the "mythology of the modern man" that proclaimed the natural inferiority of the African American race. Redfield argued that this was nonsense and that "a child of one skin color starts even with a child of any other skin color, if you let him. We don't let him, and we entertain a false biology which seems to justify us. I say again that race is of consequence because of what men think and feel about it and not

[56] *Missouri ex re. Gaines v. Canada*, 305 U.S. 337 (1938).

[57] Tushnet, *NAACP's Legal Strategy* (cit. n. 52), pp. 87–88.

[58] Maslow to Thurgood Marshall, 14 Apr. 1947, 28 Apr. 1947, NAACP Papers, Series IIB, Box 204, Folder: "University of Texas, Sweatt v. Painter, Correspondence, Jan.–June 1948."

[59] George W. Stocking, Jr., "Robert Redfield," in *Dictionary of American Biography,* Suppl. 6: *1956–1960,* ed. John A. Garraty (New York: Scribner's, 1980), pp. 532–534.

because of anything that race is of itself." Between 1947 and 1950 Redfield served as the director of the American Council on Race Relations (ACRR), an organization founded in 1944 to "bring about full democracy in race relations through the advancement of the knowledge concerning race relations." The NAACP had been working extensively with two ACRR sociologists, Robert Weaver and Louis Wirth, to amass sociological data on the effects of racially restrictive housing covenants. It is undoubtedly through this work that Redfield came to the attention of the NAACP. While Redfield had no formal connection to CCI, he was aware of the research on integrated sales counters and interracial housing.[60]

Redfield's testimony demonstrates how the social scientific ideas that were at the heart of CCI's research program could be used in a court of law to argue against segregation. Many of Redfield's arguments were precisely those that CCI's researchers were attempting to make in the social science literature. Redfield testified that segregation was inimical to "public security and general welfare." Drawing on his experiences at the desegregated University of Chicago, he argued that "segregation policy, and the stigma which segregation attaches to the segregated increases prejudice, mutual suspicion between Negroes and Whites and contributes to the divisiveness and disorder of the national community, contributing to crime and violence." Redfield's third and final point was that desegregation could be expected to proceed smoothly, that the "abolition of segregation in education, is likely to be accomplished with beneficial results to public order and the general welfare."[61]

When cross-examining Redfield, attorney general Price Daniel turned to an issue that would prove to be one of the dominant themes in litigation concerning educational segregation: whether desegregation could be "imposed" on a community that did not desire it. In other words: Could legal change precede attitude change in race relations? Daniel pressed Redfield to admit that it was "impossible to force the abolition of segregation upon a community that has had it for a long number of years." Redfield refused to admit the point, arguing that "segregation in itself is a matter of law, and that law can be changed at once." Daniel refused to give up and questioned Redfield about the speed of attitude change within the community—wasn't it true that such change could not be forced on a community? Redfield then admitted that, depending on circumstances, the attitudes of the community could resist desegregation; nonetheless, he insisted, "in every community there is some segregation that can be changed at once, and the area of higher education is the most favorable for making the change."[62]

The arguments that Redfield made in Texas underscore the importance of the research undertaken by CCI. The idea that segregation increases racial tension and that integration would decrease it was at the heart of CCI's research into the contact hypothesis. Moreover, the empirical studies that CCI undertook would add to the credibility of the witnesses in

[60] Robert Redfield, "Race and Human Nature: An Anthropologist's View" (1944), rpt. in *The Social Uses of Social Science: The Papers of Robert Redfield,* Vol. 2, ed. Margaret Park Redfield (Chicago: Univ. Chicago Press, 1963), pp. 137–145, on p. 142; and Louis Wirth, memorandum, 1 May 1947, Louis Wirth Papers, University of Chicago Archives, Regenstein Library, Chicago, Illinois. On the NAACP's work with ACRR see Clement E. Vose, *Caucasians Only: The Supreme Court, the NAACP, and the Restrictive Covenant Cases* (Berkeley: Univ. California Press, 1959), pp. 159–163. Redfield's pretrial notes for the *Briggs* trial a few years later demonstrate his familiarity with CCI research: Robert Redfield Papers, Box 23, Folder 4, University of Chicago Archives (hereafter cited as **Redfield Papers**).

[61] Pretrial notes, Redfield Papers, Box 33, Folder 7.

[62] Redfield's testimony in *Sweatt* was part of the record of *Briggs v. Elliot,* 98 F. Supp. 529 (1951). All references are to Redfield's testimony as printed in the *Briggs* transcript. See Redfield testimony, pp. 166–167, copy in Tom C. Clark Papers, Tarleton Law Library, University of Texas, Austin, Texas.

Brown who would argue, as Redfield had in the Texas case, that desegregation could be imposed on an unwilling populace.[63]

Just a few months after Redfield's testimony in Texas, the NAACP persuaded CCI director Stuart Cook to present the results of the Chein survey in an Oklahoma lawsuit against segregated graduate education. Unfortunately, because of the way the legal issues were framed Cook's testimony was disallowed by the trial judge. Nonetheless, Cook reported to American Jewish Congress executive director David Petegorsky that the NAACP was "very enthusiastic about the potential effect of our survey . . . on the psychological effects of segregation."[64]

The Texas and Oklahoma trials demonstrate that the NAACP was beginning to show considerable interest in the legal use of social science data. A good summary of the NAACP's position on social science was provided by Annette H. Peyser in 1948. Peyser was a young staffer with a sociology degree from New York University who had come to work for the NAACP in 1945 when the political scientist Harold Lasswell recommended her as a propaganda analyst. Soon after arriving she began to work with the legal staff to assemble sociological materials relating to the NAACP campaign against restrictive covenants in housing. The lone social scientist on the NAACP staff, Peyser began a "public relations attempt at compensating for the fact that the NAACP does not have a counterpart to the Commission on Law and Social Action, such as CCI." The announced purpose of her paper, which was presented at a legal conference, was an effort to "effect a better relationship between the legal expert and the social scientist." Peyser recounted the use of social scientific data and expert witnesses in *Westminster*, the graduate school cases, and other cases. She noted that the NAACP had emphasized sociological data to prove that segregation leads to inequalities in facilities because there simply were no studies regarding psychological damage. The NAACP had, "in the absence of specific scientific studies relating to the psychological effects of segregation, emphasized the factual aspects of segregation." Peyser looked to the AJC to bridge this gap. While the NAACP had neither the funds nor the personnel to undertake original research, the same was not true of the AJC, with its two divisions of social scientists and lawyers: "The CCI works independently of but in cooperation with the Commission on Law and Social Action of the American Jewish Congress. It may be that AJC, because of its physical structure, will be able to perform some of the necessary research on the psychological effects of segregation." Peyser also noted that CCI had already been an invaluable source of social scientific materials for the NAACP. In an obvious reference to the social science survey, she claimed that the "American Jewish Congress has been instrumental in reprinting articles that have appeared in scientific journals as well as preparing and writing these articles for the purpose of eventually having them 'planted' in such journals." Peyser closed her address with a call for further cooperation between social scientists and lawyers.[65]

[63] See, e.g., the trial testimony of M. Brewster Smith in *Davis et al. v. County School Board of Prince Edward County,* 1952, Case File 1333, Box 126, Vol. 2, pp. 292–293, Civil Case Files, 1938–1958, Richmond Division, Records of the U.S. District Court to the Eastern District of Virginia, Record Group 21, National Archives— Mid Atlantic Region, Philadelphia, Pennsylvania.

[64] Cook to David Petegorsky, 24 May 1948, AJC Papers, Box 19, Folder: "CCI, 1948–49."

[65] Annette H. Peyser to M. Brewster Smith, 25 Mar. 1952, M. Brewster Smith Papers, Box M605, Folder: "NAACP," Archive for the History of American Psychology (hereafter cited as **Smith Papers**); and Annette H. Peyser, "The Use of Sociological Data to Indicate the Unconstitutionality of Racial Segregation," Clark Papers, Box 63, Folder: "Background Reports, Undated." This document is undated, but internal evidence suggests that it is the 1948 address. On Peyser's work more generally see Jack Greenberg, *Crusaders in the Courts* (New York: Basic, 1994), p. 35; and Tushnet, *Making Civil Rights Law* (cit. n. 55), p. 89.

The NAACP's need for social scientists would intensify after 1950, when the Supreme Court ruled that segregated graduate education was unconstitutional. In both the Texas and Oklahoma graduate school cases, the Court based its decision on ineffable and intangible factors, benefits denied African Americans by segregation. In the case of Heman Sweatt, such an intangible was a thing like the opportunity to attend a law school with the reputation of the University of Texas. In the case of George McLaurin in Oklahoma, it was the opportunity to exchange ideas with his fellow graduate students. While the Supreme Court stopped short of directly overturning the "separate but equal" doctrine, the door was now opened wide to do so.[66] Moreover, the way through that door could be through pointing out "intangible factors." When they turned their attention to elementary and secondary education in the cases that eventually were decided in *Brown,* the attorneys of the NAACP would attempt to use social science to help them articulate some of those factors. And social scientists were waiting for them, armed with studies identifying lines of evidence that supported the notion that segregation was psychologically damaging. The NAACP could not merely turn to CCI to find social scientists, however, for on the eve of the campaign against elementary and public school segregation the social scientists at the AJC were dispirited and in disarray.

CONCEDING DEFEAT: DISSOLVING THE SCIENCE/LAW COLLABORATION

Civil rights attorneys and socially minded social scientists were dedicated to the same end: the elimination of prejudice and discrimination. Their styles and the methods they employed were fundamentally different, however. The law worked on a strict timetable, and lawyers had to adhere to that timetable. To be persuasive to judges and juries, a good legal argument had to be forceful and unambiguous, leaving no room for other interpretations. On the other hand, social science proceeded at a much more leisurely pace, since professional publications usually enforced no strict deadlines. To be persuasive to social scientific peers, a good social scientific argument had to be provisional and filled with careful caveats. The possibility of other interpretations would often be left open. The differences in the processes by which each community produced its arguments against discrimination eventually doomed the institutionalized merger of law and social science at the AJC.

Isidor Chein's survey of social science opinion was by far the most successful example of social science research designed to be used as a legal argument. More typically, CCI's social scientific research did not fit in with the AJC's larger campaign against discrimination. Despite the fact that the CCI social scientists quite deliberately aimed their research priorities in a manner that would be useful to CLSA and other activists at the AJC, social science never made a significant contribution to the AJC's struggle against discrimination.

A tension always existed between CCI and the rest of the American Jewish Congress. On the one hand, CCI was to be an objective, scientific research agency. On the other hand, it was funded by and worked within the American Jewish Congress—an avowedly political and activist organization. This tension was recognized by Stuart Cook, who told a colleague that "first, CCI must be an active functioning participant in the general [American Jewish] Congress program. Second, within the framework of [the American Jewish] Congress—which to the outsider's eye is a partisan, political organization—it must be a scientific research group holding the complete confidence of non-Congress organizations

---

[66] *McLaurin v. Oklahoma State Regents for Higher Education,* 339 U.S. 637 (1950); *Sweatt v. Painter,* 339 U.S. 629 (1950); and Tushnet, *NAACP's Legal Strategy* (cit. n. 52), pp. 130–132.

and individuals. As you know, we have waged an up-hill fight for this dual objective." Despite the optimistic tone of much that was written by CCI social scientists, the organization never fit comfortably within the AJC framework. The concerns raised by Ronald Lippitt in his 1945 memorandum came to haunt CCI in the course of the next five years: the production of social scientific materials was too slow to meet the needs of the other branches—the real world, it was argued, moved too fast for the social scientists of CCI. The result was that in 1948, soon after Cook assumed control, he faced a series of budget and staff cuts that eventually decimated CCI. Cook tried desperately to hold the organization together, but by 1950 he had had enough and accepted an offer from New York University. In his resignation letter he wrote that he was unhappy with the AJC leadership. "What made me accept [the NYU offer] was that no meaningful assurances were really possible; that [CCI's budget of] $240,000 had not really shrunk to a stable $78,000 but rather that no one could really know where the end of the path was to be," he explained.[67]

With Cook's resignation, Isidor Chein became CCI director. He faced further staff reductions and budget cuts. By 1952 the situation had come to a "crisis," and Marie Jahoda, a distinguished social psychologist, was called in to take stock. Jahoda surveyed both CCI staffers and American Jewish Congress leaders in an attempt to discover the source of the friction. CCI staff members argued that the AJC neither understood nor respected what they were attempting to accomplish. One, John Harding, complained that the AJC's position on any issue was decided "ideologically." That is to say, the AJC had its agenda; and if social science happened to agree with that agenda, well and good. But social science could never actually guide the AJC's position. "Empirical research *is* seen by Congress leadership as serving a useful function in providing evidence from time to time of the correctness of the Congress stand on various issues," Harding wrote to Jahoda. "However this use of research is a dispensable luxury, since the correctness of the Congress position is always clearly evident . . . before the research is done."[68]

Chein took the opportunity provided by Jahoda's assessment to announce that the situation at CCI was "inherently unstable and similar crises, with all the incident demoralization, must inevitably recur. In other words, regardless of the outcome, I shall be looking for another position."[69] Soon Chein would join Cook at NYU.

Jahoda's survey revealed that, while CCI was roundly criticized by the other branches of the American Jewish Congress, there was no clear consensus as to what exactly was wrong. In her final report, she noted that the criticisms leveled at CCI by the other branches often were contradictory: "CCI's scientific standards were said by some to be too high, by others too low. On the one hand, CCI was presented as being too perfectionist and too much concerned with meeting scientific requirements when the needs of Congress might have been satisfied with a less thorough job. On the other hand, jobs whose usefulness were recognized . . . were criticized as not really representing a scientific contribution." All these criticisms could be true, she argued, because *"there exists . . . no generally recognized and defined standard against which the functioning of CCI could be measured and judged adequate or inadequate. CCI's function as a social science department within Congress is undefined."*[70]

---

[67] Cook to Alfred Marrow, 5 June 1947, Marrow Papers, Box M1938, Folder 14: "1946–1947 CCI Correspondence"; and Cook to Marrow, 11 Aug. 1950, AJC Papers, Box 70, Folder: "Dr. Stuart Cook."

[68] John Harding to Marie Jahoda, 9 Apr. 1952, Marrow Papers, Box M1938, Folder 17: "1952–1953 CCI Correspondence."

[69] Chein to Jahoda, 9 Apr. 1952, Marrow Papers, Box M1938, Folder 17: "1952–1953 CCI Correspondence."

[70] Marie Jahoda, "The Commission on Community Interrelations of the American Jewish Congress," 1952, pp. 14, 15, Marrow Papers, Box M1938, Folder 21: "CCI Papers."

The fundamental problem confronting CCI, Jahoda claimed, was that "science proceeds at a lower speed, with less flexibility in tackling new problems and with different standards of success than [other] organization activities." Social scientists had little to contribute to an attack on discrimination until they had completed a study on the specific problem in question. Until then, all a social scientist could report was that "the analysis of the data continues." Jahoda noted that such a report could be regarded as "unsatisfactory and even annoying," but she also noted: "A more exact content report . . . might have been even more annoying: it might well have read thus: 'a code was developed and applied. Reliability checks proved that it was inadequate. The code was revised with better reliability checks.' Many of the processes which enter into research are . . . boring and unrevealing."[71]

Jahoda recommended that the American Jewish Congress continue to fund CCI but define more clearly what it expected from social scientific work. CCI did continue, as a shadow of its former self—it conducted no more serious research into intergroup relations. Many of CCI's social scientists followed Cook and Chein to NYU, where they continued their research in a more academic setting.

CONCLUSION

The tension between CCI and the larger AJC demonstrated the difficulties of combining the disparate cultures of social science and legal action. Social scientists were definitely interested in being part of the activist AJC, and the AJC was interested in receiving whatever aid the social scientists could offer. But even given the socially conscious nature of the social science work done at CCI, the merging of the two groups worked only poorly. CCI's social scientists realized that they had to conduct careful, scholarly research in order to be useful, but the activism of the AJC could not endure the natural pace of scientific research. Nor was the AJC happy with the provisional nature of social scientific research— the uncertain conclusions, the meticulous caveats, and the calls for further study that are the hallmarks of careful scientific work. All of these characteristics limited the usefulness of social science for the larger AJC. The processes by which social science argued against discrimination and prejudice were incompatible with the processes by which the law attacked the same foes. Hence the AJC's attempt to combine social science research and legal action can be viewed as a failure.

And yet, in another sense, the merger was a triumph, for the work done by CCI was appreciated from afar. The NAACP greatly appreciated the social scientific research generated by the American Jewish Congress. CCI had, for all intents and purposes, come to an end by 1952, just as the NAACP was beginning to litigate against segregated education in public elementary and secondary schools. Yet CCI left behind a body of research that would prove useful. Its research on the separation of attitudes and behavior and on the contact hypothesis would emerge time and again in the *Brown* litigation. Moreover, the social scientists who had worked at CCI, most notably Cook, Chein, and Saenger, would become key players in the later litigation campaign. Chein's survey of social science opinion would be cited twice by Earl Warren in the *Brown* opinion.[72] Despite Kenneth Clark's protestations to the contrary, the social scientific community had indeed prepared for the legal battle of *Brown* for a decade.

---

[71] *Ibid.,* pp. 18, 20.

[72] *Brown v. Board of Education of Topeka,* 347 U.S. 483 (1954). See also John P. Jackson, Jr., "Creating a Consensus: Psychologists, the Supreme Court, and School Desegregation, 1952–1955," *J. Soc. Issues,* 1998, *54*:143–177.

We are left, then, with what appears to be a paradox: CCI, the very organization that failed to integrate law and science under the auspices of the American Jewish Congress, was the organization responsible for the research that played a central role in the dismantling of legal segregation. In other words, the products of CCI's research proved valuable in the legal battles against segregation, though the processes by which those studies were produced were only an annoyance to the lawyers of the CLSA. Maurice Rosenberg wrote that "in general any meeting [between the legal and scientific styles of thought] that occurs will take place after the scientists have gathered their data. . . . The methods are far apart until the scientist has made his findings."[73]

The studies produced by CCI were very useful in the courtroom; however, lawyers did not need—and did not really desire—to be involved in their creation. In fact, the attempt at close collaboration between social scientists and lawyers produced frustration for both groups. The dream of Kurt Lewin and Alexander Pekelis that lawyers and social scientists would work side by side to battle prejudice and discrimination did not come to fruition. But the attempt led to the dismantling of legalized segregation and forever changed the landscape of the United States.

[73] Maurice Rosenberg, "Comments," *Journal of Legal Education,* 1970, 23:199–204, on p. 204.

Visions of a Cure

Visualization, Clinical Trials, and Controversies in Cardiac Therapeutics, 1968–1998

By David S. Jones

ABSTRACT

In the early 1970s physicians engaged in fierce debates over the most appropriate method of evaluating the efficacy of coronary artery bypass grafting (CABG). With millions of patients and billions of dollars at stake, CABG sparked fierce controversy. Skeptics demanded that randomized controlled trials (RCTs) be performed, while enthusiasts argued that they already had visual proof of CABG's efficacy. When RCTs appeared, they did not settle the controversy. Participants simply reasserted their preconceptions, defending a trial's strengths or exploiting its flaws. The debate centered on standards of knowledge for the evaluation of therapeutic efficacy. Specifically, cardiologists and cardiac surgeons struggled to assess the relevance of different measures of therapeutic success: physiological or clinical, visual or statistical. Many factors contributed to participants' decisions, including disciplinary affiliation, traditions of research, personal experience with angiography, and assessments of the history of cardiac therapeutics. Physicians had to decide whether angiography provided a meaningful representation of the disease and its treatment or whether demonstrations of therapeutic success could come only from long-term statistical evaluation of mortality data.

I N THE EARLY 1970s physicians struggled to determine the efficacy of coronary artery bypass grafting (CABG), a new surgical treatment for coronary artery disease. The debate polarized both cardiology and cardiac surgery. Some surgeons believed that the efficacy of CABG could be shown only "by a large-scale, prospective randomized study of a homogenous group of patients with a definable syndrome of symptomatic coronary artery disease." The cardiologist Eugene Braunwald argued that such randomized clinical

Isis 91, no. 3 (September 2000): 504–41.

I am indebted to Robert Martensen for the initial inspiration for this work. Many others have provided valuable advice, especially Allan Brandt, Harry Marks, Scott Podolsky, Deborah Weinstein, Elizabeth Caronna, and eight anonymous referees. Bruce Fye, Eugene Braunwald, and David Spodick provided additional insight and guidance. A portion of this essay was presented at the 69th Annual Meeting of the American Association for the History of Medicine, Buffalo, New York, 12 May 1996. This research was supported by grants from the Office of Enrichment Programs and the Department of Social Medicine, Harvard Medical School, and by a scholarship from the Medical Scientist Training Program Grant, National Institutes of Health.

trials (RCTs) had to be done before CABG, like a genie escaped from its bottle, became an uncontrollable force. Others, such as the pioneering cardiac surgeon Michael DeBakey, disagreed: "The insistence on the use of prospective randomized studies for the evaluation of surgical diagnostic and therapeutic techniques reflects a naive obsession with this research tool." René Favaloro, the "father" of CABG, warned against the "almost religious sanctification" of the ideology of RCTs: "If relied on exclusively they may be dangerous."[1]

How could RCTs—a research methodology—be dangerous?[2] Powerful forces converged in the debates over the RCTs of CABG. This was a battle over fundamental questions of medical epistemology, a battle over the production of medical knowledge. In the 1960s RCTs had been established as the gold standard for evaluating the efficacy of new treatments. In theory, they would replace traditional, biased methods of evaluation, including both retrospective cases series and the vagaries of individual physicians' clinical judgments. As CABG spread in the 1970s, many cardiologists and cardiac surgeons argued that only an RCT could prove its value. Against them stood the advocates of CABG, who argued that the physiological logic and the immediate, visible results of the operation proved, to them and to their patients, that the surgery had been a success. Underlying their hostility to RCTs was faith in the power of visualization. RCTs typically use statistical evidence of decreased mortality to demonstrate the efficacy of a therapy. But in the context of a model of coronary artery disease that identified the pathology as obstructed blood flow, there was an alternative method for demonstrating efficacy: visualization of restored blood flow, through the technique of coronary angiography.

Thus protagonists in the debate had to make choices about the relationships between definitions of a disease and definitions of a cure, about the comparative value of statistical and visual evidence. Did relief of symptoms, restoration of blood flow, or improved survival most meaningfully represent a successful treatment? Compelling arguments could be made for each case. This opened a space for judgment and preconception, grounded in tensions between cardiology and cardiac surgery, tensions within cardiology over the value of physiological and clinical research, and personal experiences with the reliability of angiography. Faith in angiography proved to be crucial, determining not only judgments about the relevance of visual evidence of revascularization but also physicians' assessments of the legacy of past, failed treatments of coronary artery disease. Skeptics, citing the recurring cycle of therapeutic enthusiasm and disillusionment, demanded that the burden of proof for CABG be set very high. Enthusiasts, citing the ability of angiography to provide accurate diagnosis and postoperative assessment, argued that CABG deserved the benefit of the doubt.

The epistemological debates did not occur in a vacuum. The controversy over RCTs for

---

[1] Herbert N. Hultgren, Timothy Takaro, Noble Fowler, and Elizabeth G. Wright, "Evaluation of Surgery in Angina Pectoris," *American Journal of Medicine,* 1974, *56*:1–3, on p. 2; Eugene Braunwald, "Direct Coronary Revascularization . . . A Plea Not to Let the Genie Escape from the Bottle," *Hospital Practice,* 1971, *6*:9–10, on p. 9; Michael E. DeBakey and Gerald M. Lawrie, "Aortocoronary-Artery Bypass," *Journal of the American Medical Association,* 1978, *239*:837–839, on p. 838; and René G. Favaloro, "Critical Analysis of Coronary Artery Bypass Graft Surgery: A Thirty-Year Journey," *Journal of the American College of Cardiology,* 1998, *31*(4 [Suppl. B]):1B–63B, on pp. 37B–38B. The operation has been known by many names: direct myocardial revascularization, saphenous vein autograph replacement, coronary artery bypass surgery, and many variations on these themes. I use coronary artery bypass grafting (the current usage) throughout.

[2] This episode was an opening salvo in a continuing debate between advocates of RCTs (whose view is now embodied in the outcomes movement of evidence-based medicine) and those who see such probabilistic methods as a threat to physician knowledge, judgment, and authority. See Sandra J. Tanenbaum, "Evidence and Expertise: The Challenge of the Outcomes Movement to Medical Professionalism," *Academic Medicine,* 1999, *74*:757–763.

CABG was also a battle for professional authority and financial resources. After all, CABG had the potential to transform the treatment of coronary artery disease, one of the most prevalent diseases in the United States and the leading cause of death.[3] Uncertainty about its ability to increase the life expectancy of its recipients had enormous stakes: CABG involved millions of patients and billions of dollars. Powerful groups—cardiologists, cardiac surgeons, patients, and insurers—all had interests in the process of evaluating the efficacy of the operation. Moreover, these concerns have not been confined to CABG in the 1970s. Instead, the tensions between visual and statistical evidence reemerged in cardiology and cardiac surgery during debates over the efficacy of angioplasty, minimally invasive bypass surgery, and gene therapy. This is not to imply that cardiac therapeutics has been plagued by hostile confrontation. Instead, as Peter Galison's notion of the disunity of science would suggest, there has been a productive tension between competing traditions of therapeutic evaluation.

TRIALS' TRIBULATIONS

Since the pioneering work of Bruno Latour, analysis of knowledge production has been a major interest in social studies of science. One leading area of investigation has been randomized controlled trials (RCTs). In an RCT, clinical investigators evaluate the efficacy of a new treatment by randomly assigning patients to receive either the new treatment or the existing, established treatment. Ideally, the trial is double-blinded: neither the patients nor the researchers know which treatment a patient received until after the results have been analyzed. Done properly, RCTs can provide unbiased results. This contrasts to traditional methods of clinical evaluation, such as retrospective case series, in which researchers compare the results of a new treatment against existing results with the old treatment. For medical researchers, RCTs have become the route to objective, rigorous knowledge of therapeutic efficacy. For historians, they are a crucial window into the practices and values of modern medicine.

As Harry Marks notes, RCTs emerged during the contingencies of antibiotic research during World War II. By 1970 the Food and Drug Administration required evidence not only of drug safety but also of efficacy, acquired through "well-controlled investigations" with "appropriate statistical methods." This requirement had been implemented as the expectation of double-blind testing and RCTs. Researchers hoped that RCTs would "neutralize the investigators' belief about the value of novel therapies" and create an "impersonal standard of scientific integrity."[4]

However, RCTs have not fulfilled these high expectations. As early as the mid 1970s, Marks and other observers of medicine became fascinated by the failure of RCTs to resolve therapeutic uncertainty. They have analyzed the failure of RCTs to change the conventional

---

[3] In 1968, diseases of the heart were the leading cause of death in the United States, responsible for 38.6 percent (1,040,292) of all deaths; 65 percent of these deaths were from ischemic heart disease: U.S. Department of Health and Human Services, *Vital Statistics of the United States, 1968*, Vol. 2: *Mortality* (Hyattsville, Md.: Center for Health Statistics, 1972), Tables 1-5, 1-26.

[4] Harry Marks, *The Progress of Experiment: Science and Therapeutic Reform in the United States, 1900–1990* (Cambridge: Cambridge Univ. Press, 1997), pp. 126, 3, 144. On the requirements of the Food and Drug Administration see Henk J. H. W. Bodewitz, Henk Buurma, and Gerard H. de Vries, "Regulatory Science and the Social Management of Trust in Medicine," in *The Social Construction of Technological Systems*, ed. Wiebe E. Bijker, Thomas P. Hughes, and Trevor J. Pinch (Cambridge, Mass.: MIT Press, 1987), pp. 243–259, on pp. 252–253; Marks, *Progress of Experiment*, pp. 11, 129, 133; and Steven Epstein, *Impure Science: AIDS, Activism, and the Politics of Knowledge* (Berkeley: Univ. California Press, 1996), pp. 185, 196.

wisdom about the value of low-fat diets, the refusal of researchers to pursue RCTs of vitamin C as a treatment for cancer, and the ability of politically empowered research populations to shape the conduct of RCTs for treatments of AIDS. RCTs are not an uncomplicated, objective technique but a social and political process, "the product of a negotiated social order." They "cannot be pried apart from the vested interests and social objectives which they embody." RCTs have such "interpretative flexibility" that, "rather than settling controversies, [they] may instead reflect and propel them." In sum, they are part of broader processes of the "social management of trust."[5]

The debates over CABG provide an opportunity to extend this analysis of RCTs to the field of surgery, where their role has been fiercely contested. RCTs of CABG have already attracted some analysis, as examples of the process of therapeutic evaluation, of the success of this process, and of obstacles to this process. For instance, Jochen Schaefer has argued that by the time surgeons had perfected CABG enough to allow a meaningful RCT, the "cardiological and cardiosurgical industry" was too committed to the procedure to perform such an "exact and critical analysis of its own work." These analyses leave crucial questions unanswered. An examination of the debates that followed the early RCTs of CABG shows that concerns with the technical details of the trials were far less important than the preconceptions held by protagonists.[6] The presence of a substantial debate before the RCTs were published allows an analysis of the origins of these preconceptions. This essay will explain these preconceptions and trace their impact on the ensuing debates. It will show how disciplinary tensions, disciplinary histories, and traditions of physiological and clinical research all shaped the evaluation of cardiac therapeutics. It will demonstrate how visual validation of CABG provided an alternative to the statistical ideal of RCTs.

NOTORIOUS TREATMENTS OF CORONARY ARTERY DISEASE

To understand the ferocity of the debates that surrounded coronary artery bypass surgery, it is necessary to appreciate the magnitude of the problem surgeons sought to treat. During the 1970s coronary artery disease (also known as ischemic heart disease) affected millions of people in the United States. By various estimates, it killed over 600,000 people each year, incapacitating another 3.5 to 5.0 million people. Leading cardiologists, such as Harvard Medical School's Eugene Braunwald, considered coronary artery disease "to be the greatest scourge of Western man." Its burden of suffering, disability, and deaths "dwarfs all of the other critical problems that face us in this turbulent era." Donald Effler, the chief

---

[5] Marks, *Progress of Experiment*, pp. 8, 240, 134 ("negotiated social order"); Evelleen Richards, "The Politics of Therapeutic Evaluation: The Vitamin C and Cancer Controversy," *Social Studies of Science,* 1988, *18*:653–701, on p. 685 ("vested interests and social objectives") (see also Richards, *Vitamin C and Cancer: Medicine or Politics?* [London: Macmillan, 1991], pp. 216–217); Epstein, *Impure Science*, p. 33 ("interpretative flexibility"); Steven Epstein, "Activism, Drug Regulation, and the Politics of Therapeutic Evaluation in the AIDS Era," *Soc. Stud. Sci.*, 1997, 27:691–726, on p. 716 (relation to controversies); and Bodewitz *et al.*, "Regulatory Science and the Social Management of Trust," p. 257.

[6] Jochen Schaefer, "The Case against Coronary Artery Surgery: A Paradigm for Studying the Nature of a So-Called Scientific Controversy in the Field of Cardiology," *Metamedicine,* 1980, *1*:155–176, on p. 162. On the process see John B. McKinlay, "From 'Promising Report' to 'Standard Procedure': Seven Stages in the Career of a Medical Innovation," *Millbank Memorial Fund Quarterly,* 1981, *59*:233–270, esp. p. 250; on its successes see J. C. Baldwin, M. J. Reardon, and R. J. Bing, "Coronary Artery Surgery," in *Cardiology: The Evolution of the Science and the Art*, ed. Bing, 2nd ed. (New Brunswick, N.J.: Rutgers Univ. Press, 1999), pp. 166–182, esp. pp. 160–161; on its obstacles see Walsh McDermott, "Evaluating the Physician and His Technology," *Daedalus,* Winter 1977, pp. 135–157, esp. p. 151. Marks and Epstein have addressed the impact of preconceptions in their own studies: Marks, *Progress of Experiment*, pp. 193–194; and Epstein, *Impure Science*, p. 343.

of cardiac surgery at the Cleveland Clinic, emphasized how coronary artery disease struck down "the most active and responsible" members of society: it was "so common among professional people, executives and men in public office" that it could "cripple the nation." It was "so prevalent and ominous that heroic measures to deal with it seem not only appropriate but essential."[7]

Unfortunately, through the 1960s doctors had found no satisfactory treatments. The work of James Herrick in the early twentieth century had provided a simple model for coronary artery disease.[8] The coronary arteries supply blood flow to the muscles of the heart (see Figure 1). If flow through these vessels is obstructed, by atherosclerosis or spasm, the demand of the muscle for oxygenated blood exceeds the supply. This imbalance initially produces angina pectoris, "the wailing of an anguished heart during that period of stress when it is getting inadequate perfusion." If the imbalance is severe enough the muscle cells start to die, producing a myocardial infarction, or heart attack.[9]

With this understanding of coronary artery disease, physicians recognized two treatment strategies: increase the supply of blood or reduce the demand for it. Many creative techniques had been tried along both lines.[10] Early efforts to reduce demand, such as operations in the 1930s to remove the thyroid gland to slow the heart's rhythm, were abandoned because of unacceptable complications. Efforts to increase supply showed more promise. Sympathectomy, first performed in 1916, cut the sympathetic nerves to dilate coronary arteries. While it did not actually improve blood flow, it inadvertently eased the symptoms of angina by interrupting sensory transmission from the heart. In the 1930s the Cleveland Clinic surgeon Claude Beck developed many techniques to provide alternative blood supplies for the heart. He connected a variety of noncardiac tissues—including pericardium, fat, muscle, skin, lung, omentum, stomach, intestines, liver, and spleen—to the heart, hoping that blood would flow from the donor tissue to the cardiac muscle. Other researchers

---

[7] Braunwald, "Direct Coronary Revascularization" (cit. n. 1), p. 9; and Donald B. Effler, "Surgery for Coronary Disease," *Scientific American*, Oct. 1968, *210*:36–43, on p. 43. For the figures see note 3, above. See also F. C. Spencer, O. W. Isom, E. Glassman, A. D. Boyd, R. M. Engelman, G. E. Reed, B. S. Pasternak, and J. M. Dembrow, "The Long-Term Influence of Coronary Bypass Grafts on Myocardial Infarction and Survival," *Annals of Surgery*, 1974, *180*:439–451, esp. p. 451; and Effler, "Surgery for Coronary Disease," p. 36.

[8] Christopher Lawrence, "Moderns and Ancients: The 'New Cardiology' in Britain, 1880–1930," in *The Emergence of Modern Cardiology*, ed. W. F. Bynum, Lawrence, and Vivian Nutton (London: Wellcome Institute for the History of Medicine, 1985), pp. 1–33, esp. pp. 26–27; Joel D. Howell, "Concepts of Heart-Related Diseases," in *The Cambridge World History of Disease*, ed. Kenneth F. Kiple (Cambridge: Cambridge Univ. Press, 1993), pp. 91–102, on p. 91; Robert A. Aronowitz, "From the Patient's Angina to the Cardiologist's Coronary Heart Disease," in *Making Sense of Illness: Science, Society, and Disease* (Cambridge: Cambridge Univ. Press, 1998), pp. 84–110; R. J. Bing, "Arteriosclerosis," in *Cardiology*, ed. Bing (cit. n. 6), pp. 118–132; and O. Paul and Bing, "Coronary Artery Disease," *ibid.*, pp. 133–153.

[9] René G. Favaloro, Donald B. Effler, Laurence K. Groves, F. Mason Sones, and David J. G. Ferguson, "Myocardial Revascularization by Internal Mammary Artery Implant Procedures: Clinical Experience," *Journal of Thoracic and Cardiovascular Surgery*, 1967, *54*:359–370, on p. 370. On coronary artery disease more generally see Michael Lesch, Richard S. Ross, and Eugene Braunwald, "Ischemic Heart Disease," in *Harrison's Principles of Internal Medicine*, 8th ed., ed. George W. Thorn, Raymond D. Adams, Braunwald, Kurt. J. Isselbacher, and Robert G. Petersdorf (New York: McGraw-Hill, 1977), pp. 1261–1271, on pp. 1264, 1262; and Effler, "Surgery for Coronary Disease" (cit. n. 7), p. 36.

[10] J. Hilario Murray, R. Porcheron, and W. Roschlau, "Surgery of Coronary Heart Disease," *Angiology*, 1953, *4*:526–530; John H. Vansant and William H. Muller, "Surgical Procedures to Revascularize the Heart: A Review of the Literature," *Circulation*, 1960, *100*:572–578; David C. Sabiston, "Role of Surgery in the Management of Myocardial Ischemia," *Modern Concepts of Cardiovascular Disease*, 1966, *35*:123–127, esp. p. 123; Herbert N. Hultgren and Edward J. Hurley, "Surgery in Obstructive Coronary Artery Disease," *Advances in Internal Medicine*, 1968, *14*:107–150, esp. pp. 107–108; Norman V. Richards, *Heart to Heart: A Cleveland Clinic Guide to Understanding Heart Disease and Open Heart Surgery* (New York: Atheneum, 1987), p. 107; and Baldwin *et al.*, "Coronary Artery Surgery" (cit. n. 6), pp. 154–157.

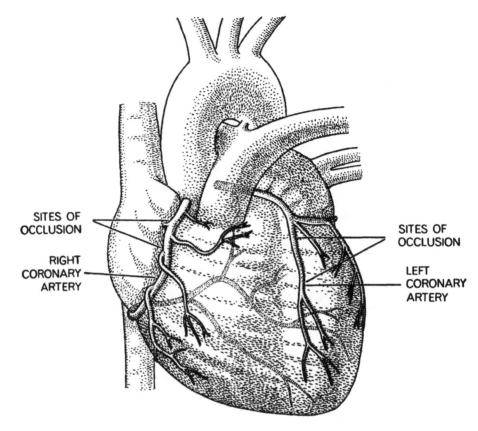

Figure 1. *The heart and its coronary arteries. Reprinted with permission from the estate of Eric Mose. (From Donald B. Effler, "Surgery for Coronary Disease," Scientific American, October 1968, 210:36–43, on page 38.)*

attempted direct manipulation of blood vessels. In 1939 Italian surgeons claimed to increase coronary blood flow by ligating (tying off) the internal mammary artery, a vessel that carries blood to the rib cage. In 1946 the Quebec surgeon Arthur Vineberg began to implant the cut end of arteries—the subclavian, intercostal, carotid, or splenic—directly into the wall of the heart. Meanwhile, Beck connected the coronary veins to the aorta, reversing their flow of blood, turning veins into arteries to bring new blood to heart. Across this range of techniques, surgeons typically reported success—relief of pain and ability to return to work—in 80 to 90 percent of patients.

Evaluation of these new techniques was complicated by the elusive nature of angina pectoris. Observers had long recognized that "symptoms of cardiac ischemia may come and go in a random fashion, seemingly unrelated to therapeutic interventions." As a result, skeptics argued that relief of angina did not equal therapeutic success: "the notorious unreliability of such data should preclude any important inferences." Even surgeons agreed: "Relief of angina pectoris by an operative procedure is not a certain index of surgical success." Long experience had also taught physicians that angina was remarkably susceptible to placebo effects. In a classic study, Henry Beecher showed that angina—

The heart and its coronary arteries labels:

SITES OF OCCLUSION

RIGHT CORONARY ARTERY

SITES OF OCCLUSION

LEFT CORONARY ARTERY

and many other symptoms—responded to irrelevant treatments, consistently improving in 35.2 percent of patients.[11]

Placebo effects of surgery had been strikingly demonstrated in the case of internal mammary artery ligation. Many observers, noting that the internal mammary artery had no connections to the heart, could not believe surgeons' claims of success. Two groups conducted controlled studies in which thirty-five unwitting patients were assigned to receive either ligation or a sham operation. Remarkably, patients in both groups reported increased exercise tolerance and reduced need for pain-relieving medication. Struggling to explain this dramatic demonstration of the power of surgical placebos, the researchers credited the many aspects of cardiac surgery that contributed to the perceived effects: "The frightened, poorly informed man with angina, winding himself tighter and tighter, sensitizing himself to every twinge of chest discomfort, who then comes into the environment of a great medical center and a powerful positive personality and sees and hears the results to be anticipated from the suggested therapy is not the same total patient who leaves the institution with the trademark scar."[12]

By the late 1960s these experiences had left physicians frustrated. New drugs showed promise for reducing cardiac oxygen consumption, but the cardiologists Edward Orgain and Henry McIntosh feared that these would fail because of complications that included heart failure, shock, and death. Cardiologists frequently confessed their impotence. For instance, in 1972 David Spodick, from Tufts University School of Medicine, concluded that medical management "had failed." Surgeons were quick to criticize, with Donald Effler characterizing cardiologists' efforts as "not terribly impressive, either to the patient or to the cardiac surgeon." But surgeons themselves had little to celebrate. Effler admitted that by the early 1960s coronary artery surgery had fallen into "virtual disrepute": "papers on this subject were viewed with frank skepticism and the authors looked upon with suspicion." The cardiologist Mason Sones saw promise in Effler's work but nonetheless chastised his surgical colleagues: "Gentleman, I suggest you surgeons get with it!"[13]

[11] Irwin J. Schatz, "Commentary on 1971 Reflection of 1970 Statistics," *Chest*, 1972, *61*:477–479, on p. 478; Favaloro et al., "Myocardial Revascularization by Internal Mammary Artery Implant Procedures" (cit. n. 9), p. 367 (surgeons' view); and Henry K. Beecher, "The Powerful Placebo," *J. Amer. Med. Assoc.*, 1955, *199*:1602–1606, on p. 1606.

[12] E. Grey Dimond, C. Frederick Kittle, and James E. Crockett, "Comparison of Internal Mammary Artery Ligation and Sham Operation for Angina Pectoris," *American Journal of Cardiology*, 1960, *5*:483–486, on p. 486. On placebo effects of internal mammary artery ligation see Henry K. Beecher, "Surgery as Placebo: A Quantitative Study of Bias," *J. Amer. Med. Assoc.*, 1961, *176*:1102–1107; and Ernest M. Barsamian, "The Rise and Fall of Internal Mammary Artery Ligation in the Treatment of Angina Pectoris and the Lessons Learned," in *Costs, Risks, and Benefits of Surgery,* ed. John P. Bunker, Benjamin A. Barnes, and Frederick Mosteller (New York: Oxford Univ. Press, 1977), pp. 213–220. For the controlled studies see Leonard A. Cobb, George I. Thomas, David D. Dillard, K. Alvin Merendino, and Robert A. Bruce, "An Evaluation of Internal-Mammary-Artery Ligation by a Double-Blind Technic," *New England Journal of Medicine*, 1959, *260*:1115–1118; and Dimond et al., "Comparison of Internal Mammary Artery Ligation and Sham Operation for Angina Pectoris."

[13] Edward S. Orgain and Harry D. McIntosh, "Editorial: Coronary Artery Disease: Physiological Aspects and Surgical Therapy," *Circulation*, 1967, *35*:1–2, on p. 2; David H. Spodick, "Letter: Surgery in Coronary Artery Disease—I," *Amer. J. Cardiol.,* 1972, *29*:581–582, on p. 582; Favaloro et al., "Myocardial Revascularization by Internal Mammary Artery Implant Procedures" (cit. n. 9), p. 370 (quoting Effler, "not terribly impressive"); Donald B. Effler, Laurence K. Groves, F. Mason Sones, and Earl K. Shirey, "Increased Myocardial Perfusion by Internal Mammary Artery Implant: Vineberg's Operation," *Ann. Surg.*, 1963, *158*:526–536, on p. 526 ("virtual disrepute") (see also W. Bruce Fye, *American Cardiology: The History of a Specialty and Its College* [Baltimore: Johns Hopkins Univ. Press, 1996], p. 257); René G. Favaloro, Effler, Groves, Mehdi Razavi, and Tail Lieberman, "Combined Simultaneous Procedures in the Surgical Treatment of Coronary Artery Disease," *Annals of Thoracic Surgery*, 1969, *8*:20–29, on p. 29 (skepticism and suspicion); and Sones, quoted in Effler, Sones, Favaloro, and Groves, "Coronary Endarterotomy with Patch-Graft Reconstruction," *Ann. Surg.*, 1965, *162*:590–601, on p. 601.

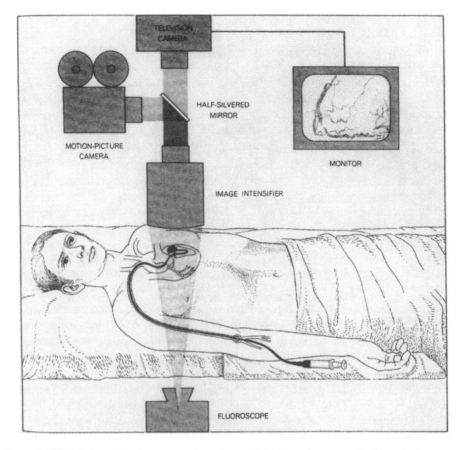

Figure 2. *"Visual diagnosis of coronary artery disease." Angiography uses a flexible catheter, threaded through a peripheral vein and into a coronary artery, to inject a radio-opaque dye, which is then visualized with fluoroscopy. The technique "makes possible diagnosis with an unprecedented degree of accuracy." Reprinted with permission from the estate of Eric Mose. (From Donald B. Effler, "Surgery for Coronary Disease,"* Scientific American, *October 1968, 210:36–43, on page 39.)*

THE ORIGIN AND SPREAD OF CORONARY ARTERY BYPASS GRAFTING

The situation changed dramatically in 1968, when a Cleveland Clinic team led by René Favaloro published their first report about coronary artery bypass grafting. Their enthusiasm quickly spread among physicians and patients, transforming the treatment of ischemic heart disease.

The Cleveland Clinic story began in 1958, with Mason Sones's accidental discovery of selective coronary arteriography (angiography). This technique, which used a catheter to inject contrast dye into coronary arteries, gave cardiologists their first opportunity to visualize these arteries, and their obstructions, in a living patient (see Figure 2). Visualization transformed their understanding of coronary artery disease. Instead of seeing the expected diffuse obstructions from coronary atherosclerosis, Sones found "remarkable localization"

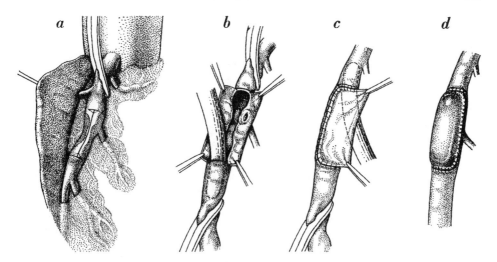

Figure 3. The "Vista Dome": Effler's pericardial patch-graft reconstruction. An incision is made into the narrowed coronary artery (a). The vessel is dilated with a probe (b). A patch of pericardial membrane is sewn into place along the edges of the incision (c). When complete, "a gentle bulge replaces the previous narrowing" (d). Reprinted with permission from the estate of Eric Mose. (From Donald B. Effler, "Surgery for Coronary Disease," Scientific American, October 1968, 210:36–43, on page 41.)

of the disease process.[14] This suggested new surgical techniques: instead of creating new conduits to bring blood to the heart, surgeons needed only to find a way to fix the focal obstructions of the coronary arteries.

Effler and his team first tried endarterectomy, removing the obstructing plaque from the coronary artery. They abandoned this technique when they realized that fragments broke off the plaque and produced new obstructions further along the arterial tree. In 1962 Effler attempted a new operation: he left the plaque in place but used a patch of pericardium to expand the diameter of the coronary artery. This procedure, named the "Vista Dome" for its resemblance to the passenger cars of railroads, allowed blood to flow past the obstruction (see Figure 3). The operation, a "direct, surgical attack," provided "instant revascularization." Combining this direct approach with Vineberg's internal mammary artery implants allowed Effler's team to treat patients who had both focal and diffuse obstructions. Earlier beliefs that only a limited set of patients could benefit from surgery were "restrictive and unimaginative"; surgery could be "applicable to a rather wide group of coronary artery problems."[15] The future seemed full of promise.

---

[14] Claude Bernard first performed cardiac catheterization in an animal. Werner Forssmann performed the first human catheterization, on himself, in 1929. During the 1940s Richard Bing used catheterization of the coronary sinus to study cardiac metabolism and a group in Sweden performed nonselective coronary angiography (visualizing the coronary arteries with the aorta). Sones inadvertently performed the first selective coronary angiography in 1958 when the tip of his ventricular catheter slipped into a coronary artery. See D. Baim and R. J. Bing, "Cardiac Catheterization," in *Cardiology,* ed. Bing (cit. n. 6), pp. 1–15. The "remarkable localization" found by Sones is reported in Effler *et al.*, "Coronary Endarterotomy with Patch-Graft Reconstruction," p. 590.

[15] On the problems with coronary endarterotomy see Effler, "Surgery for Coronary Disease" (cit. n. 7), p. 42. On the "Vista Dome" see Donald B. Effler, Laurence K. Groves, Ernesto L. Suarez, and René G. Favaloro, "Direct Coronary Artery Surgery with Endarterotomy and Patch-Graft Reconstruction," *J. Thorac. Cardiovasc. Surg.,* 1967, *53*:93–101, on p. 99; and John E. Connolly, "The History of Coronary Artery Surgery," *ibid.,* 1978,

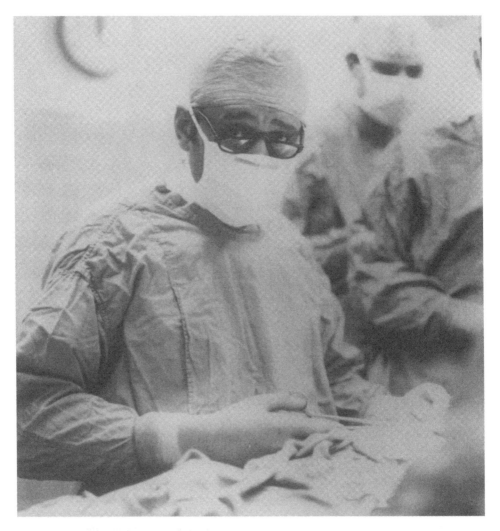

Figure 4. *René Favaloro in the operating room, Cleveland Clinic, circa 1970. Reprinted with permission from the Cleveland Clinic Archives.*

In this setting René Favaloro began his work at the Cleveland Clinic (see Figure 4). Favaloro had been a general surgeon in rural Argentina when, in 1962, he committed himself to becoming a heart surgeon and flew to Cleveland with only a letter of introduction. He arrived soon after Effler's first successful patch-graft repair and Sones's demonstration of blood flow through Vineberg implants. Although he shared in the excitement of Effler's early successes, Favaloro was not satisfied with the existing techniques. Impressed by the work of vascular surgeons, who used vein grafts to bypass obstructions in renal arteries, he adapted the bypass technique for the heart (see Figure 5). Favaloro's

76:733–744, on p. 739. See also Effler, "Surgical Treatment of Myocardial Ischemia," *Clinical Symposia*, 1969, *21*:3–17, on p. 6 ("instant revascularization"); and Effler *et al.*, "Coronary Endarterotomy with Patch-Graft Reconstruction," p. 591.

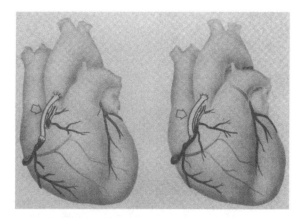

Figure 5. *Favaloro's technique of saphenous vein bypass grafting (CABG). Two variations of CABG, using a piece of saphenous vein as a conduit between the aorta and the right coronary artery: end-to-end anastomosis (left) and end-to-side anastomosis (right). Reprinted with permission from Mosby, Inc. (From Donald B. Effler, René G. Favaloro, and Laurence K. Groves, "Coronary Artery Surgery Utilizing Saphenous Vein Graft Techniques: Clinical Experience with 224 Operations,"* Journal of Thoracic and Cardiovascular Surgery, *1970, 59:147–154, on page 149.)*

ideal candidate for his new operation arrived in May 1967: a fifty-one-year-old woman with an obstructed right coronary artery. Favaloro used a piece of saphenous vein to bypass the obstruction. He described the operation's immediate success: when the team removed the clamps, they saw that "blood flowed rapidly"; "we could see the branches of the right coronary artery fill with blood." Postoperative angiography showed "excellent function of the graft" (see Figure 6).[16]

Favaloro was immediately convinced of the value of the operation; only Sones's conservatism constrained his "Latin enthusiasm." They performed only a few bypass operations during the next months, waiting to see if the vein grafts would remain patent (i.e., open to blood flow during normal physiological conditions). Within a year, they had what they considered convincing evidence. Favaloro and his colleagues began to perform CABG with increasing frequency: 37 in 1967, nearly 200 in 1968, and 1,500 in 1969. The Cleveland Clinic team was full of enthusiasm. CABG "appeals to the imagination by its very simplicity": it provided a direct and immediate solution to the fundamental problem of obstructed flow. The team reported "complete symptomatic relief and excellent postoperative angiographic verification": the "majority of patients are fully recovered, enjoy a normal life, and work full time." Although they could not yet prove that CABG would prevent heart attacks and prolong life, by 1973 they claimed that "experience over the past 5 years in the Cleveland Clinic suggests that this is true."[17]

[16] René G. Favaloro, *The Challenging Dream of Heart Surgery: From the Pampas to Cleveland* (New York: Little, Brown, 1994) (hereafter cited as **Favaloro, *Challenging Dream of Heart Surgery***), pp. 1–27, 70–86, 96 (quotations); and Favaloro, "Saphenous Vein Autograft Replacement of Severe Segmental Coronary Artery Occlusion," *Ann. Thorac. Surg.*, 1968, 5:334–339, on p. 337 ("excellent function"). On the "ideal candidate" see Favaloro, "Critical Analysis of Coronary Artery Bypass Graft Surgery" (cit. n. 1), p. 2B. Favaloro's use of this technique was not based on animal experimentation but grew out of his experience with other coronary artery operations on hundreds of patients: Favaloro, *Challenging Dream of Heart Surgery*, p. 98. Favaloro's choice of a female patient is discussed later in this essay; see note 42, below.

[17] Donald B. Effler, "Myocardial Revascularization—Direct or Indirect?" *J. Thorac. Cardiovasc. Surg.*, 1971, 61:498–500, on p. 498 (simplicity); René G. Favaloro, Effler, Laurence K. Groves, William C. Sheldon, and

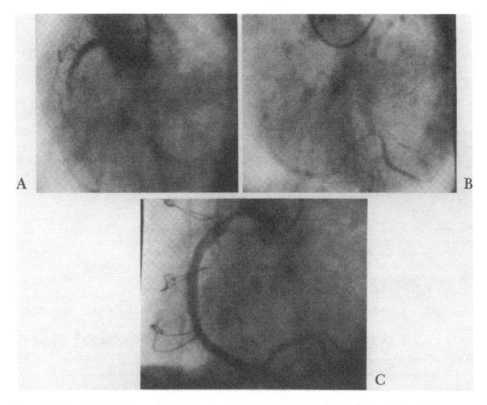

Figure 6. *Visual proof of success? Coronary angiogram showing total occlusion of the right coronary artery (A), perfusion of the distal right coronary artery by collateral branches from the left coronary artery (B), and reconstruction of right coronary artery with a saphenous vein graft (C). Reprinted with permission from the Society of Thoracic Surgeons. (From René Favaloro, "Saphenous Vein Autograft Replacement of Severe Segmental Coronary Artery Occlusion," Annals of Thoracic Surgery, 1968, 5:334–339, on page 338.)*

Their enthusiasm quickly infected other groups of surgeons. Many vascular surgeons—especially those, like David Sabiston and Michael DeBakey, who had prior experience with bypass surgery—immediately recognized Favaloro's accomplishment.[18] By 1972, the

---

Mohammed Riahi, "Direct Myocardial Revascularization with Saphenous Vein Autograft: Clinical Experience in 100 Cases," *Diseases of the Chest*, 1969, 56:279–283, on p. 282 (symptomatic relief); Favaloro, "Direct Myocardial Revascularization," *Surgical Clinics of North America*, 1971, 51:1035–1042, on p. 1042 (full recovery); and Sergio V. Moran, Robert C. Tarazi, Jorge U. Urzua, Favaloro, and Effler, "Effects of Aorto-Coronary Bypass on Myocardial Contractility," *J. Thorac. Cardiovasc. Surg.*, 1973, 65:335–342, on p. 335 (five years' experience). For the statistics see Baldwin *et al.*, "Coronary Artery Surgery" (cit. n. 6), p. 158; Favaloro, "The Present Era of Myocardial Revascularization—Some Historical Landmarks," *International Journal of Cardiology*, 1983, 4:331–344, esp. pp. 334–337; Katherine M. Detre, "Non-Randomized Studies of Coronary Artery Bypass Surgery," *Statistics in Medicine*, 1984, 3:389–398, esp. pp. 389–390; and Sheldon, Favaloro, F. Mason Sones, and Effler, "Reconstructive Coronary Artery Surgery: Venous Autograft Technique," *J. Amer. Med. Assoc.*, 1970, 213:78–82, esp. p. 78.

[18] Experiments with coronary bypass grafts in animals began in 1910 but were abandoned because of unacceptable operative mortality: Favaloro, "Present Era of Myocardial Revascularization," p. 340; Vansant and Muller, "Surgical Procedures to Revascularize the Heart" (cit. n. 10), p. 578; Lester R. Sauvage, Stephen J. Wood, Kenneth M. Eyer, Alexander H. Bill, and Robert E. Gross, "Experimental Coronary Artery Surgery: Preliminary Observations of Bypass Venous Grafts, Longitudinal Arteriotomies, and End-to-End Anastomoses,"

cardiologist Richard Ross believed that CABG might become "the most significant advance in the therapy of heart disease in our time." Use of CABG spread to many medical centers and community hospitals. The range of its applications quickly expanded; by 1968 it had been used to treat patients during acute heart attacks. CABG received such favorable media publicity that many patients came to hospitals seeking the surgery. By 1974, 100,000 CABGs had been performed. Surgeons and cardiologists believed that the procedure relieved pain, prevented heart attacks, and prolonged lives. Effler was triumphant: "There seems to be little doubt today that the surgeon will continue to play a dominant role in the treatment of the patient who suffers from ischemic heart disease."[19] His forecast was accurate: by 1980 roughly 150,000 procedures were performed each year (see Figure 7). With a typical cost of $15,000 to $20,000 for each operation, the annual cost of CABG exceeded $2 billion, roughly 1 percent of the total national health expenditure.[20]

SKEPTICS' CRITICISMS AND CALLS FOR TRIALS

Not everyone shared the excitement about CABG. Favaloro encountered skepticism as soon as he presented his results. His talks at meetings of the American Heart Association and the American College of Cardiology "produced a sour taste because it was difficult to convince the cardiologists in spite of the evidence available."[21] What were his audiences upset about?

Some critics did not trust the surgeons. The cardiologist Henry Zimmerman argued that surgeons' conflicting advocacy of different techniques, from omental grafts to bypass, had eroded their credibility: coronary artery surgery was "in a state of almost total chaos." Some wanted to see data, not optimism: "enthusiasm is not a substitute for evidence." Specific data caused alarm. While teams at elite centers like the Cleveland Clinic achieved

---

J. Thorac. Cardiovasc. Surg., 1963, *46*:826–836, on p. 835. David Sabiston performed the first human bypass in 1962; the patient died three days later. Sabiston did not attempt another such repair until after he had heard of Favaloro's success. Edward D. Garrett and DeBakey performed a successful bypass in 1964, but they did not report their results until 1973: Connolly, "History of Coronary Artery Surgery" (cit. n. 15), pp. 738, 740; Michael E. DeBakey, "The Development of Vascular Surgery," *American Journal of Surgery*, 1979, *137*:697–738, on pp. 710–711; and Favaloro, *Challenging Dream of Heart Surgery*, pp. 154–155. A Russian team performed a bypass in 1965; this was reported in English in 1967 and commented on by Effler before Favaloro's first attempt: Igor Konstantinov, "The First Coronary Artery Bypass Operation and Forgotten Pioneers," *Ann. Thorac. Surg.*, 1997, *64*:1522–1523. Favaloro was the first to apply the technique systematically and publicize his results.

[19] Richard Ross, "Surgery for Coronary Artery Disease Placed in Perspective," *Bulletin of the New York Academy of Medicine*, 1972, *48*:1163–1178, on p. 1163; Favaloro, "Direct Myocardial Revascularization" (cit. n. 17), p. 1041 (range of applications) (see also note 85, below); Eugene Braunwald, "Editorial: Coronary-Artery Surgery at the Crossroads," *New Engl. J. Med.*, 1977, *297*:661–663, on p. 663 (patients seeking CABG); Edwin L. Alderman, Harvey J. Matlof, Lewis Wexler, Norman E. Shumway, and Donald C. Harrison, "Results of Direct Coronary-Artery Surgery for the Treatment of Angina Pectoris," *ibid.*, 1973, *288*:535–539, esp. p. 536 (surgeons' and cardiologists' views); and Donald B. Effler, comment following Spencer *et al.*, "Long-Term Influence of Coronary Bypass Grafts on Myocardial Infarction and Survival" (cit. n. 7), p. 449.

[20] Eldred D. Mundth and W. Gerald Austin, "Surgical Measures for Coronary Heart Disease," *New Engl. J. Med.*, 1975, *293*:13–19, 75–80, 124–130, esp. p. 128; Gina Kolata, "Coronary Bypass Surgery: Debate over Its Results," *Science*, 1976, *194*:1263–1265, on p. 1263; Eugene Braunwald, "Coronary Artery Bypass Surgery—An Assessment," *Postgraduate Medical Journal*, 1976, *52*:733–738, on p. 737; Braunwald, "Editorial: Coronary-Artery Surgery at the Crossroads," p. 663; Henry D. McIntosh and Jorge A. Garcia, "The First Decade of Aortocoronary Bypass Grafting, 1967–1977: A Review," *Circulation*, 1978, *57*:405–431, on pp. 405–406; and McIntosh, "Commentaries on the Consensus Conference of Coronary Artery Bypass Surgery," *Journal of the Florida Medical Association*, 1981, *68*:833–835, on p. 835.

[21] Favaloro, *Challenging Dream of Heart Surgery*, pp. 113–114. As one observer described, "You could virtually palpate his frustration": Thomas J. Ryan, "Revascularization: Reflections of a Clinician," *J. Amer. Coll. Cardiol.*, 1998, *31*(4[Suppl. B]):89B–96B, on p. 89B.

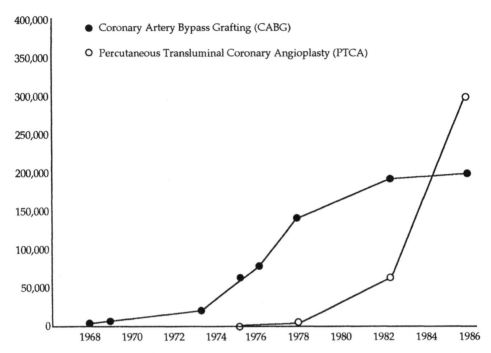

Figure 7. *The spread of CABG and PTCA, 1968–1988; number of procedures performed each year. Data compiled from Henry D. McIntosh and Jorge A. Garcia, "The First Decade of Aortocoronary Bypass Grafting, 1967–1977: A Review," Circulation, 1978, 57:405–431, on pages 405–406; Jay L. Hollman, "Myocardial Revascularization: Coronary Angioplasty and Bypass Surgery Indications," Medical Clinics of North America, 1992, 76:1083–1097, on pages 1084–1085; and Charles Landau, Richard A. Lange, and L. David Hillis, "Percutaneous Transluminal Coronary Angioplasty," New England Journal of Medicine, 1994, 330:981–993, on page 981.*

operative mortality rates as low as 1.4 percent, a 1970 survey found average rates between 7.2 and 11.8 percent. Other "disquieting facts" had appeared in the literature: between 8 and 30 percent of grafts obstructed and between 5 and 29 percent of post-CABG patients experienced myocardial infarctions within one year. One group of Veterans Administration physicians wondered whether CABG could prolong life and whether the improvement in quality of life counterbalanced the risks and costs of the surgery.[22]

Other critics were concerned with financial issues. Surgeons collected one-quarter of the annual cost of CABG. Writing in 1972, Ross noted that there were "economic factors at work which make it difficult to be objective." Such concerns captured the attention of the Senate, which held hearings about resource allocation and the uncertain benefits of CABG. Critics also cited an "even more insidious problem": the growth of an industry of facilities and training programs with a "momentum and constituency of its own." These

[22] Henry A. Zimmerman, "Editorial: The Dilemma of Surgery in the Treatment of Coronary Artery Disease," *American Heart Journal*, 1969, 77:577–578, on p. 577; Jerome Cornfield, "Approaches to Assessment of the Efficacy of Surgical Revascularization," *Bull. N.Y. Acad. Med.*, 1972, 48:1126–1134, on p. 1126 (enthusiasm vs. evidence); Schatz, "Commentary on 1971 Reflection of 1970 Statistics" (cit. n. 11), pp. 477–478 (mortality rates); Eliot Corday, "Status of Coronary Bypass Surgery," *J. Amer. Med. Assoc.*, 1975, 231:1245–1247, on p. 1245 ("disquieting facts"); and Hultgren *et al.*, "Evaluation of Surgery in Angina Pectoris" (cit. n. 1), p. 2.

factors led observers to conclude that "financial incentives for performing the operation are enormous, and there is no balancing economic disincentive to restrain the operation."[23]

These critics and skeptics were always quick to state that they did not oppose CABG—everyone hoped that it would work. Cardiologists had long struggled to provide relief to patients suffering from angina, and CABG brought the prospect that a cure might finally be at hand. Tufts cardiologist David Spodick, a leading skeptic, believed that CABG was "a quantum jump ahead of its predecessors in concept and execution"; he agreed "with Favaloro that bypass surgery holds the greatest promise for definitive management." Braunwald held a similar position. The concern was not so much substantive as epistemological. Spodick stated this most clearly: "My criticism of the surgical enthusiasts is not that they are wrong (or even probably wrong), but rather that they have not attempted to really prove themselves right"; "the professional quality and technical skill of the disputants is not in question. The basis of their evidence and beliefs is."[24]

By identifying advocates of CABG as "enthusiasts," Spodick implied that their trust in the procedure resembled religious faith, not scientific certainty.[25] Where had CABG enthusiasts gone wrong? Typically, surgical researchers compiled the results from several years of their own experience and compared them to previously published reports of medical treatments. Such case series abounded, with groups at Stanford, New York, Boston, and Houston reporting up to five-year follow-up results on thousands of patients showing that their CABG patients had better survival rates than medically treated groups. Such retrospective studies dominated the field: by 1977, 250,000–300,000 patients had received CABG; fewer than 1,300 had been enrolled in randomized trials.[26]

Unfortunately for CABG advocates, many physicians in the early 1970s considered retrospective case series to be a "time-dishonored approach" that produces "misleading results and is clearly wasteful of time, resources—and lives." Spodick dismissed case series as "peep-show reports of a handful of chosen survivors." Audiences at national meetings were "beguiled by beautiful slides and motion pictures." Case series were so problematic because many surgical groups compared their results with those in a group of

---

[23] McIntosh and Garcia, "First Decade of Aortocoronary Bypass Grafting" (cit. n. 20), p. 406 (surgeons' share); Ross, "Surgery for Coronary Artery Disease Placed in Perspective" (cit. n. 19), p. 1170; Fye, *American Cardiology* (cit. n. 13), p. 259 (Senate hearings); Braunwald, "Editorial: Coronary-Artery Surgery at the Crossroads" (cit. n. 19), p. 663 (growth of an "industry"); and Thomas A. Preston, "The Hazards of Poorly Controlled Studies in the Evaluation of Coronary Artery Surgery," *Chest*, 1978, *73*:441–442, on p. 441.

[24] David H. Spodick, "Letter: Coronary Bypass Operations," *New Engl. J. Med.*, 1971, *285*:55–56; Spodick, "Letter: Surgery in Coronary Artery Disease—I" (cit. n. 13), p. 582; Braunwald, "Direct Coronary Revascularization" (cit. n. 1), p. 9; Spodick, "Letter: Need for Controlled Study," *J. Amer. Med. Assoc.*, 1970, *213*:1344; and Spodick, "Editorial: Revascularization of the Heart—Numerators in Search of Denominators," *Amer. Heart J.*, 1971, *81*:149–157, on p. 156.

[25] Marks has noted that "enthusiasts" was a term frequently used by therapeutic reformers and RCT advocates after World War II: Marks, *Progress of Experiment* (cit. n. 4), pp. 149–150. Spodick might have had this meaning of "enthusiast" in mind. However, he made the claim about religion explicitly. When Effler offered to open the doors of his operating room to anyone who wanted to come learn his techniques, Spodick responded derisively: "Apparently none can become members of the Elect until they make the prescribed pilgrimage (Mecca; Lourdes?)": Donald B. Effler, "Myocardial Revascularization," *Chest*, 1973, *63*:79–80, on p. 79; and David H. Spodick, "Aortocoronary Bypass," *ibid.*, pp. 80–81, on p. 81. See also note 38, below.

[26] For retrospective studies see Alderman *et al.*, "Results of Direct Coronary-Artery Surgery for the Treatment of Angina Pectoris" (cit. n. 19); Spencer *et al.*, "Long-Term Influence of Coronary Bypass Grafts on Myocardial Infarction and Survival" (cit. n. 7); Lawrence H. Cohn, Caryl M. Boyden, and John J. Collins, "Improved Long-Term Survival after Aortocoronary Bypass for Advanced Coronary Artery Disease," *Amer. J. Surg.*, 1975, *129*:380–385; and George J. Reul, Denton A. Cooley, Don C. Wukasch, E. Ross Kyger, Frank M. Sandiford, Grady L. Hallman, and John C. Norman, "Long-Term Survival Following Coronary Artery Bypass: Analysis of 4,522 Consecutive Patients," *Archives of Surgery*, 1975, *110*:1419–1424. For the statistics see McIntosh and Garcia, "First Decade of Aortocoronary Bypass Grafting" (cit. n. 20), p. 412.

medically treated patients from the Cleveland Clinic in the 1960s, patients treated before the advent of cardiac care units and many valuable drugs, "at a chronologically different and therapeutically not comparable time." Without data from RCTs, the CABG debates were "a battle of wits between unarmed opponents."[27]

Moreover, the retrospective case series generally relied on unconvincing assessments of therapeutic efficacy. Surgeons celebrated the extent to which CABG could relieve angina. As noted earlier, however, angina was well known to be susceptible to placebo effects. Furthermore, even if CABG did relieve angina, it might have done so by damaging the coronary nerves or by inducing an intraoperative heart attack rather than by improving the underlying coronary artery disease. Some surgeons celebrated their patients' ability to return to work. But since many patients had avoided work because of doctor-induced fear that work would cause angina, they could be "cured" by suggestion: "A patient who is not working because of iatrogenic prophylaxis and who later returns to work because of surgical charisma may be falsely designated as improved due to a bypass graft whose main effect was to evoke enthusiastic iatrotherapy." Finally, demonstration of graft patency with angiography was also inconclusive: patency did not prove that significant blood flowed under ordinary conditions.[28]

Instead, critics wanted definitive answers about the impact of CABG on mortality. Braunwald stated this firmly in 1971: "Many questions must be answered. First and foremost: How do the survival rates and symptoms compare in operated and nonoperated patients?" Ross, Spodick, and Schatz all agreed. Since medical treatments could provide three-year survival in 80 to 90 percent of patients, surgical treatments had only a small window in which to show improvement. This had to be balanced against the risk of the operation. These skeptics all believed that only an RCT had the necessary statistical rigor to manage these statistical subtleties. Admittedly, it would not be an ideal trial: since a sham operation for the medical control group was not considered ethical, the trial could not be blinded. However, by randomly assigning patients to medical or surgical treatment groups, the trial would minimize bias, provide a meaningful control group, and determine whether CABG could improve survival, "the most unequivocal and definitive" of all outcomes.[29]

Spodick—who had trained under Thomas Chalmers, one of the chief advocates of

[27] Spodick, "Letter: Coronary Bypass Operations" (cit. n. 24), p. 56 ("time-dishonored approach"); Spodick, "Editorial: Revascularization of the Heart" (cit. n. 24), p. 149 ("peep-show report"); Spodick, "Aortocoronary Bypass" (cit. n. 25), p. 81 (audiences "beguiled"); Timothy Takaro, "The Controversy over Coronary Arterial Surgery: Inappropriate Controls, Inappropriate Publicity," *J. Thorac. Cardiovasc. Surg.*, 1976, 72:944; and David H. Spodick, "The Randomized Controlled Clinical Trial: Scientific and Ethical Bases," *Amer. J. Med.*, 1982, 73:420–425, on p. 425 ("battle of wits").

[28] Alvan R. Feinstein, "The Scientific and Clinical Tribulations of Randomized Clinical Trials," *Clinical Research*, 1978, 26:241–244, on p. 244. For less benign explanations of how CABG could relieve angina see Louis A. Soloff, "Letter: Effects of Coronary Bypass Procedures," *New Engl. J. Med.*, 1973, 288:1302–1303, on p. 1303; and M. H. Frick, "An Appraisal of Symptom Relief after Coronary Bypass Grafting," *Postgrad. Med. J.*, 1976, 52:765–769, on p. 765. On the inconclusiveness of patency see Masasyoshi Yokoyama, "A Critical Examination of the Validity of the Use of Vein Grafts in Treating Ischemic Heart Disease," *Amer. Heart J.*, 1972, 84:61–65, on p. 64.

[29] Braunwald, "Direct Coronary Revascularization" (cit. n. 1), p. 9; Ross, "Surgery for Coronary Artery Disease Placed in Perspective" (cit. n. 19), p. 1172; Spodick, "Letter: Surgery in Coronary Artery Disease—I" (cit. n. 13), p. 582; Irwin J. Schatz, "The Need to Know," *Chest*, 1973, 63:82–83, esp. p. 83; and Marvin L. Murphy, Herbert N. Hultgren, Katherine Detre, James Thomsen, Timothy Takaro, and Participants of the Veterans Administration Cooperative Study, "Treatment of Chronic Stable Angina: A Preliminary Report of Survival Data of the Randomized Veterans Administration Cooperative Study," *New Engl. J. Med.*, 1977, 297:621–627, on p. 627. Mortality had long been the measure of choice for therapeutic reformers and RCT advocates. It was convenient, reliable, and objective: Marks, *Progress of Experiment* (cit. n. 4), p. 246.

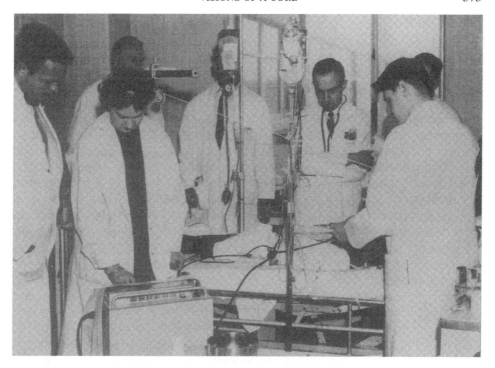

Figure 8. *David Spodick at the bedside, Lemuel Shattuck Hospital, 1960s. From left to right: Ed Moore (hepatologist), Constance Dorr (technician), Thomas Chalmers (chief of medicine), [man blocked by I.V.], David Spodick (chief of cardiology), Joseph Cohen (resident), [man blocked]. Reprinted with permission of David Spodick.*

RCTs—led the way (see Figure 8). Beginning in 1970, he waged a campaign of letters, editorials, and articles stressing that RCTs were valuable because they "tackle head-on the immediate problem of whether it works and, through stratification, for whom." Since researchers and regulators, particularly the Food and Drug Administration, demanded RCTs of medical therapy, they should also require them of surgical therapy: "We abandon our patients along with our intellect if surgical treatment is immune to the high standards demanded of other kinds of treatment." Accepting CABG on the basis of case series would establish an irrational double standard: "Somehow, the mystique of surgery—the presumed efficacy of a mechanical rearrangement of tissue"—made researchers and regulators "suspend disbelief in a way that no pill could." Calling surgery a "Sacred Cow, faith in which continues to ensure immunity from disbelief," Spodick argued that "we should demand quality control of the Sacred Cowboys who milk them and market the products."[30]

Other cardiologists, such as David Schatz and Eugene Braunwald, shared Spodick's demand that RCTs be used to replace "common consent" with "rational analysis of data." Chalmers asserted that RCTs were an ethical imperative, "in the best interest of the patient

[30] David H. Spodick, "Letter: Surgery for Coronary Artery Disease," *Amer. J. Cardiol.*, 1972, *30*:449 (on whether surgery works); Spodick, "Letter: Surgery in Coronary Artery Disease—I," p. 582 (standards); Spodick, "Letter: Coronary Revascularization," *Circulation*, 1971, *44*:302 ("mystique of surgery"); and Spodick, "Editorial: The Surgical Mystique and the Double Standard," *Amer. Heart J.*, 1974, *85*:579–583, on p. 582 (Sacred Cows and Cowboys).

entering the trial as well as of all mankind." Finally, RCTs provided the best means of defending surgeons against the specter of financial conflict of interest that loomed behind their case series.[31]

Some surgeons accepted these calls for trials. In 1968 Timothy Takaro and other Veterans Administration surgeons and cardiologists had begun an RCT of Vineberg implants. They ended this trial prematurely in 1970, with fewer than one hundred patients enrolled, when "attention and interest shifted almost completely" to CABG. Responding to this shift, they modified their trial protocol and began an RCT of CABG.[32] At the 1970 meeting of the American Association of Thoracic Surgery Takaro admitted that this trial might seem "a little heretical" to surgeons, but he defended its importance: "Why don't we require of ourselves the same degree of objectivity in assessing new operative procedures as we require of drugs?" He acknowledged that trials might be difficult to conduct but suggested that if surgeons demonstrated the "same high degree of boldness and ingenuity" that they brought to designing new operations, they would quickly produce results "that the agnostics among us, as well as among our medical colleagues, could accept." Trials presented no threat: if CABG really was superior, it would be validated.[33]

Results of randomized trials of CABG soon began to appear. One small trial was published in 1975 but provoked little interest.[34] Most observers awaited the results of the larger V.A. Cooperative Study, led by Takaro and his colleagues and published in 1977. The researchers sought to answer a basic question: Does CABG improve survival in patients with chronic stable angina? The results—reported for 310 patients treated medically and 286 treated surgically—showed that in most cases CABG did not provide a significant benefit: 87 percent of the medically treated patients were alive after three years, compared to 88 percent among the surgical group. However, early results had shown that 113 patients with obstruction of the left main coronary artery clearly benefited from surgery. This

[31] Schatz, "Commentary on 1971 Reflection of 1970 Statistics" (cit. n. 11), p. 478; Braunwald, "Direct Coronary Revascularization" (cit. n. 1), p. 9; Thomas C. Chalmers, "Randomization and Coronary Artery Surgery," *Ann. Thorac. Surg.*, 1972, *14*:323–327; and Richard D. Sautter, "Reply to Effler," *J. Thorac. Cardiovasc. Surg.*, 1974, *6*:977–978, esp. p. 978. Braunwald recalls that cardiologists believed that cardiac surgeons "were making a fortune" inappropriately: Eugene Braunwald, personal communication, 17 June 1998.

[32] On the RCT of Vineberg implants see Timothy Takaro, comment following Charles H. Dart, Yutaka Kano, Stewart M. Scott, Robert G. Fish, William M. Nelson, and Takaro, "Internal Thoracic (Mammary) Arteriography: A Questionable Index of Myocardial Revascularization," *J. Thorac. Cardiovasc. Surg.*, 1970, *59*:117–127, on p. 127; on its premature termination see Takaro, "The Enigma of the Vineberg-Sewell Implant Operation," *Chest*, 1973, *64*:150–151. On the RCT of CABG see Herbert N. Hultgren, Takaro, and Elizabeth C. Wright, "Veterans Administration Study of Coronary Bypasses," *New Engl. J. Med.*, 1973, *289*:105; and Hultgren, Takaro, Katherine Detre, and Participants in the Veterans Administration Cooperative Study, "Medical and Surgical Treatment of Stable Angina Pectoris: Progress Report of a Large Scale Study," *Postgrad. Med. J.*, 1976, *62*:757–764, esp. p. 759. The V.A. system had long been a favorite site for RCTs, starting with studies of streptomycin following World War II and continuing with studies on psychopharmacology and hypertension in the 1950s. At first glimpse, the V.A. system seems an ideal site—large, centralized, and organized: William G. Henderson, "Some Operational Aspects of the Veterans Administration Cooperative Studies Program," *Controlled Clinical Trials*, 1980, *1*:209–226. However, it is also a site of many conflicting interests: Marks, *Progress of Experiment* (cit. n. 4), pp. 12, 116–120, 132.

[33] Timothy Takaro, comment following René G. Favaloro, Donald B. Effler, Laurence K. Groves, William C. Sheldon, Earl K. Shirey, and F. Mason Sones, "Severe Segmental Obstruction of the Left Main Coronary Artery and Its Divisions: Surgical Treatment by the Saphenous Vein Graft Technique," *J. Thorac. Cardiovasc. Surg.*, 1970, *60*:469–482, on pp. 479–480.

[34] Virendra S. Mathus, Gene A. Guinn, Lakis C. Anastassiades, Robert A. Chahine, Ferenc L. Korompai, Alfredo C. Montero, and Robert J. Luchi, "Surgical Treatment for Stable Angina Pectoris: Prospective Randomized Study," *New Engl. J. Med.*, 1975, *292*:709–713. Two other large trials were also under way at this time. Organized by the National Heart, Lung, and Blood Institute, they studied patients with exertional angina and with unstable angina: Kolata, "Coronary Bypass Surgery" (cit. n. 20), p. 1265.

subgroup was partitioned from the remainder of the study and the results published separately.[35]

The V.A. study, published in the *New England Journal of Medicine* with tremendous publicity, was celebrated by those who had been skeptical of CABG all along. In a strongly supportive editorial, Braunwald celebrated the trial's results and refuted anticipated criticisms. In a special correspondence section published by the *Journal,* Spodick praised the V.A. study as a "meticulously controlled trial" and Braunwald argued that the results merited "the most careful consideration."[36]

OPPOSITION TO TRIALS

How did CABG enthusiasts respond to the damaging results of the V.A. study? Many surgeons, especially those at the Cleveland Clinic and the Texas Heart Institute, had opposed the trials before the study was published, criticized it afterward, and continued to publish traditional case series. Members of the Cleveland Clinic team had been particularly outspoken in their opposition to the need for trials. In 1969 Effler expressed his belief that the obvious mechanical evidence justified his faith in CABG: "Arteriographic proof that the pathological ligature has been removed effectively and myocardial perfusion has been restored is prima facie evidence that the needs of that particular individual have been met. The cardiologist who would loudly deny the existence and validity of such factual evidence by refusal to examine it is, in my opinion, allowing emotion to prevail over scientific evaluation." In 1974 he dismissed the "constant harping" of those who demanded RCTs. In 1976 he espoused a different ethic than that of Chalmers, arguing that performing an RCT for CABG would be unethical: no patient who had been adequately informed would consent to go without surgery.[37] Even after the V.A. study was published, many surgeons continued to be categorical in their dismissals of RCTs, mocking "the almost religious fervor of those who would sanctify randomized studies as the only means of learning the truth." DeBakey agreed: to insist that an RCT "is the only scientific basis for assessment is, itself, unscientific."[38]

Surgeons acknowledged that RCTs had value in principle but held that they were inappropriate for surgery. In 1975 Jack Love argued that Spodick's call for routine surgical

[35] For the general results see Murphy *et al.,* "Treatment of Chronic Stable Angina" (cit. n. 29), p. 621; for the subgroup results see Timothy Takaro, Herbert N. Hultgren, Martin J. Lipton, Katherine M. Detre, and Participants in the Veterans Administration Study Group, "The VA Cooperative Randomized Study of Surgery for Coronary Arterial Occlusive Disease, II: Subgroup with Significant Left Main Lesions," *Cardiovascular Surgery,* 1976, *54*:III-107–III-116, esp. p. III-107.

[36] Braunwald, "Editorial: Coronary-Artery Surgery at the Crossroads" (cit. n. 19); and "Special Correspondence: A Debate on Coronary Bypass," *New Engl. J. Med.,* 1977, *297*:1464–1470, on pp. 1465–1466 (Spodick), 1469–1470 (Braunwald).

[37] Donald B. Effler, "The Role of Surgery in the Treatment of Coronary Artery Disease," *Ann. Thorac. Surg.,* 1969, *8*:376–379, on p. 377; Effler, "Revascularization Surgery—Letter," *J. Thorac. Cardiovasc. Surg.,* 1974, *6*:977; and Kolata, "Coronary Bypass Surgery" (cit. n. 20), p. 1264 (RCT for CABG unethical). By the mid 1970s, the original Cleveland Clinic team had split up. In 1971 Favaloro returned to Buenos Aires to establish a cardiovascular surgery center; by 1976 Effler had moved to St. Joseph's Hospital in Syracuse, New York, where he established a CABG center in a community hospital.

[38] Lawrence I. Bonchek, "Are Randomized Trials Appropriate for Evaluating New Operations?" *New Engl. J. Med.,* 1979, *301*:44–45, on p. 45; and De Bakey and Lawrie, "Aortocoronary-Artery Bypass" (cit. n. 1), p. 839. See also Ralph Berg, "Reply to Spodick," *J. Thorac. Cardiovasc. Surg.,* 1982, *83*:150–151. Hywel Davies, a V.A. surgical chief, mocked Chalmers as "the high priest among the randomizer theologians" and was skeptical of the "cult nature" of RCT advocates: Davies, "Letter: More on Coronary Bypass Surgery—I," *Amer. J. Cardiol.,* 1979, *43*:1060–1061.

RCTs was "unrealistic and naive. It fails completely to take into account some important differences between drugs and operations." After all, every patient presented a unique challenge, every surgeon had different skills, and each operation could utilize a bewildering range of procedures. Since sham operations were not considered ethical, the study could not be blinded: patients and physicians would know which treatment each patient received, reintroducing a bias RCTs were designed to eliminate. And, as many surgeons noted, since the studies required many years of follow-up, they faced the problem of evolving techniques: "Just when we have accumulated enough data over a sufficient time period, we find that surgical technique has improved or medical therapy changes, or both, and conclusions no longer apply." Surgeons used these excuses to avoid RCTs. Since the results of RCTs would not be "completely *objective*," David Sabiston argued in 1971, the studies should not be done. In 1979 Hywel Davies described how he had feared that the V.A. study would be "an expensive and time-consuming effort without valid conclusions"; as a result, the V.A. hospital at which he was chief of surgery did not participate.[39]

As soon as the main report of the V.A. study was published, CABG enthusiasts responded with a firestorm of critique. The *New England Journal of Medicine* received so many letters that it published a special correspondence section in a subsequent issue. Cardiac surgeons highlighted weaknesses in the design and conduct of the study (complaining that it was "marred by very serious flaws"), suggested that the study's surgeons had performed CABG poorly, and provided their own superior (and retrospective) results to demonstrate the real value of CABG. Many surgeons were so eager to contest the findings of the V.A. study that they arrived at academic conferences armed with slides illustrating their own mortality data.[40]

Debate about the V.A. study continued for years. The *American Journal of Cardiology* published a typical exchange. The Cleveland Clinic team critiqued the V.A. trial, described their own superior results (95 percent survival after CABG, better than both the medical and surgical groups in the V.A. trial), and protested, "Why should we judge a mode of therapy on the basis of mediocre performance?" The V.A. group responded that their results were consistent with those from other centers and that the study's weaknesses were insignificant. Braunwald praised the study as "at least a step in the direction of rationally attempting to compare the results of two methods of management of chronic stable angina pectoris." In a similar symposium in *Clinical Research*, Chalmers defended the rigor of the V.A. trial, while the Cleveland Clinic cardiologist William Proudfit argued that RCTs are "not the only route to wisdom." Leading surgical groups fueled the controversy by publishing ever larger retrospective case series. For example, the Texas Heart Institute team insisted that their experience, with more than 10,000 operations, demonstrated the superiority of CABG.[41]

---

[39] Jack W. Love, "Drugs and Operations: Some Important Differences," *J. Amer. Med. Assoc.*, 1975, *232*:37–38; Daniel J. Ullyot, Judith Wisneski, Robert W. Sullivan, and Edward W. Gertz, "Reply to Takaro," *J. Thorac. Cardiovasc. Surg.*, 1976, *72*:945; David C. Sabiston, "Reply to Spodick," *Circulation*, 1971, *44*:302; and Davies, "Letter: More on Coronary Bypass Surgery—I," pp. 1060–1061.

[40] See, e.g., Floyd D. Loop, comment following Raymond C. Read, Marvin L. Murphy, Herbert N. Hultgren, and Timothy Takaro, "Survival of Men Treated for Chronic Stable Angina Pectoris: A Cooperative Randomized Study," *J. Thorac. Cardiovasc. Surg.*, 1978, *75*:1–16, on pp. 13–14. For the special section see "Special Correspondence: A Debate on Coronary Bypass" (cit. n. 36).

[41] Floyd D. Loop, William L. Proudfit, and William C. Sheldon, "Coronary Bypass Surgery Weighed in the Balance," *Amer. J. Cardiol.*, 1978, *42*:154–156, on p. 155 (Cleveland Clinic team); Herbert N. Hultgren, Timothy Takaro, Katherine M. Detre, and Marvin L. Murphy, "Evaluation of the Efficacy of Coronary Bypass Surgery—I," *ibid.*, pp. 157–160; Eugene Braunwald, "Evaluation of the Efficacy of Coronary Bypass Surgery—II," *ibid.*,

Amidst this controversy, one aspect of the V.A. study went unquestioned. Critics neither noted that all 596 patients in the V.A. study were men nor criticized the authors for not indicating the racial composition of the patient groups. When CABG appeared, coronary artery disease was defined as a problem of affluent men, though, ironically, Favaloro's first CABG patient seems to have been a woman.[42] What was going on here? On the surface, race and gender were not relevant categories for the surgeons. In choosing patients for CABG surgeons focused on angiographic details of their coronary arteries. Favaloro chose his first patient because she had an obstructed right coronary artery with good collateral flow: if the experiment failed, she would have been no worse off than before. Three years and 228 patients later, Favaloro still chose patients according to the angiographic state of their vessels.[43] After all, once the patient was draped for surgery only the heart and its obstructed arteries remained visible.

Indifference to race and gender does not explain the selection of patients for CABG, however. When gender was specified, the patients were overwhelmingly male. When race was specified, the patients were overwhelmingly white. White men were over-represented among patients chosen for CABG given the distribution of heart disease in the population. Did the surgeons at elite centers, such as the Cleveland Clinic and the Texas Heart Institute, see only "professional people" with the resources to reach those centers?[44] Did racial and gender biases skew both the diagnosis of coronary artery disease and the choice of surgical treatment? Careful analysis of patient selection in the surgical series and randomized trials might unearth the forces at work in the 1970s. When these questions began to receive adequate scrutiny in the 1990s, researchers confirmed the importance of gender and racial

---

pp. 161–162, on p. 162; Thomas C. Chalmers, Harry Smith, Alexander Ambroz, Dinah Ritman, and Biruta J. Schroeder, "In Defense of the VA Randomized Control Trial of Coronary Artery Surgery," *Clin. Res.,* 1978, *26*:230–235; Proudfit, "Criticisms of the VA Randomized Study of Coronary Bypass Surgery," *ibid.,* pp. 236–240, on p. 236; and Frank M. Sandiford, Denton A. Cooley, and Don C. Wukasch, "The Aortocoronary Bypass Operation: Myth and Reality: An Overview Based on 10,000 Operations at the Texas Heart Institute," *International Surgery,* 1978, *63*:83–89.

[42] For the study see Murphy *et al.,* "Treatment of Chronic Stable Angina" (cit. n. 29), p. 621. On Favaloro's first CABG patient see Favaloro, "Present Era of Myocardial Revascularization" (cit. n. 17), p. 334; and Favaloro, "Critical Analysis of Coronary Artery Bypass Graft Surgery" (cit. n. 1), p. 2B. Elsewhere, however, Favaloro writes that his first CABG patient was a man: Favaloro, *Challenging Dream of Heart Surgery,* p. 95. Braunwald's discussion of risk factors for coronary artery disease compared men with a history of high cholesterol, high blood pressure, and smoking to men free of those diseases: Lesch *et al.,* "Ischemic Heart Disease" (cit. n. 9), pp. 1262–1263. As mentioned earlier, Effler emphasized the impact of coronary artery disease on "professional people, executives and men in public office": Effler, "Surgery for Coronary Disease" (cit. n. 7), p. 43.

[43] Favaloro, "Present Era of Myocardial Revascularization," p. 334; and Favaloro *et al.,* "Severe Segmental Obstruction of the Left Main Coronary Artery and Its Divisions" (cit. n. 33), p. 475. Explicit discussions of race or gender rarely appeared in the cardiac surgery literature before the 1990s. A keyword search on Medline for "race" or "gender" in the *Annals of Thoracic Surgery* and the *Journal of Thoracic and Cardiovascular Surgery* revealed no matches from 1966 to 1974; "race" had four matches and "gender" three matches between 1975 and 1986.

[44] Effler provided one clue: "The Cleveland Clinic experience is based upon private patients who are referred from all parts of the world." Effler, "Myocardial Revascularization" (cit. n. 25), p. 79. Of the Cleveland Clinic's first 1,000 patients, 87.4 percent were men: W. C. Sheldon, G. Rincon, D. B. Effler, W. T. Proudfit, and F. M. Sones, "Vein Graft Surgery for Coronary Artery Disease: Survival and Angiographic Results in 1,000 Patients," *Circulation,* 1973, *47–48* (Suppl. 3): III-184–III-189, on p. III-185. Of 4,522 patients at the Texas Heart Institute, 86 percent were men: Reul *et al.,* "Long-Term Survival Following Coronary Artery Bypass" (cit. n. 26), p. 1419. Neither the early Cleveland Clinic trials nor the V.A. study mentions race. The first mention of race in a study of CABG that I found was the CASS trial, published in 1983. Of the 780 patients, 90.3 percent were male and 98.3 percent were white; CASS Principal Investigators and Their Associates, "Coronary Artery Surgery Study (CASS): A Randomized Trial of Coronary Artery Bypass Surgery: Survival Data," *Circulation,* 1983, *68*:939–950, on p. 942. Death rates per 100,000 from "diseases of heart" in 1968: white male, 362.9; white female, 180.5; other male, 391.4; other female, 271.4. See U.S. Department of Health and Human Services, *Vital Statistics of the United States, 1968,* Vol. 2: *Mortality* (cit. n. 3), Table 1-6.

analysis in cardiac trials; the role of bias in referral for cardiac procedures remains contested.[45]

The V.A. trial was not the last word on the problem of CABG. RCTs, case series, and consensus reports continued to appear for decades. But the debate over the V.A. study was the defining moment in the history of RCTs for CABG. Many physicians wished the whole affair had never happened. Cardiologists at Emory feared that the controversy over the RCTs had "blunted the enthusiasm for such an approach." As Braunwald wrote in 1981, everyone would have been better off without the "major controversy, even acrimony. The conflicts among cardiologists and cardiovascular surgeons spilled into the lay press, confusing patients and physicians alike."[46] What had produced such a contentious and intractable debate?

WAS IT ALL IN THE DETAILS?

As Harry Marks, Steven Epstein, and Evelleen Richards have shown, RCTs are never simple to conduct. Problems with patient selection, adherence to assigned treatments, and interpretation of results all fuel fierce debate. There is much evidence that this happened with RCTs of CABG. Critics immediately attacked perceived weaknesses of the V.A. study: selection of a low-risk group of patients (resulting in remarkably high survival in the medical group), poor surgical results (high operative mortality, low graft patency), and poor compliance with the treatment specified (many patients initially assigned to the medical group subsequently had surgery).[47]

Consider operative mortality. Supporters of the study defended its 5.6 percent operative mortality rate as consistent with rates for most surgery performed between 1972 and 1974. But opponents argued that the results were unacceptably poor. Both groups were right, depending on what standards (elite, national average, community hospital) were seen as most relevant. As the controversy continued, some contestants resorted to rhetorical chicanery. Critics from the Cleveland Clinic and the Texas Heart Institute argued that the V.A. data—399 patients operated on in thirteen hospitals over three years—proved that the participating surgeons had inadequate experience with CABG, performing less than one operation per hospital per month. The V.A. group was quick to defend their experience:

[45] Women had different cardiac outcomes than men: E. S. Tan, J. van der Meer, P. Jan de Kam, P. H. Dunselman, B. J. Mulder, C. A. Ascoop, M. Pfisterer, and K. I. Lie, "Worse Clinical Outcome but Similar Graft Patency in Women versus Men One Year after Coronary Artery Bypass Graft Surgery Owing to an Excess of Exposed Risk Factors in Women," *J. Amer. Coll. Cardiol.*, 1999, *34*:1760–1768. Minorities were less likely to receive invasive cardiac procedures than whites: M. B. Wenneker and A. M. Epstein, "Racial Inequalities in the Use of Procedures for Patients with Ischemic Heart Disease in Massachusetts," *J. Amer. Med. Assoc.*, 1989, *261*:253–257; and J. Whittle, J. Conigliaro, C. B. Good, and R. P. Lofgren, "Racial Differences in the Use of Invasive Cardiovascular Procedures in the Department of Veterans Affairs Medical System," *New Engl. J. Med.*, 1993, *329*:621–627. On questions of gender and racial bias see H. J. Geiger, "Race and Health Care—An American Dilemma?" *ibid.*, 1996, *335*:815–816; and L. M. Schwartz, S. Woloshin, H. G. Welch, and the V.A. Outcomes Group, "Misunderstandings about the Effects of Race and Sex on Physicians' Referrals for Cardiac Catheterization," *ibid.*, 1999, *341*:279–283.

[46] J. Willis Hurst, Spencer B. King, R. Bruce Logue, Charles R. Hatcher, Ellis L. Jones, Joe M. Craver, John S. Douglas, Robert H. Franch, Edward R. Dorney, B. Woodfin Cobbs, Paul H. Robinson, Stephen D. Clements, Joel A. Kaplan, and James M. Bradford, "Value of Coronary Bypass Surgery: Controversies in Cardiology: Part I," *Amer. J. Cardiol.*, 1978, *42*:308–329, on p. 310; and Eugene Braunwald, "Let's Not Let the Genie Escape from the Bottle—Again," *New Engl. J. Med.*, 1981, *304*:1294–1296, on p. 1295.

[47] See, e.g., "Special Correspondence: A Debate on Coronary Bypass" (cit. n. 36).

during the study period they operated on over 1,300 patients who were not enrolled in the study.[48]

But were such methodological criticisms really at the core of critics' evaluation of the V.A. study? Their own actions indicate that this cannot be the case. First, Favaloro and many other CABG enthusiasts, though they attacked the main V.A. study, accepted the data subset showing the benefit of CABG in treating left main coronary artery obstruction. The V.A. group was quick to mock the illogic of such a position: "this subset was part of the Cooperative Study and was treated by the same surgeons, in the same institutions, under the same conditions, and in the same time frame as the remaining 88% of the patients in the study." Second, enthusiasts continued to publish and rely on their own case series, which they admitted were even more methodologically flawed than the V.A. study.[49]

Alvan Feinstein, professor of medicine and epidemiology at Yale, had a valuable insight. He had observed the competing claims about the quality of the V.A. study and concluded that both sides were right: V.A. surgery was worse than that at the Cleveland Clinic but comparable to surgery in most institutions and better than that in some. The V.A. study was flawed, but so were retrospective case series. Reasonable arguments could support either position; reasonable criticisms could undermine them. For Feinstein, this was the "essence of tragedy": "the destructive collision of two protagonists holding opposing positions, each of which is right."[50]

Such a conclusion suggests that responses to the trial were underdetermined by the available information. In fact, the responses depended on contestants' prior commitments—a fact that is made particularly clear by how observers reported the results of the V.A. study. While skeptics emphasized the negative results (no benefit in most patients), enthusiasts emphasized the positive results (benefit for patients with left main disease). Many observers noted the crucial role played by preconceptions. Spodick complained that too many people had "their minds made up in advance as to the outcome." Braunwald and McIntosh agreed. One review concluded that responses to the trials seemed "related to how such results compare with preconceived views." Chalmers decried the "reluctance of physicians to accept the results of clinical trials when the conclusions are contrary to conventional wisdom."[51]

[48] On the issue of standards see Braunwald, "Evaluation of the Efficacy of Coronary Bypass Surgery—II" (cit. n. 41), p. 161; and Feinstein, "Scientific and Clinical Tribulations of Randomized Clinical Trials" (cit. n. 28), p. 243. For the charges of inadequate experience see Don C. Wukasch, Denton A. Cooley, Robert J. Hall, George J. Reul, Frank M. Sandiford, and Sherri L. Zillgitt, "Surgical versus Medical Treatment of Coronary Artery Disease: Nine Year Follow-up of 9,061 Patients," *Amer. J. Surg.*, 1979, *137*:201–207; and Loop, comment following Read *et al.*, "Survival of Men Treated for Chronic Stable Angina Pectoris" (cit. n. 40), p. 13. For the V.A. group's defense see Timothy Takaro and the Participants in the V.A. Cooperative Study of Surgery for Coronary Arterial Occlusive Disease, "Results of a Randomized Study of Medical and Surgical Management of Angina Pectoris," *World Journal of Surgery*, 1978, *2*:797–807, on p. 803.

[49] For an essay that accepted the results regarding left main coronary artery obstruction see Favaloro, "Critical Analysis of Coronary Artery Bypass Graft Surgery" (cit. n. 1), p. 9B. The V.A. authors mocked this logic in Takaro and Participants in V.A. Cooperative Study, "Results of a Randomized Study of Medical and Surgical Management of Angina Pectoris," pp. 800–801; on reliance on methodologically flawed case series see Thomas A. Preston's comment following this essay: p. 809. See also the contributions of Spodick and Braunwald to "Special Correspondence: A Debate on Coronary Bypass" (cit. n. 36), pp. 1465–1466, 1469–1470.

[50] Feinstein, "Scientific and Clinical Tribulations of Randomized Clinical Trials" (cit. n. 28), p. 241.

[51] David H. Spodick, "Coronary Bypass: How Good and for Whom?" *Modern Medicine*, 2 Apr. 1973, *41*: 80–81; Braunwald, "Editorial: Coronary-Artery Surgery at the Crossroads" (cit. n. 19), p. 662; Henry D. McIntosh and Robert A. Buccino, "Value of Coronary Bypass Surgery: Controversies in Cardiology (Continued)," *Amer. J. Cardiol.*, 1979, *44*:387–389, on p. 388; Robert L. Frye, Lloyd Fisher, Hartzell V. Schaff, Bernard J. Gersh, Ronald E. Vliestra, and Michael B. Mock, "Randomized Trials in Coronary Artery Bypass Surgery," *Progress in Cardiovascular Diseases*, 1967, *30*:1–22, on p. 15; and Thomas C. Chalmers, "The Clinical Trial,"

The debates over the V.A. study and about RCTs for CABG in general, therefore, did not grow out of technical concerns with the studies. Those who attacked RCTs most fiercely were those who had previously denied the need for them, and vice versa. Participants' positions were not the products of rational analysis of study data. Instead, the debates grew out of faith, or lack of it, in CABG itself. Preconception and rationalization drove the technical debates. But what was the source of participants' faith? What generated their allegiances, the preconceptions through which they perceived the trial?

SPODICK AS PSYCHOLOGIST

One explanation for the faith of CABG enthusiasts appeared very early in the debates. In 1971 Spodick wrote an editorial in which he described factors that influenced the psychological disposition of cardiologists and cardiac surgeons toward CABG.[52] He sought to explain why so many people in both groups (for nearly every cardiac surgeon performing CABG, there was a diagnosing cardiologist who recommended the procedure) did not see a need for trials. He traced this "credulity" to human, statistical, and professional factors.

At the human level, Spodick believed that the Cleveland Clinic team and other surgical pioneers were blinded by hopes that their innovation would succeed: "Few treatments succeed as well as they do in the hands of their originators. Here, Invention too often becomes the mother of Necessity." At the statistical level, Spodick argued that enthusiasts succumbed to classic statistical delusions, demonstrating both a remarkable willingness to be convinced by anecdotal reports and an uncritical awe of massive case series. At the professional level, Spodick characterized cardiac specialists as "prima donnas." Their professional stature had given them a sense of "olympianism." "The loftily self-sufficient doctor is convinced that he is a uniquely qualified judge of his own decisions, which therefore require little or no outside assistance." Finally, surgeons were motivated by ideals of activism. Like climbers facing a new mountain, surgeons performed CABG "simply 'because it's there.' "[53]

Spodick's analysis, however, begs the question. These psychological factors might have operated. But where did they come from? Why were some people outside of this psychological atmosphere calling for RCTs? Enthusiasm and olympianism must be contextualized.

DISCIPLINARY WARFARE?

An effort to explain the origins of controversy and enthusiasm must be grounded in an understanding of the disciplinary perspectives of cardiology and cardiac surgery. Before exploring this background, however, I must stress that a simple framework of interdisciplinary hostility does not explain the actions of the participants in the debates. To begin with, cardiologists themselves were fundamentally involved in the CABG industry, often performing the diagnostic angiography and providing pre- and postoperative care. Fur-

---

Millbank Mem. Fund Quart., 1981, 59:324–339, on p. 328. For skeptics' views of the V.A. study see Braunwald, "Editorial: Coronary-Artery Surgery at the Crossroads," p. 661; and McIntosh, "Commentaries on the Consensus Conference of Coronary Artery Bypass Surgery" (cit. n. 20), p. 835. For enthusiasts' views see Hurst *et al.*, "Value of Coronary Bypass Surgery" (cit. n. 46), p. 327; and James Jude, "Comment," *J. Florida Med. Assoc.*, 1981, 68:835.

[52] Spodick, "Editorial: Revascularization of the Heart" (cit. n. 24).

[53] *Ibid.*, pp. 152, 154, 155.

thermore, the controversy did not split cleanly along disciplinary lines. Spodick blamed not surgeons, but journal editors who did not demand RCTs from surgical researchers. Braunwald collaborated with cardiac surgeons, including Nina Braunwald, his wife. Favaloro had nothing but praise for Mason Sones, his cardiologist collaborator. Effler, who rarely restrained his attacks against cardiologists, admitted that the problem was not with the whole specialty, just its older and more conservative members.[54] And some surgeons, like Takaro and his V.A. collaborators, were willing to perform trials.

Letters and editorials provide ample evidence that the debate was perceived by some as a disciplinary battle, however. Military metaphors abounded, with Effler and Edward Diethrick, who had trained with DeBakey, writing about "battle lines," "gladiators," and the "resistance movement." Participants had a clear sense of right and wrong. Henry McIntosh contrasted rational skeptics and surgical zealots. Spodick earned the accolade "the conscience of cardiology" for his advocacy of RCTs. Each side accused the other of being old-fashioned. Diethrick decried the "atmosphere of notorious conservatism" that pervaded cardiology; Effler accused cardiologists of "wallowing in the glory that came with the development of the electrocardiogram (which happened in my childhood)." Cardiologists, in turn, criticized surgeons for their reliance on case series, the method "used to support previous (now discredited) operations that produced a similar relief of angina"; it was "no more scientific today than it was 40 years ago."[55]

Such rhetoric is not surprising. CABG involved the health of·millions of patients and billions of dollars in physician fees. At some hospitals there was considerable tension between the two groups. Further controversy appeared in debates about whether CABG should be confined to elite academic medical centers or allowed into local community hospitals.[56]

This competitiveness was exacerbated by the different histories of the cardiologists and cardiac surgeons over the previous three decades. As described by Bruce Fye, traditional taboos against cardiac surgery dissolved between 1940 and 1970, transforming the field from helplessness to hopefulness. New technologies, especially the development of cardiopulmonary bypass in the 1950s, greatly expanded surgeons' therapeutic abilities, as demonstrated by DeBakey's successful repairs of previously fatal aortic aneurysms. This work reached its climax with the first successful heart transplant, by Christiaan Barnard, in December 1967. The media celebrated surgeons' heroics and chronicled the feuds between their "quite enormous egos."[57] Such dramatic progress created a specialty of people

[54] For criticism of journal editors see David H. Spodick, "More on Coronary Bypass Surgery—II," *Amer. J. Cardiol.,* 1979, *43*:1061–1062, on p. 1061. On Eugene Braunwald's collaboration with his wife see Harvard University Press Release, 7 Jan. 1972, Countway Library Archives, Harvard Medical School, Boston, Archives GC File: Eugene Braunwald, M.D. New York U., 1952. Sones is praised in Favaloro, *Challenging Dream of Heart Surgery,* p. 115. Effler criticizes conservative cardiologists in Effler, "Myocardial Revascularization—Direct or Indirect?" (cit. n. 17), p. 499.

[55] Donald B. Effler, "Myocardial Revascularization at the Community Hospital Level," *Amer. J. Cardiol.,* 1973, *32*:240–242; Effler, "Myocardial Revascularization—Direct or Indirect?" pp. 498–499; Edward B. Diethrick, ". . . And the War Goes On," *Chest,* 1973, *63*:83–85; McIntosh, comment following Read *et al.,* "Survival of Men Treated for Chronic Stable Angina Pectoris" (cit. n. 40); introduction given to Spodick's Ewart Angus Lecture at the Wellesley Hospital, Univ. Toronto, 14 May 1981: Spodick, personal communication, 3 June 1998; Diethrick, ". . . And the War Goes On," p. 83; Effler, comment following Effler *et al.,* "Coronary Endarterotomy with Patch-Graft Reconstruction" (cit. n. 13), p. 601; and Preston, "Hazards of Poorly Controlled Studies in the Evaluation of Coronary Artery Surgery" (cit. n. 23), p. 441.

[56] Braunwald remembers tensions at Boston's Brigham and Women's Hospital in the 1970s: Eugene Braunwald, personal communication, 17 June 1998. On the debate regarding the spread of CABG see Effler, "Myocardial Revascularization at the Community Hospital Level."

[57] On the transformation of cardiac surgery see Fye, *American Cardiology* (cit. n. 13), pp. 164–175. DeBakey's

with tremendous confidence in themselves, as manifested in the olympianism described by Spodick. Cardiac surgeons had learned that direct repair of damaged hearts cured patients. Why should CABG be different?

As noted earlier, cardiologists had been feeling substantially less triumphant in 1968, especially about coronary artery disease. This might have made them particularly sensitive to surgeons' enthusiastic claims about the powers of CABG. But while cardiologists might have been overshadowed in the media, they did make dramatic progress in many areas during the first decade of CABG. New understanding of risk factors, such as diet, lack of exercise, and smoking, enabled them to suggest better preventive care for their patients. New drugs enabled them to treat anemia, hypertension, and other conditions that exacerbated coronary artery disease. New technologies, including echocardiography and radionucleotide imaging, improved diagnostic accuracy. Better methods of resuscitation and intensive care enabled them to save the lives of patients whose coronary artery disease culminated in a heart attack. While cardiologists might have lacked the boundless confidence of their surgical colleagues, they had faith that continued work would eventually lead to definitive medical treatments of coronary artery disease. By 1978 Braunwald could declare that the "golden age" was at hand.[58] With success seemingly within their reach, cardiologists found the heroics of surgery unnecessary.

DISCIPLINARY STANDARDS OF KNOWLEDGE

Disciplinary allegiances were not simply a medium through which disagreement was voiced. The different disciplinary histories left other traces as well. By 1970 RCTs had become enshrined—in principle—as the standard for evaluating the efficacy of drug-based therapeutics. But as the debates about CABG show, the expansion of this standard into cardiac surgery met substantial resistance.

Traditionally, surgeons had relied on three types of research: animal experiments, case reports, and case series. Many groups (though not the Cleveland Clinic team) had experimented with CABG in animal models before attempting it in humans. Similarly, Favaloro's first two reports about CABG involved only small groups of patients. His evaluation focused on operative technique and technical feasibility: the patients survived and revascularization was achieved; clinical concern was secondary. Many surgeons believed that

---

work is treated in Michael E. DeBakey, "Developments in Cardiovascular Surgery," *Cardiovascular Research Center Bulletin,* 1980, *19*:12–20; DeBakey, "Development of Vascular Surgery" (cit. n. 18); and Francis D. Moore, "Perspectives, Surgery," *Perspectives in Biology and Medicine,* 1982, *25*:698–721. Barnard's transplant is discussed in Fye, *American Cardiology,* p. 297; R. Dewall and R. J. Bing, "Cardiopulmonary Bypass, Perfusion of the Heart, and Cardiac Metabolism," in *Cardiology,* ed. Bing (cit. n. 6), pp. 54–83; and J. C. Baldwin, S. A. Lemaire, and Bing, "Transplantation of the Heart," *ibid.,* pp. 104–117. For cover stories on these heroics see "The Ultimate Operation," *Time,* 15 Dec. 1967, *90*:64–72; and Matt Clark, "New Hearts for Old," *Newsweek,* 18 Dec. 1967, *70*:86–90. Reporting on the media circus see "Surgery and Show Biz," *Time,* 15 Jan. 1968, *71*:49. For references to "quite enormous egos" see Thomas Thompson, "The Texas Tornado vs. Dr. Wonderful: Houston's Two Master Heart Surgeons Are Locked in a Feud," *Life,* 10 Apr. 1970, *68*:62B–74; and "Transplants: An Act of Desperation," *Time,* 18 Apr. 1969, *93*:58.

[58] Eugene Braunwald, "The First Twenty Years of the *American Journal of Cardiology:* A Chronicle of the Golden Age of Cardiology," *Amer. J. Cardiol.,* 1978, *42*:5–7. See also Fye, *American Cardiology,* p. 249. On progress in cardiology during the first decade of CABG see David H. Spodick, "Current Treatment of Angina: What Do We Know?" *Louisiana State Medical Journal,* 1972, *124*:99–104; McIntosh and Garcia, "First Decade of Aortocoronary Bypass Grafting" (cit. n. 20), p. 415; Aronowitz, "From the Patient's Angina to the Cardiologist's Coronary Heart Disease" (cit. n. 8); Robert A. Aronowitz, "The Social Construction of Coronary Heart Disease Risk Factors," in *Making Sense of Illness* (cit. n. 8), pp. 111–144; Fye, *American Cardiology,* p. 176; and Ryan, "Revascularization" (cit. n. 21), p. 92B.

such simple methods were sufficient to confirm the efficacy of CABG, just as they had proven the value of penicillin and appendectomies.[59]

Critics had a simple response. Howard Hiatt, an oncologist at the Harvard School of Public Health, agreed that therapeutic efficacy could be self-evident in some cases, when the disease was "uniformly fatal in outcome and often devastating in manifestations." This had been the case with appendicitis and aortic aneurysms. But coronary artery disease was not such a case: CABG "does not lead to speedy and uniform improvement," and the symptoms of coronary artery disease "are subject to inexplicable remissions and exacerbations."[60] Even when surgeons operated following the acute drama of a heart attack, they focused on preventing a second heart attack, a probabilistic phenomenon that could be demonstrated only with well-controlled trials. Surgeons thus faced a chronic condition that did not allow dramatic, definitive demonstrations of therapeutic efficacy. Furthermore, by moving into an area traditionally managed medically, they had to confront medical standards of knowledge—the RCT. In this model, the ensuing controversy reflected the growing pains of a new standard of knowledge introduced into surgery.

However, as already noted, the antagonists in the controversy over CABG did not split cleanly along disciplinary lines. To begin with, the status of RCTs within cardiology was complicated: RCTs did not have a monopoly on knowledge production. Instead, different standards of knowledge coexisted. Some, like Spodick, who had trained with RCT guru Thomas Chalmers, remained thoroughly committed to trials as the surest route to knowledge. Others, like Braunwald, experienced in the instrumental traditions of cardiac physiology, moved freely between advocacy of RCTs for CABG and the use of less rigorous protocols for other research questions.[61]

There were many reasons for cardiologists' continuing affinity for the traditional methods of cardiac physiology. Spodick claims that cardiologists were the last of the medical subspecialists to join the RCT bandwagon because of their unique ability to measure and modify the heart's function and dysfunction: "we can make the heart perform tricks, with everything from simple bedside maneuvers to sophisticated pharmacologic and physiologic interventions. It reacts promptly, with responses we can measure in milliseconds, and even our treatments often produce rapid and quantifiable responses."[62]

This ability to make the heart perform tricks had long shaped the traditions of cardiac research. Into the late 1960s cardiologists, like cardiac surgeons, maintained active research programs in cardiac physiology. Braunwald's 1966 review of progress in cardiology reveals a veritable menagerie of animal models and organ preparations: cows, rabbits, dogs, cats, frogs, and humans; normal hearts, isolated hearts, trypsin-digested embryonic hearts,

---

[59] On use—or not—of animal models see Favaloro, *Challenging Dream of Heart Surgery,* p. 98; and Connolly, "History of Coronary Artery Surgery" (cit. n. 15), pp. 734–741. Favaloro's focus is clear in Favaloro, "Saphenous Vein Autograft Replacement of Severe Segmental Coronary Artery Occlusion" (cit. n. 16). For another surgeon's view see Sir John Loewenthal, "Comment," *Australian and New Zealand Journal of Medicine,* 1979, 9:118–120. Ellen Koch has described the empirical focus of surgical research in the 1950s: Koch, "In the Image of Science? Negotiating the Development of Diagnostic Ultrasound in the Cultures of Surgery and Radiology," *Technology and Culture,* 1993, 34:858–893, on p. 875.

[60] Howard H. Hiatt, "Lessons of the Coronary-Bypass Debate," *New Engl. J. Med.,* 1977, 297:1462–1464, on p. 1462.

[61] See, e.g., Peter R. Maroko, Peter Libby, and Eugene Braunwald, "Effect of Pharmacologic Agents on the Function of the Ischemic Heart," *Amer. J. Cardiol.,* 1973, 32:930–936. As Braunwald describes it, he was not an RCT "crusader" like Spodick. He believed that they were crucial for some questions but that other research designs still had tremendous value: Braunwald, personal communication, 17 June 1998.

[62] David H. Spodick, "Controlled Clinical Trials of Cardiac Surgery—What Happens Next Time?" *Cardiovascular Medicine,* 1978, 3:871–876, on p. 871.

transplanted hearts, and acutely failing dog hearts. This physiological tradition shaped cardiologists' clinical research. In 1967 Nina and Eugene Braunwald developed a new method of treating angina, electrical stimulation of the carotid sinus. Two reports—case series based on only two, and then seventeen, patients—were published in the prestigious *New England Journal of Medicine*. A controlled trial of this new technique was only in the planning stage when Favaloro and Effler's "landmark report" on CABG made carotid sinus stimulation obsolete.[63]

The central presence of physiological research in cardiology was strengthened, throughout the 1960s and 1970s, by a series of new technologies. As Fye has described, cardiac care units, catheterization, angiography, cardiac ultrasound, nuclear cardiology, and pacemakers all captured the attention of cardiologists. This affinity for physiological and instrumental approaches persisted as RCTs became the new standard for clinical research. Cardiologists were left suspended between ideals of research, between the rigorous power of RCTs and the simpler, more accessible appeal of case series, between statistical analyses of mortality and physiological assessments of cardiac blood flow and perfusion. While large studies and the abstracted experience of hundreds of patients provided the surest evidence of the impact of CABG on life expectancy, their generalized results could not be easily applied to the specific circumstances of individual patients.[64] More narrowly defined subgroups could improve the clinical applicability of RCTs, but they would require larger, more complicated trials. All the while cardiologists remained ambivalent about the relevance of clinical (angina, work capacity) and statistical (life expectancy) outcomes, about prioritizing the quality or the quantity of life.

IS SEEING BELIEVING?

Such ambivalence was exacerbated by disagreement about the persuasiveness of visualization. CABG, as its supporters celebrated, had direct, immediate, and visible mechanical effects. When Favaloro's team completed their first saphenous vein graft, they "could see the branches of the right coronary artery fill with blood." Subsequent publications always described the visual evidence of graft patency provided by angiography. Patient and surgeon could see that the pathological obstruction had been removed and that coronary circulation had been restored: the patient was cured. As one historian of CABG has noted, "the morphological aspect seemed so convincing and self-evident that functional proof seemed unnecessary."[65]

---

[63] Eugene Braunwald, "Heart," *Annual Review of Physiology,* 1966, 28:227–266; Eugene Braunwald, Stephen E. Epstein, Gerald Glick, Andrew S. Wechsler, and Nina S. Braunwald, "Relief of Angina Pectoris by Electrical Stimulation of the Carotid-Sinus Nerves," *New Engl. J. Med.,* 1967, 277:1278–1283; Epstein, David Beiser, Robert E. Goldstein, David Redwood, Douglas R. Rosing, Glick, Wechsler, Morris Stampfer, Lawrence S. Cohen, Robert L. Reis, Nina S. Braunwald, and Eugene Braunwald, "Treatment of Angina Pectoris by Electrical Stimulation of the Carotid-Sinus Nerves," *ibid.,* 1969, 280:971–978; and Eugene Braunwald, "Myocardial Ischemia, Infarction, and Failure: An Odyssey," *Cardioscience,* 1994, 5(Suppl. 3):139–144, on p. 140 ("landmark report").

[64] Fye, *American Cardiology* (cit. n. 13), pp. 250–273; and Kolata, "Coronary Bypass Surgery" (cit. n. 20), p. 1265.

[65] Favaloro, *Challenging Dream of Heart Surgery,* p. 96 ("fill with blood"); Favaloro, "Saphenous Vein Autograft Replacement of Severe Segmental Coronary Artery Occlusion" (cit. n. 16), p. 337 (description of visual evidence); René G. Favaloro, "Saphenous Vein Graft in the Surgical Treatment of Coronary Artery Disease: Operative Technique," *J. Thorac. Cardiovasc. Surg.,* 1969, 58:178–185, on pp. 184–185 (description of visual evidence); Effler, "Role of Surgery in the Treatment of Coronary Artery Disease" (cit. n. 37), p. 377 (obstruction removed); and Schaefer, "Case against Coronary Artery Surgery" (cit. n. 6), p. 159 (functional proof unnecessary).

The appeal of visualization is easy to understand. Humans are visual creatures. Vision dominates our experience of the world, our study of nature, and our scientific epistemology. As Peter Galison has observed, "Vision and visuality have come to be culturally super-valued, not only but markedly in the history and philosophy of science."[66] In cardiology, visualization has been a central element of research since William Harvey's demonstrations of the circulation of the blood. Electrocardiograms became popular in the 1920s because of their ability to turn the electrical activity of the heart into a visible tracing that the cardiologist could read. In the 1940s and 1950s cardiologists learned to use contrast agents in ventricular and coronary angiography to make the soft tissues of the heart visible. The study of blood flow, especially in the coronary circulation, was a major area of research in cardiology in the 1960s; it depended on techniques of visualizing this flow, including electromagnetic flowmeters and radioisotopes.[67]

Visualization of flow was crucial for treatments of coronary artery disease. Effler and his team believed that obstructed flow caused both angina and heart attacks: "Atherosclerotic obstructions that produce major myocardial perfusion deficits constitute a threat to the myocardium and, thereby, to the life of the patient." Therefore: "Any surgical procedure that could remove, or circumvent, a significant arterial occlusion and relieve a myocardial perfusion deficit would have theoretic value."[68] To put it simply:

$$\text{flow} = \text{health}$$
$$\text{no flow} = \text{disease}$$
$$\text{restored flow} = \text{cure.}$$

CABG cured this disease, as demonstrated by postoperative angiography.

The Cleveland Clinic team tried to document these assertions in a series of studies. In 1969 they reported that graft patency predicted relief of angina: "There was a direct correlation between the angiographic findings and the clinical evaluation." In 1971 they used flowmeters to show that CABG increased myocardial perfusion, which increased oxygen consumption and improved cardiac output.[69] These studies confirmed the link between visualization of flow and restoration of function. If the flow could be visualized, the obstruction had been repaired and the patient had been cured. For the Cleveland Clinic team, seeing was believing.

However, while the Cleveland Clinic team found visual evidence compelling, others were less convinced. There has long been anxiety among scientists about the extent to which visualization technologies, such as angiography, accurately represent living tissue. Ellen Koch, Nicolas Rasmussen, and Peter Galison have shown that scientists often struggle to determine whether images accurately portray the objects they study. Cardiologists had debated the extent to which angiography accurately reflected the state of the coronary arteries. In a 1973 review McIntosh and his colleagues found that, in certain circumstances,

[66] Peter Galison, *Image and Logic: A Material Culture of Microphysics* (Chicago: Univ. Chicago Press, 1997), pp. 463–464. Similarly, only the visual power of electron microscopes could have resolved such crucial issues as the structure of cells and the nature of viruses: Nicolas Rasmussen, *Picture Control: The Electron Microscope and the Transformation of Biology in America, 1940–1960* (Stanford, Calif.: Stanford Univ. Press, 1997).

[67] Braunwald, "Heart" (cit. n. 63).

[68] Donald B. Effler, René G. Favaloro, and Laurence K. Groves, "Coronary Artery Surgery Utilizing Saphenous Vein Graft Techniques: Clinical Experience with 224 Operations," *J. Thorac. Cardiovasc. Surg.*, 1970, *59*:147–154, on p. 152.

[69] Favaloro *et al.*, "Direct Myocardial Revascularization with Saphenous Vein Autograft" (cit. n. 17), p. 283 (quotation); and René G. Favaloro, "Surgical Treatment of Coronary Arteriosclerosis by the Saphenous Vein Graft Technique: Critical Analysis," *Amer. J. Cardiol.*, 1971, *28*:493–495, esp. p. 494.

angiography could be "accurate": "High quality, selective coronary arteriography can define the presence or absence of significant occlusive disease of the coronary arteries with greater than 90 percent accuracy in the hands of an experienced angiographer." However, "the degree of correlation between what is seen on the film and what is actually present in the vessel" depended on the quality of the film and the experience of its interpreter. An analysis of the angiography in the V.A. study revealed enough problems with intra- and interobserver reliability that the researchers instituted second evaluations of all films in the study.[70]

Most of the debate about coronary angiography centered on a different question of visual representation, however. Most people accepted that angiography accurately represented the state of the arteries. But they debated whether the state of the arteries represented the disease. As I have described, coronary artery disease had many facets: blocked flow both injured cells (angina) and killed cells (heart attack). Each facet could be evaluated: flow, symptoms, and mortality. Did relief of obstruction, as evidenced by the visual method of angiography, provide the most meaningful indicator of disease treatment? Or was relief of symptoms and improvement in survival, as evidenced by RCTs, more significant? Could revascularization be a meaningful "surrogate marker" for mortality?[71]

In an ideal world, the methods of angiography and RCT should have produced compatible data. Galison has demonstrated such a productive tension between visual and statistical traditions in physics. However, those who were skeptical about angiography saw a discontinuity in the chain of representation between restoration of flow and decreased mortality. In the 1960s Takaro and other members of the V.A. study group had learned to doubt the "functional significance of the morphological findings of angiography." They performed a series of animal studies and concluded that angiographic findings did not reliably reflect blood flow under physiological conditions. Subsequent clinical research showed that angiographic findings did not reliably predict relief of symptoms or mortality rates. Instead, angiographic results depended on the exact position of the angiographic catheter and the pressure used when injecting the contrast media.[72] The researchers' critical disillusionment with angiography motivated their decision to begin the RCT of the Vineberg implants that would become the RCT of CABG.[73]

Spodick had become similarly disillusioned with the visual evidence offered by angiography. His own experience taught him that angiographic measures of coronary artery dimensions and blood flow often did not correlate with relief of angina. The immediate

[70] Koch, "In the Image of Science?" (cit. n. 59), pp. 873, 876; Rasmussen, *Picture Control* (cit. n. 66), pp. 245, 253; Peter Galison, "Judgment against Objectivity," in *Picturing Science, Producing Art*, ed. Caroline A. Jones and Galison (New York: Routledge, 1998), pp. 327–359; James S. Cole, Rudolf F. Trost, Kinsman E. Wright, and Henry D. McIntosh, "Who Should Have Coronary Arteriography," *Geriatrics*, May 1973, pp. 125–129, on pp. 125, 125–126; and Katherine M. Detre, Elizabeth Wright, Marvin L. Murphy, and Timothy Takaro, "Observer Agreement in Evaluating Coronary Angiograms," *Circulation*, 1975, 52:979–986.

[71] Steven Epstein has reviewed the debates over "surrogate markers" in HIV trials: Epstein, "Activism, Drug Regulation, and the Politics of Therapeutic Evaluation in the AIDS Era" (cit. n. 5), pp. 693, 699–700.

[72] Galison, *Image and Logic* (cit. n. 66); and T. Takaro, C. H. Dart, S. M. Scott, R. G. Fish, and W. M. Nelson, "Coronary Arteriography: Indications, Techniques, Complications," *Ann. Thorac. Surg.*, 1968, 5:213–221, on p. 219. See also Dart et al., "Internal Thoracic (Mammary) Arteriography" (cit. n. 32), p. 127; and Takaro, "Enigma of the Vineberg-Sewell Implant Operation" (cit. n. 32), p. 150.

[73] Takaro and his colleagues did not demand, and perform, RCTs of CABG as a knee-jerk response. RCTs were needed in this case because the two short-term indicators, angina (clinical) and angiography (physiological), were unreliable. Like Braunwald, they believed that other, less rigorous research protocols could have value in specific circumstances, as in their own small case series of carotid sinus nerve stimulation: C. H. Dart, S. M. Scott, W. M. Nelson, R. G. Fish, and T. Takaro, "Carotid Sinus Nerve Stimulation Treatment of Angina Refractory to Other Surgical Procedures," *Ann. Thorac. Surg.*, 1971, 11:348–359.

and visible mechanistic success of CABG was reassuring but unimportant. Only evidence of long-term benefit, provided by RCTs, could demonstrate a successful treatment.[74]

Takaro and Spodick did not doubt that angiography produced accurate images of coronary arteries. Rather, they doubted that angiography, as a proxy for flow, provided a meaningful representation of the disease. Since restoration of flow did not necessarily yield relief of symptoms, it could not reliably indicate that the patient's problem had been fixed. Visual evidence of revascularization did not necessarily prove that life expectancy would be increased. Since angina, in its unpredictability, was suspect as well, mortality was left as the only meaningful measure of success. And demonstration of subtle changes in mortality rates required an RCT.

But why were Effler and his colleagues so enthusiastic about the visual power of angiography, while Spodick, Takaro, and others remained unconvinced? Effler himself had noted significant limitations of angiography. In his work with endarterectomy and patch-graft repairs, he encountered a number of cases in which "the localized obstruction proved to be far more extensive than anticipated": the team had been "misled by the preoperative angiograms." Other evidence, however, suggests that the Cleveland Clinic team had access to higher quality angiography than other centers; perhaps this strengthened their faith in the technique's results.[75] Surgeons' individual clinical experiences with angiography must have contributed to their assessments of the value of its visual evidence.

VISUALIZATION AND THE GHOSTS OF TREATMENTS PAST

I have described how participants in the debates over CABG had contrasting assessments as to whether angiography could represent coronary artery disease. These views in turn determined whether they held angiography to be a reliable measure of therapeutic efficacy. Assessments of angiography had one further crucial impact: in evaluating the relevance of the legacies of the history of cardiac therapeutics, specifically the checkered history of surgical treatments of coronary artery disease.

For many skeptics, the strongest argument against CABG was the history of cardiac therapeutics. Braunwald saw a clear lesson: "even the most casual student of medical history will acknowledge the frequency of noncritical, overenthusiastic acceptance of newly developed modes of therapy, whether medical or surgical." The cycle of enthusiasm and disillusionment was painfully familiar for those concerned with the treatment of coronary artery disease. Surgeons had produced a long series of failures, "a long, chequered and, until recently, undistinguished history." Spodick and McIntosh had both learned that few therapies are "obviously efficacious." Everyone in the field was aware of the classic work of Leonard A. Cobb, E. Grey Dimond, and others on sham trials of internal mammary artery ligation. These experiments demonstrated the necessity of controlled trials.[76] Recent experience with Vineberg implants had reinforced many observers' caution.

[74] Spodick, "Current Treatment of Angina" (cit. n. 58), p. 99; Spodick, "Editorial: Revascularization of the Heart" (cit. n. 24), p. 156; and Spodick, "Letter: Surgery for Coronary Artery Disease" (cit. n. 30), p. 449.

[75] Donald B. Effler, "Myocardial Revascularization Surgery since 1945 A.D.: Its Evolution and Its Impact," *J. Thorac. Cardiovasc. Surg.*, 1976, 72:823–828, on p. 826; and Effler, "Surgical Treatment of Myocardial Ischemia" (cit. n. 15), p. 8. By the time of Favaloro's first CABG, Sones had the world's most extensive experience with angiography. In contrast, during an earlier visit to the Texas Heart Institute (1964), Favaloro had been struck by the "poor quality of their angiography": Favaloro, *Challenging Dream of Heart Surgery,* p. 60.

[76] Braunwald, "Evaluation of the Efficacy of Coronary Bypass Surgery—II" (cit. n. 41), p. 161; Braunwald, "Coronary Artery Bypass Surgery—An Assessment" (cit. n. 20), p. 733; David H. Spodick, "Letter: Surgical

Skeptics saw CABG as just the most recent entrant in this series. As Spodick noted—and Ross and Takaro agreed—"Well founded optimism for the effectiveness of coronary bypass surgery cannot be divorced from the knowledge that previous attempts at revascularization were proclaimed and hotly pursued with equal optimism." Charles Bailey, who had experienced his own cycle of enthusiasm and disillusionment after developing endarterectomy in 1957, felt obliged "to pour cold water" on the enthusiasm for CABG generated by a Cleveland Clinic presentation at the 1970 meeting of the Society for Thoracic Surgery. He reminded the audience that in 1967 Effler had described patch-graft repairs "as the next best thing to sexual intercourse. Today he will tell you it wasn't so good."[77] Skeptics saw little reason to expect the fate of CABG to be different.

How did CABG enthusiasts respond? They did not deny the notorious history of coronary revascularization. After the failures of sympathectomy, omentopexy, and Vineberg implants, Effler could understand the "widespread disillusionment" of cardiologists, who saw persistent surgeons "as dubious characters, if not true charlatans." He regretted that these early efforts had ever been made, "as the rewards were meager and the heritage of medical resentment and suspicion remains today." Instead of denying the lessons of history, Effler and his colleagues denied their relevance: the clear moral of past failures did not apply to CABG. Why? Not because of surgical superiority, but because of the new diagnostic power of angiography. Effler argued that all previous operations had violated "the basic principle of therapy": "treatment was undertaken before adequate diagnosis." Traditional diagnostic methods—looking at symptoms, age, weight, occupation, ethnicity, and EKG—had yielded notoriously unreliable diagnoses. Healthy patients were given diagnoses of serious disease, while others suffered heart attacks "shortly after they have been given a clean bill of health." Because of inaccurate diagnosis, surgeons had in the past operated on many patients who did not actually have coronary artery disease: "it is little wonder that the early era of coronary artery surgery was destined to end in disrepute."[78]

For Effler and his colleagues, this historical pattern was shattered by the advent of selective coronary angiography. As the technique spread rapidly in the 1960s, physicians—at least those who had faith in angiography—could for the first time directly visualize the coronary arteries of their patients. Sones's "monumental work" transformed the world for surgeons at the Cleveland Clinic. It provided "visual diagnosis," a "leap forward in our ability to read coronary disease that can be fairly likened to the impact of the invention of the printing press on the written word." It gave surgeons "a literal 'road map' of the heart's

---

Treatment of Aortic Stenosis," *New Engl. J. Med.,* 1970, *282*:340; and Henry D. McIntosh, "Benefits from Aortocoronary Bypass Graft," *J. Amer. Med. Assoc.,* 1978, *239*:1197–1199, on p. 1198. On the demonstrated need for controlled trials see, e.g., Hultgren *et al.,* "Evaluation of the Efficacy of Coronary Bypass Surgery—I" (cit. n. 41), pp. 159–160; and Barsamian, "Rise and Fall of Internal Mammary Artery Ligation in the Treatment of Angina Pectoris and the Lessons Learned" (cit. n. 12), pp. 213, 217–218.

[77] Spodick, "Letter: Coronary Bypass Operations" (cit. n. 24), pp. 55–56; Ross, "Surgery for Coronary Artery Disease Placed in Perspective" (cit. n. 19), pp. 1167, 1163; Herbert N, Hultgren, Timothy Takaro, Katherine M. Detre, and Marvin L. Murphy, "Aortocoronary-Artery-Bypass Assessment after Thirteen Years," *J. Amer. Med. Assoc.,* 1978, *240*:1353–1354; and Charles Bailey, comment following René G. Favaloro, Donald B. Effler, Laurence K. Groves, William C. Sheldon, and F. Mason Sones, "Direct Myocardial Revascularization by Saphenous Vein Graft: Present Operative Technique and Indications," *Ann. Thorac. Surg.,* 1970, *10*:97–111, on p. 111.

[78] Donald B. Effler, "A New Era of Coronary Artery Surgery," *Surgery, Gynecology, and Obstetrics,* 1966, *123*:1310–1311, on pp. 1311, 1310; Effler, "Myocardial Revascularization Surgery since 1945 A.D." (cit. n. 75), p. 824; Effler, "Current Era of Revascularization Surgery," *Surg. Clin. North Amer.,* 1971, *51*:1009–1013, on p. 1009; Effler, "Surgery for Coronary Disease" (cit. n. 7), pp. 36, 38; and Effler, "New Era of Coronary Artery Surgery," p. 1311.

blood supply, with the obstructions clearly visible."[79] It provided preoperative diagnoses and postoperative evidence of successful revascularization.

Acknowledging the transformative power of visualization, Cleveland Clinic surgeons divided their knowledge of coronary artery disease into two eras: "that before and that after coronary angiography." Other CABG advocates agreed about the pivotal contribution of Sones's technique, "the seminal event that prepared the way for the development of coronary revascularization." This new diagnostic power made angiography "the *sine qua non* of revascularization surgery." Postoperative demonstration of graft patency, again provided by angiography, validated the era of CABG. Effler hoped that it was "an era that may never end."[80]

So while critics argued that the legacy of past failures required that the burden of proof for CABG be set very high, enthusiasts disagreed. Revascularization surgery in the era of angiography bore no relation to what had come before. For Effler, this meant that CABG should not be constrained by the failures of the past: "Whatever surgical efforts were expended before are of historical interest only, and it does little good to dwell on past failures; besides, the statute of limitations for an earlier era should have expired by now."[81]

HISTORY REPEATS ITSELF?

The debates over CABG showed that faith in angiography, or lack thereof, shaped not only protagonists' evaluations of physiological and clinical data but also their evaluations of the legacy of the history of cardiac therapeutics. In subsequent controversies over treatments of coronary artery disease, visualization remained crucial as cardiologists and cardiac surgeons struggled to apply the lessons of CABG.

By the end of the 1970s, the controversy over CABG began to diminish. Continuing study and consensus panels essentially confirmed the findings of the V.A. study.[82] Spodick and Braunwald conceded that while survival was similar in most patients treated medically or surgically, surgery produced longer survival in some groups and better quality of life in most. Consensus panels from the American Medical Association and the National Institutes of Health agreed. For many physicians, the fact that trials had been conducted was as important as the findings themselves. As Spodick was pleased to note in 1977, "wholesale application of the procedure finally is being channeled by appropriate studies of what it accomplishes and for whom."[83]

---

[79] Effler, "Surgery for Coronary Disease," p. 38; and Richards, *Heart to Heart* (cit. n. 10), p. 99. On the changes wrought by angiography see Fye, *American Cardiology* (cit. n. 13), pp. 112, 175; and F. M. Sones, E. K. Shirey, W. T. Proudfit, and R. N. Wescott, "Cinecoronary Arteriography," *Circulation,* 1959, *20*:773–774.

[80] Favaloro, *Challenging Dream of Heart Surgery,* p. xv (see also Favaloro, "Critical Analysis of Coronary Artery Bypass Graft Surgery" [cit. n. 1], p. 1B); Ryan, "Revascularization" (cit. n. 21), p. 90B ("seminal event"); and Effler, "Myocardial Revascularization Surgery since 1945 A.D." (cit. n. 75), p. 828.

[81] Effler, "Myocardial Revascularization at the Community Hospital Level" (cit. n. 55), p. 240.

[82] The V.A. researchers refined their analysis for many years: Katherine Detre, Peter Peduzzi, Marvin Murphy, Herbert Hultgren, James Thomsen, Albert Oberman, Timothy Takaro, and the Veterans Administration Cooperative Study for Surgery for Coronary Arterial Occlusive Disease, "Effect of Bypass Surgery on Survival in Patients in Low- and High-Risk Subgroups Delineated by the Use of Simple Clinical Variables," *Circulation,* 1981, *63*:1329–1338, esp. p. 1336. Other large trials were conducted: Kolata, "Coronary Bypass Surgery" (cit. n. 20), p. 1265; CASS Principal Investigators and Their Associates, "Coronary Artery Surgery Study" (cit. n. 44); and Baldwin *et al.,* "Coronary Artery Surgery" (cit. n. 6), p. 160. The basic finding of the V.A. study, as described by one cardiologist, remains true today: "the sicker the patient clinically and angiographically and the poorer the heart function, the greater my enthusiasm for surgical therapy." Ryan, "Revascularization" (cit. n. 21), p. 94B.

[83] David H. Spodick, "Aortocoronary Bypass Surgery: Emerging Triumph of Controlled Clinical Trials," *Chest,*

However, while the RCTs of CABG eventually came to be seen as a success, it had taken ten years for adequate evaluation of CABG to emerge. The experience of the RCTs of CABG became a story that no one wanted to repeat. Time, effort, money, and even patients' lives had been wasted while the controversy lingered. Participants in the CABG debates committed themselves to doing a better job the next time around. As early as 1973, Spodick argued that although "prejudice has now made it too late to do properly designed, controlled trials of bypass operations, we should at least be mindful of the need in the next procedure to come along." In 1978 Braunwald expressed the hope that after the experiences with CABG physicians would insist on "careful, objective assessment, by prospective randomized trials when necessary." These needed to be done as early as possible, "before the genie escapes from the bottle."[84]

These dreams did not come true. Since the 1970s, new treatments for coronary artery disease have continued to appear and spread without trials, generating the same post hoc calls for trials. CABG was applied to the treatment of acute heart attacks as early as April 1968. Favaloro, Effler, and fellow enthusiasts quickly accepted its value: postoperative angiography showed that "the vast majority of heart muscle can be saved." Although they lacked long-term follow-up data, they believed that the operation prevented impending heart attacks and preserved heart muscle in patients experiencing heart attacks. Chalmers, McIntosh, and others demanded long-term data and called for trials: "Can we learn from our mistakes of the past?"[85]

When cardiologists developed drugs, such as streptokinase and other fibrinolytic agents, that could dissolve the blood clots implicated in heart attacks, Braunwald immediately called for trials "to prevent a decade or more of confusion about the powers of this latest genie." Angiography did indeed show that streptokinase could restore blood flow through an acutely occluded vessel. But did streptokinase really prevent the progression of a heart attack? Braunwald warned that the old ideal of restoring blood flow might actually create a risk: experience from animals and patients had shown that reperfusion of myocardium during an infarction could lead to serious hemorrhage. Controversy lingered for years.[86]

The desired lesson of CABG—that all subsequent treatments should be evaluated with RCTs immediately—had been inverted. CABG had demonstrated that certain kinds of techniques, particularly those supported by physiological common sense and visual dem-

1977, 71:318–319; Braunwald, "First Twenty Years of the *American Journal of Cardiology*" (cit. n. 58), p. 5; American Medical Association, "Report on Aortocoronary Bypass Graft Surgery," *J. Amer. Med. Assoc.*, 1979, 242:2701; Council on Scientific Affairs, American Medical Association, "Indications for Aortocoronary Bypass Graft Surgery," *ibid.*, 1979, 242:2709–2711; "National Institutes of Health Consensus Development Conference Statement on Coronary Artery Bypass Surgery: Scientific and Clinical Aspects," *Circulation*, 1982, 65(Suppl. 2):II-126–II-129; and Spodick, "Aortocoronary Bypass Surgery," p. 318.

[84] Spodick, "Coronary Bypass" (cit. n. 51), p. 81; and Braunwald, "Evaluation of the Efficacy of Coronary Bypass Surgery—II" (cit. n. 51), p. 162.

[85] Favaloro, "Direct Myocardial Revascularization" (cit. n. 17), p. 1041 (heart muscle saved); René G. Favaloro, Donald B. Effler, Chalit Cheanvechai, Robert A. Quint, and F. Mason Sones, "Acute Coronary Insufficiency (Impending Myocardial Infarction and Myocardial Infarction): Surgical Treatment by the Saphenous Vein Graft Technique," *Amer. J. Cardiol.*, 1971, 28:598–607; Chalmers, "Randomization and Coronary Artery Surgery" (cit. n. 31), p. 326 (learning from mistakes); and Henry D. McIntosh and Robert A. Buccino, "Editorial: Emergency Coronary Artery Revascularization of Patients with Acute Myocardial Infarction: You Can . . . But Should You?" *Circulation*, 1979, 60:247–250.

[86] Braunwald, "Let's Not Let the Genie Escape from the Bottle—Again" (cit. n. 46), pp. 1296 (quotation), 1294–1295. Trials sponsored by the National Heart, Lung, and Blood Institute and other groups did eventually appear. See Michael B. Mock, Guy S. Reeder, Hartzell V. Schaff, David R. Holmes, Ronald E. Vliestra, Hugh C. Smith, and Bernard J. Gersh, "Percutaneous Transluminal Coronary Angioplasty versus Coronary Artery Bypass: Isn't It Time for a Randomized Trial?" *New Engl. J. Med.*, 1985, 312:916–919, esp. p. 918.

onstration, could be incorporated into medical practice without trials. The history of angioplasty provides the most striking example. In 1977 cardiologists introduced percutaneous transluminal coronary angioplasty (PTCA) as a less invasive alternative to CABG for relieving obstructed coronary arteries. In this procedure a balloon-tipped catheter is threaded into the coronary arteries and inflated within the narrowed atherosclerotic region. By cracking the plaque and stretching the vessel walls, PTCA increases the functional lumen of the vessel, allowing new pathways for blood flow.[87]

PTCA shared the aesthetic and mechanistic appeals of CABG. It modified the plaques perceived to be the cause of coronary artery disease. Its effects were direct and immediate, visualizable with angiography and real-time fluoroscopy. Furthermore, PTCA required a shorter hospital stay than CABG and was much cheaper to perform. As a result, PTCA experienced an even more spectacular spread in the 1980s than CABG had in the 1970s. The first PTCA was performed in 1977; 2,000 were performed in 1979 (compared to 144,000 CABGs). More than 80,000 PTCAs were performed annually in the mid 1980s (compared to roughly 205,000 CABGs). By the late 1980s, the number of PTCAs done by cardiologists surpassed the number of CABGs done by cardiac surgeons (see Figure 7). This growth continued into the 1990s, with more than 300,000 PTCAs performed each year.[88]

As with CABG, the early spread of PTCA occurred in the absence of rigorous statistical data about its efficacy. Calls for trials came early. Spodick again led the way. In 1979 he expressed his frustration that the Food and Drug Administration did not hold new procedures to the same standards as drugs. He called on cardiologists not to repeat the mistakes surgeons had made with the Vineberg procedure and CABG: "we must not prematurely let this new genie out of its bottle." Although PTCA seemed promising, cardiologists had not demonstrated that it provided long-term benefits. Since medical therapy already offered excellent survival rates for most patients, the main question was whether PTCA gave better relief of symptoms. Spodick hoped that hospital committees and journal editors would be "professional guardians of scientific integrity" and demand RCTs.[89] But no trials appeared.

Consensus was quickly reached that PTCA worked best for single vessel disease and CABG for left main disease. However, indications for patients with intermediate disease—"the vast majority of patients requiring revascularization"—remained ambiguous. But still no trials appeared. Calls for trials continued throughout the 1980s and early 1990s, citing RCTs as the most reliable way of comparing the symptomatic relief, the survival benefit, and the cost of PTCA and CABG. Trials comparing PTCA to CABG did not begin to be

[87] Andreas Grüntzig, "Transluminal Dilatation of Coronary-Artery Stenosis," *Lancet*, 1978, *1*:263; Donald S. Baim, "New Devices for Coronary Revascularization," *Hospit. Pract.*, 15 Oct. 1993, pp. 41–52, esp. p. 41; Howell, "Concepts of Heart-Related Diseases" (cit. n. 8), pp. 92–93; and Charles Landau, Richard A. Lange, and L. David Hillis, "Percutaneous Transluminal Coronary Angioplasty," *New Engl. J. Med.*, 1994, *330*:981–993, on p. 981.

[88] On the visibility of PTCA's effects see Landau *et al.*, "Percutaneous Transluminal Coronary Angioplasty," p. 982. For the statistics see *ibid.*, p. 981; and Jay L. Hollman, "Myocardial Revascularization: Coronary Angioplasty and Bypass Surgery Indications," *Medical Clinics of North America*, 1992, *76*:1083–1097, on pp. 1084–1085. On the spectacular spread of PTCA see Fye, *American Cardiology* (cit. n. 13), pp. 301–304.

[89] David H. Spodick, "Letter: Percutaneous Transluminal Coronary Angioplasty," *Annals of Internal Medicine*, 1979, *90*:850–851 ("new genie"); Spodick, "Letter: Percutaneous Transluminal Coronary Angioplasty," *Mayo Clinic Proceedings*, 1981, *56*:526 (question of symptom relief); Spodick, "Editorial: Percutaneous Transluminal Coronary Angioplasty: Opportunity Fleeting," *J. Amer. Med. Assoc.*, 1979, *242*:1658–1659, on p. 1659 ("guardians of scientific integrity"); and Spodick, "PTCA: Need for Prospective Randomized Controlled Trials," *Amer. J. Cardiol.*, 1983, *51*:1467–1468.

published until 1992.[90] Cardiologists, who had aggressively criticized the epistemological standards of cardiac surgeons in the 1970s, thus accepted and performed coronary angioplasty for fifteen years without data from RCTs.

Like CABG enthusiasts before them, PTCA enthusiasts offered many reasons why RCTs were too difficult to conduct and too limited in their results. They cited many methodological complications: inadequate criteria for characterizing each patient's degree of atherosclerosis; variations in how the procedure is performed and in how success and complications are evaluated; statistical problems in analyzing small patient populations and rare adverse outcomes. They also complained that the trials were both too time consuming and too expensive to conduct. As Andreas Grüntzig, the developer of PTCA, stated: "the call for randomization is easily made but difficult to follow."[91] Meanwhile, supporters found solace in the compelling evidence provided by angiography: as they deflated the balloon and removed the catheter, they could see blood flowing where none, or not enough, had flowed before.

Eventually, some RCTs of PTCA were completed. The parallels with CABG are striking. The results of two long-awaited trials were published, with an accompanying editorial, in the *New England Journal of Medicine* in 1994. Both trials found that, in most cases, PTCA and CABG produced equivalent long-term outcomes. In the absence of definitive answers about therapeutic efficacy, the choice was left to individual patients and doctors.[92]

The pattern did not end with angioplasty. Starting in 1995, new techniques of minimally invasive CABG became increasingly popular in the United States. Instead of requiring a 30-cm incision through the patient's sternum, these procedures used an 8-cm "keyhole" incision and a series of small ports, like those used in laparoscopic abdominal surgery, to gain access to the heart. In some versions cardiopulmonary bypass was not used: the surgeon operated on a slowed but beating heart. Early results—from case series—showed that minimally invasive CABG caused less pain, required shorter hospital stays, and cost less than traditional CABG. Formal evaluation of its efficacy, however, did not appear.[93]

---

[90] Hollman, "Myocardial Revascularization" (cit. n. 88), p. 1088. For a call for trials see Mock *et al.*, "Percutaneous Transluminal Coronary Angioplasty" (cit. n. 86), p. 917; and Tom Treasure, "Angioplasty: Where's the Proof?" *British Journal of Hospital Medicine*, 1990, *43:*95, 99. For a discussion of these trials see L. David Hillis and John D. Rutherford, "Coronary Angioplasty Compared with Bypass Grafting," *New Engl. J. Med.*, 1994, *331:*1086–1087, on p. 1086.

[91] Andreas R. Grüntzig and Jay Hollman, "Reply to Spodick," *Amer. J. Cardiol.*, 1983, *51:*1468. For examples of PTCA enthusiasts' complaints see Ronald E. Vliestra, David R. Holmes, Hugh C. Smith, Geoffrey O. Hartzler, and Thomas A. Orszulak, "Reply to Spodick," *Mayo Clin. Proc.* 1981, *56:*526–527; Baim, "New Devices for Coronary Revascularization" (cit. n. 87), pp. 51–52; and Joseph Lindsay, Ellen E. Pinnow, Jeffrey J. Popma, and Augusto D. Pichard, "Obstacles to Outcomes Analysis in Percutaneous Transluminal Coronary Revascularization," *Amer. J. Cardiol.*, 1995, *76:*168–172.

[92] C. W. Hamm, J. Reimers, T. Ischinger, H.-J. Rupprecht, J. Berger, and W. Bleifeld, "A Randomized Study of Coronary Angioplasty Compared with Bypass Surgery in Patients with Symptomatic Multivessel Coronary Disease," *New Engl. J. Med.*, 1994, *331:*1037–1043; S. B. King, N. J. Lembo, W. S. Weintraub, A. S. Kosinski, H. X. Barnhard, M. H. Kutner, N. P. Alazraki, R. A. Guyton, and X.-Q. Zhao, "A Randomized Trial Comparing Coronary Angioplasty with Coronary Bypass Surgery," *ibid.*, 1994, *331:*1044–1050; and Hillis and Rutherford, "Coronary Angioplasty Compared with Bypass Grafting" (cit. n. 90), pp. 1086–1087.

[93] For descriptions of "keyhole" procedures see Tea E. Acuff, Rodney J. Landreneau, Bartley P. Griffith, and Michael J. Mack, "Minimally Invasive Coronary Artery Bypass Grafting," *Ann. Thorac. Surg.,* 1996, *61:*135–137; Landreneau, Mack, James A. Magovern, Acuff, Daniel H. Benckart, Tamara A. Sakert, Lynda S. Fetterman, and Griffith, " 'Keyhole' Coronary Artery Bypass Surgery," *Ann. Surg.,* 1996, *224:*453–462; and James A. Magovern, Benckart, Landreneau, Sakert, and George J. Magovern, "Morbidity, Cost, and Six-Month Outcome of Minimally Invasive Direct Coronary Artery Bypass Grafting," *Ann. Thorac. Surg.,* 1998, *66:*1224–1229. Preliminary reports have focused on initial safety, cost, and morbidity and cannot yet answer questions of long-term efficacy. See Aubrey C. Galloway, Richard J. Shemin, Donald D. Glower, Joseph H. Boyer, Mark A. Groh, Richard E. Kuntz, Thomas A. Burdon, Greg H. Ribakove, Bruce A. Reitz, and Stephen B. Colvin, "First Report of the Port Access International Registry," *ibid.*, 1999, *67:*51–58.

Surgeons assumed that if the immediate revascularization was comparable to that with traditional CABG, then the long-term results should be as good. But even this had not been well studied: intra- or postoperative angiography was not consistently performed.[94] How did the technique prosper despite such lack of validation? Like traditional CABG, minimally invasive CABG provided a direct, mechanical fix for the perceived cause of coronary artery disease. When the clamps are released, the surgeon can see the blood flow. This sight was so convincing that angiography seemed unnecessary.

As cardiac surgeons continued to refine their techniques, cardiologists introduced a fundamentally new approach: gene therapy. In November 1998 Jeffrey Isner—who was first exposed to cardiology as a medical student working with Mason Sones in 1967— reported the successful use of vascular endothelial growth factor to induce the formation of new blood vessels in ischemic myocardium. Within thirty days the gene therapy had relieved angina in all five patients, each of whom had had crippling, intractable angina despite multiple previous revascularization procedures. All had evidence of improved perfusion, as visualized with single photon emission computed tomography. Isner acknowledged the ideal of an RCT and hoped that one would be done soon. But since his technique required an operation for administering the genes, a proper RCT would require a sham operation for the control group, which he, the National Institutes of Health, and the Food and Drug Administration were unwilling to allow.[95] It seems likely that Isner's treatment will be held to the high standard of an RCT once less invasive methods of gene administration are developed, both because of widespread cultural anxieties over gene therapy and because gene therapy lacks the direct, immediately visualizable appeal of both CABG and PTCA.

Meanwhile, the evaluation of these new techniques is no longer simply a matter of physicians debating their efficacy. Instead, in the financially constrained contexts of managed care, physicians must not only convince themselves, but also their insurers, not only of efficacy, but also of cost efficiency. To complicate matters further, these new technologies have become major growth areas for commercial enterprise. Physicians have formed alliances with private companies. Millions of dollars have poured in from venture capital firms. This raises serious concerns about financial conflicts of interest as physicians claim to generate objective knowledge of therapeutic efficacy.[96]

As historians such as Harry Marks, Steven Epstein, and Norman Richards have shown, the ability of RCTs to resolve questions of therapeutic efficacy will always be contested. The specific case of CABG demonstrates not only the decisive role played by precon-

[94] See, e.g., Acuff et al., "Minimally Invasive Coronary Artery Bypass Grafting," p. 136; Landreneau et al., " 'Keyhole' Coronary Artery Bypass Surgery," p. 461; Robert R. Lazzara and Francis E. Kidwell, "Minimally Invasive Direct Coronary Bypass versus Cardiopulmonary Technique: Angiographic Comparison," Ann. Thorac. Surg., 1999, 67:500–503; and Mohammad Bachar Izzat, Kim S. Khaw, Wassim Atassi, Anthony P. C. Yim, Song Wan, and H. Hazem El-Zufari, "Routine Intraoperative Angiography Improves the Early Patency of Coronary Grafts Performed on the Beating Heart," Chest, 1999, 115:987–990. The assumption that long-term results should be similar if early revascularization is comparable appears in Robert H. Jones, commentary following Landreneau et al., " 'Keyhole' Coronary Artery Bypass Surgery," p. 460.

[95] Douglas W. Losordo, Peter R. Vale, James F. Symes, Cheryl H. Dunnington, Darryl D. Esakof, Michael Maysky, Alan B. Ashare, Kishor Lathi, and Jeffrey M. Isner, "Gene Therapy for Myocardial Angiogenesis: Initial Clinical Results with Direct Myocardial Injection of phVEGF$_{165}$ as Sole Therapy for Myocardial Ischemia," Circulation, 1998, 98:2800–2804 (on unwillingness to allow sham operations see p. 2804); N.S., "Gene Therapy Advances Go to the Heart," Science News, 1998, 154:346; and William Clifford Roberts, "Jeffrey Michael Isner, M. D.: A Conversation with the Editor," Amer. J. Cardiol., 1999, 82:78, 83, 86.

[96] On money from venture capital firms see Joan O'C. Hamilton, "Bypassing the Trauma," Business Week, 4 Sept. 1995, pp. 32–34; on financial conflicts of interest see Stephen Klaidman, "Should Smart Operators Mix Business and Surgery?" Ann. Thorac. Surg., 1998, 66:1119–1120.

ceived opinions but also the origins of such preconceptions. Personal experiences—from Spodick's training under Chalmers to Takaro's disillusionment with angiography—played a role. Interdisciplinary hostility, though present, was overshadowed by intradisciplinary differences in standards of knowledge, specifically the relevance of physiological and clinical measures of coronary artery disease. Assessments of the persuasiveness of visual evidence were crucial. Did angiographic demonstration of restoration of blood flow represent a successful treatment? Did angiographic diagnosis and postoperative assessment make CABG different from all the treatments that had come before? Individuals' answers to these questions guided their evaluation of CABG and its RCTs.

Fundamentally, these cases demonstrate the consequences of the coexistence of multiple representations of a single disease. Each representation, whether physiological or clinical, visual or statistical, allows different modes of assessing therapeutic efficacy. In the case of cardiac therapeutics, the traditions of visual demonstration will always stand as an alternative to the statistical ideals of RCTs. But as Galison and others have shown, such disunity in science need not be feared.[97] The cost of pluralism might be therapeutic confusion at best and the infliction of untested treatments on patients at worst. Nonetheless, it continues to spark imaginative efforts against the "greatest scourge of Western man."

---

[97] Galison concludes: "Science is disunified, and—against our first intuitions—it is precisely the disunification of science that brings strength and stability." Galison, *Image and Logic* (cit. n. 66), p. 781. See also Koch, "In the Image of Science?" (cit. n. 59), p. 892.

Making Dollars Out of DNA

The First Major Patent in Biotechnology and the Commercialization of Molecular Biology, 1974–1980

By Sally Smith Hughes

ABSTRACT

In 1973–1974 Stanley N. Cohen of Stanford and Herbert W. Boyer of the University of California, San Francisco, developed a laboratory process for joining and replicating DNA from different species. In 1974 Stanford and UC applied for a patent on the recombinant DNA process; the U.S. Patent Office granted it in 1980. This essay describes how the patenting procedure was shaped by the concurrent recombinant DNA controversy, tension over the commercialization of academic biology, governmental deliberations over the regulation of genetic engineering research, and national expectations for high technology as a boost to the American economy. The essay concludes with a discussion of the patent as a turning point in the commercialization of molecular biology and a harbinger of the social and ethical issues associated with biotechnology today.

O N 2 DECEMBER 1980 THE U.S. PATENT OFFICE issued the first major patent in the new biotechnology, one of three patents subsequently known as the Cohen-Boyer recombinant DNA cloning patents.[1] The first patent is based on the 1973–1974 development by Stanley N. Cohen of Stanford and Herbert W. Boyer of the University of California, San Francisco (UCSF), of a fundamental process of molecular biology that came

Isis 92, no. 3 (September 2001): 541–75.

This research was supported in part by an Othmer Fellowship at the Chemical Heritage Foundation. I am grateful for the comments of Mary Ellen Bowden, Kenneth Carpenter, Stanley N. Cohen, Seymour Mauskopf, Marcia Meldrum, Niels Reimers, Margaret Rossiter, and the anonymous referees for *Isis*.

[1] Stanley N. Cohen and Herbert W. Boyer, "Process for Producing Biologically Functional Molecular Chimeras," United States Patent 4,237,224, 2 Dec. 1980. As Robert Bud has illustrated, industry based on biotechnology, broadly described as the use of living organisms for commercial purposes, is ancient: Bud, *The Uses of Life: A History of Biotechnology* (Cambridge: Cambridge Univ. Press, 1993). Although I recognize that the present science and industry of biotechnology have antecedents, in this essay I use "biotechnology" in the more recent and narrower sense of a science and industry based on the manipulation of DNA.

to be known as recombinant DNA technology.[2] The concept arose in November 1972 as the two scientists brainstormed on a late-evening walk after attending a scientific conference in Hawaii. Sketching out a research collaboration over deli sandwiches, Cohen and Boyer agreed to pool their respective expertise in plasmid biology and bacterial restriction enzymes in experiments that they began soon after returning to the mainland. By March 1973, at most four months after initiating the experiments, Cohen, Boyer, and their co-workers at Stanford and UCSF knew that their DNA splicing and cloning (amplifying) procedure worked. In November they published a paper entitled "Construction of Biologically Functional Bacterial Plasmids *In Vitro*." Years later, Boyer recalled his emotions at the instant of success: "I think the most exciting moment was when we did the first experiments with the recombined plasmid DNA. [Colleague] Bob Helling and I looked at the first [electrophoresis] gels, and I can remember tears coming to my eyes, it was so nice. I mean, there it was. You could visualize your results in physical terms, and after that we knew we could do a lot of things."[3]

What became apparent as recombinant DNA technology was rapidly adopted in molecular biology laboratories around the world was that it could indeed "do a lot of things." It gave scientists a simple method for isolating and amplifying any gene or DNA segment and moving it with controlled precision, allowing analysis of gene structure and function in simple and complex organisms.[4] The process was revolutionary for molecular biology. But it was more than that: it was soon found to have great potential in commercial applications as well. In November 1974 Stanford and the University of California (UC) applied for a patent on recombinant DNA. What was expected to be a routine patent application turned into a six-year ordeal. The history of the effort to secure the patent has never been told in detail. There are two existing accounts, a cursory three-page paper by Stanley Cohen and a somewhat longer paper by Niels Reimers, who as director of Stanford's Office of Technology Licensing (OTL) managed the patenting and licensing of the Cohen-Boyer invention. Both are narrow, virtually unreferenced accounts written for scientific or technical audiences.[5]

---

[2] Cohen and colleagues introduced the term "recombinant DNA" in early 1974, intending it to be synonymous with DNA cloning. The term is also used in a narrower sense to refer only to the composite molecules that result from the joining of DNA fragments in a test tube. See Stanley Cohen's presentation on "The Science" at a conference entitled "The Emergence of Biotechnology: DNA to Genentech," sponsored by the Chemical Heritage Foundation, Philadelphia, 13 June 1997, transcripts, pp. 23–32, on p. 23. The wider meaning of "recombinant DNA" and "recombinant DNA technology" is intended in this essay. The process consists of isolating and inserting a gene or segment of DNA into a plasmid (a circle of DNA found in bacteria) and transferring the "recombinant" DNA to bacteria, where it is reproduced or "cloned" in identical genetic copies. The foreign gene can be transcribed and expressed in the bacteria as a protein, such as a hormone.

[3] Stanley N. Cohen, Annie C. Y. Chang, Herbert W. Boyer, and Robert B. Helling, "Construction of Biologically Functional Bacterial Plasmids *In Vitro*," *Proceedings of the National Academy of Sciences, USA,* 1973, *70*:3240–3244; and Herbert W. Boyer, interview by Sally Smith Hughes, 28 Mar. 1994, draft, p. 56. (Interviews with Boyer cited in this essay are steps toward an oral history, as yet incomplete.) For a contemporary account of the splicing and cloning procedure see Cohen, "The Manipulation of Genes," *Scientific American,* 1975, *233*:24–33.

[4] One indication of the speed with which the recombinant DNA method was adopted in science is Stanley Cohen's list, dated September 1974, of the names of the thirty-two recipients of his plasmid, pSC101, at the time the only vector or vehicle available for transfer of the selected DNA segment into bacteria for cloning. Stanley N. Cohen to Dick Roblin, 5 Sept. 1974, Cohen Personal Correspondence, Biohazard Collection, Stanford Univ., Stanford, California. For contemporary accounts of the technical usefulness and early applications of recombinant DNA in basic research see Jean L. Marx, "Molecular Cloning: Powerful Tool for Studying Genes," *Science,* 1976, *191*:1160–1162; Bob Williamson, "First Mammalian Results with Genetic Recombinants," *Nature,* 1976, *260*:189–190; and Eleanor Lawrence, "Nuts and Bolts of Genetic Engineering," *ibid., 263*:726–727.

[5] Stanley N. Cohen, "The Stanford DNA Cloning Patent," in *From Genetic Engineering to Biotechnology—*

A full history of the patent merits telling, not least because of its importance as a turning point in the commercialization of molecular biology. Historians have emphasized the continuities between modern DNA-based biotechnology and previous commercially oriented biological research that likewise had industry sponsorship and proposed to use science to manipulate living organisms for commercial purposes.[6] While I agree that some continuities exist, commercial activity in molecular biology was until recently episodic and generally unrepresentative of a discipline overwhelmingly focused on basic problems. With its sweeping claims on the recombinant DNA process, the patent became an agent in the transformation of molecular biology from a predominantly academic discipline to one with pervasive practical applications and industry connections. This patent history sets the stage for a shift in attitude and research emphasis in molecular biology in the 1980s toward a focus on applications, extensive interaction with the commercial world, and a resultant blurring of boundaries between academia and industry.[7]

Three sets of actors—universities and academic scientists, government officials, and entrepreneurs and established corporations—followed the course of the Cohen-Boyer patent application, watching for clues to the future of recombinant DNA as an industrial process. But in addition to the patent's status as first a barometer and then a landmark of commercial biotechnology, its history has its own intrinsic significance. It was played out during a critical period of debate in the 1970s about the role of science and technology in the national economy. Who should benefit from public support of academic science? Should science and technology be regulated and constrained? If so, how and by whom? These questions were being discussed in many areas of science and technology in the 1970s, but nowhere with more heat and publicity than in the controversy over the research and development of recombinant DNA.[8] The effort to patent this breakthrough in molecular science and technology became a major focus in the debate.

In this essay I detail the six years of concerted effort by Stanford to obtain a patent on the Cohen-Boyer process and describe its relationship to the politics of the recombinant DNA controversy and shifts in national policy and priorities. Early anxieties over the social

---

The Critical Transition, ed. W. J. Whelan and Sandra Black (New York: Wiley, 1982), pp. 213–216; and Niels Reimers, "Tiger by the Tail," *Chemtech*, Aug. 1987, pp. 464–471. Cohen's paper references only the patent, and the two references in Reimers's article, probably added by an editor, are to previously published articles in *Chemtech* on patenting in biotechnology.

[6] Angela Creager, "Biotechnology and Blood: Edwin Cohn's Plasma Fractionation Project, 1940–1953," in *Private Science: Biotechnology and the Rise of the Molecular Sciences*, ed. Arnold Thackray (Philadelphia: Univ. Pennsylvania Press, 1998), pp. 39–62; Lily Kay, *The Molecular Vision of Life: Caltech, the Rockefeller Foundation, and the Rise of the New Biology* (New York: Oxford Univ. Press, 1993); Jean-Paul Gaudillière, "The Molecularization of Cancer Etiology in the Postwar United States: Instruments, Politics, and Management," in *Molecularizing Biology and Medicine: New Practices and Alliances, 1910s–1970s*, ed. Soraya de Chadarevian and Harmke Kamminga (Amsterdam: Harwood, 1998), pp. 139–170; and Nicholas Rasmussen, "The Forgotten Promise of Thiamin: Merck, Caltech Biologists, and Plant Hormones in a 1930s Biotechnology Project," *Journal of the History of Biology*, 1999, *32*:245–261.

[7] Almost simultaneous with the patenting of the Cohen-Boyer procedure was Cesar Milstein and Georges Kohler's development and publication in 1975 of the monoclonal antibody/hybridoma technique. For an ethnography of its path from discovery to commercialization, which took a different course from the Cohen-Boyer method, see Alberto Cambrosio and Peter Keating, *Exquisite Specificity: The Monoclonal Antibody Revolution* (New York: Oxford Univ. Press, 1995).

[8] The regulation of recombinant DNA research and the complex politics surrounding it have been documented by a number of authors. See, e.g., Susan Wright, "Recombinant DNA Technology and Its Social Transformation, 1972–1982," *Osiris*, 2nd Ser., 1986, *2*:303–360; Wright, *Molecular Politics: Developing American and British Regulatory Policy for Genetic Engineering, 1972–1982* (Chicago: Univ. Chicago Press, 1994); Sheldon Krimsky, *Genetic Alchemy: The Social History of the Recombinant DNA Controversy* (Cambridge, Mass.: MIT Press, 1985); and Herbert Gottweis, *Governing Molecules: The Discursive Politics of Genetic Engineering in Europe and the United States* (Cambridge, Mass.: MIT Press, 1998).

and environmental implications of science and technology and their commercial develop-ment, while they did not altogether disappear, were overwhelmed by the economic concerns and pro-business policies that came to the forefront in the late 1970s. The Carter and Reagan administrations turned to high-technology areas such as semiconductors and recombinant DNA for help in restoring the nation's technological leadership and economic strength. Yet this transition was neither simple nor seamless, particularly for universities and their science faculties. Professors and institutions had to reconcile traditional academic ideals with the quest for commercial profit in biology. Could universities and their scientists maintain the tradition of free and disinterested academic inquiry when confronted with the pressures of the marketplace? Would patenting and other forms of commercial activity in academia blem-ish the image of the university as a bastion of pure research, education, and public service? Who, in short, should own and control and profit from scientific knowledge?

The patent history, encompassing these wider issues, provides fresh insight into the national and academic debates over science and technology policy in the 1970s and the broader economic and political contexts in which biomedical scientists and institutions shifted from ambivalence about to acceptance of commercial science and technology. I will argue that the patent set legal precedent and provided reassurance at a critical moment in the movement toward extensive commercialization of recombinant DNA and other molecular technologies. In so doing, it served as an agent and symbol of institutional and attitudinal change that encouraged American universities and business interests to become more actively involved in the commercial exploitation of basic biomedical research, par-ticularly in molecular biology.

STANFORD CONSIDERS A PATENT APPLICATION

A front-page story in the 20 May 1974 edition of the *New York Times* provoked Stanford's interest in applying for a patent on the Cohen-Boyer cloning method. The article, by Victor McElheny, covered research that Cohen, Boyer, and their collaborators had just published on the cloning of genes of a higher organism, the South African clawed frog. The work represented an exciting extension of two earlier publications on the use of the method in lower organisms. Entitled "Animal Gene Shifted to Bacteria; Aid Seen to Medicine and Farm," the *Times* article described a "practical" method for "transplanting" and amplifying animal genes in bacteria. Although noting the method's scientific value as a means for studying gene expression, McElheny focused on its potential practical uses: "The new method, its discoverers say, gives promise of meeting some of the most fundamental needs of both medicine and agriculture, such as supplies of now scarce hormones, and nitrogen-fixing micro-organisms growing near the roots of wheat and corn plants, thus reducing requirements for fertilizers."[9]

A Stanford University Medical Center news release, timed to appear on the day the paper on the frog DNA was published, also touted the method's likely applications in the

---

[9] Victor McElheny, "Animal Gene Shifted to Bacteria," *New York Times*, 20 May 1974, p. 1. Three publica-tions, only two of which were written by both Cohen and Boyer, describe the Cohen-Boyer research on gene cloning. The first is Cohen *et al.*, "Construction of Biologically Functional Bacterial Plasmids *In Vitro*" (cit. n. 3); the second, coauthored by Cohen and his laboratory technician, is Annie C. Y. Chang and Stanley N. Cohen, "Genome Construction between Bacterial Species *In Vitro:* Replication and Expression of *Staphylococcus* Plas-mid Genes in *Escherichia coli*," *Proc. Nat. Acad. Sci. USA*, 1974, *71*:1030–1034; the third, which prompted McElheny's article, is John F. Morrow, Cohen, Chang, Herbert W. Boyer, Howard M. Goodman, and Robert B. Helling, "Replication and Transcription of Eukaryotic DNA in *Escherichia coli*," *ibid.*, pp. 1743–1747.

synthesis of pharmaceuticals and the correction of hereditary defects. It quoted Joshua Lederberg, a recent Nobel laureate and chairman of the Stanford genetics department, who called the method "a major tool in genetic analysis" and suggested that "it may completely change the pharmaceutical industry's approach to making biological elements such as insulin and antibiotics." Although not himself a practitioner, Lederberg had communicated two of the three DNA cloning papers of which Cohen was a coauthor to the U.S. *Proceedings of the National Academy of Sciences* and was well informed about the scientific and commercial possibilities of recombinant DNA technology. The next day the *San Francisco Chronicle* headlined an article "Getting Bacteria to Manufacture Genes" and referred to the possibility of bacteria being transformed into "factories" for the production of drugs such as insulin. (The bacteria-as-factories metaphor was to be endlessly repeated in the years ahead in the commercial promotion of recombinant DNA.) The research also caught the attention of *Newsweek,* which carried an article on the "gene transplanters" and their possible contributions to medicine and agriculture.[10] The message of the procedure's possibly wide commercial utility was hard to miss.

Stanford's Office of Public Affairs routinely sent articles on faculty research of potential commercial interest to Niels Reimers, who directed the university's patenting and licensing effort. (See Figure 1.) Upon receiving the *Times* article, Reimers immediately recognized the discovery as a patenting and licensing opportunity. Reimers was a former engineer who had arrived at Stanford in 1968 with experience in directing contracts at technology-based companies. Noting the university's meagre profits from patents administered by the Research Corporation of New York, in 1969 he had launched a pilot technology licensing program. Faculty members were asked to propose inventions, from which the most commercially promising were to be chosen as the basis for filing patent applications with the United States Patent and Trademark Office in Washington, D.C. To encourage submissions, Stanford structured royalties so that one third was awarded to the inventor, one third to the inventor's department, and one third to university unrestricted funds.[11]

PATENTING AND ENTREPRENEURIALISM IN AMERICAN ACADEMIA

The pilot program with its license marketing focus generated $55,000 in licensing fees in its first partial year, almost ten times more than Stanford had earned in the previous ten years or so. On 1 January 1970 the program was formally established as the Stanford University Office of Technology Licensing, with Reimers as administrator. Its goals were to provide a more efficient mechanism for bringing Stanford's research discoveries forward to public use and benefit and to generate an additional and unrestricted source of income

[10] [News release], Stanford University News Service, 20 May 1974, S74-43, Correspondence 1974–1980, Archives, Office of Technology Licensing, Stanford Univ., Stanford, California (hereafter cited as **OTL Archives**); "Getting Bacteria to Manufacture Genes," *San Francisco Chronicle,* 21 May 1974, p. 6; and "The Gene Transplanters," *Newsweek,* 17 June 1974, p. 54. Lederberg was a scientific advisor to Cetus, a small life science research and development firm founded in the San Francisco Bay Area in 1971. He was thus oriented toward commercialization of discoveries in biomedicine.

[11] Reimers describes his reaction in "Tiger by the Tail" (cit. n. 5). On the pilot licensing program see Niels Reimers, "Stanford's Office of Technology Licensing and the Cohen/Boyer Cloning Patents," an oral history conducted in 1997 by Sally Smith Hughes, Regional Oral History Office, Bancroft Library, Univ. California, Berkeley, 1988 (hereafter cited as **Reimers oral history**), p. 12. Also online: http://www.lib.berkeley.edu/BANC/ROHO/ohonline. On the division of royalties see "Patents" [Stanford patent policy], Oct. 1974, Arthur Kornberg Papers, SC 359, Box 5, Folder: "1980," Green Library, Stanford Univ. For a discussion of patenting and licensing in the university-industry relationship see Roger G. Ditzel, "Patent Rights at the University/Industry Interface," *Society of Research Administrators Journal,* Winter 1983, pp. 13–20.

Figure 1. *Niels Reimers, ca. 1980. Photograph courtesy of Office of Technology Licensing, Stanford University.*

to support university education and research. This was a proactive and innovative program of commercialization that few if any other American universities at the time could equal. Most universities of the day lacked the capacity to evaluate, let alone exploit, the commercial potential of faculty research findings.[12]

Stanford's technology licensing program was one aspect of the institution's history of

---

[12] For the goals of OTL see "Patents"; on Stanford's unusual program see Robert Rosenzweig, seminar on university-industry relationships, 26 Oct. 1999, Center for Studies in Higher Education, Univ. California, Berkeley.

fostering markets for its science. Stanford has long encouraged close interactions with companies in the region, beginning with Frederick Terman's indefatigible promotion of industrial patronage in the 1930s. The university's contribution to the genesis of the electronics industry in Silicon Valley and the creation of the Stanford Research Institute and Stanford Industrial Park are examples of the business orientation of its faculty and administration. If more deliberate, Stanford was far from alone among American research universities in its investment in the world beyond the ivory tower.[13] Beginning early in the twentieth century, many sought patent rights on faculty research discoveries and contracts of one kind or another with government and industry. Stanford, the University of California, Columbia, Wisconsin, and MIT, for instance, have patent policies dating from the 1920s, 1930s, and 1940s and have long held substantial numbers of patents on faculty inventions. Because patenting in academia has a history of controversy, universities have offered a variety of justifications for their engagement in patenting and licensing. Among them are the wish to return patent royalties to research support, to prevent outsiders from profiting at the expense of the university and individual scientists, and to control the uses to which patented inventions are put.[14]

Patenting in academic biomedicine was by tradition especially suspect on ethical grounds, having been explicitly decried in the American Medical Association's Code of Ethics of 1847. Although surgical instruments were freely patented, medical inventions for the relief of the sick were felt to be in a unique ethical category to which those in need should have free and unrestricted access. Harvard, for example, decided in the 1920s to refuse to profit from faculty research in public health and therapeutics and in 1934–1935 formalized the practice in a policy dedicating such patents to the public. In 1975 the institution abandoned this commitment and permitted royalties to be shared among the inventor, the university, and outside entities.[15] Harvard's initial restraint was at one extreme; a number of North American universities from at least the 1910s on did take out patents in biomedicine—but with accompanying publicity that sought to persuade the public that the patent was to its benefit.

Many members of American university faculties, Stanford and UC included, were un-

[13] On Stanford see Rebecca S. Lowen, *Creating the Cold War University: The Transformation of Stanford* (Berkeley: Univ. California Press, 1997); and Stuart W. Leslie, *The Cold War and American Science: The Military-Industrial-Academic Complex at MIT and Stanford* (New York: Columbia Univ. Press, 1993). For an account of American industry's increasing dependence in the twentieth century on university science and engineering see David F. Noble, *America by Design: Science, Technology, and the Rise of Corporate Capitalism* (Oxford: Oxford Univ. Press, 1974), esp. pp. 128–147. For a general treatment of the control of scientific information and the issues it raises see Dorothy Nelkin, *Science as Intellectual Property: Who Controls Scientific Research?* (AAAS Series on Issues in Science and Technology) (New York: MacMillan, 1984).

[14] Richard Shryock, *American Medical Research Past and Present* (New York: Commonwealth Fund, 1947), p. 141. For example, Harry Steenbock at the University of Wisconsin argued that patents on his method for producing vitamin D would allow him and the Wisconsin Alumni Research Foundation, created in 1925 to manage university patents, to protect the public against unscrupulous manufacturers and ensure the quality of commercial products made by licensees: Rima Apple, "Patenting University Research: Harry Steenbock and the Wisconsin Alumni Research Foundation," *Isis*, 1989, *80*:375–394, on p. 376. The University of Toronto made similar arguments regarding its patent on insulin: Michael Bliss, *The Discovery of Insulin* (Chicago: Univ. Chicago Press, 1982), pp. 131–134. On the issues raised by patenting at one university see Henry Etzkowitz, "Knowledge as Property: The Massachusetts Institute of Technology and the Debate over Academic Patent Policy," *Minerva*, Autumn 1994, pp. 283–421.

[15] On medical inventions as a "unique ethical category" see Charles Weiner, "Universities, Professors, and Patents: A Continuing Controversy," *Technology Review*, Feb./Mar. 1986, pp. 33–43; and Shryock, *American Medical Research*, pp. 140–144. On Harvard's policy change see Daniel J. Kevles, "*Diamond v. Chakrabarty* and Beyond: The Political Economy of Patenting Life," in *Private Science*, ed. Thackray (cit. n. 6), pp. 66–79, on p. 67.

Figure 2. *Stanley N. Cohen, ca. 1975. Photograph courtesy of Office of News and Public Affairs, Stanford University Medical Center.*

comfortable with patenting in academia. Until recently, most professors—except those in scientific disciplines with obvious practical applications, such as engineering, agriculture, and chemistry—were unfamiliar with the requirements of the patent system and gave little consideration to the possibility of commercializing their research discoveries. It was common practice beginning early in this century for activities pertaining to intellectual property to be handled by an outside organization somewhat or entirely distanced from the university, such as the Research Corporation and the Wisconsin Alumni Research Foundation.[16] Thus Stanford in 1970 was an exception among American research universities in having a strong tradition of entrepreneurialism and a dynamic, campus-based technology-transfer operation.

COHEN'S AND BOYER'S REACTION TO PATENTING

Reimers's first step after recognizing the Cohen-Boyer method as a patenting opportunity was to contact Cohen. (See Figure 2.) His suggestion to file a patent application caught Cohen by surprise. As Cohen commented recently, up to this point he had "not . . . dreamed of the notion of patenting any of this." But he had recognized the method's possible practical applications well before the *New York Times* article and spelled out some of them in a paper published earlier in 1974. The opportunity Reimers presented was very different from Cohen's previous experience with patenting. During his postdoctoral years (1965–1967) at Albert Einstein College of Medicine, Cohen had developed a filter box for use in his own research. A salesperson from New Brunswick Scientific had seen him using it and had persuaded Cohen to let New Brunswick patent and market the device. The patent

---

[16] Reimers oral history, p. 12 (unfamiliarity with patent system); and Martin Kenney, *Biotechnology: The University-Industry Complex* (New Haven, Conn.: Yale Univ. Press, 1986), pp. 74–75 (role of outside organizations). For the role of the Research Corporation as an intermediary between university and industry see Charles Weiner, "Patenting and University Research: Historical Case Studies," *Science, Technology, and Human Values*, 1987, *12*:50–62.

was subsequently assigned to New Brunswick, and Cohen received modest royalties for several years. But the notion of patenting a basic process of laboratory science such as recombinant DNA was new and alien to him. "I suppose," he remarked in an interview, "my framework was that one patents devices, not basic scientific methodologies."[17] As Cohen told Reimers, he was not certain that patenting was appropriate for recombinant DNA, a basic science technique that he felt required many years of development before it reached a point of significant commercial application.[18] Cohen's opinion reflected the academic biologist's unfamiliarity, common at the time, with a prime purpose of the patent system: to spur industry to develop inventions into socially useful products by providing a period of protection.

Cohen's initial resistance to Reimers's suggestion may also have been related to an incongruity between the patenting process and perceptions of science as a communal and cumulative endeavor.[19] Although the patent application procedure requires citation of the research upon which an invention is based, it also aims to reduce the number of inventors to the one or few deemed responsible for the conceptual, rather than merely technical, contribution. Singling out one or a few inventors from a scientific team, as patenting protocol requires, diverged from the custom in scientific publications of assigning individual credit through coauthorship to everyone who had directly contributed to the research being reported. Moreover, compared to scientific convention, the legal definition of inventorship seemed to slight the full dimensions of a scientific discovery, leaving out some collaborators, institutional and personnel resources, and the background of research upon which a discovery is based.[20]

Recalling that he had to talk to Cohen "like a Dutch uncle" in obtaining his permission to file a patent application, Reimers maintained that the commercial development of penicillin had been delayed for eleven years for want of patent protection. This simplification of penicillin history was meant to suggest that a patent would encourage the industrial application of recombinant DNA technology through the sale of license rights. As for Cohen's concern to credit the scientists upon whose research recombinant DNA was based,

---

[17] Stanley N. Cohen, interview by Sally Smith Hughes, 28 Jan. 1999, draft, pp. 28, 32. (Interviews with Cohen cited in this essay are steps toward an oral history, as yet incomplete.) Having succeeded in splicing and cloning DNA molecules from two different species of bacteria, Chang and Boyer observed: "The replication and expression of genes in *E. coli* that have been derived from a totally unrelated bacterial species . . . now suggest that interspecies genetic transfer may be generally attainable. Thus, it may be practical to introduce into *E. coli* genes specifying metabolic or synthetic functions (e.g. photosynthesis, antibiotic production) indigenous to other biological classes." Chang and Cohen, "Genome Construction between Bacterial Species *In Vitro*" (cit. n. 9). Lederberg communicated the paper in November 1973; it was published in April 1974.

[18] Reimers maintains that Cohen initially believed that commercial development of recombinant DNA might take twenty years: Reimers, "Tiger by the Tail" (cit. n. 5), p. 466. Cohen's recollection is that he projected that the first commercial products would be available in less than ten years: Cohen, personal communication, 19 Jan. 2000.

[19] Jean-Paul Gaudillière and Ilana Löwy point out the transformation in access to knowledge that occurs through patenting: "Patents, and other forms of legal appropriation, mark the symbolic passage from the production of what is viewed as freely circulating knowledge to the production of saleable commodities: thus, they regulate science by guaranteeing legally restricted access to materials, processes, and know-hows." Gaudillière and Löwy, eds., *The Invisible Industrialist: Manufactures and the Production of Scientific Knowledge* (New York: St. Martin's, 1998), "Introduction" to Pt. 3, p. 298.

[20] See Phillippe Ducor, "Coauthorship and Coinventorship," *Science*, 2000, 289:873, 875. For a discussion and examples of the textual and perceptual distinctions between knowledge claims in science and patent claims in the legal and business sphere see Gregg Myers, "From Discovery to Invention: The Writing and Rewriting of Two Patents," *Social Studies of Science*, 1995, 25:57–105; for discussion of a patent's creation of a whiggish, internalist version of history see Geof Bowker, "What's in a Patent?" in *Shaping Technology/Building Society: Studies in Sociotechnical Change*, ed. Wiebe E. Bijker and John Law (Cambridge, Mass.: MIT Press, 1992), pp. 53–74.

Reimers responded that "no invention is made in a vacuum"; all are dependent on previous work of other scientists.[21] Patents and the licenses arising from them, he continued, were not only a mechanism for encouraging commercial development of basic science discoveries but also potential generators of unrestricted funds that Stanford could then apply to support university research and education.

Despite his reservations, on 24 June 1974 Cohen completed and signed an invention disclosure on a standard OTL form, the first formal step in the patent application procedure. The one-page disclosure of "A Process for Construction of Biologically Functional Molecular Chimeras" outlined the conceptual genesis of the scientific discovery (Cohen and Boyer's conversation in Hawaii in November 1972), listed the three cloning papers of 1973 and 1974 as the publication base, and noted "many" oral disclosures at seminars and symposia.[22] Only Cohen and Boyer were named as inventors, despite the fact that four other individuals—Annie Chang, Robert Helling, John Morrow, and Howard Goodman— were coauthors of one or more of the papers on which the invention was based. As it turned out, limiting inventorship to Cohen and Boyer would cause problems with Helling and Morrow.

Cohen told Reimers that he would renounce any future royalties to which he was entitled, despite Stanford policy awarding a third to the inventor. He intended the decision to emphasize that his motivation in permitting a patent application to go forward was not the hope of personal gain or aggrandizement but, rather, to provide a possible source of income for the university. The decision was a measure of Cohen's unease concerning the propriety of patenting, but it also turned out to have political advantages. Some years later, after he had been involved in the public debate over regulation of recombinant DNA research, he commented that "the fact that I had already turned over all royalties to Stanford enabled me to speak out in ways which would not have been possible if my motives were being questioned."[23]

Reimers was accustomed to dealing with faculty members of Stanford University Medical School concerning intellectual property matters and in fact had previously worked with Cohen in regard to a copyright on a software-based drug interaction program. At about this time, the medical faculty accounted for approximately 35 percent of OTL's licensed inventions. Thus the stigma historically associated with patenting in medicine does not appear to have deterred some members of the Stanford medical faculty from pursuing patents on their discoveries, perhaps because the inventions were not directly related to patients' health.[24] Nonetheless, there is evidence, to be described later in this essay, of campus dissension about the propriety of patenting in a university environment.

[21] [News release], Stanford University News Service, 3 Aug. 1981, S74-43, Correspondence 1980–1982, OTL Archives ("Dutch uncle"); and Cohen, "Stanford DNA Cloning Patent" (cit. n. 5), p. 215 (Reimers's response).

[22] Stanford University, Invention Disclosure, "A Process for Construction of Biologically Functional Molecular Chimeras"; Inventors: S. N. Cohen, H. W. Boyer; signed and dated: Stanley N. Cohen, 24 June 1974, 74-134-1, Folder: "Cohen, S. et al.," Univ. California Office of Technology Transfer, Oakland (hereafter cited as UC OTT Archives). In Greek mythology, a chimera is a creature composed of parts of three different animals. Cohen's use of the term refers to an entity made up of different genetic components.

[23] Niels Reimers to Josephine Opalka, 26 June 1974, 74-134-1, Folder: "Cohen, S. et al.," UC OTT Archives (renouncing royalties); and Nicholas Wade, "Cloning Gold Rush Turns Basic Biology into Big Business," Science, 1980, 208:688–692, on p. 689. At a later date, Cohen reversed his decision and began to take his share of the licensing income so that he had some control over its use: Cohen, personal communication, 10 July 2000.

[24] Reimers oral history, p. 3 (earlier work with Cohen); and Debby Fife, "The Marketing of Genius," Stanford Magazine, n.d., [1975 or 1976], pp. 48–53 (35 percent). Of the six medical patents generating royalties for Stanford in 1977, the most lucrative was an instrument, the fluorescence-activated cell sorter produced by Becton Dickinson Electronics. See Bill Snyder, "Discoveries: Finding the Way to the Marketplace," Stanford MD, Spring 1977, pp. 18–23.

Shortly after his conversation with Reimers, Cohen had telephoned Boyer to broach the possibility of filing a patent application. Up to this point, Boyer had not considered trying to patent the cloning procedure. He commented recently: "It was certainly not Stanley's idea or my idea to [patent] it. Very few molecular biologists [at the time] knew anything about patents, their place in science and business, and even in the vitality of our country." Nonetheless, Boyer had already considered possible commercial applications of recombinant DNA. The work with frog DNA indicating that genetic material from a higher organism could be cloned in bacteria had suggested the intriguing possibility of using the method to clone human genes in bacteria and express them as human proteins: "we were thinking about [commercial applications] at that time [of the frog experiments, 1974]. Not in any grandiose ways, but sitting around the laboratory talking about, Well, now if you can clone eukaryotic [higher organism] DNA, you can clone the gene for human insulin or human growth hormone or whatever you can think of, and you should be able to make it in a bacterium."[25]

Boyer reacted to Cohen's call by questioning whether Stanford and UC had the right to patent research that had been partially funded by the National Institutes of Health (NIH); only the federal government, he mistakenly believed, could hold title to a patent arising from research it had supported. Cohen then explained to Boyer what Reimers had explained to him: their method was potentially patentable under Stanford's institutional patent agreement (IPA) with NIH. Like many other American research universities, Stanford had negotiated an IPA with the Department of Health, Education, and Welfare (DHEW). Until passage of the Bayh-Dole Act of 1980 simplified patent policy regarding federally funded research, a university wishing to patent the results of work supported by a federal agency was required to petition DHEW for transfer of ownership of each invention it sought to patent to the university.[26] This cumbersome procedure accomplished, the university could then seek patent protection and hold title if the patent issued.

Satisfied with Cohen's clarification concerning Stanford's IPA and attracted by the commercial possibilities of recombinant DNA, Boyer had no further hesitation in proceeding with a patent application. He was unwilling, however, to renounce his personal share of any future royalties, despite Cohen's decision to do so. Of the two inventors, Boyer was to be less involved in the patenting and licensing processes, in part because Stanford, with UC consent, took the lead in all aspects of patent prosecution. Unlike Cohen, Boyer was quite willing to let others handle the patenting process, stepping in only when his input was critical: "I must admit, I didn't have a lot of patience with patent law and trying to figure it out. So I just told [Bertram Rowland, the patent attorney] everything I knew, and the guy went ahead and did it [i.e., filed a patent application]. Stanley helped him out quite a bit; they were always working on it together. I tossed in a few ideas."[27]

The readiness with which Boyer accepted Stanford's proposal to file a patent application may have reflected his institutional environment. While UCSF in the 1970s did not have

[25] Boyer interview by Hughes, 22 July 1994, 28 Mar. 1994, draft, pp. 168, 60.

[26] Boyer interview by Hughes, 27 July 1994, draft, p. 170. For a description of DHEW procedure for patenting prior to and after Bayh-Dole see Nelkin, *Science as Intellectual Property* (cit. n. 13), pp. 13–15. Nelkin states that of some thirty thousand government-owned patents, only 4 percent had been marketed as of 1979.

[27] Boyer interview by Hughes, 27 July 1974, draft, pp. 167–169, on pp. 168–169; and William Carpenter to Reimers, 18 Sept. 1974, S74-43, Correspondence 1974–1979, OTL Archives (on Boyer's unwillingness to renounce royalties). Cohen agrees that Boyer was not eager "to spend the time" on patenting and that "most of the details of the patents . . . were left to me": Cohen interview by Hughes, 28 Jan. 1999, draft, pp. 39–40.

a strong history of commercial interests and ties with industry, it was a health sciences campus in which basic research and clinical application were occurring side by side. The School of Medicine, in which Boyer was an associate professor of microbiology working in the arcane field of bacterial restriction enzymes, was at this time strengthening its basic science enterprise by hiring young faculty skilled in the latest molecular techniques. It was a culture focused on the synergism of basic biomedical science and clinical medicine in which practical application was a clear goal, even though patenting played only a minor role. Intent on extending the basic science of recombinant DNA—and also engaged in Vietnam War protests and other liberal causes—Boyer regarded the patent application as a peripheral matter, one that Stanford and Stanley Cohen seemed well capable of handling. This viewpoint was in keeping with the "laid-back" image he projected, with his cap of brown curls and jeans-and-running-shoes style of dress—"a baroque angel in blue jeans," as the *Wall Street Journal* later described him.[28]

In contrast, Cohen was and is a more cautious and reticent personality. He immediately saw and was preoccupied by the ethical dilemmas that patenting presented. His concern that a patent did not acknowledge the body of research on which any invention is built was perhaps related to his own care to ensure that he and his laboratory received proper scientific credit.[29] At one point Cohen had considered becoming a rabbi but instead had gone to the University of Pennsylvania Medical School; he subsequently decided to combine a career in clinical medicine with basic biological research. He arrived at Stanford in 1968 with an appointment as an assistant professor in the Department of Medicine. In the early 1970s he was balancing clinical responsibilities and commitment to his molecularly based research on bacterial plasmids—circles of DNA independent of the nucleus—as bearers of antibiotic resistance, a basic science problem with obvious clinical relevance.

THE UNIVERSITY OF CALIFORNIA: A SECONDARY PARTNER

On 26 June 1974 Reimers sent Cohen's invention disclosure to Josephine Opalka, an administrator at the University of California patent office in Berkeley that handled patenting and licensing activities for the sprawling University of California system. Reimers had first contacted Opalka on 20 May, the day the *New York Times* article appeared, to alert her to the possibility of a patentable invention for Stanford and UC. Since the cloning method was a joint discovery by professors at the two institutions, both universities would have to agree to proceed with a patent application. Opalka reported that UC supported an application and was amenable to having Stanford administer the invention under its IPA with NIH. But she objected to Reimers's proposal that UC share the expense of filing the

[28] Quoted in William Boly, "The Gene Merchants," *California Magazine*, 1982, 7:76–79, 170–172, 174–176, 179, on p. 170. Restriction enzymes, made by bacteria, "restrict" or cut DNA molecules at specific nucleotide sequences; they are an important component of the recombinant DNA process. In 1975 Boyer and his laboratory left the Department of Microbiology to join the Department of Biochemistry, where he was given space in newly opened laboratory towers. For an account of the rise of molecular biology at UCSF see William J. Rutter, "The Department of Biochemistry and the Molecular Approach to Biomedicine at the University of California, San Francisco," an oral history conducted in 1992–1993 by Sally Smith Hughes, Regional Oral History Office, Bancroft Library, Univ. California, Berkeley, 1998. Also online: http://www.lib.berkeley.edu/BANC/ROHO/ohonline.

[29] Both Cohen and Boyer speak in the interviews toward the preparation of their oral histories of the debate over the order in which the authors should be listed on the 1973 and 1974 papers: Cohen interview by Hughes, 1 Mar. 1995, draft, p. 28; and Boyer interview by Hughes, 22 July 1994, draft, pp. 153–155.

patent application. Not wanting to relinquish a chance for commercialization, Reimers agreed to proceed at Stanford's expense. Opalka, who understood the scheme to involve no initial cost or risk to UC, decided that there was thus no need to solicit the customary opinions as to the invention's commercial viability from scientists in the field.[30]

The UC patent office, according to Reimers, was in the mid 1970s "overloaded with technology that they didn't move. They did not market [licenses]; they'd wait for a company to come to them, instead of taking the very proactive way we did it at Stanford. So this invention of Cohen and Boyer was just one more thing to do." Although Reimers can hardly be considered an unbiased observer, some members of the UC faculty today agree that in the 1970s the university's patenting and licensing activities did not adequately capitalize on the inventiveness of its faculty. The university owned a number of royalty-producing patents but did not invest sufficient financial or personnel resources to make licensing a priority until the early 1980s. As late as 1977 the UC Board of Patents, which oversaw activities of the university patent office, though aware of the commercial potential of UC discoveries in genetic engineering, recommended a low-key approach to the commercialization of university inventions: "Inventions and patents should continue to be considered as fortuitous by-products of research, nothing more." Many on the Berkeley faculty, known for its liberal cast, felt that the university should not be in "the invention business."[31] In short, until the last two decades of the century, the focus of UC patent policy was on making faculty inventions available to the private sector; marketing and making money from them were secondary concerns. It was a far cry from the situation at Stanford.

In August 1974 Reimers wrote to NIH to set the IPA approval process in motion. Noting "the tremendous significance of this invention," he doubtless expected patent prosecution to proceed more or less routinely.[32] Such was not to be the case.

THE RECOMBINANT DNA CONTROVERSY: A COMPLICATION IN THE PATENTING EFFORT

The politically fraught context of biology in the mid 1970s was to be a formidable factor in the effort to patent the Cohen-Boyer procedure, influencing how Stanford and UCSF conducted and publicly presented the patenting process and raising the question of whether a patent should be sought at all. The event precipitating concerns about the safety and implications of recombinant DNA research had occurred five months before the first Cohen-Boyer paper appeared in print. Despite his agreement with Cohen to keep the discovery quiet until after publication, an ebullient Boyer had revealed to some 130 scientists attending the Gordon Conference on Nucleic Acids in June 1973 that he and his colleagues

[30] Reimers to Opalka, 2 May 1975, 74-134-1, Folder: "Cohen-Boyer Exploitation," UC OTT Archives; Opalka to Reimers, 19 Aug. 1974, 74-134-1, Folder: "Cohen, S. et al.," UC OTT Archives; Reimers, personal communication, 22 Dec. 1999 (Opalka's unwillingness to share filing expense); Opalka to Dr. Robert N. Hamburger, 14 May 1975, 74-134-1, Folder: "Cohen, S. et al.," UC OTT Archives (Reimers to proceed at Stanford's expense); and Josephine Opalka, telephone interview by Sally Smith Hughes, 3 June 1999. Both university patent offices routinely relied on evaluations from scientists in the field to help them decide whether to proceed with the expensive patenting process.

[31] Reimers oral history, p. 4; for corroboration of Reimers's view from a UC faculty member see William J. Rutter, interview by Sally Smith Hughes, 31 May 1999, draft, pp. 1–2, 10–12, passim. On the policy in 1977 see John A. Perkins to UC President David Saxon, 4 Apr. 1977, CU-6, UC (System), Office of Technology Transfer, Box 61, Folder: "Board of Patents, Priority Administration, file #1, 7/24/57–12/31/76," Bancroft Library Univ. California, Berkeley. The laissez-faire policy was to change in the early 1980s, when for the first time experts trained in patent law and licensing were hired to staff UC's technology transfer office. The phrase "the invention business" was used by a participant who prefers anonymity.

[32] Reimers to Norman J. Latker, 20 Aug. 1974, S74-43, Correspondence 1974–1979, OTL Archives.

Figure 3. *Paul Berg, 1972. Photograph courtesy of Office of News and Public Affairs, Stanford University Medical Center.*

had developed a method for splicing and cloning DNA. A participant's comment—"Well, now we can put together any DNA that we want to"—brought home the significance and to some the potential danger of the new technical capability.[33]

To summarize history that has been told in detail elsewhere, conference participants were concerned that genetically engineered molecules might prove hazardous for laboratory workers and the public. They voted to have Maxine Singer and Dieter Soll, cochairs of the Gordon Conference, write letters to the National Academy of Science (NAS) and the Institute of Medicine requesting the formation of a committee to investigate the potential risk of recombinant DNA research and the possible need for research guidelines. The academy subsequently formed the Committee on Recombinant DNA to study the problem, with the Stanford biochemist Paul Berg at its head. (See Figure 3.) In July 1974, while the patenting issue was being considered at Stanford, the committee published a letter in *Science* calling for a voluntary moratorium on certain kinds of recombinant DNA research until a conference could be convened to assess risks and develop research guidelines.[34] Berg's name was first on the list of ten scientists who signed the letter; Cohen and Boyer were also signatories.

Stanford and UCSF were at the time at the forefront of the science of recombinant DNA.

---

[33] John Lear, *Recombinant DNA: The Untold Story* (New York: Crown, 1978), p. 70.

[34] For commentary on and reproductions of key documents associated with the controversy over and the commercialization of recombinant DNA see James D. Watson and John Tooze, *The DNA Story: A Documentary History of Gene Cloning* (San Francisco: Freeman, 1981). The Singer-Soll letters are reproduced on pp. 5 and 6; the letter published in *Science* is reproduced on p. 11. For a detailed history see, e.g., Krimsky, *Genetic Alchemy* (cit. n. 8), pp. 70–96.

Because Cohen's plasmid was at first the only vector suitable for use in recombinant DNA research, he received requests for it from other scientists even before the November 1973 paper had appeared. Although initially insistent on refusing such requests until the paper was published, about six weeks before publication Cohen gave in to the pressure and began to send the plasmid to scientists if they agreed to follow safety precautions of his own construction. For their part, Boyer and his laboratory were further developing restriction enzyme technology and exploring the potential of DNA cloning as a general procedure in molecular genetics.[35] But the politics of the recombinant DNA debate could not be divorced from the science, and it was the Stanford campus that was its political center. Berg, as head of the NAS committee, was arguably the most visible figure in the controversy and a forceful spokesman for the need to assess risks and devise research guidelines. Among his many other activities in 1974, including the chairmanship of the Stanford biochemistry department, he began to organize an international conference, the now-famous Asilomar Conference on Recombinant DNA Molecules, which would convene on California's Monterey Peninsula in February 1975. Its purpose was to review progress in recombinant DNA research and to discuss biosafety guidelines so that the voluntary moratorium on recombinant DNA experiments could be lifted.[36]

The high-profile conjunction of recombinant DNA research and politics on the Stanford campus presented a major problem for those interested in commercializing the Cohen-Boyer discovery. Stanford and UC documents concerning patenting of the procedure begin in the summer of 1974 to reflect a pervasive theme of the next few years: how to accommodate the patenting and licensing effort and the biohazards controversy. Could the process of commercialization be prevented from exacerbating the political problem? Reimers assured NIH that Stanford intended "to exercise great care in the administration of this invention, insofar as is feasible within the constraints of the patent grant which may be issued, to ensure against misuse of the invention." He believed, however, that the appropriate mechanism for controlling recombinant DNA research was not the power of the patent holder but, rather, "the restraint that individuals of the international scientific community must exercise against uncontrolled [recombinant DNA] experimentation."[37] In so saying, he was departing from the traditional argument of American universities that owning a patent enabled a university to monitor use of the invention it protected.

[35] Lear, *Recombinant DNA* (cit. n. 33), pp. 83–84 (sharing the plasmid); and Howard M. Goodman, Herbert W. Boyer, Department of Biochemistry and Biophysics, UCSF, Annual Report 1974, Library, UCSF Department of Biochemistry, [n.p.]. Goodman and Boyer had agreed at this time to collaborate in research and include each other's names as authors of publications. The arrangement soon disintegrated.

[36] Paul Berg, David Baltimore, Sydney Brenner, Richard O. Roblin III, and Maxine Singer, "Asilomar Conference on Recombinant DNA Molecules," *Science*, 1975, *188*:991–994. Robert Bud points out that, rather than marking the beginning of concerns about genetic engineering, Asilomar represented the culmination of almost two decades of such concerns: Bud, *Uses of Life* (cit. n. 1), p. 175. Berg was chairman from 1969 to 1974, having followed Arthur Kornberg to Stanford in 1959 when the latter became the first chairman of the new biochemistry department. Both Kornberg and Berg, longtime colleagues and friends, recount in their oral histories the story of how almost the entire Kornberg laboratory migrated from Washington University, St. Louis, to Stanford: Arthur Kornberg, "Biochemistry at Stanford and Biotechnology at DNAX," an oral history conducted in 1998 by Sally Smith Hughes, Regional Oral History Office, Bancroft Library, Univ. California, Berkeley, 1998; and Paul Berg, "A Stanford Professor's Career in Biochemistry, Science Politics, and the Biotechnology Industry," an oral history conducted in 1998 by Sally Smith Hughes, Regional Oral History Office, Bancroft Library, Univ. California, Berkeley, 2000. The Kornberg oral history is online: http://www.lib.berkeley.edu/BANC/ROHO/ohonline.

[37] Reimers to Latker, 20 Aug. 1974, S74-43, Correspondence 1974–1979, OTL Archives.

ASSESSING THE COMMERCIAL POTENTIAL OF RECOMBINANT DNA

The political controversy was only one of Reimers's problems. He also had the patent bar to consider. U.S. patent law requires a patent application to be filed within the year following the first public disclosure of the invention. The first paper on the cloning method had been published in November 1973; Reimers was all too aware that the deadline of November 1974 was fast approaching. He had not heard about the work until May 1974 and was now hard pressed to make a final decision as to whether a patent application was in Stanford's best interest and, if so, to file an application before the bar. Yet it was not known whether industrial application of recombinant DNA technology was practical. Facing an uncertain market, Reimers assigned William Carpenter, one of the Stanford business school master's degree candidates whom he habitually hired for summer work at OTL, to investigate its commercial potential. This was routine practice: Reimers would proceed with the costly patenting procedure only if corporations expressed interest in developing the invention and the university had a reasonable chance of recouping the financial outlay through licensing fees and royalties.

Carpenter's first step was to schedule separate meetings with Cohen and Boyer for the purpose of exploring possible practical uses of the cloning method. Although no record has been found of his meeting with Cohen, there are handwritten notes of his discussion with Boyer. Elated about the method's expanding use in basic research, Boyer also saw immediate commercial applications in the synthesis of certain hormones and enzymes. He predicted that the new procedure, by increasing the yield of these substances from bacterial sources, would be of interest to commercial concerns in microbiology and suggested that it could also be used to produce antigens for the synthesis of antibodies. Boyer mentioned insulin, a drug with a highly profitable worldwide market, as another target but cautioned that there was "much work to be done before such applications can be made."[38] Thus, more than eighteen months before he cofounded the biotechnology company Genentech in April 1976, Boyer had definite ideas about specific commercial uses of recombinant DNA technology.

Carpenter's report to Reimers, based on his discussions with Cohen and Boyer, described the "plasmid gene transplantation technique" as a means of turning bacteria into "genetic factories" for the production of "hard-to-make" substances such as insulin, viral proteins for vaccine synthesis, hormones, and other proteins. The assessment provided the justification Reimers sought for proceeding with a patent application. But he had to move quickly if the university was to meet the patent bar, now only a few weeks away. As it was, Stanford and UC had lost the chance to obtain foreign intellectual property rights on recombinant DNA; the "first-to-file" patent system operating abroad precluded filing a patent claim on a previously published invention. Reimers engaged Bertram Rowland, a patent attorney whose San Francisco law firm had an office in Palo Alto, to work with Cohen and Boyer to draw up a patent application. On 4 November, just one week before the bar, Stanford in coordination with the University of California filed a patent application entitled "Process and Composition for Biologically Functional Molecular Chimeras."[39] As

---

[38] Carpenter to Cohen, 4 Sept. 1974; and BC [Bill Carpenter] to File S74-43, re: Dr. Herbert Boyer, 18 Sept. 1974: S74-43, Correspondence 1974–1979, OTL Archives.

[39] Carpenter to Reimers, 18 Oct. 1974, S74-43, Correspondence 1974–1979, OTL Archives; and "Cohen-Boyer Application," Serial No. 520, 691, filed 4 Nov. 1974, CU-6, UC (System), Office of Technology Transfer, Box 52, Exhibit Book II, Bancroft Library, Univ. California, Berkeley. Many countries require a patent application to be filed before the invention is published. One estimate is that Stanford and UC lost 50 percent of

the title states, the application was for a patent on both the recombinant DNA process and product (or "composition," in patenting jargon).

In August 1975 OTL again approached Boyer about commercial development of the recombinant DNA process. Practical applications had to be demonstrated, Boyer pointed out, before the pharmaceutical industry could be expected to make any significant capital investment in developing the technology. To provide just such a demonstration, Boyer planned to establish "a general synthesis procedure for small polypeptide hormones."[40] He proposed to make the hormones by synthesizing DNA, using chemicals off the shelf rather than employing the natural genes. Using recombinant DNA technology, the artificial gene would then be cloned in bacteria. One considerable advantage of the synthetic DNA approach was that it would not be subject to the NIH guidelines. Then being formulated in the wake of the Asilomar conference, these guidelines were to set forth physical and biological containment standards designed to reduce the chance of biohazard arising from recombinant DNA research involving living organisms and natural genes; synthetic genes made in a test tube, as Boyer envisioned, would not fall under their purview.

Because of their limited supply, Boyer believed the production of human hormones to be the most immediately promising field for commercial development of recombinant DNA technology. His proposed target was angiotensin-2, a hormone that causes vasoconstriction, which Boyer chose for its small size and resulting ease of synthesis. He told OTL that he expected the synthesis to be completed by unnamed "collaborative German biochemists" on 1 September 1975. "Once the [synthetic DNA] sample is received," the OTL memo continued, "approximately three months of basic test work will be required to determine whether synthesis of the hormone can be achieved by the recombinant DNA process. If this procedure is successful, the next logical extension would focus on the production of insulin."[41]

Thus angiotensin was merely the first step in Boyer's scheme to establish the commercial viability of the recombinant DNA process in the production of drugs for human use. He chose insulin as a target because it was a product with far greater medical significance and commercial value than angiotensin. Human insulin, he projected, should have none of the adverse side effects sometimes occasioned by use of the bovine and porcine insulin currently on the market. Its relatively small size for a biological molecule also held promise that the synthetic process might be less laborious than that for some larger molecules. Commercial development could proceed more quickly, Boyer commented, if he had the help of "organic chemists" in synthesizing DNA molecules. He asked OTL to contact firms to determine their interest in supporting development work of this nature. Boyer's ideas had advanced a step: by the summer of 1975 he not only had aspirations for commercial development of recombinant DNA and DNA synthesis but had also devised a research approach and selected drug targets.[42]

---

potential licensing revenue from the Cohen-Boyer discovery because of the lack of foreign patent rights: Kenneth S. Dueker, "Biobusiness on Campus: Commercialization of University-Developed Biomedical Technologies," *Food and Drug Law Journal,* 1997, *52:*453–509, on p. 493 n 256.

[40] Ken Imatani to Reimers, memo, "Discussion with Dr. Boyer for Future Development Work for Recombinant DNA Process," 6 Aug. 1975, S74-43, Correspondence 1974–1979, OTL Archives.

[41] *Ibid.*

[42] *Ibid.* In a May 1975 interview Boyer mentioned his current work on the expression of genes of higher organisms in bacteria and went on to remark: "I think this has a lot of implications for utilizing the technology in a commercial sense, that is, could one get bacteria to make hormones, etc., etc." Herbert Boyer, interview by Rae Goodell, 20 May 1975, Recombinant DNA Controversy Oral History Collection, Institute Archives and Special Collections, MIT Libraries, Cambridge, Massachusetts, p. 35.

No evidence has been found that research by "German biochemists" ever transpired. Instead, Boyer began to contemplate forming a commercial venture of his own. In January 1976 he was visited by the venture capitalist Robert Swanson, who was likewise interested in creating a company based on the recombinant method. Their discussions led to the formation of Genentech, Inc.—*Genetic Engineering Technology*—in April 1976. Boyer and his UCSF laboratory group immediately began collaborating in work financed by Genentech on the synthesis of somatostatin, another small-molecule hormone and the company's first research project. He and Swanson held the work to be a critical test both of recombinant DNA technology as a new commercial approach to drug synthesis and of the future viability of Genentech, the first company founded to exploit that technology.[43]

The discovery that industrial research was being conducted and funded in a public institution divided the UCSF Department of Biochemistry and prompted a campus investigation. The personal hostility directed at Boyer left enduring scars and suggested the novelty and precarious state of corporate relationships in academic biology at the time. He was an early target of the tensions that were to erupt throughout academia in coming years as universities and scientists at the forefront of molecular biology sought to capitalize on commercial opportunities.[44]

DISSENSION REGARDING THE PATENT APPLICATION

In January 1975, after confirming that the fundamental concept for the recombinant DNA discovery originated with Cohen and Boyer alone, Rowland wrote to the four coauthors of the 1973 and 1974 papers on which the patent application was based, asking them to disclaim inventorship. This was a precaution, not a requirement of patent law, which Rowland hoped would deter the U.S. Patent Office examiner from questioning Cohen's and Boyer's status as sole inventors. Rumors circulated that Helling and Morrow refused to sign the affidavits. Cohen was distressed that Rowland had not consulted him before sending the letter and by his colleagues' reaction to it. The incident served to heighten his lingering discomfort with Stanford's patenting effort, particularly in light of mounting tensions associated with the recombinant DNA controversy. He wrote Rowland that the

---

[43] On the work financed by Genentech see Herbert Boyer, Associate Professor of Biochemistry, Department of Biochemistry and Biophysics, University of California, San Francisco, Annual Report 1975, published May 1976, Library, UCSF Department of Biochemistry; and "Sponsored Research Agreement between Genentech, Inc., and the Regents of the University of California," 1 Aug. 1976, 77-064-1, Folder: "Goodman et al., Rutter et al., Deposit of Microorganisms," UC OTT Archives. For publicity on the somatostatin project see "Synthetic DNA Put to Work in Living Cells" [press release], UCSF News Services, 28 Oct. [1977]; and "A Commercial Debut for DNA Technology," *Business Week,* 12 Dec. 1977, pp. 128–129.

[44] On the campus investigation see James E. Cleaver, Chairman UCSF Biosafety Committee, to Chancellor Francis A. Sooy, 4 Nov. 1977, Sooy to Cleaver, 9 Nov. 1977, AR86-7, Carton 2, Folder 76, Archives and Special Collections, UCSF Library; and "UC Funds Aid Private Groups," *Los Angeles Times,* 7 Sept. 1978, p. 1. On the hostility directed at Boyer see Boyer interview by Hughes, 30 Apr. 1994, draft, pp. 139–143. In a newspaper article triggered by Genentech's successful production of somatostatin, Boyer gave a simple explanation for his participation in the company: "I wanted to see that the technology gets transferred to private industry so that the public benefits come out as soon as possible." Charles Petit, "The Bold Entrepreneurs of Gene Engineering," *San Francisco Chronicle,* 2 Dec. 1977, p. 2. The science media of the late 1970s and 1980s is rife with articles on the promise and problems of greater university-industry interaction in biomedicine. See, e.g., the response of the president of Harvard to the public furor when the university contemplated investing in a company formed by a Harvard professor to commercialize DNA research: [Derek C. Bok], "Business and the Academy," *Harvard Magazine,* May–June 1981, pp. 23–35. The controversy over the proper boundary between academic and industrial research continues. See, e.g., the reaction to UC Berkeley's alliance in 1998 with Novartis Agribusiness Discovery Institute: "UC Finalizes Research Deal with Biotech Firm; Pie Tossers Leave Taste of Protest," *San Francisco Chronicle,* 24 Nov. 1998, p. A17.

letter had given recipients the impression that he and Boyer were personally filing the patent application and sought to profit financially if a patent was issued. He was anxious to clarify for his colleagues that it was Stanford as an institution, rather than he and Boyer as individuals, that chose to pursue a patent application. He then stated his reason for succumbing to Reimers's request to file a patent application: "I can accept the view that it is more reasonable for any financial benefits derived from this kind of scientific research carried out at a non-profit university with public funds to go to the university, rather than be treated as a wind-fall profit to be enjoyed by profit-motivated businesses; I agreed to cooperate with Stanford for that reason."[45]

In an effort to quell any discussion that he was pursuing a patent for personal financial gain, Boyer at this point reversed his position and announced that he would not accept royalties from any future patent.[46] His decision indicated the sensitivity of this particular situation but was reinforced by the ambivalent position patenting occupied in academia at this time, even at a university as entrepreneurial as Stanford.

In early February 1975, less than three weeks before the Conference on Recombinant DNA Molecules was to convene at Asilomar, Cohen and his Stanford colleagues David Hogness, Charles Yanofsky, Ronald Davis, and Paul Berg, who were using recombinant DNA in research on a variety of organisms, met with Reimers in Berg's office in the biochemistry department of the medical school. Berg—and perhaps Hogness, Yanofsky, and Davis as well—had only recently learned of the patent application. The meeting had been called to discuss whether the scientists due to assemble at Asilomar might accuse Stanford of conflict of interest. It was common knowledge in molecular biology that Berg was spearheading the assessment of the risks and benefits of recombinant DNA research.[47] Yet his own university was simultaneously seeking patent ownership of the technology for the purpose of encouraging industrial development.

Those gathered in Berg's office were exquisitely aware that there could be adverse political fallout from Stanford's patenting effort. A false step ran the danger of igniting the biohazards controversy, tarnishing the public image of the two universities, and adversely affecting the future course of recombinant DNA science and its commercial application. The meeting was strained. Berg was opposed to any endeavor that would dilute his influence as a spokesman for creating federal regulatory policy so that the moratorium could be lifted, research guidelines formulated, and the application of recombinant DNA in basic research fully exploited. He was also concerned about the breadth of the Cohen-Boyer patent application, which claimed title to "production of all possible recombinants, joined in all possible ways, cloned in all possible organisms, using all possible vectors." In his view, the method Cohen and Boyer had devised was a basic building block of genetic engineering that should not be privatized. As he explained three years later: "I had certainly not been in support of the patent application. I thought it was the wrong time, the wrong kind of effort that the university should be making."[48]

[45] Reimers, personal communication, 22 Dec. 1999; and Cohen to Bertram I. Rowland, 22 Jan. 1975, S74-43, Correspondence 1974–1979, OTL Archives.

[46] Reimers to William Massy, 10 Feb. 1975, S74-43, Correspondence 1974–1979, OTL Archives. Boyer later donated his share of net royalties to a fund in the UCSF Department of Biochemistry that is currently known as the Herbert W. Boyer Fund.

[47] On the meeting see Reimers to Massy, 10 Feb. 1975. Although associated in the public mind with the speculative risks of recombinant DNA technology, Berg was not only aware of but also, as early as 1974, a spokesman for the projected benefits of the technology in producing antibiotics, hormones, and food sources: Bud, *Uses of Life* (cit. n. 1), p. 178.

[48] Paul Berg, interview by Charles Weiner, 17 Apr. 1978, Recombinant DNA Controversy Oral History Col-

Cohen's response reflected his informal tutorials with Reimers on the patenting process: broad patent claims and narrow determination of inventorship were common objectives of the patent system, however inimical they might seem to traditional scientific values. The group meeting in Berg's office debated whether Stanford should proceed as planned with patent prosecution, turn the invention over to the Research Corporation for patenting, or abandon the patent application entirely—without resolution.[49]

What is not reflected in surviving documentation of this and other encounters between Cohen and Berg is the undercurrent of personal animosity. In 1972, a year before the first Cohen-Boyer publication, Berg and colleagues had published their method for joining DNA segments and transferring the recombinant molecules to bacteria. It was a sophisticated procedure requiring special enzymes and expertise that only the Stanford biochemistry department at the time could provide. Since his arrival at Stanford Cohen had been given access to the biochemistry department's resources, intellectual and material, that were not available in his own Department of Medicine. Now he and Boyer, outside the inner core of Stanford biochemistry, had developed a technique so simple that virtually any molecular biologist could master it.[50] Although Cohen repeatedly acknowledged the groundwork scientists at Stanford and elsewhere had provided, Berg felt that his own role and that of others in the development of recombinant DNA had been slighted. The patent application, with only Cohen and Boyer listed as inventors, added a personal grievance to Berg's political objections.[51]

Joshua Lederberg was another opponent of patenting the Cohen-Boyer work, even though he appreciated and even expounded its commercial promise. He saw patenting in academia as a barrier to open scientific communication. Arthur Kornberg had similar reservations. The opposition of Lederberg, Kornberg, and Berg—the first two Nobel laureates, the third a leading scientist in the media limelight, and all three former or current Stanford department chairmen—was another problem to be negotiated by those interested in patenting the Cohen-Boyer procedure.[52] Although these three were not the only academics to find problems with the Stanford-UC patenting effort, their power and prestige and the fact that they were able to meet face to face with those pursuing patent prosecution made it all but impossible for Stanford to ignore their viewpoint. This was one of many situations occurring throughout the patenting process in which Stanford was compelled to

---

lection (cit. n. 42), pp. 71, 70 (quotations); and Reimers to Distribution [Cohen, Daniel Federman, Robert Lehman, Clayton Rich, Robert Rosenzweig], 15 Nov. 1976, S74-43, Correspondence 1974–1979, OTL Archives.

[49] John Poitras to File S74-43, 21 May 1976, S74-43, Correspondence 1974–1979, OTL Archives.

[50] David Jackson, Robert Symons, and Paul Berg, "Biochemical Method for Inserting New Genetic Information into DNA of Simian Virus 40," *Proc. Nat. Acad. Sci. USA,* 1972, *69*:2904–2909. The complex history, much simplified here, of the simultaneous development at Stanford of various approaches to recombining DNA is told in detail in the Berg oral history (cit. n. 36). Only the Cohen-Boyer method enabled cloning of the recombinant DNA.

[51] Reimers remarked recently: "I was *so* glad when [Berg] got that [Nobel Prize]. . . . I was feeling bad before then, because Paul was a distinguished scientist, and he had been very upset with the patenting process and not being recognized." Reimers oral history, p. 38. For an example of Cohen's acknowledgment of earlier work see Stanley N. Cohen, "The Transplantation and Manipulation of Genes in Microorganisms," in *The Harvey Lectures* (Series 74) (New York: Academic, 1979), pp. 173–204.

[52] For Lederberg's position see Joshua Lederberg, "DNA Research: Uncertain Peril and Certain Promise," *Prism,* Nov. 1975, *3*:33 (rpt. in Watson and Tooze, *DNA Story* [cit. n. 34], pp. 56–60); for Kornberg's views on patenting see Arthur Kornberg, *The Golden Helix: Inside Biotech Ventures* (Sausalito, Calif.: University Science Books, 1995), pp. 231–254. The Nobel Prize that Berg would share in 1980 for "his fundamental studies of the biochemistry of nucleic acids with particular regard to recombinant DNA" is one measure of his standing in science. See Gina Bari Kolata, "The Nobel Prize in Chemistry," *Science,* 1980, *210*:887–889. Lederberg was chairman of genetics and in 1969 Berg followed Kornberg as chairman of biochemistry.

weigh the benefits of commercialization against its political liabilities and to adjust activities in light of faculty dissension and broader social and political developments.

On 7 April 1975, more than a month after the Asilomar conference had concluded, the same group, with the addition of a Stanford provost, met for a second time. Reimers reported that tensions within the group regarding the Cohen-Boyer patent application had "defused" since the previous meeting; no one at the Asilomar conference had made an allusion to it. The presence of the provost suggests that top Stanford administrators were by this time aware that the intersection of the patent question and the recombinant DNA debate could be politically explosive and required high-level administrative oversight. They were also beginning to appreciate that the patent, if granted, might become the impressive royalty generator that the university had thus far never had. Berg and Yanofsky were sufficiently concerned about the propriety of the patenting effort that they had consulted officials at two federal science agencies. Reimers noted in a memo: "NIH and NSF [National Science Foundation] individuals with whom Drs. Berg and Yanofsky talked did not see anything amiss in Stanford seeking such [patent] development. (In fact, the people at NSF had good words to say about our licensing program.) It was thus concluded that Stanford should proceed with development in its normal fashion."[53]

But the patent application had for some time not been "normal"; it had already moved beyond usual OTL routine. NSF was to take a back seat to the National Institutes of Health on the patent issue, and the NIH was soon to hedge on its affirmation of Stanford's patenting effort. From this point on, high-level Stanford officials were to be instrumentally involved in the patent issue; patenting in the area of recombinant DNA was seen as politically sensitive and at the same time as critical to the commercial development of the technology. If not handled astutely in regard to the biohazards controversy, the effort to commercialize the technology could turn into a political liability for Stanford and UC and for the NIH as administrator of the recombinant DNA guidelines and provider of research funds.

UNREST IN THE ACADEMIC COMMUNITY

In June 1976 questions regarding Stanford's patent application were openly voiced in a public forum. The occasion was a symposium at MIT on genetic engineering attended by eight hundred people, including Cohen and Boyer, the press, and those opposed to recombinant DNA research. It was one of many contentious events that summer, including the Cambridge City Council's fiery deliberations on recombinant DNA experimentation at Harvard and MIT. In this politically charged atmosphere, participants remarked on persistent rumors of Stanford's attempt to patent recombinant DNA technology and asked whether anyone in attendance could speak to the matter. Boyer remembers the tension of the situation: "[Stanley and I] gave talks, and people were jumping up and yelling, 'Is it true you're patenting recombinant DNA? How can you do that?'" Sensing all eyes focused on him, Cohen made the agonizing walk to the podium while trying to gather his thoughts for a response. Intensely uncomfortable, as he later privately acknowledged, he nonetheless managed to assuage anxiety by arguing, contrary to Reimers's contention, that a patent on recombinant DNA technology would provide a means to control its commercial devel-

[53] Reimers to "File—Gene Transplant," 9 Apr. 1975, S74-43, Correspondence 1974–1979, OTL Archives. An anonymous *Isis* referee remarked that the National Science Foundation began to emphasize technology transfer in this period. Hence it is no surprise that the agency supported Stanford's position.

opment.[54] This was a position that Reimers and other university adminstrators were soon to reject explicitly.

Cohen was in the spotlight for another reason. In 1975 he had become a scientific advisor to Cetus Corporation, a young Bay Area company that wanted to acquire expertise in recombinant DNA. It was not the propriety of Cohen's consultant position per se that was at issue: Stanford policy in the 1970s described consulting as a privilege beneficial to both individual faculty members and the university as long as it did not interfere with academic responsibilities. What raised concern in Cohen's case was the fact that he was an inventor on a Stanford patent application and at the same time a paid consultant for a company seeking a license on the invention being patented. He tried to reassure critics by arguing that he expected to be able "to effectively separate my relationship with Stanford as the Inventor, from my relationship with Cetus as a scientific consultant."[55] It was a fine—if not impossible—line that Cohen attempted to draw between Cohen the scientist, Cohen the inventor, and Cohen the corporate consultant. Patenting and licensing efforts constituted a process in which science, invention, and business were intertwined. Attempts to draw boundaries between the three interlocking realms were artificial and ultimately futile, as the emergence of a DNA-based biotechnology industry in the next decade was repeatedly to demonstrate. The problem that Cohen sought to solve on an individual basis and that Stanford and UC administrators faced at the institutional level was how to uphold academic standards of free inquiry and open communication while at the same time supporting the privatization and commercial development of university discoveries. This question would be at the heart of academic interaction with commercial biotechnology from the 1980s on.

THE POTENTIAL FOR PROFITS AND PROBLEMS

In May 1976 Reimers forwarded a draft of his plan for licensing the Cohen-Boyer technology to Mark Owens, his counterpart at the UC Board of Patents. The plan outlined potential commercial applications in a number of industrial areas, including human and animal pharmaceuticals, industrial enzymes, and agriculture. The draft contained sizable subsections devoted to "the biological hazard" and "the public relations hazard." It made manifest how closely the process of commercialization and the politics of the biohazard controversy were intertwined. Yet in July—contrary to historic arguments in university patenting—Reimers made it clear that the patent was not regarded as a means to control in the public interest the invention's use by industry:

> Stanford and the University of California . . . cannot . . . by a license agreement, legislate morality, nor prevent a licensee from conducting research in an area of potential hazard, nor prevent an accident by a licensee in releasing a biologically hazardous substance. It does appear reasonable, however, to seek from licensees, prior to issuing a license, an expression of their

[54] Boyer interview by Hughes, 27 July 1994, draft, p. 169 (quotation); and unattributed memo to file, 11 June 1976, S74-43, Correspondence 1974–1979, OTL Archives. For an account of the Cambridge, Massachusetts, and other local initiatives for regulation see Krimsky, *Genetic Alchemy* (cit. n. 8), pp. 294–311; and the Recombinant DNA Controversy Oral History Collection at MIT (cit. n. 42).

[55] "Policy on Consulting by Members of the Academic Council: Principles and General Standards," 11 Mar. 1977, and Richard W. Lyman, [Stanford University] President, to Members of the Academic Council, 18 Mar. 1977: Kornberg Papers, SC359, Box 5, Folder: 1977, Green Library, Stanford; and Cohen to Reimers, 14 June 1976, S74-43, Correspondence 1974–1979, OTL Archives (quotation).

understanding of the potential hazards involved and their agreement to take precautions to conform with both law, good sense, common ethics, and the NIH guidelines.[56]

Neither Stanford nor NIH had the legal authority, the capacity, or the wish to regulate industry compliance with the NIH guidelines—which, after all, did not apply to industry in the first place.

There was yet another complication: Reimers learned from the U.S. Patent Office that claims on the recombinant organisms themselves, the so-called product claims, were for the time being unlikely to be allowed. As a consequence, the two universities filed a new patent application on 17 May 1976 that sought to protect only the basic Cohen-Boyer process and omitted the product claims. Second and third patent applications for the biological transformants—the cells genetically modified by the introduction of recombinant DNA—were being held in abeyance. "Product claims on new 'bugs' [microorganisms] are doubtful," Reimers explained, "although GE is pursuing coverage for their oil-eating bug in the courts." He alluded to the now-famous *Diamond v. Chakrabarty* case, in which General Electric sought to patent a living entity, a microorganism constructed to degrade crude oil. The case eventually reached the Supreme Court. Until the justices came to a decision, the U.S. Patent Office refused to consider patent applications claiming living organisms. Ananda Chakrabarty, the scientist who had developed the microorganism, did not use recombinant DNA technology to construct it. Nonetheless, *amicus curiae* briefs filed in the case, including one written by Genentech legal counsel, made reference to the implications of the decision for the commercial application of genetic engineering and the future viability of young biotechnology companies.[57] The question of the patentability of living organisms was of obvious significance to those interested in commercial exploitation of recombinant DNA.[58]

GENENTECH: CONTENDER FOR A PATENT LICENSE

Beginning in the spring of 1976, when Genentech was being formed, company president Robert Swanson became a persistent contender for license rights to the Cohen-Boyer

[56] Reimers to Mark Owens, 27 May 1976, and "Licensing Plan," draft, 14 May 1976: CU-6, UC (System), Office of Technology Transfer, Box 52, Folder: "Science & Technology, rDNA research," Bancroft Library, UC Berkeley; and Reimers to File S74-43, "Recombinant DNA Process," 17 July 1976, S74-43, Correspondence 1974–1979, OTL Archives (quotation).

[57] "Cohen-Boyer Application" [draft], serial no. 687,430, filed: 17 May 1976, CU-6, UC (System), Office of Technology Transfer, Box 52, Exhibit Book III, Bancroft Library, UC Berkeley (new patent application); Reimers to "Distribution," 15 Nov. 1976, S74-43, Correspondence 1974–1979, OTL Archives; and *Diamond v. Chakrabarty*, 447 U.S. (*United States Supreme Court Reports*), 303, 100 S.Ct. (*Supreme Court Reporter*) 2204 (1980), pp. 2204–2214. For a history of this case see Kevles, "*Diamond v. Chakrabarty* and Beyond" (cit. n. 15), pp. 65–79. *Amicus curiae* briefs were filed by nine organizations, including the University of California, the American Patent Law Association, the Pharmaceutical Manufacturers Association, the American Society of Microbiology, and Genentech, all of which had a stake in industrial application of the new genetic technologies. As Sheldon Krimsky has implied, the justices must have been aware of the commercial context and impact of their decision: Krimsky, "The Profit of Scientific Discovery and Its Normative Implications," *Chicago-Kent Law Review*, 1999, *75*:15–38. I thank an anonymous referee for calling my attention to this article.

[58] In October 1981 academic biologists and patent attorneys concerned with intellectual property rights in biotechnology met to explore the practical impact of the *Chakrabarty* decision on biological science and patent law. The resulting publication is David W. Plant, Niels J. Reimers, and Norton D. Zinder, eds., *Patenting of Life Forms* (Banbury Report 10) (New York: Cold Spring Harbor Laboratory, 1982). The impact of *Chakrabarty* and other intellectual property legislation and decisions, as well as the new tools provided by biotechnology, were of prime interest to the seed industry. See Frederick H. Buttel and Jill Belsky, "Biotechnology, Plant Breeding, and Intellectual Property," *Sci., Technol., Hum. Val.*, 1987, *12*:31–49.

process. He insisted in a flurry of letters and visits to Reimers and Opalka that license rights were critical to the young company's survival. Genentech needed the value represented in owning relevant license rights as a means to attract corporate investors. On 19 April, not two weeks after Genentech had been incorporated, Swanson proposed to Stanford and UC officials that the universities grant the company nonexclusive worldwide rights to use the licensed technology to produce "any product covered by the Licensed Technology" and exclusive rights to produce polypeptide hormones. Swanson's request for worldwide rights was moot because, as already noted, Stanford was precluded by law from filing for foreign rights to the invention. In exchange for an exclusive license, Swanson offered the two universities equity interest in Genentech in the form of four thousand shares of common stock. It was an early sign of the intimate relationship that commercial biotechnology was to create with academic institutions.[59] Reimers appreciated the damaging effect that a decision to issue only nonexclusive licenses might have: "This of course means that Genentech will not obtain its desired exclusive, that we forego equity and a possible substantial front payment for an exclusive, and it may mean that Genentech as a viable company cannot survive." Nonetheless, Reimers could not give Swanson the answer he wanted: "the jury was still out," he told him, in regard to a final decision about Stanford's licensing plan. Nonexclusive licensing was generally held to be viable only when inventions were revolutionary or broadly enabling.[60] It was becoming increasingly apparent that the Cohen-Boyer invention was in precisely this category.

CONSULTING THE NIH DIRECTOR

In 1976 Stanford officials sought counsel at the highest levels of university governance and the federal biomedical hierarchy by bringing the patent issue to the attention of Stanford's board of trustees and the director of NIH. In June Robert Rosenzweig, Stanford Vice President of Public Affairs, wrote to NIH Director Donald Fredrickson asking for advice as to whether Stanford should continue to seek patent protection on the Cohen-Boyer work. Rozenzweig acknowledged the political and ethical problems tied to the Cohen-Boyer patent application but couched his argument mainly in economic terms:

> It is a fact that the financing of private universities is more difficult now than at any time in recent memory and that the most likely prediction for the future is that a hard struggle will be required to maintain their quality. . . . To put the point as precisely as I can, we cannot lightly discard the possibility of significant income [from patent licenses] that is derived from activity that is legal, ethical, and not destructive of the values of the institution.[61]

---

[59] Boyer and Robert A. Swanson to "Gentlemen, Stanford University, University of California," 19 Apr. 1976; Swanson to Opalka, 28 Apr. 1976; and Opalka to Swanson, 10 May 1976: 74-134-1, Folder: "Cohen-Boyer Exploitation," UC OTT Archives. The long-standing collaboration between university scientists and the pharmaceutical industry predates the similarly close relationship in DNA-based biotechnology. For a history of the former see John P. Swann, *Academic Scientists and the Pharmaceutical Industry: Cooperative Research in Twentieth-Century America* (Baltimore: Johns Hopkins Univ. Press, 1988). For histories of biotechnology see Kenney, *Biotechnology* (cit. n. 16); and Sheldon Krimsky, *Biotechnics and Society: The Rise of Industrial Genetics* (New York: Praeger, 1991).

[60] Reimers to Robert Augsburger, 19 July 1976, and Reimers to File S74-43 Recombinant DNA, 2 Aug. 1976, S74-43, Correspondence 1974–1979, OTL Archives. On nonexclusive licensing see Dueker, "Biobusiness on Campus" (cit. n. 39), p. 497.

[61] For Stanford's attempts to solicit guidance see Reimers to Rodney Adams, 8 Apr. 1976, S74-43, Correspondence 1974–1979, OTL Archives; and Reimers to Owens, 27 May 1976, CU-6, UC (System), Office of Technology Transfer, Box 52, Folder: "Science & Technology, rDNA," Bancroft Library, UC Berkeley. Rosenzweig's letter to Donald Fredrickson is reproduced in Watson and Tooze, *DNA Story* (cit. n. 34), p. 499; for the quotation see Robert M. Rosenzweig to "Those Interested in Recombinant DNA," 4 June 1976, S74-43, Correspondence 1974–1979, OTL Archives.

Rosenzweig's focus on patent-as-moneymaker was a new notion in university patenting in the United States, one that the Cohen-Boyer patent was to usher in and establish as a prime justification for patenting in academia.

Pursuant to its Institutional Patent Agreement with NIH, Stanford did not need additional permission from the government to file for a patent. Nonetheless, because of this patent's political sensitivity, Rosenzweig had decided to seek counsel in Washington. Fredrickson was fully enmeshed in problems related to recombinant DNA. In light of the biohazards issue and concern that patents in the field might inhibit scientific communication, an important question was whether NIH patent policy should be revised and more rigorous procedures established for patenting in the area of recombinant DNA. The fact that the NIH guidelines were scheduled for official release in late June, ending the sixteen-month research moratorium, served to increase tensions.[62] Could Fredrickson, as head of the federal agency issuing the guidelines, also support Stanford's efforts to obtain proprietary rights on recombinant DNA technology? Critics held that the patenting and licensing of recombinant DNA risked expanding to industry the sphere of action in which biohazards might arise.

Fredrickson had learned that the University of Alabama and a number of other universities were following Stanford's lead in filing for patents on discoveries in molecular biology. It was therefore imperative to resolve the political, ethical, and intellectual property protection issues raised by the Cohen-Boyer patent application and others to follow in biotechnology. To garner advice, Fredrickson began to consult various government and professional organizations. These deliberations were occurring during a period of extensive congressional reappraisal of federal patent and technology transfer policies prompted by concern over the flagging American economy and declining technological competitiveness. Because questions relating to Stanford's patent application were interwoven with unsettled federal policy regarding intellectual property rights in government-sponsored research, it would be almost two years before Fredrickson gave Stanford an answer.[63] Meanwhile, the Cohen-Boyer patent application and others involving recombinant DNA were stalled in the U.S. Patent Office, awaiting resolution of federal patent policy deliberations and a Supreme Court case.[64]

By March 1977, the California Assembly was considering a tough bill to regulate all laboratories in the state that practiced recombinant DNA technology. A number of mea-

---

[62] Fredrickson to Rosenzweig, 2 Mar. 1978, S74-43, Correspondence 1980–1982, OTL Archives. For a personal view of the problems facing Fredrickson at this time see D. S. Fredrickson, "A History of the Recombinant DNA Guidelines in the United States" (1979), in Watson and Tooze, *DNA Story*, pp. 396–399.

[63] For Fredrickson's efforts to solicit advice see, e.g., Fredrickson to Robert Carow, Association of American Medical Colleges, 7 Sept. 1976, CU-6, UC (System), Office of Technology Transfer, Box 52, Folder 2, Bancroft Library, UC Berkeley; and Fredrickson to Members of the Recombinant Advisory Committee, 27 Aug. 1976, CU-6, UC (System), Office of Technology Transfer, Box 52, Folder: "Science & Technology, rDNA Research." According to Nelkin, before 1980 patent policy regarding federally funded research was "in a state of confusion," with twenty-six different patent policies in effect at government agencies. The Bayh-Dole Act of 1980 provided a uniform federal patent policy designed to encourage commercial development of government-supported research. Nelkin, *Science as Intellectual Property* (cit. n. 13), pp. 13–14.

[64] In an effort to encourage rapid dissemination of information concerning recombinant DNA Research, the U.S. Patent Office in January 1977 instituted procedures intended to speed up review of related patent applications. In April the procedures were withdrawn as untimely in view of ongoing congressional consideration of recombinant DNA research legislation. Pending a decision on the patentability of recombinant organisms and plasmids, the patent office subsequently suspended review of applications claiming them until the issue was resolved. Thomas Kiley, "Patent and Political Shock Waves of the Biological Explosion," in *Proceedings of the Southwestern Legal Foundation Patent Law 17th Annual* (New York: Bender, 1979), pp. 253–285, esp. pp. 264–265.

sures with similar intent were before Congress, other state legislatures, and local government bodies. Although Stanford as an institution took no official position on the pending
legislation, Berg, Cohen, Kornberg, and others on campus, as well as scientists elsewhere,
were mounting an active opposition. The lobbying effort, one of the largest ever waged
over a technical issue before Congress, helped to persuade legislators that the scientific
and commercial benefits of genetic engineering outweighed its potential risks. Berg, now
convinced that genetic engineering posed no real risk and opposed to giving government
control of the science, argued forcefully against regulatory legislation at any governmental
level and for "the tremendous opportunities afforded by the recombinant DNA methodology."[65]

COMMERCIAL PROSPECTS BEGIN TO CARRY THE DAY

A critical event in turning the tide of congressional opinion was Genentech's demonstration
in the fall of 1977 of the feasibility of using recombinant DNA methodology to create in
bacteria human proteins virtually identical to the naturally occurring forms. The successful
production of the hormone somatostatin, which Genentech announced at a press conference
in November, was presented as validating commercial application of the technology. "Molecular biology," Boyer declared to the media, "has reached the point where it can become
involved in industrial applications." Boyer, whose image was to appear on the cover of
Time Magazine in 1981, was one of the most visible proponents of capitalizing on molecular biology to make products of commercial use. (See Figure 4.) Cohen, more circum-
spect, continued to portray himself as respecting the traditional view of science as a communal endeavor.[66]

Berg and Philip Handler, president of the National Academy of Sciences, immediately
announced the somatostatin achievement at U.S. Senate hearings on genetic engineering,
calling it "a scientific triumph of the first order." The announcement was interpreted both
in the Senate and in the media as indicating the exciting commercial prospects of applied
biotechnology. As *Chemical and Engineering News* commented: "The success of the
California researchers, both Handler and Berg declare, is a vindication of the utility of
recombinant DNA research which should further defuse a tiny group of scientific critics
who claim that the technique is potentially dangerous to laboratory workers and the
public."[67]

Participants in the Senate hearings not only witnessed the biohazard issue being down-

[65] Paul Berg to Sen. Harrison Schmitt, 5 Jan. 1979, in Watson and Tooze, *DNA Story* (cit. n. 34), pp. 389–
391. For the proposed California legislation see Assembly Bill No. 757, 3 Mar. 1977, introduced by the Committee on Health, copy in Cohen Personal Correspondence, Biohazard Collection, Stanford; and David Perlman,
"Tough Rules on Creating New Forms of Life," *San Francisco Chronicle,* 4 Feb. 1977, p. 1. For scientists'
federal lobbying efforts see "Testimony by Paul Berg," Subcommittee on Science, Technology, and Space, [U.S.
Senate], 2 Nov. 1977, draft, Paul Berg Correspondence, SC 358, Accn. 90-020, Box 1, Folder: "Senate Testimony," Green Library, Stanford; Cohen's Senate testimony, in *Hearing before the Subcommittee on Health of
the Committee on Labor and Public Welfare,* U.S. Senate, 94th Cong., 22 Apr. 1975 (Washington, D.C.: Government Printing Office, 1975), pp. 3–12; and Arthur Kornberg to the Honorable Paul N. McCloskey, 8 Aug.
1977, Kornberg Papers, SC 359, Box 3, Folder: "Genetic Engineering," Green Library, Stanford.

[66] "A Commercial Debut for DNA Technology," *Bus. Week,* 12 Dec. 1977, pp. 128, 132. The 9 Mar. 1981
cover of *Time Magazine* reads: "Shaping Life in the Lab: The Boom in Genetic Engineering: Genentech's Herbert
Boyer." On Cohen's self-portrayal see David Dickson, "Inventorship Dispute Stalls DNA Patent Application,"
Nature, 1980, *284:*388.

[67] "Testimony by Paul Berg," 2 Nov. 1977; and "Human Gene in *E. Coli:* It Works!" *Chemical and Engineering
News,* 7 Nov. 1977, p. 4.

Figure 4. *Herbert W. Boyer on the cover of* Time Magazine, *9 March 1981. At the time Boyer was Professor of Microbiology and Biochemistry at the University of California, San Francisco, and cofounder of Genentech. Photograph courtesy of TimePix.*

played; they saw the ground being prepared for further industrial exploitation. The official report on the hearings was billed as "the first systematic effort to examine the issues that are likely to arise in the regulation of future large-scale commercial applications of recombinant DNA techniques." Although safety was still a discussion point, government and private-sector support of industrial expansion was described as a necessity if the United States had any hope of maintaining its leadership in the new genetic technologies.[68]

In the end, all federal bills to regulate recombinant DNA research died in Congress, in large part because legislators, heavily lobbied by a scientific coalition working through the American Society for Microbiology, recognized the value of the technology as a boost for the American economy. In July 1978 NIH issued a revised set of guidelines that eased restrictions on experiments; further relaxations were shortly to follow. In September Genentech and the City of Hope Medical Center announced to the media the production of recombinant human insulin, an accomplishment that was interpreted as the dawning of a new era of commercial biotechnology.[69] The waning of the recombinant DNA controversy loosened a constraint that had dominated Stanford's patenting and licensing strategy and complicated the process of commercialization. Although the biosafety question did not disappear from the public debate and has recently been revived with regard to genetically modified crops, it was fast losing force in the United States as a serious obstacle to mounting commercial pressures.[70]

Despite Stanford's repeated requests for guidance, it was not until March 1978 that Fredrickson finally announced the NIH position on intellectual property rights in recombinant DNA research. After consulting "a broad range of individuals and institutions on this matter," he had decided that no special procedure should be instituted for patent applications on such inventions. Fredrickson also gave Stanford permission to proceed with its licensing effort, though with the proviso that the licensed company "provides assurance of compliance with the physical and biological containment standards set forth in the [NIH] Guidelines in any production or use of recombinant DNA molecules under the license." Compliance, however, was to be voluntary; neither NIH nor Stanford had or wanted legal regulatory authority over industry.[71] There was no mention of an alternate enforcement mechanism.

Although the patent application was still stalled at the U.S. Patent Office pending the

---

[68] *Recombinant DNA Research and Its Applications,* Oversight Report by the Subcommittee on Science, Technology, and Space of the Senate Committee on Commerce, Science, and Transportation, August 1978 (Washington, D.C.: Government Printing Office, 1978), pp. iii (quotation), 37.

[69] For accounts of the derailing of U.S. federal legislation, including the role of Stanford and UCSF scientists, see Wright, *Molecular Politics* (cit. n. 8), pp. 256–278; and Krimsky, *Genetic Alchemy* (cit. n. 8), pp. 312–337. For an assessment of NIH guideline revision at this time see Wright, *Molecular Politics,* pp. 281–311. On the production of recombinant human insulin see "First Successful Laboratory Production of Human Insulin Announced," 6 Sept. 1978, Corporate Communication Division Files, Genentech, Inc., South San Francisco, California; "Human Insulin: Seizing the Golden Plasmid," *Science News,* 16 Sept. 1978, pp. 195–196; and Matt Clark with Joseph Contreras, "Making Insulin," *Newsweek,* 18 Sept. 1978, p. 93.

[70] Internationally, political concern regarding recombinant DNA waxed and waned at different rates in different countries. Maureen McKelvey notes that as the biohazard issue waned in the United States after mid 1977, the media in Sweden increasingly described the risks of the technology: McKelvey, *Evolutionary Innovations: The Business of Biotechnology* (Oxford: Oxford Univ. Press, 1996), pp. 109–113. For the situation in Great Britain see Wright, *Molecular Politics.* Although perhaps most noticeable in biotechnology, other academic sciences in the United States, such as those related to the electronics industry, were also experiencing greater commercial activity in the 1980s. For differences in industry growth patterns at this time see "A Comparison of the U.S. Semiconductor Industry and Biotechnology," Appendix C, *Commercial Biotechnology: An International Analysis* (Washington, D.C.: Office of Technology Assessment, Congress of the United States, 1984), pp. 531–540.

[71] Fredrickson to Rosenzweig, 2 Mar. 1978, S74-43, Correspondence 1980–1982, OTL Archives; and Reimers, "Tiger by the Tail" (cit. n. 5), p. 39 (on compliance).

Supreme Court decision in *Chakrabarty,* Reimers wasted no time in proceeding to finalize a licensing plan. A definitive decision had been made: Stanford would issue nonexclusive licenses on the Cohen-Boyer technology. Since Cohen had designated his royalties for research and education, Reimers pointed out that Stanford could claim that two-thirds of any future net income from the patent would be used for these purposes. This strategy was intended to deflate critics of university patenting and to mollify the coauthors of the original DNA cloning papers. Helling and Morrow nonetheless continued to refuse to sign a release of claim to inventorship. Like Paul Berg, Helling felt that no university should be allowed to claim rights on a technology as fundamental and broadly applicable as recombinant DNA. Their protest in the end was fruitless. Not only was the first patent to issue with only Cohen and Boyer as inventors; the second and third patents did likewise, even though the question of inventorship cropped up again in the 1980s.[72]

In June 1980 the Supreme Court held in *Chakrabarty* that living organisms engineered by man were potentially patentable under existing statutes. In a landmark 5-4 decision, the justices ruled that Chakrabarty's microorganism was not a product of nature but, rather, a novel invention of his own ingenuity and thus patentable subject matter. This was a critical ruling for commercial biotechnology: the patent system henceforth was to be used for securing property rights on all manner of living organisms and their components. Partly on the strength of the decision, Genentech went public on 14 October. Within minutes after the opening bell, frantic investors on the New York Stock Exchange purchased one million shares and the stock price rose from $35 to $89. Without an immediately marketable product, Genentech had in a few short hours raised $38.5 million, and Boyer and Swanson had gained a paper profit of $60 million on an initial investment of $500 each.[73] A period of speculative frenzy over genetic engineering had begun.

The decision in *Chakrabarty* released for review a stream of patent applications involving living organisms that had been held up in the patent office awaiting the opinion.[74] The Cohen-Boyer application was among them. On 2 December 1980 the U.S. Patent Office issued patent 4,237,224 on a "Process for Producing Biologically Functional Molecular Chimeras."[75] In early August 1981 Stanford announced the availability of licenses for the Cohen-Boyer invention, describing it as the start of "an unprecedented effort to license the entire genetic engineering industry for use of its basic scientific technique." By 15 December seventy-two companies had licensed the Cohen-Boyer technology, which Reimers described in his promotional material as "the basic tool needed in genetic engineering."[76] Each company had paid a license issue fee of $10,000, with another $10,000 due

[72] On the decision about licensing see Reimers to Massy, 26 Jan. 1979, 74-134-1, Folder: "Cohen-Boyer, Exploitation," UC OTT Archives; for Helling's views see Joel Gurin and Nancy E. Pfund, "Genetic Engineering: Bonanza in the Bio Lab," *Nation,* 1980, 22:542–548, on p. 544; on later questions about inventorship see Marjorie Sun, "Stanford's Gene Patents Hit Snags," *Science,* 1982, 218:868–869.

[73] Kevles, "*Diamond v. Chakrabarty* and Beyond" (cit. n. 15), pp. 70–71; and Boly, "Gene Merchants" (cit. n. 28), p. 79.

[74] By one account, 114 patent applications involving living organisms were held up in the patent office awaiting the Supreme Court's decision in *Chakrabarty:* Krimsky, "Profit of Scientific Discovery" (cit. n. 57).

[75] The legal status of the 1980 patent and the two pending patents was repeatedly debated. See, particularly, Albert P. Halluin, "Patenting the Results of Genetic Engineering Research: An Overview," in *Patenting of Life Forms,* ed. Plant *et al.* (cit. n. 58), pp. 67–126; and Jorge Goldstein, "A Footnote to the Cohen-Boyer Patent and Other Musings," *Recombinant DNA Technical Bulletin,* Dec. 1982, pp. 180–188.

[76] Stanford University News Service, advance for release Monday, 3 Aug. 1981, S74-43, Correspondence 1980–1982, OTL Archives. In an effort to encourage early licensing, Stanford offered credit to firms taking out licenses by 15 Dec. 1981 of five times the initial payment of $20,000 against future royalties on resulting products.

on 1 February 1982. By the latter date total license income would be $1,420,000, a sign of the financial windfall to come.[77]

CONCLUSION

This essay has shown that the universities' hard-won success in securing the recombinant DNA patent was tied to events at the national level. By the time the patent was issued, the cautionary approach to science and technology of the early 1970s was giving way under the Reagan administration to a view that science-based technology was central to economic growth. Congress was taking steps to encourage technological development through such means as deregulation, corporate financing, and tax relief, and federal agencies such as the National Science Foundation were developing programs designed to promote collaborations between universities and industry. Biotechnology figured among the high-technology areas that the government sought specifically to encourage for commercial development.[78] The field's promise as a new and profitable sector of the American economy was among the reasons for passage of the Bayh-Dole Act of 1980, which gave universities the right and incentive to hold patents on innovations arising from federally funded research. The increase in university patenting after the act went into effect in 1981 is one measure of its success in encouraging academic entrepreneurialism. In 1980 universities held title to approximately 150 patents; by 1990 the number had grown to 1,600. According to one historian of biotechnology, the new economic and political climate in Washington was "the crucial factor" in intensifying corporate competition after 1979 to apply recombinant DNA commercially.[79]

In tracing the patent history, we have seen interest in the industrial application of the new technology overtake earlier safety and regulatory concerns. The course patent prosecution took reflects a shift in national political focus from protectionism to technological and industrial expansionism. At first a prime consideration for those formulating patenting and licensing strategy, the biohazard issue was by the end of the decade losing out to forces in government, academia, and industry propelling recombinant DNA technology

[77] Roger G. Ditzel, [UC] Patent Administrator, to [UC] Vice President William B. Fretter, 18 Dec. 1981, 74-134-1, Folder: "Cohen, S. et al.," UC OTT Archive ($1,420,000). Media coverage of Stanford's announcement of the licensing plan focused on its moneymaking potential and its importance for the production of new biological products. See, e.g., Charles Petit, "Stanford, UC Plan to Get Rich on Gene Patent," *San Francisco Chronicle,* 1 Aug. 1981, p. 7; Victor Cohn, "Stanford Moves to Speed Use of Gene Splicing," *Washington Post,* 3 Aug. 1981, p. A6; and David Dickson, "Stanford Sells Gene-Splicing Licenses," *Nature,* 1981, *292:*405. The three patents expired as a unit on 2 Dec. 1997, having earned over $250 million in royalties and license fees: Floyd Grolle, former licensing officer of the Cohen-Boyer patents, Stanford OTL, personal communication, 8 Oct. 1998.

[78] On the new attitude under the Reagan administration see David Dickson, *The New Politics of Science* (New York: Pantheon, 1985), p. 5. On efforts to encourage collaborations see Lois S. Peters and Herbert I. Fusfield, "Current U.S. University/Industry Research Connections," in *University-Industry Research Relationships: Selected Studies* (Washington, D.C.: National Science Foundation, 1982), pp. 1–162; and Dorothy Nelkin and Richard Nelson, "Commentary: University-Industry Alliances," *Sci., Technol., Hum. Val.,* 1987, *12:*65–74 (part of a special issue called "Private Appropriation of Public Research"). On governmental interest in biotechnology see *Commercial Biotechnology* (cit. n. 70), esp. pp. 8–21.

[79] Wright, "Recombinant DNA Technology" (cit. n. 8), p. 345. Wright's article considers additional factors contributing to the intensification of competition in commercial biotechnology in this period. On the Bayh-Dole Act see "Technology Transfer Comes of Age: Bayh-Dole Fifteen Years Later," summary report, Annual Meeting of the Association of University Technology Managers, Ottawa Carleton Research Institute, 1995, Sect. 3: "The Bayh-Dole Act and What It Means," pp. 4–6; the 1980 and 1990 figures are on p. 6. The report can also be found online: http://strategis.ic.gc.ca/SSG/tf00159e.html. Between 1980 and 1990 the number of patent applications on NIH-funded inventions increased by almost 300 percent: Krimsky, "Profit of Scientific Discovery" (cit. n. 57).

toward privatization and commercial exploitation. By late 1980, when the patent was issued, the NIH guidelines had been weakened to the point of virtual dismantling. The single brief provision in the patent license agreement asking licensees voluntarily to comply with the NIH guidelines is symbolic of the national shift away from social control of technology.[80]

In regard to the commercialization of molecular biology, it is clear from the historical record that the Cohen-Boyer patent was not critical to the initial diffusion and earliest industrial exploitation of recombinant DNA. The technology's potential for tailoring life processes more precisely to mankind's practical needs had been quickly apparent to Cohen and Boyer and others familiar with its capabilities. Hence the protection and exclusivity traditionally provided by a patent had not been required to induce companies to take up the technology and develop it for commercial purposes. As this essay has described, Genentech and a number of large corporations and small genetic engineering firms were utilizing recombinant DNA technology in commercial processes well before the first patent was issued in December 1980. For the entrepreneurs and business executives of these enterprises, what was paramount was the availability of a revolutionary new tool with a variety of exciting practical applications.[81] Despite the murkiness of patent law as applied to genetic engineering, individuals founded biotechnology companies, academic scientists and others became corporate executives and employees, established corporations developed capacity in recombinant DNA technology, and investors sank capital into the new endeavors. Although those running the first companies knew the special importance of owning and controlling intellectual property in biotechnology, the viability and legal standing of patents in the new field was generally seen as a problem to be worked out over time as legal experience and precedents were established. In this vein, the president of Biogen, one of the new biotechnology companies, concluded: "We're going to be aggressive in applying for patents, but we don't know what they're worth and we won't depend on them. We'll depend on fast running."[82]

What then is the importance of the first Cohen-Boyer patent for commercial biotechnology? The answers lie not so much in the realm of hard facts—although there were some, as the patent was widely licensed and gave signs of becoming a real moneymaker for Stanford and UC—as in the realm of business psychology, legal precedent, and agency.

---

[80] For the circumstances of the political decline of the recombinant DNA controversy in the United States and Britain see Krimsky, *Genetic Alchemy* (cit. n. 8), pp. 283–293; and Wright, *Molecular Politics* (cit. n. 8), pp. 337–437. On the weakened NIH guidelines see Wright, "Recombinant DNA Technology," p. 338. In June 1980 the Recombinant DNA Advisory Committee, responding to corporate concern over possible disclosure of trade secrets during the review process, revised the guidelines to provide for protection of proprietary information at this stage: Nelkin, *Science as Intellectual Property* (cit. n. 13), p. 29. On the advice of NIH, the clause in the Cohen-Boyer patent license agreement concerning the NIH guidelines was further softened. Instead of "Licensee agrees to comply with the physical and biological containment standards set forth in the NIH Guidelines . . . ," the wording was changed to "Licensee specifically expresses its intent to comply . . .": William P. O'Neill to Bernard Talbot, M.D., Associate Director, NIH, 24 Apr. 1981, S74-43, Correspondence 1980–1982, OTL Archives.

[81] Another lure was of course the chance to make a personal fortune. In a parody of the so-called central dogma of molecular genetics—DNA (makes) RNA (makes) protein—one wag remarked: DNA (makes) RNA (makes) money: Usher Fleising and Alan Smart, "The Development of Property Rights in Biotechnology," *Culture, Medicine, and Psychiatry*, 1993, *17*:43–57, on p. 53.

[82] Hal Lancaster, "Profits in Gene Splicing Bring the Tangled Issue of Ownership to the Fore," *Wall Street Journal*, 3 Dec. 1980, p. 1. Eighteen biotechnology firms were founded in 1980, thirty-three in 1981: Kenney, *Biotechnology* (cit. n. 16), p. 140. For a quantitative assessment of linkages between biological faculty at American universities and the biotechnology industry between 1985 and 1988 see Sheldon Krimsky, James G. Ennis, and Robert Weissman, "Academic-Corporate Ties in Biotechnology: A Quantitative Study," *Sci., Technol., Hum. Val.*, 1991, *16*:275–287.

Stanford's well-publicized prosecution of the patent in the face of major political and legal obstacles had attracted a diverse audience in academia, government, and industry that regarded its fate as suggestive of the commercial prospects of recombinant DNA. The mere fact that the patent was issued, despite formidable setbacks, provided a psychological boost for those interested in commercial biotechnology. But its effects on the field were more than momentary. Its broad claims on the recombinant DNA process required any company practicing the technology in the United States to license it or risk litigation.[83] Following within six months of the *Chakrabarty* decision affirming the patentability of living organisms, issuance of the Cohen-Boyer patent served to reinforce confidence that commercial biotechnology had a future and was a sound investment opportunity. The patent's wide scope and dominant position and the immediate success of its licensing plan were stabilizing and reassuring factors in the heady but uncertain business and legal environment of the pioneering years of an emerging biotechnology industry.

Investors and other interest groups had few economic indicators by which to estimate the potential success of the biotechnology start-up firms formed from the late 1970s on and often touted with considerable hyperbole. What these companies mainly offered in lieu of immediate products were scientific and technological concepts and expertise whose merit the investment community had difficulty in assessing. Consequently, a firm's ability to instill confidence in its capacity to generate, sustain, and protect intellectual property was critical to its ability to attract investors and fund its research and development. The encouragement and reassurance that the Cohen-Boyer patent and, to a lesser extent, other early patents in biotechnology contributed was significant in the formative stage of a commercial field in which proprietary rights to scientific processes and products were central and critical. In this vein, a cofounder and former CEO of a biotechnology company remarked: "all the early patents were viewed as positive, because if you couldn't protect this intellectual property, then people were not going to invest in the field. So it was the fact that patents would issue, even if they were in your way, that gave people confidence that the field would be able to create value."[84]

The patent history also shows the continuity of concerns that had accompanied commercial ventures in academia for much of the twentieth century. Lederberg's and Kornberg's fear that Stanford's patenting effort would induce secrecy in science was far from new.[85] The university's new president, Donald Kennedy, worried in 1980 that growing

[83] In late 1981 Reimers set up a $200,000 fund from royalties for use if litigation ensued: KRP [Kent R. Peterson] to DXK [Donald Kennedy], "End-Game on Cohen-Boyer," 2 Dec. 1981, S74-43, Correspondence 1980–1982, OTL Archives. The Cohen-Boyer patents were never litigated. By August 1982, a year after Stanford advertised the availability of licenses, Stanford reported income to the two universities of $1.4 million: "Genetic Engineering Patent Delayed," News Bureau, Stanford University Medical Center, 5 Aug. 1982, S74-43, Correspondence 1980–1982, OTL Archives.

[84] Edward E. Penhoet, quoted in "Regional Characteristics of Biotechnology in the United States: Perspectives of Three Industry Insiders," oral histories with Hugh A. Andrade, David P. Holveck, and Edward E. Penhoet conducted in 1998 and 1999 by Sally Smith Hughes, Regional Oral History Office, Bancroft Library, Univ. California, Berkeley, 2001, p. 102. In addition to its role as a confidence booster, the Cohen-Boyer patent and its licensing plan have served as models for structuring other broad patents and their licensing plans. For a recent example see Alan Dove, "Opinions Evolve on Kauffman Patent," *Nature Biotechnology*, 2000, *18*:373.

[85] Kornberg's reservations about commercial ventures in biotechnology were and are restricted to university patenting. In 1980 he, Paul Berg, and Charles Yanofsky cofounded the private DNAX Research Institute of Molecular and Cellular Biology, which was acquired by Schering-Plough in 1982. For histories of DNAX see the Kornberg oral history (cit. n. 36); and Kornberg, *Golden Helix* (cit. n. 52). For a study of specific patent claims in the monoclonal antibody field and their implications for "a subtle but significant shift in the political economy of science and technology" see Michael Mackenzie, Peter Keating, and Alberto Cambrosio, "Patents and Free Scientific Information in Biotechnology: Making Monoclonal Antibodies Proprietary," *Sci., Technol., Hum. Val.*, 1990, *15*:65–83, on p. 65.

campus interest in patenting and other commercial activities threatened to create "a proprietary atmosphere" that could jeopardize the academic tradition of free scientific inquiry and open communication.[86] It was a concern to be argued with new immediacy in the 1980s as a biotechnology industry highly dependent on basic academic science began to coalesce, enticing increasing numbers of academic scientists and their universities to capitalize on the new opportunities. (See Figure 5.) With the increase in patenting in biotechnology, new power relationships were established through intellectual property rights that determined who had access to specific information and who did not.[87] The public and scientific media in the 1980s featured countless stories on the benefits and risks of increased patenting in academia and the various research arrangements being created between universities and industry, with biotechnology in most cases providing the illustrative examples. Policy at American research universities was reformulated after 1980 largely to accommodate greater faculty and institutional involvement in outside business concerns and efforts to secure proprietary rights.[88]

In the practical realm, the patent served as inspiration and model for other universities, setting a pattern for and spurring growth of more efficient and profitable patenting and licensing operations on American campuses.[89] The instability of government research support, which Rosenzweig had noted in the mid 1970s, was to continue in the next decade to motivate universities to commercialize their research inventions and seek greater private-sector investment. Just as Cohen overcame his initial hesitations over patenting, so would many other biomedical scientists and institutions reinterpret their contract with society in ways that allowed the growing commercial movement in American biomedicine to proceed.

The patent and its two companions of 1984 and 1988 were instruments in the transformation of perceptions and policy regarding commercial activity in academia. No longer was a university patent perceived primarily as a means for technology transfer or for

[86] [Robert] Beyers, "Free Inquiry Must Be Rule in Research," [Stanford] *Campus Report*, 5 Nov. 1980, *13*(7):1, 18. The implications of the concept of science as property were of wide interest in science in the 1980s. An examination of the subject by a committee of the American Association for the Advancement of Science resulted in Nelkin's 1984 summary in *Science as Intellectual Property* (cit. n. 13).

[87] Although close university-industry collaborations in chemistry and other applied fields have occurred since the nineteenth century, the two spheres remained largely distinct and interaction was restricted. A recent development is the creation of an "interphase" in which academic, industrial, and sometimes third parties intermingle. For discussion and examples see Gaudillière and Löwy, eds., *Invisible Industrialist* (cit. n. 19): "Introduction" to Pt. 3, pp. 298–299; Vivien Walsh, "Industrial R&D and Its Influence on the Organization and Management of the Production of Knowledge in the Public Sector," *ibid.*, pp. 301–344; and Nelly Oudshoorn, "Shifting Boundaries between Industry and Science: The Role of the WHO in Contraceptive R&D," *ibid.*, pp. 345–368, esp. p. 361.

[88] One of many occasions at which university policy regarding interactions with industry was considered was the 1982 Pajaro Dunes, California, conference arranged by Stanford president Donald Kennedy and attended by corporate leaders and university presidents. For a contemporary account see Barbara Culliton, "Pajaro Dunes: The Search for Consensus," *Science*, 1982, *216*:155–156, 158. Among 1980s articles on patenting in academia, with a focus on biotechnology, see Jeffrey L. Fox, "Can Academia Adapt to Biotechnology's Lure?" *C&EN* [*Chem. Eng. News*], 12 Oct. 1981, pp. 39–81; Culliton, "The Academic-Industrial Complex," *Science*, 1982, *216*:960–962; and David Blumenthal, Michael Gluck, Karen Seashore Louis, Michael A. Stoto, and David Wise, "University-Industry Research Relationships in Biotechnology: Implications for the University," *ibid.*, 1986, *232*:1361–1366. As anyone in current touch with the media knows, the debate about the effect of increased patenting and other forms of commercial activity in academia continues. See, e.g., Michael A. Heller and Rebecca S. Eisenberg, "Can Patents Deter Innovation? The Anticommons in Biomedical Research," *ibid.*, 1998, *280*:698–701.

[89] "On loan" from Stanford, Reimers spent 1985–1986 as director of the Technology Licensing Office at MIT, where he "reformed existing licensing office and developed staff." A major impetus for the appointment was the success of the Cohen-Boyer patents. Resumé in Reimers oral history, p. 50.

Figure 5. *Illustration on the cover of* Chemical and Engineering News, *1981, depicting an academic scientist in a DNA helix being tugged in two different directions by a university scientist and a businessman.*

controlling use of an invention; in the 1980s patenting became a way for universities and scientists to make money. Biological knowledge was being privatized, commodified, and given economic value in addition to its value as a cultural good. But for those willing to look beyond their resounding commercial success, the Cohen-Boyer patents also carried subtle seeds of concern regarding the effect of patenting and licensing on the culture and norms of academic research. They were thus simultaneously agents of attitudinal and institutional change and harbingers of ethical issues that were to accompany the accelerating commercialization of academic biomedicine and the rise of the biotechnology industry in the years ahead.[90]

[90] For an account of subsequent commercialization of the biological sciences and its effect on the culture and ethos of academic biology see Krimsky, "Profit of Scientific Discovery" (cit. n. 57).

The Polonium Brief

A Hidden History of Cancer, Radiation, and the Tobacco Industry

By Brianna Rego

ABSTRACT

The first scientific paper on polonium-210 in tobacco was published in 1964, and in the following decades there would be more research linking radioisotopes in cigarettes with lung cancer in smokers. While external scientists worked to determine whether polonium could be a cause of lung cancer, industry scientists silently pursued similar work with the goal of protecting business interests should the polonium problem ever become public. Despite forty years of research suggesting that polonium is a leading carcinogen in tobacco, the manufacturers have not made a definitive move to reduce the concentration of radioactive isotopes in cigarettes. The polonium story therefore presents yet another chapter in the long tradition of industry use of science and scientific authority in an effort to thwart disease prevention. The impressive extent to which tobacco manufacturers understood the hazards of polonium and the high executive level at which the problem and potential solutions were discussed within the industry are exposed here by means of internal documents made available through litigation.

> The American public is exposed to far more radiation from the smoking of tobacco than they are from any other source.
> —Reimert Thorolf Ravenholt (1982)

> [Publishing our research] has the potential of waking a sleeping giant. This subject is rumbling . . . and I doubt we should provide facts.
> —Paul A. Eichorn of Philip Morris, cautioning against publishing internal industry research on polonium (1978)

Isis 100, no. 3 (September 2009): 453–84.

Many thanks to Robert Proctor for first hinting to me that polonium might make an interesting story and for his advice and guidance during the writing of this essay. Vilma Hunt, Jack Little, and Ann Kennedy read the manuscript and spoke with me about their experiences researching polonium. The editor and anonymous referees for *Isis* offered excellent comments that improved the analysis tremendously. A special thanks to David Sepkoski, Jonathan Payne, David Holloway, Dick Zare, Linda Huynh, Geoffrey Sea, Angela Creager, Spencer Weart, and family and friends who heard a lot about polonium, read drafts, and offered invaluable comments and suggestions. Parts of this essay were presented at the History of Science Society annual meeting in 2007.

O N 1 NOVEMBER 2006, former KGB operative Alexander V. Litvinenko fell ill after a meeting at a London sushi restaurant. Over the next three weeks his illness took several unexpected turns for the worse, ultimately prompting doctors to announce that he had been poisoned by an unknown radioactive substance. Litvinenko died on 23 November, and the next day the cause of death was determined to be radiation poisoning from polonium-210. In subsequent days, as concern grew in London that others had been exposed, the British Health Protection Agency reassured the population that there was "no radiation risk" to the general public, stressing that polonium-210 "can only contaminate if it is ingested, inhaled, or taken in through a wound." Shortly thereafter, a *New York Times* op-ed by Robert N. Proctor pointed out that this statement was misleading, given that more than a billion people worldwide expose themselves daily—most often unknowingly—to polonium-210. The vector of this mass irradiation is not a vengeful government, nor an adversary in the style of Cold War espionage, but, rather, something far more common and available, something that is in fact quite ordinary: cigarettes.[1]

That tobacco contains polonium comes as a surprise to many. A 2001 survey by K. Michael Cummings of the Roswell Park Cancer Institute found that 86 percent of 1,046 adult smokers polled were unaware that there are radioactive isotopes in tobacco.[2] The scientific evidence, however, dates back more than four decades. The radiochemists Edward Radford and Vilma Hunt first published on this topic in the 17 January 1964 issue of *Science*, only a few days after the appearance of the surgeon general's report on smoking and health on 11 January.[3] Their paper launched an extensive and often impassioned debate, as public and industry researchers alike pursued the problem with great interest.

The presence of polonium-210 in tobacco is of particular concern because it is one of the most radioactive isotopes, and even a very small dose can have devastating consequences. An extremely rare metalloid that occurs naturally in uranium ores, polonium was discovered in 1898 by Marie and Pierre Curie and named for her homeland of Poland. It has several radioactive isotopes, the most common of which is polonium-210 (^{210}Po), an alpha emitter, which means that it releases a high-energy helium ion (^{4}He) in its radioactive decay. Alpha radiation is innocuous so long as it remains outside the body, but internally alpha particles are, generally speaking, the most hazardous form of radiation. Polonium-210 is a product of the natural uranium-238 decay series; the parent isotopes include radium-226, radon-222, and lead-210. Polonium-210 itself decays to the stable isotope lead-206 (plus an alpha particle). The isotope has a half-life of 138 days, which is short when compared to the 4.5 billion–year and 1,600-year half-lives of uranium-238 and radium-226, respectively. Its short half-life means that the isotope is very "hot," emitting alpha particles at a rapid rate. And since smokers are exposing themselves to new doses with each cigarette, a constant cycle of exposure and decay is played out in the lung.

---

[1] "'No Radiation Risk,' Public Told," *BBC News*, 24 Nov. 2006, http://news.bbc.co.uk/go/pr/fr/-/2/hi/health/6181190.stm (accessed 1 Dec. 2006); Jill Lawless, "Putin Faces Barrage in Death of Ex-Spy," *Washington Post*, 26 Nov. 2006; and Robert N. Proctor, "Puffing on Polonium," *New York Times*, 1 Dec. 2006.

[2] K. Michael Cummings, Andrew Hyland, Gary A. Giovino, Janice L. Hastrup, Joseph E. Bauer, and Maansi A. Bansal, "Are Smokers Adequately Informed about the Health Risks of Smoking and Medicinal Nicotine?" *Nicotine and Tobacco Research*, 2004, 6(S3):S333–S340, on p. S336. My own experience discussing polonium in tobacco suggests that this statistic is too high and that far fewer than 14 percent of smokers are in fact aware that there is polonium in their cigarettes. Polonium is often confused with plutonium, and only one smoker and a handful of nonsmokers I've spoken with have been aware that there are radioactive particles in tobacco.

[3] Edward P. Radford and Vilma R. Hunt, "Polonium-210: A Volatile Radioelement in Cigarettes," *Science*, 1964, *143*(3603):247–249.

Although polonium-210 is quite rare in the earth's crust, it is extremely toxic. Estimates judge it to be about 250 million times more toxic than cyanide (by weight), and the isotope emits five thousand more radioactive particles per unit time than an equal amount of radium-226. Before its recent use as a spy-killing poison, polonium's most common industrial use was as an antistatic agent for large machinery and in brushes to remove dust from photographic film. The isotope was also used by the military as an initiator for nuclear weapons. In addition, it has been employed as a lightweight heat source for artificial space satellites. Considering its industrial and military uses, polonium-210 does not seem the sort of isotope wanted in one's lungs, yet the number of those exposed every day is staggering. About 5.7 trillion cigarettes are smoked annually worldwide—lined up end to end, they would reach to the sun and back, with enough left over for several trips to Mars.[4] Each of those cigarettes contains about 0.04 picocuries (pCi) of polonium-210.[5] Although that might seem like a very small dosage, puff by puff the radioactive particles concentrate in a smoker's lungs to the equivalent, by one estimate, of three hundred chest x-rays per year. That is a nontrivial exposure to radiation over the course of a lifetime—and a massive aggregate exposure that in terms of public health effects is the most important source of radiation to which humans are exposed.[6]

Exposure to polonium has been as wide ranging and long standing as the cigarette itself, which has permeated every aspect of society, adapting to the smoker's every imaginable mood, need, and image. The love affair with smoking, however, has been filled with contradictions—"smoking is patriotic yet rebellious, risky yet safe, calming yet exciting." But the history of tobacco is about much more than the history of smoking. Recent scholarship has documented the increasing popularity of tobacco in developing countries, the unquestioned brilliance and success of tobacco advertising, and smoking's leading role in Hollywood. We know a lot about the marvel that is the ephemeral yet persistent cigarette: the triumph of popular protests against secondhand smoke, the indignant responses of smokers to their increasing marginalization, and the trials of the tobacco industry. As Allan Brandt summarizes in *The Cigarette Century*, "there are few, if any, central aspects of American society that are truly smoke-free."[7] Despite a shift in the

---

[4] "Polonium," *Human Health Fact Sheet, August 2005*, Argonne National Laboratory, EVS, http://www .ead.anl.gov/pub/doc/polonium.pdf (accessed 5 Mar. 2007); and Robert N. Proctor, "Tobacco and the Global Lung Cancer Epidemic," *Nature Reviews Cancer*, 2001, *1*:82–86, on pp. 85–86.

[5] Several units for radiation are used in the United States. The curie (Ci) measures the actual amount of radiation: 1 Ci = 3.7 × 10^{10} decays/second (or the activity of 1 gram of radium-226). A picocurie (pCi) is one trillionth (10^{-12}) of a curie. The international equivalent of the curie is the becquerel (1 Bq = 1 decay/sec and 1 Bq = 2.7 × 10^{-11} Ci). A rad (radiation absorbed dose) is a unit of radiation dose equivalent to 0.01 joule of energy absorbed per kilogram of tissue. It measures the amount of radioactive energy actually in the tissue. In order to measure the physiological effect of a dose, the amount in rad is multiplied by the relative biological effectiveness (RBE, a "quality factor" often represented by the letter Q) of the radioactive element in question to get the dose in rem (röntgen equivalent in man), which measures the biological effectiveness (the degree to which the actual tissue is affected and destroyed) of an absorbed dose of radiation. The RBE differs from radioisotope to radioisotope, depending on many factors, including the type of radiation and the location and type of tissue affected. Alpha particles can have a Q as high as 20 (beta particles, x-rays, and gamma particles usually have $Q = 1$). Given its chemical properties as an alpha emitter, the amount of polonium-210 that is in cigarettes is, according to Ann Kennedy, who did research on polonium in the 1970s, "quite capable of promoting damage that could lead to cancer": telephone conversation with Ann Kennedy, 24 May 2007.

[6] Thomas H. Winters and Joseph R. Difranza, "Letter to the Editor," *New England Journal of Medicine*, 1982, *306*:364–365 (chest x-ray equivalence); and Reimert Thorolf Ravenholt, cited by the Associated Press, "Smokers Said to Risk Cancers beyond Lungs," *New York Times*, 29 July 1982.

[7] Robert N. Proctor, "Agnotology: A Missing Term to Describe the Cultural Production of Ignorance (and Its Study)," in *Agnotology: The Making and Unmaking of Ignorance*, ed. Proctor and Londa Schiebinger (Stanford, Calif.: Stanford Univ. Press, 2008), pp. 1–33, on p. 11 ("patriotic yet rebellious"); and Allan M. Brandt, *The

United States in recent years toward limiting and marginalizing smoking, Brandt was right to stress the extent to which tobacco permeated the twentieth century. The cigarette became an essential prop in hand and a cornerstone of the American economy.

The World Health Organization has stressed that smoking is the most avoidable cause of death.[8] The destructive nature of smoking (as described by historians and scientists alike) has driven much of the scholarship on cigarettes, which until internal industry documents became available in the mid-1990s was largely focused either on quantifying disease consequences or on tracing the cultural responses to the rise of the industry. Following the release of millions of industry documents, scholars have been able to gain insights into the industry side of tobacco's history as well—and thus to offer a more nuanced and multifaceted story.

One aspect that has especially interested scholars is how the tobacco companies have understood and taken advantage of scientific authority, consensus, and controversy. The primary goal of cigarette manufacturers has been to influence and even minimize the extent to which the smoking public has understood the dangers of smoking. Emphasizing any wisp of uncertainty in science or any whisper of disagreement among scientists—if only with regard to what methods or tools to use in their research—has been a long-standing tactic of the tobacco industry.[9] In response to mounting evidence that smoking is harmful, the industry launched a campaign in the mid-twentieth century to stress that, despite what had been published in the scientific literature, the dangers of smoking had not been "proven" and the results were little more than "statistics." Any lack of consensus among researchers has therefore been spun by the tobacco industry in its continued insistence that even the experts "don't really know" what makes smoking harmful.[10] Emphasizing the argument that science cannot offer "proof" but only a "suggestion" that smoking is dangerous, the tobacco industry has attempted to undermine any kind of evidence demonstrating the harms of smoking.

The tobacco companies have been successful in their use of science, influencing and even manipulating their customers by enlisting experts to defend their product in the face of increasing evidence that smoking is harmful: witness the advertisements proclaiming that "More Doctors Smoke Camels" and that L&M filters are "Just What the Doctor Ordered." The industry used its own scientific authorities to counter and even undermine the scientific and medical legitimacy of countless researchers and doctors who stressed the dangers of tobacco. Smokers could choose which doctor they wanted to believe, and—as was the aim of the industry's advertising campaign—that doctor was often the one who

---

Cigarette Century: The Rise, Fall, and Deadly Persistence of the Product That Defined America (New York: Basic, 2007), p. 3.

[8] World Health Organization, *WHO Report on the Global Tobacco Epidemic, 2008: MPOWER Package* (Geneva: World Health Organization, 2008), p. 8.

[9] On the goals of the cigarette manufacturers see Robert N. Proctor, "Tobacco and Health" [expert witness report submitted in United States of America v. Philip Morris], *Journal of Philosophy, Science, and the Law*, 2004, *4*:2–32, esp. pp. 9–18. On the tactic of emphasizing uncertainties and disagreements see Brandt, *Cigarette Century* (cit. n. 7); Richard Kluger, *Ashes to Ashes: America's Hundred-Year Cigarette War, the Public Health, and the Unabashed Triumph of Philip Morris* (New York: Knopf, 1997); Stanton Glantz, *The Cigarette Papers* (Berkeley: Univ. California Press, 1998); and Proctor, *Cancer Wars: How Politics Shapes What We Know and Don't Know about Cancer* (New York: Basic, 1995).

[10] Proctor, "Agnotology" (cit. n. 7), p. 11. In her work on climate science Naomi Oreskes has discussed a similar stress on a lack of consensus among scientists. Some corporations have emphasized uncertainty in science in an effort to undermine scientific knowledge about global climate change, essentially echoing the tobacco industry's strategy. See Naomi Oreskes, "The Scientific Consensus on Climate Change," *Science*, 2004, *306*:1686; and Oreskes and Erik M. Conway, "Challenging Knowledge: How Climate Science Became a Victim of the Cold War," in *Agnotology*, ed. Proctor and Schiebinger (cit. n. 7), pp. 55–89, on p. 78.

smoked Camels. Thus, as Brandt discusses, the industry's "strategic campaign to obscure and confuse the ongoing scientific enterprise" resulted in disordered responses by the public and delayed acceptance of the dangers of smoking.[11]

In the case of polonium, the tobacco industry went so far as to initiate a research program of its own, staffed by industry scientists whose knowledge, training, awareness, and expertise were similar to those of scientists outside the industry. Cigarette manufacturers could therefore be selective about what science they promoted and used in their own defense, what science they hushed, and what science they questioned, either publicly or privately. So while the external scientific community stressed that the concentration of polonium in tobacco was significant and perfectly capable of causing cancer, tobacco industry scientists worked in silence to stress just as firmly that this had not been proven with absolute certainty. Despite extensive research, industry results were never published. Nor—underlining the unpublicized nature of this research—have I been able to find a single instance in which the industry issued a public denial on polonium. This silence is in stark contrast to the thousands of denials released by the industry on other hazards, such as nicotine. With polonium, the industry did not even engage in a public discussion— neither admitting nor denying the hazards—remaining silent rather than risk inspiring a debate that could only be harmful to the tobacco companies.

Despite the extensive scope of Richard Kluger's *Ashes to Ashes* and Brandt's *Cigarette Century*, neither mentions either polonium or the history of radiation research on tobacco. Aside from the 2006 *New York Times* op-ed by Proctor and a brief discussion in his 1995 book *Cancer Wars*, the presence of radioisotopes in cigarette smoke and the research conducted on alpha radiation as a cause of lung cancer in smokers have been ignored by historians. However, the topic of polonium in tobacco has had an extensive and somewhat tumultuous career within the tobacco industry, the external scientific community, and the wider public sphere. A recent article published in the *American Journal of Public Health* by a group from the Mayo Clinic and the Stanford University School of Engineering discussed the tobacco industry's response to the polonium issue. Quoting at length from internal documents, the authors stress that the industry has known about the presence of polonium in cigarettes for forty years and has done nothing to reduce the danger. This paper concluded that cigarette packs should be labeled with radiation warnings.[12] It is remarkable, however, how often such concerns have been raised, only to be quickly forgotten.

The story of polonium in cigarettes lies at the junction of tobacco, cancer, risk, policy, and radiation, and, as evidenced by the wide-ranging scholarship on these subjects, there are many angles from which to study their intricate and multifaceted stories. In contrast to previous histories treating tobacco, the purpose of this essay is to chronicle the largely untold story of a single hazardous element and to use this to expose the depravity of the tobacco industry in its use of science.[13] The story of polonium-210 offers an opportunity

---

[11] Brandt, *Cigarette Century* (cit. n. 7), p. 4.

[12] Proctor, *Cancer Wars* (cit. n. 9), pp. 174, 306 n 1; and Monique E. Muggli, Jon O. Ebbert, Channing Robertson, and Richard D. Hurt, "Waking a Sleeping Giant: The Tobacco Industry's Response to the Polonium-210 Issue," *American Journal of Public Health*, 2008, *98*(9):1–8.

[13] This study is part of a larger "elemental historiography" I am exploring in my doctoral thesis. My sources for the project include papers from scientific journals, press releases, and articles from newspapers and magazines, as well as the Legacy Tobacco Documents Library, maintained by the University of California at San Francisco. This database has made available internal tobacco industry memoranda, meeting notes, research reports, and correspondence pertaining to the polonium issue, as well as many other aspects of the internal workings of the tobacco industry. The documents are online at http://legacy.library.ucsf.edu, and users can

to look into the relationship between science and the tobacco industry and to explore the sophisticated and well-funded scientific investigations conducted by the industry. Despite their extensive and long-term research on polonium, the tobacco companies never published their results nor even admitted the existence of their research program. Nor did the manufacturers ever make any serious effort to remove polonium from tobacco, despite the availability of several different techniques for doing so. Instead, the policy recommended by company scientists and supported by executives and attorneys was to stay abreast of external knowledge on the topic and to be prepared, if the need ever arose, to face the issue and deal with the problem.[14] The impressive extent to which the industry understood the hazards of polonium and the high executive level on which the problem and potential solutions were discussed are exposed here by means of internal documents that have become available through litigation.[15]

HOT SPOTS

The discovery of polonium in tobacco was the result of chance and curiosity stemming from increased radiation research and nuclear fear during the 1950s. Like other labs across the country at the time, the radiochemical group at Harvard was involved in Atomic Energy Commission–sponsored research on the effects of radioactive fallout. Key figures in the discovery were Edward Radford and Vilma Hunt. Radford (1922–2001) was educated in biology and medicine at MIT and Harvard before he joined the Air Force and became involved in nuclear arms testing following World War II; Hunt (b. 1926), a native Australian, was originally trained in dentistry (after serving in the Royal Australian Air Force) before moving to the United States and refocusing her research on radiochemistry and occupational health. Collaborating at the Harvard School of Public Health, Radford and Hunt (and others in the radiochemical group) measured tooth and bone samples from across New England to determine the concentration in human tissue of naturally occurring alpha-emitting isotopes—specifically radium and polonium. The unexpected detection in 1962 of a particular radium isotope in one skeletal sample would divert Radford and Hunt from the fallout studies and lead to the publication a few months later of their discovery of polonium in cigarette smoke.[16]

The Harvard group's analyses of teeth and bones had shown the consistent presence of radium-224 and radium-226, but in one skeleton from a three-year-old child the group observed a new phenomenon: a pattern of decay that indicated the possible presence of radium-223. Hunt recalls this being "of interest, to say the least!" as it indicated that a different chain of uranium was responsible for the radium in this child's skeleton. As she was contemplating these results—and wondering what might have caused this pattern—Hunt's eye glided over cigarette ash left behind by a smoking colleague, and she thought to test it for radium and polonium. Exactly why, she doesn't know; as she now recalls, it was the way her "brain worked that day." She knew it was "a long shot that radium-223 might be a constituent with a different pattern of radioactive decay," but she recognized

---

search by keyword or browse through different collections. Documents are identified by a Bates number stamped on each page. I have followed the citation format used and recommended by Robert N. Proctor.

[14] Gonzalo Segura to Abraham Bavley, "Polonium in Tobacco and Smoke" (memo), 27 Oct. 1964 (Philip Morris), Bates 100188168.

[15] To give a sense of the often-overwhelming magnitude of the Legacy Tobacco Documents Library: in early July 2009 a search for "polonium" yielded more than thirteen thousand results in the entire archive and a search for "radiation" more than eighty thousand results.

[16] Telephone conversation with Vilma Hunt, 2 Feb. 2009.

that such a finding could improve the analytical method the Harvard group was using at the time for very low concentrations of radium isotopes in biological materials.[17]

The results of her spontaneous study would stun her: no polonium was detected in the ash. No other biological material the group had tested in their laboratory (including plants) had lacked polonium when radium was found. As Hunt recalled in a recent conversation, she must have had the volatile temperature of polonium-210 in the back of her mind while looking over her lab notes, because it suddenly made sense to her that if there were no polonium in the ash it must have gone into the smoke.[18] And so Hunt diverted her studies to focus more intensively on polonium and worked with Radford to measure the radioactivity in smoke.

Studies in the 1950s had looked into the possibility that radioactive isotopes might be responsible for lung cancer in smokers. But the particles measured—potassium-40 and radium isotopes—were quickly dismissed as insufficiently volatile at the temperature of a burning cigarette (600° to 800° C), which means they could not be responsible for irradiating the lungs of smokers. Radford and Hunt undertook their studies with the intent of reevaluating the presence of alpha emitters in tobacco, determined not only to measure the concentration of radioisotopes but also to explore whether the levels were significant enough to cause lung cancer in smokers.[19]

Before Radford and Hunt could hypothesize how much polonium was actually deposited on the bronchial epithelium (the membrane lining the respiratory tract), they had to measure the minimal dose that would result from the smoke simply passing through the lung, without accumulating in any one place. Radford and Hunt artificially "smoked" cigarettes, drawing air through a filter designed to collect all tobacco smoke–sized particles. Cigarettes were "puffed" in this manner at a rate of a two- to three-second puff every fifty seconds—the average rate Radford and Hunt observed while spying on their smoking colleagues. The smoke was passed through a trap to capture both mainstream and sidestream smoke and treated with hydrogen chloride. The polonium was then plated on silver and counted for alpha activity in windowless gas-flow proportional counters, following radiochemical procedures developed by the Harvard group during their research into radium and polonium in teeth and bones. They found that for a person who smoked two packs of cigarettes a day for twenty-five years, the total-lung minimum dose was about 36 rem, or seven times the normal background radiation exposure.[20]

Radford and Hunt also realized, however, that this minimum dose was not a meaningful representation of the way polonium-210 behaves when inhaled through a cigarette. Because of the way the lung branches, the radioisotopes settle and concentrate at the points of bifurcation, forming so-called hot spots of intense radioactivity. Radford and Hunt focused their original study on these areas of greatest concentration. Accurately

[17] *Ibid.*; and email correspondence with Hunt, 27 Feb. 2009. Radium-223 decays from uranium-235, radium-226 decays from thorium-232, and radium-224 (a parent isotope of polonium-210) decays from uranium-238.

[18] Email correspondence with Hunt, 27 Feb. 2009; and telephone conversation with Hunt, 2 Feb. 2009.

[19] Radford and Hunt, "Polonium-210" (cit. n. 3), p. 247. For the 1950s studies see F. W. Spiers and R. D. Passey, "Radioactivity of Tobacco and Lung Cancer," *Lancet*, 1953, *265*:1259–1260; R. C. Turner and J. M. Radley, "Naturally Occurring Alpha Activity of Cigarette Tobaccos," *ibid.*, 1960, *1*:1197–1198; and E. S. Harlow and Charles R. Greene, "Some Comments on 'Temperature Profiles throughout Cigarettes, Cigars, and Pipes,'" *Science*, 1956, *123*(3189):226–227.

[20] Radford and Hunt, "Polonium-210," pp. 247–248. For the radiochemical procedures see Edward P. Radford, Jr., Vilma R. Hunt, and Dwyn Sherry, "Analysis of Teeth and Bones for Alpha-Emitting Elements," *Radiation Research*, 1963, *19*(2):298–315, on pp. 300–301; they were also discussed in my telephone conversation with Hunt, 2 Feb. 2009.

measuring the precise dosage of radiation at these hot spots was not easy, as it was difficult to reproduce a proper smoking technique in the lab. Rising to this challenge, they estimated a range of several hundred to more than 1,000 rem for somebody smoking two packs a day for twenty-five years.[21] It was already known that, in addition to the high doses of radiation at such hot spots, certain pathological consequences of heavy smoking (such as ciliastasis) made the epithelium even more vulnerable to the effects of polonium. Because of the delicacy of the bronchial system and the well-known physiological damage caused by cigarette smoke, Radford and Hunt proposed that even minimal radiation from polonium-210 could cause lung cancer by initiating mutations in cellular development.[22]

Radford and Hunt's research was focused on measuring radioactivity in smoke, and they hypothesized about its deposition in the lungs. It was their colleague at Harvard, John ("Jack") Little, who conducted a study of human lung tissue and showed that polonium was "indeed deposited and retained in specific regions of the bronchi." Little, a radiobiologist and physician, was able to collect samples of human lung tissue within a few hours of death from pathologists in the Boston area (a project that, he recalls, involved many late-night dashes to the hospital). Little dissected specimens of the epithelium from multiple areas in the bronchial tree and showed that the highest concentration of polonium was found in segmental bifurcations, just as Radford and Hunt had proposed.[23]

These papers on polonium in tobacco provoked an impressive response from radiochemists, cancer biologists, and tobacco researchers both inside and outside the industry. Radford and Hunt's first paper inspired numerous letters to the editor of *Science*, and more papers on the subject were published within the year. Much of the research that followed Radford and Hunt's original publication and Little's 1965 study focused on the biology and chemistry of the tobacco plant itself. During the late 1960s, researchers measured concentrations of polonium in various tobacco crops and cigarette brands in an attempt to determine the origin of polonium-210 in the tobacco plant and the stage of the cigarette manufacturing process during which it could be most effectively and efficiently removed.

In November 1964, T. C. Tso of the U.S. Department of Agriculture, Naomi Hallden of the U.S. Atomic Energy Commission (who would stay involved in polonium research, later publishing under her married name Harley), and L. T. Alexander, also of the Department of Agriculture, published a paper suggesting that there were significant differences in the concentrations of polonium in tobacco grown in different regions of the country. Sampling tobacco crops from various localities, Tso, Hallden, and Alexander found that North Carolina tobacco contained about three times more polonium than Maryland tobacco. These results were viewed as particularly significant, since more than a third of American tobacco was grown in North Carolina, accounting for a billion dollars per year in state income. The authors suggested that these variations in polonium levels could be caused by the natural radiation content of the soil and phosphate fertilizer used, as some phosphate rocks naturally contain more uranium than others.[24]

Research by the New Zealand–born scientist L. P. Gregory in 1965 expanded Tso,

[21] Radford and Hunt, "Polonium-210," p. 248.

[22] *Ibid.*, pp. 248–249. Ciliastasis is the stiffening of the cilia, the hair-like organelles, similar to flagella, that line the surface of the lung. In a healthy lung the cilia wave back and forth, cleansing the lung.

[23] Email correspondence with John Little, 4 Mar. 2009; telephone conversation with Little, 5 Mar. 2009; and J. B. Little, E. P. Radford, Jr., H. L. McCombs, and V. R. Hunt, "Distribution of Polonium-210 in Pulmonary Tissues of Cigarette Smokers," *N. Engl. J. Med.*, 1965, *273*(25):1343–1351.

[24] T. C. Tso, N. A. Hallden, and L. T. Alexander, "Radium-226 and Polonium-210 in Leaf Tobacco and Tobacco Soil," *Science*, 1964, *146*(3647):1043–1045, esp. pp. 1043–1044.

Hallden, and Alexander's work, showing that there were significant variations in the polonium-210 concentration of tobacco grown in different countries. Gregory found that the radioactivity of New Zealand tobacco was less than half that of American tobacco. South African tobacco had concentrations similar to the American, while Rhodesian was slightly more radioactive.[25] Despite such dramatic differences in tobacco crops, Gregory, along with other researchers, found no significant difference between various brands of cigarettes or between filter and nonfilter varieties.[26] The lack of variation among brands suggested that the isotope was naturally present in the tobacco leaf, not something added during the curing process.

In their original report, Radford and Hunt had speculated briefly on the origins of polonium in tobacco. They had suggested two possibilities: either daughter isotopes of natural atmospheric radon-222 (a polonium precursor) settled on the leaves, decaying to polonium once attached to the tobacco plant, or lead-210 (polonium-210's parent isotope) decayed to polonium-210 after being absorbed through the plant's roots.[27] In order to test the first hypothesis—that polonium was absorbed by the tobacco plant from nuclear fallout—Tso, Harley, and Alexander conducted an experiment growing tobacco plants inside a greenhouse pumped full of radon to about five hundred times the normal background atmospheric concentration, as determined by German tests.[28] Although there was an increase in the polonium measured in the tobacco plants grown in this radioactive atmosphere, the concentration was not significant enough for fallout to be the primary source of radioactive particles in tobacco. To test the second hypothesis—that lead-210 in phosphate fertilizer entered the tobacco plant through the roots—Tso, Harley, and Alexander conducted a soil experiment, testing two different kinds of fertilizers: a commercial superphosphate and a specially mixed fertilizer made from chemically pure secondary calcium phosphate. The differences between the two were remarkable. The commercial fertilizer had about thirteen times more radon-226 than the specially mixed fertilizer, resulting in polonium levels in the leaf that were nearly seven times higher.[29]

Tso, Harley, and Alexander felt they had established that most of the polonium in tobacco originated in the phosphates used to fertilize the plant, but subsequent research by Chester Francis, Gordon Chesters, and Wilfred Erhardt, of the University of Wisconsin, suggested that phosphate fertilizers were not a significant enough source of polonium to eliminate fallout as the "principal mechanism" of polonium-210 entry into the tobacco plant. In order to determine the amount of polonium resulting from fallout, Francis, Chesters, and Erhardt measured the concentration of lead-210 in rainwater they collected in Wisconsin during the summer of 1966.[30] Although their results were intriguing, they did

---

[25] Gregory hypothesized that the lower levels of polonium-210 in New Zealand tobacco could be due to an "insular effect" that could cause most natural nuclear fallout to be dispersed over the ocean, rather than over crops; see L. P. Gregory, "Polonium-210 in Leaf Tobacco from Four Countries," *Science*, 1965, *150*(3692): 74–76, esp. pp. 74–75.

[26] Thomas F. Kelley, "Polonium-210 Content of Mainstream Cigarette Smoke," *Science*, 1965, *149*(3683): 537–538, on p. 537.

[27] Radford and Hunt, "Polonium-210" (cit. n. 3), p. 248.

[28] The "normal" atmospheric radon concentration cited by Tso, Harley, and Alexander came from work by the German radiation physicist Wolfgang Jacobi. See W. Jacobi, "Die natürliche Radioaktivitfit der Atmosphäre," *Biophysik*, 1963, *1*:175–188.

[29] T. C. Tso, Naomi Harley, and L. T. Alexander, "Source of Lead-210 and Polonium-210 in Tobacco," *Science*, 1966, *153*(3783):880–882, esp. p. 881. Superphosphate fertilizer is made by treating phosphate rock with sulfuric acid.

[30] Chester Francis, Gordon Chesters, and Wilfred Erhardt, "²¹⁰Polonium Entry into Plants," *Environmental Science and Technology*, 1968, 2:690–695, on p. 690. Although tobacco was apparently the first plant to be

not sample rainwater in other tobacco-producing areas—notably North Carolina—and their data were limited to a single summer. It is therefore not known if the results were an accurate representation of average annual fallout over tobacco crops.

As research on the origins of polonium-210 in the tobacco plant continued through the 1960s, and as the physiological effects began to be understood, cigarette manufacturers became increasingly concerned. Internal tobacco industry documents reveal a flurry of activity and correspondence in reaction to the 1964 paper by Radford and Hunt. The same day that this paper was published in *Science*, for example, Philip Morris researcher Ted Katz scrawled a note to Abraham Bavley, manager of the company's research division, commenting on the discovery. Katz wrote that polonium is "reportedly an alpha emitter" that, it seemed to him, "would be a most dangerous material once inside the body." Bavley was sufficiently concerned to commission Philip Morris radiochemist Gonzalo Segura to do a literature review on the topic, the results of which he received ten months later. Segura confirmed that "it is true" that polonium could cause cancer and stressed that both dose and rate of intake should be considered. He immediately recognized that the study of polonium in tobacco had the potential to become a "major project" but did not think that Philip Morris should worry about pursuing the issue. He did, however, strongly recommend that management keep a "particularly sharp look-out" for any research developments that might unfavorably affect the industry. The company should strive to keep "ahead of adverse publicity and be in a position to counter it quickly" if the problem of polonium in tobacco ever became critical.[31]

Management took Segura's advice and continued to keep a close eye on relevant scientific literature. In 1965, Philip Morris senior research chemist Robert Carpenter wrote to his supervisors Helmut Wakeham (vice president of research and development) and Robert Seligman (assistant director of tobacco research and development), pointing out that polonium was receiving increased attention from the biomedical field. He suggested, therefore, that it might be time for Philip Morris to look into the "polonium situation" or "problem," as it was often called.[32]

"THE INDUSTRY MIGHT BE INTERESTED . . ."

In February 1965 the Department of Research and Development at the American Tobacco Company released its budget for the following year, earmarking $95,725 for its radio-chemical section. Twenty percent of this was to go to development, leaving 80 percent for applied research. Philip Morris also decided to pursue its own research on polonium and by the end of 1965 had talked with Edward Radford (who was described, in something of an understatement, as being "actively interested in the polonium in smokers").[33] Despite

---

studied as a source of polonium, by the late 1960s researchers were beginning to realize that everything from fruit trees to lettuce contained traces of polonium. As Radford and Hunt noted, what makes polonium dangerous in tobacco and not in, say, broccoli, is that polonium-210 is volatile above 500° C, well below the temperature of a burning cigarette (about 600° to 800° C), which allows it to adhere strongly to the smoke particles and to gain direct access to the lung. According to a 2005 fact sheet from Argonne National Laboratory, the cancer risk from inhaling polonium-210 is about six times greater than the risk from ingesting it, a determination that highlights the danger of the presence of polonium-210 in tobacco. See Radford and Hunt, "Polonium-210" (cit. n. 3), p. 248; Francis *et al.*, "²¹⁰Polonium Entry into Plants," p. 690; and "Polonium," *Human Health Fact Sheet, August 2005* (cit. n. 4).

[31] Ted Katz, "Polonium in Smoke," 17 Jan. 1964 (Philip Morris), Bates 1001896995; and Segura to Bavley, "Polonium in Tobacco and Smoke" (memo), 27 Oct. 1964 (Philip Morris), Bates 1001896688–6689.

[32] R. D. Carpenter, "Polonium in Tobacco," 16 Dec. 1965 (Philip Morris), Bates 1001881339.

[33] For the American Tobacco Company budget figures see "Department of Research and Development New

the companies' interest in the matter and increasingly sophisticated radiochemical programs, the industry did not publicize its internal research or interest in this area. The closeted nature of their research, however, did not stop industry scientists from discussing methods and materials with outside experts.

Following the publication of his 1965 paper in the *New England Journal of Medicine*, Jack Little was visited by several scientists from the large tobacco companies who were interested in learning in detail the techniques used by Little's group, as well as their results. As he recalls, they "seemed very pleasant and interested, were clearly actively investigating the question themselves, and seemed open in discussing their findings." However, he was unable to establish any further contact with the scientists; and despite the visits he was not aware of the extent of the research being conducted by the industry. In 1967, American Tobacco sent Ronald Davis to the University of Massachusetts at Lowell to confer with Professor of Radiological Sciences Kenneth Skrable on various techniques involved in measuring polonium in tobacco. Applying information gleaned through such interactions with outside scientists, American Tobacco confirmed in its own laboratories earlier reports that existing filters had no effect on the concentration of polonium in cigarette smoke. By November, Philip Morris had also measured and reconfirmed polonium concentrations in cigarettes. The results, like those of the other tobacco companies, were never published.[34]

Over the next few years, industry scientists would become increasingly involved in pursuing polonium. The industry's scientific effort, however, had goals that differed from those of researchers like Radford and Hunt, who were hoping to prevent cancer. Industry researchers pursued the same problem—polonium in tobacco—but they tweaked parameters and methods so as to suggest that Radford and Hunt's measurements might be exaggerated. Philip Morris (and perhaps other tobacco companies) measured the concentration of polonium over the *entire* bronchial area and did not take the varying concentrations in different parts of the lung into account. The Harvard researchers, in contrast, had focused their measurements on the branching points of the bronchial epithelium (the so-called hot spots), which account for only 2 to 3 percent of the weight of the lung. As Little and Radford stressed in a 1967 letter to the editor of *Science*, much higher concentrations of polonium are found there than in the rest of the bronchial area.[35] Industry results were therefore diluted, showing lower concentrations of polonium-210 than were found by Radford, Hunt, and Little.

As research on polonium in tobacco continued, and as work by external scientists strengthened the argument that radioactive particles in cigarette smoke could pose a health hazard, one might have anticipated that the public media would pick up the story. The issue would seem to offer the makings of a full-blown press frenzy, along the lines of the thalidomide and asbestos stories of the 1960s and 1970s. However, the press never seized

Products Division Budget—1966," 24 Nov. 1965 (American Tobacco Company), Bates 966013836–3838. On the talks with Radford see R. D. Carpenter, "Polonium in Tobacco," 16 Dec. 1965 (Philip Morris), Bates 1001881339.

[34] Email correspondence with Little, 4 Mar. 2009; Ronald W. Davis, "Visit to Industrial Reactor Laboratory," 14 Aug. 1967 (American Tobacco Company), Bates 950113464–3465 (conferring with Skrable); E. C. Cogbill and R. W. Davis, "Progress Report—Month of February," 6 Mar. 1969 (American Tobacco Company), Bates 950282988–2989; Cogbill and Davis, "Progress Report—Radiochemical Section March 1969," 11 Apr. 1969 (American Tobacco Company), Bates 950282986–2987; and J. L. Charles to R. B. Seligman, "Meeting with Mr. Alex Holtzman" (interoffice correspondence), 14 Nov. 1980 (Philip Morris), Bates 2060534987 (on confirmations of polonium concentrations).

[35] John B. Little and Edward P. Radford, "Polonium-210 in Bronchial Epithelium of Cigarette Smokers," *Science*, 1967, *155*(3762):606–607, esp. p. 606.

on and pursued the story of radioactive material in tobacco to the extent one might have expected. Though a small number of pieces were published in a few newspapers and journals following Radford and Hunt's original report, sustained coverage of the issue was minimal.

Indicative of the lack of public concern about the subject is the surprising absence of consumer letters written to the tobacco industry on this topic. One might expect anxious smokers to have written the tobacco companies, wondering about the radioactivity levels of their favorite brand of cigarettes. However, among the many thousands of letters on countless topics (the surgeon general's report, for example, inspired more than five thousand queries), I could find only a couple from the 1960s that mentioned polonium at all. On 27 January 1964, only ten days after Radford and Hunt's first publication on polonium, Charles J. Smyth of Staten Island, New York, wrote to the director of research at R. J. Reynolds Tobacco Company:

> The Surgeon General's recent report linking cigaret smoking to lung cancer and the subsequent report by the Harvard scientists, which indicated the presence of the radioactive heavy metal polonium in tobacco leaves has done little to enhance the future of the cigaret industry.
> . . . It is with the finding of polonium in the tobacco leaves that I concern myself and it is after some thought and urging of others that I forward the following suggestion in an effort to provide a "safe" cigaret.

Smyth's suggestion—developing a "porous filter" that he believed would remove the risk of inhaling polonium—was acknowledged in a response from Reynolds to the effect that his letter was "appreciated" and that the company felt "complimented" that he had turned to them with his suggestion.[36] And then the matter was dropped. The tobacco companies clearly did not have much to worry about from their customers on this issue. Once the press coverage immediately following the 1964 paper died down, the tobacco industry simply refrained from comment and waited for the research and reporting to abate.

Even though there were not many papers on polonium published in the late 1960s, and therefore not much public awareness, there was a significant amount of scientific research during this time. Following Radford and Hunt's 1964 paper, the industry received several calls and even visits from scientists interested in pursuing polonium studies. During the first few months after the discovery, the tobacco manufacturers were not particularly interested in funding or supporting any research on the issue.[37] By 1967, however, as the industry was strengthening its own internal research programs, the Council for Tobacco Research (the industry's external research arm) began to receive more serious proposals from reputable scientists. One of these in particular caught their attention.

In July 1967 John E. Noakes of Oak Ridge Associated Universities (now professor of geology and director of the Center for Applied Isotope Studies at the University of Georgia) submitted an application for a research grant to the Council for Tobacco Research (CTR). The application proposed a three- to four-year project in two phases. The first phase, estimated to last about a year, was to be a geochemical study of the origins of polonium in the tobacco plant, including research to identify the parent isotopes (radium-226, radon-222, or lead-210) responsible for its presence. The second phase, to be

---

[36] Charles J. Smyth, "Letter to R. J. Reynolds," 27 Jan. 1964 (R. J. Reynolds), Bates 11294070; and Robert D. Rickert, "Response to Charles J. Smyth," 31 Jan. 1964 (R. J. Reynolds), Bates 11294071.
[37] Robert C. Hockett, "Proposals for Study of Radiation in Tobacco," 20 July 1964 (Council for Tobacco Research), Bates 01127511.

completed during the second and third (and, if necessary, fourth) years of research, would be a medical study of the levels of polonium needed to cause cancer in humans. The funding requested by Noakes was $30,900 for the first phase and $61,200 per year for the second. Citing Gregory's 1965 study of varying levels of polonium in different regions, Noakes stressed that the polonium content of U.S. tobacco was among the highest in the world. He then referenced Tso's soil studies and proposed to focus his own research on phosphate fertilizer as a source of polonium in the tobacco plant. Noakes hypothesized that the high polonium content of American-grown tobacco could be due to the extensive use of Florida phosphate fertilizer, which has a naturally high level of uranium and therefore high levels of its daughter isotopes, including lead-210 and polonium-210. During the 1960s the majority of American phosphate used in fertilizer was from the central Florida region, and Noakes suggested that using phosphate from a different source might lower the levels of polonium in tobacco.[38]

Noakes's proposal was submitted to the Council for Tobacco Research after he had made "two or three visits" to the office of Robert Hockett (associate scientific director of the CTR), during which he discussed his past work and future research goals. Noakes impressed Hockett, who remarked in a CTR memo that his research had the potential to demonstrate "relatively simple and inexpensive means of reducing the [polonium] content materially." Hockett went on to say that, regardless of the actual hazard posed by polonium, "the industry might be interested in . . . minimizing this contamination, as a matter of public policy." In August 1967 Hockett wrote to Helmut Wakeham, vice-president and director of research and development at Philip Morris, asking him to attend a meeting of the CTR's science advisory board and offer his input on Noakes's grant application. According to Hockett, the general opinion of the board was that polonium in tobacco did not "constitute any appreciable hazard." However, if Noakes's proposed research were to deliver the results it promised, then the board "would probably agree that 'the less the better' if it can be reduced easily."[39]

In Hockett's view, if Noakes's hypothesis that polonium levels in tobacco could be reduced simply by switching fertilizer sources proved correct, it would give the tobacco companies a chance to "benefit chiefly in terms of public relations." While all this sounded perfectly fine to the board, the main question was the "economic feasibility" of such a large-scale change. As Hockett said, "If the 'solution' to the problem is not really practical or practicable, the value of pointing a finger at the source of the polonium (or unknown hazard) would vanish." Convinced that this might be the case, the CTR ultimately rejected Noakes's proposal. The board had not found any fault in the proposed research plan itself. However, in their response to Noakes the CTR focused on the basis of his proposed project, questioning the "evidence implicating polonium in tobacco as a health hazard," despite well-documented internal concern at the time. The CTR was clearly worried that highlighting the dangers of radioisotopes in tobacco might harm the industry. What's more, by funding such research the industry would be admitting that polonium was a

---

[38] Noakes, "Application for Research Grant," submitted to the Council for Tobacco Research, 11 July 1967, Bates 1003546978–6995.

[39] Robert C. Hockett to Leon O. Jacobson, Clayton G. Loosli, and Stanley P. Reimann, "New Grant Application from John E. Noakes, Ph.D.—No. 624" (memo), 29 Aug. 1967 (Council for Tobacco Research), Bates 01188040–8041; and Hockett to Helmut Wakeham, 31 Aug. 1967 (Philip Morris), Bates 001609361–9362.

problem. With all this in mind, the board rejected Noakes's proposal, telling him they felt funds should be "committed in other directions."[40]

Despite the CTR's rejection of Noakes's proposal, his idea of investigating fertilizers clearly appealed to the tobacco companies. Apparently willing to give Noakes's solution a try, despite not funding his research, the industry experimented with North Carolina phosphate in the early 1970s. T. C. Tso believed that the North Carolina phosphate was "free" of radium (and therefore of radium's daughter isotopes), and he suggested conducting several studies to determine if phosphate were, in fact, the origin of polonium in tobacco. Tso also stressed to the Council for Tobacco Research that even if the cost of removing polonium from tobacco were high, it would ultimately prove to be a "self-supporting" investment.[41]

During the late 1960s, the tobacco men were focusing a significant amount of attention and resources on the polonium problem behind closed doors, yet there were not many publications during this time. Some isolated papers appeared in the late 1960s and early 1970s, but for the most part work on the topic remained quiet. In May 1974, a paper published in *Nature* by Edward Martell of the National Center for Atmospheric Research in Boulder, Colorado, revived interest in the subject. Martell, a graduate of West Point, earned his doctorate in radiochemistry from the University of Chicago. Following World War II he was involved in researching radiation effects at the Nevada Test Site and at Bikini Atoll. After witnessing the destructive power of nuclear energy he made a switch in his professional life, dedicating himself to researching radiation-induced cancers.[42] In the 1970s he became interested in radioactive particles as a cause of lung cancer and focused his research on the physiological effects of polonium and the biology of the tobacco plant.

Much of Martell's 1974 paper discussed the presence of lead-210 in cigarette smoke. Stating that previous research on the origins of polonium in tobacco had been "contradictory" (with Tso, Harley, and Alexander finding that most of the polonium was from root uptake of soil enriched with phosphate fertilizers, while Francis, Chesters, and Erhardt believed that the source of polonium was radioactive fallout), Martell began his study by looking closely at the tobacco leaf in order to determine how polonium behaves in the plant itself.[43]

He noted that the surface of the tobacco leaf is covered by tiny hairs called trichomes, about 85 percent of which have a glandular head coated with a sticky organic substance. A mature tobacco leaf can have up to nine hundred trichomes per cm[2] on each side of the leaf. Martell suggested that the stickiness of these trichomes attracts settling particles and provides a surface on which radioactive fallout can collect while the tobacco leaf is growing. Because the sticky coating on the glandular head is hydrophobic, he postulated

[40] Hockett to Wakeham, 31 Aug. 1967; and W. T. Hoyt to John Noakes, 19 Dec. 1967 (Council for Tobacco Research), Bates 50083248.

[41] Noakes's proposed research appealed to the tobacco companies: see E. S. Harlow to Wakeham, 14 Sept. 1967 (American Tobacco Company), Bates 1001609331. Regarding industry work on this topic see R. C. Hockett, "Visit by Dr. T.C. Tso of the U.S. Department of Agriculture, to Discuss Martell's Concepts of Lead-210 and Polonium-210 as Contaminants of Tobacco, August 21, 1975," 25 Aug. 1975 (Council for Tobacco Research), Bates 1000279828–9830.

[42] Bob Henson, "NCAR Mourns the Death of Ed Martell, Its Only Radiochemist," *Staff Notes Monthly* [University Corporation for Atmospheric Research], July 1995, *30*(7).

[43] Edward Martell, "Radioactivity of Tobacco Trichomes and Insoluble Cigarette Smoke Particles," *Nature*, 1974, *249*:215–217. He was referring to Tso *et al.*, "Source of Lead-210 and Polonium-210 in Tobacco" (cit. n. 29); and Francis *et al.*, "210Polonium Entry into Plants" (cit. n. 30).

that even soluble particles would not be washed away by rain. Measuring the polonium-210 content of two thousand trichomes from North Carolina–grown tobacco leaves, he found that the average concentration of lead-210 was $3.2 \pm 0.6 \times 10^{-6}$ pCi per trichome. In keeping with earlier research indicating that North Carolina tobacco has a higher concentration of radioactive particles than tobacco grown in most other places, tobacco grown in Turkey and Kentucky showed lead-210 concentrations that were a half and a quarter, respectively, of that in the North Carolina tobacco.[44]

Martell agreed with Noakes and Tso, Harley, and Alexander that the geographic variations in tobacco concentration were due to variations in the uranium concentration in the rock used to make phosphate fertilizers. Rather than believing that polonium was taken up by the roots, Martell suggested that soils with a concentration of uranium-rich phosphate fertilizer would release increased amounts of radon-222 into the surrounding atmosphere; it would then be deposited in the form of lead-210 on the leaves of the growing tobacco crop.[45]

Like Radford and Hunt, Martell was concerned with the buildup of polonium-210 in certain areas of the lung. It had been generally accepted for some time that exposure to radiation from radon daughters was the cause of elevated cancer risk among uranium miners, so Martell reasoned that radiation should also be accepted as the agent of cancer in smokers. He concluded that, given the chronic exposure to low doses of insoluble radioactive particles that were concentrated in specific areas of the lungs, polonium-210 was likely the primary cause of smokers' lung cancer and perhaps, as he suggested in a later paper, other types of cancer as well.[46] Martell shifted his studies from polonium-210 to lead-210, which is present in tobacco in a volatile state and as soluble and insoluble particles. He suggested that it was the *insoluble* lead-210 particles that were most likely the primary agents of lung cancer in smokers. The volatile lead-210 could disperse easily and be exhaled, and the soluble particles would dissolve into the bloodstream and ultimately would be excreted, resulting in the higher levels of polonium measured in the urine of smokers.[47]

On the basis of comparisons with radiation exposures of uranium miners, Martell

[44] Martell, "Radioactivity of Tobacco Trichomes and Insoluble Cigarette Smoke Particles," p. 215. The earlier work on polonium concentrations included Gregory, "Polonium-210 in Leaf Tobacco from Four Countries" (cit. n. 25); and Little *et al.*, "Polonium-210 in Bronchial Epithelium of Cigarette Smokers" (cit. n. 35).

[45] For an industry report on Martell's findings see "Polonium-210 Entry into Plants from Superphosphate Fertilizers," 5 Dec. 1975 (American Tobacco Company), Bates 962004691–4693.

[46] In the 1960s, researchers found elevated levels of polonium in the lungs and also in the blood, urine, bronchial lymph nodes, and even the skeletons of smokers. In addition to their original measurements of the levels of polonium-210 in the bronchial epithelium, Radford and Hunt also measured the concentration in urine. Heavy smokers (two packs a day) had a urine concentration of 0.065 pCi of polonium per twenty-four hours, nearly six times as much as the 0.011 pCi of nonsmokers. The presence of elevated levels of polonium in urine suggested an increased level in the bladder as well, indicating that the radioactive polonium-210 isotope, originally inhaled into the lung, can be traced throughout the body. Because of this, Martell suggested that polonium-210 in cigarette smoke could be tied to other radiation-induced cancers, such as osteosarcoma and leukemia, in addition to lung cancer. Hannes Eisler of the University of Stockholm suggested in a letter to the editors of *Science* that the presence of polonium in the urine of smokers could be an indication that the increased risk of bladder cancer among smokers might be attributed to radiation from tobacco. See Radford and Hunt, "Polonium-210" (cit. n. 3); Richard B. Holtzman and Frank H. Ilcewicz, "Lead-210 and Polonium-210 in Tissues of Cigarette Smokers," *Science*, 1966, *153*(3741):1259–1260, esp. p. 1260; Hannes Eisler, "Polonium-210 and Bladder Cancer," *ibid.*, 1964, *144*(3621):952–953; Martell, "Radioactivity of Tobacco Trichomes and Insoluble Cigarette Smoke Particles" (cit. n. 43), p. 217; and Edward Martell, "Tobacco Radioactivity and Cancer in Smokers," *American Scientist*, 1975, *63*(4):404–412.

[47] Martell, "Radioactivity of Tobacco Trichomes and Insoluble Cigarette Smoke Particles," p. 217. Lead-210 (^{210}Pb) is a precursor of polonium-210 in the natural uranium-238 decay series. A beta emitter, it decays to

suggested that the danger came not with the volume dose at any given time but, rather, with exposure over time. He argued that lead-210 (which has a half-life of twenty-two years) entered the lungs along with the polonium, settling in the lower bronchial lobes. The high exposure associated with a lifetime of smoking would presumably give the lead enough time to decay to polonium-210. Basing his findings on the average North Carolina flue-cured tobacco concentration (3×10^{-6} pCi of lead-210 per smoke particle), Martell found that the alpha radiation dose to cells in the immediate vicinity of a radioactive particle (about six cell diameters) would be about 0.5 rad, or 5 rem per year. Over a twenty-five-year period, the total radiation dose would exceed 200 rem, with much higher doses possible in hot spots of concentrated activity.[48] A lifelong smoker, therefore, could be at a high risk for cancer despite the relatively low dose of polonium-210 per cigarette.

"NO COMMERCIAL ADVANTAGE"

Like Radford and Hunt's paper a decade earlier, Martell's inspired a flood of research and new publications that expanded polonium research from botany to animal studies. In the years since the publication of Radford and Hunt's paper, the polonium scene at Harvard had shifted away from radiochemistry and toward radiobiology. Radford had left Harvard for the University of Cincinnati and Hunt had moved to Yale; polonium research at Harvard was taken up by the radiobiologist John Little, who had worked with Radford and Hunt in the 1960s. Under Little, the focus was on biological studies and animal experiments that would help researchers show just how damaging even low doses of polonium could be.

In a 1974 paper, Little and fellow Harvard scientist William O'Toole showed astonishing results after forcing polonium into the tracheas of hamsters in an effort to determine whether extremely low doses could cause cancer: 94 percent of hamsters in the highest exposure group developed lung tumors with doses so small that there was no inflammation. A similar study conducted a year later by Little, Ann Kennedy, and Robert McGandy exposed hamsters to a very low dose of polonium-210 aerosols over a period of several weeks.[49] The results showed that 10–36 percent of the animals developed malignant tumors in their lungs, compared to the 15 percent of lifelong smokers who develop lung cancer. Little, Kennedy, and McGandy continued their research for several years, and their experiments demonstrated that lung cancer *could* be caused by relatively small amounts of radioactive polonium (and alpha particles more generally); this concerned the tobacco companies a great deal.[50]

Much as in the 1960s, however, the tobacco industry did not respond publicly to this

---

bismuth-210, which itself decays to polonium-210 via beta decay. Lead-210 has a half-life of twenty-two years; see Holtzman and Ilcewicz, "Lead-210 and Polonium-210 in Tissues of Cigarette Smokers," p. 1260.

[48] Martell, "Radioactivity of Tobacco Trichomes and Insoluble Cigarette Smoke Particles," p. 216.

[49] John B. Little and William F. O'Toole, "Respiratory Tract Tumors in Hamsters Induced by Benzo(α)pyrene and-210Po α-Radiation," *Cancer Research*, 1974, *34*:3026–3039; Little, Ann R. Kennedy, and Robert B. McGandy, "Lung Cancer Induced in Hamsters by Low Doses of Alpha Radiation from Polonium-210," *Science*, 1975, *188*(4189):737–738; and Little, Kennedy, and McGandy, "Effect of Dose Rate on the Induction of Experimental Lung Cancer in Hamsters by α Radiation," *Radiation Res.*, 1985, *103*(2):293–299. The Syrian golden hamster was chosen for Little's studies because it tends to be resistant to pulmonary infections and rarely develops spontaneous lung tumors.

[50] Although Little, Kennedy, and McGandy's work on low doses was the most prominent, their studies were not the first on the subject. In 1967, C. L. Yuile, H. L. Berke, and T. Hull of the University of Rochester had found lung cancer growths in rats exposed to a single dose of polonium aerosol; see C. L. Yuile, H. L. Berke, and T. Hull, "Lung Cancer Following Polonium-210 Inhalation in Rats," *Radiation Res.*, 1967, *31*(4):760–774.

latest wave of research. Internally, on the other hand, there was a significant amount of activity. There are over two thousand documents from the 1970s in the UCSF Legacy Tobacco Documents online library that mention polonium, and more than a third of these date from 1974 and 1975. They show a flurry of interdepartmental correspondence, research reports, and meetings between top-level representatives from several tobacco manufacturers. In many of these memos one can sense a certain level of apprehension and urgency as the industry realized that it might eventually have to face this potentially embarrassing problem.

Only a couple of months after Martell's paper was published in *Nature*, Walter Gannon, director of new product development at Philip Morris, wrote a memo recapping a phone conversation he had had with Martell on 2 July 1974. Martell had mentioned that he was studying lung samples sent to him by Edward Radford and that he (Martell) expected the polonium-210/lead-210 ratio of these samples to support his previous research and hypotheses. Martell also said that he would be in touch with Gannon before he presented a paper to the Tobacco Working Group of the National Cancer Institute in September of that year.[51]

This was not the first time Philip Morris representatives had spoken with Martell. The company's contact with him reached back at least to March 1973, when Tibor Laszlo met with Martell at the National Center for Atmospheric Research. During his visit to the "new and unusually beautiful" research building at the center, Laszlo was able to spend quite some time speaking with Martell about his research on polonium and lead in tobacco. Martell displayed an interest in staying in touch with the R&D department at Philip Morris; he noted that he traveled back and forth to Washington quite regularly and would be pleased to stop in and discuss his research, as he was "very anxious" to gauge the general attitude of the department concerning the validity of his work. In response to Martell's entreaties for his opinion, Laszlo "gave a noncommittal answer."[52] Cigarette manufacturers were not yet ready, it would seem, to come out in the open on the matter of polonium.

Despite the amount of work on the subject and the accumulating evidence that polonium was indeed present in cigarette smoke and likely one of its carcinogens, there was still a sense among some in the industry that the science was wrong—or at the very least exaggerated. In a memo to Tim Cahill of the R. J. Reynolds corporate public relations department, Alan Rodgman, head of the analytical section of the Reynolds research department, wrote:

> While the biological results presented by the Harvard group [i.e., Little, Kennedy, and McGandy's 1975 paper] are suggestive of a relationship between polonium and cancers observed in hamsters so treated, it should be realized that the experiment conducted was unrealistic in terms of dose of polonium in an artificial way not related to the cigarette smoke inhalation process. The smoker probably receives his exposure in small incremental doses during the puffs with ample opportunity for his lung clearance mechanism to function either between exposures to individual puffs or between the smokings of successive cigarettes.[53]

[51] W. F. Gannon to Wakeham, F. E. Resnik, Thomas Osdene, D. A. Lowitz, T. S. Laszlo, and Robert Jenkins, "Call to Dr. E. A. Martell on July 2, 1974" (interoffice memo), 10 July 1974 (Philip Morris), Bates 2012601880.

[52] Laszlo to Wakeham, "IRI Meeting, March 26–27, 1973" (interoffice memo), 29 Mar. 1973 (Philip Morris), Bates 000016756–6758.

[53] Alan Rodgman to Tim Cahill (interoffice memo), 15 July 1975 (R. J. Reynolds), Bates 501016384. Rodgman is referring to Little, Kennedy, and McGandy's 1975 paper, "Lung Cancer Induced in Hamsters by Low Doses of Alpha Radiation from Polonium-210" (cit. n. 49), published four months before this memo was

Rodgman was being somewhat economical with the truth, however, as the challenges of replicating a "proper" smoking technique in the lab were well known and discussed by researchers. Moreover, his statement that radiation posed no long-term health hazards because the lung will cleanse itself of any inhaled polonium reveals a lot of unfounded faith in the weakened bronchial "clearance mechanism" of smokers, which could have been damaged by such physiological effects of long-term smoking as ciliastasis.

In several of the industry minutes and memos written on the subject in the late 1970s, one can sense that tobacco manufacturers were aware that they might eventually be called on to address this problem and had begun exploring various techniques to remove at least some of the polonium from tobacco. The industry knew that polonium could cause problems as a carcinogenic "additive," and Philip Morris was concerned that tobacco might fall under "something akin" to the Delaney proviso of the 1958 Food and Drug Administration Food Additives Amendment, which prohibited any known carcinogens from being added to food.[54] If such a proviso were ever extended to tobacco, the companies could be forced to remove polonium from their product (or at least reduce levels "below existing methods of detection"). In a memo to the president of the American Tobacco Company in 1975, R&D manager R. M. Irby wrote, "If the future should dictate that steps would have to be taken to ensure that zero amounts of polonium were present, work would certainly have to include treatment of tobacco as well as filtration."[55] Both of these potential solutions would be brought up repeatedly throughout the following decade by external and internal scientists and by high-level industry executives. But it seems that in most cases research on the removal of polonium went no further than brainstorming or preliminary experimentation, and no industry results were ever published.

Regarding a proposed solution that would presumably also remove many of the other criticized constituents of tobacco, Rodgman wrote to Cahill,

> It should be noted that the members of the Harvard group [i.e., Little, Kennedy, and McGandy] have little, if any, knowledge about the ease of removal of specific components from tobacco smoke. While polonium probably does not contribute to the "flavor" of tobacco smoke, it is a component of the so-called particulate phase of smoke. The particulate phase consists of the aerosol particles—small liquid spheres—each containing many thousands of components. To remove polonium selectively from these spherical balls by filtration is virtually impossible. Since the polonium is present in the particulate phase, one way to decrease its level in smoke is to reduce the level of "tar" delivered by the cigarette during smoking.[56]

---

written. In "The Golden Holocaust: A History of the Global Tobacco Plague" (ms draft, forthcoming) Robert N. Proctor calls Tim Cahill "Cold Facts Cahill" for his consistent use of the phrase "the cold facts are . . ." in response to consumer inquiries.

[54] On 11 June 2009 the U.S. Senate passed the Family Smoking Prevention and Tobacco Control Act, which was signed by President Obama on 22 June. The bill gives the Food and Drug Administration regulation over tobacco and the power to set standards, such as requiring graphic images on cigarette packs, banning certain cigarette flavors, limiting or lowering chemicals in tobacco, and banning the use of marketing terms such as "light" or "low-tar" by cigarette manufacturers. Although the legislation does not allow the federal government to outlaw cigarettes, it does for the first time bring tobacco under the jurisdiction of the FDA. As has concerned Philip Morris and other tobacco companies, polonium is likely to be one of the constituents of cigarettes that the FDA will choose to regulate.

[55] "Radiochemistry—Polonium," 15 Apr. 1977 (Philip Morris), Bates 1003372188; and R. M. Irby, Jr., to Robert Heimann, "Polonium-210" (memo), 19 May 1975 (American Tobacco Company), Bates 950113500. Polonium is not, strictly speaking, an "additive," but in this memo the industry called it such, interestingly placing it in the same category as other substances that are added to tobacco during the curing process.

[56] Rodgman to Cahill (interoffice memo), 15 July 1975 (R. J. Reynolds), Bates 501016384.

Despite differing opinions as to the most efficient and straightforward way of removing polonium, industry scientists had determined that if the isotope *were* in tobacco, it would show up in the particulate phase of smoke. This determination allowed research on the removal of polonium to focus on a filter. On 22 July 1975 the Council for Tobacco Research held an Industry Technical Committee meeting with representatives from all the tobacco companies, including the CTR's Hockett, Philip Morris's Wakeham, Osdene, Seligman, and Jenkins, and R. J. Reynolds's Nystrom and Rodgman. Much of the meeting was spent discussing polonium, and the attendees were given a thorough literature review as well as photocopies of selected references. To the high-powered industry personnel involved in this discussion, a filter was an attractive solution, as it might remove particulate matter (and a proportional amount of polonium) without altering the chemistry of the cigarette or the process of growing and curing the tobacco leaf.[57]

The idea of a filter would resurface the following month in a meeting on polonium held by the Tobacco Working Group of the National Cancer Institute in Bethesda, Maryland. The August 1975 meeting, unlike the CTR meeting in July, was attended not only by representatives of the tobacco industry but also by figures from the National Cancer Institute and the U.S. Department of Agriculture and external polonium researchers, including T. C. Tso, Naomi Harley, and Ann Kennedy (of the Little *et al.* group from Harvard that caused such a stir with its paper on the consequences of low doses of polonium in hamsters).[58] Kennedy, then a graduate student in radiation biology working with Jack Little at Harvard (and today a professor of research oncology at the University of Pennsylvania School of Medicine), represented the position of the Harvard group. She stood by their research indicating that polonium posed a health hazard and stated that "something should be done about its removal." But there was outspoken disagreement among the other attendees, many of whom felt that the risks were being exaggerated. After Kennedy raised the possibility of removing polonium from cigarette smoke by means of an ion-exchange resin filter, industry representatives questioned its likely effectiveness. The position of the tobacco manufacturers at this meeting is remarkable, given that in their own meeting only a month earlier they seem to have favored exploring the possibility of using a filter to remove polonium from cigarette smoke.[59]

Had the tobacco industry changed its stance on removing radioactive isotopes in the time between the two meetings? More likely, the opinions expressed during the meeting with Kennedy fit with the tobacco industry's lawyered public stance on polonium, while

[57] "Minutes from Industry Technical Committee Meeting, Council for Tobacco Research," 25 July 1975 (CTR), Bates 950149527–9529. As suggested by these meeting minutes, the industry continued to shy away from admitting the presence of radioactive materials in cigarettes (at least publicly), saying that "polonium-210 in cigarette smoke, if any," would exist in the particulate phase of smoke. The attendees at this meeting included Helmut Wakeham (vice president for research and development), Thomas Osdene (director of research), Robert Seligman, and Robert Jenkins from Philip Morris, as well as Charles Nystrom and Alan Rodgman from R. J. Reynolds. Following the discussion on polonium, Wakeham gave a brief demonstration on the presence of carbon monoxide in cigarette smoke. He claimed that its dangers are not as grave as some have suggested because "the smoke is diluted first by the air already in the smoker's lungs as well as by that taken in with the puff."

[58] "National Cancer Institute Smoking and Health Program: Workshop on the Significance of Po210 in Tobacco and Tobacco Smoke," 25 Aug. 1975 (Lorillard), Bates 01421854–1857. In a telephone interview on 24 May 2007, Ann Kennedy recalled that her attendance at this meeting was funded by the National Institutes of Health. She stressed that she would not have attended a meeting funded by the tobacco industry, as her research focuses on cancer prevention—something that is clearly not the top priority of the tobacco industry!

[59] On Kennedy's position see Wakeham to Resnik, "Meeting on Polonium" (cc'd to Seligman, Osdene, and Jenkins), 26 Aug. 1975 (Philip Morris), Bates 1003728418–8419.

the memoranda, notes, and minutes from the July 1975 CTR meeting that were exclusive to industry personnel reveal private and internal views of the matter. In the years since Radford and Hunt's first paper on polonium, the industry had been careful not to draw attention to the issue by keeping results unpublished and avoiding public debate with researchers or health officials. At the same time, however, top-level industry managers and executives had kept abreast of ongoing external research. Industry scientists had worked on their own parallel experiments and had recorded their results and measurements, always keeping their work secret and private. Thus, there had been a wide discrepancy and inconsistency between what the industry was admitting publicly about polonium and what it was saying and doing privately.

Despite their outspoken disagreement during much of the joint industry-NIH meeting of the Tobacco Working Group, and despite the fact that it was decided that an ion-exchange filter would not be pursued, Kennedy and the industry representatives were able to come up with a research plan recommending three areas of further study. First, the amount of polonium in cigarettes would be determined for both filter and nonfilter commercial brands. The polonium concentration remaining in the filter after a cigarette had been smoked would also be measured to determine the effectiveness of the filter and how much polonium had actually been inhaled. Second, laboratory dogs would be exposed to polonium-210 particulates in smoke, and their lungs would be examined for any radio-activity and ensuing damage. (Dogs were often used in radiation research because "people believe dogs," as they are much more similar to humans than rodents.)[60] Special attention would be focused on how polonium was distributed through the lung and on whether there were indeed areas of concentration, or hot spots, as first suggested by Radford and Hunt in 1964. Third, the amount of polonium in tobacco would be "monitored," as it varied with the use of different fertilizers. Work led by T. C. Tso of the Department of Agriculture would consider various sources of fertilizer.[61]

Concern about polonium remained strong within the tobacco industry into the late 1970s, and there was considerable worry that more scientists and researchers would become interested in the problem. In 1977, Robert Jenkins of Philip Morris traveled to Arizona to meet with John McKlveen, a professor of nuclear engineering at Arizona State University, to dissuade him from studying polonium. McKlveen had shown an interest in beginning research similar to that of Martell, but Jenkins cautioned against it, convincing him that "there are areas of unknown science that are more important and are necessary before" he should commence research on polonium.[62]

The industry's own researchers, on the other hand, had been addressing the issue since 1964, and following Martell's publication in 1974 industry scientists conducted several projects and experiments on removing polonium from tobacco. By 1975, 80 percent of the personnel in Philip Morris's radiochemical section were involved in the polonium project.

---

[60] Telephone conversation with Kennedy, 24 May 2007.

[61] It was suggested that the industry should look into a new uranium-free fertilizer developed in the 1970s by the Uranium Recovery Corporation, which was by then being prepared commercially. In the mid-1970s, several companies researched the possibility of producing uranium-free fertilizers, among them Uranium Recovery Corporation, Gulf Oil, Westinghouse Electric Company, and Freeport Minerals. According to Gulf Oil, 96 percent of the uranium could be removed from phosphate fertilizers, with a loss of only 0.01 percent of the phosphate. See Wakeham to Resnik, "Meeting on Polonium" (cc'd to Seligman, Osdene, and Jenkins), 26 Aug. 1975 (Philip Morris), Bates 1003728418–8419; and "Uranium from Phosrock," Chemical Week, 9 July 1975, p. 24.

[62] Jenkins to Osdene, "Visit with Dr. J. McKlveen" (interoffice correspondence), 23 Nov. 1977 (Philip Morris), Bates 000016590–6591.

Their work included measuring the concentration of polonium in tobacco, but the majority of industry scientists focused on ways to remove or reduce it. In June 1977 Robert Jenkins of Philip Morris completed a three-month study titled "Nuclear and Radiochemistry of Smoke." In keeping with Martell's findings, Jenkins reported that about 50 percent of the soluble polonium-210 could be removed from the bottom leaves of the tobacco plant by simple washing. The amount of polonium removable by washing decreased toward the top of the stalk, leading Jenkins to conclude that the technique was not effective enough to be worth implementing, despite the intimation that any reduction of polonium would reduce radioactive exposures.[63]

By 1980 there seems to have been a certain acceptance within the tobacco industry that the issue was not going to fade away as it had in the 1960s. In a meeting at Philip Morris on 11 November of that year it was noted that the "key point here is that interest is continuing" on the issue, with the principal concern being that future publications and research might draw a new wave of attention to the isotope. In a memo written the next day, Roger Comes of Philip Morris followed up with Alex Holtzman, the company's vice president and associate general counsel, stressing that the polonium problem "will not be leaving us."[64] Although Philip Morris acknowledged the advantage of not engaging publicly on polonium, the company also recognized the necessity of monitoring the issue very closely so it could respond to any new developments. In his memo to Holtzman, Comes was careful to urge that "the entire subject of low level radiation effects on public health from whatever source (Mt. St. Helens, Three Mile Island, Chinese Nuclear Testing, Tobacco, etc, etc) is one we must be aware of and must be addressing."[65]

The tobacco industry was in fact mulling over new methods for reducing the polonium in cigarettes. Several years earlier, Ramsey Campbell of the Stauffer Chemical Company had submitted a patent application for a treatment of tobacco leaves that would reduce the levels of radioactive lead and polonium in tobacco. The patent was granted on 25 March 1980, and it launched several years of correspondence and deliberations between the Stauffer Chemical Company and several of the major tobacco manufacturers, including Philip Morris and R. J. Reynolds.[66] The patent proposed to remove radioactive lead and polonium from tobacco by washing the leaves with a dilute acid solution of hydrogen peroxide. The process would involve either spraying the leaves with the solution while they were still growing or dipping them in the acid solution after they were harvested. The leaves would then be rinsed with water before being allowed to dry.

Philip Morris had been in contact with Campbell since the mid-1970s, and Robert Jenkins in particular had spoken with him on several occasions. In January 1976 Jenkins

[63] R. W. Jenkins, "Nuclear and Radiochemistry of Smoke" (report), 7 June 1977 (Philip Morris), Bates 1001925327–5328. On work in the Philip Morris radiochemical section see Jenkins to W. R. Johnson, "Projected Six-Month Research Plans for-210Pb-^{210}Po" (interoffice correspondence), 24 Sept. 1975 (Philip Morris), Bates 2012614839–4843.

[64] Charles to Seligman, "Meeting with Mr. Alex Holtzman—210 Polonium Briefing—November 11, 1980" (interoffice correspondence) (cc'd Osdene, Alexander Holtzman, Jenkins, Roger A. Comes, and Edward B. ["Ted"] Sanders), 14 Nov. 1980 (Philip Morris), Bates 000016574–6575; and Comes to A. Holtzman, "Follow-up to Discussion of November 11, 1980" (interoffice correspondence) (cc'd to Seligman, Osdene, Sanders, Charles, and Jenkins), 12 Nov. 1980 (Philip Morris), Bates 1000083336.

[65] Comes to A. Holtzman, "Follow-up to Discussion of November 11, 1980." Regarding the importance of following all developments closely see Marian DeBardeleben to Comes (memo), 28 Feb. 1980 (Philip Morris), Bates 2012600443; Osdene to A. Holtzman (memo), 11 Apr. 1980, Bates 000016577; and Charles to Seligman (interoffice correspondence), 14 Nov. 1980 (Philip Morris), Bates 2060534987.

[66] Ramsey G. Campbell, "United States Patent: Removal of Radioactive Lead and Polonium from Tobacco," 25 Mar. 1980 (R. J. Reynolds), Bates 501522607–2611.

traveled to California to visit with Campbell to discuss his interest in polonium. Jenkins offered a brief review of the "open literature," including an explanation of "just what Martell was saying." Campbell then asked (with what seemed to Jenkins to be "previous knowledge") about any polonium research being conducted by the tobacco industry. All Jenkins revealed was that, "in general terms, Martell's findings were essentially accurate in their radiochemical determinations." Jenkins and Campbell followed up this visit with a phone conversation on 16 February, during which (in response to questioning) Jenkins recommended certain laboratory equipment that Campbell might find useful for his own polonium research. He also referred Campbell to T. G. Williamson of the University of Virginia as someone who might be able to help him with polonium analyses. In recommending Williamson, Jenkins felt he was ensuring that Philip Morris would have full access to any of the resulting data and could control their release; as Jenkins put it, "Dr. Williamson 'knows where his bread is buttered.'"[67]

Between 1980 and 1985, several tobacco companies reviewed the Stauffer patent, considering whether they should adopt the acid washing procedure to reduce the polonium in tobacco leaves. Philip Morris had been experimenting with washing leaves since the 1970s, and Jenkins had obtained results similar to Campbell's (perhaps Jenkins pursued his own research following his discussion of this technique with Campbell in January 1976?). On 14 April 1980, Philip Morris patent agent Susan Hutcheson wrote to Seligman that she thought Jenkins was quite unhappy about the Stauffer patent; but, as she reasoned, not much could be done about it, as "Stauffer is *not* in the tobacco business—which makes a difference!" Several years later, Philip Morris was still discussing whether it should adopt the process described in the Stauffer patent. In March 1985 Jenkins described the research Philip Morris had conducted in the 1970s on washing. He stated that it is "well known that, under the right chemical conditions," soluble lead-210 and polonium-210 can be dissolved. These results were supported not only by the work leading to Campbell's technique but also by research conducted by Jenkins himself. Jenkins's data showed that about 60 percent of all soluble polonium-210 could be removed by washing tobacco leaves.[68]

Jenkins himself said in this same 1985 memo that "Mr. Campbell utilizes accepted technology and at this time no fault can be found with his radiochemistry." Despite the "scientific validity" and promising results of Campbell's washing process, however, both Philip Morris and R. J. Reynolds conclusively decided not to adopt it. According to Reynolds's "resident expert on polonium," Charlie Nystrom, the greatest challenge of the Stauffer patent was the impracticality of implementing the procedure on a "commercial scale basis." Numbered first among the company's reasons for rejecting the patent was that "complete removal . . . would have no commercial advantage."[69]

Philip Morris, like R. J. Reynolds, also seemed most concerned with the "practicality of this patent," rather than with the potential health benefits of washing. Because the tobacco leaf would have to be washed before curing, it was likely the farmer who would

---

[67] Jenkins to Osdene (interoffice correspondence), 25 Feb. 1976 (Philip Morris), Bates 2012614498.

[68] For the Philip Morris patent expert's comment see S. Hutcheson to Seligman, "Re: Washing Tobacco" (memo), 14 Apr. 1980 (Philip Morris), Bates 1003725586. Regarding Jenkins's data see Jenkins to Tom Goodale, "Removal of Radioactive Lead and Polonium from Tobacco, U.S. Patent 41 94 541" (interoffice correspondence), 15 Mar. 1985 (Philip Morris), Bates 2012615307.

[69] Jenkins to Goodale, "Removal of Radioactive Lead and Polonium from Tobacco, U.S. Patent 41 94 541"; and C. W. Nystrom to Rodgman, "Comments on the Stauffer Patent No. 4,194,514 for Removal of Radioactive Lead and Polonium from Tobacco" (interoffice memo), 5 Mar. 1982 (R. J. Reynolds), Bates 504970288.

apply Campbell's technique.[70] Both Philip Morris and R. J. Reynolds were rightly worried about the added expense for the tobacco farmer, but they listed other concerns about the impracticality of the Stauffer patent process, including the problem of disposing of the acid solution once the leaves had been washed and the fact that a large portion of cured tobacco leaves are water soluble. If, however, the washing took place *before* the tobacco leaf were cured, the second concern would be a nonissue, leaving only the matter of disposing of the used acid solution, a problem that could probably have been resolved. Ultimately, Reynolds opted to not pursue the patent because it was felt that, given the lack of consumer concern, there would be "no commercial advantage in providing a tobacco product with reduced quantities of these constituents."[71]

Despite the companies' decision not to wash tobacco leaves, Robert Jenkins noted that if it were ever "deemed desirable" to remove or reduce the amount of radioactive material in tobacco, there were procedures and methods in addition to those patented by the Stauffer Chemical Company that promised to do just that. He concluded, however, that the "real question" was whether it would be of any commercial value to the industry to remove polonium from tobacco. In concert with Jenkins, Alan Rodgman of R. J. Reynolds noted that the polonium issue had "appeared and disappeared periodically" since 1964, suggesting that there was really no need to invest in resolving the problem as it would certainly disappear once again.[72]

"THE WORST PART . . . THERE MAY BE SOME DEGREE OF VALIDITY"

By 1980 the wave of polonium-related research that followed Martell's papers had subsided somewhat, but the tobacco industry correctly anticipated that this lull would be short lived. Interest in the issue was revived in early 1982 by a letter written to the editor of the *New England Journal of Medicine* by Thomas Winters and Joseph Difranza, both of the University of Massachusetts Medical Center. Winters and Difranza felt that there had not been nearly enough research on polonium as a carcinogen in tobacco, and they stressed that in the seventeen years since the surgeon general's original report work on radiation had been "conspicuous because of its absence."[73] In their brief review of the relevant research, however, Winters and Difranza overlooked most of the papers that had been published on the topic, citing only Radford and Hunt's "Polonium-210: A Volatile Radioelement in Cigarettes" (1964), Little, Radford, McCombs, and Hunt's "Distribution of Polonium-210 in Pulmonary Tissues of Cigarette Smokers" (1965), and Martell's "Radioactivity of Tobacco Trichomes and Insoluble Cigarette Smoke Particles" (1974).

Their limited references embittered many of the other scientists who had published on the topic, and the 29 July 1982 issue of the *New England Journal of Medicine* published seven responses to Winters and Difranza, including letters from Martell, Cohen, and

---

[70] Jenkins to Goodale, "Removal of Radioactive Lead and Polonium from Tobacco, U.S. Patent 41 94 541"; and Hutcheson to Seligman, "U.S. Patent 4,194,514 Assigned to Stauffer" (law department memo), 14 Apr. 1980 (Philip Morris), Bates 1003725585.

[71] Nystrom to G. R. Di Marco, "Evaluation of Idea from Ramsey G. Campbell for Removal of Radioactive Lead and Polonium from Tobacco" (interoffice memo), 22 July 1985 (R. J. Reynolds), Bates 504205144–5146. For other company concerns with regard to impracticality see Jenkins to Goodale, "Removal of Radioactive Lead and Polonium from Tobacco, U.S. Patent 41 94 541"; and Hutcheson to Seligman, "U.S. Patent 4,194,514 Assigned to Stauffer."

[72] Jenkins to Goodale, "Removal of Radioactive Lead and Polonium from Tobacco, U.S. Patent 41 94 541"; and Rodgman to Roy Morse, "Stauffer Patent" (memo), 8 Mar. 1982 (R. J. Reynolds), Bates 501522602.

[73] Winters and Difranza, "Letter to the Editor" (cit. n. 6).

Harley. Although a couple of these letters criticized Winters and Difranza's suggestion that there had been little research on polonium as a tobacco carcinogen and offered extensive lists of references as proof of earlier work, most applauded the renewed attention they had drawn to the subject. In the same issue, Winters and Difranza wrote a letter responding to the enormous reaction they had provoked. It had become clear to them that although there had certainly been an extensive amount of work on polonium in tobacco, only a few people were aware of this research.[74]

Unlike the tobacco industry memoranda written following the publication of Martell's paper, several of the industry's internal documents from the months after Winters and Difranza's letter display a surprising lack of knowledge on the subject of polonium in tobacco, not to mention polonium itself. Some industry employees in 1982 were surprised and shocked to learn that there is polonium in tobacco and flatly denied that radioactive particles could be a cause of cancer in smokers. In a Brown & Williamson report written just after Winters and Difranza's original letter, Senior Field Manager Arthur Flynn wrote,

> The N.E. Journal of Medicine reports this week that two scientists working for the University of Massachusetts report that after extensive testing, found that cigarette smoke produces an extremely high amount of a Radioactive ingredient called "Polonium."
> Websters Definition: Polonium—So named by its co-discoverer, Marie Curie, after her native land, Poland. A Radioactive chemical element formed by the disintergration [sic] of Radium.
> It was further stated that Polonium in cigarette smoke is absorbed in the tissue of the lungs and that a cigarette smoker that smokes a pack and a half a day receives the Radioactive equivelent [sic] of 300 chest X-Rays during a given year!
> Our R&D Dept. will just love to hear this!

But of course the company's R&D department *already* knew all about it—or at least had known about it only a few years before. The initial shock of learning about radioisotopes in tobacco aside, in 1982 manufacturers were clearly concerned with the potential consequences of this newest wave of interest in polonium research, especially after Winters and Difranza noted in the 29 July issue of the *New England Journal of Medicine* that they were "gratified to receive hundreds of phone calls from smokers who quit on learning about the alpha radiation in cigarette smoke." This evidence that "more smokers are encouraged to quit as they learn of the presence of radiation" was striking, and the industry realized that it could lose many customers because of the recent attention drawn to the issue.[75]

From several documents produced following Winters and Difranza's letter, it is evident that the industry was now focusing on the fact that there was no way of knowing for certain whether radioactivity in tobacco could cause cancer in smokers. That is, instead of acknowledging that polonium *could* be a carcinogen and taking precautions, the industry was drawing attention to any doubt and disagreement there might be among researchers.[76]

---

[74] Letters were sent by Edward A. Martell, Jeffrey I. Cohen, Beverly S. Cohen and Naomi H. Harley, C. R. Hill, Walter L. Wagner, R. T. Ravenholt, and Dietrich Hoffman and Ernst L. Wynder: "Letters to the Editor," *N. Engl. J. Med.*, 1982, *307*(5):309–313. For the rejoinder see Winters and Difranza, "Letter to the Editor," *ibid.*, p. 313.

[75] Arthur J. Flynn, "Senior Field Manager's Report," Feb. 1982 (Brown & Williamson), Bates 670915637; Winters and Difranza, "Letter to the Editor" (cit. n. 74), p. 313; and Andrew A. Napier to PM Munich, Brussels, Paris, Amstelveen, Benelux BOX, Athens, and UK Feltham, "Polonium-210" (interoffice correspondence), 11 Oct. 1982 (Philip Morris), Bates 2501025243.

[76] See, e.g., Napier to PM Munich, Brussels, Paris, Amstelveen, Benelux BOX, Athens, and UK Feltham,

In an R. J. Reynolds memorandum with the subject line "With Friends Like This, We NEED Enemies," Frank G. Colby of the legal department reacted to a German paper written in response to Winters and Difranza's letter by Franz Adlkofer, director of the scientific division of the German Cigarette Industry Trade Association. Adlkofer had suggested that polonium, and its role as a carcinogen in tobacco, "has not received sufficient attention to date by researchers." He went on to say that, in his opinion, "everything should be done to avoid introducing polonium into tobacco through fertilizers" and, presumably, through other sources as well. In his memorandum, Colby—clearly concerned by Adlkofer's statement—said, "It is glaringly obvious that instead of making this appalling and scientifically erroneous statement," Adlkofer should instead have drawn attention to the fact that there is not complete agreement among physicians and scientists on this issue.[77]

In a June 1982 memo to Thomas Osdene, director of research at Philip Morris, Robert Jenkins expressed his concern that future papers on polonium would only feed the current frenzy. He was particularly worried about Martell, whom he called "sensationalism at its best" and who, he felt, would "receive wide acclaim from the anti-smoking foes and the press media." Jenkins strongly urged Osdene to consider making Philip Morris's radio-chemical research public, as the company could not properly counter such "hypothetical papers" by "anti-smoking foes" if it did not become actively involved in the scientific debate on the subject:

> At present, the major funding support for any research along these lines is from the anti-smoking forces. The tobacco industry has chosen not to answer these types of studies with well conducted scientific research, but has chosen to remain quiet in hopes "it too shall pass." As we have constantly seen since 1964, it continues to make news. The worst part being that there may be some degree of validity amonst [*sic*] the many assumptions that are grossly incorrect.[78]

Jenkins wrote to Osdene again in July 1982, stressing that it was time for Philip Morris to publish some of its research on polonium (research that had been largely conducted by Jenkins himself in the 1970s), as the results "would serve to offer an alternative inter-pretation to the world's scientific community." Jenkins felt that by publishing its own work Philip Morris "would cause the public to realize that this issue is indeed just an unproven controversy, not a fact."[79] The best way to approach the matter was for the tobacco manufacturers to be open about the research they had done on polonium and to fund private scientists working on the question. Publishing would draw attention to the fact that the industry had been working on the problem since the 1960s, perhaps lending some legitimacy to its point of view. However, it could also backfire, as it would leave the

---

"Polonium-210"; and Frank G. Colby to Samuel B. Witt III, "RE: With Friends Like This, We NEED Enemies—Part VI. Polonium—Lung Cancer—Prof. Adlkofer" (memo), 20 Apr. 1982 (R. J. Reynolds), Bates 511221904.

[77] Colby to Witt, "RE: With Friends Like This, We NEED Enemies—Part VI. Polonium—Lung Cancer—Prof. Adlkofer" (this includes the quotation from Adlkofer). At a 1988 joint meeting of the worldwide tobacco industry held in London, Adlkofer "deviated from the agenda" to discuss the direction of future secondhand smoke research. He stated that nothing was likely to come from continuing present research and suggested that, rather than using marketable science in public relations campaigns against the secondhand smoke issue, the industry should use its resources to develop a safe threshold for secondhand smoke exposure. This notion met with great disagreement from the other meeting attendees, who felt that it was dangerous to set a threshold as that "provides *a priori* proof of causation for anti-smoking advocates": "Joint Meeting on ETS—London, England" (memo), 15 July 1988 (Philip Morris), Bates 2021548222–8235.

[78] Jenkins to Osdene, "Review of Manuscript by E. A. Martell" (interoffice correspondence), 11 June 1982 (Philip Morris), Bates 1000083314–3319.

[79] Jenkins to Osdene, "Significant Scientific Accomplishments of Our Past-210Po Research Studies" (inter-office correspondence), 2 July 1982 (Philip Morris), Bates 1000083334–3335.

manufacturers vulnerable to criticism that in spite of more than twenty years of unpublished research, there were as yet no concrete advances in reducing the polonium in cigarettes.

Despite such fears—and fortunately for the industry—the matter would once again receive little public attention. Polonium was mentioned briefly in an April 1985 article in *Reader's Digest*, titled "Deadly Mixers"; following this, a concerned smoker named Chris Heimerl wrote to R. J. Reynolds to ask whether there was any polonium in his Salem Lights 100's. He also wanted to know how a radioactive material could come to be in cigarettes. Unaccustomed to answering such questions, Miriam Adams of the company's public relations office forwarded the letter to Alan Rodgman, saying that she had "no information in file to offer" in answer to Heimerl's questions. Rodgman, in turn, forwarded Adams's memo to Charlie Nystrom. In the R. J. Reynolds "Quarterly Status Report on Smoking and Health," dated 16 July 1985, it was mentioned in the "miscellaneous" section that a memo had been sent to Adams outlining appropriate responses by public relations representatives to consumer concerns about polonium.[80] This correspondence shows how rarely the industry was forced to confront such worries, despite the significant research that had been conducted since the 1960s.

The issue was reignited in February 1986 by a paper published in the *Southern Medical Journal* by Jerome Marmorstein, a physician and medical writer from Santa Barbara, California. In this paper, titled "Lung Cancer: Is the Increasing Incidence Due to Radioactive Polonium in Cigarettes?" Marmorstein, citing most of the research published since the 1960s, suggested that the increased incidence of lung cancer among smokers in recent decades could be due to an increase in the amount of polonium in tobacco. Despite the fact that by the mid-1980s 90 percent of American cigarettes were filtered and 15 percent of the population (nearly thirty million Americans) had quit smoking, the incidence of lung cancer among smokers had actually risen since the 1960s, and twice as many American men and three times as many women had died of the disease in 1980 as in 1960. As 85 percent of lung cancers were in smokers, it was clear to Marmorstein that whatever was responsible for this increase would be found in changes to tobacco and cigarette design since the mid-twentieth century.[81]

Marmorstein laid out four features of tobacco carcinogens that would be necessary for the higher incidence of cancer he was trying to explain: the carcinogen must be "inadequately filtered" by existing cigarette filters; it must cause cancer even at a very low dose; smokers' lungs must have a greater concentration of the carcinogen than the lungs of nonsmokers; and there must be a reason for an increase in the levels of this carcinogen since the 1960s.[82] After considering more than a hundred known carcinogens in tobacco, Marmorstein found only three that caused cancer by inhalation: benzopyrene, nitrosamines, and polonium-210.[83]

---

[80] Lowell Ponte, "Deadly Mixers: Alcohol and Tobacco," *Reader's Digest*, Apr. 1986, *126*:53–56; Chris Heimerl, "Consumer Inquiry," 27 Mar. 1985 (R. J. Reynolds), Bates 504974166–4167; Miriam G. Adams, "Consumer Inquiry: Chris Heimerl" (interoffice memo), 4 Apr. 1985 (R. J. Reynolds), Bates 504974165; and Anthony V. Colucci, "Quarterly Status Report on Smoking and Health, 2nd Quarter, 1985," 16 July 1985 (R. J. Reynolds), Bates 504974078–4088.

[81] Jerome Marmorstein, "Lung Cancer: Is the Increasing Incidence Due to Radioactive Polonium in Cigarettes?" *Southern Medical Journal*, 1986, *79*(2):145–150.

[82] *Ibid.*, p. 145. Polonium had been shown by both industry and external scientists to be unaffected by existing cigarette filters, and so it seemed likely to Marmorstein that it was a leading cause of cancer among smokers.

[83] For a while benzopyrene had been considered the leading candidate for the dubious honor of being the cancer-causing constituent of tobacco, but its concentration in tobacco had been shown to decline dramatically

Testing his hypothesis that the rising incidence of lung cancer was due to a change in the tobacco itself, Marmorstein looked at shifts in fertilizer use since the 1960s. Tobacco samples from 1938 were measured for radioactivity and the measurements compared to those made by Tso, Hallden, and Alexander in the 1950s and 1960s. The tobacco from 1938 had one-third to one-sixth the concentration of polonium, indicating that there had been a significant rise in radioactive particles in tobacco crops throughout the middle part of the century. Marmorstein then asked *why* there was more radioactive polonium-210 in tobacco grown in the 1960s, which would presumably account for the higher incidence of lung cancers developing twenty years later. He pointed to the increasingly widespread usage of artificial high-phosphate fertilizers in developing countries. In the United States, fertilizer manufacturing had begun in earnest with the establishment of the Tennessee Valley Authority in the 1930s, and phosphate fertilizers gained further popularity through the "Big Agriculture" movement of the postwar era. Marmorstein also noted that the quality of a tobacco crop and the resulting flavor of the cigarette were adversely affected by high nitrogen concentrations. In order to reduce the amount of nitrogen in their tobacco plants, therefore, farmers had been saturating their land with exceptionally large amounts of phosphate fertilizer. This, of course, resulted in even higher levels of polonium in tobacco grown in areas of high fertilizer use. Citing a 1965 paper by C. R. Hill, Marmorstein also noted the lower concentration of polonium found in tobacco grown in developing countries such as Turkey, India, and Indonesia, where organic fertilizers (such as manure and guano) were used instead of phosphates.[84]

Marmorstein's paper depicted the hazards of polonium rather dramatically, but other researchers—industry and otherwise—had shown essentially the same results since 1964. And yet, despite early evidence that the tobacco industry had at various times been interested in investigating ways to reduce the levels of radioisotopes in cigarettes, nothing had been done along these lines. This lack of action is all the more incredible given the fact that—as revealed by internal industry documents—several methods for reducing radioactivity in cigarette smoke had been evaluated and considered by the manufacturers.

Since the 1960s the industry had flirted with several potential solutions, among them developing a strain of tobacco that did not have trichomes and adding materials to tobacco that would react with lead and polonium to prevent their transfer to smoke.[85] One option that significantly intrigued the manufacturers involved developing an ion-exchange resin filter, as suggested in the late 1960s by Erick Bretthauer and Stuart Black of the U.S. Public Health Service (and recommended once again in the mid-1970s). Such a filter had been shown by Bretthauer and Black to "markedly reduce" exposure to polonium. Ann Kennedy had pushed this option in 1975, stressing that an ion-exchange filter could

---

with the advent of filter cigarettes. Furthermore, the type of cancer caused by benzopyrene, squamous cell carcinoma, had become less prevalent among smokers, replaced in the preceding twenty years by adenocarcinoma as the most common type of lung cancer among smokers. Similarly, approximately 80 percent of nitrosamines were removed by filters, so they, too, were unlikely causes of the increased prevalence of lung cancer. See Ronald G. Vincent, John W. Pickren, Warren W. Lane, Irwin Bross, Hiroshi Takita, Loren Houton, Alberto C. Gutierrez, and Thomas Rzepka, "The Changing Histopathology of Lung Cancer: Review of 1682 Cases," *Cancer*, 1977, *39*:1647–1655.

[84] Marmorstein, "Lung Cancer" (cit. n. 81), p. 148; and C. R. Hill, "Polonium-210 in Man," *Nature*, 1965, *208*(5009):423–428.

[85] Regarding potential solutions see P. D. Schickedantz to H. J. Minneveyer, "Comments on Recent Articles Concerning Polonium-210 as a Tobacco Smoke Carcinogen" (memo), 5 Sept. 1975 (Lorillard), Bates 01092297–2299; and J. D. Mold to R. W. Tidmore, "Radioactive Particles in Cigarette Smoke" (memo), 3 Dec. 1975 (Liggett & Myers), Bates 81151936–1938.

remove up to 92 percent of the polonium in cigarette smoke.[86] Bretthauer and Black gave a rough estimate that it would cost the tobacco industry 0.5 cents per pack of cigarettes to incorporate the 0.12 grams of resin needed for each cigarette, making the ion-exchange resin filter a relatively cheap fix that could be used until a more effective solution could be found.

Another straightforward option, following Martell's research in the 1970s, had been to wash the tobacco leaves to remove polonium that had collected on the trichomes that covered the leaf's surface, per the discarded Stauffer patent and Robert Jenkins's research at Philip Morris. Yet a third solution would have been to remove trichomes from the cured tobacco leaf mechanically. T. C. Tso had estimated in 1975 that 30–50 percent of polonium could easily be removed from fertilizer and that washing could eliminate another 25 percent.[87] Adding to that the effects of an ion-exchange filter, the polonium content of tobacco could have been significantly reduced using techniques that were well known and repeatedly discussed by both external and industry scientists and by high-level industry executives and attorneys. But the tobacco companies were clearly focused on other priorities.

A "SLEEPING GIANT"

Although research on polonium has slowed in recent years, the tobacco industry has continued to monitor relevant literature, keeping abreast of advances by compiling bibliographies and extensive reviews of scientific papers and news articles.[88] In 1999, polonium, along with other potentially hazardous smoke constituents, was reviewed for Philip Morris by scientists at INBIFO (Institut für Biologische Forschung [Institute for Biological Research]), a bioresearch laboratory in Germany acquired by Philip Morris in 1971. The first of the two reviews was written by INBIFO Manager of Bioresearch Support Helmut Schaffernicht. Work at INBIFO focused on "quantitative biological product evaluation," but Schaffernicht's review of radioisotopes in tobacco was brief and uninformed: he had not read any of the literature on the topic, nor did he scan internal industry research and reports. Schaffernicht gave polonium the lowest priority level (1 on a scale of 1 to 5) for contributing to potential Philip Morris programs in relation to health and safety, potential need to alter the product, and company credibility. The second review, however, saw polonium as a more serious threat. This reviewer, INBIFO Manager of Cell Biology Jan Oey, had taken quite an extensive look at the scientific literature since 1964 but did not examine the internal industry literature. Oey gave polonium a rank of 3 for "Potential Contribution to a PM Smoking/Health Program," describing it as an issue

---

[86] Erick W. Bretthauer and Stuart C. Black, "Polonium-210: Removal from Smoke by Resin Filters," *Science*, 1967, *156*(3780):1375–1376; and Resnik to Wakeham, "Meeting on Polonium—August 26, 1975" (memo), 26 Aug. 1975 (Philip Morris), Bates 1003728418–8419 (Kennedy's view).

[87] Schickedantz to Minneveyer, "Comments on Recent Articles Concerning Polonium-210 as a Tobacco Smoke Carcinogen" (memo), 5 Sept. 1975 (Lorillard), Bates 01092297–2299 (mechanical removal); and "National Cancer Institute Smoking and Health Program: Minutes of the Workshop on the Significance of Po^{210} in Tobacco and Tobacco Smoke," 25 Aug. 1975 (Philip Morris), Bates 1000268053–8056 (Tso's estimates).

[88] Ronald Davis, Radiochemical Section, "Radionuclide Concentration: A Literature Survey of the Principal Radionuclides Found in Tobacco, Tobacco Smoke, Soil, Other Agricultural Products, Air, Water, Human, and Other Animals," June 1967 (American Tobacco Company), Bates 962004628–4682; Robert Jenkins, "Polonium-210," 9 Nov. 1976 (Philip Morris), Bates 1002977456–7473; Roger Comes, "Investigations of Polonium-210 in Cigarette Tobacco and Whole Smoke Condensate," 12 May 1977 (Philip Morris), Bates 1000365379–5414; and "Polonium," 1986 (American Tobacco Company), Bates 950008179–8208.

of "moderate" priority to Philip Morris. Both reviewers felt that the only future step necessary on the part of the company was to monitor any relevant literature.[89]

External researchers, in contrast, have given far more dire assessments. Reimert Thorolf Ravenholt, a career epidemiologist with the U.S. Centers for Disease Control, the U.S. Agency for International Development, and the U.S. Food and Drug Administration, said in a 1982 interview that "the American public is exposed to far more radiation from the smoking of tobacco than they are from any other source."[90] Nonetheless, and despite forty years of research, there is little awareness of the problem outside of a small group of scientists and tobacco industry personnel. The serial irradiation of smokers' lungs by polonium-210 remains a repeatedly exposed and then forgotten story.

In a sweeping call for awareness, Monique Muggli, Jon Ebbert, Channing Robertson, and Richard Hurt of the Mayo Clinic and Stanford's School of Engineering suggested in a recent issue of the *American Journal of Public Health* that all cigarette packs should come with a radiation warning.[91] Yet another admonition, however, is unlikely to make a dent in the smoking population significant enough to concern the tobacco industry. After all, warnings pertaining to the smoker's own health, the health of the smoker's unborn child, and the health of those around the smoker are already displayed on packs of cigarettes. The grotesque images of foul teeth and gums, cancerous lungs, and open-heart surgeries on European packs go so far as to enshroud cigarettes in the very images of their consumers' potential futures. And yet people still smoke. Would one more warning— even one highlighting the specific dangers posed by radiation to the smoker's lungs— really make a difference?

Although it is certainly striking that the tobacco manufacturers have not made a definitive move to reduce the concentration of radioisotopes in cigarettes, it is equally striking that, despite forty years of research suggesting that polonium is a leading carcinogen in tobacco, they have felt no pressure from the public, the government, or the medical and public health communities to do so. So long as there continues to be only episodic awareness of the issue, and no pressure from powerful entities (such as the Food and Drug Administration, the surgeon general, or public opinion) to remove polonium from cigarettes, the tobacco industry does not need to worry about designing new filters or washing leaves.

No matter how simple and straightforward some of the proposed solutions to reducing polonium may seem, implementing them would cost the industry money and manpower. As tobacco manufacturers have never shown much concern about the health hazards of cigarettes (which they didn't even publicly admit until the late 1990s), it is no wonder that they have remained passive on the issue of polonium as well. It is unlikely, too, that the latest wave of interest, provoked by the death of Alexander Litvinenko in November 2006, will have much of an impact on their thinking or behavior.[92] The tobacco industry no doubt kept a close eye on press releases and commentaries written in the months following Litvinenko's death. Statements such as the one by the British Health Protection Agency that polonium poses no risk to the general public only continue to limit awareness of the presence and health hazards

---

[89] H. Schaffernicht, "WSA Categorization Form: Smoke Constituents," 27 Oct. 1999 (Philip Morris), Bates 2074168043; and J. Oey, "WSA Categorization Form: Smoke Constituents," 27 Oct. 1999 (Philip Morris), Bates 2074168044. On the INBIFO mission see Institut für Biologische Forschung, "History and Capabilities of INBIFO," 1988 (Philip Morris), Bates 2505235055–5088.

[90] Ravenholt, quoted in "Smokers Said to Risk Cancers beyond Lungs" (cit. n. 6).

[91] Muggli *et al.*, "Waking a Sleeping Giant" (cit. n. 12).

[92] *Ibid.*; and Proctor, "Puffing on Polonium" (cit. n. 1).

of polonium in tobacco. What's more, smokers do not see themselves as victims of radiation poisoning, and there is no group of polonium sufferers in solidarity with each other. Most do not even know that their lungs have been infiltrated by a highly radioactive isotope, and they certainly do not identify with the sickly images of Alexander Litvinenko.

As historians of tobacco have illustrated, a smoker's death is not the dramatic and sensational one generally associated with radiation poisoning but, rather, a quiet and lonely one consuming its victims slowly, decades after their smoking habit began. This is a marked contrast with the perceived injustice and sin against youth and vitality attributed to other cancers. The physician and historian of science Robert Aronowitz has argued that over the past two hundred years breast cancer has evolved from a matter of private suffering and endurance on the part of the individual victim to a collective focus for societal fear and concerns about risk.[93] This shift has resulted in public pleas for more studies on risk and treatments and has motivated walks and other events to raise money for research. Exhibiting breast cancer awareness and activism has become a part of modern society; a simple pink ribbon graces the clothing, jewelry, and bumper stickers of victims, loved ones, survivors, and supporters. Strikingly, Aronowitz notes that the mortality rate from breast cancer has remained more or less steady through the second half of the twentieth century. The number of deaths from lung cancer, in contrast, has shot up dramatically. It is the *awareness* of breast cancer, not the actual number of cases, that has grown over the past hundred years; this sort of public awareness and solidarity has not become a part of the story of lung cancer.

More than a general lack of awareness, however, the story of polonium is marked by cycles of forgetting and remembering—or, better said, of the dying down and reigniting of awareness. Such a waxing and waning of interest and of gaining, then losing and forgetting, knowledge is discussed in a volume on "the making and unmaking of ignorance" edited by Robert N. Proctor and Londa Schiebinger. To describe this phenomenon, a new term was introduced as the title of the book: agnotology. Defined broadly as the study of ignorance, agnotology pays homage to the lost and forgotten, the never known and the carefully concealed.[94] Addressing the importance of ignorance throughout history and the impressive extent to which it has been disregarded by scholars, Proctor discusses three categories: "ignorance as *native state* (or resource), ignorance as *lost realm* (or selective choice), and ignorance as a deliberately engineered and *strategic ploy* (or active construct)." I would add a fourth: ignorance (or forgetting) as default. It takes *something* to keep a memory going: momentum, reignition, power, emotion. And certainly the press is all too ready to leap ahead to the next big story, leaving yesterday's headlines as kindling. For a story to continue beyond a single news cycle is extraordinary, a phenomenon the tobacco industry understands well and has taken full advantage of in its quest to minimize the polonium-210 problem. In discussing agnotology in relation to the tobacco industry, Proctor compares the ways in which cigarette companies have censored knowledge to the classifications of military secrecy: in both cases "we don't know what we don't know" because "steps have been taken to keep [us] in the dark!"[95]

In this spirit, Big Tobacco has been careful to avoid drawing attention to what Paul Eichorn of Philip Morris, in a handwritten 1978 memo to his boss, Robert Seligman (then vice president of research and development), called the "sleeping giant" of polonium. (See

---

[93] Robert A. Aronowitz, *Unnatural History: Breast Cancer and American Society* (Cambridge: Cambridge Univ. Press, 2007).

[94] Proctor and Schiebinger, eds., *Agnotology* (cit. n. 7), p. vii.

[95] Proctor, "Agnotology" (cit. n. 7), pp. 3, 11.

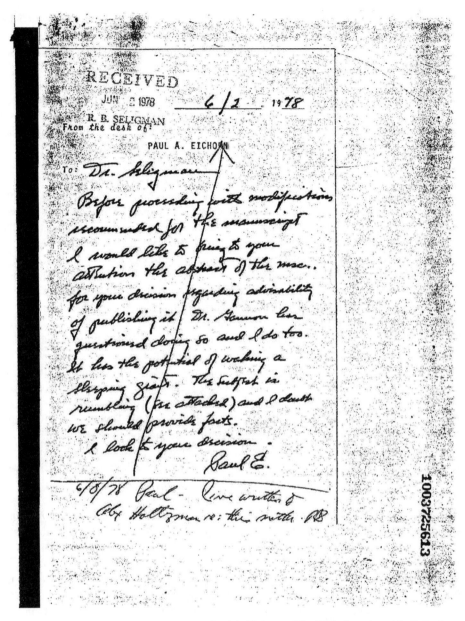

Figure 1. *Philip Morris memorandum from Paul A. Eichorn of the R&D department to Robert Seligman, vice president for R&D, cautioning against publishing the company's research on polonium. Eichorn was concerned that providing any facts at all had the potential to wake the "sleeping giant" of polonium. Seligman responded that articles related to this matter should be given to the company's vice president and general counsel, Alexander Holtzman. See Paul A. Eichorn to Robert Seligman (memo), 2 June 1978 (Philip Morris), Bates 1003725613.*

Figure 1.) The giant was to be kept quiet by keeping results unpublished and by avoiding public debate with researchers or health officials. At the same time, tobacco manufacturers have continuously stayed abreast of the research, conducting their own parallel experiments and recording their own results and measurements, always careful to keep their work secret and private. The industry got so far as to single out and debate the drawbacks and benefits of several potential solutions to the polonium problem; however, as Charlie Nystrom of R. J. Reynolds wrote to Alan Rodgman, "removal of these materials would have no commercial advantage."[96]

When I told Ann Kennedy of the extensive radiochemical research programs at Philip Morris and R. J. Reynolds, she was "astonished" to hear that the tobacco industry had been conducting its own polonium research throughout the 1960s, 1970s, and 1980s.[97] That someone as deeply involved in researching radioactivity in tobacco as Kennedy had no idea that the manufacturers themselves were spending hundreds of thousands of dollars investigating the same topic—always in secret—highlights the discrepancy between what the industry was admitting publicly and what it was saying and doing privately.

The polonium story reveals a dark chapter in the history of science and scientific authority. Tobacco manufacturers have long used the persuasive powers of science to their advantage, hindering disease prevention. Here in the annals of a single isotope is this microcosm of deceit and silence, with disease and death as its result. And although the story of polonium has been repeatedly forgotten, the stakes have remained consistently high: in 2008, the National Cancer Institute estimated that there were 162,000 deaths from lung cancer in the United States, 90 percent of them due to smoking. It is impossible to know how many of these cancers were caused by alpha-emitting isotopes in tobacco, but if the polonium had been reduced through methods known to the industry a certain fraction could have been avoided. The industry made the conscious choice not to act on the results of its own scientific investigations; but it is the customers who have had to live with—and die from—that decision.

[96] Paul Eichorn to Seligman (memo), 2 June 1978 (Philip Morris), Bates 1003725613; and Nystrom to Rodgman, "Comments on the Stauffer Patent No. 4,194,514 for Removal of Radioactive Lead and Polonium from Tobacco" (interoffice memo), 5 Mar. 1982 (R. J. Reynolds), Bates 504970288.
[97] Telephone conversation with Kennedy, 24 May 2007.

CONTRIBUTORS

Joshua Blu Buhs is an independent scholar living in California. His research on the fire ants culminated in *The Fire Ant Wars: Science and Public Policy in Twentieth-Century America* (University of Chicago Press, 2004). More recently, he published *Bigfoot: The Life and Times of a Legend* (University of Chicago Press, 2009).

Ellen Herman is professor of history at the University of Oregon. She is the author of *The Romance of American Psychology: Political Culture in the Age of Experts* (University of California Press, 1995), and *Kinship by Design: A History of Adoption in the Modern United States* (University of Chicago Press, 2008). She maintains a website, The Adoption History Project, and is currently working on a history of autism.

Sally Smith Hughes is a historian of science at the Bancroft Library, University of California, Berkeley. She has conducted over one hundred in-depth, web-based oral histories on bioscience, medicine, and biotechnology. Her current focus is on the scientific, economic, legal, social, and political context for the rise of the earliest biotechnology companies in the 1970s. She is the author of *The Virus: A History of the Concept* (Heinemann Books, 1977), and *Genentech: The Beginnings of Biotech* (University of Chicago press, 2011).

John P. Jackson, Jr., is associate professor in the Department of Communications at the University of Colorado at Boulder. He published *Science for Segregation: Race, Law, and the Case against Brown v. Board of Education* (New York University, 2005), and *Social Sciences for Social Justice: Making the Case against Segregation* (New York University, 2001).

David S. Jones is the A. Bernard Ackerman Professor of the Culture of Medicine at Harvard University. Trained as a historian and a psychiatrist, he studies how doctors think about disease and its treatments. He is author of *Rationalizing Epidemics: Meanings and Uses of American Indian Mortality since 1600* (Harvard University Press, 2004), and *Heart Attacks: Historical Reflections on Blindspots in Cardiac Therapeutics* (Johns Hopkins University Press, 2012).

David Kaiser is the Germeshausen Professor of the History of Science at the Massachusetts Institute of Technology, and also senior lecturer in MIT's Department of Physics. His books include *Drawing Theories Apart: The Dispersion of Feynman Diagrams in Postwar Physics* (University of Chicago Press, 2005), which received the History of Science Society's Pfizer Prize, and *How the Hippies Saved Physics: Science, Counterculture, and the Quantum Revival* (W. W. Norton, 2011).

Scott G. Knowles is the author of *The Disaster Experts: Mastering Risk in Modern America* (University of Pennsylvania Press, 2011), and editor of *Imagining Philadelphia: Edmund Bacon and the Future of the City* (University of Pennsylvania Press, 2009). He is associate professor of history at Drexel University, where he also directs the Center for Interdisciplinary Inquiry.

Sally Gregory Kohlstedt is professor and director of the Program in the History of Science and Technology at the University of Minnesota. She studies aspects of science in American culture, especially institutional development, women and gender in science, and education in and dissemination of science. Her most recent book is *Teaching Children Science: Hands-On Nature Study, 1890–1930* (University of Chicago Press, 2010).

Stuart W. Leslie is professor of the history of science and technology at Johns Hopkins University. "'Industrial Versailles'" was the first in a series of essays on laboratory design and architecture, culminating in *The Architects of Modern Science*, to be published by the University of Pittsburgh Press in 2013. With Robert Kargon, Leslie is also working on a project, *Spaces of Inquiry*, that explores the intersection of the laboratory, the studio, and the clinic in late nineteenth-century Philadelphia, 1930s Detroit, and postwar San Diego.

W. Patrick McCray is professor of history at the University of California, Santa Barbara. His essay in this volume comes from his book *Keep Watching the Skies! The Story of Operation Moonwatch and the Dawn of the Space Age* (Princeton University Press, 2008). His latest book is *The Visioneers: How a Group of Elite Scientists Pursued Space Colonies, Nanotechnologies, and a Limitless Future* (Princeton University Press, 2012); it focuses on people who used their expertise to design, develop, and promote visions of the technological future.

Philip J. Pauly was professor of history at Rutgers University with a strong interest in exploring the relationship between culture and science. His publications include *Controlling Life: Jacques Loeb and the Engineering Ideal in Biology* (Harvard University Press, 1987), *Biologists and the Promise of American Life* (Princeton University Press, 2000), and *Fruits and Plains: The Horticultural Transformation of America* (Harvard University Press, 2008). He died in 2008.

Brianna Rego is a PhD candidate in the Department of History at Stanford University. She is currently writing a dissertation entitled "Behind a Veil of Smoke: Research and Development at Philip Morris from 1952 to 1985." She is interested in the many internal research laboratories built by the tobacco industry to study behind closed doors the health effects of smoking. Following the completion of her dissertation she hopes to apply her historical understanding of the industry's deep scientific knowledge of the hazards of its product toward current and future tobacco control initiatives.

Michael Rossi is a postdoctoral researcher at the École normale supérieure de Cachan. His research focuses on the historical ways in which science has functioned as a tool for judging matters of truth, rectitude, and aesthetic worth. He is currently revising his manuscript, *The Rules of Perception*, a history of color science in the United States in the late nineteenth and early twentieth centuries.

Daniel W. Schneider is professor in the Department of Urban and Regional Planning and an ecologist for the Illinois Natural History Survey at the University of Illinois at Urbana/Champaign. He has recently published an environmental history of the sewage treatment plant, *Hybrid Nature: Sewage Treatment and the Contradictions of the Industrial Ecosystem* (MIT Press, 2011), that examines the simultaneous naturalization and industrialization of microbial processes.

Paul S. Sutter is associate professor of history at the University of Colorado at Boulder and the author of *Driven Wild: How the Fight against Automobiles Launched the Modern Wilderness Movement* (University of Washington Press, 2002). His current book project, from which the article in this volume is drawn, is tentatively titled *Pulling the Teeth of the Tropics: Environment, Disease, Race, and the U.S. Sanitary Program in Panama, 1904–1914.*

Alex Wellerstein is an associate historian at the American Institute of Physics, College Park, Maryland. He received his PhD from the Department of History of Science at Harvard University in fall 2010. His dissertation, "Knowledge and the Bomb: Nuclear Secrecy in the United States, 1939–2008," looked at the evolution of information control in the American nuclear-weapons complex from the Manhattan Project through the War on Terror.

Catherine Westfall is visiting professor at Lyman Briggs College at Michigan State University. She has written books and articles on how science and technology are produced at the U.S. national laboratories. Her most recent book, with Lillian Hoddeson and Adrienne Kolb, is *Fermilab: Physics, the Frontier, and Megascience* (University of Chicago Press, 2008).

INDEX